化学英語［精選］文例辞典

CD-ROM付

松永義夫
編著

朝倉書店

はじめに

化学の分野そのものが多様化するうえ，異なる学問分野にもかかわる学際的研究が求められる今日，その成果を化学用語に基づいて文法上正しく明快な英文にまとめることは，必要な英語の知識の多様化をも意味する．本辞典は広く化学全般に関する文例を提供し，英文執筆の便を図らんとするものある．

特に論文に関しては，緒言，実験，結果，考察，謝辞，正誤表作成に至るまでの文例を蒐集してある．主語と述語の一組だけを含む単文はもちろん，それを組み合わせた重文，複文の文例を取り上げるとともに，煩雑さを避け，読みやすい文に仕上げるに有効な句や節を活用した文例も多く取り入れた．単文を並べるだけで意図を伝えられるものではないから，主語，目的語などの入れ替えや不要部分の削除によって，複雑な文例も目的の英文に到達する途となることを期待したい．

また，論議の展開に必要な知識を英文で執筆者でもある読者に提供することにも努めた．用語の定義，解説の範囲は関連の深い数学，材料科学，結晶学，鉱物学，生物学の表現に及ぶ．また，図の解説に必要な白丸，黒丸，陰影，実線，点線，破線，鎖線のほか，丸括弧，角括弧，中括弧などの表現にも配慮してある．

見出し語は編集の便宜上選ばれたものであるから，使用頻度の高い語ほど多くの他の項目にも文例が存在する．したがって，検索を試みれば一層好ましい文例に出会えることであろう．また，類義語は多少とも意味が異なるものであるから，添えられた類義語を検討することで，より相応しい単語を見出せる可能性があるし，和訳では違いが明らかではない場合でも，類義語の比較によって意味の隔たりを知ることもできよう．

本辞典は，筆者が十年余にわたって日本化学会速報誌掲載論文の最終チェックを行う製作顧問を務める傍ら，その体験を反映しつつ蒐集した文例を編集したものである．

刊行にあたって大変お世話になった朝倉書店編集部の方々に厚く御礼申し上げる．

2015年5月　松永義夫

例文作成の参考とした文献

例文は次の英米の単行本，総説誌ならびに論文誌を参考とし，現行の米式文法に従って作成した．

単行本

N. L. Allinger, M. P. Cava, D. C. de Jongh, C. R. Johnson, N. A. Lebel, C. L. Stevens, Organic Chemistry, Worth Publishers, Inc., New York, 1971.

P. W. Atkins, Physical Chemistry, Oxford University Press, Oxford, 1978.

P. Atkins, Concepts in Physical Chemistry, Freeman & Co., New York, 1995.

J. W. Baker, Hyperconjugation, Oxford University Press, London, 1952.

G. M. Barrow, Physical Chemistry, McGraw-Hill Book Co., New York, 1961.

F. Basolo, R. G. Pearson, Mechanisms of Inorganic Reactions, John Wiley & Sons, New York, 1958.

R. S. Berry, S. A. Rice, J. Ross, Physical Chemistry, John Wiley & Sons, New York, 1980.

A. Bondi, Physical Properties of Molecular Crystals, Liquids, and Glasses, John Wiley & Sons, New York, 1968.

A. N. Campbell, N. O. Smith, The Phase Rule and its Applications, Dover Publications, New York, 1951.

R. G. Compton, G. H. W. Sanders, Electrode Potentials, Oxford University Press, Oxford, 1996.

M. C. Day, Jr. , J. Selbin, Theoretical Inorganic Chemistry, 2nd ed., Reinhold Book Corporation, New York, 1962.

M. J. S. Dewar, The Electronic Theory of Organic Chemistry, Oxford University Press, London, 1950.

B. E. Douglas, D. H. McDaniel, J. J. Alexander, Concepts and Models of Inorganic Chemistry, 3rd ed., John Wiley & Sons, New York, 1994.

A.B. Ellis, M. J. Geselbracht, B. J. Johnson, G. C. Lisensky, W. R. Robinson, Teaching General Chemistry, Materials Science Companion, ACS, Washington, DC, 1993.

H. J. Emeleus, A. G. Sharpe, Modern Aspects of Inorganic Chemistry, 4th ed., Routledge & Kegan Paul, London, 1973.

R. C. Evans, An Introduction to Crystal Chemistry, 2nd ed., Cambridge University Press, London, 1966.

L. F. Fieser, Experiments in Organic Chemistry, 3rd ed., D. C. Heath and Co., Boston, 1957.

R. Foster, Organic Charge-Transfer Complexes, Academic Press, London, 1969.

D. Fox, M. M. Labes, A. Weissberger, Physics and Chemistry of the Organic Solid State, Interscience Publishers, New York,
vol. 1, 1963.
vol. 2, 1965.

A. A. Frost, R. G. Pearson, Kinetics and Mechanism, John Wiley & Sons, New York, 1953.

B. S. Furniss, A. J. Hannaford, V. Rogers, P. W. G. Smith, A. R. Tatchell, Vogel's Textbook of Practical Organic Chemistry, 4th ed., Longman, London, 1978.

W. E. Garner, Chemistry of the Solid State, Butterworths Scientific Publications, London, 1955.
G. W. Gray, Thermotropic Liquid Crystals, John Wiley & Sons, Chichester, 1987.
N. N. Greenwood, A. Earnshaw, Chemistry of the Elements, Pergamon Press, Oxford, 1984.
W. T. Hall, F. P. Treadwell's Analytical Chemistry, John Wiley & Sons, New York,
vol. 1, Qualitative Analysis, 1937.
vol. 2, Quantitative Analysis, 1935.
K. B. Harvey, G. B. Porter, Introduction to Physical Inorganic Chemistry, Addison-Wesley Publishing Co., Reading, 1965.
R. B. Heslop, K. Jones, Inorganic Chemistry, Elsevier Scientific Publishing Co., Amsterdam, 1976.
J. H. Hildebrand, R. L. Scott, The Solubility of Nonelectrolytes, Dover Publications, New York, 1964.
J. E. Huheey, E. A. Keiter, R. L. Keiter, Inorganic Chemistry, Principle of Structure and Reactivity, 4th ed., Harper-Collins College Publishers, 1993.
C. Kittel, Introduction to Solid State Physics, 2nd ed., John Wiley & Sons, New York, 1956.
I. M. Klotz, R. M. Rosenberg, Chemical Thermodynamics, Basic Theory and Methods, John Wiley & Sons, New York, 1994.
H. R. Kruyt, Colloid Science, Elsevier Publishing Co., Amsterdam, 1952.
P. W. Kuchel, Biochemistry, 3rd ed., McGraw-Hill Co., New York, 2009.
K. J. Laidler, Chemical Kinetics, 3rd ed., Harper-Collins Publishers, New York, 1987.
K. J. Laidler, The World of Physical Chemistry, Oxford University Press, Oxford, 1993.
K. J. Laidler, J. H. Meiser, Physical Chemistry, 2nd ed., Houghton Mifflin Co., Boston, 1995.
R. J. Lewis, Sr., Hawley's Condensed Chemical Dictionary, 12th ed. Van Nostrand Reinhold Co., New York, 1993.
G. R. Luckhurst, G. W. Gray, The Molecular Physics of Liquid Crystals, Academic Press, London, 1979.
W. F. Luder, S. Zaffanti, The Electronic Theory of Acids and Bases, 2nd ed., Dover Publications, New York, 1961.
D. A. MacInnes, The Principles of Electrochemistry, Dover Publications, New York, 1961.
J. W. McBain, Colloid Science, D. C. Heath, Boston, 1950.
E. Martell, M. Calvin, Chemistry of the Metal Chelate Compounds, Prentice-Hall, New York, 1952.
S. H. Mauskopf, Chemical Sciences in the Modern World, University of Pennsylvania Press, Philadelphia, 1993.
F. M. Menger, D. J. Goldsmith, L. Mandell, Organic Chemistry, A Concise Approach, W. A. Benjamin, Menlo Park, 1972.
R. T. Morrison, R. N. Boyd, Organic Chemistry, 5th ed., Allyn and Bacon, Newton, 1987.
N. F. Mott, R. W. Gurney, Electronic Processes in Ionic Crystals, 2nd ed., Oxford University Press, London, 1953.
J. N. Murrell, The Theory of the Electronic Spectra of Organic Molecules, Methuen & Co., London, 1963.
R. D. Noble, P. A. Terry, Principles of Chemical Separations with Environmental Applications, Cambridge University Press, 2004.
B. Norden, A. Rodger, T. Dafforn, Liner Dichroism and Circular Dichroism, A Textbook on

Polarized-Light Spectroscopy, RSC Publishing, Cambridge 2010.
Opportunities in Chemistry, National Academy Press, Washington, D.C. 1985.
J. R. Partington, An Advanced Treatise on Physical Chemistry, John Wiley & Sons, New York,
volume 1, Fundamental Principles, The Properties of Gases, 1962.
volume 2, The Properties of Liquids, 1955.
volume 3, The Properties of Solids, 1957.
volume 4, Physico-Chemical Optics, 1962.
volume 5, Molecular Spectra and Structure, Dielectrics and Dipole Moments, 1962.
L. Pauling, General Chemistry, Dover Publications, New York, 1988.
M. J. Pilling, P. W. Seakins, Reaction Kinetics, Oxford University Press, Oxford, 1996.
C. S. G. Phillips, R. J. P. Williams, Inorganic Chemistry, Oxford University Press,Oxford,
vol. 1, Principles and Nonmetals, 1965.
vol. 2, Metals, 1966.
J. E. Ricci, The Phase Rule and Heterogeneous Equilibrium, Dover Publications, New York, 1966.
J. D. Roberts, M. C. Caserio, Modern Organic Chemistry, W. A. Benjamin, New York, 1967.
H. Rossotti, Diverse Atoms, Profiles of the Chemical Elements, Oxford University Press, Oxford, 1998.
R. T. Sanderson, Chemical Periodicity, Reinhold Publishing Corporation, New York, 1960.
D. W. Smith, Inorganic Substances, Cambridge University Press, Cambridge, 1990.
J. C. Slater, Introduction to Chemical Physics, Dover Publications, New York, 1970.
W. D. Stansfield, J. S. Colomé, R. J. Cano, Molecular and Cell Biology, McGraw-Hill, 1996.
A. Streitwieser, Jr., Molecular Orbital Theory for Organic Chemists, John Wiley & Sons, New York, 1962.
W. J. Thiemann, M. A. Palladino, Introduction to Biotechnology, 2nd ed., Pearson Education, San Francisco, 2009.
W. A. Waters, The Chemistry of Free Radicals, 2nd ed., Oxford University Press, London,, 1948.
A. F. Wells, The Third Dimension in Chemistry, Oxford University Press, London, 1966.
A. R. West, Basic Solid State Chemistry, John Wiley & Sons, Chichester, 1984.
G. W. Wheland, Advanced Organic Chemistry, 2nd ed., John Wiley & Sons, New York, 1957.
A. N. Winchell, The Microscopic Characters of Artificial Inorganic Substances or Artificial Minerals, John Wiley & Sons, New York, 1931.
M. J. Winter, Chemical Bonding, Oxford University Press, Oxford, 1994.
P. J. Wheatley, Determination of Molecular Structure, 2nd ed., Dover Publications, New York, 1981.
E. A. Wood, Crystals and Light, An Introduction to Optical Crystallography, 2nd ed., Dover Publications, New York, 1977.

総説誌および論文誌

Accounts of Chemical Research
Angewandte Chemie, International Edition
Chemistry in Britain
Chemical Communications
Chemical Society Reviews

Dalton Transactions
Journal of the American Chemical Society
Journal of Chemical Education
Journal of the Chemical Society
Journal of Chemical Society, Dalton Transactions
Journal of Chemical Society, Faraday Transactions
Journal of Chemical Society, Perkin Transactions
Journal of Materials Chemistry
Journal of Organic Chemistry
Journal of Physical Chemistry
Organic & Biomolecular Chemistry
Philosophical Transactions of the Royal Society of London
Physical Chemistry Chemical Physics
Pure and Applied Chemistry
Quarterly Reviews, Chemical Society

その他参考にした書籍

R. E. Allen, The Concise Oxford Dictionary of Current English, 8th ed., Oxford University Press, Oxford, 1990.

D. Baker, G. Chantrell, Oxford Learner's Grammar, Oxford University Press, Oxford, 2005.

A. M. Coghill, L. R. Garson, The ACS Style Guide, 3rd ed., Oxford University Press, New York, 2006.

P. Hanks, Collins Dictionary of the English Language, Collins, London, 1979.

M. Lester, L. Beason, The McGraw-Hill Handbook of English Grammar and Usage, McGraw-Hill, New York, 2005.

N. Lewis, The Dictionary of Good English, A Guide to Grammar and Correct Usage, Penguin, New York, 1987.

P. Peters, The Cambridge Guide to English Usage, Cambridge University Press, Cambridge, 2004.

J. P. Pickett, The American Heritage Guide to Contemporary Usage and Style, Houghton Mifflin Co., Boston, 2005.

J. Sinclair, Collins Cobuild English Usage, HarperCollins, London, 1992.

M. Swan, Practical English Usage 3rd. ed., Oxford University Press, Oxford, 2005.

L. Urdang, The Book of Synonyms and Antonyms, New Revised Edition, Penguin, New York, 1986.

L. Urdang, The Oxford Thesaurus, American Edition, Oxford University Press, New York, 1992.

Webster's Seventh New Collegiate Dictionary, Merriam, Springfield, 1969.

小稲義男, 新英和大辞典, 第 5 版, 研究社, 1980.

竹林滋, 東信行, 諏訪部仁, 市川泰男, 新英和中辞典, 第 7 版, 研究社, 2003.

日本化学会, 文部省学術用語集, 化学編, 増訂 2 版, 南江堂, 1986.

日本化学会, 標準化学用語辞典, 丸善, 1991.

日本化学会, 化学英語のスタイルガイド, 朝倉書店, 2007.

荻野博, 山本学, 大野公一, 英和化学用語辞典, 東京化学同人, 2008.

中野運, 生化学分子生物学英和用語集, 化学同人, 2002.

D

a.見出し語
b.品詞
c.和訳

damage n **1** 物の損傷 syn *destruction, harm, impairment*

▶ Damage to plants from pollution ranges from adverse effects on foliage to destruction of fine root systems.

vt **2** 損傷を与える syn *harm, impair*

▶ Any material or substance which in normal use can be damaging to the health and well-being of man is said to be hazardous.

damp adj **1** 湿気のある syn *moist* ☞ dry

▶ The induction period could be eliminated by the addition of a small drop of slightly damp solvent.

vt **2** 減じる syn *lessen, reduce, suppress*

d.類義語

▶ The parameter r_D determines how strongly the potential is damped from its pure Coulomb value.

e.関連のある見出し語

damping n 制動

f.文例

▶ A new mechanism in heavy-ion-induced reactions termed deeply inelastic collisions is characterized by damping of large amounts of collective nuclear energy through interactions with nucleonic modes of excitation.

danger n 危険 syn *hazard, risk*

▶ Esters of tertiary alcohols are prone to acid-catalyzed elimination, and the alkoxide method precludes danger from this reaction.

dangerous adj 危険な syn 用語の解説 *rious*

▶ Potassium perchlorate and other perchlorates are oxidizing agents, somewhat less vigorous and less dangerous than the chlorates.

dangerously adv 危険なほどに syn *badly, severely*

▶ Diazomethane is highly toxic, dangerously explosive and cannot be stored without decomposition.

dangling bond n ダングリングボンド syn *unsaturated bond*

▶ The chemical inertness of alkali halide nanocrystals, based on their composition of closed-shell atomic ions, contrasts sharply to the reactivity of metal or semiconductor clusters which have unsaturated, or dangling, bonds at their surfaces.

dark n **1** 暗所 syn *blackness, darkness*

▶ Phosphorus seems a lively element, which glows in the dark and is essential for biological energy transfer and for heredity.

adj **2** 暗い syn *dim, shadowy*

▶ When viewed down an optic axis, anisotropic crystals appear to be isotropic, i.e., they are dark between crossed polarizers.

darken vi 薄黒くなる syn *blacken*

▶ Hypophosphoric acid is not reduced by zinc and dilute sulfuric acid and gives with silver nitrate a white precipitate which does not darken in the light.

dash n ダッシュ

▶ Dashes are inserted when the reagent causes no oxidation.

dashed line n 破線 syn *discontinuous line, broken line* なお，鎖線は dashed and dotted line ☞ dotted line, solid line

▶ There is a gradual, overall decrease in radius as the d shell is filled, as shown by the dashed line that passes through Ca^{2+}, Mn^{2+}, and Zn^{2+}.

dashpot n ダッシュポット

▶ The elastic elements are represented by springs, and the viscous by dashpots, the motion of which is retarded by a viscous liquid.

data n データ datum の複数形であるが，data を動詞の単数形と組み合わせる扱いも広く行われる．しかし，科学論文では複数形として扱うことが多い． syn *facts, record*

▶ The spectroscopic properties of this compound were consistent with the data available in the literature.

database n データベース

▶ The standard procedure is to use classical scattering theory with refractive index data as input data available from different databases.

g.成句
h.成句の和訳
i.成句の類義語
j.成句の文例

date n **1** to date 現在まで syn *until now*

▶ To date, all applications of liquid crystal technology involve organic molecules, and in

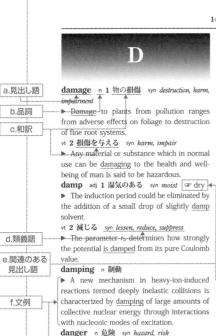

使用の手引き

1. **項目の構成**　次の配列による.

a. 見出し語

b. 品詞（略号）

名詞	n	助動詞	aux
代名詞	pron	前置詞	prep
形容詞	adj	接続詞	conj
副詞	adv	接頭辞	pref
自動詞	vi	接尾辞	suff
他動詞	vt	略語	abbr

c. 和訳

d. 類義語（略号 syn を付け，斜字体で示す）

e. 関連のある見出し語（☞印）

f. 文例（見出し語には下線）

g. 成句（品詞の指定があれば，括弧つきの品詞）

h. 成句の和訳

i. 成句の類義語（略号 syn を付け，斜字体で示す）

j. 成句の文例（成句には下線）

主体は文例であるから，成句があっても見出し語に直接対応する文例がない場合は，c-f を欠く．例えば

accord n **1 in accord with 調和して** syn *consistent*
▶ The results are in complete accord with the postulated molecular arrangement.
vi **2 accord with と一致する** syn *agree, conform, coincide*
▶ In an ideal solution the vapor pressure of the component present in small abundance accords with Raoult's law.

成句がなければ g-j を欠く．

2. **見出し語**　化学用語は重点的に収録した．それ以外の語は品詞に軽重を付けることなく取り上げてある．配列はアルファベット順によった．この配列は

見出し語がもつ複数の意味を示せること，簡単な説明を付けられること，類義語を記載できるなどの利点がある．和訳から適当な見出し語を探すには索引を使用されたい．

動詞は原形で示す．一般の辞書にも形容詞と記載されている現在分詞形や過去分詞形は動詞とは別項目として扱った．動詞の原形は不定詞句として，現在分詞形，過去分詞形は形容詞として用いられるほか，動名詞句，現在分詞句，過去分詞句に使用される．これら場合の見出し語も動詞の原形であるので，本来の動詞としての使用と区別するため，文例には白抜きの記号▷を文頭に使用した．この場合，本来の動詞としての文例は割愛されたが，検索すれば他の見出し語の文例中にその使用例は見出されるであろう．

3. 和訳　文例を考慮に入れて訳を限定した．同じ見出し語に複数の意味があり，それぞれに文例があれば，和訳に番号を付けた．品詞として異なるとなるものも，番号は通しで付けてある．

化学用語の場合は基準となる

「文部省 学術用語集化学編」（増訂 2 版，南江堂，1986）

に定められたものを継承，記載してある

荻野博・山本学・大野公一編「英和化学用語辞典」（東京化学同人，2008）

に従った．

4. 成句　研究社「新英和大辞典」，「新英和中辞典」を参考にして採択した．この中には英米の辞典，文法書に品詞の指定があるものが含まれる．その場合も独立の見出し語とすることなく，括弧入りで品詞を示すに止めた．他の成句と同様に見出し語と同格に扱い，通し番号を付けてある．自動詞と前置詞または副詞，他動詞と前置詞または副詞の 4 種類の組み合わせも同様に成句として配列した．

例えば

［項目 account］

account n **1 説明**
2 on account of (prep) **…の理由で**
3 on no account 決して…ない
4 take account of …を考慮に入れる
5 take into account 考慮に入れる

vi **6 account for**　**…の理由を説明する**

［項目 addition］

addition　n **1 付加**
2 添加
3 in addition　（adv）　**さらに**
4 in addition to　（prep）　**…に加えて**

［項目 as］

as　adv **1 as…as**　**だけ**
2 as…as possible　**出来るだけ**
prep **3 として**
conj **4 ように**
5 …すると
6 …だから
7 概念を制限して
8 as for　（prep）　**に関する限り**
9 as if　（conj）　**まるで…であるかのように**
10 adj + as it is　**それは…であるけれども**
11 as…so　**と同じように**
12 as though　（conj）　**まるで…であるかのように**
13 as to　（prep）　**…について**
14 so as to　**…するように**
15 so…as to　**…するほどに**

［項目 bring］

bring　vt **1 持ってくる**
2 bring about　**もたらす**
3 bring in　**導入する**
4 bring out　**引き出す**
5 bring together　**まとめる**

5. 成句の和訳　さらに，一つの成句に複数の意味がある場合は，これらを(a)，(b)で区別した．例えば

according　adv **according to**　（prep）　（a）**…にしたがって**
（b）**…によれば**

6. 類義語　同義語ともいわれるが，意味が全く同じ語はないという立場から，また辞典で類義語として扱われる語には幅があって，見出し語と必ず交換できるとは限らないから，ここでは類義語と呼ぶことにした．略号 syn を用い，見出し語と意味が近いものを選んで，斜字体で示した．類義語には見出し語として

存在しないものも含まれる．

化学用語には，見出し語としては異なっても，和訳は全く同じものがある．この場合も類義語として，特に区別はしていない．用語として推奨されているもの一つに絞って，他の用語を類義語として斜字体で示すに止めた．例えば，

decay constant syn *disintegration constant*

化学用語のうちどれが推奨されているかは

日本化学会編「標準化学用語辞典」(丸善，1991)

および前記の英和化学用語辞典に記載がある．

異なる和訳が存在する場合，例えば，孤立電子対の場合，非結合性電子対，非共有電子対は和英索引に残すため，すべて別項目として扱った．その際も，同義の見出し語は斜字体で示してある．例えば

lone pair syn *nonbonding electron pair, unshared electron pair*

7. 関連のある見出し語 化学用語には，関連の深いもの，紛らわしいもの，反対の意味をものが多くある．それらの見出し語を一括して，☞印に続けて示した．意味が同じ化学用語でも，その一つは他を簡略化した表現である場合は，類義語ではなく関連のある見出し語として扱った．例えば

electric charge ☞ charge

関連のある見出し語の一例をあげると

filled circle n **黒丸** ☞ hatch, open, shaded

から，斜線を引いた丸，白丸，陰影を付けた丸の表現に到達できる．

8. コンマの使用法 米式文法と英式文法とで異なる．出典にかかわらず，現行の米式文法によることとした．その詳細は本辞典の著者の手になる

日本化学会監修「化学英語のスタイルガイド」(朝倉書店，2007)

を参照されたい．

9. ハイフンと二分ダッシュ ハイフンは一般の複合形容詞，例えば

transition-metal complex, first-order reaction, rate-determining step, five-membered ring, head-to-tail configuration

などに使用される．ただし，名詞と名詞を組み合わせに関するハイフンの使用は著者次第で未だ統一されていないのが現状である．

二分ダッシュは二つの名詞が対等な関係にあるとき，例えば

acid–base indicator, keto–enol tautomerism, spin–spin coupling, sol–gel process, Maxwell–Boltzmann distribution

などに使用される．より詳しくは，「化学英語のスタイルガイド」の解説を参照されたい．

10. 索引 各項目中の和訳を五十音順に配列し，英語の単語または成句を対応させ，そのページ数が記載したものである．検索を行う場合には，和英辞典として利用されたい．

11. 検索による文例の活用 検索を行うと見出し語の文例以外から多くの望ましい例を見出す可能性がある．例えば，赤外スペクトルに関して数値を扱った記述を求めたいとしよう．まず，赤外に相当する infrared で検索を行うと（下線は本辞典の見出し語）

▶ In the **infrared** spectrum of a hydrogen–bonded alcohol, the most conspicuous feature is a strong, broad band in the 3200–3600 cm^{-1} region due to O–H stretching.
▶ In the **infrared** spectrum of 1-butene the absorption band near 1650 cm^{-1} is characteristic of the stretching vibration of the double bond.

が得られる．

次に，これら二つの文例に共通な band で検索を行うと

▶ Another strong, broad **band** due to C–O stretching appears in the 1000–1200 cm^{-1} region, the exact frequency depending on the nature of the alcohol.
▶ The –CHO group of an aldehyde has a characteristic C–H stretching band near 2720 cm^{-1}; this, in conjunction with the carbonyl **band**, is fairly certain evidence for an aldehyde.
▶ The C≡N stretching **band** in cyano compounds not only appears in the range 2040–2170 cm^{-1} but is also identifiable by its sharpness and high intensity.
▶ **Bands** due to carbon–carbon stretching may appear at about 1500 and 1600 cm^{-1} for aromatic bonds, at 1650 cm^{-1} for double bonds, and at 2100 cm^{-1} for triple bonds.
▶ Nitro compounds exhibit two very intense absorption **bands** in the 1560−1500 cm^{-1} and 1350−1300 cm^{-1} regions of the spectrum arising from asymmetric and symmetric stretching vibrations of the highly polar nitrogen–oxygen bonds.
▶ The absorption at 2750 cm^{-1} in the carbon–hydrogen stretching region is indicative of the aldehyde C–H bond, and the **band** at 1730 cm^{-1} suggests a carbonyl group.
▶ The feature which enables one to distinguish a carboxylic acid from all the other carbonyl compounds is a broad absorption **band** which extends from 3300 cm^{-1} to 2500 cm^{-1}.

さらに，band は absorption band を簡略化した表現であることを考慮して

absorption で検索を行うと

▶ Aliphatic **absorption** is strongest at higher frequency and is essentially missing below 900 cm^{-1}.

▶ For aromatic rings, out-of-plane C-H bending gives strong **absorption** in the 675-870 cm^{-1} region, the exact frequency depending upon the number and location of substituents.

▶ We may rule out the presence of a hydroxyl group in a material that does not exhibit an **absorption** in the 3600 cm^{-1} region.

▶ **Absorption** from about 5000 to 1250 cm^{-1} is generally associated with changes in the vibrational states of the various bonds and is relatively characteristic of the types of bonds present.

▶ If a compound has a hydroxyl group, there will be an **absorption** in the 3600 cm^{-1} region.

が得られ，absorption band は band のみならず absorption と呼んでも差し支えないこともあると知れる．文例を眺めると，stretching や vibration が赤外スペクトルの記述に重要な表現で，検索の手掛かりとなることも知れよう．

これら文例から話題によって何を記載すべきかが知れるから，手持ちのデータを考慮して文例を選び，あるいはそれらをさらに組み合わせたのち，必要な語や数値の入れ替えを行えば，希望する文に容易に到達することができよう．なお，文例を選択するに当たっては前後の文との調和，接続に考慮を払うことが望ましい．

付録 CD-ROM について

本書の付録 CD-ROM には本文の PDF データが入っています．「使用の手引」の 11 のような検索にご活用下さい．

A

a, an article **1 一つの**
▶ An example of a system which forms compounds with congruent melting points is that of water and iron(Ⅲ) chloride which forms four stable hydrates.
不定冠詞 a, an は初めて話題としたときに使用し，同一の名詞を話題とすることが読者にとっても明らかであれば，次からは定冠詞 the を付ける．
▶ Alkyl halide formation from an alcohol and a hydrogen halide affords an important example of a reaction wherein the C–O bond of the alcohol is broken.
2 …というもの その種類全体に通じる一般的なことを述べるとき
▶ An isothermal process is a process that occurs at constant temperature.
同じ目的で，the を付けた単数形，the を付けない複数形が用いられる． ☞ the 語順は一般に a＋名詞，a＋形容詞＋名詞，a＋副詞＋形容詞＋名詞であるが，so＋形容詞＋a＋名詞のような特殊な場合がある．
▶ In dilute system, ultrahigh purity is not so crucial a factor since short-range, exciton-type energy transfer does not take place.

abandon vt 決定的に**放棄する** syn *give up*
▶ The classical definition of oxidation and reduction in terms of gain or loss of oxygen has been abandoned in favor of the concept of electron loss and electron gain.

abbreviate vt 語句の一部を残して**短縮する** syn *shorten*
▶ The name for this type of reaction is abbreviated S_N, S for substitution and N for nucleophilic.

abbreviated adj **短縮した** syn *brief, concise, short*
▶ The amino acid abbreviated on the left is always the one that has a free amino group, and the one abbreviated on the right has the free carboxyl function.

abbreviation n **省略形** syn *contraction, shortening*
▶ SI is an abbreviation in many languages for Système International d'Unités.

ability n **能力** syn *capability, capacity*
▶ For certain bacteria, their ability to cause disease requires that they attach to human tissues.

ab initio calculations n **アブイニシオ計算**
▶ In ab initio calculations, no experimental data are employed other than the electronic and nuclear charges, the nuclear masses, and the value of the Planck constant.

able adj **…することができて** syn *capable*
▶ NaCl is able to dissolve a small amount of $CaCl_2$, and the mechanism of solid solution formation involves the replacement of two Na^+ ions by one Ca^{2+} ion; one Na^+ site, therefore, becomes vacant.

abnormal adj **異常な** syn *peculiar, unusual, uncommon*
▶ The addition of hydrogen bromide to an olefin may be abnormal if oxygen or peroxides are present.

abnormal addition n **異常付加**
▶ The abnormal addition of hydrogen bromide to an olefin can be more or less completely inhibited by traces of such antioxidants as hydroquinone or diphenylamine.

abnormality n **異常性** syn *peculiarity*
▶ Acquired mutations can cause abnormalities in cell growth leading to cancerous tumor formation, metabolic disorders, and other conditions.

abnormally adv **並外れて**
▶ If the ions are of the same sign, the preexponential factors are abnormally low.

abolish vt **無効にする** syn *annihilate, destroy, eliminate*
▶ Modification of an interface cysteine found in a number of parasitic triosephosphate isomerases abolishes activity.

abound vi **たくさんある** 対応する名詞は

abundance syn *prevail*

▶ Nonstoichiometry abounds particularly for compounds with the transition elements, and many of chalcogenides can be considered as metallic alloys.

about adv **1** およそ syn *almost, approximately, around, nearly*

▶ X-rays are electromagnetic radiation with a wavelength of about 1 Å (10^{-10} m), the approximate size of an atom.

prep **2** …に関して syn *concerning*

▶ Intramolecular hydrogen bonds in five- and seven-membered rings are also common, but we do not have data at this point about competition between these bonds and intermolecular hydrogen bonds.

3 …の周囲に syn *around*

▶ Clustered about each ion is a group of polar solvent molecules, oriented with their negative ends toward the carbocation and their positive ends toward the anion.

4 …の周りの syn *around*

▶ H_2O_2 is the smallest molecule known to show hindered rotation about a single bond.

above prep **1** 位置が…より上に ☞ below

▶ In some complexes stacking angle is about 30°, while in others the donor and acceptor molecules are vertically above one another.

2 基準，数量など…を越えて syn *more than, over*

▶ Anhydrous $NaClO_2$ crystallizes from aqueous solutions above 37.4 ℃, but below this temperature the trihydrate is obtained.

adv **3** 上のほうに

▶ The results given above show that the relative amounts of the various isomers differ markedly depending upon the halogen used.

abrasive n 研磨材

▶ Abrasive in powder form may be affixed to paper or textile backing after the particles have been coated with an adhesive.

adj 研磨する

▶ Glass grinding is done by rubbing together two pieces of glass with an abrasive powder such as silicon carbide or fused aluminum oxide.

abrupt adj 急激な syn *sudden*

▶ Many materials undergo abrupt changes in structure or property on heating.

abruptly adv 急に syn *suddenly*

▶ By carefully monitoring and varying the concentration of reactant gases, the refractive index can be changed abruptly or gradually from the center to the outside of the fiber.

abscissa n 横座標 ☞ ordinate

▶ The abscissa ϕ is the angle of rotation as measured from some arbitrary starting point.

absence n 存在しないこと syn *deficiency, non-existence*

▶ One striking feature of the 1:2:3 structure is the complete absence of oxygen ions in the *y* plane.

absent adj 存在しない syn *lacking, missing*

▶ Vitamin B_{12} has a reduced porphyrin ring in which one methine bridge is absent.

absolute adj 絶対的 syn *complete, flawless, perfect*

▶ The absolute accuracy of the results obtained with this apparatus can only be estimated by a comparison of the derived thermodynamic quantities with those determined by other methods.

absolute alcohol n 無水アルコール ☞ anhydrous

▶ Absolute ethanol is made by adding benzene to 95% alcohol and removing the water in the volatile azeotrope.

absolute configuration n 絶対配置 ☞ relative configuration

▶ Bijvot determined using X-ray diffraction the actual arrangement in the space of the atoms of the rubidium sodium salt of (+)-tartaric acid and thus made the first determination of the absolute configuration of an optically active substance.

absolutely adv 完全に syn *completely, entirely, fully*

▶ Whenever the word insoluble is used in this text, it is with the understood limitation that no substance is absolutely insoluble in water.

absorb vt **1** 気体を吸収する syn *take in* ☞ adsorb, sorb

▶ Benzene vapors are absorbed to a considerable extent by water and all aqueous salt solutions.

2 光を吸収する

▶ An isolated carbon-carbon double bond absorbs near 180 nm ($\varepsilon \fallingdotseq 15000$) as a result of a $\pi \rightarrow \pi^*$ electronic transition.

3 緩和する syn *buffer, lessen, reduce*

▶ When pressure is applied to a system at equilibrium, Le Chatelier's principle asserts that it will adjust to absorb the effect of the pressure increase.

absorbance n 吸光度

▶ The quantity $A = \log_{10}(I_0/I)$ is known as the absorbance. It was formerly called the extinction or the optical density, but IUPAC now discourages these terms.

absorbent n 吸収剤

▶ In an absorption method, the mixture of gases is treated with a series of absorbents.

absorption n **1** 物質の吸収 ☞ adsorption, sorption

▶ In chemical terminology, absorption is referred to the penetration of one substance into the inner structure of another, as distinguished from adsorption, in which one substance is attracted to and held on the surface of another.

2 光の吸収

▶ There is an intense absorption in the spectrum of methyl vinyl ketone which is due to the excitation of one of the π electrons of the conjugated system.

absorption band n 吸収バンド

▶ A single, structureless and very broad absorption band was found in the range 3000–12000 Å.

absorption cross section n 吸収断面積

▶ Compounds with donor-π-donor, donor-acceptor-donor, and acceptor-donor-acceptor structural motifs could exhibit exceptionally large value of two-photon absorption cross section.

absorption edge n 吸収端

▶ A lighter element, such as iron, would absorb Cu Kα radiation as well as Kβ, because it absorption edge is displaced to higher wavelength.

absorption maximum n 吸収極大

▶ One may correlate the position of the ultraviolet absorption maximum with the degree of conjugation within a molecule.

absorption spectrum n 吸収スペクトル

▶ The absorption spectrum of H_2 shows bands corresponding to the ${}^1\Sigma_u^+ \leftarrow {}^1\Sigma_g^+$ and ${}^1\Pi_g \leftarrow {}^1\Sigma_g^+$ transitions.

absorptive power n 吸収力

▶ The dehydrated zeolites, both natural and synthetic, have a considerable absorptive power for gases other than water vapor.

absorptivity n 吸光率 syn *absorption factor*

▶ The spectrum of carboxypeptidase is irregular and has a high absorptivity, indicating that a regular tetrahedron is not present.

abstract n **1** 抄録 syn *outline, summary*

▶ Chemical Abstracts is the most indispensable information source in chemical literature and is the largest scientific abstract journal in the world.

adj **2** 抽象的な syn *conceptual, intellectual, theoretical*

▶ A concept is an abstract idea generalized from particular instances.

vt **3** 引き抜く syn *cut, shorten*

▷ Triphenylmethyl can act as a powerful reducing agent by abstracting an electron from another molecule.

abstraction n 引き抜き syn *extraction, removal*

▶ After allowance is made for differences in the probability factor, the rate of abstraction of hydrogen atoms is always found to follow the sequence tertiary＞secondary＞primary.

abundance n **1** 存在率 ☞ natural abundance, isotopic abundane

▶ ^{15}Nitrogen occurs naturally in the air at an abundance of 0.3663 percent.

2 多量 syn *plenty*

▶ There is an abundance of data describing vapor-liquid equilibrium for many systems.

abundant adj 豊富な 最上級の後では，しばしば名詞は省略される. syn *ample, plentiful*
▶ Cerium is the twenty-sixth most abundant of all elements, being half as abundant as Cl and five times as abundant as Pb.
academic n **1** 教員
▶ It is often asserted by university academics, to justify the time and public money they spend on research, that the most effective teachers at this level are those actively involved in research.
adj **2** 学術的な syn *intellectual, scholatic*
▶ Rubidium and cesium are of considerable academic interest but so far have few industrial applications.
accelerate vt 促進する syn *advance, speed*
▶ According to our general principles, an increase in dielectric constant will accelerate the S1 reaction greatly but will retard the S2 reaction, though to a smaller extent.
acceleration n 加速
▶ The acceleration of rate of racemization by hydroxide ion is apparently a result of base hydrolysis leading to a decomposition of the complexes.
accelerator n 促進剤
▶ Organic accelerators greatly reduce the time required for vulcanization of natural and synthetic rubbers, at the same time improving the aging and other physical properties.
accentuate vt 強調する syn *emphasize, stress*
▶ The oxo acids of P are clearly very different structurally from those of N, and the difference is accentuated when the standard reduction potentials and oxidation-state diagrams for the two sets of compounds are compared.
accept vt **1** 受容する syn *receive*
▶ The porphyrin can accept two hydrogen ions to form the +2 diacid or donate two protons and become the −2 dianion.
2 認める syn *admit, allow, recongnize*
▶ It is generally accepted that the interaction between a simple cation like Na^+ and water is mainly of an ion-dipole nature, the water molecules nearest to the ion having their oxygen atoms adjacent to it.
acceptable adj **1** 容認できる syn *adequate, agreeable*
▶ The commonly used abbreviations for organic groups, such as Me, Bu, Ph etc., are acceptable in inorganic formulas.
2 満足な syn *adequate, satisfactory, tolerable*
▶ It is not acceptable to estimate geometries using bond lengths and bond angles of similar species: even small changes in geometry can result in substantial changes in calculated values of molecular properties.
acceptably adv 満足に
▶ Eventually, the preponderance of the data will be most acceptably represented by the best hypothesis, and the majority of the scientific community will come to accept it.
acceptance n 採用 syn *prevalence, recognition*
▶ The recommended names for these compounds, phosphinic acid and phosphinates, have not yet gained wide acceptance for inorganic compounds but are generally used for organophosphorus derivatives.
accepted adj 一般に承認された syn *current, well-established*
▶ The commonly accepted mechanism of this alkylation is based on the study of many related reactions.
acceptor n **1** 受容体 syn *electron acceptor, electron-pair acceptor*
▶ π Acceptors include aromatic systems containing electron-withdrawing substituents such as nitro, cyano and halo, also acid anhydrides, acid chlorides, and quinones.
2 半導体の分野におけるアクセプター ☞ donor
▶ Impurities of the simple acceptor or donor type are generally considered to be electrically neutral.
acceptor level n アクセプター準位 ☞ donor level
▶ The acceptor levels are discrete if the concentration of gallium atoms is small, and it is not possible for electrons in the acceptor

levels to contribute directly to conduction..

access n **1** 立ち入り syn *admission, approach, entry, passage*

▶ Access of direct sunlight was prevented as far as possible by using black cloth wrapping on the vessel.

2 到達できる方法 syn *approach, road*

▶ Sulfonation is of special importance in the chemistry of naphthalene because it gives access to the β-substituted naphthalenes.

vt **3** 利用する syn *get*

▶ Several decay pathways can be accessed simultaneously.

accessibility n **1** 入手可能なこと

▶ The utility of any method of synthesis is limited by the accessibility of the starting materials.

2 接近できること

▶ Generally, accessibility and rate of diffusion were better for ordered mesoporous materials than for amorphous ones, although in the former case it is important to avoid blocking the entrances to the pores.

accessible adj **1** 近づきやすい syn *open*

▶ The π system of a carbon–carbon double bond is readily accessible to a variety of oxidizing agents.

2 入手しやすい syn *available, obtainable, ready*

▶ The most accessible compounds of bismuth are those in the +3 oxidation state with an inert pair.

accessory pigment n 補助色素

▶ Some of the energy of the light not absorbed by the chlorophyll itself is captured by accessory pigments.

accident n **1** 事故 syn *misfortune, mishap*

▶ The source materials used in metal organic chemical vapor deposition are extremely hazardous; several accidents have occurred in the preparation and purification of some of the alkyls.

2 偶然の出来事 syn *chance*

▶ It is a fortunate physiological accident that eye reacts to the variable intensity as though the pattern did consist of maxima and minima.

accidental adj 偶発的な syn *chance, fortuitous, unintentional*

▶ When it is desired to inhibit unwanted explosions or detonations, it is particularly important to prevent the accidental heating up of small regions of the solid.

accidentally adv 偶発的に syn *by chance, incidentally*

▶ Many of the complexes of S_2 were first obtained accidentally.

accommodate vt **1** 収容する syn *lodge, put*

▶ Second phase can precipitate at the grain boundaries, where the imperfect nature of the atomic packing can accommodate them more easily.

2 順応させる syn *adapt, fit, suit*

▶ There is an extensive mosaic structure of small coherent crystalline regions within each particle that is better able to accommodate the volume change.

accommodating adj 与しやすい syn *cooperative, helpful*

▶ Nature is rarely so accommodating as to provide simple, well-defined systems.

accommodation n **1** 収容

▶ The Frenkel defect involves the removal of an ion from its normal site in the structure and its accommodation elsewhere in an interstitial position.

2 適応 syn *adaptation, adjustment, modification*

▶ Adaptation is the accommodation of a living organism to its environment.

accommodation coefficient n 適応係数

▶ Other information concerning chemisorption on metals may be obtained from the measurement of accommodation coefficients.

accompany vt 伴う syn *attend*

▶ The transition of an electron from the ground state to an excited electronic state is accompanied by vibrational and rotational changes in the molecule.

accompanying adj 伴う syn *consequent, following, related*

▶ Energy changes accompanying chemical reactions have been measured with a high degree of accuracy and by a variety of experimental methods.

accomplish vt なしとげる syn *achieve, complete*

▶ Oxidation of primary alcohols to carboxylic acids is usually accomplished by use of potassium permanganate.

accord n **1 in accord with** …と調和して syn *consistent*

▶ The results are in complete accord with the postulated molecular arrangement.

vi **2 accord with** …と一致する syn *agree, conform, coincide*

▶ In an ideal solution the vapor pressure of the component present in small abundance accords with Raoult's law.

accordance n **in accordance with** (prep) …と一致して syn *consistent*

▶ Ozone is absorbed quantitatively by means of sodium arsenite solution in accordance with the following reaction.

according adv **according to** (prep) (a) …に従って syn *consistent with, in conformity with*

▶ We classify a carbon atom as primary, secondary, or tertiary according to the number of other carbon atoms attached to it.

(b) …によれば syn *consistent with, in agreement with*

▶ According to Brønsted's definition, the species HA, H_3O^+, and BH^+ are acids, while B, OH^-, and A^- are bases.

accordingly adv それゆえ syn *consequently, hence, therefore, thus*

▶ Accordingly a large cell area is required for generating even moderate amounts of electric energy.

account n **1** 説明 syn *description, explanation*

▶ The mathematical treatment of conductivity is somewhat complicated, and only a very general accounts of the main ideas can be given here.

2 on account of (prep) …の理由で syn *because of*

▶ Manganese is of interest on account of its polymorphism and the complexity of some of its structures.

3 on no account 決して…ない syn *never*

▶ Every precaution must be taken in using sodium for drying purposes or as a reagent; on no account must the metal come in contact with water.

4 take account of 考慮に入れる syn *consider*

▶ The BET theory attempts to take account of higher layer formation by removing the Langmuir restriction to monolayer adsorption.

5 take into account 考慮に入れる syn *consider, take into consideration*

▶ To compare the acidities of substituted aromatic acids, resonance and inductive effects have to be taken into account.

vi **6 account for** …の理由を説明する syn *explain, rationalize*

▶ To account for the diamagnetism of $[Re_3Cl_{12}]^{3-}$, it was suggested that the rhenium–rhenium bonds were double rather than single bonds.

accountable adj 説明できる syn *responsible*

▶ The observed electronic absorption spectra, characteristic of the complex as a whole, are accountable as intermolecular charge-transfer transitions.

accounting n 計算 syn *bookkeeping*

▶ The so-called transfer of electrons is an accounting device for effecting the changes in oxidation states and for balancing the equations.

accumulate vi **1** 蓄積する syn *collect, gather, store*

▶ Over the past few years, evidence has accumulated which shows that cation formation is a fairly common occurrence among nonmetals, at least insofar as polycations of the type A_m^{n+} are concerned.

vt **2** 蓄積する syn *assemble, pile up*

▷ The presence of a detergent molecule reduces the surface tension by accumulating in the interface, and the decrease in Gibbs function stabilizes the emulsion.

accumulation n 蓄積 syn *collecting, gather-*

ing

▶ Although the reaction probably proceeds in two stages no evidence could be found for the accumulation of the presumed intermediate imidate esters in the solid state.

accuracy n 正確さ syn *correctness, precision*

▶ The true value is not usually known so that it is not always possible to obtain a numerical measure of the absolute accuracy of a measurement or analysis.

accurate adj 精密な syn *correct, exact, precise*

▶ More accurate experimental data and theoretical calculations are necessary before the relative contributions of electrostatic and charge-transfer effects to changes in the infrared spectrum can be unraveled.

accurately adv 正確に syn *exactly, perfectly, truly*

▶ Susceptibility anisotropies have been very accurately determined for small molecules by molecular beam techniques and less accurately determined for large aromatic molecules by single crystal measurements.

accustom vt 慣らす syn *familiarize*

▶ Chemists are accustomed to use different grades of filter paper, some which make rapid filters and others which make slow filters but are required for precipitates that are likely to pass through the pores of a rapid filter paper.

acetylate vt アセチル化する

▶ Aniline is acetylated either with acetic anhydride in aqueous solution or by refluxing the amine with acetic acid.

acetylation n アセチル化

▶ Acetylation, in the case of a reducing sugar, is complicated by the possibility of formation of the α- or the β-anomeric form, or a mixture of both.

acetyl coenzyme A n アセチル補酵素 A

▶ Acetyl coenzyme A is the building block from which the long chains of fatty acids are synthesized.

acetylenic linkage n アセチレン結合 ☞ triple bond

▶ Alkynes in the presence of palladium or platinum and hydrogen gas are reduced to alkenes by addition of hydrogen across the acetylenic linkage.

achievable adj なしとげられる syn *feasible, possible*

▶ A much wider range of micelle-tempelated silicas can be formed, allowing for better control over surface chemistry as well as a major extension in the pore size achievable.

achieve vt 達成する syn *accomplish, complete, execute*

▷ The various stoichiometries reflect differing ways of achieving charge balance, preferred coordination polyhedra, and hydrogen bonding.

achievement n **1** 業績 syn *attainment*

▶ Tremendous achievements in chemistry and other sciences have resulted from fundamental research.

2 達成 syn *attainment, completion, realization*

▶ The achievement of equilibrium with melting ice exposed to 1 atm pressure serves to establish the value of the length of the mercury column.

achiral adj アキラル ☞ chiral

▶ Glycine is achiral because it contains two hydrogen atoms bonded to the α-carbon.

acid n **1** 酸 ☞ Lewis acid

▶ Acids are referred to as strong or weak according to the concentration of H^+ ion that results from ionization.

adj **2** 酸性の syn *acidic*

▶ Phosphoric acid really has three replaceable hydrogens, but only the first is acid toward Methyl Orange and two hydrogen atoms are acid toward phenolphthalein.

acid-base catalysis n 酸塩基触媒作用

▶ Acid-base catalysis is defined as the process in which there is transfer of a proton in the transition state of the reaction.

acid-base indicator n 酸塩基指示薬 ☞ indicator

▶ An acid-base indicator is normally some large organic molecule which can exist as either a protonated or an unprotonated form, and of which one form is colored.

acid-base reaction n 酸塩基反応

▶ In addition to simple dissolution, ionic dissociation, and solvolysis, two further classes of reaction are of importance in aqueous solution chemistry, namely, acid-base reactions and oxidation-reduction reactions.

acid-base titration n 酸塩基滴定 ☞ acidimetry

▶ Acid-base titrations are very conveniently performed by monitoring the pH of the solution.

acid catalysis n 酸触媒作用

▶ After the reaction, the ketone could be regenerated by acid catalysis.

acid catalyst n 酸触媒 ☞ base catalyst

▶ Dehydration of cyclohexanol to cyclohexane can be accomplished by heating the cyclic secondary alcohol with an acid catalyst at a moderate temperature or by distillation over alumina or silica gel.

acid constant n 酸定数 syn *acid dissociation constant*

▶ A polyprotic acid has several acid constants, corresponding to dissociation of successive hydrogen ions.

acid dissociation n 酸の解離

▶ The behavior of a buffer can be understand from the equilibrium equation for the acid dissociation.

acid dissociation constant n 酸解離定数 syn *acid constant*

▶ Although a precise value of the acid dissociation constant of pyrrole is not available, it is estimated at about 10^{-15}.

acid hydrolysis n 酸加水分解

▶ There is an increase in acid hydrolysis at the lower pH due to acid catalysis and another increase at the higher pH due to base catalysis.

acidic adj 酸性の syn *acid* ☞ alkaline, basic

▶ The iodine molecule may be regarded as acidic in its reaction with iodide ion, a base or donor, to form I_3^-.

acidic oxide n 酸性酸化物 ☞ basic oxide

▶ MoS_2 is converted, via the acidic oxide MoO_3, to ammonium molybdate, which can be reduced to the metal with hydrogen.

acidic solution n 酸性溶液

▶ The acid HOBr disproportionates in acidic solution, while the element itself disproportionates in alkaline solution.

acidification n 酸性化

▶ Acidification of a solution of hexaamminecobalt(III) results in no immediate change, and several days are required at room temperature for degradation of the complex despite the favorable thermodynamics.

acidify vt 酸性にする

▷ Continuous silica gel is formed by acidifying the solution of an alkali-metal silicate.

acidimetry n 酸滴定 ☞ acid-base titration

▶ In acidimetry, acids are determined either by titration with standard alkali solution or a known amount of the latter is added and the excess titrated with standard acid.

acidity n 酸性度 ☞ alkalinity, basicity

▶ In aqueous solution, the acidity and basicity of an acid and its conjugate base are related by the expression $K_a \times K_b = 10^{-14}$.

acidolysis n アシドリシス

▶ The inertness of hexaamminecobalt(III) results from the absence of a suitable low-energy pathway for the acidolysis reaction.

acid rain n 酸性雨

▶ If the sulfur is not removed before combustion of petroleum, it contributes to acid rain.

acid soap n 酸性せっけん

▶ In concentrated solutions, acid soaps are solubilized by the soap and leave the soap solution completely clear and transparent.

acid strength n 酸強度

▶ If a different solvent is chosen, then the acid strengths will be different.

aci form n アシ形

▶ The aci and nitro forms may be distinguished from one another by the fact that the aci, but not the nitro, forms give characteristic colors with iron(III) chloride and rapidly absorb bromine.

acoustic adj 音響の ☞ sound wave

▶ With the assumption that the velocity of propagation of the acoustic waves is independent of wavelength and direction, Debye obtained the expression for the heat capacity of an isotropic solid at constant volume.

acquainted adj **aquinted with** …に精通して syn *familiar with*

▶ Of the varied chemical properties of alcohols, there is just one pair that we should become acquainted with at this point: their acidity and basicity.

acquire vt 取得する syn *gain, get, obtain, secure*

▷ Many atoms are readily ionized, either by losing electrons to form cations or by acquiring electrons to form anions, and sometimes several alternative states of ionization are possible.

acquired adj 後天性の syn *obtained*

▶ Prolonged exposure to ultraviolet light can cause acquired mutations in skin cells leading to skin cancer.

acquisition n 取得 syn *gain*

▶ The distribution of bond lengths in TCNQ is consistent with the acquisition of 0.2 units of charge.

acronym n 頭字語 syn *abbreviation, symbol*

▶ Laser is an acronym for light amplification by stimulated emission of radiation.

across adv **1** さしわたし syn *over*

▶ The macropores are 1 μm in diameter, and the mesopores are 50 nm across.

2 横切って

▶ The electrons of one metal may find it energetically favorable to spill across into the other.

prep **3** 横切って syn *over*

▶ Ionization potentials gradually increase from left to right across the periodic table.

4 …にまたがって syn *over*

▶ Hydrogen chloride, hydrogen bromide, and hydrogen iodide readily add across multiple bonds.

5 …のいたるところで syn *over*

▶ With no crystallographic constraints, metallic glasses can be produced across wide composition ranges, giving great flexibility for optimizing material properties.

act n **1** 行為 syn *action, operation*

▶ If a measurement is made on a particle such as an electron, the act of making the measurement disturbs the particle, so that there is inevitably an uncertainty in the measurement of the quantity.

2 in the act of 行動中

▶ An electron transfer between a semiconductor and a gas molecule in the act of chemisorption is reflected in a change in the electric conductivity of the semiconductor.

vi **3** 役を務める syn *function, operate, work*

▷ In the presence of acid the hemiacetal, acting as an alcohol, reacts with more of the solvent alcohol to form the acetal, an ether.

4 作動する syn *perform, play*

▷ The force acting on a nucleus in a molecule is the sum of the electrostatic forces of repulsion due to the other nuclei and of attraction due to the electron density which is treated as a continuous distribution of negative charge.

5 act upon 作用する syn *alter, affect, attack*

▶ The insoluble substance may form a protective coating over particles of material that have not been acted upon.

actin n アクチン

▶ The binding of Ca^{2+} to troponin displaces the complex and allows actin and myosin to interact and ATP to power the relative movement of one protein across the other.

actin filament n アクチンフィラメント

▶ Calcium ions binding to calmodulin stimulate the movement of caldesmon so that myosin heads can associate with the actin filaments.

actinide n アクチニド syn *actinoid*

▶ The actinides are base metals and have to be made by methods such as the reduction of their oxides or fluorides with calcium or barium.

action n **1** 作用 syn *activity, effect, function, reaction*

▶ Chlorobenzene has solvent action similar to

that of benzene and toluene but considerably greater and is used for crystallizing sparingly soluble substances.

2 in action 実行中で

▶ Producing human insulin in recombinant bacteria remains an outstanding example of microbial biotechnology in action.

activate vt **1** 作動させる syn *mobilize, trigger*

▶ The many types of gauges are activated by mechanical, ultrasonic, electronic, magnetic, and pneumatic means.

2 活性化する

▷ A group is classified as activating if the ring it is attached to is more reactive than benzene and is classified as deactivating if the ring it is attached to is less reactive than benzene.

activated adj 活性化された

▶ The rate of decomposition of the activated complex, which takes place in a time of the order of 10^{-13} seconds, must not be confused with the rate of decomposition of the activated molecule, which is much slower.

activated alumina n 活性アルミナ

▶ High surface area per unit mass alumina, either amorphous or crystalline, which has been partially or completely dehydrated is termed activated alumina.

activated carbon n 活性炭

▶ Activated carbons are distinguished by their enormous surface area which is typically in the range 300–2000 m^2/g.

activated complex n 活性錯体

▶ The term activated complex refers to the configuration of reactants and products at a peak in the reaction profile energy curve, i.e., at the transition state.

activated sludge n 活性汚泥

▶ The activated sludge process for deriving sewage digestion is a form of fermentation.

activating group n 活性化基

▶ In some cases, it may be necessary to have an activating group to facilitate substitution, which would otherwise be very difficult.

activation n 活性化 ☞ deactivation

▶ A photochemical activation of a molecule insufficient to produce complete dissociation may, nevertheless, bring about a molecular rearrangement.

activation analysis n 放射化分析

▶ In activation analysis, an element E is determined by bombarding the sample with neutrons and measuring the intensity of radioactivity induced.

activation energy n 活性化エネルギー

▶ Since adsorption is usually associated with only a small activation energy, it occurs rapidly, so that a perfectly clean surface will at once become covered if it is exposed to the atmosphere.

activator n 活性化物質

▶ The emission spectra consist of one or more broad bands whose position frequently depends markedly on the nature of the activator.

active adj **1** 活発な syn *brisk*

▶ Research into crystal defects is an active and rapidly advancing area of solid-state science.

2 活性の

▶ A knowledge of the active surface area is of value and, for quantitative considerations, essential.

3 能動の syn *effective, operative, working*

▶ The development of molecular electronics depends upon the synthesis and development of active molecules and molecular systems, and this represents a great challenge for chemists.

active center n 活性中心

▶ The active centers of metalloenzymes often involve distortion of the normal stereochemistry of the metal ion with changes in stereochemistry occurring during the reaction cycle.

actively adv 活発に syn *vigorously*

▶ The study of the kinetics of enzyme reactions is one that has been actively pursued during the past few years, and the subject has now reached a very interesting phase.

active methylene n 活性メチレン

▶ The Knoevenagel reaction embraces a number of base-catalyzed condensations between a carbonyl compound and a component having an active methylene group.

active site n 活性サイト

▶ Although only one active site per supercage is modified, it is many orders of magnitude more active than the remaining unmodified sites.

active species n 活性種

▶ Laser materials often contain a transition metal ion as the active species, e.g., the ruby laser is essentially Al_2O_3 doped with a small amount of Cr^{3+}.

active state n 活性状態

▶ It remains to be determined which electronic transition leads to the photochemically active state for these reactions.

activity n **1 活動** syn *endeavor, interest, pursuit*

▶ The synthesis of coordination polymers has been a field of recent intense activity.

2 活性 syn *function*

▶ The resulting material shows good activity in the aerial oxidation of alkylaromatics.

3 活量

▶ For ions there may be serious deviations from ideality, and activities must be used.

activity coefficient n 活量係数

▶ The emf measurements over a range of concentrations lead to values for the activity coefficients.

actual adj 実際の 名詞の前でのみ用いる. syn *real, true*

▶ The actual value of the resistivity of silicon depends markedly on purity but is ca. 40 ohm cm at 25 ℃ for very pure material.

actuality n 現実 syn *reality*

▶ In actuality, the collision frequencies of chlorine and bromine atoms toward methane differ by only a few percent.

actually adv 実際に syn *in fact, really*

▶ A reversible change is an ideal change, which cannot actually occur, although an actual change may approach closer and closer to an ideal change as the pressure and temperature differences approach zero.

acute adj **1 重大な** syn *critical, crucial, intense, severe*

▶ Where the solubility differences are not so acute, we may make use of repeated phase equilibria manually as in fractional crystallization or repeated solvent extraction.

2 急性の

▶ Acute toxicity refers to exposure of short duration, i.e., a single brief exposure.

3 鋭角の syn *sharp* 鈍角のは obtuse.

▶ Small single crystals appeared as thin parallelograms with a measured acute profile angle of 57°.

acutely adv 激しく syn *profoundly, very*

▶ These V^{IV} complexes are stable to the atmosphere in the solid state but acutely sensitive to oxygen and moisture in solution.

acyclic adj 非環式の ☞ cyclic

▶ The addition of a variety of nitroalkanes to both cyclic and acyclic enones was examined.

acylating agent n アシル化剤

▶ When an acylating reagent such as carboxylic anhydride is used, still more catalyst is required since some is consumed in converting the acyl compound to the acyl cation.

acylation n アシル化

▶ Unlike alkylation, acylation is easily controlled to give monosubstitution because, once an acyl group is attached to a benzene ring, it is not possible to introduce a second acyl group into the same ring.

adapt vt 適応させる syn *conform, fit, modify, suit*

▶ Pauling introduced the concept of resonance to adapt the simple valence bond notation to situations in which electrons are delocalized.

adapted adj 適応した

▶ The naturally adapted moorland plants have evolved a high degree of tolerance to aluminum.

adaptability n 適応性 syn *flexibility*

▶ The specific adaptability of gluten to bread making is due to its elastic, cohesive nature that enables it to retain the bubbles of carbon dioxide evolved by leavening agents.

adaptation n **1** 改作 syn *alteration, change, modification*

▶ Most dicarboxylic acids are prepared by adaptation of methods used to prepare monocarboxylic acids.

2 適応 syn *adjusting, fitting, suiting*

▶ The hydrophobic nature of triglycerides means that special molecular adaptations have evolved for their digestion, absorption, and intertissue transport.

adaptive adj 適応性のある

▶ Mass gain will generally lead to an adaptive rise in basal metabolic rate.

adatom n 吸着原子

▶ The cinchona molecule is adsorbed through the quinoline N atom adjacent to a Pt adatom.

add vi **1** 付加する syn *combine, join, unite*

▶ All evidence on the addition of free radicals to styrene indicates that process by which $X^{\cdot}$ adds to the CH_2 end of the double bond is greatly favored over addition at the CH end.

vt **2** 加える syn *supplement*

▷ On adding an arsenite to a solution of a hypochlorite, the former is oxidized to arsenic acid, while the latter is reduced to chloride.

3 add in 含める syn *include*

▷ The structures of many solid materials are often considered as being held together by ionic forces with some covalent character added in.

4 add on 付け足す さらに to を伴う

▶ The reaction has produced another radical, which in turn can add on to another molecule of monomer.

added adj 追加の syn *extra, extraneous, plus*

▶ Superoxide formation is an added difficulty since this reacts explosively with K metal.

addend n 付加剤

▶ Many of the most important synthetic reactions of carbonyl compounds involve enolate anions, either as addends to suitably activated double bonds or as participants in nucleophilic substitutions.

addition n **1** 付加

▶ In general, we predict that the direction of addition of an unsymmetrical monomer will be such as to give always the most stable growing-chain radical.

2 添加 syn *adding, putting together*

▶ The addition of the cations Ag^{I}, Ba^{II}, or Pb^{II} to aqueous dichromate solutions causes immediate precipitation as insoluble chromates rather than their more soluble dichromates.

3 in addition （adv）さらに syn *additionally, furthermore, moreover*

▶ When an X-ray beam is incident on powdered material, the usual diffraction pattern is formed by the scattered rays at relatively wide angles; in addition, there is intense scattering at low angles.

4 in addition to （prep）…に加えて syn *as well as, besides*

▶ If an external pressure is applied in addition to the saturated vapor pressure, the vapor pressure becomes a function of pressure as well as of temperature.

1, 4-addition n 1,4-付加

▶ Conjugated dienes generally react with dienophiles by 1,4-addition to form six-membered ring compounds.

additional adj 追加の syn *extra, extraneous, further*

▶ The Hofmann rearrangement provides a good way to prepare amines from acids of one additional carbon atom, since the latter may be readily converted into amides.

additionally adv さらに syn *moreover*

▶ Additionally, a variety of catalyst systems have been developed for the asymmetric conjugate addition of nitroalkanes.

addition compound n 付加化合物

▶ One of the simplest types of chemical reactions which is fundamental to a large number of more complex reactions is the formation of addition compounds by Lewis acids and bases.

addition elimination n 付加脱離 複合形容詞として使用

▶ Esters are reduced by lithium aluminum

hydride to primary alcohols via addition-elimination mechanism.

addition polymerization n 付加重合

▶ The most important type of addition polymerization is that of the simple vinyl monomers such as ethylene, propylene, styrene, etc.

addition reaction n 付加反応

▶ To discuss addition reactions in a comprehensive manner, one needs to compare the behavior of compounds with different functional groups.

additive n **1** 添加剤

▶ Studies have shown that the active additives to the metallic glass are highly concentrated in the surface film produced in a corrosion experiment.

adj **2** 加成性の

▶ Most sets of radii are additive and self-consistent, provided one does not mix radii from different tabulations.

additivity n 加成性

▶ The Wiedemann additivity law was applied to determine the diamagnetic susceptibility of a solute from that of the solution.

address vt **1** 提出する syn *deliver*

▶ The effect of the metal ion on water molecules was also addressed.

2 処理する syn *approach*

▶ There remain other, as yet largely unexplored, issues regarding DNA structure and replication that might well be addressed by such molecular probes.

addressable adj 処理可能な

▶ Addressable carboxylate groups on wild-type mosaic virus particles were utilized as anchor groups for an organic, redox-active viologen derivative.

adduct n 付加物

▶ Urea adducts are inclusion complexes of the channel or canal type.

adenosine diphosphate n アデノシン二リン酸

▶ Enough free energy is supplied to drive the combination of inorganic PO_4^{3-} with adenosine diphosphate to form adenosine triphosphate, a conversion called oxidative phospholylation.

adenosine triphosphate n アデノシン三リン酸

▶ Whenever an energy-requiring process is reversed, it is likely that the adenosine triphosphate concentration will decline.

adequate adj **1** 適切な syn *proper, suitable*

▶ The fugacity of a gas or a gas in a mixture can be evaluated if adequate *P–V–T* data are available.

2 十分な syn *enough, satisfactory, sufficient*

▶ Photons of wavelengths in the far-ultraviolet region correspond to energies greater than 143 kcal/einstein, which is more than adequate to break almost any bond in a typical organic molecule.

adequately adv 適切に syn *satisfactorily, well*

▶ Some molecule or ions cannot be represented adequately by a single valence bond formula.

adhere vi 付着する syn *attach, stick*

▶ The ionic bond is formed when one or more electrons are transferred from one atom to another, and the resulting ions adhere by direct Coulombic attraction.

adherence n 付着 syn *adhesion*

▶ Most electrodes can be pre-etched in situ in the ionic liquid such that subsequent deposition leads to better adherence to the substrate.

adherent adj 付着した

▶ One theory of the cause of passivity is the formation of a closely adherent oxide film.

adhesion n 接着 syn *adherence*

▶ The adhesion of aluminum to mild steel is greatly enhanced by etching the electrode in situ in the ionic liquid prior to deposition.

adhesive n **1** 接着剤 syn *glue*

▶ Hot-melt types of adhesives offer the possibility of almost instantaneous bonding, making them well-suited to automated operation.

adj **2** 付着力のある syn *sticky, tacky*

▶ The silanol present in typical silicone adhesives provides ample sites for crosslinking in the adhesive structure.

ad hoc adj その場限りの
▶ The resulting description of the bonding is an ad hoc mixture of four oversimplified limiting models and should more logically be replaced by a generalized MO approach.

adiabatic adj 断熱的な
▶ The correlations of ionization potential with the energy of the charge-transfer band have involved mainly adiabatic values.

adiabatic expansion n 断熱膨張
▶ By definition, an adiabatic expansion is one that is not accompanied by a transfer of heat.

adipocyte n 脂肪細胞
▶ Adipocytes, being storage cells for fat, respond to variations in supply from bloodstream.

adipose tissue n 脂肪組織
▶ A great deal of energy is expended in maintaining ion gradient in resting muscle whereas adipose tissue is relatively inactive, being mainly lipid droplets and possessing little cytoplasm.

adjacent adj 隣接する syn *close, near, nearby, neighboring*
▶ The decrease in distance between adjacent basal planes makes it much more difficult for basal slip to occur.

adjunct n 補助物 syn *attachment, supplement*
▶ The process known as electroosmosis has some practical application as an adjunct to ordinary osmosis.

adjust vt 調節する syn *alter, modify, regulate*
▶ If the material is readily soluble in one member of a solvent pair and sparingly soluble in the other, a solution of the sample in the first solvent can be adjusted to condition suitable for crystallization by dilution with the second solvent.

adjustable adj 調節できる syn *flexible*
▶ MO treatments of metal complexes incorporating metal–ligand overlap are more notable for the rationalization of known quantities, often with the aid of a more or less extensive set of adjustable parameters.

adjustment n 調節 syn *regulation, tuning*
▶ Fine temperature control is made by adjustment of the distance of the flame from the flask.

adjuvant n アジュバント
▶ The use of adjuvants, covalent attachment of carbohydrates to immune-stimulants or integration into complex multicomponent vaccines has helped in the progress of carbohydrate vaccine development leading to a renewed interest in the field.

adlayer n 吸着層
▶ Normally, an adlayer will agglomerate to form three-dimensional islands, but if an interaction exists it may spread.

administer vt 薬を投与する syn *give, provide*
▶ Many substances that are intensely poisonous are actually beneficial when they are administered in micro amounts.

administration n 投与 syn *funishing, supplying*
▶ Short-term imaging does not seem to damage the tissue locally at the site of administration.

admirable adj 見事な syn *excellent, marvelous, splended*
▶ Alchemy is an admirable illustration of the proper meaning of the term natural philosophy.

admission n 導入 syn *access, entry*
▶ The weight of the aqueous sample was determined by weighing before and after its admission to the flask.

admit vi **1** 導く syn *introduce*
▶ Many reactions do not admit to the assignment of an order.
vt **2** 認める，許す syn *allow, permit, tolerate*
▶ First we establish the meaning of a standard state of a solute that obeys Henry's law, and then we admit deviations from ideal dilute behavior.
3 入れる syn *accept, allow to enter, receive*
▶ When hydrogen is gradually admitted to nickel at room temperature, the adsorption is essentially isothermal and reaches a steady state.

admittance n アドミタンス
▶ Conductivity experiments are most often

carried out by using the complex impedance or complex admittance techniques, which permit bulk conduction to be distinguished from grain boundary or surface contributions.

admittedly adv 明らかに

▶ This example of a structure proof is admittedly an idealized one.

admix vt 混ぜ合わせる syn *mix*

▷ Chlorobenzene and other aryl chlorides are usually unreactive unless added to the magnesium admixed with a more reactive halide.

adopt vt 取る syn *accept, acquire, assume, take*

▶ The antifluorite structure is adopted by Li_2O, for which the ionic radii are 0.73 and 1.28 Å for Li^+ and O^{2-}, respectively.

adsorb vt 吸着する ☞ absorb, sorb

▷ The most frequently encountered adsorption experiment is the measurement of the relation between the amount of gas adsorbed by a given amount of adsorbent and the pressure of the gas.

adsorbable adj 吸着される

▶ Isothermal adsorption is a good assumption only when the adsorbable component concentration is low and/or the heat of adsorption is low.

adsorbability n 吸着能

▶ Separation by chromatography is not feasible, because the two isomers have the same degree of adsorbability on alumina.

adsorbate n 吸着質 ☞ sorbate

▶ At any surface coverage, it is subject to the influence of adsorbate-adsorbate interactions.

adsorbed water n 吸着水

▶ The highly lipophilic amino acids travel almost as fast as the organic solvent, whereas the very hydrophilic ones are largely retained by the adsorbed water and make little progress.

adsorbent n 吸着剤 ☞ sorbent

▶ No general principles can be given for the nature of the adsorption bond of any gas on any adsorbent.

adsorption n 吸着 ☞ absorption, sorption

▶ Adsorption is always more or less selective in nature, and precipitates tend to adsorb some substances in solution to a much greater extent than they do others.

adsorption complex n 吸着錯体

▶ Between certain detergent-to-protein ratios, adsorption complexes separate as coacervates.

adsorption indicator n 吸着指示薬

▶ A precipitate of silver chloride will turn red in a solution containing even a minute excess of silver ion, if fluorescein is present. In this example, fluorescein is the adsorption indicator.

adsorption isotherm n 吸着等温式 ☞ isotherm

▶ The relation between the amount adsorbed and the pressure in the gas, the adsorption isotherm, is a complicated one and depends on the nature of the solid surface, on its homogeneity, and of course on the forces between gas molecules and solid.

adsorption site n 吸着サイト

▶ The surface is considered to consist of a number of adsorption sites, and the rate of desorption is assumed to be proportional to the fraction of the surface covered.

adsorptive capacity n 吸着容量

▶ The activated carbon in equilibrium with the solution will have reached its adsorptive capacity and cannot adsorb any more pollutant under the current conditions.

advance n **1** 進歩 syn *development, progress*

▶ It is well recognized that advances in technology owe much to basic research, and there are many examples that illustrate this.

2 in advance 前もって, あらかじめ syn *ahead of time, beforehand*

▶ The only condition one can specify in advance is that molecules must conform to those most fundamental laws of energy, the laws of thermodynamics, which provide the boundary conditions of all physical change.

adj **3** 前もっての

▶ Advance justification for the research may be equally difficult to provide.

vt **4** 進める syn *extend, improve*

▶ Present theories of chemical bonding are not sufficiently advanced to treat definitively molecules as large as heptacoordinate compounds.

5 提出する syn *propose, suggest*

▶ Several reasons have been advanced for this.

advanced adj **進歩した** syn *refined, sophisticated*

▶ The discovery of high-temperature oxide superconductors directed scientific and public interest to advanced ceramic materials.

advancing adj **前進する** syn *progressive*

▶ The advancing liquid zone accumulates the contaminants as it passes, for they are preferentially partitioned into the liquid phase.

advantage n **1 利点** syn *benefit, improvement, usefulness*

▶ Despite these advantages, so far the commercial application of microemulsions has been rather limited.

2 take advantage of …を利用する syn *use*

▷ The reaction was followed by taking advantage of the fact that pyridine hydrochloride is only slightly soluble in benzene and precipitates out as the reaction proceeds.

advantageous adj **便利な** syn *convenient, favorable, profitable, useful*

▶ For drying precipitates an electric oven with automatic temperature control is most advantageous.

advantageously adv **好都合に** syn *favorably*

▶ A wide variety of theoretical and empirical methods have been used for evaluating atomic charges, and most of these have served advantageously in the correlation of atomic-charge-related properties.

advent n **出現** syn *appearance*

▶ The advent of X-ray and UV photoelectron spectroscopy in the early 1970s presented the chemists interested in the surfaces of solids with remarkable insights.

adverse adj **不利な** syn *negative, unfavorable*

▶ Pressure has an adverse effect on rates if there is a volume increase when the activated complex is formed.

adversely adv **不利に** syn *badly, unfavorably*

▶ Substitution for Ba or Cu adversely affects superconductivity.

advisable adj **当を得て** syn *expedient, practical, proper, suitable*

▶ To assist hydrolysis, boiling is advisable, because water is much more dissociated at this temperature than when cold.

aerial adj **空気の**

▶ The aerial oxidation of olefins to aldehydes or ketones using palladium(II) as a catalyst is the Hoechst-Wacker process.

aerobe n **好気生物** ☞ anaerobe

▶ Microbes that use oxygen for ATP production are called aerobes because they undergo oxygen-dependent metabolism.

aerobic adj **好気性の** ☞ anaerobic

▶ Most aerobic microorganisms have compounds called siderophores which solubilize and transport iron(III).

aerobic bacterium n **好気性菌** ☞ anaerobic bacterium

▶ Aerobic bacteria in the water can further modify polychlorinated biphenyls, and others can then convert them into water, carbon dioxide, and chloride.

aerobic metabolism n **好気代謝** ☞ anaerobic metabolism

▶ Microbes can convert many chemicals into harmless compounds either through aerobic metabolism or through anaerobic metabolism.

aerobic respiration adj **好気呼吸** ☞ anaerobic respiration

▶ Aerobic respiration occurs when the terminal electron acceptor is molecular oxygen.

aerogel n **エーロゲル**

▶ Xerogels are denser than aerogels, have high surface areas, and are often microporous.

aerosol n **エーロゾル**

▶ Aerosols are always unstable, tend to aggregate, and then precipitate.

affair n **状況** syn *issue, matter, topic*

▶ The electric field that is conveniently applied to a solution is not an overwhelming factor in the affairs of ions.

affect vt 影響を及ぼす syn *change, influence, modify*

▶ Side chains affect the properties of proteins not only by their acidity or basicity but also by their other chemical properties and even by their sizes and shapes.

affinity n 親和力

▶ Quinone has a large affinity for electrons and so is a powerful oxidizing agent.

n 親和性 syn *alliance*

▶ Often an inhibitor of an enzyme can be used as the selective affinity ligand.

affinity chromatography n アフィニティークロマトグラフィー

▶ In affinity chromatography the gel matrix has been chemically modified with a ligand that specifically interacts with the protein of interest.

affirm vt 肯定する syn *confirm, establish, prove, settle*

▶ From an electrostatic viewpoint it is necessary to affirm that atomic and ionic sizes are not constants but are simply a result of balanced forces of attraction and repulsion.

affix vt 固定する syn *fix, pin*

part ▶ Arora et al., by affixing a methyl group to the central ring of one of their phenylene bis (benzoates), have prepared a material with a nematic range of 72-156 ℃.

affluent adj 豊富な syn *rich*

▶ On average, ca. 100 g of triglyceride is consumed by an adult on an affluent Western diet; this constitutes 30-50% of the daily energy intake.

afford vt もたらす syn *furnish, give, offer, provide*

▶ Certain solutions of iodine afford interesting examples of acid-base interaction in the electron donor-acceptor sense.

after prep **1** 時に関して…の後に

▶ After about 10-15 minutes, the level of liquid in the pycnometer is adjusted to the mark by touching the short arm with a filter paper.

2 順序に関して…の次に

▶ Silicon is the most abundant element in the earth's crust after oxygen.

3 …にちなんで

▶ Olefins are named after the corresponding paraffins by adding ene or ylene to the stem.

4 after all つまり syn *eventually*

▶ After all, one of the chief uses of soaps, detergents, and emulsifiers is to exert protective action on droplets and suspensions.

afterglow n 残光

▶ The afterglows are normally very short, of the order of 10^{-4} to 10^{-5} sec, indicating that the phosphorescence is due to a forbidden transition.

afterward adv その後 syn *subsequently*

▶ For less accurate analyses, the gases can be collected over water which has been previously saturated with the gas to be analyzed, and the analysis must be made immediately afterward.

again adv **1** 再び syn *further*

▶ For nucleophilic carbon, we turn again to the carbanion-like organic group of an organometallic compound.

2 again and again 幾度も syn *repeatedly*

▶ As a cell enters mitosis, each extended chromosome begins to shorten and thicken by supercoiling on itself again and again.

against prep **1** …に反対して syn *in opposition to*

▶ The second piece of evidence against the four-center mechanisms is that addition reactions carried out in the presence of nucleophilic reagents often give mixtures of products.

2 …に逆らって

▶ If a proton is placed in an external magnetic field, its magnetic moment can be aligned in either of two ways: with or against the external field.

3 …に競って

▶ Against the mild action of solvents other forces can compete, and this competition can be observed and measured.

4 …に対して syn *versus*

▶ Acids and bases may be titrated against each other by the use of substances, usually colored, known as indicators.

5 …に備えて syn *resistant*

▶ From the applied standpoint, scavengers have been widely considered as possible protective agents against damage due to radiation.

agate mortar n めのう乳鉢 ☞ mortar, pestle

▶ All platinum compounds, when heated with soda on charcoal, are reduced to the gray, spongy metal, which assumes a metallic luster on being rubbed with pestle in an agate mortar.

age n **1** 年代 syn *period, time*

▶ Radiocarbon dating is a method of determining quite accurately the age of carbon-bearing material derived from living plants or animals within the last 70,000 years.

vt **2** 熟成させる syn *mature*

▶ In order to obtain crystals with more uniform behavior, it was found helpful to age a batch of crystals at room temperature for 12 h to allow the more reactive ones to rearrange and to study only those which survive this test unchanged.

agency n 作用 syn *action, activity, operation*

▶ Practically all biochemical reactions are carried on through the agency of enzymes.

agent n 試薬 syn *reagent*

▶ Frequently antifoaming agents lose their efficiency at higher temperatures.

agglomerate n **1** 集積による塊 syn *stack*

▶ The main panel of Figure 1 demonstrates on a larger scale that these agglomerates form very long branched chains.

vi **2** 塊になる syn *accumulate, aggregate, pile up*

▶ The small primary drug particles often agglomerate and coagulate, thus reducing the active surface area.

agglomeration n 集合 syn *ensemble, mass, pile*

▶ The food industry uses agglomeration in the sense of increasing the particle size of powdered food products.

aggregate n **1** 集合体 syn *cluster, ensemble*

▶ Considerable current research is directed toward understanding and utilizing molecular recognition in chemical aggregates.

vi **2** 凝集する syn *accumulate, stack*

▶ Surfactants aggregate into a wide variety of structural forms in aqueous solution, and interactions between aggregates can result in very high viscosity, viscoelasticity, or even gel formation.

aggregation n 凝集 syn *accumulation, pile, stack*

▶ The aggregation of chain-like molecules is a fascinating phenomenon; it is of importance in such areas as liquid crystals and micelle formation in soaps and detergents and in biological membrane phenomena.

aggressive adj 活動的な syn *active, vigorous*

▶ Aqua regia is more aggressive than concentrated nitric acid, due to the formation of free Cl_2 and ClNO and the superior complexing ability of the chloride ion.

aging n **1** 熟成

▶ Attenuated vaccines involve using live bacteria or viruses that have been weakened through aging or by altering their growth conditions to prevent their replication after they are introduced into the recipient of the vaccine.

2 老化

▶ Most human body cells can divide a maximum of 50–90 times before they show sign of aging, a process called senescence, which eventually leads to cell death.

agitate vt かきまぜる syn *shake, stir*

part ▶ Flotation is a method of separating minerals from waste rock or solids of different kinds from one another by agitating the pulverized mixture of solids with water, oil, and special chemicals.

agitation n **1** かきまぜ syn *disturbance, stirring*

▶ The rates of nucleation and transformation are usually dependent as much on temperature, particle size, purity, and physical agitation as on the nature of the structural change involved.

2 運動 syn *motion, movement*

▶ At temperatures below 100 ℃, it is very rare indeed that thermal agitation alone can supply sufficient energy to break any significant number of bonds stronger than 30 to 35 kcal/mole.

ago adv 現在より過去にさかのぼって前に syn *back*

▶ The two diagnostics of superconductivity, resistanceless current flow and perfect diamagnetism, were appreciated even many years ago as having tremendous technological implications.

agonist n アゴニスト ☞ antagonist

▶ Signals that stimulate are called agonists, whereas those that inhibit are known as antagonists.

agree vi 意見が一致する syn *accord, coincide*

▶ Different workers do not agree on the magnitude of the viscosity effect.

agreement n 合意 syn *approval, harmony*

▶ This peak results from an absorption of photons by the electron as it is excited to a higher energy level, but not all workers are in agreement as to the nature of the excited state.

ahead adv **1** 前へ syn *forward*

▶ In principle, one could go ahead and calculate each molecule listed above by an appropriate computational method.

2 ahead of (prep) …より前に syn *in advance*

▶ Supply consistently ran ahead of demand.

aid n **1** 助け syn *assistance, help, support*

▶ Our present theoretical knowledge is of great aid in selecting the substances which may reasonably be expected to conform a particular equation.

2 with the aid of …の助けを借りて syn *by means of, through*

▶ Determination of the molecular weights and dimensions of proteins has been made with the aid of an impressive array of physical techniques.

vi **3** 促進する syn *assist, help, promote, support*

▶ Photochemical activation sometimes can aid in breaking the M–CO bond, but many reactions are carried out thermally.

aim n **1** 目的 syn *goal, purpose*

▶ Most work on peptide synthesis has had as its aim the preparation of compounds identical with naturally occurring ones.

vt **2 aim at** 明確な目標とする syn *focus on, plan for, seek*

▷ This structural study should facilitate future designs aimed at exploiting hydrogen-bonding environments in dioxygen activation chemistry or tuning the redox behavior of metal centers.

air n **1** 空気

▶ Nitrogen prepared entirely from air has a density 0.5% greater than nitrogen prepared from ammonia or in any other chemical way.

2 the air 大気 syn *atmosphere*

▶ The carbonate, nitrate, and oxalate of zinc are readily and quantitatively changed to zinc oxide by ignition in the air.

airborne adj 風で運ばれる

▶ Prolonged exposure to airborne suspensions of asbestos fiber dust can be very dangerous.

airborne infection n 空気伝染

▶ Airborne infection is a source of cross infections in hospitals and may be a factor in epidemics.

air oxidation n 空気酸化

▶ The white CuCl is left covered with the reducing solution for protection against air oxidation until it is to be used and then dissolved in hydrochloric acid.

air pollution n 大気汚染

▶ Photosynthesis has been shown to be substantially inhibited by air pollution to the extent of 20% in rural locations and 33% in urban areas.

akin adj 類似して syn *alike, related, similar*

▶ Mössbauer or γ-ray spectroscopy is akin to NMR spectroscopy in that it is concerned with transitions that take place inside atomic nuclei.

alarming adj 大変な syn *seious*

▶ Population growth has been accelerating at an alarming pace.

alarmingly adv 大いに syn *dangerously*

▶ Classification and analysis of innovations

indicate that chemical product and process advances have fallen off alarmingly over the past ten years.

albeit conj とはいえ syn *although, though* 用法が異なることに注意せよ.

▶ Any compound which has a neutral oxygen atom with an unshared pair of electrons is also a base, albeit a very weak one in most cases.

alcoholic adj アルコール性の ☞ methanolic, ethanolic

▶ 1,1-Dichloroethane treated with alcoholic potassium hydroxide affords acetylene by the elimination of two moles of HCl.

alcoholysis n アルコーリシス

▶ Both hydrolysis and alcoholysis may be considered as forms of solvolysis.

aldol addition n アルドール付加 syn *aldol condensation*

▶ The ingredients in the key step in aldol addition are fundamentally an electron-pair donor and an electron-pair acceptor.

aldol condensation n アルドール縮合 syn *aldol addition*

▶ The aldol condensation results in the formation of a new carbon–carbon bond and is , therefore, very useful in the synthesis of carbon chains and rings.

algebraic adj 代数学的な

▶ The measurable magnetic susceptibility of any substance can be represented as the algebraic sum of its diamagnetic and paramagnetic contributions.

algebraically adv 代数学的に

▶ A maximum value of a function for a value $x = x_1$ is a value algebraically greater than all values in the immediate neighborhood of x_1.

algorithm n アルゴリズム

▶ The primitive algorithm does indeed generate all structural isomers of C_nH_{2n+2}.

alicyclic adj 脂環式の

▶ The rings of crown ethers can exist in much the same conformations as the alicyclic rings.

align vt 配列させる syn *arrange, array, line, order*

▶ Ammonium fluoride has the wurtzite lattice, in which the fluoride ions are aligned through hydrogen bonds with the tetrahedrally dispersed hydrogen atoms on the ammonium ions.

alignment n 一列になること syn *arrangement, orientation*

▶ Formation of a smectic phase requires not only parallel alignment of rod-shaped molecules but also weak end-to-end intermolecular attraction so that the layers of the laminar structure are free to slide by one another.

alike adj よく似ていて 名詞の前では, similar, identical を用いる.

▶ There are many pairs of molecular species so nearly alike in their attractive forces that their liquids mix with little or no heat effect.

aliphatic hydrocarbon n 脂肪族炭化水素

▶ Aliphatic hydrocarbons undergo chiefly addition and free-radical substitution.

aliquot n アリコート, 試薬の一部分

▶ It is best to make up the solution with precipitate to a definite volume and use an aliquot part of the filtered solution.

alkali n **1** アルカリ

▶ Any substance which in water solution is bitter and is irritating or caustic to the skin and mucous membranes, turns litmus blue, and has a pH value greater than 7.0 is called an alkali.

2 アルカリ金属 syn *alkali metal*

▶ The salts of the alkalis are more or less volatile and impart characteristic colors to the nonluminous flame.

alkaline adj アルカリ性の

▶ Salts of weak acids and strong bases have an alkaline reaction.

alkalinity n アルカリ度 ☞ basicity

▶ pH is a value taken to represent the acidity or alkalinity of an aqueous solution.

alkylating agent n アルキル化剤

▶ The Friedel–Crafts reaction cannot be used for the preparation of long-chain aromatic derivatives since isomerization may occur when the alkylating agent contains more than two carbon atoms.

alkylation n アルキル化

▶ The alkylation reaction is by no means

hopeless as a practical method for the preparation of amines because usually the starting materials are readily available and the boiling-point differences between mono-, di-, and trialkylamines are sufficiently large to make for easy separations by fractional distillation.

all pron **1 全部** syn *everything*

▶ Evidence of other kinds may already have limited the number of possible structures, and all that is required of the NMR spectrum is that it let us choose among these.

2 all at once すべて一時に syn *suddenly*

▶ Rapidly mixed reactions can be achieved simply by pouring the initiator solution into the aniline solution all at once and rapidly mixing them.

3 all of 次に定冠詞または限定詞を必要とする．続く名詞は複数形または不加算名詞で，もし動詞が続くときは，名詞に対応させて単数形または複数形を用いる．

▶ A variety of substituted benzenes are known with one or more of the hydrogen atoms of the ring replaced with other atoms or groups, and, in almost all of these compounds, the special stability associated with the benzene nucleus is retained.

4 at all (adv) (a) 否定文に用いて，**少しも** syn *never, scarcely*

▶ There is no very practical route at all for conversion of phenol to chlorobenzene.

(b) 条件文に用いて，**少しでも** syn *conceivably, possibly*

▶ Because of considerable strain present in small ring, the synthesis of cyclopropanes and cyclobutanes can never be carried out using reactions which are at all reversible.

adj **5** 個々の構成要素を念頭に置いて，**すべて，全部の**

加算名詞の複数形の前に置いたときは，動詞は複数形を使用する．三つ以上のものを対象とする．対象が二つしかないときは，both を用いる．the などの限定詞があるときは，その前に置かれる．

▶ All forms of carbon, apart from diamond, have been found to be graphitic in character.

不可算名詞の前に用いたときは，動詞は単数形を使用する．

▶ In the oxidation of aqueous $H_2{}^{18}O_2$ by chlorine and MnO_4^- and in the catalytic decomposition of hydrogen peroxide by Fe^{3+} and MnO_2, all the molecular oxygen that is released comes from the H_2O_2, none from the water.

adv **6 すっかり** syn *entirely*

▶ The available experimental magnetic susceptibilities of the carboxylic acids are all listed in Table 1.

7 all over いたるところ syn *everywhere, throughout*

▶ An important question is whether resistive heating of graphite gives the same products in the same ratio all over the world.

alliance n **組み合わせ** syn *association, combination*

▶ The alliance of group theory with chemistry has proved even more rewarding than that of combinatorial theory.

allied adj **同類の** syn *associated, coupled, related*

▶ Closely allied to the problem of autoxidation is that of the selective hydrogenation of systems of the same linolenate type.

allocate vt **割り当てる** syn *assign, apportion, distribute*

▶ In molecular orbital theory, the electrons are allocated to orbitals which in general extend over the whole molecule.

allocation n **割り当て** syn *assignment, distribution*

▶ The determination of the dimensions of the unit cell involves measurement of the Bragg angles, allocation of the correct indices, and evaluation of the interplanar spacings.

allosterically adv **アロステリックに**

▶ Elevation of glucose 6-phosphate is sufficient to allosterically activate glycogen synthase, even in the absence of any insulin-induced dephosphorylation of the enzyme.

allosteric effect n **アロステリック（分子変容）効果**

▶ Allosteric effects play a critical role in the control and integration of molecular events in biological systems.

allosteric site n アロステリック部位

▶ The control is not likely to be due to binding at the active site but at an alternative site, called an allosteric site.

allotrope n 同素体

▶ The prototypical example of a covalently bonded solid is the diamond allotrope of carbon.

allotropic adj 同素体の

▶ Ozone is an allotropic form of oxygen produced by the absorption of ultraviolet light or electric energy.

allotropic phase transformation n 同素相変態

▶ Iron undergoes several allotropic phase transformation, from a body-centered cubic phase, called α-iron or ferrite, at room temperature, to a face-centered cubic phase, called γ-iron or austenite, at temperatures between about 900 and 1400 ℃.

allotropy n 同素

▶ The term allotropy has been used to cover all cases where an element exists in more than one form, the word form including differences in crystal structure and also differences such as that between the molecule of oxygen and that of ozone.

allow vi **1 allow for** …を考慮に入れる syn *account, consider*

▶ It is not possible to allow experimentally for any interaction between solute and solvent molecules, and the presence of this solvent effect may sometimes be serious.

vt **2** 可能にする syn *enable, permit*

▷ By allowing the temperature to rise in a potentially reactant system, both pyrolysis and explosive processes can be studied.

3 見込む syn *make allowance, set aside*

▶ A liquid is contained in a glass tube. The tube may allow space for expansion when the liquid becomes warm.

4 be allowed to …に構うことなく置かれる syn *let, permit*

▶ The furnace was turned off and allowed to cool to room temperature.

allowable adj 許容される syn *permissible, tolerable*

▶ The allowable range of solvents is large, from hydrocarbons to concentrated sulfuric acid, and for most compounds it is possible to find a suitable solvent.

allowance n **1** しんしゃく syn *consideration*

▶ Orange II separates from aqueous alcohol with two molecules of water of crystallization, and allowance for this should be made in calculation of the yield.

2 make allowance for …を酌量する syn *consider, rationalize*

▶ It is possible to make allowance for these other effects and thus to deduce from dipole moments the ionic character of a bond.

allowed adj 許容された syn *permissible*

▶ For most molecules, the spacing of these allowed rotational-energy levels is quite small compared with a room temperature value of kT.

allowed transition n 許容遷移 ☞ forbidden transition

▶ The d^2 configuration in a cubic field is expected to give rise to three spin-allowed transitions.

alloy n **1** 合金 syn *blend, composite, mixture*

▶ Alloys involve a distribution of atoms of two or more different metals over one set of crystallographically equivalent sites.

vt **2** 合金にする

▷ Either alone or alloyed with palladium, gold often appears to possess unique activities.

alloying n 合金にすること

▶ The alloying of isoelectronic species introduces disorder which scatters phonons without much disruption of electron transport.

allylation n アリル化

▶ C-allylation of sodium phenoxide as observed in nonpolar solvents is not the result of O-allylation followed by rearrangement.

almost adv **1** ほとんど syn *about, approximately, nearly*

▶ The length of a given bond is found to be almost identical in different compounds, provided that the bond is not incorporated into a conjugated system.

2 almost all ほとんどすべて syn *majority*

▶ Phosphorus forms a great variety of binary compounds with almost all elements, often by direct action; the exceptions are antimony, bismuth, and the noble gases.

alone adj **1** ただ…だけ 名詞の後に置く syn *only*

▶ Silver forms all four monohalides. The fuoride alone is soluble in water.

adv **2 単独に** syn *exclusively, merely, only, solely*

▶ Boron stands alone among the group 13 elements in forming a large number of hydrides and hydride derivatives.

along prep **1 細長いものに沿って**

▶ The so-called square complexes contain not only four firmly bonded ligands in the *xy* plane but also two somewhat more weakly held groups along the *z* axis.

2 along with …と一緒に 形容詞または副詞の働きをする前置詞句をつくる. syn *together with*

▶ 0.1 M aqueous $FeCl_3$ contains only 10% Fe^{3+}, along with 42% $FeCl^{2+}$, 40% $FeCl_2^{+}$, 6% $FeOH^{2+}$, and 2% $Fe(OH)_2^{+}$.

alongside adv **alongside of** (prep) **…の横に並んで** syn *adjacent to, beside*

▶ The chains with trans double bonds are seen to be able to lie alongside of each other, ordered to form a semicrystalline array.

alpha n アルファ

▶ Halogenation of saturated aldehydes and ketones usually occurs exclusively by replacement of hydrogens alpha to the carbonyl group.

alphabetical adj アルファベット順の

▶ The order in which the standard heats of formation are given is alphabetical according to the symbol of the principal element.

already adv すでに syn *earlier, formerly, previously*

▶ HCl is an acid by virtue of the fact that it can donate a hydrogen ion, not because it might have already done so in an aqueous solution.

also adv 事実を並べてまた 一般に also は動詞の前に置く. ただし, be 動詞の場合はその後に置く. syn *as well, besides, too*

▶ The intermolecular factors that are important in determining how a gas deviates from ideal behavior are also the quantities that govern the critical-state constants.

alter vt 部分的に変更する syn *change, modify, revise, transform, vary*

▶ Small quantities of surface contamination or a fraction of a monolayer of a different metal can drastically alter a catalyst's reactivity.

alteration n 変化 syn *change, modification, transformation*

▶ Alterations in the structure of the polymer, such as crosslinking produced by irradiation of the material with high-energy radiation, produce molecular changes that lead to marked changes in the physical properties.

alternant hydrocarbon n 交互炭化水素

☞ nonalternant hydrocarbon

▶ Most of the electron-spin resonance studies of aromatic systems have been concerned with radical anions and cations of alternant hydrocarbons.

alternate adj **1** 交互に現れる syn *every other, successive*

▶ In layered CdX_2-type structures, the metal ions occupy layers of octahedral interstices between alternate pairs of close-packed halide layers.

vi **2** 交互する syn *be in rotation, be in succession,*

▶ There is a tendency for the enthalpies of formation of compounds of the group 14 elements to alternate from C–Si–Ge–Sn–Pb.

alternately adv 交互に syn *by turns, in rotation, successively*

▶ Iron and molybdenum form cubane sulfides, M_4S_4, with atoms alternately at the corners of a cube.

alternating adj 交互する syn *periodic*

▶ Borazines are six-membered hetrocycles with alternating boron and nitrogen atoms.

alternation n 交互 syn *exchange, rotation, succession*

▶ The melting points for various series of *cyclo*-$(NPX_2)_n$ frequently show an alternation, with values for *n* even being greater than those for adjacent *n* odd.

alternative n **1** 代わり syn *alternate, selection, substitute, variant*

▶ Molybdenum is a chemically suitable alternative but is much harder to fabricate.

adj **2** 代わりとなる syn *alternate, another*

▶ There are excellent alternative ways to make alkyl halides, conveniently and from readily available precursors.

alternatively adv **1** 代わりに syn *instead of, preferably, rather than*

▶ In applying the substitution technique to N_2O, one might prepare a sample of NNO^{18} or, alternatively, replace one or more of the N^{14} atoms with N^{15}.

2 交互に

▶ Although the structure of **2** is dominated by the zig-zag chain of four phosphorus atoms like in **1**, the position of the two naphthalene rings in phosphonium environments is very different, these lying alternatively below and above the mean plane defined by the four phosphorus atoms.

although conj であるが syn *despite, even though, though, while*

▶ Although the ionic metal oxides are usually basic, ZnO and Al_2O_3 are amphoteric.

altitude n 高度 syn *elevation, height*

▶ At some altitude, the temperature becomes so low that the vapor becomes thermodynamically unstable with respect to the liquid, and a cloud should form as the water vapor liquefies.

altogether adv **1** 完全に syn *completely, entirely, totally, utterly, wholly*

▶ The cooling must stop altogether until the transition is complete.

2 全体で syn *in all*

▶ Individually, each of these ion–dipole bonds is relatively weak, but altogether they provide a great deal of energy.

aluminosilicate n アルミノケイ酸塩

▶ Further complications arise in other aluminosilicates in which Al^{3+} ions may occupy octahedral sites as well as tetrahedral ones.

always adv **1 almost always** たいてい syn *usually*

▶ For solutions, and particularly for solutions of electrolytes, deviations from ideal behavior are almost always very large.

2 nearly always たいてい

▶ Alkyl halides are nearly always prepared from alcohols.

3 not always いつも…とは限らない

▶ In general, although not always, the molecule is more polarizable when the electron is in the antibonding orbital since it is less tightly bound when in the excited state.

amalgam n アマルガム syn *alloy, composite*

▶ When one of the constituents of an alloy is mercury, the alloy is referred to as an amalgam.

amalgamate vt アマルガムにする syn *blend, integrate, mix*

▶ Zinc can be amalgamated so that it will not react with acid and yet will reduce iron(III) ions to the iron(II) condition.

amalgam electrode n アマルガム電極

▶ The particular value of amalgam electrodes is that one can use active metals such as sodium in such electrodes.

amazing adj 驚くべき syn *extraordinary, remarkable, surprising*

▶ The basic units of life have an amazing complexity that until recently seemed to be such as to remain beyond man's understanding.

amazingly adv 驚くばかりに syn *extremely, particularly, surprisingly*

▶ Cyclodextrins and their inclusion compounds have found amazingly diverse uses.

ambidentate ligand n 両座配位子 ☞ bridging ligand

▶ The thiocyanate ion has been much studied as an ambidentate ligand, in which either S or N is the donor atom; it can also act as a bidentate bridging ligand.

ambient adj 外界の

▶ Complex **1** was treated with Ag［PF_6］in dichloromethane at ambient temperature and stirred for 20 min.

ambiguity n あいまいさ syn *indefiniteness, uncertainty*

▶ A problem with this approach is ambiguity about the origin to be assigned to a bond.

ambiguous adj あいまいな syn *unclear, uncertain*

▶ Microwave heating of solids primarily occurs via dipolar or conduction losses, but it should be noted that terminology can be ambiguous when identifying the physical phenomena responsible for microwave heating.

amenable adj されることが可能で syn *easy*

▶ In the case of nickel, the oxide ores are not generally amenable to concentration by normal physical separations and so the whole ore has to be treated.

amide linkage n アミド結合 ☞ peptide bond

▶ Each protein chain is synthesized on ribosomes and consists of an unbranched, linear sequence of amino acid monomers connected by covalent amide linkages called peptide bonds.

amino acid residue n アミノ酸残基

▶ A tetrahedrally coordinated zinc center is attached to the protein backbone by three amino acid residues, with the fourth site being occupied by a water molecule.

amino acid sequence n アミノ酸配列

▶ The amino acid sequence determines the manner in which a polypeptide folds upon itself to yield a biologically active protein.

ammoniacal adj アンモニア性の

▶ By means of ammoniacal CuCl, acetylene is absorbed and forms red copper acetylide(Cu_2C_2・H_2O).

ammonolysis n アンモノリシス

▶ Ammonolysis of halides is generally carried out either by allowing the reactants to stand together at room temperature or by heating them under pressure.

among prep **1** 集団をなす三つ以上のものを対象として，…の間の syn *in the middle of, surrounded by*

▶ The positrons were found among the particles produced by the interaction of cosmic rays with matter.

2 among other 数ある…の中で

▶ For solid complexes, in addition to the symmetry effects discussed above, molecular packing, lattice forces, ligand flexibilities, metal–ligand bond distances, d–d electronic transition energies, orbital overlap and orientation effects, and hydrogen bonding among other factors must also be considered.

amorphous adj 無定形の

▶ If a substance has neither the smooth external faces nor the regular internal arrangement of atoms, it is said to be amorphous.

amorphous alloy n 非晶質合金

▶ Amorphous alloys for use in transformer coils are made by quick-quenching the melt.

amount n **1** 量 syn *mass, quantity*

▶ Gram amounts of $XeOF_4$ have been isolated and studied.

vi **2 amount to** …に等しい syn *add up to, come to*

▶ The overall synthesis thus amounts to the conversion of alkenes into ketones and esters.

3 達する syn *become, develop into*

▷ The effect of molecular thermal motion leads to an apparent shortening of bond lengths amounting to several hundredths of an angstrom unit.

amperometric adj 電流測定の

▶ In an amperometric enzyme electrode, the current is a direct measure of the rate of enzyme-catalyzed reaction and, therefore, under suitable circumstances, is directly related to the concentration of analyte.

amphiphile n 両親媒性物質

▶ A self-assembling amphiphile with $n \geq 8$ appeared to be prerequisite for vesicle templating.

amphiphilic adj 両親媒性の syn *amphipathic*

▶ The surfactant is adsorbed because of the amphiphilic nature of its molecules, which usually contain a hydrophobic alkyl chain and a hydrophilic head group that by itself would be highly soluble in water.

amphoteric adj 両性の

▶ Alcohols, like water, are amphoteric and are usually neither strong bases nor strong acids.

amphoterism n 両性

▶ Aluminum in the form of an amalgam demonstrates amphoterism by dissolving in liquid ammonia solution containing an ammonium salt or an amide.

ample adj 十分な syn *abundant, enough, sufficient*

▶ In the case of solution in water, there is ample evidence that the ions exist coordinated by water molecules.

amplification n 増幅

▶ When the aim is to produce large amounts of transgene DNA or protein, the transgene vector is transformed into bacteria for amplification.

amplifier n 増幅器

▶ The p–n–p or n–p–n junction acts as a current or voltage amplifier.

amplify vt 拡大する syn *expand, extend*

▶ The endoplasmic reticulum compartmentalizes the cell and amplifies the surface area available for specialized biochemical reactions that occur at or across membranes.

amplitude n 振幅

▶ When a solid is heated, the molecules, atoms, or ions that constitute it vibrate with a greater amplitude and at some temperature, the melting point, are able to move from site to site in the lattice.

amply adv 十分に syn *fully, substantially, well*

▶ It has since become amply clear that some, but by no means all, of the infinite structures were uniquely stabilized by common impurity atoms such as hydrogen or carbon acting in an interstitial role.

ampule n アンプル

▶ Ampules were rotated end-over-end in a thermostat at 25.0° until the solution attained equilibrium.

amusing adj 面白い syn *exciting, interesting*

▶ $(F_5SO)_2$ is an amusing example of a compound accidentally prepared as a by-product of SF_6 and S_2F_{10} due to the fortuitous presence of traces of molecular oxygen in the gaseous fluorine used to fluorinate sulfur.

amylase n アミラーゼ

▶ Several bacterial enzymes are used to manufacture foods, such as carbohydrate-digesting enzymes called amylases that are used to degrade starches for making corn syrup.

anabolic reaction n 同化反応 ☞ catabolic reaction

▶ The energy released during catabolism is captured in a form that drives anabolic reactions.

anabolism n 同化作用 ☞ catabolism

▶ The processes that involve the synthesis of complex molecules from simpler ones are referred to as anabolism.

anaerobe n 嫌気生物 ☞ aerobe

▶ Microbes that use fermentation are called anaerobes.

anaerobic adj 嫌気性の ☞ aerobic

▶ Rice plants with roots in anaerobic soils transport dioxygen to the periphery of the roots where it oxidizes the iron(II) to iron(III).

anaerobic bacterium n 嫌気性細菌 ☞ aerobic bacterium

▶ Some anaerobic bacteria are involved in the first step of breaking down polychlorinated biphenyls by cleaving off chlorine and hydrogen groups.

anaerobic glycolysis n 嫌気解糖

▶ Human red blood cells contain no mitochondria so they drive their energy from glucose purely on the basis of anaerobic glycolysis.

anaerobicity n 嫌気性

▶ All of these solutions and solids are excellent reducing agents and react rapidly with air and moisture so that complete anaerobicity must be maintained during synthesis and handling.

anaerobic metabolism n 嫌気代謝 ☞ aerobic metabolism

▶ Anaerobic metabolism yields lactic acid which is subsequently metabolized by the liver.

anaerobic respiration n 嫌気呼吸 ☞ aerobic repiration

▶ Anaerobic respiration occurs when the terminal electron acceptor is an inorganic molecule other than oxygen.

analogous adj **analogous to …に類似の** syn *akin, like, parallel, similar*

▶ We see here a situation exactly analogous to one we have encountered several times before.

analogously adv **同じように，類似して**

▶ One of the important reasons that organosulfur compounds have properties which differ from analogously constituted oxygen compounds is that sulfur is substantially less electron-attracting than oxygen.

analogue n **類似物** syn *analog, counterpart, equivalent*

▶ Aminoborane, H_2NBH_2, and iminoborane, HNBH, are the very reactive BN analogues of ethylene and acetylene.

analogy n **1 類似** syn *resemblance, similarity*

▶ Similar analogies can be established for weak bases in water and weak donors in the semiconductor.

2 類推 syn *correspondence, parallelism*

▶ Carrying the analogy further, the cations correspond to acids, while the anions are bases.

analysis n **1 分析** syn *assay, separation* ☞ chemical−, gravimetric−, quantitative−, volumetric−

▶ Supercritical-fluid chromatography is a developing area of analysis in which supercritical ammonia, carbon dioxide, and hexane have all been used as the mobile phase in a hybrid of gas and liquid chromatography.

2 解析 syn *examination, investigation, study* ☞ crystal structure−, structure−

▶ The analysis of line widths in NMR spectra has provided valuable information about processes having half-lives of 10^{-5} s.

analyte n **被検体**

▶ Binding of analytes to receptors immobilized on the pore walls results in an increase in the average refractive index of the material and hence an increase in the optical thickness.

analytical adj **1 分析の**

▶ Chromatography is now a standard operation in the organic chemistry laboratory for both analytical and preparative separations.

2 解析の syn *theoretical*

▶ An analytical method gives the answer exactly as an equation or a set of equations.

analytically adv **1 分析的に**

▶ Analytically pure crystals were obtained after two recrystallizations.

2 解析的に

▶ For the Nernst equation to be useful analytically the potential responses to changes in activity must be fast; the equilibrium must be readily attained, and the charge transfer must be rapid.

analytical reagent n **分析用試薬**

▶ Tin(II) chloride dihydrate (25 g) of analytical reagent grade is dissolved in concentrated hydrochloric acid (20 mL) and water (10 mL)

analyze vt **1 分析する** syn *assay, separate test*

▶ If the solution to be analyzed contains sulfate, it is advisable to remove it.

2 解析する syn *examine, investigate, study*

part ▶ Van't Hoff's work on osmotic pressure was mainly directed toward analyzing the results obtained by Pfeffer.

analyzer n **検光子**

▶ The analyzer is held in the barrel of the microscope, usually somewhere up near the middle on a sliding carriage.

anchor vt **固定する** syn *affix, attach, fix, pin*

▶ The molecule-substrate interaction must be strong enough to anchor the molecule so that it can be imaged but not so strong that the molecule is altered by the anchoring.

and conj **1 と，および** 名詞に限らず，文法上，対等の関係にある語，句，節どうしを連結するのに用いる．列挙するときは，A, B, C, and D の形をとる．同じ名詞であっても，測定方法と機器は対等の関係にない．

▶ Eley and Pepper and Eley and Richards have shown that no cocatalyst is necessary in the tin(IV)-chloride-induced polymerization of 2-ethylhexyl vinyl ether.

▶ The ability to form solid solutions between solids having the same zinc blende structure and possessing atoms of comparable size is of great technological importance.

▶ Krypton can be trapped into ice and be contained as a clathrate inside the hydrogen-bonded but rather openwork cage of β-quinol.

▶ We consider that the molecules behave as rigid spheres in elastic collisions and that there are two kinds of molecules, A and B, with diameters d_A and d_B.

複数の名詞が一つの概念を表すときは，動詞は単数形を用いる．

▶ The reversible formation and decomposition of transition metal hydrides has been an area of active research, as these compounds can be used as hydrogen storage materials.

2 文頭に用いて，**それでは，さて**

▶ And attention should be called once again to the fact that these wonderful materials occur in what are often considered simple binary metal-halogen systems.

3 and so forth …など

▶ Many detailed problems of fibrous protein structure remain: the effect of bulky side groups, the role of disulfide cross links, and so forth.

4 and so on …など

▶ Some natural ionophores for alkali-metal ions have O donor atoms from carboxylates, cyclic peptides, and so on.

and/ or conj **およびまたは** one or the other or both の意味である and と or のいずれかで十分ではないか，検討してから使用する．

▶ Every organic reaction involves the making and/or the breaking of chemical bonds.

anesthetic n **1 麻酔剤**

▶ N_2O has been much used as an anesthetic.

adj **2 麻酔の**

▶ Halogenation increases the anesthetic and narcotic action of aliphatic hydrocarbons.

angle n **1 角**

▶ The C(2)-C(1)-C(6) angle in the picryl ring is much less than 120°.

2 角度 syn *inclination*

▶ When viewed from another angle, SiO_4 tetrahedra possess twofold rotation axes which pass through the central silicon and bisect the O-Si-O bonds.

3 at an angle 斜めに

▶ The faces of crystals lie at angles to one another that have definite characteristic values.

vt **4 ある角度に曲げる** syn *bend, tilt, twist*

▶ The two benzene rings are not in the same plane but angled so as to form a butterfly arrangement.

angle bending n **変角**

▶ The interactions fall broadly into two classes: nonbonded interactions such as electrostatic and van der Waals intersactions and bonded interactions such as bond stretching, angle bending, and torsional twisting.

angled adj **角をなす** syn *inclined, oblique*

▶ The I_8^{2-} anion is found with an acute-angled planar Z configuration in its Cs^+ salt.

angle of diffraction n **回折角**

▶ From the Bragg equation, it can be seen that the angle of diffraction is related to the interplanar spacing.

angle of incidence n **入射角**

▶ The light ray meets the surface at an angle of incidence of 40°as measured between the light beam and the normal to the surface.

angle strain n **角ひずみ**

▶ Any deviations from the normal bond angles are accompanied by angle strain.

angular adj **1 角のある**

▶ ICl_2^+, BrF_2^+, and ClF_2^+ ions have been found to have angular structures in accordance with the fact that their valence shells have four electron pairs, which would be expected to have a tetrahedral arrangement.

2 角度の

▶ Subgrain boundaries involve a difference in the relative angular orientation between two parts of essentially the same crystal.

anharmonic adj **非調和な** ☞ harmonic

▶ When there are significant deviation from the parabolic curve, the vibrations are no longer harmonic and are said to be anharmonic.

anharmonicity n 非調和性

▶ Anharmonicity causes a slight curvature in a potential energy surface.

anharomonicity constant n 非調和定数

▶ If we can measure the progressive diminution in spacing, we can obtain a value for the anharmonicity constant and from this a value for the spectroscopic heat of dissociation.

anhydrous adj 無水の ☞ absolute alcohol

▶ Anhydrous LiCl is soluble in anhydrous amyl alcohol, while KCl and NaCl are difficultly soluble in this liquid.

anion n 陰イオン ☞ cation

▶ Most cations are Lewis acids and most anions are bases. Hence, salts are automatically acid–base complexes.

anion exchange n 陰イオン交換

▶ Different anions can subsequently be introduced by anion exchange.

anion-exchange membrane n 陰イオン交換膜 ☞ cation-exchange membrane

▶ Anion-exchange membranes have restored interest in alkaline fuel cells and in catalytic properties of materials in highly alkaline environments.

anionic adj 陰イオンの ☞ cationic

▶ In an attempt to stabilize the highly conducting α-AgI phase at lower temperatures, various anionic and cationic substitutions have been tried.

anionic polymerization n アニオン重合

▶ Anionic polymerization of alkenes is quite difficult to achieve, since few anions are able to add readily to alkene double bonds.

anionic surfactant n 陰イオン活性剤

▶ With the correct level of electrolyte added to a solution of an anionic surfactant, cylindrical micelles can be induced to grow to lengths equal to many times their diameter.

anionoid reaction n アニオノイド反応 syn *nucleophilic reaction*

▶ A number of anionoid replacement reactions of allyl derivatives are known where rearrangement accompanies replacement.

anionoid reagent n アニオノイド試薬 syn *nucleophile*

▶ In the salts of ketones, the electron affinity of oxygen is increased and so likewise the reactivity to anionoid reagents.

anion radical n 陰イオンラジカル syn *radical anion*

▶ Apart from salts of anion radicals made by treatment of an organic compound with the free metal, there are few alkali-metal compounds soluble in nonpolar solvents.

anion water n 陰イオン水 ☞ lattice water, water of crystallization

▶ When copper sulfate pentahydrate is heated, the water of the complex ion is lost in two stages giving first $CuSO_4{\cdot}3H_2O$ and then $CuSO_4{\cdot}H_2O$; the anion water remains tenaciously held up to 520 K.

anisometric adj 等軸でない ☞ isometric

▶ The lattice vibrations of anisometric molecules are generally anisotropic.

anisotropic adj 異方性の ☞ isotropic

▶ Anisotropic substances transmit light with unequal velocities in different directions; they include all crystals, except those of the isometric system which are unstrained.

anisotropically adv 非等方的に ☞ isotropically

▶ For all of the other structures, all nonhydrogen atoms were refined anisotropically.

anisotropy n 異方性

▶ While experimental determination of magnetic susceptibility anisotropy in liquid crystals is a formidable task, its calculation from local values is quite straightforward if the structure of the molecule is known.

anneal vt 焼きなます syn *temper*

▶ Further heating anneals product $BaTiO_3$ at the microwave-induced plasma temperature of ca.1100 ℃ giving the single tetragonal phase.

annealing n 焼きなまし

▶ Annealing of a crystalline mass causes growth of the larger crystals at the expense of adjacent smaller ones.

annihilate vt 消滅させる syn *destroy, elimi-*

nate, extinguish

▶ When a mobile conduction-band electron recombines with a mobile valence-band hole, both carriers are annihilated and no longer contribute to electric conductivity in the solid.

annihilation n 消滅 syn *destruction*

▶ The term exciton is applied to the neutral entity produced by the partial combination, without annihilation, of an electron with a hole.

annotation n 注釈付け syn *comment, note*

▶ Annotation can also involve identifying the regulatory elements of a gene such as promoter and enhancer sequences.

annulated adj 環状になった

▶ Backbone metallation, observed in some imidazolylidene-based systems, is eliminated in the annulated frameworks.

anode n **1** 電解のアノード，陽極 ☞ cathode

▶ The instability of anodic materials will present difficulties for systems that require inert anodes, e.g., for alloy deposition.

2 電池のアノード，負極 syn *negative electrode*

▶ Lithium is a good anode because of its high potential and its low equivalent weight.

3 X 線管の陽極

▶ In the generation of X-rays, the electron beam, provided by a heated tungsten filament, is accelerated toward an anode by a potential difference of ca. 30 kV.

anode slime n アノード泥

▶ The main source of Se and Te is the anode slime deposited during the electrolytic refining of Cu.

anodic adj 陽極の ☞ cathodic

▶ The distribution of cathodic and anodic areas is controlled by the ionic conductivity of the electrolyte.

anodically adv 陽極で

▶ The Cl_2 produced by anodic oxidation of Cl^- reacts with cathodic OH^- to give hypochlorite which then either disproportionates or itself further anodically oxidized to ClO_3^-.

anodic protection n アノード防食

▶ Anodic protection is achieved by connecting to the object a metal with a more negative electrode potential, such as magnesium.

anodic reaction n 陽極反応

▶ Central to the study of corrosion are the electrode processes, in particular the anodic reactions of metals.

anodize vt 陽極酸化する

▷ Anodized coatings are hard and have good electric insulating properties.

anodizing n 陽極処理

▶ The protection of metals by coherent oxide films is deliberately brought about in the process of anodizing.

anolyte n アノード液，陽極液 ☞ catholyte

▶ Because the membrane is impermeable to Cl^- or OH^- ions, the anolyte does not became basic, and the caustic product remains uncontaminated by chloride ions.

anomalous adj 異常な syn *abnormal, exceptional, peculier*

▶ Some ferroelectrics show an anomalous and large increase in resistivity as the temperature approaches the ferroelectric–paraelectric transition temperature.

anomalous dispersion n 異常分散 ☞ normal dispersion

▶ Some media, in certain parts of the spectrum, refract longer waves more than shorter and are said to show anomalous dispersion.

anomalously adv 異常に

▶ The anomalously large internuclear distance implied for LiI is readily explained if this distance is determined by anion–anion rather than anion–cation contact.

anomaly n 異常 syn *abnormality, exception, irregularity*

▶ A variety of anomalies are associated with this phase transition, and some are discussed below.

anomer n アノマー

▶ In the chair conformation, the hydroxyl substitutent at C1 is equatorial in the β-anomer and axial in the α-anomer; hence the β-anomer should be more stable

anomeric adj アノマーの

▶ Like D-(+)-glucose, other monosaccharides exist in anomeric forms capable of mutarotation and react with alcohols to yield anomeric glycosides.

another pron **1** もう一つのもの

▶ Substitution reactions include the replacement of one ligand by another in a complex or one metal ion by another.

adj **2** もう一つの 可算名詞の単数形の前に置く. syn *different, further, other, second*

▶ Another aspect of stoichiometry involves the determination of empirical formulas for compounds having extended structures.

answer n **1** 答え syn *response*

▶ Attempts to find answers have made use of chemical and enzymatic methods.

vt **2** 答える syn *respond*

▶ Questions to be answered include whether nonpolar nucleoside analogues occupy the same orientation and geometry as their natural counterparts.

antagonist n アンタゴニスト ☞ agonist

▶ A particular chemical or physical signal may be an agonist under some conditions and an antagonist in others.

antenna n アンテナ

▶ Novel polyelectrolytes with antenna chromophores, photozymes, have been synthesized, which can solubilize sparingly water-soluble organic compounds and act as catalysts in their phototransformation.

anthropogenic adj 人間の活動に由来する

▶ The sources of the dioxins are predominantly anthropogenic, in particular resulting from the use of organochlorine chemicals and combustion processes.

anti adj 反対の

▶ Groups which have a dihedral angle relationship of 180° are said to be anti with respect to each other.

anti addition n アンチ付加 ☞ syn addition

▶ To describe the kinds of stereochemistry possible in addition reactions, the concepts of syn addition and anti addition are used.

antibacterial adj 抗菌性の

▶ The sulfonamides are antibacterial compounds that are sulfur-containing analogs of *p*-aminobenzoate.

antibiotic n 抗生物質

▶ Like crown ethers antibiotics wrap around the cation, holding it through ion-dipole bonds.

antibiotic resistance n 抗生物質耐性

▶ Antibiotic resistance has become one of the main concerns in bacterial disease treatment.

antibody n 抗体 ☞ antigen

▶ Some blood proteins function to form antibodies, which provide resistance to disease.

antibonding adj 反結合性 ☞ bonding, nonbonding

▶ The energy of the state Ψ_A is always above the energy of two separate atoms and is, therefore, called an antibonding state.

antibonding orbital n 反結合性軌道 ☞ bonding orbital

▶ Since the electronically excited states of organic molecules contain electrons in antibonding orbitals, we cannot represent these species by means of the usual structural formulas.

anticancer adj 抗がん性の

▶ While DNA has been implicated as the major target related to anticancer activity, the identification of the active species and mechanism of action has been poorly understood.

anticancer drug n 抗がん剤

▶ Stimulated by the discovery of the guanine-crosslinking properties of anticancer drug cisplatin, metal complexes have traditionally been targeted to the nucleophilic guanine bases of DNA.

anticipate vt 予想する syn *forecast, predict*

▶ Another development is the study of interactions which are anticipated to be more akin to the drug-receptor complex.

anticipation n **1** 予想 syn *expectation*

▶ With the anticipation of finding additional examples of materials exhibiting such properties and improving the physical properties as well as discovering new phenomena, new donors and acceptors continues to be the subject of intense study.

2 in anticipation of …を予想して

▶ In anticipation of the introduction of the internal-energy property, the energy of the system for all molecules in their lowest allowed energy level was denoted by E_0.

anticlockwise adj 反時計回りの，左回りの syn *counterclockwise* ☞ clockwise

▶ The S_n chains form helices, either clockwise or anticlockwise.

anticoagulant n 抗凝血剤 ☞ coagulant

▶ Pharmaceutical heparin is one of the oldest drugs currently still widespread clinical use as an anticoagulant.

antiferroelectric n **1** 通常複数形で，反強誘電体 ☞ ferroelectric

▶ Examples of antiferroelectrics, with their Curie temperatures, are lead zirconate, $PbZrO_3$, 233 ℃; sodium niobate, $NaNbO_3$, 638 ℃; and ammonium dihydrogenphosphate, $NH_4H_2PO_4$, −125 ℃.

adj **2** 反強誘電性の ☞ ferroelectric

▶ In antiferroelectric materials, individual dipoles occur but they generally arrange themselves so as to be antiparallel to adjacent dipoles.

antiferromagnetic adj 反強磁性の ☞ ferrimagnetic, ferromagnetic

▶ The unpaired electrons may be aligned in antiparallel fashion, giving zero overall magnetic moment and antiferromagnetic behavior.

antiferromagnetism n 反強磁性 ☞ ferromagnetism

▶ It is convenient to consider a three-dimensional antiferromagnetism as consisting of at least two interpenetrating lattices on each of which all the spins are aligned similarly but oppositely to those on the other sublattice.

antifluorite structure n 逆ホタル石型構造 ☞ fluorite structure

▶ The antifluorite structure contains tetrahedrally coordinated cations and eight-coordinated anions.

antifoaming agent n 泡止め剤

▶ It is well to remember that many of these antifoaming agents are of influence only when they are freshly presented to the foam.

anti form n アンチ形 ☞ syn form

▶ Butane will exist as an equilibrium mixture of the anti and gauche forms.

antifreeze n **1** 不凍剤

▶ Ethylene glycol is an excellent permanent antifreeze for automotive cooling systems, since it is miscible with water in all proportions and a 50 percent solution freezes at −34 ℃.

adj **2** 抗凍結の

▶ Some coolants provide antifreeze protection.

antigen n 抗原 ☞ antibody

▶ An antigen is an infective organism with molecular weight of at least 10,000; it is able to induce formation of an antibody in an organism into which it is introduced.

antigenic adj 抗原性の

▶ A bacterial cell has many different antigenic molecules on its surface.

antigenicity n 抗原性

▶ Other strategies to improve antigenicity involve the design of conjugates incorporating chemical modifications on the carbohydrate epitopes.

antiknock n アンチノック

▶ The action of antikocks such as tetraethyllead can be studied by kinetic spectroscopy.

antiknock agent n アンチノック剤

▶ The antiknock agents used in lead-free gasolines are nonmetallic compounds such as *tert*-butyl methyl ether or a mixture of methanol and *tert*-butyl alcohol.

antimalarial n 抗マラリア剤

▶ New strains of malaria have recently appeared which are more or less resistant to some or all of the synthetic antimalarials.

anti-Markovnikov addition n 逆マルコフニコフ付加

▶ The hydroboration-oxidation process gives products corresponding to anti-Markovnikov addition of water to the carbon–carbon double bond.

antimatter n 反物質 ☞ matter
▶ Most of the fundamental particles can be described as constituting either matter or antimatter.

antioxidant n 酸化防止剤 ☞ oxidant, oxidizing agent
▶ The inhibition of the radical-chain reaction by an antioxidant is presumably due to the ability of such a substance to act as a chain breaker.

antiparallel adj 逆平行の ☞ parallel
▶ The magnitude of energy difference between the parallel and antiparallel spin states is varied by varying the applied magnetic field strength.

antiparasitic adj 寄生虫に対して作用する
▶ The field of antiparasitic vaccines has experienced less scientific impetus since the appearance of the first synthetic carbohydrate vaccine against malaria.

antiparticle n 反粒子 ☞ particle
▶ Particles of antimatter are created by collisions, and the antiparticles are then rapidly destroyed as they react with particles of ordinary matter with which they collide.

antiprism n アンチプリズム ☞ prism
▶ The symmetrically bicapped trigonal antiprism or puckered hexagonal bipyramid is derived from a cube by a trigonal distortion.

antiprismatic adj アンチプリズムの ☞ prismatic
▶ For common coordination numbers of 2, 4, 6, and 8 the configurations would be linear, tetrahedral, octahedral, and square antiprismatic, respectively.

antiseptic n 防腐剤
▶ Tincture of iodine is a useful antiseptic.

antisymmetric adj 反対称の ☞ symmetric
▶ The molecular orbitals of naphthalene must be either symmetric or antisymmeric with respect to reflection in the *yz* and *xz* planes.

antisymmerization n 反対称化
▶ Neither the use of a self-consistent field nor antisymmetrization provides for another feature of electron repulsion, namely, electron correlation.

antisymmerized adj 反対称化された
▶ The principal application of the antisymmetrized molecular orbital with configuration interaction procedure has been to the calculation of the relative energy levels of excited states for comparison with experimental spectra.

antitumor adj 抗腫瘍性の
▶ A resurgence of interest in these seemingly simple complexes of platinum started when B. Rosenberg and co-workers discovered the antitumour activity of *cis*-$[PtCl_2(NH_3)_2]$.

antiviral adj 抗ウイルスの
▶ Many polyoxometallates show promising antiviral activities in vitro and in vivo.

any pron **1** どれも
▶ A state of equilibrium exists when there is no change with time in any of the system's macroscopic properties.
adj **2** どんな
▶ For any isoelectronic series of cations, the radius decreases with increasing charge.

anything pron **anything but** (adv) …のほかは何でも
▶ Some compounds such as SiO_2 are so prone to supercool that it is difficult to obtain anything but a glass upon cooling the molten material.

anyway adv とにかく syn *in any case*
▶ Since the s → d excitation energy decreases anyway in going from Ca to Ba, the influence of the halides on the sd^x hybridaization should be more important for Ca compared to Sr and Ba.

anywhere adv どこでも
▶ For $Pb_{1-x}Sn_xTe$, emission can occur anywhere between 6.5 and 32 μm (1690–312 cm^{-1}).

apart adj **1** 離れて syn *isolated, separate*
▶ Benzene is known, from X-ray diffraction and spectroscopic measurements, to be a flat molecule with six carbon atoms, 1.397 Å apart, in a hexagonal ring.
adv **2** 離れて syn *separately*
▶ Cleavage is the tendency of the crystal to

come apart between planes of atoms when struck or pulled apart.

3 apart from (prep) (a) …は別にして syn *aside from, except for*

▶ Apart from very light atoms, the height of the electron density peak is approximately equal to the atomic number of that atom.

(b) …のほかに syn *besides*

▶ Apart from the provision of raw materials of adequate quality, there are many chemical aspects to the preparation of ferroelectric materials.

aperture n 孔 syn *hole, gap*

▶ Gaseous uranium hexafluoride is allowed to diffuse through a series of microporous barriers, whose apertures are of the molecular dimensions.

apex n 頂点 複数形は apexes syn *peak, vertex*

▶ The apex of this tetrahedron is the anion at the top corner with coordinates 0, 0, 1.

apical adj 頂点の

▶ In a trigonal bipyramidal cluster, the apical metal atoms are three-coordinated and the equatorial atoms four-coordinated.

apoenzyme n アポ酵素

▶ Enzymes are composed of a protein structure called an apoenzyme and a small prosthetic groups which may be either a simple metal ion or a complexed metal ion.

apparatus n 装置 syn *device, equipment, instrument, tool*

▶ The distillation apparatus shown in Figure 1 consists of a 50-mL distilling flask equipped with a thermometer with its bulb in the vapor space, a condenser, a receiving flask, and a microburner.

apparent adj **1** 名詞の前において見かけの syn *superficial*

▶ The most effective way of creating a large apparent viscosity with a small amount of suspended material is for the particles to link up to large loose ramifying aggregates extending through and effectively immobilizing much of solvent.

2 明白な 推論を含む表現である. syn *clear, evident, obvious*

▶ The importance of heterocyclic compounds is apparent from the wealth and variety of such compounds that occur naturally or are prepared on a commercial scale by the dye and drug industries.

apparently adv **1** どうも…らしい syn *it appears that, it would appear that, seemingly*

▶ The niobium-sulfur system is quite complex, having apparently homogeneous compositions from Nb_2S to NbS_4, but there is a stable phase corresponding Nb_2S_3.

2 明白に syn *clearly, evidently, obviously*

▶ This new class of clathrate formers exhibits a sharp selectivity for forming crystals containing certain organic molecules, the selectivity being based, apparently, on the shape rather than on the molecular volume occupied by the organic moiety.

appeal n 要請 syn *application, request*

▶ Only one fixed parameter is needed to give the lattice energies of a large class of crystals, without appeal to experimental data on the crystal structure or dimensions.

appealing adj 魅力的な syn *attractive*

▶ A valence bond description of ferrocene requires 560 contributing structures. In this situation the MO description is more appealing.

appear vi **1** 思われる syn *look, seem*

▶ On nickel, twice as much carbon monoxide appears to be adsorbed as hydrogen.

2 現れる syn *arise, emerge*

▷ The electrons appearing as products in one electrode reaction are used up as reactants in the other.

appearance n **1** 外見 syn *form, manner, shape*

▶ Nickel oxide, when prepared at temperatures below 500 ℃, is black in appearance and contains slightly more oxygen than nickel.

2 出現 syn *existence, manifestation, occurrence*

▶ The versatility of the coordination of sodium accounts for its appearance in ferroverdin.

3 掲載 syn *publication*

▶ Figures should be numbered according to

the sequence of their appearance in the text.

append vt 添える syn *add, attach*

▶ In naming an alcohol by the IUPAC system, the ending -ol is appended to the name of the parent hydrocarbon.

appendix n 付録 複数形は appendixes syn *addition, supplement*

▶ A brief summary of this nomenclature and symbolism is given in the appendix.

applicability n 適用の可能性 syn *feasibility, relevance, usefulness*

▶ The applicability of the above two conditions to the first term of eq 1 has been thoroughly discussed.

applicable adj 適用できる syn *fitting, proper, suitable, usable*

▶ Raoult's law is not applicable to most solutions but is often approximately applicable to a mixture of closely similar substances.

application n **1** 印加 syn *imposition*

▶ The electrolysis of HCl can be effected by the application of a sufficient voltage to the electrodes of a cell.

2 応用 syn *operation, use, utilization*

▶ Many potential applications of superconductors are precluded by the critical magnetic field at which the conversion from superconducting to normal behavior occurs.

applied adj **1** 応用の syn *practical, technical*

▶ Applied research has produced the multitude of new materials that have revolutionized industry, agriculture, and medicine in the last 50 years or so.

2 印加された syn *imposed*

▶ In a given molecule, protons with the same environment absorb at the same applied field strength; protons with different environments absorb at different applied field strengths.

applied voltage n 印加電圧

▶ There will be some applied voltage which will keep the reaction at equilibrium.

apply vt **1** 適用する syn *employ, use, utilize*

▶ This approach of in situ formation of stable adducts has been successfully applied to the growth of binary, ternary, and quaternary compounds of the Ga-In-As-P system.

2 …に当てる syn *commit, concentrate, focus*

▶ When a powerful flash created by a discharge of 2,000-4,000 J through an inert gas is applied to a photochemically responsive system, very high momentary concentrations of atoms and free radicals are produced.

3 使用する syn *employ, use*

▶ If excessive heat is applied and the vapor becomes superheated, the drop will disappear, the liquid-vapor equilibrium will be upset, and the temperature will rise above the boiling point.

4 言葉を…に用いる syn *assign, use*

▶ The term nucleotide is applied to a fundamental unit of nucleic acids.

apportion vt 割り当てる syn *distribute, divide, split*

▷ The data show that the calculated charges are strongly dependent on the recipe used for apportioning electron density among atoms.

appraisal n 評価 syn *evaluation*

▶ The charge data can be useful in demonstrating the existence of other factors and helping in the appraisal of their relative significance.

appraise vt 見積もる syn *estimate, evaluate, measure*

▶ It is desirable to be able to appraise, even qualitatively, the effects of dipoles upon the fugacities of the components of solutions.

appreciable adj かなりの syn *considerable, significant*

▶ If a solution contains a sufficiently large number of ions, it is only reasonable to expect their electrostatic interactions to be appreciable.

appreciably adv 認められるほどに

▶ Since chromium(III) is kinetically inert, its complexes often exist in different isomeric forms which do not change appreciably from one to the other during the time needed for a laboratory preparation.

appreciate vt 認識する syn *perceive, real-*

ize, recognize, understand

▶ It is well appreciated that the crystal structure is the key to understanding both the thermodynamic and kinetic behavior of a material as well as its physical and chemical properties.

appreciation n 評価 syn *awareness, recognition*

▶ With an appreciation of the close relationship between cubic and tetragonal symmetry, we can now understand the suggestion of Linnett that hybridization may play a role in determining these structures.

approach n **1** 接近 syn *access, entry*

▶ It may be supposed that hydration prevents the close approach of the anions by making the smaller cations larger.

2 研究方法 syn *manner, method, procedure, technique*

▶ Experimental approaches to inorganic synthesis are often limited by conventional ideas based on the limits imposed by hydrolysis occurring in the narrow window of acidities and basicities available in the common solvent water or by the redox limits of that solvent.

vt **3** …に近づく syn *advance, come close to*

▶ The relaxed structures of the nanocrystals are very stable; their cohesive energy per atom quickly approaches a large fraction of the energy of the perfect crystalline solid.

appropriate adj 適切な syn *fitting, pertinent, proper, suitable*

▶ Such reactions do not occur in the absence of an appropriate proton sink or source, that is, in the absence of appropriate basic or acidic catalysts.

appropriately adv 適切に syn *correctly, properly, suitably*

▶ Equations must be modified appropriately to account for these unequal atomic contributions.

approximate adj **1** 近似の syn *imprecise, loose, rough*

▶ In systems of interest to organic chemists, the most common procedure has been to find approximate solutions for approximate Hamiltonian.

vi **2** 接近する syn *approach, come close to*

▷ Aryl halides are slightly polar substances and accordingly have boiling points approximating those of hydrocarbons of the same molecular weights.

approximately adv おおよそ syn *about, nearly*

▶ The overall dipole moment of the dichlorobenzenes is approximately the vector sum of the C–Cl moments.

approximation n 近似 syn *estimate, estimation, evaluation*

▶ To a first approximation, the dipole moment of the crystal can be considered to be the vector sum of the dipoles of the individual molecules.

a priori adv 推測的に

▶ In inorganic compounds, it is not so easy to assign bond valences a priori since they are usually nonintergral and range widely in value.

aprotic solvent adj 非プロトン性溶媒 ☞ protic solvent

▶ Aprotic solvents dissolve ionic reagents chiefly through their bonding to the cation; they leave the anion relatively free and highly reactive.

apt adj 適切な syn *appropriate, fitting, proper, suitable*

▶ The analogy to a *PV* isotherm is so apt that in gaseous monolayer regime the two-dimensional analogue of the ideal gas law pertains; $\Pi A = nRT$.

aptitude n 素質 syn *ability, capability, propensity, tendency*

▶ In many rearrangements, the problem of relative intrinsic migratory aptitudes does not enter, since the intermediate with the open sextet contains only a single atom or group that could conceivably migrate.

aptly adv 適切に syn *appropriately, properly*

▶ An elastic jelly is commonly, although not very aptly, defined as one that when immersed in a suitable liquid swells either to an apparent limit or indefinitely to form a sol.

aqueous adj 水に溶けた

▶ To the retained filtrate and acid washings was added 3.5 mL of 10% aqueous sodium hydroxide, and the clear solution was extracted with 75 mL of ether in three portions.

aqueous ammonia n アンモニア水

▶ Solutions of cobalt(II) chloride in aqueous ammonia gradually turn brown in air and then wine-red on being boiled.

aqueous solution n 水溶液 ☞ nonaqueous solution

▶ When the concentration of H_2O_2 reaches about 5.5 g/L, the organic phase is extracted with water to give an 18% aqueous solution.

aquo complex n アコ錯体

▶ Ammine complexes are generally less acidic than are aquo complexes, just as ammonium ion is a weaker acid than hydronium ion.

arbitrarily adv 任意に syn *at random, fortuitously, randomly*

▶ Crystal growth from the vapor can be subdivided rather arbitrarily into several different methods.

arbitrary adj 任意の syn *inconsistent, random, uncertain, unpredictable*

▶ A helpful, though arbitrary and imperfect, classification of solid electrolytes is presented in Table I.

arc n **1** 円弧 syn *circle, curve*

▶ All the arc-marked angles in Figure 1 are θ.

2 電弧, アーク

▶ In a hydrogen arc lamp, molecules are produced in the $^3\Sigma_g^+$ state, and the characteristic emission is due to the $^3\Sigma_g^+ \rightarrow {}^3\Sigma_u^+$.

archetypal adj 典型的な syn *representative*

▶ The experiment is the archetypal form of many enzyme modification procedures that are now used routinely to identify active site constituents.

architecture n 構成 syn *construction, fabrication*

▶ The basic structural characteristic of a surfactant is a diblock architecture composed of immiscible components.

area n **1** 地区 syn *region, section, zone*

▶ Dust suspensions in enclosed industrial areas are a serious fire hazard regardless of the chemical nature of the dust.

2 分野 syn *discipline, field, specialty, subject*

▶ My objective is to illustrate the major concepts and areas of interest, the difficulties that need to be overcome, and the potential for future research.

3 面積 syn *extent, size*

▶ The area under an NMR signal is directly proportional to the number of protons giving rise to the signal.

arene n アレーン syn *aromatic hydrocarbon*

▶ The arenes themselves are generally quite toxic; some are even carcinogenic and inhalation of their vapors should be avoided.

argentation n 銀化

▶ Detailed information concerning the argentation of aromatic hydrocarbons is lacking.

argue vi **1** 議論する syn *debate, discuss, reason*

▶ Van't Hoff argued on the basis of the variation of the equilibrium constant with the temperature and pointed out that a similar relationship should hold for the rate constant of a reaction.

vt **2** 主張する syn *contend, debate*

▶ It is true that this is negative evidence; one might argue that isomers exist which have never been isolated or detected simply because the experimental techniques are not good enough.

argument n 議論, 論証 syn *debate, disagreement, discussion, dispute*

▶ From thermodynamic arguments, the bond energies can be calculated from heats of formation of the original compound and the various atoms and radicals, and it is often these values that are tabulated.

arise vi **1** ことが起こる 抽象的な主語について用いられる. syn *happen, occur*

▶ The colligative properties are frequently used for molecular-weight determinations, but special difficulties arise with macromolecules.

2 arise from …から起こる syn *originate, stem*

▷ Once identified as arising from an aldehyde or ketone, its exact frequency can give a great deal of information about the structure of the molecule.

arithmetically adv 算術的に syn *mathematically*

▶ Forces are added arithmetically only if they have the same point of application and the same direction.

arithmetic mean n 算術平均 ☞ wighted average

▶ In a series of observations, all of which possess an equal degree of probability, the most probable value of the quantity is the arithmetic mean.

aroma n 香り syn *fragrance, odor, perfume, smell*

▶ Many organosulfur compounds with higher molecular weights or sulfur in a higher oxidation state do not have unpleasant aromas.

aromatic adj **1** 芳香のある

▶ The term aromatic is often used in the perfume and fragrance industries to describe essential oils which are not aromatic in the chemical sense.

2 芳香族の

▶ Naphthalene is classified as aromatic because it properties resemble those of benzene.

aromatic amino acid n 芳香族アミノ酸

▶ Chymotrypsin preferentially attacks peptide bonds whose carbonyl function is furnished by an aromatic amino acid.

aromaticity n 芳香族性 syn *aromatic character*

▶ This method of determining relative aromaticities offers a clear-cut comparison free of the ambiguities surrounding other criteria of aromaticity.

aromatic ring n 芳香環

▶ Replacement of the diazonium group by $-H$ provides a method of removing an $-NH_2$ or $-NO_2$ group from an aromatic ring.

aromatization n 芳香族化

▶ One of the best methods of aromatization is catalytic dehydrogenation, accomplished by heating the hydroaromatic compound with a catalyst like platinum, palladium, or nickel.

aromatize vt 芳香族化する

▶ Acyl derivatives of β-phenylethylamine are cyclized by treatment with acids to yield dihydroisoquinolines, which can then be aromatized.

around adv **1** 周囲に syn *about, in the neighborhood, in the vicinity*

▶ Except for a very few special cases, substances which contain six groups around a central ion have an octahedral configuration.

2 あちこち syn *about, everywhere, here and there*

▶ The flexibility of a cyclopentane ring is such that these deformations move rapidly around the ring.

3 数量についておよそ syn *about, almost, approximately, nearly, roughly*

▶ Finding new and more efficient materials is important because the substances currently in use require fields of around several thousand volts per centimeter to produce a sufficient change in the refractive index.

4 出回って syn *everywhere*

▶ The conventional aqueous battery systems have been around for many years.

arouse vt 喚起する syn *encourage, provoke*

▷ Points arousing much current interest are the geometry of the complexes formed and the factors responsible for complexation in aqueous solutions.

arrange vt **1** 整える syn *adjust*

▶ The conditions are arranged so that two gases are above their critical temperature.

2 配列する syn *array, dispose, place, position*

part ▶ The molecule of dinitrogen monoxide is known to be linear with the atoms arranged in the sequence N–N–O.

arrangement n 配置 syn *array, disposition, order, organization*

▶ In normal hexagonal ice, each O is surrounded by a nearly regular tetrahedral arrangement of four other O atoms (three at 276.5 pm and one, along the *c* axis, at 275.2 pm).

array n **1** 規則的な配列 syn *arrangement, order*

▶ The lattice of gallium dichloride is based on

an array of Ga^+ and $GaCl_4^-$ ions.
2 ものの集まり syn *combination*
▶ Silicates provide an interesting array of structural types that show greater variety than usually encountered for compounds of other elements.
vt **3** 配列する syn *arrange, dispose*
▷ The larger etch pits on sucrose crystals frequently occur arrayed in rows.
arresting adj 人目に付く syn *attractive*
▶ The spheroidal geometry of the fullerenes is their most arresting feature.
Arrhenous equation n アレニウスの式
▶ The conductivity data for the β-alumina fit the Arrhenius equation very well over large ranges of temperature and conductivity.
Arrhenius plot n アレニウスプロット
▶ A curved Arrhenius plot, meaning a temperature-dependent E_a, requires the pre-exponential factor to be temperature-dependent.
arrive vi **arrive at** 到達する syn *attain, reach*
▶ Chemical kinetics provides some of the information needed to arrive at the mechanism of a reaction.
arrow n 矢印 syn *indicator, sign*
▶ In formulas, evolution may be indicated by an upward vertical arrow.
art n 技巧 syn *aptitude, ingenuity, skill, technique*
▶ Under these circumstances, the initial synthesis of totally new types of phases sometimes depends a good deal more on art, and luck as well, than on the qualities usually associated with scientific exploration.
article n 論文 syn *paper*
▶ Invited articles are critical and concise accounts of topics which are deemed to be of particular significance.
artifact n **1** 人為結果 syn *product*
▶ There are some cases in the literature where high-pressure effects reported for chiral substances are almost certainly artifacts due to strain birefringence of the anvil windows.
2 装飾品
▶ As silver is stable in damp air, it was used for coinage and for jewelry and small artifacts.
artifice n 工夫 syn *device, maneuver*
▶ As a useful artifice, Debye considered a crystal to be continuous, elastic medium and treated the oscillation frequencies of the atoms as acoustic waves, the nodes of which coincide with the boundaries of the crystal.
artificial adj 人造の syn *man-made, synthetic, unnatural*
▶ Colors in topaz, smoky quartz, and amethyst are induced by natural and artificial irradiation.
artificially adv 人工的に
▶ Crosslinking can be effected artificially, either by adding a chemical substance and exposing the mixture to heat or subjecting the polymer to high-energy radiation
artificial radioisotope n 人工放射性同位体 ☞ radioisotope
▶ Artificial radioisotopes are made by neutron bombardment of stable isotopes in a nuclear reactor.
as adv **1 as…as** だけ
▶ If equimolar amounts of methane and ethane are allowed to react with a small amount of chlorine, about 400 times as much ethyl chloride as methyl chloride is obtained, showing that ethane is 400 times as reactive as methane
2 as…as possible できるだけ
▶ The flask and condenser should be as dry as possible to begin with.
prep **3** …として
▶ Pinacol is a liquid but can be isolated as the crystalline hexahydrate.
conj **4** …のように
▶ As discussed in Part I, we believe that there is only one temperature at which the high- and low-temperature forms can coexist in true equilibrium.
5 すると 節を文の後半におくとき，コンマは用いない． syn *when, while*
▶ It is not enough to establish that the composition of the system is not changing as time goes on.
6 …だから 節を文の後半におくとき，その前にコンマを用いる． syn *because of, for, owing to,*

since

▶ The hydrogen atom is the simplest possible, as its nucleus consists usually of just one proton whose single positive charge is balanced by the negative charge of the one associated electron.

7 概念を制限して

▶ Reagents were obtained from commercial sources and used as received unless otherwise noted.

8 as for (prep) …に関する限り syn *as to, concerning*

▶ As for any equilibrium, the concentrations of the components are related by the expression

$$K_{eq} = [RCOO^-][H_3O^+]/[H_2O][RCOOH]$$

in which K_{eq} is the equilibrium constant.

9 as if (conj) まるで…であるかのように 仮定法過去を用い，be 動詞は were とする． syn *like*

▶ On an atomic level, absorbed heat causes the atoms of a solid to vibrate, much as if they were bonded to one another through springs.

10 形容詞 + **as it is** それは…であるけれども

▶ Valid as they are, these deductions concerning the properties of interatomic forces still leave unanswered the question as to their origin.

11 as…so …と同じように

▶ The 9,10-bond in phenanthrene is quite reactive; in fact, almost as much so as an alkene double bond.

12 as though (conj) まるで…であるかのように syn *as if, like*

▶ The kinetic response of electrons to electric forces indicates that electrons behave as though they have a mass, called the effective mass, that is usually somewhat smaller than the value of their mass in vacuum.

13 as to (prep) …について syn *as for, concerning*

▶ In bromination, there is a high degree of selectivity as to which hydrogen atoms are to be replaced.

14 so as to …するように

▶ In the layered perovskite structure, the layers remain separated and are stacked in a staggered way so as to form a body-centered tetragonal structure.

15 so…as to …するほどに

▶ Lithium carbonate is so sparingly soluble as to be precipitated from solutions of lithium salts by sodium carbonate.

asbestos n アスベスト

▶ Asbestos is a carcinogen and is highly toxic by inhalation of dust particles.

ascend vt 上がる syn *climb, rise*

▶ As the series of straight-chain alkanes is ascended, each additional CH_2 group contributes a fairly constant increment to the boiling point and density and to a lesser extent to the melting point.

ascertain vt 確かめる syn *determine, discover, find out, see*

▶ Work is currently in progress to ascertain whether the proposed model offers significant advantages over existing schemes for hydrocarbon derivatives.

ascribe vt …に帰する syn *assign, attribute, credit*

▶ A trace of water added to a dilute solution of HCl in ethanol causes a marked drop in the electric conductivity, which is ascribed to the partial suppression of proton jumps resulting from the capture of proton from $C_2H_5OH_2^+$ ions.

ash n 灰

▶ In analytical chemistry, the term ash refers to the residue remaining after complete combustion of a material.

ash content n 灰分

▶ The ash content of charcoals varies widely in nature and amount and is another major factor in the sorptive behavior.

aside adv aside from prep は別にして syn *apart from*

▶ Aside from structural studies, NMR is now being used to give information about the equilibria and rates of rapid chemical reactions in which there is a change in the magnetic environment of a nucleus.

ask vt 尋ねる syn *inquire, question*

▶ If three molecules A, B, and C are close to

one another, we can consider the energies of the individual pairs A–B, B–C, and A–C and ask whether the total energy is the sum of these three energies.

aspect n 見地 syn *point of view, viewpoint*

▶ Sulfur demonstrates its diversity in many aspects of its chemistry.

aspect ratio n 縦横比

▶ Filamentous bacteriophages are a subset of these viruses that have a high aspect ratio.

assay n **1** 検定 syn *analysis, test*

▶ Biochemical assays may now be performed in volumes of a few microliters in a microtiter plate rather than a few milliliters in a test tube.

vt **2** 分析する syn *analyze, test*

▶ Pharmaceutical products are assayed to validate the amount of drug present in a given unit.

assemblage n 人, 物の集合 syn *assembly, ensemble*

▶ Nitrites of less basic metals such as Co(II), Ni(II), and Hg(II) are often highly colored and are probably essentially covalent assemblages.

assemble vi **1** 集合する syn *congregate, get together,*

▶ Supramolecular assemblies of amphiphilic molecules, in the form of micelles, lamellae, and biocontinuous phases, are able to cooperatively assemble with inorganics, like silica, to form mesostructured inorganic relica materials.

vt **2** 集合させる syn *collect, gather*

▶ One hundred or more filaments are assembled into a cable which has extremely high data-carrying capacity.

assembly n 集合体 syn *assemblage, cluster, group*

▶ Curved surfaces with regions of high surface charge and reactivity can be generated from supramolecular assembly, aggregation, and controlled polymerization.

assert vt 主張する syn *claim, emphasize, support*

▶ Despite some misleading observations, there now appears to be sufficient evidence to assert that many compounds used as drugs can participate in charge-transfer complex formation, usually as the donor species with simple acceptor molecules.

assess vt 評価する syn *estimate, evaluate, judge*

▶ The charge on the polymer can be assessed from electrophoretic measurement, from ion-exchange studies, or from the study of the pH of their solutions.

asset n 長所 syn *advantage*

▶ The fact that triethylammonium dichlorocuprate(I) can be both oxidized and reduced is an asset in the above reactions, but it is a handicap when the system is used merely as a solvent.

assign vt **1** 指定する syn *fix, prescribe, specify*

▶ The emf corresponding to the standard hydrogen electrode is arbitrarily assigned the value of zero.

2 …に帰する syn *ascribe, attribute*

▶ Pyridine exhibits an absorption spectrum very similar to that of benzene with an additional absorption band at 270 nm which is assigned to the transition involving the nitrogen lone pair.

assignable adj 帰せられる

▶ The expansion of the outer orbitals is measured experimentally in the probe ions Tl^{+}, Pb^{2+}, and Bi^{3+} as the frequency of the $^{3}P_{1} \leftarrow {}^{1}S_{0}$ transition and is proportional to Jorgensen's *h* function, a number assignable to a ligand set.

assignment n **1** 帰属 syn *specification*

▶ The assignment of structure by this method cannot be accepted as proof because of the assumptions which had to be made.

2 指定 syn *designation*

▶ On the basis of the collision theory, these reactions require an assignment of a probability factor of 10^{-4} or 10^{-5}.

assist vt 助長する syn *aid, help, support*

▶ To assist a chemical reaction, it is necessary to increase the concentration of one of the original reacting substances or lessen the concentration of one of the substances formed.

assistance n 助力 syn *aid, help, support*

▶ Formation of tertiary cations is relatively

easy and needs little nucleophilic assistance; in any case, crowding would discourage such assistance.

associate vi **1** 会合する syn *combine, link*

▶ In a less polar solvent, such as isopropanol, the polymer may associate better with itself than with the solvent and, therefore, form compact aggregates.

vt **2 associate with** (a) …と連携する syn *mingle, mix*

▶ Cadmium is usually associated with zinc in its ores.

(b) …と関係づける syn *connect with, involve with*

▶ Cation formation has normally been associated with the metallic elements.

associated adj 関連した syn *related*

▶ Biomineralization is another example of how inorganic chemistry can drive the development of associated disciplines.

associated liquid n 会合液体

▶ An associated liquid has a boiling point that is abnormally high for a compound of its molecular weight and dipole moment.

association n **1** 会合 syn *assembly*

▶ Hyrodgen fluoride is strongly hydrogen-bonded but, with only one hydrogen atom per molecule, association can take place only in two dimensions.

2 提携 syn *combination, mixture, union*

▶ Chemotherapy can be characterized as a permanent cooperative association of two or more disciplines around certain problems.

assume vt **1** 仮定する syn *imagine, presume, suppose, surmise*

▷ By assuming that the C–C bond-energy term is constant and equal to the bond energy found experimentally in diamond, the number of adjustable parameters is minimized.

2 執る syn *acquire, adopt, take*

▷ When double bonds are cis, steric hindrance prevents the chains from assuming an ordered structure and the bulk of the material exists in an amorphous state with randomly oriented chains.

assumed adj 仮定された syn *hypothetical, presumed, supposed, theoretical*

▶ The assumed configurations of the glyceraldehydes and, hence, the assumed configurations of all compounds related to them were indeed the correct ones.

assumption n **1** 仮定 syn *guess, inference, premise, supposition*

▶ The partial molar volumes in Table 1 are based on the widely used assumption that the partial molar volume of H^+ is zero.

2 on the assumption that …という仮定の下に

▶ The data have been interpreted on the assumption that from each hydrocarbon two water-soluble complexes, $AgAr^+$ and Ag_2Ar^{2+}, are formed.

assurance n 保証 syn *guarantee*

▶ Even if the data point to one mechanism in the case of a certain compound or group of compounds, there is no assurance that such a mechanism operates for all related systems.

assure vt 保証する syn *confirm, ensure, guarantee*

▶ If the capacity to store lithium is high, then a high gravimetric energy density will be assured provided the molar mass of intercalation host is small.

asterisk n 星印

▶ The carbons labeled with an asterisk in **1** are asymmetric.

astonishing adj 目覚しい syn *amazing, remarkable, striking*

▶ An alloy of nickel and titanium exhibits several astonishing properties, which identify it as a smart material capable of responding to external stimuli.

asymmetric adj **1** 非対称性の syn *unsymmetrical*

▶ The acetone–chloroform system is slightly asymmetric, but the asymmetry is in the opposite direction to that predicted by the volume ratios.

2 不斉の syn *chiral, dissymmetric*

▶ A carbon atom is asymmetric when it has

four nonequivalent substituents.

asymmetrically adv 非対称に

▶ Optical isomers of an asymmetrically substituted amine are very rapidly interconverted by an inversion process involving a planar transition state.

asymmetric carbon atom n 不斉炭素原子

▶ The most common form of dissymmetry found in organic compounds is associated with the asymmetric carbon atom or asymmetric center.

asymmetric center n 不斉中心 syn *center of asymmetry, chiral center*

▶ Mirror-image substances which contain an asymmetric center are called enantiomers or optical anipodes.

asymmetric polymer membrane n 非対称高分子膜

▶ Asymmetric polymer membranes consist of a very dense top layer or skin with a thickness of 0.1 to 0.5 μm supported by a porous sublayer with a thickness of about 50 to 150 μm.

asymmetric synthesis n 不斉合成

▶ Asymmetric syntheses and decompositions are of especially great importance in biological reactions, since the reactions which occur in living systems are so frequently brought about by optically active reagents or are catalyzed by optically active catalysts.

asymmetric vibration n 非対称振動

▶ Nitro compounds exhibit two very intense absorption bands in the 1560–1500 cm^{-1} and 1350–1300 cm^{-1} regions of the spectrum arising from asymmetric and symmetric stretching vibrations of the highly polar nitrogen–oxygen bonds.

asymmetry n 非対称 syn *dissymmetry*

▶ The asymmetry of the atmosphere will have the effect of dragging the central ion back. This is the relaxation or asymmetry effect.

asymptote n 漸近線

▶ A nonzero electron tunneling rate occurs if the total energy of the system is above the electron-detachment asymptote.

asymptotic adj 漸近的の

▶ It is doubtful that modern theories of atomic structure would admit the definition of ionic radii because of the asymptotic fall-off of electron density at large distances from the nucleus.

asymptotically adv 漸近的に

▶ With soap solutions, the interfacial tension is steadily and asymptotically reduced to zero.

at prep **1** 場所，位置の一点を表して…において

▶ The maximum rate of coagulation is at the isoelectric point.

2 時の一点を表して…に

▶ We are not yet at a stage at which the structures of transition metal compounds can often be predicted.

3 温度を表して…で

▶ Anhydrous α-maltose decomposes at about 100–120 ℃, and anhydrous α-lactose at 200–220 ℃.

4 数量を表して…で

▶ The absorption at 2750 cm^{-1} in the carbon-hydrogen stretching region is indicative of the aldehyde C–H bond, and the band at 1730 cm^{-1} suggests a carbonyl group.

atactic adj アタクチック

▶ Polystyrene, made by free-radical polymerization in solution, is atactic, which means that, if we orient the carbons in the polymer chain in the form of a regular zigzag, the phenyl groups will be randomly distributed on one side or the other when we look along the chain.

atactic polymer n アタクチックポリマー

▶ Atactic polymers have randomly arranged side groups.

atmosphere n **1** 大気 syn *the air*

▶ The amount of dioxygen in the atmosphere has probably been roughly constant for the last 500 million years.

2 雰囲気 syn *environment, medium*

▶ Unless otherwise stated, all manipulations were performed under an oxygen-free nitrogen atmosphere.

3 気圧

▶ Sodium phenoxide absorbs carbon dioxide at room temperature to form sodium phenyl carbonate and, when this is heated to 125 ℃ under a pressure of several atmospheres of carbon dioxide, it rearranges to sodium salicylate.

atmospheric adj 大気の，空気による

▶ In the massive state, none of Ni, Pd, and Pt is particularly reactive, and they are indeed very resistant to atmospheric corrosion at normal temperatures.

atmospherically adv 大気圏で

▶ An atmospherically important association reaction involves the generation of stratospheric ozone.

atmospheric pressure n 大気圧 複合形容詞として使用

▶ The first organic atmospheric-pressure superconductor, $(TMTSF)_2ClO_4$, was discovered by Bechgaard and co-workers in 1981.

atom n 原子

▶ The metallic elements forming the central atoms of complexes can be divided roughly into two classes, according to trends in stability of their complexes with different ligand atoms.

atomic adj 原子の

▶ Atomic dimensions are on the order of angstroms; therefore, unraveling the atomic positions of a solid requires a physical technique that operates on a similar spatial scale.

atomically adv 原子の単位で

▶ The luminescence is supposed to be associated with impurity ions atomically dispersed in the lattice.

atomic force microscope n 原子間力顕微鏡

▶ The atomic force microscope operates by sensing the force between surface atoms and the atom in a tip.

atomic force microscopy n 原子間力顕微鏡法

▶ The formation of polyaniline-DNA nanowires was followed by atomic force microscopy.

atomic hydrogen n 原子状水素

▶ Atomic hydrogen can be conveniently prepared in low-pressure glow discharges.

atomic orbital n 原子軌道 ☞ molecular orbital

▶ It is reasonable to suppose that in the regions close to the nuclei the molecular orbitals look very like atomic orbitals.

atomic pile n 原子炉 ☞ pile

▶ One of the main limitations on the use of neutron diffraction is the need for an atomic pile to produce sufficient neutrons for diffraction purposes.

atomic polarizability n 原子分極率 ☞ molecular polarizability

▶ Atomic polarizabilities are the differential change in electron density with change in the energy of an electron in an isolated atomic orbital.

atomic polarization n 原子分極 ☞ distortion polarization, electronic polarization, orientation polarization

▶ The atomic polarization may be expected to occur in alternating electric fields the frequency of which is very small in comparison with that of light.

atomic radius n 原子半径 複数形は atomic radii ☞ covalent radius, ionic radius, van der Waals radius

▶ Atomic radii vary considerably for a particular atom depending on bond type and coordination number, and many tabulations of radii are available.

atomic refraction n 原子屈折 ☞ molar refraction

▶ No single value of the atomic refraction of fluorine will, when combined with the standard refractions of other atoms, produce good agreement with the observed molar refractions of all fluorine-containing compounds.

atomic scattering factor n 原子散乱因子

▶ At higher angles of diffraction, the different scattered rays will interfere, the scattering will be reduced, and the proportionality factor, which is known as the atomic scattering factor, will become less than the number of electrons.

atomistic adj 原子論的な

▶ Density measurements do not, of course,

give any atomistic details of the vacancies or interstitials involved but only a bulk mechanism.

atomization n 原子化

▶ While bond energies are necessarily sensitive to the values taken for the heats of atomization of the elements, some of which are often not known accurately.

atomize vt 霧にする syn *spray*

▷ An aerosol bomb contains liquids which are emitted as an atomized spray on release of pressure.

atop prep …の上に syn *above*

▶ Although the overall unit cell of Cu_2HgI_4 is tetragonal, with square bases and rectangular sides, it can be viewed as two fcc unit cells of iodide ions, with one cube atop the other.

ATP synthetase n ATP 合成酵素

▶ The ATP synthetases are channel proteins that span the membrane and conduct ions through the membrane to drive the synthesis of ATP from ADT and inorganic phosphate.

atropisomer n アトロプ異性体

▶ Over the last few years, there has been a growing interest in the preparation of non-biaryl atropisomers.

atropisomeric adj アトロプ異性の

▶ Provided rotation about the Ar–CO bond is slow enough, atropisomeric chirality results from this perpendicular arrangement.

atropisomerism n アトロプ異性

▶ In general, atropisomerism is observed when an anilide carries two different non-hydrogen ortho-substituents.

attach vi **1** 伴う syn *be attracted*

▶ Considerable theoretical and stereochemical interest attaches to phosphorus pentahalides because of the variety of structures they adopt.

2 取り付ける syn *bond, connect*

▶ The molecules making up the liquid-crystalline phase are characterized by an aromatic or alicyclic core structure to which groups are usually attached at both ends.

attached adj 結合した syn *connected, joined*

▶ Aryl halides are compounds containing halogen attached directly to an aromatic ring.

attachment n **1** 付属物 syn *adjunct*

▶ The total number of attachments to the central atom is called the coordination number.

2 取り付け syn *attaching, connection*

▶ In view of the basic nature of alkenes and the acidic nature of BH_3, the principal driving force of the reaction is almost certainly attachment of boron to carbon.

attack n **1** 攻撃 syn *incursion*

▶ The general chemical inertness of boron at lower temperatures can be gauged by the fact that it resists attack by boiling concentrated aqueous NaOH.

2 着手 syn *onset*

▶ Many physical and chemical attacks have been made on the problem of protein structure and behavior.

vt **3** 冒す syn *destroy*

▷ These substitution reactions are called electrophilic because the species attacking the aromatic rings are electron-deficient.

attain vt 達成する syn *accomplish, acquire, get, obtain*

▶ Purely ionic bonding in crystalline compounds is an idealized or extreme form of bonding which is rarely attained in practice.

attainable adj 達成できる syn *accessible, feasible, possible*

▶ It is predictable that Mn(II) is the most stable and that Mn(VII) is attainable, although highly oxidizing.

attaining n 達成 syn *achievement*

▶ Neutralization does not mean the attaining of pH 7.0; rather it means the equivalent point for an acid-base reaction.

attainment n 達成 syn *achievement*

▶ The choice of solvent was important for the attainment of high enantioselectivity.

attempt n **1** 試み syn *endeavor, effort, trial, try*

▶ The first attempt to calculate the preexponential factor was based on the kinetic theory of collisions, the assumption being that the molecules are hard spheres.

vt **2 試みる** syn *seek, try, undertake*

▷ Before attempting to optimize ionic liquids for electrodeposition applications, it is important to appreciate the ways in which ionic liquids differ from aqueous solutions.

attend vt 事が結果として…に伴う syn *provide, serve*

▶ 5-Amino-2,3-dihydro-1,4-phthalazinedione is commonly known as luminol because oxidation of the substance is attended with a striking emission of light.

attendant adj **付随する** syn *accompanying, following, related*

▶ In a concentrated solution, the probability of an ion and its attendant solvent environment being hindered in its movement by other moving species becomes greater than in more dilute solutions.

attention n **1 注目** syn *concentration, consideration*

▶ Considerable attention has been given to the possible ways in which peptide chains can be arranged so as to give stable conformations.

2 attract attention 注目を引く syn *notice*

▶ As carbonaceous polymers are of complex origin, their chemical structure is complex, ill-defined, and imperfectly understood so that they have attracted little attention from organic chemists.

3 call attention 注目を促す syn *emphasize, mention, point*

▶ We call attention, first, to peaks in the region, 280–400 mμ.

4 pay attention to …に注目を払う syn *look, mark, observe*

▶ In an ionic solution, if we pay attention to ions bearing charges of the same sign, we are likely to find fewer pairs in such situations.

attenuate vt **減じる, 弱める** syn *reduce*

▶ Dielectric constants represent a molecule's ability to attenuate an electric field generated between macroscopically distant electrodes relative to vacuum ($\varepsilon = 1$ by definition).

attenuation n **希薄化**

▶ The attenuation of Ti^{3+} was not a simple masking of the substrate, since the amounts of deposited metal were far too small.

attest vi **物事が…の証明となる** syn *affirm, assert, confirm, verify*

▶ Tropolone has many properties which attest to its aromatic character.

attitude n **態度** syn *manner, position*

▶ The chemical industry people took the attitude that they devised and made materials which they believed to be beneficial.

attract vt **引き付ける** syn *draw*

▶ Any portion of matter on the earth is attracted toward the center of the earth by the force of gravity; this attraction is called the weight of the portion of matter.

attraction n **引力**

▶ Dipole–dipole interaction is the attraction of the positive end of one polar molecule for the negative end of another polar molecule.

attractive adj **1 興味をそそる** syn *appealing, interesting*

▶ Room temperature ionic liquids exhibit many properties which make them potentially attractive media for homogeneous catalysis.

2 引力の

▶ We conclude that the majority of short C–H⋯O, C–H⋯N, and C–H⋯Cl contacts are attractive interactions, which can reasonably be described as hydrogen bonds.

attractive force n **引力**

▶ The attractive force is found to be inversely proportional to the sixth power of the separation distance between molecules and is, therefore, proportional to the square of the density of the gaseous molecules.

attractive surface n **引力面**

▶ In between the two theoretical extremes of attractive and repulsive surfaces, we can envisage a whole range of mixed energy release surfaces.

attributable adj **起因する**

▶ The necessity for heating finely ground and well-mixed precursor compounds in this synthesis is attributable to the structures of the starting materials and products.

attribute n **1** 特質，属性 syn *character, characteristic, property, quality*

▶ The most important attribute of high polymers arises from their peculiar physical properties, which are largely a function of unusual molecular length, shape, and interaction with neighboring molecules.

vt **2** ある原因に帰する syn *ascribe, assign, credit, impute*

▶ The small resistance to electric current in a metal can be attributed to the vibrations of the metal atoms which scatter the mobile electrons.

atypical adj 異常な syn *peculiar, unusual*

▶ A reason why N is atypical of group 15 is that its covalency is limited to a maximum of four.

Auger electron n オージェ電子

▶ The energy spectrum of the Auger electrons can be used to determine the types of atoms present to a depth of about 100 nm from the surface.

Auger electron spectroscopy n オージェ電子分光法

▶ Higher energy techniques, such as X-ray photoelectron spectroscopy and Auger electron spectroscopy have been used to identify surface species.

augment vt 強化する syn *enlarge, increase, intensify, supplement*

▶ If the induced magnetic field reinforces the applied field, then the field felt by the proton is augmented, and the proton is said to be deshielded.

authentic adj 本物の syn *factual, genuine, real, true*

▶ A variety of authentic cyclodiazomethanes, or more properly diazirines, have been prepared, and these have found to have very different properties from the diazoalkanes.

authenticate vt 本物であることを証明する syn *substantiate, verify*

▷ $[Y(NCS)_6]^{3-}$ and $[La(NCS)_6]^{3-}$ are among the few well-authenticated, six-coordinate complexes of Y and La.

authenticity n 真正性 syn *actuality, reality*

▶ The corresponding Tl^{III} sesquichalcogenides, Tl_2X_3, are either nonexistent or dubious authenticity.

autocatylysis n 自触媒現象

▶ Catalysis brought about by one of the products of a reaction is referred to as autocatalysis.

autocatalytic reaction n 自触媒反応

▶ The autocatalytic reaction is a reaction induced by a product of the same reaction.

autoclave n オートクレーブ

▶ An autoclave provides a safe method of conducting large-scale pressure reactions

autoignition n 自然発火

▶ Dried peat is susceptible to autoignition, storage conditions must be such as to minimize this risk.

autoionization n 自動イオン化

▶ Autoionization of sulfuric acid results in the formation of the hydrogensulfate ion and a solvated proton.

autoionize vi 自動イオン化する

▶ In addition to water, many other solvents autoionize with the formation of cationic and anionic species that are regarded as the strongest acidic and basic species that exist in the medium.

autolyse vi 自己分解する

▶ In the absence of the peptidoglycan layer, the cell autolyses and dies due to its internal osmotic pressure.

automated adj 自動化した

▶ Fluorescent nucleosides were incorporated into oligonucleotides by automated synthesis.

automatically adv **1** 自動的に

▶ Thermogravimetry automatically records the change in weight of a sample as a function of either temperature or time.

2 必然的に syn *neccessarily*

▶ On univariant curves, if one variable, say pressure, is fixed, then the other, temperature, is automatically fixed.

automatic control n 自動制御

▶ Maintenance of desired process conditions by means of sensing devices which function either electromechanically or electronically is called automatic control.

autoprotolysis n 自己プロトリシス

▶ In any pure liquid, the transfer of a proton from one molecule to another molecule has been named autoprotolysis.

autoprotolysis constant n 自己プロトリシス定数

▶ Formic acid has a dielectric constant almost 10 times greater than that of acetic acid, it shows a particularly high autoprotolysis constant.

autoradiography n オートラジオグラフィー

▶ Autoradiography is a technique for locating radioactive compounds within cells.

autoxidation n 自動酸化

▶ Nowadays, an autoxidation process is used in the manufacture of hydrogen peroxide.

auxiliary adj 従属的な syn *extra, secondary*

▶ Several auxiliary relationships are convenient for practical calculations.

avail n **of no avail** 効果がない syn *useless, vain*

▶ If the substance is pure CeO_2, the heating with concentrated hydrochloric acid is of no avail; CeO_2 will not dissolve.

availability n 入手の可能性

▶ Because of availability of alcohols, each of these reactions is one of the best ways to make the particular kind of product.

available adj 利用できる syn *accessible, convenient, obtainable, ready*

▶ Several techniques are available for growing thin layers of inorganic materials on crystalline substrates.

average n **1** 平均値 syn *mean*

▶ The number of molecules with a particular kinetic energy is greatest for an energy near the average and decreases as the energy becomes larger or smaller than the average.

2 on average 概して syn *normaly, usually*

▶ On average, DNA polymerase III alone inserts the incorrect nucleotide one every ca. 10^5 additions, but after proofreading via the exonuclease, this error rate is reduced to 1 in every ca. 10^7 incorrect base.

adj **3** 平均の syn *mean*

▶ It is assume that the ionic interactions are less than the average thermal energy.

4 普通の syn *general, ordinary, usual*

▶ Average healthy human male has ca. 20% of body mass as triglycerides with the proportion being higher in females (ca. 25%).

vt **5** 平均する

▶ Rotation about the single bond is sufficiently fast to average out the differences between the protons, and an average resonance line position is observed.

avoid vt 避ける syn *escape, prevent*

▶ To avoid further oxidation to the corresponding acid, the aldehyde is removed as rapidly as possible by distillation through a fractionating column.

avoidance n 回避 syn *prevention*

▶ The motivation is essentially the same, the avoidance of soluble metal complexes which may be difficult to separate from reaction mixtures and are likely to lead to metal-contaminated products.

await vt 待ち受ける syn *expect, hope, look*

▶ The effect of Lewis acid strength is a facet of this reaction that awaits systematic experimental study.

aware adj **aware of** …に気付いて syn *familiar, sensitive*

▶ Since the completion of this work, we became aware of a report of dimorphism of a complex quinone with some features similar to those observed here.

awareness n 気付いていること syn *knowledge, realization, recognition*

▶ Recently, there has been an increased awareness that heterogeneous catalysts could significantly improve a large number of fine chemical processes.

away adv **1** 離れて syn *distant, off*

▶ The vibration of a diatomic molecule consists of the motion of the atoms toward and away from each other.

2 far and away 断然 比較級，最上級を強め

て断然 syn *clearly, definitely*

▶ Light absorption is far and away the most selective technique for producing excited molecules.

awkward adj 厄介な syn *delicate, difficult, inconvenient, tricky, troublesome*

▶ Functionalization of the dienes resulting from enyne ring-closing metathesis proved to be rather awkward.

axial adj **1** アキシアル位の ☞ equatorial

▶ In the pentagonal bipyramidal arrangement, the two axial sites are less crowded than the five equatorial sites.

2 軸方向の

▶ For M = Mo, the axial interactions are weakened enough to be overcome at temperatures before decomposing but strong enough to allow access to the mesophase.

axial bond n アキシアル結合 ☞ equatorial bond

▶ The bond holding the hydrogen atoms that are above and below the plane of a cyclohexane ring are pointed along an axis perpendicular to the plane and are called axial bonds.

axial chirality n 軸性キラリティー

▶ Axial chirality in ortho-substituted anilides was recognized more than fifty years ago.

axially adv 軸方向に

▶ While the Lagevin treatment is applicable to axially symmetrical systems, the more general theory of Van Vleck must be applied in other cases, particularly in the cases of polyatomic molecules.

axial symmetry n 軸対称

▶ A object has axial symmetry when it is invariant to rotation by some fraction of 2π radians.

axis n 軸 複数形は axes

▶ The edges of the unit cell are used as reference axes, and such a choice gives rise to simpler coordinates.

axis of rotatory inversion n 回映軸

▶ An axis of fourfold rotatory inversion combines the process of rotation through $360°/4$ with simultaneous inversion through a center of symmetry.

axis of symmetry n 対称軸

▶ We can generalize by saying that if an object has an *n*-fold axis of symmetry, it occupies the same position in space after each angular rotation of $360°/n$ around that axis.

azeotrope n 共沸混合物 syn *azeotropic mixture*

▶ An azeotrope occurs when the vapor and liquid compositions of a system become identical such that, without altering the system, no further separation is possible.

azeotropic adj 共沸の

▶ If a mixture shows azeotropic behavior, either maximum or minimum, it cannot be separated completely by fractional distillation into its components but only into one component and the azeotropic mixture.

azo compound n アゾ化合物

▶ The coupling of diazonium salts with aromatic phenols and amines yields azo compounds, which are of tremendous importance to the dye industry.

azo dye n アゾ染料

▶ The resulting diazo reagents undergo a wide variety of reactions including those of interest in manufacture of azo dyes and pharmaceuticals.

B

β-alumina n ベータアルミナ

▶ β-Alumina is a member of a class of solid materials with high ionic conductivities.

back adj **1** 逆の syn *backward*

▶ The amount of water formed in the esterification is too small, compared to the amount of methanol, to cause any serious back reaction.

adv **2** 逆戻りして syn *backward*

▶ Acid chlorides, esters, and amides can be hydrolyzed back to carboxylic acids, the process being fastest for chlorides and quite slow for amides.

3 back and forth 行ったり来たり

▶ Secretory vacuoles shuttle back and forth between the Golgi apparatus and the plasma membrane.

backbone n **1** 中枢 syn *main support*

▶ Silicon chips form the backbone of modern computing.

2 主鎖 ☞ main chain

▶ The active sites of these enzymes feature a zinc center attached to the protein backbone by three or four amino acid residues.

back-donate vt 逆供与する

▶ In the species containing a metal in the lowest oxidation state, more of the electron density is back-donated to the ligand.

back donation n 逆供与

▶ Back donation in carbonyl complexes is monitored conveniently by IR spectroscopy.

backfill vt 埋め戻す

▶ An oven-dried Schlenk flask equipped with a magnetic stir bar was evacuated and backfilled with nitrogen.

background n **1** 背景 syn *experience, qualifications*

▶ With this background in mind, we set out to synthesize an analogue of bone that comprised a porous hydroxyapatite–octacalcium phosphate–calcium dodecylphosphate composite film grown on titania–titanium substrate.

2 バックグラウンド

▶ In the presence of heavy atoms, errors due to absorption and uncertainty in the atomic scattering factors become so great that hydrogen atoms are lost in the general background of the Fourier and difference maps.

adj **3** バックグラウンドとなる

▶ There was only minimal background reaction observed in the absence of base and no background reaction in the absence of catalyst.

backside n 背面 syn *rear*

▶ If an optically active form of an organometallic compound were available, we could ascertain that S_E2 reactions involve front-side or backside attack, depending on whether the configuration of the products is retained or inverted.

backside displacement n 背面置換

▶ The actual product is the trans alcohol, from which we infer that reaction occurs by backside displacement.

back titration n 逆滴定

▶ Unchanged dichromate is determined by back titration against standard ammonium iron (II) sulfate solution.

backward adj **1** 逆の syn *reverse*

▶ For systems close to equilibrium, forward and backward rates are comparable, and the analysis of opposing reactions becomes important.

adv **2** 逆に syn *in reverse*

▶ The most useful way to plan a multistep synthesis is to work backward from the product to starting material.

backward scattering n 後方散乱 syn *back scattering* ☞ forward scattering

▶ Low-energy electron diffraction is the coherent backward scattering of low-energy electrons from the uppermost few atomic layers of a solid surface.

bacterial adj **1** 細菌の

▶ Bacterial lectins may bind to the ends of surface carbohydrates or to more internal parts of these molecules.

2 細菌による

▶ Bacterial mineralization is generally associated with cell wall processes in which extruded metabolic products coprecipitate with extraneous metal ions in the surrounding environment.

bacteriophage n バクテリオファージ

▶ Bacteriophages are viruses that infect bacteria.

bacterium n **細菌，バクテリア** 通常，複数形 bacteria が用いられる．

▶ Bacteria are referred to as prokaryotic cells or simply prokaryotes, named from the Greek words meaning before nucleus because they do not have a nucleus, an organelle that contains DNA in animal and plant cells.

bad adj **1 不利な** syn *unfavorable, unfortunate*

▶ Of all these reagents, the least bad results were obtained with sulfuryl chloride which gave sulfone in 19% yield.

2 ひどい syn *nasty, severe*

▶ The difficulties involved in the measurement of the solubilities of crystalline substances are bad enough, but those for amorphous or gelatinous substances are much worse.

badly adv **1 不都合に** syn *inadequately, insufficiently, poorly*

▶ Cis unsaturated acid chains have a bend at the double bond and fit each other badly.

2 大いに syn *greatly, very much*

▶ Predictive schemes for reactivity of polycyclic aromatic hydrocarbons based on whole molecule properties such as the HOMO energy or ionization potential can fail badly for large polycyclic aromatic hydrocarbons.

baffle n **じゃま板**

▶ The mixing chamber will contain an elaborate set of baffles to promote turbulent flow and mixing of the two reactant solutions.

balance n **1 均衡** syn *equilibrium, stability*

▶ The concentration of any metabolite in a cell is the result of the balance between its synthesis and its degradation.

2 残り syn *difference, remainder, rest*

▶ The composition of nodules is approximately 55% manganese and 35% iron, the balance being cobalt, copper, and nickel.

3 てんびん

▶ The balance was sensitive to a change of density of 0.0002 g per liter.

vt **4 釣り合う** syn *equalize, level, match*

▶ For the system to be at equilibrium, the force exerted on the piston must exactly balance the pressure of the gas.

ball-and-stick model n **玉棒模型**

▶ Ball-and-stick models of many organic substances with normal carbon valence angles of 109.5° are easily constructed.

ball mill n ボールミル

▶ The grinding efficiency of a ball mill depends on the number of contacts between any two balls; thus, the greater the number of balls, the more effective the grinding action.

ban n **禁止** syn *exclusion*

▶ Both sodium and calcium cyclamates（cyclohexanesulfamates）were commonly used as sweeters before 1969 ban.

band n **1 色の縞** syn *line, strip*

▶ The band of colored permanganate ion is seen to move uniformly away from the cathode and toward the anode.

2 吸収バンド ☞ absorption band

▶ Another strong, broad band due to C—O stretching appears in the 1000–1200 cm^{-1} region, the exact frequency depending on the nature of the alcohol.

3 バンド ☞ conduction band, valence band

▶ The energetically nearly continuous set of electronic states is called a band and accounts for many of the common macroscopic properties of bulk solids.

band gap n バンドギャップ syn *energy gap, forbidden band*

▶ The energy separation between the top of the valence band and the bottom of the conduction band is a fundamental property of the solid, called its band gap.

band model n バンドモデル

▶ The band model has been used widely not only for metals, which are characterized by a partly filled band of appreciable width, but for semiconductors and insulators, both of which have a lower (valence) and an upper (conduction) band.

band spectrum n バンドスペクトル

▶ The band spectrum of SiH_2 shows that the Si-H length is 1.521 Å, with H-Si-H angle 92.1°.

band structure n バンド構造

▶ The vast majority of molecular solids are insulators, and to create a molecular metal it is necessary to have strong intermolecular interactions to produce a delocalized electronic band structure in the solid and partial occupation of the band, brought about by electron transfer.

band theory n バンド理論

▶ Band theory is able to account satisfactorily for the properties of semiconductors such as Si and Ge.

barely adv かろうじて syn *hardly, just, only, scarcely*

▶ The physical properties of helium are mainly a result of its very low mass and the fact that its small, tightly held, electron cloud is barely distorted by neighboring atoms.

bar graph n 棒グラフ

▶ It is common practice to represent mass spectra in the form of a bar graph with a linear *m/e* scale.

barometric pressure n 気圧

▶ The density of the air varies with the temperature, moisture and carbon dioxide content, barometric pressure, and the value of gravity.

barrier n **1** 障壁 syn *block, impediment*

▶ Van der Waals strain can affect not only the relative stabilities of various staggered conformations but also the heights of the barriers between them.

2 壁 syn *boundary, limit*

▶ The greatest technical barrier lies in the discovery of energy-efficient regeneration routes.

basal adj 基底の

▶ Between the three basal phosphorus atoms are three sites for bridging atoms of oxygen. There are three more bridging sites between each of the basal phosphorus atoms and the peak.

basal plane n 底面

▶ If the basal planes of magnesium are oriented at other than 45° to the rod axis, it is found experimentally that a larger applied stress is needed in order for plastic flow to occur.

base n **1** 塩基 ☞ Lewis base

▶ Four principal bases are found in DNA: adenine (A) and guanine (G), which contain the purine ring system, and cytosine (C) and thymine (T), which contain the pyrimidine ring system.

2 対数の底

▶ Since the base e is less than 10, the logarithm to the base e is larger than the common logarithm; it is 2.3026 times the latter.

vt **3 base on** …に基づく syn *ground*

▶ Consideration so far has been based very largely on measurements obtained on thin films of pure oxides.

base catalyst n 塩基触媒 syn *basic catalyst* ☞ acid catalyst

▶ Coumarin is formed from acetic anhydride and salicylaldehyde in the presence of triethylamine as the base catalyst.

based adj **1 based on** …に基づく 形容詞であるから，base on は be 動詞か，修飾すべき名詞の後に位置する．

▶ The silicones are a group of neutral inorganic polymers based on rings, chains, or network of alternating silicon and oxygen atoms.

ただし，分詞句と見做せば，上記の制約はない．

▶ Based on theoretical considerations first given by Evans and Polanyi, Semenov has suggested that for exothermic abstraction and addition reactions of atoms and small radicals, the following approximate equation may be used:

2 複合語の第二構成素として**基礎が…の**

▶ Ammonia-borane is a promising material for transportation-based hydrogen storage.

baseline n 基準線

▶ In DTA, a horizontal baseline corresponding to $\Delta T = 0$ occurs, and superposed on this is a sharp peak due to the thermal event in the sample.

base metal n 卑金属 ☞ noble metal

▶ The base metals are removed most readily by allowing the mercury to run in a fine stream through about a meter of eight percent nitric acid.

base pair n 塩基対

▶ In the normal structure of DNA, the base pairs are aligned parallel with one another and with the plane of each base lying approximately perpendicular to the long axis of the two chains.

base peak n 基準ピーク

▶ The most intense peak in the mass spectrum is known as the base peak.

base sequence n 塩基配列

▶ Transcription factors bind and interact with specific base sequences.

basic adj 基本的な syn *central, essential, fundamental, principal*

▶ Thus, all four amorphous calcium phosphate samples contain the same basic arrangement of Ca^{2+} and PO_4^{3-} ions regardless of Ca:P ratio.

adj **塩基性の** ☞ alkaline

▶ Alkali hydroxides are the most basic of all hydroxides.

basically adv 要するに syn *largely, practically, primarily, substantially*

▶ Basically, the Born equation for the electric work can be developed by considering a uniform field of intensity E in a vacuum.

basic catalyst n 塩基性触媒 syn *base catalyst*

▶ The actual mechanism that results in this proton removal by the basic catalyst is not particularly important.

basic carbonate n 塩基性炭酸塩

▶ Dry air does not oxidize copper, but in moist air containing carbon dioxide it becomes covered with a layer of basic carbonate which protects the metal from corrosion.

basification n 塩基化

▶ Basification of the solution or of the isolated orange precipitate with either sodium hydroxide or carbonate solution yields the bright green nitrosamine base.

basicity n 塩基性度 ☞ acidity

▶ Alkoxides are powerful bases and, by varying the alkyl group, we can vary their degree of basicity, their steric requirements, and their solubility properties.

basic oxide n 塩基性酸化物 ☞ acidic oxide

▶ In the production of high-grade steel from iron relatively high in phosphorus and sulfur, use is made of a furnace lining containing basic oxides such as calcined magnesite or dolomite, so that a slag is formed containing magnesium phosphates and sulfates.

basis n **1 根拠** 複数形は bases で，base の場合と同じ． syn *ground*

▶ On an empirical basis, the thermochemical relationships among many different chemical compounds are firmly established.

2 on the basis of …に基づいて

▶ Pyridine is classified as aromatic on the basis of its properties.

3 基準 syn *relations, term*

▶ When compared on a molar basis, Li is nearly 50% more soluble than Na in liquid ammonia.

4 主成分 syn *essential*

▶ Cellulose is the basis of many plastics, fibers, coatings, lacquers, explosives, and emulsion stabilizers.

batch n バッチ

▶ We were thus able to make *trans*-stilbenes in 40-gram batches.

batch distillation n バッチ蒸留

▶ In batch distillation the entire sample of the material to be distilled is placed in the still before the process is begun, and product is withdrawn only from the condenser of the apparatus.

bath n 浴

▶ The flask is chilled in an ice bath to promote

further crystallization.

bathe vt 浸す syn *dip, rinse, soak*

▷ If the solutions bathing each side of the membrane contain equal concentrations of sodium cation, then the magnitude of the charge separation at each interface will be equal.

bathochromic shift n 深色移動 ☞ hypsochromic shift

▶ The bathochromic and hypsochromic shifts of the alkylazulenes could be explained by the inductive effect of the alkyl group.

battery n 電池 syn *cell*

▶ Much of the impetus for research on solid electrolytes has come from their possible use in new types of battery.

be vi **1** である

▷ Titanium metal resembles iron and nickel in being hard and refractory.

2 存在する syn *be found, be present, exist*

▶ If the double bond is between two carbon atoms, geometric isomers are called cis and trans.

bead n ビーズ，玉 syn *sphere*

▶ One common approach is to fill a column with plastic beads coated with molecules that will bind to the tag protein portion of the fusion protein.

beam n 線束 ☞ ray

▶ If we direct a beam of light through a mixture of hydrogen and chlorine, it explodes, forming hydrogen chloride and evolving a large amount of heat.

bear vi **1 bear on** 関連する syn *concern, relate*

▶ The question of scope is an important one that bears on the nature of the bonds involved.

vt **2** もつ syn *carry, support*

▷ The oxidation of an alcohol involves the loss of one or more hydrogens from the carbon bearing the −OH group.

3 関係をもつ syn *display, exhibit, have*

▶ Gels differ from jellies in that they bear a much closer relation to the crystal structure of the solid which is responsible for forming the gel.

4 bear in mind 覚えておく

▶ It is also necessary to bear in mind that ring flipping will most probably occur by a nonsymmetric rotation about the two torsional angles.

5 bear out 裏付ける syn *attest, substantiate, support*

▶ The expectation of a high intrinsic viscosity for polymers in good solvents compared with poor solvents is borne out by the result for polystyrene of 1.20 to 1.30 in good aromatic solvents and 0.65 to 0.75 in poor aliphatic solvents.

bearing n 関連 syn *connection, correlation, relation*

▶ The interfacial energy between two solid phases has an important bearing on the rate of reaction between two solids or the rate of decomposition of solids that result in the production of solid products.

beautiful adj すばらしい syn *appealing, attractive*

▶ More recent work with carboxylate bridging ligands revealed a wide range of exceptionally stable coordination polymers with beautiful structures, showing gas sorption properties.

because conj **1** だから 他の意味をもつ as や since よりは，because の使用が望ましいとする意見があるが，理由が既知のときには as と since を，理由が新しい知見で特に強調したいときに because を用いる慣習もある． syn *as, since*

▶ Some solids are not easily zone-refined, either because they melt with decomposition or because they sublime too readily.

2 because of (prep) …のために 同じ意味の on account of, owing to よりも，多く使用される．

▶ Rearrangements of this keto-enol kind take place particularly easily because of the polarity of the O−H bond.

Becke line n ベッケ線

▶ Using the Becke line method, it is possible to tell whether the refractive index of the sample is larger or smaller than that of the immersion liquid.

become vi 次に主語を形容する形容詞が置かれて，…になる syn *change into, transform into, turn into*

▶ The experimental values of ionization energy and electron affinity needed to evaluate electronegativity and hardness are rapidly becoming available.

bed n 層

▶ In gas chromatography, different components move through the bed of packing at different rates and so appear separately at the effluent end.

befit vt 適する syn *conform, satisfy, suit*

▶ Each absorption line corresponds to a change in electronic energy as befits the excitation of an electron from one quantum-mechanical orbital to another.

before adv **1** 以前に syn *already, earlier, formerly, previously*

▶ As before, no way is found for the determination of the individual activity coefficients; therefore, again mean-activity coefficients are all that need be considered.

prep **2** …よりも前に syn *ahead of*

▶ Before our work, there were no examples of stable species at room temperature containing covalent S−I or Se−I bonds except SeI_6.

conj **3** …よりも前に syn *preceding, previous to*

▶ Under favorable conditions, the methane-chlorination chain may go through 100 to 10,000 cycles before termination occurs by free-radical or atom combination.

begin vi **1** 始まる syn *initiate, start*

▶ It is only recently that ionic fragmentation pathways begun to be studied in depth.

2 to begin with まず第一に syn *first, initially*

▶ To begin with, it permits the determination of the standard state potential by means of a rather straightforward extrapolation procedure.

beginning n 始まり syn *onset, outset, start*

▶ From the very beginning, the boranes posed serious problems of both structure and bonding.

behave vi 振舞う syn *conduct, function, operate, perform, play*

▶ The ligand behaves as intended, but the metal center behaves much less predictably.

behavior n 挙動 syn *actions, conduct, manners*

▶ A real understanding of organic chemistry involves much more than a knowledge of the separate behaviors of classes of compounds with different functional groups.

behind adv **1** 残して

▶ One enantiomer may react more rapidly, leaving an excess of the other enantiomer behind.

prep **2** …の背後に

▶ The drawback with this method is that it does not provide insight into the reasons behind structure preference.

belief n 確信 syn *idea, opinion, view*

▶ Evidence supporting the belief that six-membered and larger rings are puckered and strain-free can be derived not only from the enthalpy of combustion but also in other ways as well.

believe vt 思う syn *assume, consider, expect, suppose*

▷ Many hydrated salts contain an odd number of water molecules, and often there are chemical grounds for believing that these molecules are not all equivalent.

belong vi **belong to** …に属する syn *be a part, be associated, relate*

▶ The nuclei of trivinylmethyl belong to point group D_{3h}, but treatment as C_3 will give optimum use of symmetry in this case.

below adv **1** 後のほうに syn *lower down*

▶ The six major preparative methods and typical examples are outlined below.

prep **2** 位置が…より下に ☞ above

▶ The dots above and below the oxygen atoms are intended to represent electrons in different orbits about the oxygen kernels.

3 未満の syn *less than, lower than, under*

▶ Temperatures below 0 ℃ are obtained with ice, water, and salt; with 1 part of sodium chloride and 3 parts of ice, temperatures to

−20 ℃ can be reached.

bend n **1** 曲がり syn *angle, curvature, curve*

▶ The rigidity of proline creates a kink or bend in the polypeptide chain.

vt **2** 曲げる syn *distort, twist*

▶ There is good evidence that some dihalides of the group 2 metals are bent whereas they would be expected to have linear AX_2 structures.

bending n 変角 syn *bend*

▶ Absorption due to various kinds of carbon–hydrogen bending, which occurs at lower frequencies, can also be characteristic of structure.

bending mode n 変角モード

▶ Two strong perpendicular-type bands with central Q branches are found in the infra-red, and their frequencies are in the right range for them to be assigned to the two bending modes.

bending vibration n 変角振動

▶ The internal stretching and bending vibrations of small molecules affect the properties of condensed phases, such as their vapor pressure, only through their effect on the zero-point energy and their comparatively small contribution to the dispersion energy.

beneath prep …の下に syn *below, under*

▶ Any acid can react with any base lying beneath it in the sequence of descending acidity.

beneficial adj 有益な syn *advantageous, favorable, useful*

▶ Unfortunately, not all immune responses are beneficial.

benefit n **1** 利益 syn *advantage, gain, help*

▶ Aquatic organisms are thought to be rich and valuable sources of new genes, proteins, and metabolic processes that may have important applications with human benefits.

vi **2** 利益を得る syn *gain, profit*

▶ Synthetic molecular chemistry has benefited from advances in methodology, and perhaps the most widespread innovation of recent years has been application of microwave technology.

vt **3** 役に立つ syn *aid, help, serve, support*

▶ We all recognize that chemicals can benefit us, just as chemicals can harm us.

benign adj 害の少ない syn *gentle, harmless, mild, soft*

▶ Manganese is approximately 1% of the cost of cobalt and significantly more environmentally benign than either cobalt or nickel.

bent adj 曲がった syn *curved, deformed, distorted, twisted*

▶ Aside from a comparison of the potential energy barriers separating the linear and the bent geometries, the relative flexibility of the structures may be assessed by the theoretical bending force constants.

bent bond n たわみ結合 syn *banana bond*

▶ On the basis of quantum mechanical calculations, Coulson and Moffitt proposed bent bonds between carbon atoms of cyclopropane rings; this idea is supported by electron density maps based on X-ray studies.

benzenoid n ベンゼノイド

▶ There are a number of compounds that possess some measure of aromatic character typical of benzene but that do not possess a benzenoid ring.

beset vt 付きまとう syn *encompass, surround, trouble*

▶ It is evident that a rigid quantitative treatment of these relations is beset with enormous difficulties.

beside prep …のそばに syn *alongside, close to, near, nearby, next to*

▶ Here a single isolated dot beside an atomic symbol represents an unpaired electron.

besides adv **1** そのうえ syn *in addition, also, as well, further, moreover, too*

▶ Because of this conjugation, α, β-unsaturated carbonyl compounds possess not only the properties of the individual functional groups but certain other properties besides.

prep **2** …のほかに syn *apart from, aside from, in addition to, other than*

▶ Besides being a valuable technique for high-molecular-mass materials, osmotic pressure effects are important in physiological systems.

best n **1 一番よいもの** syn *finest, first*
▶ With valence-bond approximations, the best that can be done is qualitatively to assay orbital overlaps from purely symmetry arguments.
2 at best いくらよく見ても syn *just, only*
▶ The literature concerning the addition of hydrogen bromide to unsymmetrical alkenes is at best confused.
3 at its best 最良の状態に syn *ideally, under the best of circumstance*
▶ Mössbauer spectroscopy is at its best in situations where other techniques are difficult or impossible, i.e., in the study of compounds which are insoluble solid powders.
4 to the best …の限り
▶ To the best of our knowledge, the strontium and barium dibromide and diiodide dimers are yet to be fully characterized experimentally or studied using reliable computational methods.
adj **5 最もよい** syn *excellent, finest, first, superior*
▶ Infrared spectroscopy is by far the best way to detect the presence of a carbonyl group in a molecule.
adv **6 最もよく** syn *in the most suitable way*
▶ Salts of amines with inorganic or organic acids are usually best named as substituted ammonium salts.

between adv **1 in between 中間に** syn *intermediate*
▶ Liquids do not exhibit the randomness of gases or the ordered arrangement of solids but lie somewhere in between.
prep **2 …の間で** 二つを対象として用いる．三つ以上のものを集団として対象とするときには among を用いる．
▶ The reaction between hydrogen and iodine is known to take place at bimolecular collisions involving a single molecule of each kind.
次の例は，二つずつを対象とする表現をまとめたものと見なされる．
▶ Other correlations between the melting point, boiling point, enthalpy of sublimation, and enthalpy of vaporization are also present in the same data set.
3 from between …の間から
▶ The extraction of lithium from between oxide ion layers is facile, and a chemical diffusion coefficient for the coupled ($Li^+ + e^-$) diffusion of $5 \times 10^{-8}\,cm^2\,s^{-1}$ at $Li_{0.65}CoO_2$ has been measured.

beyond prep **1 …を越えて** syn *exceeding, more than, over*
▶ Although ethylene does not absorb beyond 2000 Å, 1,3-butadiene absorbs intensely at 2170 Å.
2 …以上に syn *across, over*
▶ Humans cannot degrade purines beyond uric acid because we lack the enzyme uricase.

bias n **1 バイアス，偏り**
▶ The current increases exponentially with forward bias and is negligible with reverse bias.
vt **2 一方に偏らせる** syn *incline, influence, lead*
▶ In fact, our whole thinking about inorganic systems is strongly biased toward regarding solution and aqueous solution as almost synonymous.

biaxial adj **二軸性の** ☞ uniaxial
▶ Anisotropic crystals may be classed into uniaxial or biaxial, depending on whether they have one or two optic axes.

biaxial crystal n **二軸性結晶**
▶ In optically biaxial crystals for any one wavelength, there are two directions along which the light travels with zero birefringence.

bibliography n **文献目録** syn *catalogue, list, record*
▶ A very informative annotated bibliography is available which traces the course of this controversy.

bicyclic adj **二環式の**
▶ Camphor is a particularly well-known bicyclic terpene ketone which has uses as a medicinal and as a plasticizer for nitrocellulose.

bidentate adj **二座の** syn *didentate*
▶ In $Ti(NO_3)_4$, the bidentate nitrate ions are disposed tetrahedrally around the titanium which thereby attain a coordination number of eight.

bidentate ligand n **二座配位子** syn *didentate ligand*

▶ In the absence of other ligand, ClO_4^- can function as a monodentate or bidentate ligand.

bidirectional adj 二方向の

▶ Replication in *E. coli* is a bidirectional process.

bifunctional adj 二官能性の

▶ A linear polymer is obtained with a bifunctional anhydride and a bifunctional alcohol.

bifurcated hydrogen bonding n 二差水素結合 複合形容詞として使用

▷ There is crystallographic and nuclear quadrupole resonance evidence for a bifurcated hydrogen-bonding arrangement.

big adj **1** 大きな syn *great, large*

▶ The coupling of alkyl halides with organometallic compounds is the only one of these methods in which carbon–carbon bonds are formed and new, bigger carbon skeleton is generated.

2 重要な syn *important, major, notable, outstanding, significant*

▶ A big disadvantage of this method is the need for independent optical data as input.

bilayer n 二分子層

▶ An isolated bilayer cannot exist in water because exposed hydrocarbon tails would exist at the edges and ends of the sheet; however, this is precluded by the sheet curving to form a self-sealed hollow sphere.

bile acid n 胆汁酸

▶ Bile acids are synthesized from cholesterol in hepatocytes.

bile pigment n 胆汁色素

▶ The bile pigments are degradation products of hemoglobin and various porphyrin-containing molecules.

bimolecular adj 二分子の

▶ For the elementary reaction

$$Br + H_2 \rightarrow HBr + H$$

the molecularity is two, and the reaction is said to be bimolecular.

bimolecular elimination n 二分子脱離反応

▶ As the name implies, the rate of a bimolecular elimination is proportional to the concentration of both base and substrate.

bimolecular nucleophilic substitution n 二分子求核置換反応 S_N2

▶ Bimolecular nucleophilic substitution is often designated S_N2, S for substitution, N for nucleophilic, and 2 for bimolecular.

binary adj **1** 二成分の

▶ The system salt–water is binary only in the absence of hydrolysis.

2 二進法の

▶ These magnetic bubble materials can be used as memory components for binary digital computers.

binary compound n 二元化合物

▶ Silicon does not form binary compounds with the heavier members of group 14, but its compound with carbon, SiC, is of outstanding academic and practical interest.

binary system n 二成分系 syn *two-component system*

▶ Studies of binary systems that form solid solutions often are motivated by a desire to obtain insight into the nature and mechanism of phase changes occurring in one or more of the components.

bind vt **1** 結合する syn *attach, bond, connect*

▶ For many of the aromatic ligands that form complexes with Os(II), the metal can bind either the aromatic ring or a heteroatom lone pair that is not part of the aromatic system.

2 拘束する syn *constrain, hold, oblige*

▷ Alums form an obvious group of the hydrated salts of oxo acid anions, the cations such as $Al(H_2O)_6^{3+}$ being bound to anions SO_4^{2-} through a chain of hydrogen bonds.

binder n 結合剤

▶ The protein myoglobin serves as an oxygen binder in muscle.

binding energy n **1** 原子の結合エネルギー syn *bond energy*

▶ The binding energy for metals is defined experimentally as the heat of formation of the crystalline solid from the gaseous monatomic atoms.

2 核子の結合エネルギー

▶ Iron is thought to make up much of the

earth's core, possibly because its most common isotope, ^{56}Fe, has, in terms of binding energy per nuclear particle, the most stable atomic nucleus known.

binomial adj 二項式の

▶ In general, if the radical contains N equivalent protons, there are $N+1$ hyperfine lines with a binomial intensity distribution.

binuclear complex n 二核錯体

▶ In the case of high-T_c copper oxides, the required copper-copper interactions occur through oxygen bridges similar to the copper-copper interactions in antiferromagnetic binuclear copper(II) complexes.

bioactivity n 生理活性

▶ Ethylene and many other compounds are known that are not themselves hormones but that possess bioactivity of a regulatory type.

bioassay n バイオアッセイ

▶ Assays on bacteria determine their reactions to an antibiotic or insecticide. This is called bioassay.

bioceramic n バイオセラミック

▶ In many cases, the key requirement is that the bioceramic should become an intimate part of the tissue it is replacing.

biochemical adj 生化学的な

▶ The biochemical processes that occur in living systems are relatively constant over time even though there is a continual flux of energy and matter through them.

biochemically adv 生化学的に

▶ Steroids are polycyclic compounds closely related biochemically to terpenes.

biochemical reaction n 生化学反応

▶ Most of the amino acids involved in biochemical reactions have a primary amino group in the α-position with respect to the carboxylic acid function.

biocide n 生物致死剤

▶ The next major use of organotin compounds is an agricultural biocides, and here triorganotins are the most active materials.

biocompatibility n 生体適合性

▶ The biocompatibility and toxicity of dendrimers can be regulated by synthesis, particularly through judicious choice of functional groups at the periphery.

biocompatible adj 生体適合性の

▶ With ability of porous silicon to be both a photonic device and a biocompatible material, application in biosensing was almost inevitable.

biodegradation n 生物分解

▶ As polycyclic aromatic hydrocarbons are degraded, bacteria release light that can be used to monitor biodegradation rates.

bioenergetics n 生体エネルギー学

▶ A global overview of bioenergetics can be obtained by using thermodynamic ideas in the study of networks of reactions.

biofouling n 生物付着

▶ Normally a biosensor cannot be used for practical biofluid measurements without some degree of protection from background interference and biofouling.

biogenesis n 生合成 ☞ biosynthesis

▶ The biogenesis of alkaloids has been extensively studied and, although for a time, it was thought that alkaloids arose primarily from amino acid precursors, strong evidence is now available that acetate is also involved.

bioinformatics n 生命情報科学

▶ Applying computer science to the study of DNA and protein data has created an exciting new field called bioinformatics.

biological adj 生物学的な

▶ To a large degree, the primary influence of zinc in biological systems resides with its presence in ca. 300 enzymes.

biological activity n 生物活性

▶ The biological activity of a protein depends not only upon its prosthetic group and its particular amino acid sequence but also upon its molecular shape.

biological clock n 生体時計

▶ In many ways, telomeres serve as a biological clock for counting down cell divisions leading to senescence and cell death.

biologically adv 生物学的に

▶ The amide group is extremely important

biologically because proteins are composed of amino acids linked by amide bonds.

biological membrane n 生体膜

▶ Biological membranes in the active state appear to permit rapid translational and rotational motion within their lipid components, but movement across the bilayer from one monolayer to the other is much more restricted.

biological oxidation n 生体酸化

▶ Biological oxidation of tryptophan at the 5-position, followed by decarboxylation, yields an important hormone called serotonin.

bioluminescence n 生物発光

▶ Bioluminescence arises from enzyme-catalyzed oxidations which by energy accumulation give products in electronically excited states.

biomass n バイオマス

▶ In terms of energy, wood is by far the most important component of biomass.

biomedical adj 生物医学的な

▶ Synthetic analogues of bone are being actively pursued for biomedical applications in the field of bone replacement, augmentation, and repair.

biomimetic chemistry n バイオミメティッククケミストリー

▶ A notable example of biomimetic chemistry is the development of model synthetic catalysts which imitate the action of natural enzymes.

biomineral n 生体鉱物

▶ This is very apparent from studies of biominerals where the dimension, morphology, crystallographic alignment, and assembly of mineral particles are highly controlled and replicated over millions of years.

biomineralization n 生体鉱化 ☞ mineralization

▶ Biomineralization concerns the structure and synthesis of inorganic minerals in biological environments.

biomolecule n 生体分子

▶ There are four major classes of biomolecules that are synthesized by living systems: nucleic acids, proteins, lipids, and polysaccharides.

biopolymer n 生体高分子

▶ Secretion of biopolymers, such as collagen and chitin, into the extracellular space enables complex large-scale composite materials to be fabricated.

bioreactor n バイオリアクター ☞ reactor

▶ In large culturing containers called bioreactors, cells containing the DNA of interest can be mass produced.

bioremediation n バイオレメディエーション remediation は改善の意味.

▶ Bioremediation will not be able to eliminate all pollutants from the environment, but well-planned bioremediation approaches are an important component of environmental monitoring and cleanup efforts.

biosensor n バイオセンサー ☞ senser

▶ Bacteria such as *Vibrio* have been used as biosensors to detect cancer-causing chemicals called carcinogens, environmental pollutants, and chemical and bacterial contaminants in foods.

biosphere n 生物圏

▶ In the contemporary biosphere, overwhelmingly most enzymic action is carried out by proteins.

biosynthesis n 生合成 ☞ biogenesis

▶ The biosynthesis of macromolecules like protein requires a great deal of driving force both on account of the enthalpy required but also because of the huge decrease in entropy that occurs when so many small molecules are assembled in a precisely determined sequence.

biosynthesize vt 生合成する

▶ Hexadecanoic and octadecanoic acids and other natural fatty acids are biosynthesized from a number of molecules of acetic acid.

biosynthetic adj 生合成した

▶ One important example of a biosynthetic penicillin is penicillin V generated by fermentation in the presence of phenoxyacetic acid.

biotechnology n バイオテクノロジー

▶ Stem cell technologies are some of the newest, most promising aspects of medical biotechnology, but they are among the most controversial topics in all of science.

biotope n バイオトープ

▶ A mixed bacterial community isolated from a marine biotope highly contaminated with hydrocarbons has been shown to metabolize anthraquinone almost completely.

biphasic adj 二相の

▶ Polymer-forming reactions that occur at the phase boundary of an immiscible aqueous-organic biphasic system are usually referred to as interfacial polymerizations.

bipolaron n バイポーラロン ☞ polaron

▶ A bipolaron is a polaron-like state in which either two holes or two electrons are bound together by the lattice distortion.

bipyramidal adj 両錐形の ☞ pyramidal

▶ Pentagonal bipyramidal geometry is reported for the $UO_2F_5^{3-}$ ion in the crystal lattice of the potassium salt on the basis of a two-dimensional X-ray analysis.

biradical n ビラジカル

▶ The most familiar of inorganic biradicals is oxygen O_2, which has been found spectroscopically to have two unpaired electrons per molecule.

birefringence n 複屈折 syn *double refraction*

▶ The property of having two indices of refraction is known as birefringence.

birefringent adj 複屈折性の syn *doubly refractive*

▶ In birefringent material, light is propagated as two rays, ordinary and extraordinary, which have different refractive indices.

bisect vt 二等分する syn *cleave, split*

▶ The collinear OUO group is bisected by a plane of three nitrate groups.

bit n **1** 小片 syn *piece, portion, segment*

▶ A tiny bit of fine powder from the stock supply of substance may be sprinkled into the saturated solution before cooling or evaporating.

2 a bit 副詞的にちょっと syn *somewhat*

▶ Solving an unknown crystal structure is a bit like solving a set of simultaneous equations.

3 ビット

▶ Magnetic media such as computer disks use very small domains to hold binary bits.

bitter adj 苦い syn *harsh, sharp*

▶ Almost all salts of magnesium are colorless and soluble in water and have a bitter taste.

bivalent adj 二価の syn *divalent*

▶ Beryllium, although a bivalent metal of group 2 in the periodic classification, resembles aluminum in its reaction.

bivariant adj 二変の ☞ univarient

▶ Every phase of the system, constituting a bivariant equilibrium state, is defined by an area, with two degrees of freedom.

black body n 黒体

▶ A refractory body coated with iron(III) or manganese(IV) oxide behaves approximately as a black body, but by far the best experimental black body is a cavity with a small aperture.

black-body radiation n 黒体輻射

▶ The entirely new concepts came from the quantum theory, which was born in 1900 with the analysis by Max Planck of black-body radiation.

black powder n 黒色火薬

▶ Black powder deflagrates rather than detonates.

blank experiment n 空実験 ☞ control experiment

▶ A blank experiment must always be made to see how much of silver solution is necessary to produce the red shade used in the titration when no chloride is present.

blanket vt 一面に覆う syn *cover, smother, spread*

▷ For fires in flammable liquids a blanketing or smothering effect is essential.

blanketing n 掩蔽

▶ Both these observations are most easily accounted for by a simple blanketing of the metal surface by TiO_x.

bleach vt 脱色する syn *whiten*

▶ Irradiation in the F band bleaches the F centers and produces vacant anion sites and F' centers.

bleaching n 漂白
▶ The efficiency of a bleaching plant depends on the precision in determining the hypochlorite concentration.

bleaching agent n 漂白剤
▶ Sodium peroxide finds widespread use industrially as a bleaching agent for fabrics, paper pulp, wood, etc. and as a powerful oxidant.

bleaching powder n さらし粉
▶ Bleaching powder is a white, finely powdered substance that usually smells of chlorine, because of its decomposition by water vapor in the air.

blend n **1** 混合 syn *combination, mix, mixture*
▶ The mixtures for the phase studies were prepared by weighing the components accurately into a microbeaker, melting, stirring to give an intimate blend, and cooling with stirring until solidification occurred.
vt **2** 混合する syn *combine, mix*
▷ Multicomponent oxides are made by blending oxides, hydroxides, or carbonates.

blending n 混合 syn *mixture*
▶ Soaps cause blending of methanol and cyclohexane.

blinding adj 目をくらます syn *dazzling*
▶ Mixed with 1.5 volumes of oxygen, COS burns with explosive violence and blinding light.

block n **1** ブロック syn *region, section, zone*
▶ The resultant orthorhombic unit cell comprises three cubic perovskite blocks stacked along the crystallographic *c* axis.
2 遮断 syn *barrier, impediment, inhibition, interference*
▶ A block at the last complex in the electron-transport chain prevents the reduction of oxygen and causes all the electron-transport complexes to become fully saturated with electrons.
vt **3** 遮断する syn *hamper, impede, obstruct, prevent*
▷ Because of large number of steps associated with the translation of mRNA into protein, there are numerous opportunities available for blocking it with inhibitors.

blockage n 阻害 syn *impediment, inhibition, interference*
▶ Gall bladder blockage and the inhibition of pancreatic lipases decrease the breakdown of triglycerides in the small intestine.

block copolymer n ブロック共重合体 ☞ random copolymer
▶ In block copolymers, sections made up of one monomer alternative with sections of another.

block diagram n ブロック線図
▶ A block diagram of the experimental arrangement used in this work is shown in Figure 1.

blocking n ブロッキング syn *prevention, suppression*
▶ Blocking is favored by high pressures, thick membranes, high concentrations, and absence of surface-active or stabilizing agents.

block structure n ブロック構造
▶ The block or double shear structures are characterized by the length, width, and manner of connectivity of the blocks of unreduced ReO_3 structure.

blood plasma n 血漿
▶ If the concentration of low-density lipoprotein（LDL）is high and high-density lipoprotein（HDL）is low in the blood plasma, then there will be an increased risk of heart disease.

bloodstream n 血流
▶ The enzyme projects into the bloodstream and hydrolyzes the triglyceride that is contained in chylomicrons, to yield fatty acids and glycerol.

blossom vi 発展する syn *develop, flourish, grow*
▶ The literature dealing with radiation damage and radiation effects has blossomed to impressive proportions.

blow n **1** ひと吹き syn *knock, punch*
▶ Physical signals are generally restricted to energetic blows, heat, and electromagnetic radiation.
vt **2** 吹き飛ばす syn *drive*
▶ The standard method of determining the

size of the pores is to observe the pressure of air required to blow bubbles through the wet membrane.

3 吹く syn *enlarge, expand*

▶ The tube is withdrawn and blown, without pulling, until the heated section is expanded slightly beyond the normal diameter.

blowing agent n 発泡剤

▶ Typical blowing agents are baking powder, halocarbons and dichloromethane in urethane, pentane in expanded polystyrene, and hydrazine and related compounds in various types of foamed plastics.

blue shift n ブルーシフト ☞ red shift

▶ The blue shift of n–π bands is particularly noticeable on going from a hydrocarbon to a hydroxylic solvent.

bluish adj 青みを帯びた

▶ Zinc and cadmium are silvery solids with a bluish luster when freshly formed.

blurred adj 形状がぼやけた syn *indefinite, indistinct, obscure*

▶ With increasing time, the splitting of this band becomes increasingly blurred and disappears completely after 145 s.

blurring n ぼやけること

▶ Studies of the blurring of the NMR spectrum of 1, 2-difluorotetrachloroethane with temperature provide a means of evaluating the amount of energy the molecules must have to allow rotation to take place.

boat conformation n ボート形配座 ☞ chair conformation

▶ The ions $Te_3S_3^{2+}$ and $Te_2Se_4^{2+}$ have pronounced boat conformations.

boat form n ボート形 ☞ chair form

▶ The instability of the boat form of cyclohexane relative to the chair form may be ascribed to relatively unfavorable interactions between the hydrogen atoms around the ring.

bodily adj **1** 身体の syn *physical*

▶ The interconversion of triphosphates, diphosphates, and phosphates is important in many bodily processes, including the absorption and metabolism of sugar.

2 有形の syn *substantial, tangible*

▶ Bubbles, drops, and solids suspended in a liquid exhibit bodily movement when placed in an electric field.

body n **1** 主体 syn *bulk, main portion, majority, mass*

▶ A very large body of work has been carried out on transformations in solids which are, or appear to be, continuous.

2 物体 syn *substance*

▶ The concept of temperature can be evolved as soon as a means is available for determining when a body is hotter or colder.

body center n 体心 複合形容詞として使用.

▶ The unit cell of CsCl is not body-centered cubic since there are different ions at corner and body-center positions.

body-centered cubic adj 体心立方の

▶ A common structure for metallic elements is the body-centered cubic structure, with a cubic unit cell having atoms at the corners and an identical atom in the center of the cube.

body diagonal n 体対角線 ☞ face diagonal

▶ Face-centered cubic metals have four different sets of close-packed planes which are perpendicular to the body diagonals of the cubic unit cell.

body fluid n 体液

▶ The complete inertness of niobium to body fluid makes it the ideal material for surgical use.

Bohr magneton n ボーア磁子

▶ The Bohr magneton is the natural unit of magnetism and is equal to the magnetic moment of an electron spinning on its own axis.

boil vi **1** 沸騰する syn *foam, froth*

▶ A solution of hydrochloric acid of specific gravity of 1.10 boils at 110 ℃ under atmospheric pressure.

vt **2** 煮沸する syn *heat, warm up*

▶ When collagen is boiled with water, the strands come apart and the product is ordinary cooking gelatine.

boiling n **1** 沸騰
▶ If the solution contains ammonium salts, it is possible to remove them by boiling with an excess of sodium carbonate solution.
adj **2 boiling hot** 猛烈に熱い
▶ If solutions are not boiling hot during the precipitation, the calcium oxalate forms very fine crystals.

boiling point n 沸点
▶ The boiling point may be defined as the temperature at which a liquid gives off bubbles of its own vapor from all parts of its depth.

bomb n ボンベ
▶ An oxygen bomb is used for accelerated aging tests for rubber and plastic products; oxygen under high pressure is used.

bombard vt 粒子などで衝撃を与える syn *attack*
▶ On passage of an electric current through the fluorescent lamp, the atoms of mercury are bombarded by electrons and are excited into upper electronic energy states.

bombardment n 衝撃
▶ Powdered NaCl turns a greenish-yellow color after bombardment with X-rays for half an hour or so.

bond n **1** 化学結合 syn *bonding* ☞ linkage
▶ Covalent bonding gives highly directional bonds in which one or all of the atoms present have a definite preference for a certain coordination environment, irrespective of the other atoms that are present.
vt **2** 結合する syn *combine, link, unite*
▶ There are a vast number of compounds in which phosphorus is bonded to carbon, since it can replace both nitrogen atoms and CH groups in a wide range of organic compounds.
3 接着する syn *attach, stick, weld*
▶ Tungsten carbide is bonded with cobalt at 1400 ℃; from 3 to 25% of cobalt is used, depending on the properties desired.

bond angle n 結合角
▶ Probably no tetravalent carbon compound, except one with four identical substituents, has exactly tetrahedral bond angles.

bond cleavage n 結合開裂 ☞ cleavage
▶ Both metal-metal bond cleavage and CO dissociation could occur in the $M_3(CO)_{12}$ clusters on irradiation with ultraviolet light.

bond dissociation energy n 結合解離エネルギー
▶ The amount of energy consumed or liberated when a bond is broken or formed is known as the bond dissociation energy.

bonded adj 結合した
▶ $[Ag(S_4N_4H_4)_2]^+$ has a sandwich-like structure and is unique in being S-bonded rather than N-bonded to the metal ion.

bond energy n 結合エネルギー syn *binding energy*
▶ By the use of similar procedures for molecules containing different kinds of bonds, it is possible to arrive at set of bond energies that will allow us to make approximate estimates of energies of atomization and energies of formation.

bond formation n 結合形成
▶ The liberation of a definite amount of exchange energy is one essential characteristic of covalent bond formation.

bonding n **1** 化学結合 syn *bond* ☞ linkage
▶ Commonly, the bonding may be a blend of the different types, as in TiO which is ionic and metallic or CdI_2 which is ionic, covalent, and van der Waals.
2 接合 syn *association, link, linkage*
▶ Subsequent sintering promotes bonding between the grains that compose the pellet, thereby increasing the density and mechanical strength of the pellet.

bonding electron n 結合性電子
▶ The methyl group in a CH_3-C bond repels the bonding electrons more than does hydrogen in a corresponding $H-C$ bond which was taken as the reference standard.

bonding interaction n 結合性相互作用
▶ The most energetically significant bonding interaction is between e_{1g} metal orbitals and the ligand e_{1g} pair.

bonding orbital n 結合性軌道 ☞ antibonding orbital

▶ If two parallel p orbitals combine, they produce a π-bonding orbital and a higher-energy π*-antiboning orbital.

bond length n 結合距離 syn *bond distance*

▶ Tricarbon dioxide is linear, with bond lengths 1.16 Å for carbon-oxygen and 1.28 Å for carbon-carbon.

bond moment n 結合モーメント

▶ Ammonia has a dipole moment of 1.46 D. This could be accounted for as a net dipole moment resulting from the three individual bond moments.

bond order n 結合次数

▶ Metal-metal bond orders range from 8/12 = 2/3 for the $M_6X_{12}^{2+}$ (M = Nb, Ta) species, through 12/12 = 1 for $Mo_6X_8^{4+}$ to 6/3 = 2 for the Re_3X_9 species.

bond polarity n 結合極性

▶ The utility of the scale of electronegativities has led to a search for a parallel between bond polarity and electronegativity.

bond strength n 結合強度

▶ Accurately determined M−M distances provide the most readily available indication of bond strength.

bond stretching n 結合伸縮

▶ Absorption bands in the IR spectrum result from energy changes arising as a consequence of molecular vibrations of the bond stretching and bending (deformation) types.

bone n 骨

▶ A prime consideration in the preparation and performance of a bone implant material is the selection of a surface that is able to induce nucleation of octacalcium phosphate $Ca_8(PO_4)_4$-$(HPO_4)_2 \cdot 5H_2O$ and growth of new bone.

bookkeeping n 簿記 syn *accounting*

▶ The metallic character of Rb_9O_2 comes from the excess of at least five electrons above that required for simple bookkeeping.

boost n **1** 向上 syn *encouragement, help, support*

▶ Of the classical spectroscopic methods, IR has given a remarkable boost to understanding the gas-solid interface and, to some extent, the solid-liquid interface.

vt **2** 上昇させる syn *increase, raise*

▶ If the antigen is repeatedly contacted at regular intervals, each such encounter boosts the immune response to higher protective levels.

borane n ボラン

▶ Boranes are usually named by indicating the number of B atoms with a latin prefix and the numner of H atoms by an arabic number in parentheses, e.g., B_5H_9, pentaborane(9). They are referred to as electron-deficient because there are insufficient electrons with which to form all normal electron-pair bonds.

border line n 境界線 syn *division, boundary*

▶ PCl_5 is closer to the ionic-covalent border line than is PF_5.

border line adj 境界線上の syn *disputable, doubtful, questionable*

▶ In some other cases of borderline radius ratios, distorted polyhedra and/or coordination numbers of 5 are observed.

Born-Haber cycle n ボルン・ハーバーのサイクル ☞ thermochemical cycle

▶ Consideration of the terms in a Born-Haber cycle helps to rationalize the existence of certain compounds and the nonexistence of others.

borosilicate glass n ホウケイ酸ガラス

▶ Borosilicate glasses are extensively used because of their small coefficient of thermal expansion and their easy workability.

borrowing n 借用

▶ Intensity borrowing between two different electronic excitations is more complex and less susceptible to simple quantitative analysis.

both pron **1** 両方とも

▶ Both of these interactions tend to be very weak since the polarizabilities of most species are not large.

2 A and B both それぞれ この順にあるときの both は，A と B それぞれを意味する．

▶ The rock salt and nickel arsenide structures

both have octahedrally coordinated cations and differ only in the anion stacking sequence.

adj **3 両方の** 定冠詞があれば，その前に置く．

▶ The obvious feature of both types of spectrum is the appearance of radiation at a series of discrete frequencies.

conj **4 both A and B 両者の** 両者を強調するだけで，比重の差はない．冠詞，形容詞を伴うときは，その前に置く．

▶ Intramolecular hydrogen bonding between suitable ortho groups has the effect of reducing both the melting and boiling points.

A と B は名詞に限らないが，文法上，対等の関係にある語または表現であることを要する．☞ and

▶ The mechanism requires both that there be an α hydrogen and also that the center of asymmetry be located at this α carbon.

bother vi 否定文でわざわざ…する syn *concern with*

▶ This is a straightforward calculation, and we shall not bother with the details.

bottle n **瓶** syn *container, flask*

▶ To remove alkene impurities, a gaseous alkane is bubbled through several bottles of sulfuric acid, and liquid alkane is shaken with sulfuric acid in a separatory funnel.

bottleneck n **障害** syn *impediment, snag*

▶ The bottleneck in such syntheses is the need to isolate and purify the new peptide made in each cycle; the time required is enormous, and the yield of product steadily diminishes.

bottom n **1 底面** syn *base*

▶ At the bottom of the unit cell, $z = 0$, and halfway up $z = 1/2$. The top layer ($z = 1$) must be identical to the bottom layer.

adj **2 最下の**

▶ The bottom electronic states of the 3p band in Mg are lower in energy than the upper states of the 3s band.

bottom-up adj **基本的な要素から始めて複合的な要素に至る** ☞ top-down

▶ The challenge that confronts scientists and engineers fascinated by the prospect of fabricating molecular electronic devices by a bottom-up approach calls for radical departures from current attitudes and activities in many areas of science and technology.

bound n **1 複数形で，限界** syn *boundary*

▶ It was not entirely outside the bounds of possibility in such novel circumstances.

adj **2 結合した**

▶ There are protein-bound Fe_3S_4 clusters in some enzymes and bacteria.

vi **3 跳ね返る** syn *hop, jump, spring*

▶ As the temperature increases, the molecules in a crystal become more and more agitated; each one bounds back and forth more and more vigorously in the little space left for it by neighbors.

boundary n **境界線** syn *bound, limit*

▶ It is possible to define an approximate boundary for each ion and, by an integration procedure, to calculate the total number of electrons around each ion.

boundary condition n **境界条件**

▶ The selection of the set of solutions appropriate to a given problem is accomplished by applying the correct boundary conditions.

boundary line n **境界線** syn *bound, division*

▶ That the boundary line separating normal and inverse structures is a realistic one is indicated by the fact that the borderline examples are often partially disordered with varying amounts of normal and inverse character.

bound state n **束縛状態**

▶ Laser-induced fluorescence is limited to species with a bound and accessible upper electronic state, and some important radical species cannot be detected using this method.

bracket n **1 角括弧** ☞ curly bracket, square bracket

▶ The quantities within the brackets denote either concentrations for liquid reactants or partial pressures for gaseous reactants.

vt **2 挟んで位置する** syn *classify, place, rank*

▶ In the order of abundance, strontium (384 ppm) and barium (390 ppm) are bracketed by sulfur (340 ppm) and fluorine (544 ppm) in order of abundance.

branch n **1** 分枝

▶ In contrast to our study, introduction of an ethyl branch at the same position on a long alkyl chain for M = Cu increases the solid-to-liquid crystal temperature.

2 分野 syn *division, section*

▶ Although yttrium has received little attention in the past, it is now acquiring a much higher profile because of applications of its compounds in various branches of solid-state technology.

branched adj 分岐した

▶ One of the most successful polymer electrolytes is the comb-branched block copolymer.

branched-chain adj 枝分かれ鎖の

▶ The possibility of branched-chain hydrocarbons isomeric with the continuous-chain hydrocarbons begins with butane.

branched copolymer n 分枝共重合体

▶ Branched polymers can be made at will by the addition of suitable reactants to the polymerization vessel.

branching n **1** 枝分かれ

▶ The prefix *n*- has been retained for any alkane, no matter how large, in which all carbons form a continuous chain with no branching.

adj **2** 枝分かれの

▶ A branching methyl substituent was introduced by regioslective epoxide opening.

branching point n 分岐点

▶ To some extent, branching points may be confined to amorphous regions or, in favorable circumstances, may also be incorporated to a degree as defects within the crystalline phase.

breadth n 広さ syn *extent, magnitude, range, scope*

▶ The breadth and diversity of the subject has attracted the interest of researchers from every conceivable scientific background.

break n **1** 割れ目 syn *rupture, separation*

▶ A fine scratch gives a much better opportunity for a clean break than a wide groove sawed in the tube at the expense of dulling the file.

2 急変 syn *discontinuity*

▶ If the discussion is limited to those cases where the solute forms no hydrate and where the fused solute is completely miscible with water, then it follows that a break in the solubility curve must be due to a change in the crystal form of the solid solute.

vi **3** 壊れる syn *crack, fracture, fragment, shatter, split*

▶ Foaming volume is small when all bubbles break and escape upon reaching the surface.

4 break away (a) 離脱する syn *leave*

▶ The fact that the migrating group is ordinarily not racemized to any significant extent in a 1, 2-shift suggests rather strongly that this group, in the course of its migration, never breaks completely away from the rest of the molecule.

(b) 逸脱する syn *depart*

▶ Materials chemists are recognizing the need to break away from the traditional divisions of inorganic, organic, and physical chemistry toward a multidisciplinary approach.

5 break down 分解する syn *break up, decompose*

▷ NaCl and KCl do not intercrystallize in all proportions but form a series of double salts, which are stable at the temperature of formation but unstable at ordinary temperature, breaking down into simple halides.

6 break out 脱出する syn *escape*

▶ As the photon energy increases, the iodine atoms are produced with greater kinetic energy and are, therefore, more likely to break out of the reaction cage.

vt **7** 壊す syn *fragment, rupture, shatter, split*

▶ Dissociation energy is a definite quantity that refers to the energy required to break a given bond of some specific compound.

8 乱す syn *violate*

▶ The trend of increasing $D(\mathrm{X{-}X})$ with decreasing atomic number

$$D(\mathrm{I{-}I}) < D(\mathrm{Br{-}Br}) < D(\mathrm{Cl{-}Cl})$$

is broken with F_2 which has about the same bond dissociation energy as I_2 does.

9 break open こじ開ける

▶ To separate the fusion protein from other proteins normally made by the bacteria, cells are broken open and homogenized to create a bacterial milkshake of sorts known as an extract.

10 break up 分割する syn *disintegrate*

▶ Certain interaction potentials may reasonably be broken up into central forces acting between pairs of atoms, one in each of two different molecules in the crystal.

breakage n 破損 syn *fracture*

▶ A dynamic foam is one to which new bubbles are being added, and a steady state is reached when the rate of breakage equals the rate of replenishment.

breakdown n **1** 挫折 syn *failure*

▶ The breakdown of the Curie law is due to the effect of neighboring paramagnetic atoms, so that we would not expect the Curie law to be obeyed in the solid state, where the density of paramagnetic atoms will be high.

2 崩壊 syn *destruction*

▶ Another temperature of great important in the practical use of polymers is the temperature near which thermal breakdown of the polymer chains occurs.

breaking n 破壊 syn *fracture, rupture*

▶ About 25 percent soda reduces the viscosity of a glass by a factor of 10^{10}. Chemically this can be thought of as a breaking of the crosslinking between different Si—O—Si chains.

breakthrough n 大きな進歩 syn *advance, progress*

▶ One of the most exciting scientific breakthroughs in recent years has been discovery of superconductivity at relatively high temperatures in the ceramic oxides, $La_{2-x}Ba_xCuO_{4-x}$ and $YBa_2Cu_3O_7$.

breathe vt **breathe upon** …に息を吹きかける syn *blow*

▶ Many anhydrous phosphates are reduced by heating with magnesium to phosphides, which, on being breathed upon, give the peculiar odor of phosphine.

breed vt 引き起こす syn *create, generate, produce*

▶ Pressure tuning has dramatic and extraordinary consequences which, in themselves, exemplify the truth of the statement that understanding breeds exploitation.

bridge n **1** 架橋 syn *connection, link*

▶ An interesting compound having a fully conjugated cyclic structure of four pyrrole rings linked together through their 2- and 5-positions by four methine bridges is known as porphyrin.

vt **2** 架橋する syn *connect, link, unite*

▶ Copper(II) forms acetate dimers, $Cu_2(O_2CCH_3)_4(H_2O)_2$; the four acetate groups bridge the two copper ions, forming two CuO_4 squares, one on top of the other, with a water molecule attached to the outer face of each square.

3 ギャップを埋める syn *cross over, span, traverse*

▶ An important new and rapidly growing field which attempts to bridge the gap between the gas and liquid phases is that of cluster studies.

bridgehead radical n 橋頭遊離基

▶ It was surmised that successful dimerization required high concentrations of combining bridgehead free radicals, and low solubilities were characteristic of the longer rod compounds.

bridging carbonyl n 橋かけカルボニル

▶ That the extent of π bonding to terminal and bridging carbonyls is different is clearly shown by the carbonyl stretching frequencies.

bridging ligand n 架橋配位子

☞ ambidentate ligand

▶ Very frequently CN^- acts as a bridging ligand —CN— as in Prussian-blue-type compounds.

brief n **1 in brief** 要するに syn *briefly*

▶ In brief, the overall synthesis involves the formation of a stilbene such as difluorostilbene which is then dihydroxylated, protected, and a double lithiation performed before ring closure on electrophilic sulfur.

adj **2** 簡潔な syn *concise, short*

▶ Brief mention can be made of the interesting and important class of materials known as semiconductors.

briefly adv **1** 簡潔に syn *concisely, in brief, in short, in summary*

▶ We begin by briefly reviewing the various mechanisms which have been proposed to account for the transfer of energy between molecules.

2 しばらく syn *for the moment, temporarily*

▶ When a photographic film is briefly exposed to light, some of the grains of silver bromide undergo a small amount of decomposition.

bright adj **1** 明るい syn *brilliant, shining, shiny*

▶ Iodoform is a highly insoluble, bright-yellow solid.

2 有望な syn *hopeful, promising*

research and development は一つの概念として扱われる.

▶ Future research and development in these areas is especially bright.

brightener n 光沢剤

▶ Brighteners are essential to most electroplating systems and are thought to function by two mechanisms in aqueous solutions.

brightly adv 明るく syn *clear, luminously*

▶ The brightly colored solutions obtained when sulfur is dissolved in oleum are paralleled by similar behavior of Se and Te.

brightness n 明るさ syn *brilliance, shine*

▶ On rotating a sample, extinction occurs every 90°; at 45° position, the sample exhibits maximum brightness.

brilliance n 光輝 syn *brightness, luster*

▶ The brilliance of a cut diamond is made possible by its high index of refraction.

brilliant adj 光り輝く syn *bright, luminous, lustrous, shining*

▶ Hematite is a brilliant black to blackish-red or brick red mineral with brown to cherry red streak and metallic to dull luster.

brine n かん水

▶ The brines in central Michigan were richest in bromine content.

bring vt **1** もってくる syn *bear, carry, deliver, take*

▷ In FeTe, $c/a = 1.49$; hence, the iron-iron distance parallel to c is reduced to $0.745a$, thereby bringing these iron atoms into closer contact and increasing the metallic bonding in the c direction.

2 bring about もたらす syn *achieve, cause, produce*

▷ One of the classic photochemical reductions of organic chemistry is the formation of benzopinacol, as brought about by the action of light on a solution of benzophenone in isopropyl alcohol.

3 bring in 導入する syn *produce, realize, yield*

▶ The valence-bond theory brings in the delocalization by sharing electrons, usually between two atoms but a greater number can be involved if several resonance structures can be conceived of.

4 bring out 引き出す syn *display, feature*

▶ Fluorine tends to bring out the highest valence of the element with which it combines.

5 bring together まとめる syn *assemble, gather*

▶ A double decomposition of the type

$$CsF + LiI \rightarrow CsI + LiF$$

will be expected, for it brings the smaller ions together and, hence, increases the total lattice energy.

brisk adj 勢いのよい syn *active, vigorous*

▶ In the present experiment, adequate agitation is provided by introducing a boiling stone and adjusting the flame of the burner to a point of brisk refluxing, which is continued for a period of 2 hours.

brittle adj もろい syn *fragile*

▶ All glasses are known to be brittle solids, in the sense that they are weak in tension, fracturing at low strain.

brittleness n 脆性

▶ Ozone can cause unwanted crosslinking in rubbers and other polymers with residual unsaturation, thereby leading to brittleness and fracture.

broad adj **1** おおざっぱな syn *approximate,*

rough

▶ These data should be used only for broad comparisons since the estimated bond energies depend markedly on the particular compounds being studied.

2 広範囲の syn *comprehensive, extensive, general, inclusive*

▶ Our aim is to give a broad picture of recent developments.

3 広義の syn *inaccurate, inexact, loose, rough*

▶ Sometimes the word "substance" is used in a broader sense, essentially as equivalent to material.

4 幅の広い syn *extensive, wide*

▶ Many substances with broad absorption bands will have maximum absorption below 4000 Å and yet appear strongly colored because their absorption bands extend into the visible spectrum.

broaden vt 広げる syn *expand, extend, widen*

▶ The peak is somewhat broadened and, most importantly, is shifted upfield to 169 ± 2 ppm.

broadening n 広がること syn *extension, spread*

▶ By study of the broadening of the peak or of the coalescence of two peaks, it is possible to estimate the E_{act} for rotation.

broad-line NMR n 広幅 NMR

▶ Broad-line NMR has been used to detect and to characterize motional transitions in several π-π molecular complexes of naphthalene in the solid state.

broadly adv おおまかに syn *approximately, generally, largely, roughly*

▶ The fact that the order of the stabilities is broadly similar for the three series of nitro compounds indicates that the same sort of molecular interactions is operative in all cases.

broadly speaking 概して

▶ The least readily melted halides are, broadly speaking, the best conductors when fused, and usually the transportation of current is mainly by the cations.

brominate vi **1 臭素化する**

▶ Toluene brominates much faster than nitrobenzene.

vt **2 臭素化する**

▷ The compound *N*-bromosuccinimide is a reagent used for the specific purpose of brominating alkenes at the allylic position.

bromination n 臭素化

▶ The rate of reaction of bromination by bromine is much greater than iodination by iodine.

bronze n **1 青銅**

▶ Cupper is widely employed in coinage alloys as well as the traditional bronze, brass, and special alloys such as Monel.

adj **2 ブロンズ色の** syn *bronze-colored*

▶ The alkali metal solutions in ammonia are deep blue in color when dilute and bronze when concentrated.

brown adj 褐色の

▶ The amorphous form of silicon is a dark brown powder.

Brownian movement n ブラウン運動 syn *Brownian motion*

▶ Brownian movement is just a visible manifestation of the kinetic theory of heat; every independent rigid particle, small or large, has the same average kinetic energy.

brownish adj 褐色を帯びた

▶ Silver oxide, Ag_2O, is a brownish black powder, which on being heated to 300° is completely decomposed into metal and oxygen.

bubble n **1 気泡** syn *foam, froth*

▶ By bubbles we mean either ordinary bubbles, in which air or vapor is trapped by a thin film, or cavities full of vapor in a liquid.

vt **2 泡立たせる** syn *foam*

▶ A mixture of chloride ion and hypochlorite ion is formed when chlorine is bubbled through a solution of sodium hydroxide.

bubble chamber n 泡箱

▶ Liquid hydrogen is used in bubble chambers for studying high-energy particles.

bubbler n バブラー

▶ Liquid precursors with high vapor pressures were particularly suitable for volatilization in bubblers.

Buchner funnel n ブフナー漏斗

▶ The classical apparatus for collection of

crystals by suction filtration is a Buchner funnel fitted to a filtering flask with a rubber stopper.

buffer n **1 緩衝** syn *safeguard*

▶ The simplest store acts as a buffer when vital supplies are low.

2 緩衝液 syn *buffer solution*

▶ The solution of sodium hydrogenphthalate is useful as a buffer, that is, a solution of a pH that changes only slightly on addition of small amounts of either a strong acid or a strong base.

vt **3 緩衝する**

▶ Blood and other physiological solutions are buffered; the pH of blood changes only slowly from its normal value on addition of acid or base.

buffer action n **緩衝作用**

▶ Between the two inflections corresponding to NaH_2PO_4 and Na_2HPO_4 the pH changes relatively slowly with addition of NaOH, and this is an example of buffer action.

build vt **1 組み立てる** syn *assemble, construct*

▶ The sequence in which the amino acids are built into the protein chain depends upon the sequence of bases along the messenger RNA chain.

2 build up 築き上げる syn *accumulate, develop, enlarge, increase*

▷ Colloids may be prepared by building up particles from molecular species.

builder n **ビルダー**

▶ It is common practice to add electrolytes to detergents and soap powders as builders.

building n **構築** syn *fabrication, production, structure*

▶ Model building shows that structures based on four- and five-membered rings are compatible with many of the observed properties of the larger silicate species found in solution; the proportion of five-membered rings increases as the size of the particle increases.

building block n **構成要素** syn *material*

▶ The basic five-carbon-atom building blocks of the terpenoids are formed from acetic acid via a series of condensations and decarboxylations.

building-up principle n **構成原理** syn *Aufbau principle*

▶ The building-up principle is a set of rules that helps to rationalize the experimentally determined ground-state electron configurations of atoms of the elements.

build-up n **1 蓄積** syn *accumulation*

▶ The NiAs structure may be regarded as built-up of layers of prisms.

2 強化 syn *increase*

▶ The effect of the rate of breakdown of the water skeleton on internal pressure is insufficient to counteract the build-up of repulsive forces at 150 ℃.

built-up multilayer n **累積多分子層**

▶ In explaining properties of built-up multilayer, it is often necessary to assume that the outside layer can be turned inside out by contact with specific reagents.

bulb n **球状の部分**

▶ If the thermometer bulb were immersed in the boiling liquid, it might record a temperature a little above the boiling point.

bulbous adj **丸くふくらんだ**

▶ The bulbous part is heated in a softer flame and gently pulled enough to reduce the diameter of the seal to that of the tube.

bulge n **ふくらみ** syn *bump, hump*

▶ The packing density of polyatomic molecules with their many bulges and irregular shapes is nearly unpredictable even if we could say in what array they are to crystallize.

bulk n **1 大部分** syn *majority*

▶ After using all these physical approaches to remove the bulk of the oil, millions of gallons of oil still remained attached to sand, rocks, and gravel both at the surface and below the surface of contaminated shorelines.

2 塊 syn *mass*

▶ Many properties of thin films differ from those of the same material in bulk.

adj **3 大量の**

▶ Most bulk chemical and petrochemical

processes could not be operated commercially without a heterogeneous catalyst.

4 塊の

▶ Band structure calculations have helped quantitatively correlate bulk electric and magnetic properties with structure and composition.

bulk modulus n 体積弾性率 ☞ elastic modulus,

▶ The bulk modulus requires changes in bond distances only.

bulky adj かさばった syn *massive*

▶ Many of the substances studied were too bulky to fit inside the β-cyclodextrin cavity, and it is assumed that only the side chains were complexed.

bump n **1** 隆起 syn *bulge, hump*

▶ For molecular shapes that contain bumps and hollows, a popular interpretation of the close-packing principle requires that bumps in one molecule must be fitted into the hollows of another to attain complementarity of molecular surfaces.

vi **2** 液体が激しく沸騰する

▶ Liquids bump badly when boiled at reduced pressure, and most boiling stones lose their activity in an evacuated system.

vt **3** 打ち当たる syn *collide, strike*

▷ The energy barriers resisting molecular motions in crystals arise mainly from peripheral atoms bumping together rather than being pulled apart.

bumping n 突沸

▶ To prevent superheating and bumping, bits of porous pipeclay, etc., or small capillary tubes closed at one end and containing air may be put into the liquid.

bundle n 束 syn *package, packet*

▶ The cellulose chains lie side by side in bundles, undoubtedly held together by hydrogen bonds between the numerous neighboring −OH groups.

buoyancy n 浮力

▶ The buoyancy effect for a particle which merely displaces an equal amount of liquid is simply the difference between the density of the particle and the density of the liquid.

buoyant adj 浮揚性

▶ When a substance is immersed in water, it experiences a buoyant or floating effect.

burden n 負担 syn *load, strain, trouble, weight*

▶ The formal steric enthalpy term bears the burden of all discrepancies between the sum of increments and the experimental enthalpy of formation of the single conformer of lowest energy.

burgeoning adj 急激に発展する syn *growing*

▶ There is now a burgeoning literature concerning new methods for the synthesis of trans fused polycyclic ethers and their application to the synthesis of marine polyether natural products.

burn vi **1** 燃える syn *ignite*

▶ Carbon burns to form the gases, carbon monoxide and carbon dioxide, the former being produced when there is a deficiency of oxygen or when the flame temperature is very high.

vt **2** 燃やす syn *fire, light*

▷ Practically all of carbon black is made by burning vaporized heavy-oil fractions in a furnace with 50% of the air required for complete combustion.

burning n 燃焼 syn *combustion*

▶ Burning of hydrocarbons takes place only at high temperatures, as provided by a flame or a spark.

burning velocity n 燃焼速度

▶ The methyl nitrite self-decomposition flame is suitable for study at atmospheric pressure since it has a low burning velocity and a wide reaction zone.

burst n **1** 突発 syn *explosion, flash, spark*

▶ Another type of extreme disturbance in a premixed system is produced by a sudden irradiation with a very intense burst of light energy.

vi **2** 破裂する syn *explode, rupture, shatter*

▶ The oxide skin slowed down the reaction, but above the mp globules of metal burst through.

bury vt **埋める** syn *conceal, cover*

▶ For glucose oxidase, direct electrochemical oxidation of the enzyme is slow because the active site is buried deep within the enzyme core.

but adv **1 ただ** syn *only*

▶ The nature of the bonding is but poorly understood in many cases.

prep **2 を除いて** syn *except*

▶ Page 8, last line but one in text, for "be" read "is".

conj **3 しかし** nevertheless よりは意味が弱く，however よりは強い．

▶ Azulene is isoelectronic with naphthalene, but whereas naphthalene is colorless azulene is blue.

4 …ではなくて 前に否定語がある場合

▶ The red light is emitted, not by excited lithium ions or atoms, but by the molecule LiOH.

button n **円形のもの**

▶ If a tin salt is heated with soda on charcoal, usually only a small malleable button is obtained, which, on taking away the flame, is immediately covered with a white coating of oxide.

by prep **1 よって** 本来は動作主を表す． syn *by means of*

▶ Phenols are converted into their salts by aqueous hydroxides but not by aqueous hydrogencarbonates. The salts are converted into the free phenols by mineral acids, carboxylic acids, or carbonic acid.

手段を表すときもある．

▶ The final product was purified by chromatography on alumina followed by recrystallization and identified by X-ray methods.

従来，with は測定方法や機器を表すとされていたが，近年は，この場合にも by が用いられる．

▶ Line positions were measured accurately by using field markers generated by an NMR gaussmeter, while the microwave frequency was measured by a microwave frequency counter.

2 いくらだけ

▶ The presence of a phenyl group in place of a hydrogen of methyl chloride stabilizes the cation by 61 kcal/mol.

3 …を単位にして

▶ PH_3 is an extremely poisonous, highly reactive, colorless gas which has a faint garlic odor at concentrations above about 2 ppm by volume.

4 乗除，寸法に用い**掛ける**

▶ Some of these virions can be seen with the light microscope because they generally are 300 to 450 nm by 170 to 260 nm in size.

by and large adv **概して** syn *generally, mainly, ordinarily, usually*

▶ By and large, metal complexes form much more stable radicals than organic materials, and so several groups have become interested in paramagnetic, metal-containing liquid-crystal systems.

bypass vt **飛び越えて進む** syn *avoid, ignore, jump, omit*

▶ Microreaction technology has the potential to bypass this step, attaining large volume production through the replication of unit process, enabling the direct transfer of laboratory-optimized conditions to production scale.

by-product n **副生物** syn *coproduct, side product*

▶ The by-product can be removed by crystallization from ligroin.

C

ca. abbr およそ circa の略 syn *about*

▶ By aggregating the reaction mixture in a mortar and pestle, the viscosity rapidly increases to form a sticky material after ca. 5 min.

cage n かご形のもの syn *enclosure*

▶ The simplest cage-type molecule is white phosphorus, P_4.

cage compound n かご形化合物

▶ The largest protonated cluster of water molecules yet characterized is the discrete unit $[H_{13}O_6]^+$ formed by chance when the cage compound $[(C_9H_{18})_3(NH)_2Cl]^+Cl^-$ was crystallized from a 10% aqueous hydrochloric acid solution.

cake n 平たい固い塊 syn *block, chunk, lump*

▶ When filtration requires many hours for completion, the difficulty is increased by development of cracks in the filter cake which break the vacuum.

calcination n か焼

▶ The zeolite was prepared by steam calcination at 1027 ℃ of an 80% exchanged NH_4^+ zeolite Y.

calcine vt 焼成する

▶ When gypsum is calcined at 150–165 ℃, it looses approximately three-quarters of its water of crystallization to give the hemihydrate $CaSO_4 \cdot \frac{1}{2}H_2O$.

calculable adj 計算可能な

▶ Since enthalpy data are easily calculable from those for free energies and entropies, no plot is presented.

calculate vt 計算する syn *compute, evaluate*

▶ In order to probe what factors control the degree of charge transfer from donor to acceptor in charge-transfer solids, we have calculated the Madelung energy for tetrathiafulvalene–chloranil as a function of pressure and temperature.

calculated adj 計算して確かめられた

▶ Experimentally, ammonia is found to have the pyramidal shape calculated by quantum mechanics.

calculation n 計算 syn *computation, estimation*

▶ Calculations based on the detailed structure of the solvent around an ion have met with only limited success.

calibrate vt 校正する syn *standardize*

▶ Commercial pH meters are calibrated so as to give a direct reading of the pH.

calibration n 校正 syn *standardization*

▶ IR spectra were recorded on a Perkin-Elmer FT spectrometer with internal calibration.

calibration correction n 校正補正

▶ If the calibration correction was found to be less, then the weight is too light and a negative correction should be applied in using it.

calixarene n カリックスアレーン

▶ Some advantages of calixarene receptors are evident in the complexation of UO_2^{2+} ions; in contrast to many competing heavy metal ions which have an octahedral coordination sphere, UO_2^{2+} ion requires hexagonal coordination.

call vi **1 call for** 必要とする syn *require*

▶ Advances in technology have called for materials of extreme purity.

2 call on 要求する syn *demand*

▶ Attack on an aromatic ring is probably one of the most difficult jobs a carbocation is called on to do.

3 call upon 用いる syn *mobilize*

▶ During short periods of vigorous exercise, the blood cannot supply oxygen fast enough for respiration to carry the entire load; when this happens, glycolysis is called upon to supply the energy difference.

vt **4** 名づける syn *name, designate*

▶ Naturally occurring glucose belongs to the D series and is properly called D-glucose.

5 call up 呼び寄せる syn *recruit*

▶ At ribosome, messenger RNA calls up a series of transport RNA molecules, each of which is loaded with a particular amino acid.

calomel electrode n カロメル電極

▶ In the calomel electrode, mercury is in contact with Hg_2Cl_2 immersed either in a 0.1 *m* solution of KCl or in a saturated solution of KCl.

calorimetric adj 熱量測定の

▶ It thus appears that these calorimetric methods offer significant advantages over other approaches which rely on differentiation procedures.

calorimetry n 熱量測定

▶ Calorimetry was conducted with a Perkin-Elmer DSC-7 differential scanning calorimeter equipped with a thermal analysis data station.

camera-ready adj 写真製版に利用可能な質の

▶ All figures must be submitted as camera-ready artwork and must be, therefore, be drawn to a professional standard.

can n 缶

▶ Lithium aluminum hydride is usually supplied in powdered form contained in individual plastic bags sealed within a metal can.

can aux 能力を表して，できる　否定形には cannot を用いる． syn *able*

▶ If the gas has a different molecular weight in the two phases, i.e., if it associates or dissociates in one phase but not in the other, obedience to Henry's law cannot be expected.

cancel vi 相殺する syn *counterbalance, nullify*

▶ The usefulness of bond energies for predicting heats of formation is independent of the accuracy of these values which cancel out in the overall calculation.

cancellation n 相殺 syn *elimination*

▶ Although internal pressure effects induced by the salt influence the activity coefficient of both the reactants and the activated complex, a cancellation of effects occurs during the reaction.

candidate n 候補 syn *possibility, prospect*

▶ The solid picryl ether **1a** appeared to be an interesting candidate for study as an intramolecular rearrangement in the solid state.

capability n 能力の範囲 syn *ability, capacity, potential*

▶ More work is needed to explore scanning tunneling microscopy's capabilities fully; until now these capabilities have been only scratched at.

capable adj 可能で syn *able*

▶ This method is capable of giving excellent results after a little practice.

capacitance n キャパシタンス

▶ For a given volume, a $BaTiO_3$ capacitor has 10 to 1000 times the capacitance of a dielectric capacitor.

capacitor n コンデンサー ☞ condenser

▶ The cell capacitor is calibrated with a gas of known dielectric constant, usually ammonia or carbon dioxide.

capacity n **1** 適応力 syn *ability, capability, potential*

▶ The important lead oxide is the mixed valence red lead, $Pb^{II}_2Pb^{IV}O_4$, which is used as a pigment and in various capacities in the glass, rubber, and plastics industries.

2 容積 syn *size, volume*

▶ Cylinders of F_2 are now commercially available in various sizes from 230-g to 2.7-kg capacity.

3 コンデンサーの容量

▶ The cation and the adjacent faces of its nearest-neighbor anions form a condenser, the capacity of which will to a good approximation be a constant multiple of that of a spherical condenser of radii d_1 and d_2.

capillary n 毛管 syn *capillary tube*

▶ With capillaries of less than 5-nm diameter, capillary condensation proceeds at pressures less than the saturated pressure of the adsorbate.

capillary action n 毛管作用

▶ Using conventional technology, further miniaturization is problematic; for example, evaporation becomes significant in microliter volumes, and caplillary action causes bridging of liquid between wells.

capillary condensation n 毛管凝縮

▶ Multilayer adsorption and capillary condensation are usually separable.

capped trigonal prismatic adj 面冠三角柱の

▶ Three-dimensional X-ray analysis and neutron diffraction studies established the capped trigonal prismatic structure for ReH_9^- (Re—H = 1.68 Å)

capsid n カプシド

▶ A capsid is generally rigid, forming a body that is cylindrical, spherical, or isosahedral.

capsule n **1** 色素を入れるカプセル

▶ A slurry of the capsules in a water-polymer solution gives the ink.

2 細菌を包むカプセル，莢膜

▶ Some bacteria have a capsule of slimy or gummy layers of polysaccharides or polypeptides surrounding the cell wall.

capture vt 捕捉する syn *catch*

▶ The holes may capture primary electron beam or secondary ejected electrons from the conduction band.

carbanion n カルボアニオン

▶ Heterolytic bond cleavage leads to retention of the electron pair on a carbon atom and yields species called carbanions.

carbene n カルベン

▶ Free carbenes exist in both triplet and singlet states, but those containing heteroatoms (e. g., O or N), as found in Fischer carbenes, tend to be of the latter variety.

carbenium ion n カルベニウムイオン

▶ German and French literature frequently uses the name carbenium ion for the trivalent cation: this is a logical name if we consider the latter as a protonated carbene.

carbocation n カルボカチオン syn *carbonium ion*

▶ The characteristic feature of a carbocation is, by definition, the electron-deficient carbon and the attendant positive charge.

carbohydrate n 炭水化物

▶ The term carbohydrate means hydrated carbon atoms; literally the generic formula is $(CH_2O)_n$ with as many water molecules, attached as —H and —OH groups, as there are carbon atoms in the molecule.

carbonaceous adj 炭素質の

▶ A carbonaceous polymer may contain not only functional groups but also chemisorbed species held intact by charge-transfer forces as in inorganic semiconductors.

carbonation n 炭酸化

▶ Commercial preparation of H_2NCN is by continuous carbonation of an aqueous slurry of CaNCN in the presence of graphite.

carbon fiber n 炭素繊維

▶ Carbon fibers are used for their high-strength-to-low-mass ratios as well as for their thermal and acid stability and general chemical inertness.

carbonium ion n カルボニウムイオン syn *carbocation*

▶ The alkyl groups are electron donors and inductively stabilize the transition state leading to the carbonium ion intermediate.

carbonization n 炭化

▶ All oxalates are decomposed on ignition with slight carbonization.

carbon nanotube n カーボンナノチューブ

▶ Carbon nanotubes are well-ordered hollow graphtic nanomaterials which vary in length from several hundred nanometers to several micrometers and have diameters of 0.4 to 2 nm.

carbon skeleton n 炭素骨格

▶ It is often convenient to omit the hydrogen points from the structural representations of alkanes and deal with their carbon skeleton graphs.

carbon whisker n 炭素ウィスカー

▶ Carbon whiskers are manufactured by striking a carbon arc at 3600 ℃ under 90 atm Ar.

carbonylation n カルボニル化

▶ In a 1984 patent, Garapon et al. reported the $Co_2(CO)_8$-catalyzed carbonylation of oxazolines in the presence of an alcohol to give *N*-acyl-β-amino esters.

carbonyl complex n カルボニル錯体

▶ The molecular structures adopted by simple carbonyl complexes are generally compatible with predictions based on valence shell electron-pair repulsion theory.

carcinogen n 発がん物質

▶ Carcinogens cause the development of cancerous growth in living tissue.

carcinogenic adj 発がん性の

▶ There are approximately 3500 known and suspected carcinogenic compounds, and new ones are constantly being discovered.

cardiac glycoside n 強心配糖体

▶ Cardiac glycosides constitute a group of steroids, mainly of plant origin, which have a powerful action on the heart.

care n **1 注意** syn *caution, consideration*

▶ Liquid vinyl and vinylidene compounds often polymerize on standing unless special care is taken to inhibit the reaction.

2 take care 気を付ける syn *be cautious, beware*

▶ It is desirable to use the purest possible reagents, but the chemist should always take care to test them.

careful adj 注意深い syn *cautious*

▶ With careful experimental work and data analysis, reliable equilibrium constants can be obtained.

carefully adv 注意深く syn *thoroughly*

▶ The acid solution is partly neutralized by adding carefully and in small portions 15 g of $NaHCO_3$.

carotenoid n カロテノイド

▶ The carotenoids are a special group of terpenes which are widely distributed in both the plant and animal kingdoms. They are yellow, orange, or red pigments whose color is due to a large number of conjugated double bonds.

carrier n **1 電荷担体** ☞ charge carrier

▶ Mobile electrons and holes are collectively called carriers because they are the species responsible for conductivity of a semiconductor.

2 触媒の担体 syn *support*

▶ A neutral material such as diatomaceous earth used to support a catalyst in a large-scale reaction system is called a carrier.

3 トレーサーの担体

▶ An atomic tracer carrier is a stable isotope or a natural element to which radioactive atoms of the same element have been added for purpose of chemical or biological research.

4 生化学における担体

▶ The best known function of Fe in biological systems is as an O_2 carrier in hemoglobin.

carry vt **1 運ぶ** syn *bear, transport*

▶ Mobile electrons and holes carry the electric current in the semiconductor.

2 保有する syn *bear, maintain*

▶ Charges are also carried by inorganic colloidal particles, such as AgCl, where the charge can be attributed to a preferential adsorption of Ag^+ or Cl^- ions on the surface of the particles.

3 進める syn *conduct*

▶ In aerobic respiration, the oxidation of glucose is carried to completion, and the extremely complex system of reactions is adapted to preserving as much of the energy of the molecule as possible.

4 carry out 行う syn *accomplish, perform*

▷ The high dielectric constants of the solvents most frequently used for carrying out chemical reactions greatly decrease the electrostatic work of ion separation.

5 carry over 持ち越す

▶ It is possible to carry over the standard bond lengths and bond angles from simpler molecules and to use these values to construct the geometry of the basic elements of the polypeptide or protein molecules.

cartilage n 軟骨

▶ Chondroitin sulfate is among the principal polysaccharides of cartilage and is structurally similar to hyaluronic acid.

cascade n **1 カスケード**

▶ Complement activation involves a cascade of protein interactions that leads to the formation of holes or pores in the plasma membranes of target cells, resulting in their osmotic dissolution.

vi **2 カスケードとなる**

▶ The fission may cascade to give an atomic explosion.

cascade reaction n カスケード反応

▶ The utility of radical-based methodologies is enhanced by the potential for effecting cascade reactions, where the careful design of a substrate will allow several radical reactions to proceed in sequence.

case n **1** 場合 syn *example, instance*

▶ The procedures are not only useful for the preparation of specific compounds but can serve as models for adaptation of known reactions to a new case.

2 the case 事実 syn *the fact, the reality, the truth*

▶ As is invariably the case, solution- or liquid-state geometry, the state of most general interest to chemists, is the most difficult to define.

3 in any case とにかく syn *anyway*

▶ In any case, no further report of a fluorine oxo acid transpired until 1968.

4 in case もし…なら syn *if, if it happens*

▶ In case either the alcohol or the acid involved in an esterification contains reactive double bonds or is sensitive to dehydration, boron fluoride etherate becomes the catalyst of choice.

5 in no case 決して…でない

▶ In no case, so far as the author is aware, have decomposition centers or nuclei been observed with organic crystals.

case-by-case adj その場その場の

▶ The reaction conditions usually must be worked out on a case-by-case basis.

case study n 事例研究

▶ The case studies below illustrate the evolution of precursor design from approaches that focus on volatility and stability to more recent strategies that include consideration of likely decomposition pathways.

cast vt **1** 投げかける 過去形，過去分詞ともに cast syn *throw*

▶ Doubts have been cast upon interpretations in terms of vacancy pairs.

2 形を与える syn *shape*

▶ Thin films of aromatic hydrocarbons were cast on microscopic slides by either solution evaporation or vapor deposition.

3 鋳造する syn *found*

▶ Most organic compounds stable enough to be zone-refined can be cast easily into cylindrical rods.

casual adj 軽い気持ちの syn *easy, relaxed*

▶ Casual inspection of the familiar world indicates the existence of three states of matter: solids, liquids, and gases.

catabolic reaction n 異化反応 ☞ anabolic reaction

▶ Cataboic reactions break down complex organic molecules to release energy for driving anabolic reactions that build the molecules needed by the cell.

catabolism n 異化作用 ☞ anabolism

▶ The reactions that break down large molecules into smaller units are part of what is called catabolism.

catalogue n **1** 目録 syn *record*

▶ The book "Three Dimensional Nets and Polyhedra" by A. F. Wells, published in 1977, provides a very useful catalogue of nets.

vt **2** 目録を作る syn *list, record*

part ▶ Early efforts were devoted to cataloguing the many commercial materials according to their responses under irradiation.

catalysis n 触媒作用

▶ Catalysis of chemical reactions by surfaces must proceed by chemisorption of at least one of the reactants.

catalyst n 触媒

▶ A catalyst is a substance that increases the rate of reaction without modifying the overall standard Gibbs energy change in the reaction.

catalytic adj 触媒の

▶ Tetraethyllead is frequently referred to as a catalyst, since only a trace, or catalytic amount, of it is required. This substance is, however, destroyed in the reaction, and so it is not strictly a catalyst.

catalytic activity n 触媒活性

▶ The catalytic activity and specificity in the decomposition of hydrazine are determined by the metal in the coordination polymer.

catalytically adv 触媒によって

▶ Although both the benzene ring and the carbon–carbon double bond can be hydrogenated catalytically, the conditions required for the double bond are much milder.

catalytic cracking n 接触分解

▶ The use of catalytic cracking of crude oil increased to supply high-octane aviation gasoline.

catalytic hydrogenation n 接触水素化

▶ Catalytic hydrogenation converts pyrrole and furan into the corresponding saturated heterocycles, pyrrolidine and tetrahydrofuran.

catalytic oxidation n 接触酸化

▶ Catalytic oxidation is widely used in industry and increasingly in pollution control.

catalytic polymerization n 接触重合

▶ Polyacetylene is prepared by the catalytic polymerization of acetylene in the absence of oxygen.

catalytic reaction n 触媒反応

▶ A substantial amount of hydrogen is also converted into N—H bonds in ammonia by catalytic reaction with nitrogen.

catalytic reforming n 接触改質

▶ By the process of catalytic reforming, enormous quantities of the aliphatic hydrocarbons of petroleum are converted into aromatic hydrocarbons which are used not only as superior fuels but as the starting materials in the synthesis of most aromatic compounds.

catalyze vt 触媒する

▷ The preferred point of attack depends chiefly on whether the reaction is acid-catalyzed or base-catalyzed.

catastrophic adj **1** 大変動の syn *violent*

▶ In a sequence of chlorides of isoelectronic metals such as KCl, $CaCl_2$, $ScCl_3$, and $TiCl_4$, the sudden discontinuity in physical properties of $TiCl_4$ is more a function of stoichiometry and coordination number than a sign of any discontinuous or catastrophic change in bond type.

2 大惨事の syn *disastrous*

▶ *t*-$BuAsH_2$ is a liquid supplied in small metal containers, and a spillage is not catastrophic.

catch vt 捕らえる syn *capture, trap*

▶ The distillate was caught in 60 mL of 0.5 M HCl.

categorical adj 断定的な syn *definitive*

▶ If there is a lesson to be learned from all this, it would seem to be that it does not pay to be too categorical as to what is or is not a promising line of experiment.

categorize vt 分類する syn *class, divide, label*

▷ The chemistry of organic materials is conveniently described by categorizing compounds according to their functional groups.

category n 種類 syn *class, type, variety*

▶ The one category of substitution reactions in cobalt(III) complexes which appears to be accompanied by extensive rearrangement is that of base hydrolysis.

catena n カテナ，連鎖

▶ Catena-S_n, which is obtained by pouring liquid sulfur into cold water, has $n = 2\times10^5$ if made from the melt at 180 ℃.

catenate vt 鎖状に連結する

▶ The extent to which B_3O_3 rings catenate into more complex structures or hydrolyze into smaller units such as $[B(OH)_4]^-$ clearly depends sensitively on the activity of water in the system.

catenation n カテネーション

▶ Catenation is an important feature of the chemistry of Ge, Sn, and Pb, though less so than for C and Si.

cathode n **1** 電気分解のカソード，陰極 ☞ anode

▶ The platinum anode is attacked, causing the metal to go into solution, which is deposited upon the cathode, partially, at least.

2 電池のカソード，正極 syn *positive electrode*

▶ In all these cells, the cathode reaction is the reduction of a metal oxide to a lower oxide or to a metal.

cathodoluminescence n カソードルミネ

センス

▶ Cathodoluminescence uses cathode rays or electrons to provide excitation energy.

cathode ray n 陰極線

▶ The cathode rays were discovered by J. Plücker, professor of physics in Bonn, who found that they were deflected by a magnetic field.

cathodic adj 陰極の ☞ anodic

▶ PO was obtained as a brown cathodic deposit when a saturated solution of Et_3NHCl in anhydrous $POCl_3$ was electrolyzed.

catholyte n カソード液, 陰極液 ☞ anolyte

▶ In the membrane cell, charge transport occurs by the selective movement of sodium ions across the membrane from anolyte to catholyte.

cation n 陽イオン ☞ anion

▶ As halogen cations are highly electrophilic species, they can only exist in the presence of molecules and anions of very low basicity.

cation-exchange membrane n 陽イオン交換膜 ☞ anion-exchange membrane

▶ The cation-exchange membrane may contain fixed anionic sites with which cations in the membrane are to some extent loosely chemically bonded.

cationic adj 陽イオンの ☞ anionic

▶ Polymerization by the cationic mechanism is most important for isobutylene and α-methylstyrene.

cause n 1 原因 syn *origin*

▶ The delocalization of the π system is the result of the σ framework symmetry and not the cause.

vt 2 原因となる syn *give rise to*

▶ Hydrogen sulfide reacts with iodic acid causing precipitation of sulfur and, at first, iodine is liberated, but further addition of hydrogen sulfide causes the disappearance of the iodine.

3 するようにさせる syn *induce, make*

▶ The sensitivity of the silver halides to light is caused by photoreduction of the metal ions by the halide, to give clusters of silver atoms together with halogen molecules.

caution n 1 注意 syn *care, precaution*

▶ Manipulation of any organometallic compounds should always be carried out with caution, owing to their extreme reactivity and their considerable toxicity.

vt 2 注意する syn *warn*

▶ It should be cautioned that so much heat is normally generated during oxidation of a nickel-silica catalyst that irreversible changes in specific surface and in particle size distribution may result.

cautionary adj 警戒的な

▶ Crystal field theory was used alone to view many aspects of transition metal chemistry despite the cautionary notes concerning such usage which have appeared.

cautious adj 注意深い syn *careful*

▶ This gentle, cautious heating is continued until the reaction is proceeding smoothly enough so that the mixture can be refluxed on the steam bath.

cautiously adv 注意深く

▶ If a neutral solution of a silver salt is cautiously treated with ammonia, the first drops produce a white precipitate, AgOH, which changes quickly to brown oxide, Ag_2O.

cavitation n キャビテーション

▶ Cavitation involves the formation, growth, and eventual collapse of bubbles in a liquid irradiated with ultrasound.

cavity n 1 空洞 syn *hole, pit*

▶ The laser cavity consists of the active medium, where the population inversion is generated and is bound by mirrors which reflect the light back and forth through active medium.

2 結晶中の空洞 syn *opening, space*

▶ Polyhedral frameworks, sometimes with cavities of more than one size, can be generated from unit cells containing $12H_2O$, $46H_2O$, $136H_2O$, etc.

cease vi 終える syn *stop, terminate*

▶ Many investigations on surfaces of known area have confirmed that chemisorption ceases after a unimolecular layer is formed but that physisorption may give rise to additional layers.

celebrated adj 有名な syn *famous, popular, well-known*

▶ It follows that the total number of degrees of freedom is $F = C - P + 2$. This is the celebrated phase rule.

cell n **1** 電池 ☞ battery

▶ During the last century, studies were made of the relationship between the emf of a cell and the thermodynamics of the chemical reactions occurring in the cell.

2 電解槽

▶ The latest membrane cells give chloride-free caustic at 35 percent strength at a low voltage of ca. 3.1 V.

3 胞 ☞ unit cell

▶ The mineral perovskite, $CaTiO_3$, has the larger Ca^{2+} cation in the center of the cubic cell, the smaller Ti^{4+} ions at each corner of the cube, and the O^{2-} ions bisecting the edges of the cube.

4 試料セル syn *chamber*

▶ It is possible to raise the temperature of a tiny cell containing a reaction mixture by a few degrees in less than 10^{-7} s.

5 細胞

▶ Cells are bounded by membranes, yet they are not isolated from their local environment.

cell death n 細胞死

▶ Many polymers are not suitable for drug delivery, even if cell death is the ultimate goal, because target specificity is still lacking.

cell division n 細胞分裂

▶ During cell division, chromatin is coiled into tight fibers that eventually wrap around each other to form a chromosome.

cell membrane n 細胞膜

▶ Apart from being a precursor of bile salts, cholesterol is a major component of cell membranes where it modulates membrane fluidity.

cell reaction n 電池反応

▶ The electromotive force of a cell is related to the change in Gibbs energy for the cell reaction.

cellular adj **1** 細胞の

▶ Cellular membranes behave as two-dimensional semifluid structures, allowing embedded protein molecules to constantly move about rather freely by lateral diffusion.

2 多孔性の syn *porous*

▶ A cellular metallic structure, usually of aluminum or zinc alloys, are made by incorporating titanium or zirconium hydride in the base metal.

cellulosic adj セルロースでできた

▶ The deterioration of cellulosic materials, either textiles, paper, or wood, by enzymatic degradation is an economic problem which is not yet adequately solved.

cell wall n 細胞壁

▶ The fibrous tissue in the cell walls of plants and trees contains the polysaccharide cellulose.

cementitious adj セメント質の

▶ In cements, the β polymorph of Ca_2SiO_4 has superior cementitious properties to the γ polymorph.

center n **1** 中心 syn *middle*

▶ According to the laws of electrostatics we may, with no loss of accuracy, treat a charge sphere as a point located at the center of the sphere.

2 中核 syn *core, nucleus*

▶ When fractional crystallization has occurred, the plagioclase crystals have calcium-rich centers and sodium-rich outer regions.

vt **3** 中心に置く syn *concentrate*

▶ Much of the interest in fluorophore nucleobase replacements is centered around their ability to mimic natural nucleobases.

centered adj 中心に置かれた

▶ The second decomposition process occurs over a broad temperature range centered at approximately 150 ℃.

center of gravity n 重心

▶ In nematic phase, the molecular centers of gravity are disordered as in a liquid, but we have a statistically parallel orientation of the long axes of the molecules along an axis.

center of inversion n 反転中心 ☞ center of symmetry

▶ While each molecule has a center of inversion, the arrangements of the *n*-alkyl side chains fall into two sets.

center of mass n 質量中心 複合形容詞として使用

▶ For an *n*-atom molecule, using three coordinates to specify each atom's position and subtracting three coordinates for center-of-mass translational motion and three rotational motion leaves $3n-6$ vibrational degrees of freedom.

center of symmetry n 対称心

▶ Crystals of substances in the triclinic system have no plane nor axis of symmetry but have a center of symmetry.

central adj **1** 中心の syn *inner, middle*

▶ Sodium sulfate shows a single sulfur 2p peak at the same energy as that for the central sulfur atom in $Na_2S_2O_3$.

2 主要な syn *basic, fundamental, necessary*

▶ Chemical reactions involving oxidation and reduction processes are central to metabolism.

central field approximation n 中心力場近似

▶ Although one cannot achieve spectroscopic accuracy for the total wave function and energy, it is possible to get useful representations in terms of the central field approximation and one-electron orbitals.

centrality n 重要な地位

▶ Because of the basic character of chemistry and its centrality among the sciences, introductory chemistry courses fulfill a crucial service role.

centrifugal adj 遠心の

▶ In order for smaller particles to undergo sedimentation, it is necessary to increase the effective gravitational field by subjecting the solution to centrifugal motion.

centrifugal field n 遠心力場

▶ The sedimentation of large molecules in a strong centrifugal field permits determination of both average molecular weights and the distribution of molecular weights in certain systems.

centrifugal force n 遠心力

▶ The centrifugal field is not uniform, as contrasted the gravity field, but this is easily allowed for the calculation since at any point the centrifugal force equals $\omega^2 x$ where ω is the rotational speed in radians per second, and x is the distance from the axis of rotation.

centrifugation n 遠心分離

▶ The equipment used in centrifugation is a chamber revolving at high speed to impart a force up to 17,000 times gravity.

centrifuge n **1** 遠心分離機

▶ In a centrifuge, the magnitude of the gravitational field is $r\omega^2$, and a mass m placed in this field experiences a sedimenting force of $F=mr\omega^2$.

vt **2** 遠心分離機にかける

▶ The solid was dissolved in dry, hot ethannitrile and centrifuged twice to remove the fine insoluble materials.

centrosymmetric adj 中心対称的な

▶ Neutron and X-ray diffraction studies suggest that $[Cl-H\cdots Cl]^-$ can be either centrosymmetric or non-centrosymmetric depending on the crystalline environment.

ceramic n **1** 通常複数形で，セラミックス

▶ Mixing aqueous dispersions provides a very accurate and safe method for preparing feed solutions for multicomponent ceramics.

adj **2** 陶磁器の

▶ Silica is the main component of many ceramic materials as well as being the most common oxide, apart from water, in the earth's crust.

certain pron **1** 複数扱いをして，いくらか

▶ Certain of the macroscopic properties have fixed values for a particular state of the system, whereas others do not.

adj **2** 確かな syn *confident, positive*

▶ The $-CHO$ group of an aldehyde has a characteristic C–H stretching band near 2720 cm^{-1}; this, in conjunction with the carbonyl band, is fairly certain evidence for an aldehyde.

3 一定の syn *definite, specific*

▶ A certain minimum number of properties have to be measured in order to determine the condition or state of a macroscopic system completely.

4 いくらかの syn *several*

▶ There are certain properties common to all silicones.

certainly adv 確かに syn *definitely, surely*

▶ Molybdenum(III) oxide, Mo_2O_3, is not known certainly to exist.

certainty n **1** 確信 syn *assurance, confidence*

▶ Until quite recently, the absolute configuration of no optically active compound was known with certainty.

2 確実性 syn *reality*

▶ There is no certainty about the detailed mechanism of the primary process, although this is of considerable importance in photographic science.

chain n **1** 鎖 syn *strand*

▶ In the so-called hexagonal perovskite structures, BX_6 octahedra are linked into chains by sharing opposite faces, the chains being separated by the larger A cations.

2 連鎖 syn *series*

▶ The high reactivity of Cl atoms leads to very long chain, up to 10^6.

chain carrier n 連鎖伝達体

▶ Since radicals are involved in the propagation steps consuming reactants or generating products, we need to develop expressions for these chain carrier concentrations.

chain initiation n 連鎖開始反応

▶ First, chain initiation must occur, which for methane chlorination is activation and conversion of chlorine molecules to chlorine atoms by light.

chain length n **1** 鎖長

▶ The reactions that occur when diacids are heated depend critically upon the chain length separating the carboxyl groups.

2 連鎖長

▶ The average number of propagation steps, which occur between initiation and termination, is called the chain length.

chain molecule n 鎖状分子 複合形容詞として使用

▶ Since the *n*-alkanes themselves are the simplest of these chain-molecule solids, they are a natural starting point for understanding the premelting transitions exhibited by all of these systems.

chain propagation n 連鎖成長 複合形容詞として使用

▶ In the second phase, the chain-propagation steps convert reactants to products with no net consumption of atoms or radicals.

chain reaction n 連鎖反応

▶ If a chain reaction is to provide useful energy, the heat it generates must be extracted by means of a suitable coolant and converted, usually by steam turbines, into electric energy.

chain termination n 連鎖停止反応 ☞ termination

▶ Two pathways for chain termination have been proposed for cationic dehydropolymerization of H_3NBH_3.

chain transfer n 連鎖移動

▶ If the alkene is of low reactivity and the transfer agent of high reactivity, chain transfer is so effective that there is no polymerization.

chain-transfer agent n 連鎖移動剤

▶ Even traces of certain impurities acting as chain-transfer agents or inhibitors can interfere with the polymerization process.

chair conformation n いす形配座 ☞ boat conformation

▶ The neutral S_6 ring has the expected chair conformation.

chair form n いす形 ☞ boat form

▶ The six-membered rings of decalin like those of cyclohexane are expected to be most stable when in the chair form.

challenge n 難題 syn *problem*

▶ Size characterization of particles with a complex chemical composition poses an even greater challenge than for pure particles.

challenging adj やりがいのある syn *difficult*

▶ We want a more challenging system, and crystals thought to be largely ionic certainly offer us that.

chamber n 室 syn *compartment*

▶ The crystallizer is made up of three vessels, a large outer chamber, a middle chamber, and an inner chamber with several openings in the bottom.

chance n 1 機会 syn *opportunity*

▶ The chance of finding biaxial crystals oriented with an optic axis parallel to the light beam is quite small.

2 見込み syn *possibility, prospect*

▶ If only one reasonable structure can be drawn for a molecule, the chances are good that this one structure adequately describes the molecule.

adj 3 偶然の syn *accidental*

▶ Very many colloidal particles are of the unstable type, depending for even temporary existence upon a chance stabilizing agent.

vi 4 偶然出会う syn *happen*

▶ They worked through the field of the azo compounds until they chanced upon ones with exceptional disinfectant powers.

change n 1 変化 syn *difference*

▶ Difluorotoluene molecule caused no measurable change in charge migration relative to natural thymine.

vt 2 全面的に変化させる syn *alter, modify, vary*

▶ Sometimes, relatively small changes in the reagents and conditions change the pattern of orientation.

3 change into 別のものに変える syn *convert, transform*

▶ The value of determinant is unaltered if rows are changed into columns or columns into rows.

changeover n 変換 syn *change*

▶ At sufficiently high temperatures, the intrinsic charge-carrier concentration may exceed the extrinsic value, in which case, a changeover to intrinsic behavior would be observed.

channel n 1 チャンネル syn *path*

▶ Since the benzene molecule has a thickness of ca. 3.7 Å and a diameter of ca. 6 Å, its passage through the channel is impeded unless there is appreciable distortion of the crystal.

vt 2 運ぶ syn *conduct, convey*

▶ Weak, reversible binding attracts species of interest to the selectivity filter and then channels them through the macrocyclic host.

channel-forming adj チャンネル形成の

▶ Some channel-forming proteins respond mechanically to stimuli such as phospholipid tension, others to changes in pH, where protonation disrupts intermolecular interactions that allow molecules to bridge the channel.

char n 1 黒焦げになったもの

▶ Anthracene heated under its own vapor pressure begins to form chars around 500 ℃.

vt 2 黒焦げにする

▷ Animal charcoal made by charring treated bones consists of finely divided carbon supported on calcium phosphate.

character n 1 性質 syn *characteristic, feature, quality*

▶ Unless precautions are taken, the system is thus subjected to a considerable adiabatic shock, and the reactions observed are pyrolytic rather than photochemical in character.

2 指標

▶ The sum of the diagonal elements of a matrix is its character.

characteristic n 1 特徴 syn *character, feature, property*

▶ When organic solids, whether in the fully crystalline, partly crystalline, or fully amorphous state, are subject at certain temperature ranges, to relatively small applied stresses, they exhibit many of the characteristics associated with anelastic or simple, linear, viscoelastic behavior.

adj 2 特徴的な syn *distinctive, representative, typical*

▶ The most characteristic chemical property of amines is their ability to act as bases by accepting protons from a variety of acids.

characteristic absorption band n 特性吸収帯

▶ A particular group of atoms give rise to

characteristic absorption bands; that is to say, a particular group absorbs light of certain frequencies that are much the same from compound to compound.

characteristically adv 特徴的に

▶ The chemical shifts of aromatic protons are characteristically toward lower magnetic fields than those of protons attached to ordinary double bonds.

characteristic frequency n 特性振動数

▶ When an oscillator is exposed to radiant energy, it may absorb certain characteristic frequencies.

characterization n 確認 syn *description, identification*

▶ A notable advance has recently been signaled in the isolation and X-ray structural characterization of Sb homologues of N_2 and azobenzene as complex ligands.

characterize vt 特徴付ける syn *define, identify*

▷ Crystallinity and crosslinking, by restricting certain types of motion, should cause a spread in the distribution of relaxation times characterizing the process.

charge n **1** 電荷 ☞ electric charge

▶ It is probable that the bonds have enough ionic character in the crystal to make the actual charges of the atoms nearly zero; for ZnS this would require about 50 percent ionic character.

vt **2** 充電する

▶ Comparison with data for known samples showed that freshly deposited electrodes consisted of α-$Ni(OH)_2$ and were charged to γ-NiOOH, whilst aged electrodes consisted of β-$Ni(OH)_2$ and were charged to β-NiOOH.

3 充填する syn *fill*

▶ A 250-mL flame-dried flask equipped with a magnetic stirrer bar and a septum inlet was charged with 9.5 mL of 1-octyne (64.40 mmol) in 60 mL of dry THF and then cooled with a dry ice/acetone bath.

charge carrier n 電荷担体 ☞ carrier

▶ Obviously a general answer to the question of the type of charge carrier is not available, because the predominant species depends both on the material and on the conditions of excitation.

charge cloud n 電荷雲 ☞ electron cloud

▶ Acids which are large and whose charge clouds are easily distorted interact most strongly with bases that are large and whose charge clouds are easily distorted.

charged adj 帯電した

▶ An electric current passed through a solution containing ions causes them to move to the oppositely charged electrode.

charge delocalization n 電荷非局在化

▶ Intimate information about charge delocalization with π-bonding ligands has been obtained in the series $Fe[(CN)_5L]^{n-}$.

charge density n 電荷密度

▶ The negative difference density along the O—O bond direction says that the total charge density along this line is less than the sum of the densities of spherically averaged ground-state oxygen atoms at the same nuclear positions.

charge-density wave n 電荷密度波

▶ Tantalum disulfide exhibits a periodic structure not directly associated with the atomic lattice; instead it is dominated by a charge-density wave signal.

charged particle n 荷電粒子

▶ If a charged particle, such as an ion, is introduced into the neighborhood of an uncharged, nonpolar molecule, it will distort the electron cloud of the molecule.

charge separation n 電荷分離

▶ The charge separation internally builds up a small difference in potential energy and thus produces a voltage across the junction.

charge transfer n 電荷移動

▶ For the 1,3,5-trinitrobenzene complexes of anthracene and phenanthrene, there is a mirror-image relationship between the emission and the charge-transfer absorption spectra.

charge-transfer complex n 電荷移動錯体 ☞ electron-donor–acceptor complex

▶ The designation charge-transfer complex originates from a resonance description in which the structure of the complex receives contributions from resonance forms involving transfer of an electron from the donor molecule to acceptor molecule.

charge-transfer force n 電荷移動力

▶ In many so-called charge-transfer complexes, charge-transfer forces do not provide the major contribution to the binding forces in the ground state.

charge-transfer interaction n 電荷移動相互作用

▶ The extent to which the solid matrix can be used to control the course of substitution reactions is currently being explored for a wide variety of aromatic derivatives entering into charge-transfer interactions.

charge-transfer spectrum n 電荷移動スペクトル

▶ Charge-transfer spectra of metal complexes are due to transitions which result in a possibly reversible, oxidation-reduction reaction.

charring n 黒焦げになること

▶ The procedure selected for this experiment involves catalysis by phosphoric acid; sulfuric acid is no more efficient, causes charring, and gives rise to sulfur dioxide.

chart n **1** 図 syn *diagram, graph, trace*

▶ NMR chart paper is cross-hatched, and we can conveniently estimate step heights by simply counting squares.

vt **2** 記録する syn *plot*

▶ Neutralization reactions can be established in the solid just as they are in aqueous solution and followed by shifts in the Fermi level in the same way that pH changes would be used to chart the course of titrations in water.

cheap adj 割安な syn *inexpensive*

▶ Sodium salts are cheaper than potassium salts, and for this reason they are often used when only the anion is important.

check n **1** 点検 syn *examination, inspection, test*

▶ Thorough checks on catalyst stability, the leaching of catalytically active species into solution and catalyst turnover number and reuse, are critical in all cases.

vt **2** 照合する syn *confirm*

▶ The compositions of the products were checked by chemical analysis as follows.

chelate n **1** キレート ☞ metal chelate

▶ When no hydrophilic groups are present in the chelate, precipitation from aqueous solution may be used as a test for chelation.

vt **2** キレート化する

part ▶ Sodium triphosphate acts mainly a water softener, by chelating and sequestering the Mg^{2+} and Ca^{2+} in hard water.

chelate ring n キレート環

▶ Metal ion size preferences of five- and six-membered chelate ring relate to the steric requirements of low-strain cycloalkanes.

chelating agent n キレート化剤

▶ Humic acid is an excellent chelating agent and important in the exchange of cations in soils.

chelation n キレート化

▶ In general, chelation gives a much more stable complex than one formed by binding of analogous separate ligands.

chemical activity n 化学活性

▶ The high chemical activity of organometallic compounds toward oxygen, iodine, etc., is explicable by the fact that carbon–metal bonds have low energy contents.

chemical affinity n 化学親和力

▶ In the thermodynamics, free energy was to replace chemical affinity as the driving force of chemical reaction.

chemical analysis n 化学分析 ☞ gravimetric analysis, qualitative analysis, quantitative analysis, volumetric analysis

▶ Chemical analysis showed that wüstite contained an excess of oxygen, and X-ray measurements confirmed that the crystals were a homogeneous phase.

chemical bond n 化学結合 syn *chemical bonding*

▶ The origin of the chemical bond lies in the

fact that an electron can be in a region where it simultaneously attracts two nuclei closer together.

chemical change n 化学変化
▶ Chemical changes should be distinguished from physical changes, in which only the state or condition of a substance is modified.

chemical composition n 化学組成
▶ Absorption bands are not only characteristic for the chemical composition and the phase of the particles, but they can also depend on the size, the shape, and the architecture of the aggregates.

chemical degradation n 化学分解
▶ The primary structure of the antidiabetic hormone, insulin, has been deduced by chemical degradation and synthesis.

chemical labeling n 化学ラベル
▶ Other methods of monitoring the reaction of DNA with restriction enzyme either involve extra chemical labeling of the DNA with a fluorophore or require the DNA to become single stranded after cleavage.

chemical laser n 化学レーザー
▶ In a chemical laser, the population inversion is produced by chemical reaction.

chemically adv 化学的に
▶ If replacement of either of two protons by Z would yield the same product, then the two protons are chemically equivalent in an achiral medium.

chemical milling n 化学切削法
▶ Chemical milling is used in the manufacturing of instruments and other components where exact tolerances are required.

chemical modification n 化学修飾
▶ The calixarenes are amenable to chemical modification, which make them useful neutral receptors or baskets for a range of cationic guests including transition metals.

chemical oscillation n 化学振動
▶ Chemical oscillations are driven by the overall decrease in free energy as reactants are converted into products in a far-from-equilibrium system.

chemical potential n 化学ポテンシャル
▶ Any chemical system is characterized by its electronic chemical potential and by its absolute hardness.

chemical property n 化学的性質
▶ Enantiomers have identical chemical properties except toward optically active reagents.

chemical reaction n 化学反応
▶ A chemical reaction must involve the breaking and the forming of chemical bonds whether the reaction takes place in a test tube or as part of some physiological system.

chemical reactivity n 化学反応性
▶ In general, ethers are low on the scale of chemical reactivity, since the carbon-oxygen bond is not cleaved readily.

chemical resistane n 耐薬品性
▶ When polymerized, halocarbons yield plastics characterized by extreme chemical resistance, high electric resistivity, and good heat resistance.

chemicals n 化学物質 syn *chemical substance*
▶ All other chemicals were obtained commercially and used without further purification.

chemical shift n 化学シフト
▶ Hybridization of the carbon atom has a significant effect on the chemical shift: sp^3-hybridized carbon absorbs at high field (0-60 ppm downfield from TMS), sp^2 carbon at low field (80-200 ppm) and sp carbon at intermediate values.

chemical species n 化学種
▶ Intercalation compounds consist of layers of different chemical species.

chemical transformation n 化学変換
▶ To study a chemical transformation, a reaction coordinate is chosen which varies uniformly during the reaction.

chemical vapor deposition n 化学蒸着
▶ With modern growth methods based on chemical vapor deposition and molecular beam epitaxy, junctions that are virtually atomically abrupt can be prepared.

chemiluminescence n 化学発光
▶ Chemiluminescence occurs in thousands of

chemical reactions covering a wide variety of compounds, both organic and inorganic.

chemiluminescent adj 化学発光の

▶ A beautiful demonstration of the chemiluminescent reaction is furnished by pouring a solution of *p*-chlorophenylmagnesium bromide from one flask to another in the dark.

chemisorb vt 化学吸着する ☞ physisorb

▶ Carbon monoxide chemisorbs on a low coordination number gold atom.

chemisorption n 化学吸着 ☞ physisorption

▶ Chemisorption is properly defined as adsorption involving electronic interaction between adsorbent and adsorbate.

chemistry n **1** 化学

▶ Chemistry tells us of the properties and composition of substances, of the action of substances on each other, and of the changes in composition which they undergo in a great variety of circumstances.

2 化学現象

▶ Real chemistry occurs interfacially or heterogeneously.

chemotherapeutic adj 化学療法の

▶ Much research effort has been devoted to the chemotherapeutic investigation of cancer.

chemotherapeutic agent n 化学療法剤

▶ Many compounds that were powerful disinfectants proved damaging or useless as chemotherapeutic agents.

chemotherapy n 化学療法

▶ Many drugs currently used in chemotherapy may be effective against cancerous cells but also affect normal cells.

cherish vt 心に抱く syn *preserve, sustain*

▷ The controlled delivery of a drug to its target is a long cherished dream of pharmacologists.

chief adj 主な syn *essential, main, prime*

▶ Cellulose is the chief constituent of wood and cotton.

chiefly adv 主に syn *mainly*

▶ Among hydrocarbons the factors that determine boiling point seem to be chiefly molecular weight and shape; this is to be expected of molecules that are held together chiefly by van der Waals forces.

chill vt 冷やす syn *cool, freeze*

▶ If the reaction vessel was chilled with liquid nitrogen, the unreacted fluorine could be pumped away, and then, on gradual warming, the HOF could be distilled from the vessel.

chilled adj 冷却した

▶ The crystals that have separated in this first crop are collected by suction filtration and washed free of mother liquor with a little fresh, chilled solvent.

chilling n 冷却

▶ Rapid chilling may result in coring or zoning, that is, the deposition of successive layers which are not in equilibrium with one another.

chip n 半導体の小片 syn *bit, fragment, piece*

▶ Miniaturization and increased speed of computer chips are limited by the generation of heat and the charging time of capacitors arising from the resistance of the interconnecting metal films.

chiral adj キラルな syn *dissymmetric*

▶ Both enantiomers react faster in the chiral environment than in the absence of the chiral modifier, but one reacts faster than the other.

chiral center n キラル中心 syn *asymmetric center*

▶ Addition of cyanide to the aldopentose generates a new chiral center, about which there are two possible configurations.

chirality n キラリティー syn *handedness*

▶ The shapes of many organic molecules are dissymmetric and have chirality.

chlorinate vt 塩素化する

▶ Acetic acid can be chlorinated by gaseous chlorine in the presence of red phosphorus to yield successively mono- (**1**), di- (**2**), and trichloroacetic acid (**3**); the reaction proceeds better in bright sunlight.

chlorination n 塩素化

▶ If the chlorination is stopped when approximately one molar proportion of chlorine is absorbed, the main product is monochloroacetic acid.

chlorophyll n クロロフィル

▶ Chlorophyll absorbs low-energy light in the red region. The exact frequency depends on the nature of the substituents.

chloroplast n 葉緑体

▶ Chloroplasts contain photosynthetic systems for utilizing the radiant energy of sunlight.

choice n **1** 選択 syn *preference, selection*

▶ The choice of base was found to have an impact not only on the product yield but also on the enantioselectivity.

2 of choice 一般的に好まれる

▶ Preparation of acid derivatives via acid chlorides is often the method of choice.

choking adj 窒息させる

▶ The NaBr-H_2SO_4 method is preferred to the Br_2-SO_2 method because of unpleasant, choking properties of sulfur dioxide.

cholesteric adj コレステリック液晶の

▶ Nematic phases are made cholesteric by dissolving in them optically active enantiomers.

choose vt 選ぶ syn *prefer, select*

▷ Many reactions are available for the preparation of hydrogen, and the one chosen depends on the amount needed, the purity required, and the availability of raw materials.

chromatic adj 着色した

▶ Chromatic emulsions are produced when the refractive index of the two liquids is made identical for only one wavelength of light.

chromatin n クロマチン

▶ When a cell is not dividing, DNA in the nucleus exists as an intricate combination of DNA and DNA-binding proteins called histones to form strings called chromatin.

chromatogram n クロマトグラム

▶ The chromatogram is developed by passing through a suitable solvent which washes the adsorbate down through the column.

chromatograph n **1** クロマトグラフ

▶ The analytical separation of a mixture of 17 aromatic compounds was effected by a linearly programmed chromatograph, in which column temperature was raised continuously during analysis.

vt **2** クロマトグラフにかける

▶ Benz[a]anthracene was chromatographed on alumina, using carbon tetrachloride, and then recrystallized from benzene.

chromatographic adj クロマトグラフの

▶ The purification of hexane for spectroscopic use may be readily achieved by passing through a chromatographic column having silica gel in the lower section and basic alumina in the upper section.

chromatographically adv クロマトグラフによって

▶ If all attempts to resolve a given substance chromatographically are unsuccessful, evidence is thus provided for the presence of a single pure chemical entity.

chromatography n クロマトグラフィー

☞ column chromatography

▶ Chromatography provides a critical test for homogeneity of chemicals and facilitates the comparison of similar substances, their purification or concentration, the recognition of quantitative separation in complex mixtures, and the determination of molecular structure.

chrome n クロム化合物

▶ Chrome pigments are more stable to sunlight, weathering, and chemical action than the brighter organic dyes.

chromel-alumel thermocouple n クロメル-アルメル熱電対

▶ The temperatures produced by these heating tapes were monitored with chromel-alumel thermocouples embedded between the tape and the glass tube.

chromophore n 発色団

▶ The benzene chromophore gives rise to a second band at longer wavelengths.

chromosomal adj 染色体の

▶ The long, thin strands of chromosomal DNA become coiled around clusters of histone proteins, forming structures resembling beads on a string.

chromosomal aberration n 染色体異常

▶ Two types of chromosomal aberrations can

occur in cells: changes in chromosome structure and changes in chromosome number.

chromosome n 染色体

▶ Chromosomes are the bearers of the hereditary instructions in a cell; thus they are the overall regulators of cellular processes.

chronic adj 慢性的な syn *continual, persistent*

▶ Chronic toxicity refers to exposure of long duration, i.e., repeated or prolonged exposures.

chronicle vt 記録にとどめる syn *document, record*

▷ A report of the mild hydrogenation of $[Os(NH_3)_5(\eta^2\text{-benzene})]^{2+}$ to $[Os(NH_3)_5(\eta^2\text{-cyclohexene})]^{2+}$ was followed by a string of publications chronicling acylations, alkylations, cyclizations, and tandem addition reactions of η^2-aromatic ligands.

chronologically adv 時間順に

▶ Chronologically, the first period covers the interaction of high-energy radiation with matter.

chunk n 大きな塊 syn *lump, mass*

▶ Solid chunks of material to be crystallized are introduced into the outer chamber.

chymotrypsin n キモトリプシン

▶ Chymotrypsin plays an important role in digestion by assisting in the hydrolysis of proteins in the intestinal tract.

circle n 円 ☞ filled circle, hatch inscribed circle, open, shaded

▶ The point returns to its original position once in every complete revolution on the circle, the period being the time taken to describe the circles.

circuit n 回路 syn *path*

▶ Every time 1 mol of H_2 reacts with 1 mol of Cu^{2+}, 2 mol of electrons pass through the outer circuit.

circuit board n 回路基板

▶ In the electronics industry, when Au and Al were sometimes used in close proximity on circuit boards, a purple color often appeared between strips of the two metals.

circular adj 円形の syn *round*

▶ As we moved toward the nucleus, we see a rapid increase in electron density, indicated by the close spacing of the now circular lines.

circular dichroism n 円二色性

▶ The net circular dichroism spectrum of a protein can be deconvoluted to provide an estimate of the different proportions of secondary structure.

circularly polarized light n 円偏光 ☞ plane polarized light

▶ Asymmetric molecules preferentially absorb either left- or right-handed plane or circularly polarized light.

circulate vt 循環させる

▶ When a molecule is placed in a magnetic field, its electrons are caused to circulate.

circulating adj 循環する

▶ After cooling to 160 ℃, the sulfur trioxide is adsorbed in a circulating stream of 98–99% sulfuric acid, where it unites with the small excess of water in the acid to form more sulfuric acid.

circulation n **1** 流通 syn *passage, transmission*

▶ The stirring rate in the growth chamber and the rate of circulation of fresh solvent cannot be controlled separately, since they are interdependent.

2 回転 syn *rotation*

▶ Circulation of electrons about the proton itself generates a field aligned in such a way that it opposes the applied field.

3 循環 syn *distribution*

▶ Dietary protein is hydrolyzed to free amino acids that are absorbed into the circulation.

circumferentially adv 周辺を取り囲んで

▶ Spherical liposomes become ellipsoidal in shear flow thus orienting with their long axes arranged circumferentially around the linear dichroism cell.

circumscribe vt 円を外接させる ☞ inscribe

▷ These structures may conveniently be visualized within circumscribed sphere.

circumstance n 環境 syn *condition, situation*

▶ The appearance of the same or very similar

spectra in crystals or solutions thought to contain these ions is a very direct proof of their existence in these circumstances.

circumstantial adj 状況的な syn *indirect*

▶ The most important circumstantial evidence as to reaction mechanism is the identity of the products formed.

circumvent vt 克服する syn *bypass*

▶ Various attempts have been made to circumvent these difficulties by defining pH in terms of quantities that do have thermodynamic significance.

cis adj **1** シスの

▶ The three differing Sb−Cl distances in $(SbCl_4)_n^{n-}$ reflect the influence of the lone-pair of electrons on Sb^{III}. The shortest bonds are cis to each other, whereas the intermediate bonds are trans to each other.

pref **2** シス 化学物質名の一部をなすときは，イタリック体としハイフンを用いる． ☞ trans

▶ In *cis*-2-butene, the prefix cis means that the two methyl groups are on the same side of the double bond.

cis addition n シス付加 ☞ trans addition

▶ The catalyst adsorbs the hydrogen gas and in a cis addition delivers two hydrogen atoms to the olefin.

cis form n シス形 ☞ trans form

▶ 2-Butene can exist in a cis form in which the methyl groups are on the same side of the double bond', and a trans form with the methyl groups on opposite sides of the double bond.

cis-trans isomerization n シス-トランス異性

▶ Light brings energy to the rhodopsin, energy that causes a $\pi \rightarrow \pi^*$ transformation in the retinal moiety; in effect, it opens carbon-carbon double bonds and permits the rotation that is necessary for cis-trans isomerization.

citation n 引用 syn *mention, quotation, reference*

▶ In a formula, the order of citation of symbols is based upon relative electronegativities: the more electropositive constituent(s) being cited first.

cite vt 引用する syn *mention, quote, refer*

▷ Fuller details including X-ray structures for many of the compounds are in the references cited above.

clad adj 覆われた

▶ Synthetic cryolite is manufactured in lead-clad vessels by the reaction

$$6HF + Al(OH)_3 + 3NaOH \rightarrow Na_3AlF_6 + 6H_2O$$

cladding n 被覆 syn *coating*

▶ Platinum is used in the chemical and glass industries as cladding to prevent attack by particularly corrosive materials such as hot hydrofluoric acid.

claim n **1** 資格 syn *right*

▶ The bonding in the trimer appears to be of the same order of strength and character as that in the dimer, and in this respect the trimer has every claim to considerations as a distinct molecular species.

vt **2** 主張する syn *assert, state*

▶ Although we cannot yet claim to understand the origin of such differences, it is important to recognize their existence.

clamp n **1** クランプ

▶ The clamp supporting the reflux condenser is mounted high on the ring stand and is screwed only to a loose fit so that the flask may be grasped and swirled if desired.

vt **2** 固定する syn *attach, fix, lock*

▶ Molecules in the bulk of a crystal are clamped in their places mainly by repulsions that oppose the disentanglement of interlocking molecules.

clarification n 解明 syn *explanation, interpretation, solution*

▶ Clausius introduced the concept of entropy to interpret the second law, and this brought about a great clarification of the law.

clarify vt 明らかにする syn *elucidate, explain, simplify*

▶ These examples help to clarify the relationship of neutralization to displacement mentioned at the beginning of this chapter.

clarity n 明快 syn *simplicity*

▶ For clarity, all hydrogen atoms have been omitted.

clash n 衝突 syn *collision*

▶ It is difficult to assemble six pyridine-derived donors around an octahedral metal center, on account of the clashes between hydrogen atoms adjacent to nitrogen on neighboring pyridine nuclei.

class n **1** 分類 syn *category, classification, division*

▶ Phenomenologically, it is observed that class a metals tend to be N-bonded whereas class b metals tend to be S-bonded.

2 be in a class by itself 比類がない

▶ The halogens are in a class by themselves, being deactivating but ortho, para-directing.

vt **3** 分類する syn *categorize, classify, group*

▶ Adsorptions that are classed as chemisorptions have heats of adsorption that compare with those of ordinary chemical reactions.

classic adj 典型的な syn *definitive, ideal, standard*

▶ Transition-metal hydrides are classic examples of compounds with variable stoichiometry.

classical adj 古典的な syn *established, serious, standard*

▶ The classical situation corresponds to small values of the energy-level spacing compared to kT.

classically adv 古典的に

▶ Classically, changing the sign of the time coordinate in Newton's equation of motion is equivalent to reversing the direction of all velocities.

classification n 分類 syn *category, class, division*

▶ The classification obscures the important point that there is an almost continuous gradation in properties between members of the various classes.

classified adj 機密扱いの syn *confidential*

▶ Carbaboranes burst onto the chemical scene in 1962–3 when classified work was cleared for publication.

classify vt 分類する syn *categorize, class, group*

▶ The terms insulator, semiconductor, and metal are used to classify solids based on the size of the band gap and the effect of the band-gap energy on electric conductivity.

clathrate n クラスレート

▶ The gas hydrates in which the guest molecules are not bound chemically but are retained by the structure of the host are called clathrates.

clathrate compound n クラスレート化合物

▶ The formula for any clathrate compound is determined by the ratio of available cavities to the amount of cage material.

clathration n クラスレート化

▶ The dimensions of the voids in the crystal lattice of the metal complexes must be such that molecules with a width of one benzene ring are preferred for clathration.

clay n **1** 粘土

▶ Clays form suspensions in water, acquiring a stabilization charge, often by hydrolysis or by base exchange with any salts present.

adj **2** 粘土製の

▶ Pfeffer deposited a copper hexacyanoferrate (II) membrane in the walls of a porous clay pot.

clean adj **1** きれいな syn *pure, uncontaminated*

▶ The work with clean metal surfaces has emphasized the complexities that undoubtedly occur when metal powders, chemically deposited metal films, oxides of metals, and nonmetals are used as adsorbents.

2 完全な syn *neat, simple*

▶ Nickel foil is effective in absorbing the Cu $K\beta$ radiation and most of the white radiation, leaving monochromatic, reasonably clean beam of $K\alpha$ radiation.

adv **3** きれいに syn *altogether, thoroughly*

▶ A little more fresh hot solvent is required to wash the paper clean.

vt **4** きれいにする syn *cleanse, wash*

▷ There has been growing interest in finding an efficient method for cleaning the environment of manmade pollutants.

5 clean up 浄化する syn *purify*

▶ Polluted groundwater can sometimes be very difficult to clean up because contaminated water often gets trapped in soil and rocks.

cleaner n 洗浄剤

▶ In liquid cleaners, rheological behavior is often critically important.

cleanliness n 清潔 syn *purity*

▶ Background impurity concentrations often need to be below one part in 10^9; therefore, the purity of the starting materials and cleanliness of the growth system are vital factors.

cleanly adv きれいに syn *clean*

▶ The phenomenon of cleavage, i.e., the property of splitting cleanly along certain planes, lends further emphasis to this premise.

cleanup n 浄化，清掃

▶ Because living organisms are used for the cleanup, bioremediation processes are generally cleaner than other types of cleanup strategies.

cleanse vt 化学薬品を用いて取り除く syn *clean, wash*

▶ A solid is seldom distilled with the object of effecting a separation of constituents of different degrees of volatility but rather to cleanse the solid.

clear adj **1** 明白な syn *evident, obvious, plain*

▶ That these attractions exist is clear; exact knowledge of their source and quantitative values for them are much harder to come by.

2 澄み切った syn *transparent*

▶ The resulting suspension was hot-filtered, yielding a red solid and a clear red filtrate.

vi **3** 透明になる

▶ For M = Cr, whose axial interactions are comparable to those in late transition-metal tetracarboxylates, the liquid crystal phase does not clear before decomposing.

vt **4** 透明にする

▶ A solution of tin(II) chloride dihydrate is prepared and cleared as in Roth's method, and hydrous tin(II) oxide is precipitated by addition of an aqueous solution of ammonia until the pH is about 10.

clear-cut adj 明快な syn *evident, obvious*

▶ The basis for other correlations between size, charge, and chemical properties is not so clear-cut.

clearing point n 透明点

▶ The introduction of a dopant in a liquid crystal host generally causes a shift in the clearing point that is a function of dopant-host interactions.

clearly adv 明らかに syn *certainly, evidently*

▶ Clearly charge balance must be preserved overall, but within this limitation, there is much scope for introducing ions of different charge.

cleavage n **1** へき開 syn *fracture, rupture*

▶ $(NH_4)_2PtCl_6$ is isometric and octahedral with perfect 111 cleavage.

2 開裂

▶ The cleavage of the N-substituted phthalimide is best effected by reaction with hydrazine hydrate and then heating the reaction mixture with hydrochloric acid.

cleavage face n へき開面 syn *cleavage plane*

▶ Ammonium iodide crystallizes on the cleavage surface of mica with its (111) plane in accurate alignment with the mica lattice.

cleave vi **1** へき開面に沿って裂ける syn *divide, split*

▶ Since mica cleaves so easily, pieces of different thickness are quickly and easily prepared.

vt **2** へき開面に沿って裂く

▶ Any attempt to cleave a crystal along an arbitrary plane will shatter it.

3 結合開裂させる

▶ Calcium carbide is cleaved by water to give acetylene and calcium hydroxide.

cleaving n 開裂 syn *parting*

▶ Cleaving or other fission-type processes only occur beyond a certain cluster size, corresponding to the onset of three-dimensional nanocrystal structures.

cleft n 裂け目，割れ目 syn *crack, flaw*

▶ The photopolymerization of vesicles prepared from surfactants containing styrene in their head groups results in pulling together approximately 20 monomers, thereby creating some 15 Å diameter surface clefts.

clever adj 上手な syn *elegant*

▶ These clever experiments provide definite evidence of but little resonance energy in the cyclobutadiene ring system.

cleverly adv 上手に syn *elegantly*

▶ This achievement constitutes a spectacular early example of the spontaneous, one-pot assembly of cleverly designed components to yield an intended product of some complexity.

climb n **1** 上昇 syn *rise*

▶ The process of climb is a special mechanism of dislocation motion which involves vacancies.

vi **2** 上昇する syn *rise*

▷ A climbing dislocation at the root of the whisker performs similar evolutions to a growth step at the surface.

cling vi **cling to** …にくっつく syn *adhere, attach, stick*

▷ There is formed a carbocation with the solvent molecule clinging to its back side, and it is this nucleophilically solvated carbocation that then undergoes further reaction to yield the product.

clinical adj 臨床の

▶ A further restriction on the clinical use of alumina is the response of tissues to its implantation.

clip vt 削減する syn *diminish*

▶ All of the light beam incident upon the cell must pass through the sample and not be clipped or reflected by the walls or base of the cell or the meniscus of the solution.

clockwise adj 時計回りの，右回りの ☞ anticlockwise, counterclockwise

▶ In the case of the white form, a given order of attachment of groups which is clockwise in one ring is counterclockwise in the rings above and below it in the same stack.

clog vt 詰まらせる syn *block*

▷ Suspensions do not pass through paper filters, after the pores have become clogged a little.

clogging n 目詰まり

▶ Another severe problem with electrospraying is the clogging of the capillaries.

clone vt クローンを作る syn *double, duplicate*

▶ Because the genetic code is universal, biologists can use techniques called recombinant DNA technology to clone a human gene such as the insulin gene and insert it into a bacteria so that bacterial cells transcribe and translate insulin.

close adj **1** 接近した syn *near*

▶ Gallium has a unique orthorhombic structure in which each Ga has one very close neighbor at 244 pm and six further neighbors, two each at 270, 273, and 279 pm.

2 密接な syn *intimate*

▶ NaCl forms a complete range of solid solution with the chemically less closely related AgCl on account of the close correspondence between the radii of the two cations.

3 密集した syn *dense*

▶ In metals, the direction of motion of dislocations is usually parallel to one of the directions of close packing in the crystal structure.

4 綿密な syn *careful*

▶ Closer inspection shows that some solids can exist in different crystal forms.

adv **5** 接して syn *adjacent to, nearby*

▶ Most of the electron density is concentrated close to the nuclei of the ions.

vt **6** 遮断する syn *seal*

▶ The condenser is closed by means of a tube containing soda lime.

7 完成する syn *establish, make*

▷ Most other methods for closing rings of ten or more carbons yield large amounts of intermolecular reaction in addition to the desired intramolecular process.

closed adj 閉じた syn *unopened*

▶ A gas contained within a closed vessel exerts a force on the walls of the vessel.

closed shell n 閉殻 複合形容詞として使用

▶ The valence shell electron-pair repulsion model has become the most successful and widely used model for the prediction of geometries of closed-shell molecules.

closed system n 閉じた系 ☞ open system

▶ Though most reactions in closed systems are carried out at constant volume, sometimes it is convenient to work with constant pressure.

closely adv **1** 密接に syn *intimately*

▶ A reaction closely related to acetal formation is the polymerization of aldehydes.

2 ぴったりと syn *exactly, nearly*

▶ Whereas the IR absorption spectra of weak complexes are effectively the sum of the spectra of the component molecules, in these strong interactions the IR spectra correspond closely to the sum of the spectra of the corresponding ions D^+ and A^-.

3 接近して syn *immediately*

▶ A closer look at the vibrational-absorption band of a molecule like HCl shows that it does not consist of an absorption region with a single maximum but rather has a number of closely spaced components.

4 厳重に syn *rigorously, tightly*

▶ Water pollution caused by discharge of toxic chemical wastes is closely regulated.

closeness n 近さ syn *proximity*

▶ Although the basic ideas of ion pairing is quite generally accepted, the closeness of approach necessary for ion pairs to form cannot be directly measured.

close-packed adj 最密の ☞ face-centered cubic

▶ The preferred plane of shear or slip is usually a close-packed plane.

close-packed structure n 最密構造

▶ Although it is legitimate to estimate the efficiency of packing of spheres in close-packed structures, it is incorrect to do this for space-filling polyhedra.

close packing n 最密充填 複合形容詞として使用.

▶ Modified close-packing arrangements are observed for many diatomic molecules.

closure n **1** 閉鎖

▶ The most reliable method of determining configurations by chemical means is through reactions that lead to closure of five- or six-membered rings.

2 容器の蓋

▶ To permit ready loading and unloading of samples, a vacuum-tight closure provides a suitable aperture of about 1 cm diameter in the top of the apparatus.

cloud n **1** 雲 syn *fog, mist*

▶ A vapor can be compressed beyond its saturation pressure until at a definite degree of supersaturation cloud formation suddenly occurs.

2 電子雲 ☞ electron cloud

▶ The mobile π cloud of the carbonyl double bond is pulled strongly toward the more electronegative atom, oxygen.

cloud chamber n 霧箱

▶ The cloud chamber, invented by Wilson in 1911, is a chamber containing air saturated with water vapor.

cloudiness n 濁り syn *dimness, obscurity*

▶ The optical quality of boric acid glass leaves much to be desired, water absorption leading to progressively increasing cloudiness of the specimen.

cloudy adj 濁った syn *obscure, opaque*

▶ The magnesium turnings have disintegrated, and the solution has acquired a cloudy or brownish appearance.

clue n 手がかり syn *hint, idea*

▶ It is often possible to determine repeated periods that give important clues as to the arrangement and structure of the macromolecules in the solid state.

cluster n **1** クラスター syn *collection, group*

▶ Alkali halide nanocrystals are highly ordered, even at small sizes, in ontrast to covalently bonded clusters, to metallic clusters, or to clusters held together by weak dispersion interactions.

2 クラスター，集塊

▶ Operons are essentially clusters of several related genes located together and controlled by a single promoter.

vt **3 群がらせる** syn *collect, gather*

▶ Proteins tend to fold up so that their hydrophobic residues are clustered in an interior core, away from contact with the aqueous environment, and the hydrophilic residues tend to be arranged on the exterior interacting with water.

cluster compound n **クラスター化合物**

▶ These metal cluster compounds may involve anywhere from two to twenty metal atoms connected by metal-metal bonds stabilized by appropriate ligands.

clustering n **クラスター形成**

▶ Calcination of micelle-templated silicas produces hydrophobic zones and hydrophilic zones, and grafting occurs only on the hydrophobic areas, leading to clustering of the organic functionality even at relatively low loadings.

cluster ion n **クラスターイオン**

▶ Tellurium forms cluster ions such as the square Te_4^{2+} and trigonal-prismatic Te_6^{2+}.

coacervate n **コアセルベート**

▶ The term coacervate has been broadened to include the interaction of a colloid and of an ordinary ion of opposite charge, usually polyvalent.

coagulant n **凝結剤** ☞ anticoagulant, flocculant

▶ Compounds that dissociate into strongly charged ions are normally used as coagulants.

coagulate vi **1 凝結する** syn *unite*

▷ The number of equivalents is evidently independent both of the valence of the coagulating ion and of the concentration required for coagulation.

vt **2 凝結させる** syn *set*

▶ Unstable or irreversible colloids are those that are readily coagulated by slight contamination.

coagulation n **凝結** ☞ coalescence

▶ Coagulation can often be brought about by addition of appropriate electrolytes, for example, by the addition of an acid to milk or of aluminum sulfate to turbid water.

coagulum n **凝塊** 複数形は coagula.

▶ The coagulum always contains definite amounts of the ion which was effective in producing the coagulation.

coalesce vi **合体する** syn *combine, unite*

▶ Oil droplets in contact with water tend to coalesce so that there are a water layer and an oil layer.

coalescing n **合体すること** syn *coalescence, union*

▶ Repulsion between similar charges keeps the oil droplets from coalescing.

coalescence n **コアレッセンス** ☞ coalescing, uniting

▶ Variable temperature NMR studies of **1** in toluene showed coalescence of the signals due to the two conformers of **1**, allowing us to estimate a barrier to rotation for either Ar−CO or Ar−N bond of only 63.5 kJ mol^{-1}.

coarse adj **粗い** syn *rough*

▶ As more ammonium peroxodisulfate solution is added into the aniline solution, the nanofibers become thicker and coarser, and the final reaction product contains mostly irregularly shaped agglomerates.

coat n **1 外被** syn *covering, layer*

▶ The cell walls of fungi and the coats of mold spores may contain cellulose, but chitin is their major constituent.

vt **2 覆う** syn *cover*

▶ It is usually necessary to coat the sample with a thin layer of metal, especially if the sample is a poor electric conductor, in order to prevent the build-up of charge on the surface of the sample.

coated adj **1 上塗りを施した**

▶ The electrode that liberates very readily anions is commonly a metal coated with one of its insoluble salts.

2 複合語で，…**に覆われた**

▶ The preparation of sensitizer- or catalyst-coated colloidal semiconductors is currently more of an art than science.

coating n **被膜** syn *cover, film*

▶ Dense, thick, corrosion-resistant coatings could be achieved from this electrolyte system.

coat protein n コートタンパク質

▶ For many viruses, in vitro self-assembly mechanisms of coat protein monomers into intact and stable virus-like particles have been observed.

code n **1** コード syn *notation, symbol*

▶ The hereditary message is written in a code, with each word standing for a particular amino acid.

vt **2** コード化する syn *encode*

▶ These quasi-smectic phases are coded by the letters E, J, G, H, K, and B.

coded adj コード化された

▶ If the newly coded amino acid changes the structure of the protein, then its function may be significantly changed.

codify vt 分類する syn *formulate, organize*

▶ These relationships were codified by Wade in a set of rules which have been extremely helpful not only in rationalizing known structures but also in suggesting the probable structures of new species.

codon n コドン

▶ Proof that the codon for each amino acid consists of three nucleotides was provided by genetic studies of the effects on the polypeptide product of nucleotide addition or deletion from a gene.

coefficient n 係数 syn *factor*

▶ In the so-called virial equation, *PV* is expressed as a series in which the terms involve powers of the reciprocal of the volume multiplied by coefficients which are functions of temperature:

$$PV = A + B(1/V) + C(1/V)^2 + D(1/V)^4 + E(1/V)^6 + F(1/V)^8 + \cdots$$

where *A*, *B*, *C*, *D*, *E*, *F* are the first, second, third, fourth, etc., virial coefficients, *D*, *E*, and *F* being usually very small below 1 atm.

coenzyme n 補酵素

▶ A coenzyme is not usually a protein; it is a much simpler organic molecule than an enzyme, and the enzyme–coenzyme complex is formed reversibly.

coercive field n 抗磁場

▶ In order to reduce the polarization to zero, a reverse field is required; this is the coercive field.

coercivity n 保磁力

▶ Materials that are magnetically soft are those of low coercivity.

coexist vi 共存する

▶ There may be some point where solid, liquid, and gas can all coexist in equilibrium.

cofactor n 補因子, コファクター

▶ Many enzymes require cofactors if they are to exert their catalytic effects.

coherence n コヒーレンス syn *connection, integrity*

▶ The intense irradiation clearly causes sufficient disorder and molecular crosslinking to destroy the coherence of the single crystal lattice of polymers.

coherent adj **1** 干渉性の

▶ During the transition from the ^2E level to the ground state, many Cr^{3+} ions are stimulated to decay, in phase with each other, giving an intense, coherent pulse of red light of wavelength 6934 Å.

2 密着する syn *consistent, orderly, organized*

▶ For normal samples of Al metal, a coherent protective oxide film prevents appreciable reaction with oxygen, water, or dilute acids.

cohesion n 凝集

▶ Cohesion creates a pressure within a liquid of between 10^3 and 10^4 atmospheres.

cohesive energy n 凝集エネルギー

▶ The large cohesive energies can account for the abundance of alkali halide clusters in alkali halide vapors.

cohesive force n 凝集力

▶ Boiling occurs when a temperature is reached at which the thermal energy of the particles is great enough to overcome the cohesive forces that hold them in the liquid.

coil n **1** コイル syn *spiral*

▶ The hydrogen bonds are nearly parallel to the long axis of the coil, and the spacing between the turns is about 5.4 Å.

vi **2** らせん状に巻く syn *wind*

▶ Nearly 6 feet of DNA is coiled into the nucleus of every human cell.

vt **3** ぐるぐる巻く syn *wind*

▷ The tertiary structure is the way in which the coiled polypeptide chain is folded and hydrated in the natural state.

coiling n コイル形成

▶ The principal feature of the α-helix is the coiling of peptide chains in such a way as to form hydrogen bonds between the amide hydrogens and carbonyl groups that are four peptide bonds apart.

coin vt 作り出す syn *invent, make*

▶ The terms furanose and pyranose have been coined to denote five- and six-membered rings in cyclic sugars.

coincide vi 一致する syn *accord, agree*

▶ The spherical, ordinary-ray velocity surface coincides with the extraordinary-ray velocity surface at two points only.

coincidence n 一致 syn *equality, resemblance*

▶ A symmetry operation is an operation such as rotation, performed on an object or on pattern, which brings it into coincidence with itself.

coincident adj 一致した syn *coinciding*

▶ Because both the director of the $Mo_2[O_2C(CH_2)_6CH_3]_4$ molecule and the primary component of the ^{95}Mo quadrupole coupling tensor are coincident with the metal–metal bond, molybdenum NMR should be a pertinent probe of the alignment of the liquid crystal phase in a magnetic field.

col n 鞍部 ☞ saddle point

▶ The energy corresponding to the col is from one point of view a maximum energy and from another point of view a minimum energy. The height of this col represents the activation energy of the system.

cold n **1 the cold** 冷たいところ

▶ Alkali carbonates can be titrated in the cold by using Methyl Orange as an indicator, the end point being taken as the change from yellow to reddish orange.

adj **2** 冷たい syn *cool*

▶ Since the solubility of phthalic acid in cold water is low, difficulty in the filtration can be avoided by use of a large excess of solvent.

cold drawing n 冷延伸

▶ The polymer crystallites in the surface layer are more perfectly oriented in the direction of cold drawing than are the crystallites in the film interior.

colinearity n 直線性

▶ The apparent colinearity in the higher-temperature form of many diphosphates, which was previously ascribed to a P−O−P angle of 180°, is now generally attributed to positional disorder.

collaborate vi 協力する syn *cooperate*

▶ Zsigmondy began to collaborate with Siedentopf, a physicist, on the construction of an ultramicroscope.

collaboration n 協力 syn *cooperation*

▶ We deeply appreciate collaboration with many colleagues whose names appear in the references.

collaborative adj 協力的な syn *joint*

▶ The research that produced the first sulfa drugs was a collaborative effort in which biomedical researchers and chemists played distinct and complementary roles.

collagen n コラーゲン

▶ Connective tissue and skin are indicated by X-ray diffraction photographs to be composed of collagen molecules running parallel to the long axis.

collapse n **1** 崩壊 syn *breakdown, destruction*

▶ The aggregation of anion vacancies leads to the collapse of the parent lattice and nucleation of the metal.

vi **2** 崩壊する syn *break down*

▶ By using refinements such as the MAS technique, the broad bands collapse to reveal a fine structure.

collect vi **1** 溜まる syn *accumulate, pile up*

▶ Impure *p*-benzoquinone may be purified by

placing it in a distilling flask attached to a condenser and passing a rapid current of steam into the flask: the quinone sublimes and collects in the receiver.

vt **2 集める** syn *gather*

▶ The data were collected at ambient room temperature.

collection n **1 集まり** syn *assemblage, store*

▶ The molecule is modeled as a static collection of hard spheres which completely exclude a spherical probe representing a solvent molecule.

2 収集 syn *accumulation*

▶ A prismatic crystal of dimensions 0.17×0.20×0.40 mm was selected for data collection on a Siemens R3m/V diffractometer at room temperature.

collective adj **1 集団的な** syn *cooperative, joint*

▶ From the point of view of the collective electron treatment the highest full band in alkali-halide crystals corresponds to the outermost p electrons of the halide ions.

2 包括的な syn *common, general*

▶ Nanotechnology is a collective term for a wide range of relatively novel technologies; the main unifying theme is that it is concerned with matter on the nanometer scale.

collectively adv **全体として**

▶ Collectively, these compounds have found tremendous use as ligands for modifying the performance of various transition-metal catalyst systems.

collide vi **衝突する** syn *strike*

▶ Two hydrogen atoms do not recombine whenever they collide, although the process $2H^{\cdot} \rightarrow H_2$ is very highly exothermic.

colligative property n **束一的性質**

▶ The properties of dilute solutions that depend on only the number of solute molecules and not on the type of species present are called colligative properties.

collimate vt **光線を平行にする**

▶ The output beam of the laser is directional and collimated, which means that its spectral brightness is ca. 10^{14} times as high as that of the black body.

collinear adj **同一線上の**

▶ The collinear or very nearly collinear nature of the UO_2 group appears to be on moderately sound basis.

collision n **衝突** syn *hit, impact*

▶ In an ideal experiment, we should like to prepare A and BC with very specific energies, allow them to undergo a single collision, and then analyze the product states.

collisional adj **衝突の**

▶ The observed lifetime is often less, owing to the presence of competing and faster collisional processes.

collisional deactivation n **衝突による失活**

▶ Certain types of complex molecules are peculiarly resistant to collisional deactivation and retain their ability to fluoresce even in the dissolved or solid state, where collisions are much more frequent than in the gas phase.

collision frequency n **衝突頻度**

▶ The collisions will occur at irregular intervals, but over a sufficiently long time, the average number of collisions per second approaches a definite number called a collision frequency.

collision frequency factor n **衝突頻度因子**

▶ The pre-exponential factor for a diffusion-controlled reaction in solution is a little larger than the collision frequency factor in the gas phase.

collision numer n **衝突数**

▶ For noninteracting molecules in the gas phase, the encounter frequency is simply the kinetic molecular collision number.

collision theory n **衝突説**

▶ If the reactant molecules are fairly simple, the orientation effect is small and the hard-sphere collision theory is not far from the truth.

collocate vt **一緒に並べる**

▶ These one-dimentional chains are collocated so that the Q atoms from neighboring chains cap the rectangular faces of adjacent chains to give bicapped trigonal-prismatic coordination about each metal atom.

colloid n コロイド syn *suspension*

▶ Sol-gel refers not only to the processing of aqueous oxide colloids but also to a synthesis, in which metal-organic compounds are hydrolyzed.

colloidal adj コロイド状の

▶ Gold can be obtained in red, blue, and violet colloidal forms by the addition of various reducing agents to very dilute aqueous solutions of gold(III) chloride.

colloidal dispersion n コロイド分散系

▶ It is impossible to draw a distinct line between colloidal dispersions, true solutions, and suspensions.

colloidal particle n コロイド粒子

▶ Colloidal particles may be formed in two distinct ways: by subdivision of bulk material or by growth from molecular dimensions.

colloidal solution n コロイド溶液

▶ The opalescence of gold sols is particularly prominent when a narrow beam of light passes through a colloidal solution.

colony n コロニー

▶ When grown on culture plates, each bacterial cell typically divides to form circular-shaped colonies that contain thousands or millions of cells.

color n **1** 色 syn *shade, tint*

▶ Upon treatment with chloroform and aluminum chloride, alkylbenzenes give orange to red colors.

2 in color カラーで

▶ Figures may be printed in color within the text of an article, when color is necessary to the scientific content of the paper.

vt **3** 彩色する syn *dye, stain*

▶ If a cold, neutral solution of potassium dichromate is treated with hydrogen peroxide, the solution is colored violet, owing to the formation of the potassium salt of H_3CrO_7.

colorant n 着色剤

▶ A colloidal pigment is a more effective colorant than a coarse one because of the greater surface area of its particles.

coloration n 着色

▶ Potassium chromate solution poured over a crystal of tartaric acid at room temperature gives purple-violet or black coloration and evolution of carbon dioxide.

color center n 色中心 ☞ F center

▶ Most of the color centers produced by irradiation are unstable and can be bleached thermally; therefore, coloring is generally associated with some type of reactive transient species.

color change n 変色

▶ The color change is due to a small decrease in the band gap with the change in structure.

colored adj 着色した

▶ The clear light-colored part of the crystal had an IR spectrum which showed only starting material.

colorimetric adj 比色の

▶ Colorimetric tests are often suitable for determining small quantities of a constituent.

colorimetrically adv 比色分析法で

▶ Manganese in steel can be quantitatively determined by oxidizing it by sodium bismuthate directly to permanganate and estimating the concentration colorimetrically.

coloring matter n 着色物質

▶ Samples to be purified frequently contain soluble extraneous coloring matter that gives rise to solutions and crystals that are yellowish when they should be colorless or of an off color rather than a pure color.

colorless adj 無色の syn *pale, white*

▶ Selenium does not dissolve in 95% sulfuric acid, but on addition of potassium persulfate green, yellow, or colorless solutions are obtained.

column n **1** カラム syn *pile, pillar*

▶ The molecules of the yellow form arrange themselves in stacks or columns along the *b* axis.

2 クロマトグラフのカラム

▶ The bands at the top of the column contain the most strongly adsorbed components, and the bands at the bottom the least strongly held components.

3 縦の欄 ☞ row

▶ The estimated stretching and bending frequencies for the four activated complexes are shown in columns 2 and 3 of Table 3, and column 4 gives the zero-point energies that are calculated from these frequencies.

columnar adj カラム状の

▶ Strontium chloride hexahydrate has a columnar structure with strontium atoms coplanar with three water-oxygen atoms at 2.80 Å, and the strontium atoms are connected through trigonally arranged bridging oxygen atoms of water molecules at 2.62 Å.

columnar mesophase n カラム状中間相

▶ Columnar mesophases are characterized by stacked columns of molecules, the columns being packed together to form a two-dimensional crystalline array.

column chromatography n カラムクロマトグラフィー ☞ liquid-solid chromatography

▶ Purification by column chromatography on silica gave the title compound as a white solid.

combination n **1** 結合 syn *union*

▶ When iodine and an alcohol were both added to bromoform or ethylene bromide, the molal lowering of the freezing point was considerably less than additive, indicating combination of iodine and alcohol.

2 化合 syn *synthesis*

▶ If there are two initiator fragments in each molecule, termination must have occurred by combination.

combination band n 結合バンド

▶ In polyatomic molecules, combination bands or difference bands, involving two or more modes of vibration, may occur.

combinatorial adj 組み合わせの

▶ The need to increase the throughput of laboratory assays can readily be seen by considering the vast combinatorial spaces inherent in the study of biological and macromolecular systems.

combine vi **1** 化合する syn *compound, unite*

▶ Mercury does not tarnish in air nor combine appreciably with oxygen below 350 ℃.

vt **2** 組み合わせる syn *incorporate*

▷ Study of the behavior of single crystals combining visual observation with chemical analysis and X-ray examination shows that the initial stage of the reaction of each of the oxime picryl ethers in the solid involves the formation of a solid solution of the product anilides in the parent crystal.

combined adj **1** 組み合わせた syn *united*

▶ The operation of rotatory inversion is a combined operation, and if you try to separate its parts, you destroy its character completely.

2 化合した

▶ In combined form, nitrogen is essential to all forms of life and constitutes, on average, about 15% by weight of proteins.

combustant n 燃焼剤

▶ The most important of drawbacks is the necessity to find ways to remove carbonate ions that accumulate in the liquid electrolyte over time when air is used as the combustant.

combustibility n 可燃性

▶ It is difficult to generalize about the combustibility of solids.

combustible adj 可燃性の syn *flammable, inflammable*

▶ Since any material that will burn at any temperature is combustible by definition, it follows that this word covers all such materials, irrespective of their ease of ignition.

combustion n 燃焼

▶ Carbon and hydrogen in combustible compounds can be measured by combustion of a weighed sample in a stream of oxygen and gravimetric determination of the resulting water and carbon dioxide through absorption in anhydrous magnesium perchlorate and soda lime, respectively.

combustion analysis n 燃焼分析

▶ The ability of NH_2OH to react with N_2O, NO, and N_2O_4 under suitable conditions makes it useful as an absorbent in combustion analysis.

come vi **1** 到着する syn *arrive*

▶ Collisions in which three or more molecules all come together at the same time are very unlikely.

2 生じる syn *originate, stem*

▶ The enthalpies of formation come from many sources and have been measured with variable levels of precision.

3 …になる syn *get*

▶ As ions come very close together, they repel one another because of interpenetration of electron clouds.

4 come by …を手に入れる syn *acquire*

▶ While direct experimental data are still hard to come by, we can obtain these parameters from ab initio calculations.

5 come upon 出会う syn *encounter*

▶ In our analysis of orbital coupling, we early came upon some cases whose splitting patterns were completely opposite to our expectations.

comfortably adv 何不自由なく syn *easily, well*

▶ It is only in relatively recent times that chemists have learned to work comfortably with fluorine.

coming n 来ること syn *arrival*

▶ The extent of solvation of ions and even exact interpretation of this term, in view of the rapid coming and going of solvent molecules, remain an open question.

command vt の価値がある syn *demand*

▶ Most metals form arsenides, antimonides, and bismuthides, and many of these command attention because of their interesting structures or valuable physical properties.

commence vt 始める syn *begin, initiate*

▶ Once distillation has commenced it should continue steadily without any drop in temperature at a rate not greater than 1 mL in 1.5–2 min.

commensurate adj 整合して syn *equal, proportional*

▶ In the case of a half-filled band, the periodicity of the charge-density wave is commensurate with the lattice periodicity.

comment n **1** 解説 syn *note*

▶ The reactions of arylmagnesium halides are analogous to those of alkylmagnesium halides and require little further comment.

2 批評 syn *review*

▶ We would like to thank Dr. X. for reading and making helpful comments on the manuscript.

commercial adj **1** 市販の

▶ Most reagents were obtained from commercial sources and used without further purification unless otherwise stated.

2 商業上の

▶ The reaction of chlorobenzene with sodium hydroxide solution at temperatures around 340 ℃ has been an important commercial process for the production of phenol

commercially adv **1** 商品として

▶ All other reagents were commercially obtained and subsequently purified by recrystallization or distillation until melting or boiling points were within 1 ℃ of the literature value.

2 商業上

▶ The aqueous solution of formaldehyde, known commercially as formalin, contains about 40 percent of formaldehyde.

commit vt 投入する syn *deliver*

▶ In the years past, the spent catalysts were committed to landfills.

common adj **1** ありふれた syn *familiar, usual*

▶ The elements Mg and Fe form the common silicate minerals olivine and pyroxene.

2 共通の syn *general, universal, well-known*

▶ The indole ring system is common to many naturally occurring compounds, as for example the essential amino acid, tryptophan, which is a constituent of almost all proteins.

3 in common 共通に

▶ Although the number of different compounds that are manufactured as phosphors is very large, their methods of synthesis have much in common.

4 in common with …と共通して

▶ A foreign salt is one that does not have an ion in common with any ion directly involved in the reaction.

common ion effect n 共通イオン効果

▶ The common ion effect which is involved in nearly every reaction of precipitation and of dissolution represents an application of the mass-action principle.

commonly adv 共通の概念が強い一般に syn *generally, usualy*

▶ Like calcium, strontium occurs fairly commonly, as the sparingly soluble carbonate or sulfate.

common name n 慣用名 ☞ trivial name

▶ Bis(cyclopentadienyl)iron has an extensive aromatic-type reaction chemistry which is reflected in its common name "ferrocene".

commonplace adj ありふれた syn *familiar, ordinary, prevalent, typical*

▶ Superconductivity at the temperature of liquid nitrogen is now commonplace.

common salt n 食塩

▶ Hydrochloric acid is usually made by the action of sulfuric acid upon common salt.

common sense n **1** 常識 syn *experience, intuition*

▶ Common sense suggests that the vapor should be richer in the more volatile component than the liquid.

adj **2** 常識的な 複合形容詞として使用 syn *practical, realistic*

▶ This equation expresses quantitatively the common-sense view that solutes with high boiling points and large enthalpies of fusion have low solubilities.

communal adj 共有の syn *common, general*

▶ In a molecular crystal, the stationary states of the excited system are communal states of the whole assembly.

communicate vi **1** 連絡する syn *relate*

▶ In a ferromagnet, the unpaired electrons strongly communicate with one another and align in large regions known as magnetic domains.

vt **2** 伝達する syn *transmit*

▶ The excited atom will vibrate about its new equilibrium position, and some of the absorbed energy will be communicated thermally to the lattice.

communication n **1** 学会発表論文 syn *report*

▶ This communication contains a brief summary of the thermodynamic properties of aqueous sodium chloride solutions from 0 to 40 ℃ derived principally from the measurements of cells without liquid junction containing a dilute sodium amalgam.

2 伝達 syn *transmission*

▶ The high levels of conformational control in certain benzanilides indicate that there is some degree of conformational communication between the Ar−CO and the N−CO axes, provided that the rings bear sufficiently large substituents.

community n 集団 syn *organization*

▶ Despite the increased interest in solid-state materials, there remains a relatively small community of academic researchers in chemistry involved in the syntheses of new materials.

commutative adj 可換性の

▶ Vector addition is commutative; i. e., the vector sum is independent of the order in which the vectors are added.

commute vi 交換できる syn *exchange, interchange, transpose*

▶ Matrix multiplication and the multiplication of symmetry operations do not commute, except in special cases.

compact adj **1** 密な syn *dense, heavy*

▶ Compact zirconium is hard to dissolve in acid, and alkali hydroxides react only slightly with the metal.

2 小型の syn *small*

▶ An ion with unpaired electrons is slightly less compact than if the electrons were paired.

vt **3** 圧縮する syn *contract, pack*

▷ To produce transferred LB layers, a suitable substrate, such as aluminum oxide, is normally passed vertically through the compacted monolayer.

compaction n 圧縮した状態

▶ The resistance of a polycrystalline compaction is made of four temperature-dependent components: interparticle contact resistance plus the three principal resistivities of the material.

companion n 仲間 syn *colleague*

▶ Cholesterol isolated from gallstones or from body tissues contains small amounts of the following companions: cholestanol, 7-dehydrocholesterol, and lathosterol.

comparable adj 類似の syn *equivalent, similar*

▶ Silicon halides are much more reactive than organic halides of comparable structure.

comparably adv 比較できるほどに

▶ Substitution of comparably sized group 16 element like Se or Te for As leads to an n-type material, because the dopant is a donor with one extra valence electron relative to As.

comparative adj 相対的な syn *relative*

▶ Comparative values should be more valid.

comparatively adv 比較的 syn *relatively*

▶ Only a comparatively few groups may be present in the halide molecule from which we prepare a Grignard regent.

compare vi **1 compare with** …に匹敵する syn *approximate, correspond, match, resemble*

▶ The room-temperature conductivity of $RbAg_4I_5$ may be compared with that of a 35% aqueous solution of sulfuric acid.

vt **2** …と比較する syn *contrast, relate*

▷ In comparing two materials of the same mass, the material that requires the transfer of more heat to change its temperature by a given amount has the higher specific heat.

3 compared to 類似性を指摘して，…と比べると

▶ Compared to water with an effective proton affinity range of 58 kJ, the effective proton affinity range for ammonia is more than twice as great.

4 compared with 類似性，または差異を指摘して，…と比べると

▶ The relative binding power of the zinc ion toward halide ions is reversed in the enzyme, carbonic anhydrase, compared with the free Zn^{2+} ions.

5 as compared with …と比較して

▶ The reduced electrostatic repulsion and increased van der Waals' attraction among chalcogenide ions as compared with oxide ions are commonly given as reasons for the absence of close packing.

comparison n **1** 比較 syn *contrast, equality, resemblance*

▶ Although the product is microcrystalline, comparison of its infrared spectrum and X-ray powder photographs with those of 2, 3-dichloro-1, 4-dihydoxyanthraquinone prepared by crystallization from warm toluene confirms that they are identical.

2 by comparison 比較して

▶ By comparison, the more electropositive elements of group 7 are more readily oxidized.

3 in comparison to …と比較して

▶ If the equilibrium shifts with a change in the measured variable in a time very short in comparison to the time required for the measurement, the equilibrium condition is applicable.

4 in comparison with …と比較して

▶ The electron–nuclear magnetic interaction energy is small in comparison with the electron spin resonance energy and is independent of the strength of the magnetic field.

compartment n 区画 syn *section*

▶ The top and bottom compartments of the tetragonal unit cell contain Ba^{2+} ions, and the middle compartment contains a Y^{3+} ion.

compartmentalization n 区画化 syn *segregation, separation*

▶ Compartmentalization is the key concept in the stabilization of colloidal catalyst and semiconductor particles in vesicles.

compartmentalize vt 区画に分ける syn *segregate, separate*

▶ Growth of crystals is usually compartmentalized so that there must be pumps for selected ions which move them into vesicles where the crystals form and grow.

compatibility n 適合性 syn *harmony*

▶ Sometimes hydration of either the cation or the anion is required to improve the size compatibility of the units comprising the lattice.

compatible adj **1 be compatible with** …と両立する syn *agree*

▶ Another difficulty with intermetallic compounds lies in the formulas assigned to them, which are often not compatible with any ordinary chemical valence of the metal.

2 make compatible with …と両立させる syn *reconcile*

▶ Mesoporous silicas may now be made compatible with large biomolecules such as proteins, which opens the door to heterogeneous biocatalytic applications.

compelling adj 認めざるをえない syn *powerful, strong*

▶ There is compelling evidence that similar interactions occur between a molecule in an electronically excited state and a second molecule in the ground state to form an excited complex with a finite lifetime.

compellingly adv 抗しがたいほどに

▶ Isoelectronic-isovalent principles are compellingly illustrated with the tetrahedral extended solids.

compensate vt 相殺する syn *balance, offset*

▶ In nearly every alkane, the less reactive hydrogens are the more numerous; their lower reactivity is compensated for by a higher probability factor, with the result that appreciable amounts of every isomer are obtained.

compensating adj 補償する syn *compensatory*

▶ Zeolite readily undergo ion exchange of the compensating cation in the channel.

compensation n 補償

▶ The compensation is exact when the same volume of reagent is used in the analysis as in the standardization.

compensatory adj 補償的な syn *compensating*

▶ Because the solid must remain electrically neutral, each oxide ion that is removed requires compensatory removal of two units of positive charge; this compensatory removal can be accomplished if two Cu^{3+} centers are converted to Cu^{2+} centers.

compete vi 競う syn *match, oppose*

▷ The products inhibit the reaction by competing with the reagent for the catalyst surface.

competition n **1** 競合 syn *match, opposition*

▶ If, in S_N1 reactions, more than one nucleophile is present, there will be a competition for the carbonium ion, and a mixture of products will result.

2 in competition with …と競合して

▶ Many secondary and tertiary halides undergo E1 type of elimination in competition with the S_N1 reaction in neutral or acidic solutions.

competitive adj 競合的な syn *conflicting, in opposition*

▶ For alkenes, alkynes, cycloalkanes, and benzene derivatives, the results are likewise competitive with those obtainable from much more complicated schemes which require more parameters.

competitive reaction n 競争反応 syn *competition reaction*

▶ For an exact, quantitative comparison under identical reaction conditions, competitive reactions can be carried out, in which the compounds to be compared are allowed to compete for a limited amount of a reagent.

competitiveness n 競合 syn *competition*

▶ Chemical research must now find expression in its benefit to industrial competitiveness or the quality of life.

competitor n 競争相手

▶ The similarity of a molecule to a lead compound or to competitors' products has frequently been used as a guide to synthesis.

compilation n 編集物

▶ Most of the experimental enthalpies of formation quoted in this paper were taken from the compilation of Pedley et al.

compile vt 編集する syn *assemble, collect*

▷ Having discussed some of the physical properties which are modified by inclusion of a metal, it is worth considering and compiling the evidence concerning the occurrence and significance of intermolecular metal–metal or metal–ligand interactions in the mesophases of metal-containing liquid-crystal systems.

complement n **1** 補体

▶ Antigens that are coated with these antibodies or complement factors are more efficiently phagocytosed than those without the help of these molecules.

vt **2** 補足する syn *complete, supplement*

▶ Biological deposition of silica and ice is associated with hyroxy-rich macromolecules such as polysaccharides and serine- and threonine-rich proteins. Again, the organic ligands complement the mineral chemistry.

complementarity n 相補性

▶ The specific interaction of molecules in solution as exemplified by host–guest complexes of macrocyclic ligands, enzyme-substrate reactions, and antibody–antigen coupling is determined by complementarity in size, molecular shape, and dynamics.

complementary n **1** 補足するもの

▶ Copper–chromium oxide, which may be regarded as complementary to Raney nickel, is generally useful for the reduction of oxygen-containing functions and is the catalyst of choice for converting esters into primary alcohols.

adj **2** 補足の

▶ Direct measurement of absorption spectra has been extensively used when the absorbing ion, e.g., Fe^{2+}, is a major constituent of the mineral, and to this extent it is a complementary technique.

3 補色の

▶ If we pass a beam of white light through a substance which absorbs essentially all blue light at 4800 ± 300 Å, the emergent beam appears to be yellow, which is the color complementary to blue.

complete adj **1** 完全な syn *ideal, perfect*

▶ The optically active phosphine must be heated in boiling toluene for three hours for complete racemization to occur.

2 全部の syn *entire, whole*

▶ The alkali metals form a complete range of compounds with all the common anions.

vt **3** 完成する syn *accomplish, achieve*

▶ The diamond lattice is constructed to allow every atom to complete its electron octet.

4 完了する syn *end*

▶ Drying is completed simply by filtering the ethereal solution through a layer of anhydrous sodium sulfate.

completely adv 完全に syn *fully, totally, wholly*

▶ If a paraffin hydrocarbon C_nH_{2n+2} is treated with an excess of chlorine, the completely chlorinated product C_nCl_{2n+2} is not obtained if n is greater than 3.

completeness n 完全であること syn *perfection*

▶ For the sake of completeness, it is necessary to consider other possible slip vectors on (102) -type planes.

completion n 完了 syn *perfection*

▶ The concentration for the completion of formation of micelles is ten or more times greater than the critical concentration for their initiation.

complex n **1** 複合

▶ This complex of circumstances can yield complicated relations with emulsions that have often been misunderstood.

2 錯体 ☞ charge-transfer−, metal−, molecular −

▶ The complex may be nonionic or a cation or anion depending on the charges carried by the central atom and the coordinated groups.

adj **3** 複雑な syn *complicated, elaborate, involved*

▶ Solutions of molybdic acid are very complex chemically because they show a great tendency to polymerize.

vt **4** 錯形成する

▶ The structure of hemin shows that the iron is complexed to all four of the pyrrole nitrogens.

complexation n 錯形成反応 syn *complex formation*

▶ It was important to establish unequivocally whether or not 2-dicyanomethyleneindane-1,3-

dione had rearranged to 2,3-dicyano-1,4-naphthoquinone upon complexation with the fulvalene donors.

complex compound n 錯化合物

▶ One of the characteristics of the complex compounds is that the complex ion or neutral complex that composes it will most often retain its identity in solution, although partial dissociation may occur.

complex formation n 錯形成反応 syn complexation

▶ The assumption of all activity coefficients equal to unity is common to most methods used in investigation of complex formation in nonelectrolyte systems.

complexing agent n 錯化剤

▶ Bulk deposition from ionic liquid containing just $CuCl_2$ produces black, powdery deposits whereas the addition of a complexing agent can lead to lustrous copper deposits.

complex ion n 錯イオン

▶ Many cations form complex ions with four or six molecules of water coordinated to the metal.

complexity n 複雑さ syn *complication, intricacy*

▶ The structures of MQ_3 (M = Ti, Zr, Hf, Q = S, Se) are difficult, if not impossible, to describe in terms of the anion arrangement because of their low dimensionality and inherent complexity.

complicate vt 複雑にする syn *confound, confuse*

▷ A complicating factor in some other compounds with the spinel structure is that the cation distribution may vary.

complicated adj 複雑な syn *complex, intricate, involved*

▶ Orientation of substitution in the naphthalene series is more complicated than in the benzene series.

complication n 厄介な問題 syn *complexity, problem*

▶ One of the long-standing experimental complications arises from the difficulty in finding a dielectric material suitable for a sample container.

component n 成分 syn *element, material*

▶ The number of components of a system is the number of constituents that can undergo independent variation in the different phases.

compose vt **1** …を構成する syn *constitute, construct, produce*

▷ The Cu ions occupy the corners of the cubes composing the tetragonal unit cell, and the oxide ions bisect the cube edges.

2 be composed of …からなる 後に構成材料が記される. syn *consist of, comprise, be formed*

▶ Many substances that appear to be amorphous are actually composed of microcrystalline units.

composite n **1** 複合体 syn *hybrid*

▶ Tooth-colored polymer-ceramic composites are based on difunctional methacrylate monomers and silane or graft-coated ceramic fillers.

adj **2** 複合の syn *complex, intricate, involved, multiple*

▶ The chief use of whiskers is in the manufacture of composite structures with plastics, glass, or graphite.

composite material n 複合材料 ☞ hybrid material

▶ When camphorsulfonic acid-doped polyaniline is blended in a host polymer such as poly(methyl methacrylate), the composite materials can be made electrically conductive, even when there is only 2% of polyaniline.

composite reaction n 複合反応

▶ Composite reactions are reactions that occur in more than one elementary step; they are also commonly referred to as complex reactions or as stepwise reactions.

composition n 組成 syn *proportion*

▶ Composition diagrams in organic systems can be as complex as the familiar diagrams for metals and inorganic silicates.

compositional adj 組成の

▶ The compositional homogeneity of glass may be checked by measuring the refractive index of a random selection of fragments of crushed glass.

compost n 堆肥

▶ In a compost pile, hay, straw, or other materials are added to the soil to provide bacteria with nutrients that help bacteria degrade chemicals.

compound n **1** 化合物

▶ A compound has characteristic properties quite different from those of its constituent elements.

vt **2** 結合する syn *combine*

▶ Compounds composed of the elements carbon and hydrogen are called hydrocarbons; there are several different types, depending upon how the carbon atoms are compounded.

3 混ぜて作る syn *hybrid, mix*

▶ The wave function for the F center electron can be compounded from the atomic orbitals of an alkali metal atom.

4 度を増す syn *enhance, increase, intensify*

▶ Difficulties of these sorts may be compounded by the existence of unprecedented compositions and structures for unknown phases hidden in a system.

compound semiconductor n 化合物半導体

▶ Compound semiconductors have a variety of different band gaps which cover the whole electromagnetic spectrum from IR through to the UV.

comprehend vt 理解する syn *appreciate, know, understand*

▶ The significance of the stability of the electron pair can best be comprehended from the aspect of the quantum theory.

comprehension n 理解 syn *appreciation, realization, understanding*

▶ The classification of chemical bonds is largely an empirical one and results, to some extent, from a limited comprehension of the forces involved.

comprehensive adj 包括的な syn *broad, complete, inclusive, wide*

▶ For a comprehensive understanding of particle properties, it is often important to combine many different methods.

compress vt 圧縮する syn *press, squeeze*

▷ Most of the measurements of semiconductivity have been made on compact specimens in the form of plates or disks obtained very frequently by compressing powdered materials.

compressed adj 圧縮した syn *compact, condensed*

▶ In the compressed state, poly(tetrafluoroethylene) particles lie flat and closely packed.

compressibility n 圧縮率

▶ The compressibility decreases as the pressure increases, and the values for different liquids tend to become equal at very high pressures.

compressibility factor n 圧縮因子

▶ The deviations from ideal behavior are conveniently shown by treating 1 mole of gas and plotting as ordinate the ratio PV/RT, called the compressibility factor Z.

compressible adj 圧縮できる syn *elastic, soft*

▶ One expects a solid that is physically soft to be compressible, and one that is hard to resist compression.

compression n 圧縮 syn *stress, tension*

▶ In the crystal, compression appears to have gone so far that minor changes from one crystal to another do not have any appreciable effect on the crystal radius.

compressive adj 圧縮可能な

▶ The repulsive part of the potential is not infinitely steep; i.e., the molecules are compressible.

compressive force n 圧縮力

▶ The ions in a crystal are under strong compressive forces.

comprise vt …からなる compose とは異なり能動態を用いるのが正しい. syn *compose, include, involve*

▶ Sorption comprises a number of phenomena, the most commonly recognized of which is accumulation of one or more substances in the interface.

compromise n **1** 折衷 syn *mean, middle*

▶ I prefer a compromise in which the calculations are carried out, as they have to be, in a delocalized way but are interpreted starting from a semilocalized view.

vt **2** 妥協によって解決する

▶ If the carriers are in short supply, the rapid operation of the entire pathway can be compromised.

computation n 計算 syn *calculation, evaluation*

▶ It is difficult to eliminate this hypothesis since the computations then become hard.

computational adj 計算の

▶ The two sets of computational data show very good qualitative agreement.

computationally adv 計算によって

▶ The simplest and computationally quickest method of calculating ionic or molecular volume is the analytical approximation put forward by Kitaigorodsky.

computer n コンピューター，電子計算機

▶ Computers store and manipulate data using binary numbers consisting of the digits 0 and 1.

computer graphics n コンピューターグラフィックス

▶ Recent developments in computer graphics have added a new impetus to quantitative measures of similarity.

computer simulation n コンピューターシミュレーション

▶ Unlike real experiments, a computer simulation can test the effects of adding and subtracting particular forces to the intermolecular force.

concave adj くぼんだ syn *hollow* ☞ convex

▶ The *trans*-decalin is about 2 kcal more stable than the cis isomer, largely because of relatively unfavorable nonbonded interactions within the concave area of *cis*-decalin.

conceivable adj 考えられる syn *likely, plausible, possible*

▶ Two-directional ring construction strategies involving metathesis are conceivable.

conceivably adv おそらく syn *perhaps, possibly, seemingly*

▶ The relatively sharp resonance which results upon the oxidation of chromia-alumina may conceivably be due to Cr^{+3}, Cr^{+4}, Cr^{+5}, or color centers in the oxidized chromia.

conceive vi 考える syn *imagine, speculate, think*

▶ One can conceive of four distinct mechanisms of energy transfer.

concentrate n **1** 精鉱

▶ Precious metal concentrates are obtained either from the metallic phase of the sulfide matte or as anode slimes in the electrolytic refinement of the baser metals.

vi **2** 集中する syn *center, direct, focus*

▶ Theoretical approaches in drug design have until recently concentrated on energy calculations and descriptions of molecular shape in terms of nuclear positions.

vt **3** 集中する

▶ In energy-focusing systems, the incident radiation is concentrated on a central receiver from an array of flat or parabolic mirrors.

4 濃縮する syn *condense, distill, refine*

▶ Whatever the method of preparation, ethyl alcohol is obtained first mixed with water; this mixture is then concentrated by fractional distillation.

concentrated adj **1** 濃縮した syn *strong*

▶ When an oxime is treated with strong concentrated acid, such as sulfuric or polyphosphoric acid, it is converted into an amide.

2 集中的な syn *intensive*

▶ The π^*orbital of CO, concentrated on carbon, gives better overlap than that of N_2.

concentration n **1** 集中 syn *focus*

▶ Interest in the effects of anthropogenic materials released into the atmosphere led to an unprecedented concentration of effort on the measurement of the rates of elementary reactions.

2 濃度 syn *strength*

▶ The word concentration means the amount of substance divided by the volume of the solution. The SI unit is mol m^{-3}, but the usual unit is moles per liter.

3 濃縮 syn *extraction*

▶ Storage implies concentration, and there must be either a binding mode or an energized concentration gradient relative to the external solutions.

concentration cell n 濃淡電池

▶ A simple example of a concentration cell is obtained by connecting two hydrogen electrodes by means of a salt bridge.

concentration gradient n 濃度勾配

▶ A sodium-potassium transport system exists in plasma membranes to pump Na^+ out of the cell and K^+ into the cell against concentration gradients.

concentric adj 同心の

▶ A positively charged nucleus is surrounded by electrons arranged in concentric shells or energy levels.

concept n 概念 syn *idea, image*

▶ Many metallic, ionic, covalent', and molecular crystal structures can be described using the concept of close packing.

conception n 概念 syn *idea, image, opinion*

▶ The electronic conception of oxidation and reduction makes it easy to see why an oxidation is always accompanied by a reduction.

conceptual adj 概念の syn *abstract, ideal*

▶ Compared to molecules, the theoretical study of solid-state structures is much more difficult from both the computational and conceptual viewpoint, because of the size of the structure to be considered.

conceptualization n 概念化

▶ Fortunately or not, theory is presently in no position to assess well the stability of any new phase, as such conceptualization can at best be an extrapolation of the known.

conceptually adv 概念上

▶ This methodology is conceptually different from previous efforts that focused on shaping polyaniline into nanostructures.

concern n 1 懸念 syn *worry*

▶ Environmental concerns resulted in a ban on this practice in the mid-1980s.

2 関心事 syn *interest*

▶ The concern with energy distribution among reaction products is of long standing.

vt 3 心配させる syn *disturb, trouble, worry*

▶ The zincblende and wurtzite structures are of the type AB and differ in details which need not concern us.

4 concern with …と関係している syn *relate, have relation to*

▶ Chemistry is more concerned with the properties of individual substances, in contrast to physics which is more concerned with general properties.

5 as concerns 前置詞的に，関しては

▶ When a substance is dissolved, whether in water or in acids, phenomena are often observed which may be of great importance as concerns the subsequent analysis.

concerning prep …に関して syn *about, regarding*

▶ A particularly significant observation concerning the mechanism is that bromine addition proceeds in the dark and in the presence of free-radical traps.

concerted adj 協奏的な syn *combined, joint*

▶ In the 1970s, a concerted effort was made to investigate the second-order nonlinear optical effects in polarizable noncentrosymmetric molecules.

concerted mechanism n 協奏的機構

▶ There has been some discussion of whether the cyclopropane isomerization proceeds by a concerted mechanism, in which a C—C bond is broken and a hydrogen atom is transferred at the same time, or by a diradical mechanism.

concerted reaction n 協奏反応

▶ If a concerted reaction is forbidden, the probable pathway will be one involving more than one step.

concise adj 簡潔な syn *brief, short*

▶ The discussion of the application of the resonance method to benzene may be reduced to a fairly concise prescription.

concisely adv 簡潔に syn *briefly*

▶ The structures of metal borides are concisely and conveniently discussed in terms of the disposition and environment of the boron atoms.

conclude vt **1** 結論する syn *deduce, guess, perceive*

▶ From his observations, Moseley concluded that the atomic number is a measure of the positive charge on the nucleus of an atom.

2 完結する syn *complete, end*

▷ Before concluding this section, mention should be made of apparatus designed to give continuous growth by evaporation, i. e., in which the evaporated solvent is condensed, resaturated, and recirculated.

conclusion n **1** 結論 syn *decision, deduction, inference*

▶ Conclusions from different methods applied to the same solute have not always been consistent, so additional techniques should be valuable.

2 終わり syn *end*

▶ At the conclusion of a run, or of series of runs, the sample was oxidized to obtain the zero emf reading as described above.

3 in conclusion 結論として，終わりに臨んで

▶ In conclusion, there are in the literature innumerable examples of compounds which were observed to undergo chemical change immediately on melting; in almost none of these cases has the possibility of a solid-state reaction been investigated.

conclusive adj 決定的な syn *definitive, ultimate*

▶ Completely conclusive evidence has been obtained from a study of several further rearrangements in which comparable geometrical factors are not involved but in which the configurations are still found to be retained.

concomitant n 付随するもの syn *complement, counterpart*

▶ Concomitant with the syntheses of new chalcogenides has been the discovery of new properties and new uses.

concomitantly adv 付随して

▶ Concomitantly there is a buildup of CO_2 from oxidation of fuels, such as glucose and fatty acids.

concordance n 一致 syn *accord, agreement*

▶ Such concordance is only to be expected inasmuch as errors due to imperfect estimation of the equation of state of the calibration vapor should drop out in the extrapolation to zero density.

concordant adj 合致する syn *compatible*

▶ For the region covered in this investigation, the Joffe and Berthelot equations yield entirely concordant values for the vapor density of carbon tetrachloride.

concrete adj **1** 明確な syn *definite, precise*

▶ Concrete evidence that terpene and steroid biogeneses are actually related is provided by the finding that isotopically labeled lanosterol, a tetracyclic triterpene alcohol, is converted to cholesterol by liver tissue.

2 具体的な syn *particular, specific*

▶ The thermal polymerization of cyclopentadiene by way of the Diels-Alder reaction provides a simple concrete example of how a monomer and a polymer are related.

concurrent adj 同時の syn *simultaneous*

▶ Chlorine with a trace of phosphorus reacts similarly but with less overall specificity, since concurrent free-radical chlorination can occur at all positions along the chain.

concurrently adv 同時に

▶ Concurrently with our work on porous perovskites, Prasad and co-workers have developed a large number of metathetic routes to perovskites which produce a soluble salt as a second phase.

condensability n 凝縮性

▶ The amount of the physical type of sorption is parallel with the condensability of the vapor, and heat is evolved during sorption.

condensable adj 凝縮できる

▶ It has been increasingly realized in recent years that carbonaceous chrondrites probably approximate samples of the condensable fraction of original solar-system matter.

condensate n **1** 凝縮液

▶ The rim of condensate should rise very

slowly as it approaches the top in order that the thermometer may record the true boiling point soon after distillation starts.

2 縮合物

▶ It is often difficult to distinguish catenation from condensation, for on formation of a condensate weak metal–metal interaction utilizing d electrons often occurs.

condensation n 1 凝縮

▶ The chief method of forming aerosols is by the sudden condensation of highly supersaturated vapor.

2 縮合

▶ Condensation of two phosphoric acid molecules occurs by the reaction of two hydroxyl groups to form water and an oxygen atom held by single bonds to two phosphorus atoms.

condensation polymerization n 縮合重合 syn *polycondensation*

▶ In condensation polymerization, some of the atoms of the monomer are split off in the reaction as water, alcohol, ammonia, or carbon dioxide, etc.

condensation product n 縮合物

▶ This expectation is realized in practice, and much effort has been expended to determine conditions by which practical yields of the condensation product can be obtained.

condensation reaction n 縮合反応

▶ Some polymers can be formed by either addition or condensation reactions.

condense vi 1 縮合する

▶ When orthophosphoric acid is heated, it loses water and condenses to diphosphoric acid.

vt 2 凝縮する syn *contract, diminish, shrink*

▷ Nitrogen(IV) oxide is a very toxic, brown-red gas condensing to a red-brown liquid boiling at 22.4 ℃.

3 圧縮する syn *compress*

▶ The pH scale condenses the span of 15 orders of magnitude into the range 0–14 and is a convenient measure of proton activity.

condensed adj 縮合した

▶ The various types of condensed cations, acids, and anions can all be regarded as polymeric in the sense that they contain a repeating structural unit based on the formation of hydroxyl or oxo bridges.

condensed phase n 凝縮相

▶ A fundamental effect of compression for condensed phases is to decrease interatomic and intermolecular distances and to increase overlap of the outer electronic orbitals.

condensed ring n 縮合環

▶ Naphthalene is the simplest member of the class of compounds said to possess condensed ring systems.

condensed system n 凝相系

▶ Systems are called condensed systems if the vapor phase is effectively nonexistent.

condenser n 1 冷却器

▶ The condenser can be cleaned by clamping it in a vertical position, placing under it the receiving flask emptied of its contents, and ringing the tube with a few mL of acetone.

2 コンデンサー syn *capacitor*

▶ Between the plates of the condenser, there may be either a vacuum or a dielectric.

3 集光レンズ

▶ The crystal mats were mounted in a Beckman micro solid-sample holder and beam condenser, and their infrared dichroism was studied in radiation polarized by a AgCl polarizer.

condition n 1 条件 syn *prerequisite, requirement*

▶ These are necessary, but not sufficient, conditions.

vt 2 調節する syn *adapt, modify*

▶ Flotation is further conditioned by the use of regulators, depressing agents, and activating agents.

3 制約する syn *influence*

▶ The hardness and tensile strength of metals or the bending strength of glass are not entirely bulk properties but are often conditioned by sorption of surface active materials.

conduct n 1 行為 syn *handling, operation*

▶ The conduct of these parallel calculations

serves in lieu of an error analysis.

vt **2 実施する** syn *operate, run*

▶ The vast majority of photochemical syntheses have been conducted in the liquid phase.

3 導く syn *guide*

part ▶ Platinum present in hydrochloric acid is precipitated as sulfide by conducting hydrogen sulfide into the hot solution.

4 伝導する syn *convey, transmit*

▶ Salts of larger alkaline earth cations are typically ionic and conduct electricity well in aqueous solution and in the molten state.

conductance n **コンダクタンス** syn *conductivity*

▶ When a strong acid is titrated with a strong base, the conductance of the solution at first decreases because of the replacement of hydrogen ions, which have a high mobility, by slower moving cations.

conducting polymer n **導電性高分子**

▶ The microelectrochemical transistors and diodes make use of the very large changes in the conductivity of electronically conducting polymers which accompany oxidation or reduction of the polymer backbone.

conduction n **伝導** syn *transmission* ☞ thermal conduction

▶ The mechanism of conduction in these materials is still a matter of some controversy.

conduction band n **伝導帯** ☞ valence band

▶ On the band theory, an insulator may become conducting when an electron is raised from the filled band to the empty conduction band.

conductive adj **熱や電気の伝導性の**

▶ For any molecule to be observed with the scanning tunneling microscope, it must be attached to a rigid, flat, conductive material.

conductivity n **伝導率** syn *conductance*

▶ The alkaline earth metals owe their conductivity to the overlap of their p bands with their s bands.

conductometric adj **伝導度の**

▶ The conductometric method may be used in connection with colored solutions, in which ordinary indicators cannot be used.

conductometric titration n **伝導度滴定**

▶ Conductometric titrations between these various species can be carried out, the end point being indicated by a sharp minimum in the conductivity.

conductometrically adv **伝導度によって**

▶ Neutralization reactions can be followed conductometrically, potentiometrically, or even with colored indicators such as phenolphthalein.

conductor n **熱や電気の導体** ☞ insulator, semiconductor

▶ Many chain conductors are known, including polymers such as polysulfurnitride and polyacetylene, when doped with electron acceptors.

cone n **円錐**

▶ The Tyndall effect causes the appearance of a visible cone of light through the suspension.

confer vt **confer on 授与する** syn *give, present*

▶ The small size of lithium frequently confers special properties on its compounds.

confidence n **1 信頼** syn *belief, reliance*

▶ If a decision about the molecular symmetry depends on whether the molecule has or has not a finite dipole moment, considerable confidence can be placed in the conclusion.

2 with confidence 確信をもって

▶ It is possible to use any set of radii to evaluate interionic distances in crystals with reasonable confidence.

confident adj **確信のある** syn *certain, secure, sure*

▶ At the time of this earlier investigation, no reliable approach was available to assess the presence or absence of the true C−H⋯Cl bonds in the solid state, making confident assignment of such bonds difficult.

confidently adv **確信して** syn *surely*

▶ Stringent tests had to be applied to distinguish confidently between B and C atoms in the structure of $B_{13}C_2$.

configuration n **立体配置** syn *make-up, structure*

▶ We are chiefly interested in whether the configurations of a reactant and its product are the same or different not in what either configuration actually is.

configurational adj 立体配置の

▶ The configurational stability of the phosphines resembles that of the sulfonium salts.

configurational interaction n 配置間相互作用

▶ If configurational interaction between the totally symmetric molecular orbital functions is allowed for, then the ground state is improved, its energy is lower, and it has less ionic character.

configurational isomer n 配置異性体

▶ Configurational isomers can be isolated under ordinary conditions and are thus to be distinguished from different conformations of a molecule, which are in mobile equilibrium with one another.

configurational isomerism n 配置異性

▶ Whether a species exhibits evidence of configurational isomerism or merely of different conformations depends on the activation energy for the interconversion process.

configurationally adv 立体配置の上で

▶ Other compounds could be related configurationally to one or the other of the glyceraldehydes by means of reactions that did not involve breaking bonds to a chiral center.

confine vt **1** 制限する syn *limit, restrict*

▶ The possibility of using a host-guest approach to confine the size of the reaction volume for materials synthesis is very attractive.

2 閉じ込める syn *enclose*

▷ The attraction a molecule exerts on its neighbors tends to draw them in toward itself; i. e., the attraction acts with the confining pressure to hold the molecules together.

confined adj 閉じ込められた

▶ Pressure applied to a unit area of a confined liquid is transmitted equally in all directions throughout the liquid.

confinement n 制限 syn *restraint, restriction*

▶ Confinement of mineralization reaction to the nanospace is achieved by specific molecular processes which compete successfully against nonspecific reactions occurring in the external medium.

confirm vt 確かめる syn *recognize, validate, verify*

▶ Maxwell himself later undertook some viscosity experiments, with results that did confirm the prediction, namely, that the viscosity should be independent of the pressure.

confirmatory adj 確認の

▶ Confirmatory evidence for this conclusion is derived from the X-ray powder photographs.

conflict n 不一致 syn *disagreement, discrepancy*

▶ There is no conflict with the frontier orbital theory of chemical reactivity.

conflicting adj 相反する syn *contradictory, opposing*

▶ Many conflicting data have been reported, and considerable uncertainty remains as to the exact geometry of the complexes.

confocal adj 共焦点の

▶ In the confocal setup, excitation light from a laser source is focused by a microscope objective to a diffraction-limited spot with a diameter of $\leq 1\ \mu$m.

conform vi 適合する syn *agree, correspond*

▷ In describing the electronic structure of ferrocene, we construct ligand-group orbitals conforming to the D_{5d} molecular symmetry from MOs on the two separate rings.

conformal adj 整合した

▶ Chemical vapor deposition possess an intrinsic advantage over physical deposition processes such as sputtering in its ability to produce thin layers of material with conformal coverage.

conformation n 立体配座 syn *form, shape*

☞ configuration,

▶ The stable conformation of cyclodecane is a compromise structure with slightly distorted bond angles and several very short H−H distances.

conformational adj 立体配座の

▶ The degree of conformational control in benzanilides is governed by the size of the substituents at the ortho positions and also the substituent at the nitrogen; the bulkier they are, the higher the conformational ratio.

conformational analysis n 配座解析

▶ Conformational analysis can account not only for the fact that one conformation is more stable than another but often for just how much more stable it is.

conformational isomerism n 配座異性

▶ Cis and trans isomers of cyclohexane derivatives have the additional possibility of conformational isomerism.

conformationally adv 配座に

▶ Most of conformationally sensitive bands in the IR spectra are weak; therefore, thick samples were needed.

conformer n 配座異性体 syn *conformational isomer*

▶ In the majority of the compounds, the ratio between the conformers can be established by the integration of the signals corresponding to each conformer.

conformity n 適合 syn *coincidence*

▶ Rationality is admittedly a logically weaker imposition upon acceptable explanation than conformity to any logical model.

confound vt 困惑させる syn *disturb, impede*

▶ The variability of the Si−O−Si bond angle and extent of hydroxylation confounds structural analysis.

confront vt 直面する syn *encounter, face*

▶ If an isomer is assigned to one of the many homologous series, we are then confronted with the problem of defining precisely what is meant by such a series.

confuse vt 混同する syn *tangle*

▶ The elements manganese and magnesium, as their names indicate, were often confused in the early days of chemistry, and both were called magnesium

confused adj 混乱した syn *inconsistent, indistinct*

▶ Information in the literature on the ammines of nickel cyanide is both confused and contradictory.

confusing adj **1** 紛らわしい syn *confused, obscure*

▶ The terminology of natural gums and resins is inconsistent and often confusing.

2 当惑させるような syn *involved*

▶ A confusing mixture of variously condensed and hydrated silicate ions and silicic acids seems to be formed.

confusingly adv 紛らわしく

▶ Another oxygen-containing pigment is confusingly named hemocyanin, which contains neither the heme group nor the cyanide ion; the name simply means blue blood.

confusion n 混乱 syn *ambiguity, misunderstanding*

▶ The similarity between Be and Al led to considerable confusion concerning the valence and atomic weight of Be.

congener n 周期表における同族体

▶ Magnesium, like its heavier congeners Ca, Sr, and Ba, occurs in crustal rocks mainly as the insoluble carbonates and sulfates.

congregate vi 集まる syn *assemble*

▶ Covalent molecules congregate to form liquids and solids.

congruent melting point n 調和融点

▶ If the system involves a compound with congruent melting point, the latter acts as a pure substance and thermal analysis leads to a diagram like two of the simple eutectic systems set side by side.

congruently adv 調和して ☞ incongruently

▶ At a pressure at which a compound melts congruently, the melting point of the compound is a maximum on the melting point curve, the composition of the liquid at the maximum being that of the solid compound.

conical adj 円錐形の

▶ The conical shape of the Erlenmeyer flask prevents spilling of liquid.

conjugate n **1 共役体**

▶ Chemical species that differ from each other only to the extent of the transferred proton are termed conjugates.

2 複合体

▶ A polymer moiety in a drug–polymer conjugate must be nontoxic.

vt **3 共役する**

▶ A double bond that is separated from a benzene ring by one single bond is said to be conjugated with the ring.

conjugate acid n **共役酸** ☞ conjugate base

▶ The acid HCN becomes its conjugate base CN^- when it loses a proton to the base H_2O, converting the latter to its conjugate acid H_3O^+.

conjugate addition n **共役付加**

▶ Imidazoline catalyst has been found to give good enantioselectivities for the conjugate addition of nitroalkanes to acyclic α,β-unsaturated enones.

conjugate base n **共役塩基** ☞ conjugate acid

▶ The methyl group of the phosphonium salt is weakly acidic because the negative charge on the conjugate base is stabilized by the inductive effect of the neighboring positive atom and also because the ylide is resonance stabilized.

conjugated diene n **共役ジエン**

▶ Conjugated dienes react with a variety of ethylene derivatives to give cyclohexenes, a reaction first investigated by Diels and Alder.

conjugated double bond n **共役二重結合**

▶ 1, 3-Alkadienes and other compounds with alternating double and single bonds are said to have conjugated double bonds.

conjugated protein n **複合タンパク質**

▶ Some protein molecules contain a nonpeptide portion called a prosthetic group; such proteins are called conjugated proteins.

conjugated system n **共役系**

▶ Pyridine resembles benzene in that it contains a conjugated system of six π electrons, one being contributed by each atom in the ring.

conjugation n **1 共役**

▶ If there are enough double bonds in conjugation, absorption will move into the visible region, and the compound will be colored.

2 結合

▶ Benzoate is metabolized by conjugation with glycine.

3 複合

▶ Conjugation involves the addition of a normal cellular component or a modified form of a normal component to the foreign compound.

conjunction n **in conjunction with …と連結して**

▶ A typical coenzyme is nicotinamide adenine nucleotide, which, in conjunction with the enzyme, alcohol dehydrogenase, oxidizes ethanol to acetoaldehyde with the formation of the reduced coenzyme and acetoaldehyde.

connect vt **結ぶ** syn *join, link*

part ▶ We extended this work to rutile nets in which $N(CN)_2^-$ rather than $C(CN)_3^-$ played the role of the three-connecting node.

connected adj **関係している** syn *related*

▶ There are a number of associated phenomena connected with the system where the semiconductor–metal transformation takes place.

connection n **1 関係** syn *correlation, relation*

▶ There is no simple connection between the stoichiometric equation for a reaction and the order of the reaction.

2 in connection with …に関連して

▶ One very important effect of substituent groups on aromatic substitution is the inductive effect which we have encountered in connection with the ionization of carboxylic acids.

3 in this connection この点について

▶ The amino acids, proline and hydroxyproline, are particularly important in this connection.

connective tissue n **結合組織**

▶ The widespread occurrence of siliceous biominerals as structural elements in lower plants and animals suggests a role for Si in the production and maintenance of connective tissue in higher organisms.

connectivity n 連結性

▶ Only with the close proximity of the carboxylate groups that is found in 1,2-cyclohexanedicarboxylate can the metals be sufficiently close to sustain infinite inorganic connectivity.

conrotatory adj 同旋的な ☞ disrotatory

▶ Rotation in a conrotatory fashion pushes a plus lobe at C1 of butadiene into a plus lobe at C4, a bonding interaction which allows smooth transition directly to the σ orbital of the cyclobutene

consecutive reaction n 逐次反応 syn *successive reaction*

▶ Consecutive reactions are reactions of the form A → B → C → ....

consensus n 合意 syn *accord, agreement, concordance*

▶ The general consensus of zeolite chemists is that the organic plays a templating function through the spatially controlled assembly of inorganic building blocks.

consent n 合意 syn *acceptance, agreement*

▶ The past ten years have seen, by common consent, a consolidation of the early systematization of those areas of inorganic chemistry in which the crystal field theory played such a significant role.

consequence n **1** 重要性 syn *importance, significance*

▶ The consequences of solvent effects can be seen in almost every aspect of electrolyte behavior and in virtually all solvents.

2 結果 syn *outcome, result*

▶ A practical consequence of anisotropic expansion of single crystals is the weakening of polycrystalline systems during temperature rises due to the development of stresses at the grain boundaries.

3 as a consequence ゆえに syn *consequently, therefore*

▶ Ferroelectric crystals are internally strained and, as a consequence, show unusual piezoelectric and elastic properties.

4 in consequence …の結果として syn *accordingly*

▶ In consequence the bonding is weak, the alkali metals are soft, their mp low, and they are exceptionally good conductors of electricity.

consequent adj 結果として生じる syn *eventual, subsequent*

▶ Oxygen is consumed with the consequent production of carbon dioxide and water and the release of energy.

consequently adv 結果として syn *accordingly, hence, therefore*

▶ The intermolecular complex formation is significant in many areas of chemistry and biochemistry and consequently has attracted considerable attention.

conservation n 保存 syn *maintenance, preservation*

▶ The first law is merely a statement of the principle of conservation of energy, and its application is very straightforward.

conservative adj 控えめの syn *moderate, prudent*

▶ Previous results provide some assurance that conservative lower limits for the eutectic temperatures are for **1a**–**3a** 95 ℃, for **1b**–**3b** 126 ℃, and for **1c**–**3c** 100 ℃.

conserve vt 保存する syn *keep, maintain, preserve*

▶ The energy requirement will not be as stringent in the liquid phase, because the large number of energy levels possible for a system of interacting particles will make it easier to find configurations which allow energy to be conserved.

consider vt **1** 考察する syn *take into consideration*

▶ The underlying connectivities or topologies are best considered in terms of the concept of nets.

2 みなす syn *look on, regard, think*

▶ The paraffins with an even number of carbon atoms must be considered as a series distinct from those with an odd number.

considerable adj **1** かなりの syn *appreciable,*

sizable, substantial.

▶ Depending on the raw material, carbons prepared at temperatures below 800 to 1000 ℃ often contain considerable quantities of elements other than carbon.

2 無視できない syn *notable, remarkable*

▶ On treatment with hot 50 percent sulfuric acid, compounds of the type $R\text{-}CH_2\text{-}NO_2$ give carboxylic acids and hydroxylamine, a reaction of considerable technical importance.

considerably adv ずいぶん syn *exceedingly, vastly, widely*

▶ Because of their greater molecular weights, haloalkanes have considerably higher boiling points than alkanes with the same number of carbons.

consideration n **1** 考察 syn *examination, study*

▶ A full consideration of these and many other results led to us to the following mechanism of reactions.

2 take into consideration …を考慮に入れる syn *consider, reckon*

▶ A more refined treatment takes into consideration the thermal expansion of the crystal and the changes in vibrational frequencies of the solid due to the presence of vacancies and interstitial ions.

3 under consideration 考慮中の

▶ The application of Hall coefficient measurements is limited by the nature of the material under consideration.

consist vi **1 consist in** …にある

▶ The most convenient method for preparing anhydrous hydrogen cyanide consists in allowing a mixture of equal volumes of sulfuric acid and water to drop upon sticks of 98 percent potassium cyanide.

2 consist of 部分からなる syn *be compose of, comprise*

▶ While the complex will probably exist as discrete pairs containing single donor and acceptor molecules in solution, in the solid phase the structures usually consist of infinite stacks of alternate donor and acceptor molecules.

consistency n **1** 一貫性 syn *uniformity*

▶ Experimental data from uncritical data collections or that have not been examined critically in their original publication should always be tested for internal and for external consistency.

n **2** コンシステンシー， 具体的には，濃度，密度，粘度など syn *texture*

▶ Commercial amalgams are usually of a liquid or semiliquid consistency.

consistent adj **1** 矛盾しない syn *in agreement, in harmony*

▶ The consistent observation that strong metal-titania interactions require reduction to TiO_x suggests that reduced titanium ions are central to this chemistry.

2 終始一貫した syn *preditable, regular, steady*

▶ The alkali halides represent a series in which properties dependent on the stability of the ionic lattice should vary in a consistent way.

consistently adv 矛盾なく syn *firmly, uniformly*

▶ Kinetic, thermodynamic, and statistical mechanical approaches lead consistently to the same fundamental equation or isotherm.

consolidate vt **1** 統合する syn *combine, integrate*

▶ The material deposited on the alumina rod is dried and heated to about 1600 ℃ to consolidate it into a clear glass.

2 固める syn *solidify*

▶ The surrounding lattice is consolidated by van der Waals attractive forces.

consolidation n 強化

▶ Annealing usually leads to a general mechanical consolidation of the whole mass.

consolute temperature n 共溶点 ☞ critical solution temperature

▶ It may seem quite natural that there should be upper consolute temperatures, where the more violent molecular motion overcomes the tendency of molecules of the same species to stick together in swarms and, therefore, to form two phases.

conspicuous adj 顕著な syn *outstanding, prominent, striking*

▶ In the infrared spectrum of a hydrogen-bonded alcohol, the most conspicuous feature is a strong, broad band in the 3200–3600 cm^{-1} region due to O—H stretching.

conspicuously adv 著しく syn *especially, notably*

▶ Massive quantities of corrosive liquid acids figure rather too conspicuously in the production of many core chemicals.

conspire vi 重なって働く syn *collaborate, cooperate*

▶ Together the orbital expansion and shorter Co···Co distance conspire to make possible direct overlap of Co t_{2g} orbitals across a shared octahedral edge resulting in a narrow d band and hole conduction in the mixed valence $Co^{4+/3+}$ system.

constancy n 不変性 syn *persistence, stability*

▶ In consequence of the constancy of the ratio of concentrations in the two phases, regardless of the relative volumes, extraction of a substance from an aqueous solution is more efficient if several small portions of ether are used than if the same total volume of ether is employed in one portion.

constant n **1** 定数

▶ The van der Waals constants are empirical constants; that is, their values are chosen to give the best agreement between the points experimentally observed and the points calculated from the van der Waals equation.

adj **2** 一定の syn *fixed, stable, unchanged*

▶ At constant pressure, the azeotropic mixture boils at a constant temperature without change of composition, but the azeotropic composition is different under different pressures.

3 絶えず続く syn *continuous, persistent*

▶ By means of the rapid cooling and constant stirring, the salt is obtained in the form of a crystalline powder.

constant-boiling adj 定沸点の

▶ A primary alkyl bromide can be prepared by heating the corresponding alcohol with constant-boiling hyrobromic acid (47% HBr).

constantly adv **1** 絶えず syn *consistently, permanently*

▶ Constantly appearing in these reactions is one compound, adenosine triphosphate (ATP).

2 しばしば syn *always*

▶ Names are given to certain groups that constantly appear as structural units of organic molecules.

constituent n **1** 成分 syn *element, part, unit*

▶ Zinc, as a constituent of more than 300 enzymes, plays essential roles in biological systems.

adj **2** 構成する

▶ When a large number of atoms or molecules are brought together to form a polymeric chain or a crystalline solid, an energy band will form if there is sufficient interaction of the constituent atomic or molecular orbitals.

constitute vt 構成する syn *compose, form, make*

▶ One finds that solutions of metals in mercury constitute fairly ideal solutions and that the emf's are almost correctly calculated by using concentrations instead of acitivties.

constitution n 化学構造 syn *composition*

▶ Discrete substances of unkown constitution are generally named first after their place of origin.

constitutive adj 構成的な

▶ Engineered nanoparticles have unique properties, which are different from their individual constitutive components.

constrain vt **1** 制限する syn *dictate, prescribe*

▶ The requirement for the motion of ions to accompany the oxidation and reduction of the polymer constrains the maximum rate at which these films can switch between conducting and insulating states.

2 強いてさせる syn *drive, force*

▷ Trivalent carbenium ions contains an sp^2-hybridized electron-deficient carbon atom, which tends to be close to planar in the absence of constraining skeletal rigidity or steric interference.

constrained adj 制約を受けた syn *bound*

▶ The entropy of the ordered, constrained adsorbed layer is always less than that of the gas.

constraint n 制約 syn *restraint, restriction*

▶ The relaxation of the globin–heme interaction allows the fourth heme to accept a dioxygen molecule without paying the price of the protein constraint and accounts for the affinity of $Hb(O_2)_3$ for the last dioxygen molecule.

constrict vi 収縮する syn *contract*

▶ When opposite ends of polyurethane foam are pulled apart, placing the foam under tension, it becomes longer in the direction that it is pulled and constricts in the perpendicular directions.

constriction n 収縮 syn *contraction*

▶ There are constrictions in the channel to a diameter of 4 Å.

construct n **1** 構成物 syn *organization, structure, system*

▶ We find by inspection that virtually all the biological constructs, membrane, nuclei, filaments, ribosomes, etc. are negatively charged.

vt **2** 文や理論を組み立てる syn *formulate*

▶ A detailed quantitative theory of amphoterism is difficult to construct.

3 物を組み立てる syn *compose, make, set up*

▶ The polyelectrolytes for phototransformations are constructed from hydrophobic and hydrophilic units in approximate molar ratios.

construction n 構築 syn *building*

▶ In a self-organizing system, basic construction units spontaneously associate to form a particular structure, the architecture of which is solely determined by the bonding properties and shapes of the individual components.

consult vt 参照する syn *refer to*

▶ Depending on which table of ionic radii is consulted, the diameter of a Ca^{2+} ion is in the range ca. 2.2 to 2.6 Å.

consumable adj 消費できる

▶ These reactions are applicable to a wide range of alkylaryl sulfides and proceed at room temperature using the relatively benign hydrogen peroxide as the consumable source of oxygen.

consume vt 消費する syn *exhaust, use up*

▶ There are one or more chain-propagating steps, each of which consumes a reactive particle and generates another.

consumption n 消費 syn *depletion*

▶ The consumption of bromine is obvious because the reddish color of the reagent disappears.

contact n **1** 接触 syn *touch*

▶ Assuming that the metal atoms are in contact, then the hexagonal unit cell has a definite shape given by the ratio $c/a = 1.633$.

2 in contact with …と接触して

▶ In close-packed structures, such as NaCl, the anions are either in contact with each other or are in close proximity.

3 out of contact 接触していない

▶ Alkali sulfites, when heated out of contact with the air, are changed to sulfate and sulfide.

4 電気的接点

▶ Contacts were made with gold paste to four gold wires stretched over a hole in a glass plate.

adj **5** 接触による

▶ In contact signaling it is somewhat arbitrary to say that one protein is the ligand and the other is the receptor.

vi **6** 接触する syn *touch*

▶ The fluorine–fluorine distance in CaF_2 is 2.73 Å, which indicates that the fluorines are approximately contacting.

contact angle n 接触角

▶ Surface energy is normally measured via contact angle goniometry, although this becomes difficult to perform on fine powders and rough surfaces.

contact potential n 接触電位

▶ When two metals are placed in contact, a contact potential arises due to the fact that electrons flow between the metals until the Fermi level is at the same height in both components.

contact process n 接触過程

▶ The emergent SO_2 is fed into a contact process for H_2SO_4.

contact resistance n 接触抵抗

▶ The relative significance of bulk and surface on contact resistance can be determined by deriving the conductance dispersion curve for the specimens over a range from zero frequency to approximately 100 kc/s.

contain vt **1** 成分とする syn *compose*

▷ Hemoglobin consists of a protein called globin bound to an iron-containing prosthetic group called heme.

2 含む syn *have in it, hold*

▶ Commercial chloroform contains 0.5–1% ethanol as stabilizer.

container n 容器 syn *vessel*

▶ Silica glass is so refractory that there is no container in which it could be melted without contamination.

containment n 拘束

▶ The samples were placed in a temperature-controlled aluminum block with an oval depression for sample containment.

contaminant n 汚染物質 syn *impurity*

▶ If aqueous solutions of ClO_2 is reduced by a peroxide, this adds no contaminant to the resulting chlorite solution.

contaminate vt 汚染する syn *pollut, soil*

▶ Commercial iodine is contaminated with chlorine, bromine, water, and sometimes cyanogens; it must be purified.

contamination n 汚染 syn *impurity, pollution*

▶ The deliberate contamination of a copper film with small amounts of oxygen does not promote hydrogen chemisorption.

contaminative adj 汚染する

▶ Although caution must be exhibited with all elements, it is still possible that some of the contaminative elements will yield useful dietary information in the future.

contemplate vt 予測する syn *imagine, perceive*

▶ The construction of networks based on a much wider range of structural prototypes could realistically be contemplated, in which each atom of the prototype net would be replaced by a stereochemically appropriate molecular building block and each bond of the parent would be replaced by an appropriate molecular connection.

contemporaneous adj 同時代の syn *contemporary, current,*

▶ In contemporaneous studies, Rainier et al. demonstrated that it was possible to convert the enol ethers into the bicyclic enol ethers by ring-closing metathesis mediated by the molybdenum catalyst.

contemporaneously adv 同時に syn *simultaneously*

▶ The two studies were carried out contemporaneously.

contemporary n **1** 同時代の人

▶ Thermodynamics created a physical chemistry conceived by contemporaries as a bridging science and even as a fundamental and autonomous theoretical chemistry.

adj **2** 現代の syn *current, present*

▶ A primary focus for contemporary studies of multiply charged anions is the investigation of molecular clusters of multiply charged anions with solvent molecules.

contend vi **contend with** …に対処する syn *cope*

▶ In practice, one often must contend with unwanted nuclei.

content n **1** 内容物 syn *substance*

▶ The contents of the unit cell give the chemical formula for the solid.

adj **2** 満足して syn *satisfied*

▶ Actual measurements of adsorption upon liquids have rarely been carried out; most authors are content to calculate them from the adsorption equation without experimental verification.

context n 背景 syn *circumstance, situation*

▶ The main interest in magnesium chemistry is in its organic and biochemical contexts that the element is at its most impressive.

contiguous adj 隣接する syn *neighboring*
▶ Organic molecules with protons on contiguous carbon atoms show principal resonance signals for protons of different chemical shifts.

continual adj 頻発する syn *constant, repetitive*
▶ The kinetic mechanism by which the vapor is maintained at its saturation value is the continual exchange of particles between one set of levels and the other.

continually adv 頻繁に syn *frequently, often*
▶ In the study of very fast reactions in solution, the question continually arises of the rate at which the reactant molecules encounter each other by diffusion through the solvent and whether this determines the observed rate, rather than the chemical reaction itself.

continuation n **1** 延長 syn *extention, supplement*
▶ The lines bisecting the two H−C−H angles in C_2H_4 must lie along continuations of the line joining the carbon atoms.
2 続行 syn *maintenance, preservation*
▶ Continuation of these reactions to all of available ortho and para positions of the phenol leads to a cross-linked three-dimensional polymer.

continue vi **1** 継続する syn *proceed, pursue*
▶ Substituted polyanilines continue to be of great interest since they can improve upon some of the properties of polyaniline.
vt **2** 持続する syn *keep, maintain*
▶ If the side-to-side linking of chains is continued, a sheet-like structure is produced.

continued adj 引き続きの syn *ongoing*
▶ Continued linking of silicate or aluminosilicate structures can result in the formation of complex frameworks that contain pores or channels that are large enough for water, small molecules, or cations to fit inside.

continuity n 連続 syn *constancy*
▶ There is a complete continuity of states in which the transformation from the gas to the liquid state occurs continuously.

continuous adj 連続的な syn *constant, persistent, steady, sustained*
▶ The system KNO_3−$NaNO_3$ forms a continuous series of solid solutions with a minimum point at 225.7 ℃, the composition being 55 percent KNO_3.

continuously adv 連続的に syn *constantly, continually*
▶ An essential point of difference between industrial and laboratory methods is that the most efficient industrial process is almost always a completely continuous process, in which starting materials flow continuously into a reactor and products flow continuously out.

continuum n 連続
▶ The continuum of wavelengths from the sun produces the psychological response that we call white light.

contour n **1** 等高線
▶ The contours represent lines of constant electron density throughout the structure.
2 輪郭 syn *outline, profile*
▶ It is often adequate to draw just one contour and to use that one contour simply to denote the shape of the orbital rather than to make any attempt to indicate electron density.
adj **3** 等高の
▶ The contour levels of electron density around each ion are very nearly spherically symmetric, even at the point of contact of two neighboring ions.

contract vi 収縮する syn *diminish, reduce, shrink*
part ▶ Gallium is unusual in contracting on melting, the volume of the liquid phase being 3.4% less than that of the solid.

contraction n **1** 収縮 syn *decrease*
▶ Fatty acid oxidation proceeds at its maximum rate, but it is insufficient for the supply of all the energy, so the muscle cells burn a mixture of glucose and fatty acids to sustain repeated contractions.
2 縮約語 syn *abbreviation*
▶ The word pervoration is a contraction of two words, permeation and evaporation.

contradiction n 矛盾 syn *disagreement, para-*

dox

▶ It had been reported by some workers that addition of hydrogen bromide to a particular alkene yields a product in agreement with Markovnikov's rule; by others, a product in contradiction to Markovnikov's rule; by still others, a mixture of both products.

contradictory adj 矛盾した syn *conflicting, paradoxial*

▶ The wave and particle descriptions seem literally mutually contradictory.

contradistinction n 対比 syn *difference*

▶ Sorption from solutions, in contradistinction to sorption of gases and vapors from the gas phase, is not very greatly affected by temperature.

contrary n **1** 正反対 syn *opposite*

▶ At first it might appear that the formal study of thermodynamics would not lead us very far. The contrary, however, is true.

2 on the contrary 反対に 常に文頭におく syn *however*

▶ On the contrary, single-layer graphite is accepted to have a single, characteristic HOMO level regardless of edge structure, and experiments show a single bond length.

adj **3 contrary to** （prep）…に反対の syn *opposite, reverse*

▶ A completely different effect, frequently found in biological systems, is when the flow is contrary to the concentration gradient: the solute molecules move from the low-concentration side to the high-concentration side of the membrane.

contrast n **1** 差異 syn *difference, distinction*

▶ The contrast between the reactivity of pyrrole and furan and that of pyridine is very striking.

2 by contrast 対照してみると

▶ By contrast, van der Waals adsorption requires no activation energy and, therefore, occurs more rapidly than chemisorption.

3 by contrast with …との対照によって 比較を導入する節に用いられる．文の主語を引き合いに出すものでなければならない．

▶ By contrast with the water-soluble sulfides of groups 1 and 2, the corresponding heavy metal sulfides of groups 11 and 12 are among the least-soluble compounds known.

4 contrast to 対照的に正反対なもの

▶ Being a model of both nuclear and electronic stability, helium provides a complete contrast to the chemical variety of hydrogen; indeed, much of its interest lies in its very inertness.

5 in contrast …と対照をなして syn *unlike*

▶ In contrast, in an interfacial polymerization, these nanofibers rapidly move away from the interface and diffuse into the water layer.

6 in contrast to …と対照をなして syn *unlike*

▶ In complete contrast to S_N2 reactions, the rates of S_N1 reactions of alkyl derivatives follow the order tertiary R > secondary R > primary R.

vi **7 contrast with** 対照的である syn *differ from*

▶ Conductivity measurements have shown that the anion vacancy in CaF_2 is more mobile than the interstitial F^- ion. This contrasts with AgCl in which the interstitial Ag^+ is more mobile than the cation vacancy.

vt **8** 対比する syn *compare, differentiate, distinguish*

▷ Relative stabilities of the crystallites in various polymers may be judged qualitatively by contrasting their respective melting temperatures.

contrasting adj 対照的な syn *different, distinct, dissimilar*

▶ The contrasting tendencies of benzene and borazine toward addition vs. aromatic substitution are illustrated by their reactions with bromine.

contribute vi **1 contribute to** 寄与する syn *participate*

▶ The spacings between vibrational levels are appreciable compared to the room temperature value of *kT*. Therefore, only a few of terms in the partition function will contribute significantly to the energy.

vt **2** 貢献する syn *furnish, provide*

▶ The more volatile component contributes the greater number of molecules to the vapor phase.

contribution n **1** 寄与 syn *share*

▶ In dilute solutions, individual ionic species make additive contributions to thermodynamic properties.

2 投稿原稿

▶ All contributions should include a concise, informative abstract and a maximum of six keywords.

contributor n 一因 syn *participant*

▶ At very low temperatures, the much smaller heat capacity due to freely mobile electrons in metallic solids begins to be the dominant contributor to the measured heat capacity.

contributory adj 寄与する syn *supporting, supportive*

▶ A major contributory cause for this is the rapid and successful development of the field of homogeneous catalysis where phosphine complexes feature prominently.

control n **1** 制御 syn *check, mastery, restraint*

▶ pH control is of critical importance in a large number of industrial operations.

2 対照

▶ One methodological implication of biological specificity was the need for controls, animals experimentally infected but not treated chemically, at the beginning and end of each investigation.

vt **3** 制御する syn *govern, regulate, rule*

▶ The concentration of Ca^{2+} ion controls the rates of a wide variety of functions such as muscle contraction and blood clotting.

control experiment n 対照実験 ☞ blank experiment

▶ Control experiments have shown that many technologically useful precursors including metal oxides and carbonates do not exhibit direct microwave heating; therefore, the reaction does not proceed.

controllable adj 制御可能な syn *manageable*

▶ Acid-catalyzed bromination of ketones is a controllable process, and a single hydrogen can often be substituted.

controllably adv 制御可能に

▶ Some of the Si atoms have been isomorphously replaced by Al, and the Si:Al ratio may be controllably varied from ca. 20 : 1 to 1000 : 1.

controlled adj 制御された syn *temperate*

▶ Most devices are made from thin single-crystal layers of controlled composition and electric properties, deposited onto a single-crystal substrate.

controlled atmosphere storage n 調製大気貯蔵

▶ Controlled atmosphere storage to retard post-harvest ripening is used for unprocessed fruits and vegetables.

controlled valence semiconductor n 原子価制御半導体

▶ Verway introduced the idea of controlled valence semiconductors in which the concentration of, say, Ni^{3+} ions is dependent on the addition of a controlled amount of Li_2O to NiO.

controversial adj 異論の多い syn *doubtful, questionable*

▶ Whether the intermediates in alkene addition reactions are correctly formulated with bridged structures is still a controversial matter.

controversy n **1** 論争 syn *argument, debate, dispute*

▶ The ability of carbon atoms to act as proton donors in hydrogen bonds has been the subject of controversy for many years.

2 a matter of controversy 論争の種

▶ For pure liquids, it is a matter of controversy as to whether dynamic surface tension is experimentally distinguishable from the final static surface tension.

convection n 対流 syn *convestion current*

▶ Extraordinary precautions are required to avoid convection in colloidal sols sufficiently that their sedimentation under the influence of gravity may be observed.

convenience n **1** 便利 syn *favor*

▶ The special convenience of the activity coefficient γ is that it shows explicitly the importance of the nonideality.

2 for convenience 便宜上

▶ There are systems that contain extended arrays of inorganic connectivity, which we shall refer to for convenience as extended inorganic hybrids.

convenient adj 便利な syn *suitable, useful*

▶ Thionyl chloride is a particularly convenient reagent to use since the only by-products are HCl and SO_2.

conveniently adv 都合よく

▶ Since the products formed differ in color from the reactants, the rate of dissociation was conveniently followed by measurements of optical density at predetermined wavelengths.

convention n 慣習 syn *custom, pratice, rule*

▶ As has been shown elsewhere, the convention of assigning zero entropy to aqueous hydrogen ion leads to thermodynamic inconsistencies when it is applied to some hypothetical processes.

conventional adj 通常の syn *customary, ordinary, standard*

▶ If we assume the conventional oxidation states for Y(3＋), Ba(2＋), and O(2－), then an average valence of 7/3＋ is obtained for the Cu ions in $YBa_2Cu_3O_7$.

conventionally adv 一般的に syn *generally*

▶ Since polyaniline nanofinbers are also observed in conventionally synthesized products, there should be some conditions under which these nanofibers can avoid overgrowth.

converge vi **1** 収れんする syn *coincide, merge*

▶ If the approximations used in the valence bond and molecular orbital methods are reduced, the two methods converge.

2 集中する syn *focus*

▷ On the basis of all this converging evidence, the conclusion can be drawn that the reaction between the alcohol and hydrogen chloride is accompanied by inversion.

convergence n 収束 ☞ divergence

▶ The process is repeated until a convergence between experimental observations and numerical predictions is obtained.

convergent adj 収束する ☞ divergent

▶ Signaling pathways initiated by signal transduction from different types of receptors can be convergent, having the same downstream target.

converse n **1** 逆 syn *opposite, reverse*

▶ If the *G* value accounts for an appreciable fraction of the total radiation products, the importance of the radical can be inferred, but the converse does not follow.

adj **2** 逆の syn *opposite, reverse*

▶ The X-ray powder pattern and Mössbauer spectrum of Prussian blue are the same as those of Turnbull's blue which is produced by the converse addition of $K_3[Fe^{III}(CN)_6]$ to aqueous Fe^{II}.

conversely adv 反対に

▶ Conversely, for an anion the hydrogen atoms of water are adjacent to the ion.

conversion n 転化 syn *alteration, transformation*

▶ Conversion of phenol by loss of the hydroxyl proton to phenoxide anion is expected to lead to substantially greater delocalization of the unshared pair.

convert vt 転化する syn *change, modify, transform*

▶ Toluene is more readily converted to the trinitro derivative, and this substance, on oxidation and decarboxylation, yields 1,3,5-trinitrobenzene.

converter n コンバーター

▶ Chlorophyll acts as the energy converter in photosynthesis.

convertible adj 変換可能な syn *transformable*

▶ Not all of the solar photons falling on a unit area are convertible into electric energy.

convex adj 凸状の ☞ concave

▶ The adsorption isotherms of gases at temperatures not far removed from their condensation points show two regions for most adsorbents: at low pressures the isotherms are concave, at high pressures convex toward the pressure axis.

convey vt 伝える syn *deliver, express*
▶ It is important to distinguish quite clearly between the usefulness of a model in predicting or correlating certain types of experimental observation and the general validity of the picture which this model seems to convey.

convince vt 納得させる syn *assure, satisfy*
▶ With the wisdom of hindsight, it seems remarkable that Kautsky's elegant experiments and careful reasoning failed to convince his contemporaries.

convincing adj 説得力のある syn *good, strong*
▶ A rather convincing experimental verification of ionic association is found in a series of conductance studies made by Kraus and Fuoss.

convincingly adv 説得力のあるものとして
▶ In the solid state, recent experiments have convincingly demonstrated that the electric field can influence the rate at which ion diffusion occurs.

convolution n コンボリューション
▶ Although the positions of the individual atoms in myoglobin have not been identified, the convolutions of the α-helix chain can be followed.

cool vt 冷却する syn *chill*
▶ The product was cooled, filtered, dried in the air, and then recrystallized from acetonitrile（ca. 1.2 dm^3）to yield 2-dicyanomethyleneindan-1,3-dione, 26.0 g（70%）as shining yellow plates.

coolant n 冷却剤 ☞ refrigerant
▶ One of the most effective and cheapest coolants is water, which is almost universally used in automotive and ordinary reaction equipment.

cooling n 冷却
▶ The crystals of the molecular compound which separated on slow cooling were filtered off rapidly, washed on the filter with a little acetone, and dried.

cooling curve n 冷却曲線
▶ Cooling curves are not very suitable for the study of second- and higher-order transitions.

cooperate vi 協力する syn *collaborate*
▶ The binding of one molecule of O_2 to hemoglobin cooperates in a positive way with the binding of the next molecule so that the apparent affinity of hemoglobin for O_2 increases as the degree of saturation by O_2 increases.

cooperative adj 協同的な syn *collaborative, collective, joint*
▶ Cooperative effects, in which many unpaired electrons communicate and interact with one another, can lead to more complex magnetic behavior in an extended solid than that observed for molecules in the gas or liquid states.

cooperative binding n 協同的結合
▶ The cooperative binding of O_2 by hemoglobin requires interactions between sites that are widely separated in space.

cooperatively adv 協同的に
▶ The Cooper pairs were supposed to move cooperatively through the lattice in such a way that electron–phonon collisions were avoided.

cooperative phenomenon n 協同現象
▶ Both ferromagnetism and antiferromagnetism are cooperative phenomena characterized by the existence of zones of magnetization termed domains.

cooperative process n 協同過程
▶ Addadi and Weiner have suggested that nucleation is a cooperative process involving structurally disordered sulfate groups of flexible oligosaccharide side chains and organized motifs of carboxylate ligands within β-pleated sheet surface domains of the matrix.

cooperativity n 協同作用
▶ Although hemoglobin is not an enzyme but an O_2-binding protein, the study of it has contributed a great deal to our understanding of macromolecular cooperativity.

coordinate n **1** 通常複数形で，座標
▶ It is customary to describe the structure by the use of coordinates *x* and *y*, giving the position of the atoms relative to an origin at the corner of the unit cell, with *x* and *y* taken as fractions of the edges of the unit of structure.
adj **2** 配位結合の

▶ In the clathrate compound, $Ni(CN)_2(NH_3)\cdot C_6H_6$, the six-coordinate Ni^{2+} ions are bonded to two NH_3 and the N atoms of four bridging CN^- ions; the four-coordinate Ni^{2+} ions are bonded to the C atoms of CN^- ions.

vi **3 配位する**

▶ Cyanide ion usually coordinates through C, except when it serves as a bridging group.

vt **4 配位させる**

▶ In the orthorhombic structure of $NiSO_4\cdot 7H_2O$, each nickel atom is octahedrally coordinated by six water molecules.

5 協調させる syn *organize, systematize*

▷ To achieve these objectives will require an extensive, coordinated program of fundamental chemical research.

coordinate axis n **座標軸**

▶ When a plane is parallel to one of the coordinate axes, we say it meets that axis at infinity.

coordinate bond n **配位結合** syn *dative bond*

▶ The concept of the coordinate bond as an interaction between a cation and an ion or molecule possessing a lone pair of electrons can be accepted before specifying the nature of that interaction.

coordinated water n **配位水** ☞ water of crystallization

▶ Some of the most important properties of metal ions in aqueous solution, such as the relationship between complex salts and oxo acids, the nature of hydrolytic reactions, and the formation of highly condensed basic salts, find a common interpretation in the enhanced acidity of coordinated water.

coordinate system n **座標系**

▶ The crystallographic axes provide a satisfactory coordinate system for describing the positions of atoms within the unit cell.

coordination n **1 協調** syn *arrangement, regulation*

▶ The coordination of the oxidation of fuel molecules and the synthesis of ATP can be understood by drawing on the following concepts.

2 対等関係 syn *harmony*

▶ Good coordination between the results derived from all these sources is obtained.

3 配位

▶ Lanthanum telluride, La_2Te_3, has a lanthanum coordination sphere of eight tellurium atoms with La–Te distances of 3.244 and 3.418 Å.

coordination chemistry n **配位化学**

▶ In coordination chemistry, the central atom is an electrophilic reagent and the ligands are nucleophilic reagents.

coordination compound n **配位化合物**

▶ Coordination compounds in which the coordination number of the central ion is different from either four or six are much less common.

coordination geometry n **配位構造**

▶ NO shows a wide variety of coordination geometries and sometimes adopts more than one mode within the same complex.

coordination isomer n **配位異性体**

▶ In compounds made up of both anionic and cationic complexes it is possible for the distribution of ligands between the ions to vary and so lead to coordination isomers.

coordination isomerism n **配位異性**

▶ Coordination isomerism is an extreme case of ionization isomerism which arises when both anion and cation are complexes.

coordination number n **配位数** syn *ligancy*

▶ Although the coordination number of the central ion for most compounds is either six or four, examples of compounds are known which exhibit each of the other numbers up to ten.

coordination polyhedron n **配位多面体**

▶ The oxygen dodecahedral coordination polyhedron is found in the garnet structure.

coordination polymer n **配位高分子**

▶ The term coordination polymer owes its origins to the analogy with coordination compounds, in which ligands organic or otherwise, are coordinated to monomeric metal centers.

coordination position isomer n **配位位置異性体**

▶ When two coordinating centers are not in separate ions but are joined by bridging groups, the isomers are often distinguished as coordination position isomers.

coordination structure n 配位構造

▶ Metal ions of d^0 or low d^x electronic configuration generally form more stable high-coordination structures than do metal ions of high d^x configuration.

coordinative saturation n 配位的飽和

▶ Coordinative saturation in aminoboranes can be achieved not only through partial double bond formation but also by association of the monomeric units to form $(B{-}N)_n$ rings.

cope vi cope with …をうまく処理する syn *deal with, handle*

▷ An organism must adapt to its environment, using the raw materials available to it and coping with unwanted or even toxic substances.

coplanar adj 同一平面上の

▶ Another compact ligand is the coplanar nitrate group which has a short oxygen-oxygen separation, ca. 2.1 Å.

coplanarity n 共平面性

▶ The diminished light absorption of *cis*-stilbene is attributed to interference between the two ortho hydrogen atoms of the phenyl groups, which prevents coplanarity of these rings and hence diminishes resonance.

copolymer n 共重合体

▶ If the units are nonidentical, as when different monomers are polymerized together, the product is called a copolymer.

copolymerization n 共重合

▶ The properties of polyvinyl chloride can be improved by copolymerization, as with vinyl acetate, which produces a softer polymer with better molding properties.

coprecipitate vt 共沈させる

part ▶ In many cases, the precursor is simply a finely divided, coprecipitated, solid mixture or a gelatinous mass called a gel.

coprecipitation n 共沈

▶ More elegant coprecipitations of mixtures of complex metal salts, such as carbonates, nitrates, or oxalates, also have been used in attempts to obtain finer particle dispersions and better homogeneity than possible through grinding.

coproduct n 副産物 syn *by-product, side product*

▶ Most of vanadium is obtained as a coproduct along with other materials.

copy n **1** コピー syn *replica*

▶ Recent reviews of DNA replication fidelity now reflect the idea that replicative DNA polymerase are governed largely by steric effects in making DNA copies with high fidelity.

vt **2** コピーする syn *duplicate, replicate*

▶ To synthesize proteins, genes are first copied into molecules called messenger RNA.

copying n 複製 syn *reproduction*

▶ Gene expression involves the copying, sometimes carried out many thousands of times, of small sections of the genome that are then often translated to protein.

core n **1** コア syn *center*

▶ When the core of the optical fiber has a larger refractive index than the surrounding cladding, total internal reflection results, meaning that the light is largely confined to the core and can be transmitted along the fiber with relatively little leakage.

2 中心核 syn *kernel, nucleus*

▶ The internal cavity is ca. 8 nm in diameter which sets an upper limit on the number of Fe atoms that can be accommodated in the mineral core.

adj **3** コアの

▶ Core orbitals like the 1s, 2s, and 2p orbitals of Na do not overlap as significantly.

vt **4** コアを形成する

▶ The crystals that precipitate during cooling are often cored.

corner n 隅

▶ Ferredoxins contain the Fe_4S_4 group with the atoms at alternate corners of a cube.

corner-shared adj 隅を共有した

▶ The compound Tl_4S_3, which has the same

stoichiometry as In_4X_3, has a different structure in which chains of corner-shared $Tl^{III}S_4$ tetrahedra of overall stoichiometry TlS_3 are bound together by Tl^{I}.

cornerstone n 基礎 syn *essence, fundamental*

▶ Very little attention has been paid to the development of simple molecular orbital arguments of the type which today form the cornerstone of much of molecular organic and inorganic chemistry.

corollary n 必然的な結論 syn *deduction, postulate*

▶ It is a necessary corollary that the shape of the isotherm is dependent on the form of the distribution of the sites.

corpuscle n 微粒子

▶ The energy carried by each corpuscle is $h\nu$, where ν is the frequency of light. These corpuscles are called photons.

corpuscular adj 微粒子の

▶ Both the photoelectric effect and the Compton effect indicate that light is corpuscular.

correct vt **1** 補正する syn *calibrate*

▶ Melting points are corrected.

2 訂正する syn *amend, rectify, revise*

▷ When correcting proofs, indicate in the text where the correction must be made and also write the appropriate proof-correcting symbol clearly in the margin.

correctable adj 補正できる

▶ Ionic radii are not constant but show small systematic changes, many of which are correctable but only by using ad hoc anion–anion repulsions and their associated parameters.

correction n **1** 訂正 syn *amendment, improvement, revision*

▶ Corrections must be as clear as possible in order to avoid further mistakes.

2 補正

▶ Gravimetric measurements require corrections to account for buoyancy, while volumetric measurements involve corrections for the volume of the sample.

correctly adv 適切に syn *appropriately, properly*

▶ Simple collisions between the reactants, even correctly oriented, are not generally enough to cause the reaction to take place.

correctness n 正しさ syn *accuracy, precision*

▶ It must not be forgotten that the fitting of data by proper choice of adjustable constants does not constitute proof of the correctness of the underlying theory.

correlate vi **1** 関連する

▶ In inorganic compounds, the bond valences correlate much better with bond lengths than they do in organic compounds.

vt **2** 関係づける syn *coordinate, relate*

▷ We are interested in correlating the magnitude of the donor–acceptor interactions with electronic properties of the acids and bases.

correlation n 相関 syn *connection, parallel*

▶ There is no positional correlation between molecules within the S_A or S_C layers.

correlation diagram n 相関図

▶ The simplest way to display such a correlation is with a correlation diagram.

correlation length n 相関長

▶ In traditional superconductors, the correlation length of the Cooper pairs is large, of the order of hundreds or thousands of angstroms.

correlative adj 相関的な syn *reciprocal*

▶ It is necessary to establish proof of identity by correlative experiments.

correspond vi **correspond to** …に該当する syn *fit, match*

▶ Acetone in cyclohexane solution exhibits two absorption bands; one appears at 190 nm and corresponds to the $\pi \rightarrow \pi^*$ transition, while the second is at 280 nm and corresponds to the $n \rightarrow \pi^*$ transition.

correspondence n **1** 書簡 syn *letter, note*

▶ We acknowledge with great thanks much helpful correspondence with Professor X. and also his kindness in providing us with the observed displacements for the table.

2 対応 syn *coincidence, equality, resemblance*

▶ The transition from aluminum to gallium is from an eight-shell- to an eighteen-shell-type atom, and no close correspondence of physical properties is to be expected.

corresponding adj **1** 対応する syn *equivalent, identical*

▶ Light of a frequency corresponding to an absorption of the radical reactant is passed through the reaction cell and is monitored by a suitable detector.

2 関連する syn *respective*

▶ An alkyl iodide is often prepared from the corresponding bromide or chloride by treatment with a solution of sodium iodide in acetone.

3 文通する

▶ A transfer of copyright form will be sent to the corresponding author.

correspondingly adv 相応して

▶ Many reduction processes will be hindered by the formation of mixed oxides, in which the partial pressure of the metal oxide is reduced so that the driving force of the reaction is correspondingly reduced.

corresponding state n 対応状態

▶ The most generally useful method of prediction of the volumetric properties of fluids is the hypothesis of corresponding states.

corroborate vt 確証する syn *check, confirm, verify*

▶ This conclusion is further corroborated by work on cobalt and manganese 4-cyclohexene-1, 2-dicarboxylates.

corroboration n 確証 syn *check, evidence, proof*

▶ Although preliminary work in our laboratory and some results reported elsewhere support this hypothetical concept, additional experimental corroboration is required.

corroborative adj 裏付ける

▶ The excited states of long-lived organic phosphors are triplet states, and corroborative evidence has been found by means of static magnetic susceptibility measurements.

corrodibility n 腐食性

▶ In general, the corrodibility of a metal or alloy depends upon its position in the electromotive series.

corrosion n 腐食 syn *erosion*

▶ When the corrosion system has some areas acting as anodes and others as cathodes, the current flow between them produces potential differences.

corrosion resistance n 耐食性

▶ The corrosion and oxidation resistance of steel is markedly increased by incorporation of from 15 to 18% of chromium and often a few percent of nickel.

corrosion-resistant adj 耐食性の

▶ Copper, nickel, chromium, and zinc are among the more corrosion-resistant metals and are widely used as protective coatings for other metals.

corrosive adj 腐食性の

▶ Acetic anhydride is cheap, readily available, less volatile, and more easily handled than acetyl chloride, and it does not form corrosive hydrogen chloride.

cosolvent n 共溶媒

▶ The energy lost in the disruption of water structure by added alcohol is not totally replenished in the formation of hydrogen bonds with the cosolvent.

cosolvency n 共溶媒性

▶ Blending and cosolvency are often of practical importance in industrial processes.

cosonicate vt 共に音波処理する

▶ Platinum ions were entrapped by cosonicating K_2PtCl_4 with the surfactants.

cost n **1** 価格 syn *expense, price*

▶ If the process is to be an economically feasiblle route to an improved ceramic product, the sol must be made available at a cost related to the product value.

2 at the cost of …という代償を払って

▶ By ignition, phosphorous acid is changed at the cost of its own oxygen to the higher compound.

costly adv 高価な syn *expensive, invaluable*

▶ The microtechnique is particularly appropriate because the reagent is costly and must be used sparingly.

could aux 事実に反する仮定に用いて，できるだろう

▶ Propylene could yield either of two products, the *n*-propyl halide or the isopropyl halide, depending upon the orientation of addition, that is, depending upon which carbon atoms the hydrogen and halogen become attached to. Actually, only the isopropyl halide is formed.

Coulomb barrier n クーロン障壁

▶ The term repulsive Coulomb barrier is somewhat misleading since. although the barrier is repulsive at long range, at short range the overall potential is attractive.

Coulomb force n クーロン力

▶ In water, the Coulomb forces extend only short distances, and a high yield of solvated electrons is found.

Coulomb integral n クーロン積分

▶ The Coulomb integral represents the interaction energy between each electron and its respective nucleus.

count vi **1** 重要である syn *be important, of importance*

▶ It is microscopic rigidity that counts rather than the overall solid appearance at the macroscopic level.

vt **2** 数える syn *calculate, compute, enumerate, number*

▶ To count the number of Cl^- ions within the unit cell of the NaCl structure, one is in the body center (counts as 1); twelve are at the edges (counts as $12 \times 1/4 = 3$). This gives a total of four Cl^- ions.

3 count up 数え上げる

▷ Counting up the numbers of each interstitial site, for every anion there is one octahedral site and two tetrahedral sites.

countable adj 数えられる syn *finite*

▶ Discrete molecules have easily countable numbers of atoms, and small integer relationships exist between the numbers of different kinds of atoms composing the molecule.

counter vt 逆らう syn *oppose*

▷ Countering these attractive forces are repulsive forces resulting from nucleus–nucleus repulsion and, more important, the repulsion of inner or core electrons.

counteract vt 打ち消す syn *neutralize, offset*

▶ The formation of oxygen vacancies in $BaTiO_3$ can be counteracted by the introduction of lower valency ions on the Ti sites which act as deeper traps for the electrons and prevent them from entering the conduction band.

counter anion n 対陰イオン ☞ counter cation, counter ion

▶ The development of the BEDT-TTF superconductors has been based on the role of the counter anion.

counterbalance vt 相殺する syn *cancel, counteract*

▷ Herzberg first explained chemical bonding in a simple and convincing manner in terms of bonding and antibonding electrons, an antibonding electron counterbalancing the effect of a bonding electron.

counter cation n 対陽イオン ☞ counter anion, counter ion

▶ The structures of $[M(CN)_8]^{3-}$ ($M = Mo, W$) vary, according to the environment and counter cation, between the energetically similar square-antiprismatic and dodecahedral forms.

counterclockwise adj 反時計回りの syn *anticlockwise* ☞ clockwise

▶ If a spiral proceeds clockwise, the dislocation is by definition right-handed, and counterclockwise, left-handed.

countercurrent adj 向流の

▶ The separation resulting from contact of the countercurrent streams of vapor and liquid is called fractionation

countercurrent extraction n 向流抽出

▶ Countercurrent extraction is a process in which two solvent phases flow counter to one another in a system that contains baffles to improve exchange of solutes between the phases.

counterdiffusion n 相互拡散

▶ Wagner's hypothesis states that when two solid oxides react they do so by counterdiffusion of the cations through the layer of product.

counter ion n 対イオン ☞ counter anion,

counter cation

▶ The counter ions needed for charge balance, often the smaller cations, are situated in holes or interstitial sites between the framework ions.

counterpart n 対応するもの syn *equivalent*

▶ There is no oxygen counterparts of SF_4 and SF_6.

counting n 計数 syn *calculation*

▶ Electron counting rules have proved to be extremely valuable for predicting cluster geometries and stabilities.

countless adj 数え切れない syn *abundant*

▶ Countless attempts have been made to express the periodic relationships of the elements in the form of a table, a diagram, or a space model.

couple n **1 a couple of** 同種の二つの 続く名詞は複数形，動詞も複数形を用いる．

▶ The iron is present in the +3 oxidation state and is coordinated to two or three tyrosyl residues, a couple of histidyl residues, and perhaps a tryptophanyl residue in a protein chain.

vt **2** 連結する syn *connect, join, link*

▷ Methyl Orange is prepared by coupling diazotized sulfanilic acid with *N*, *N*-dimethylaniline in a weakly acidic solution.

coupled reaction n 共役反応

▶ Interest in this type of host molecule stems from the idea of mimicking coupled reactions.

coupling n **1** 結合 syn *link*

▶ Coupling of the proton of an aldehyde group with adjacent protons has a small constant (*J* 1-3 Hz).

2 カップリング反応 ☞ coupling reaction

▶ To make an alkane of higher carbon number than the starting material requires formation of carbon-carbon bonds, most directly by the coupling together of two alkyl groups.

3 組み合わせ syn *connection*

▶ The coupling of electric and magnetic phenomena is elegantly demonstrated in a levitation experiment.

coupling constant n 結合定数 ☞ spin-spin coupling constant

▶ The distance between peaks in a multiplet is a measure of the effectiveness of spin-spin coupling and is called the coupling constant, *J*.

coupling reaction n カップリング反応 ☞ coupling

▶ The chief impurity in the Grignard reaction mixture is biphenyl, formed by the coupling reaction.

course n **1** 過程 syn *process, way*

▶ This hypothesis affords an adequate explanation of the steric course of either S_N1 or S_N2 substitutions in haloammineplatinum(II) complexes.

2 in due course そのうち syn *eventually, soon*

▶ Our findings in this area will be reported in due course.

3 of course (adv) もちろん syn *certainly, naturally, surely*

▶ This rejection can of course be enhanced by increasing the ionic strength of the aqueous phase with large additions of salt, the salting-out effect.

covalence n 共有原子価 syn *covalency*

▶ According to Fajans' rules, increasing charge results in increasing covalency, especially for small cations and large anions.

covalent adj 共有結合の

▶ The greater covalent character of the silver halide bond compared with those in the alkali halides helps stabilize discrete AgX molecules in the liquid and thus makes the melting points of the silver compounds lower than those of the potassium compounds.

covalent bond n 共有結合 syn *covalent bonding*

▶ Sulfur can form four or even six covalent bonds if another element, combining with sulfur, releases enough energy to pay for promotion of the electrons.

covalently adv 共有結合で

▶ Retinal is not only bonded covalently to the protein but is held in a lipophilic pocket.

covalent modification n 共有結合性修飾

▶ Molecular conversion, namely, reversible covalent modification of enzymes, is another method of enzyme control.

covalent radius n 共有結合半径 ☞ atomic radius, ionic radius, van der Waals radius

▶ A set of values of tetrahedral covalent radii for use in crystals of these types is given in the table.

cover vt **1** 対象とする syn *comprise, include*

▶ The term lipid covers a structurally diverse group of compounds including fatty acids, glycerolipids, sphingolipids, terpenes, steroids, and carotenoids.

2 わたる syn *extend*

▷ With sunlight, covering a wide range of wavelengths, the fraction absorbed is reduced by a factor of 10 or more, since most of the visible and near-ultraviolet radiation lies outside the absorption region of the dye.

3 覆う syn *screen, shield*

▷ The Langmuir theory suggests that the rate of evaporation can be taken to be proportional to the fraction of the surface covered.

coverage n 被覆

▶ Our interest is in the variation of adsorption heat q with coverage θ.

crack vi ひびが入る syn *fracture*

▶ Boric acid glass suffers from being brittle and hygroscopic, and thick samples readily crack unless carefully annealed.

cracked adj 砕けた

▶ In the absence of any combination reactions of cracked fragments, the intense peak that is highest in mass number corresponds to the parent molecule minus just one electron and provides a highly accurate method for measuring molecular weights.

cracking n **1** クラッキング

▶ Most cracking is directed toward the production of fuels, not chemicals, and for this catalytic cracking is the major process.

2 破砕

▶ Larger crystals undergo cracking during the initial stages of rearrangement due to strains introduced by buildup of product in the matrix of starting material.

cream n **1** クリーム

▶ In the cream, the droplets are close together, and they may even adhere or clump together, but it is important to note that they have not coalesced.

vt **2** クリーム状にする

▷ Foams are similar to creamed emulsions except that the droplets are replaced by bubbles; hence, gravity has a much greater influence.

crease vt 折り目を付ける syn *fold, rumple*

▶ The simplest way to fold a cicular cut paper is to crease it across a diameter and then, without opening the paper, make a second fold at right angles to the first one.

create vt 生み出す syn *generate, produce*

▶ Even though electrons can be created at energies well above the conduction-band edge, the electrons will rapidly come to the band edge by losing their excess energy as heat.

creation n 創作 syn *fabrication, formation, generation*

▶ The creation of a chiral active site by the adsorption of pure enantiomers onto a metal surface has proved successful in the related studies.

creatively adv 独創的に syn *ingeniously*

▶ By making new compounds or identifying suitable ones from the wide range already known, chemists are contributing creatively to a joint endeavor to find substances with new properties or combinations of properties.

credence n 信憑性 syn *acceptance, credit*

▶ The relative simplicity of the spectra lend considerable credence to the methyl radical identification.

credit n **1** 功績の帰属 syn *recognition*

▶ In spite of the importance of the earlier contributions, the major portion of the credit for the development of the periodic system must go to the Russian, Dmitrii Ivanovich Mendeléev, and to the German, Julius Lothar Meyer.

vt **2** 帰する syn *attribute*

▶ The development of the first sulfa drugs cannot be credited to any single individual.

creep n クリープ

▶ Observations of creep can readily be made at specified temperatures, but a complete or detailed temperature coverage would be most time-consuming.

creeping n 表面に広がること

▶ During evaporation many substances have the property of creeping over the edge of the crucible or dish, often causing a slight loss of the substance.

crest n 波頭 syn *peak, top*

▶ The distance from crest to crest of the wave is called the wavelength, usually indicated by lambda, λ.

crisis n 危機 複数形は crises syn *danger*

▶ Environmental toxicologists and atmospheric and oceanic modelers are discovering crises at a remarkable rate.

criterion n 判断の基準 複数形は criteria syn *measure, standard*

▶ The presence or absence of a chiral center is thus no criterion of chirality.

critical adj **1** 決定的な syn *crucial, decisive, important*

▶ Variations in the oxygen stoichiometry in the high-temperature superconductors play a critical role in determining whether a material is superconducting or not.

2 臨界の

▶ While low coordination numbers are found with low radius ratios, the correlation with the critical radius ratios is fairly poor.

critical constant n 臨界定数

▶ This scheme involves one parameter in addition to the critical constants for each substance.

critical current density n 臨界電流密度

▶ Superconducting films with critical current densities as high as nearly 10^8 A/cm^2 have been prepared.

critically adv きわどく syn *seriously, severely*

▶ Under critically controlled conditions of concentration, temperature, and reaction time, *N*-acetyl-DL-alanine can be prepared easily in high yield.

critical micelle concentration n 臨界ミセル濃度

▶ From phase rule it follows that monomers and micelles are in equilibrium only at a single monomer concentration, the critical micelle concentration.

critical phenomenon n 臨界現象

▶ In spite of its simplicity and the easily understood significance of its constants, the van der Waals equation can be used to treat critical phenomena as well as a wide range of properties.

critical point n 臨界点

▶ Since there is no distinction between liquid and gas phases above the critical point and no second phase is formed regardless of the pressure of the system, the term supercritical fluid is used instead of liquid or vapor.

critical size n 臨界寸法

▶ Any nuclei which form spontaneously by collision are intrinsically unstable up to a critical size, after which the bulk lattice energy terms dominate and crystal growth proceeds.

critical solution temperature n 臨界共溶温度

▶ The temperature at which two conjugate solutions become identical is known as the critical solution temperature.

critical temperature n 臨界温度

▶ Above a temperature characteristic of the gas, called the critical temperature, the transition from the gaseous state to the liquid state occurs without a sharp change in volume on increasing the pressure.

criticism n 批判 syn *appraisal, assessment, evaluation*

▶ While the force of these criticisms cannot be denied, we should point out that some assumptions made in the analysis are not strictly justified and that it is difficult to assess the effects of these assumptions on their results.

crop n **1** 同時に発生する集まり syn *output*

▶ The mixture was allowed to stand at room temperature overnight to afford a crop of seed crystals which were collected and washed rapidly with acetonitrile.

2 穀物 syn *harvest*

▶ Large amounts of fertilizer nitrogen are used each year by farmers to increase yields of crops.

cross n **1** 十字

▶ The arms of the cross will remain parallel to the polarization directions of the crossed polarizers.

vi **2** 交差する syn *intersect*

▶ In spite of all the work done on this system, we still do not know the thermodynamic transition point, where the two free energy curves cross.

vt **3** 越える syn *traverse*

▶ Solid-state chemistry becomes truly interdisciplinary and crosses the traditional boundaries between a number of recognizable disciplines such as physics, engineering, geology, and even biology.

cross coupling n 交差カップリング

▶ These results clearly show the phosphine-sulfone and phosphine-sulfonamide hybrid ligands to be effective ligands for palladium-catalyzed cross couplings.

cross-cut adj 横に切った

▶ Analytical chemistry is a cross-cut field in that it has application in diverse areas.

crossed adj 交差した

▶ A colorless anisotropic crystal between crossed polarizers may show bright colors.

crossed molecular beam n 交差分子線

▶ In crossed molecular beam experiments, two reactant beams are directed into a scattering region, where single reactive or unreactive collisions occur between component molecules of the beams.

cross formula n 投影式 ☞ projection formula

▶ Each of the possible stereoisomers is commonly represented by a cross formula. As always in cross formulas, horizontal lines represent bonds coming toward us out of the plane of the paper, and vertical lines represent bonds going away from us behind the plane of the paper.

cross-hair n 十字線

▶ The position of the cross-hair could be read to ±0.1 mm.

crossing n 交差

▶ An example of intersystem crossing is the singlet-to-triplet crossing that occurs as step in the mechanism of phosphorescence.

cross link n **1** 橋かけ結合 syn *crosslankage, crosslinking*

▶ Cross links are extremely important in determining physical properties because they increase the molecular weight and limit the motion of the polymer chains with respect to one another.

vt **2** 橋かけ結合される ハイフンを要する.

▶ All grades of polyethylene and most copolymers can be chemically cross-linked.

crosslinking n 橋かけ結合 syn *cross link, crosslinkage*

▶ Crosslinking increases strength, heat and electric resistance, and especially resistance to solvents and other chemicals.

crossover n 交差

▶ The actual crossover from one electronic energy level to another is adiabatic and has an equal likelihood of occurring in either direction.

cross-polarization magic angle spinning n 交差分極マジック角スピニング

▶ High resolution NMR in solids came of age with the development of cross-polarization magic angle spinning.

cross-ring adj 渡環の ☞ transannular

▶ Alternatively the structure of $Te_2Se_4^{2+}$ and $Te_3Se_3^{2+}$ cations may be described as an overall six-membered ring with a pronounced boat conformation and one cross-ring bond.

cross section n 断面 syn *section*

▶ A further restriction on the discharge flow technique is the need to retain a uniform flow velocity along the entire cross section of the tube.

cross-sectional area n 断面積

▶ The indefinite nature of the effective cross-sectional area of an adsorbate molecule scarcely merits a distinction as to the nature of the two-dimensional phase.

crowded adj 込み合った syn *dense*

▶ If there are one or more lone-pair domains we expect these domains to occupy the least crowded positions.

crowding n 混雑 syn *swarm*

▶ Crowding among the large aromatic rings tends to stretch and weaken the carbon-carbon bond joining the triphenylmethyl groups in the dimer.

crown ether n クラウンエーテル

▶ Crown ethers are cyclic ethers containing four, five, six, or more oxygen atoms. Like a cyclodextrin, a crown ether can act as a host to guest molecules.

crucial adj 決定的な syn *critical, decisive, pivotal*

▶ The choice of glycol and the amount of water in the synthesis mixture were crucial.

crucially adv 決定的に

▶ The electric and superconducting properties of the 1:2:3 materials depend crucially on the preparative conditions, such as sintering temperature, annealing, quenching rates, etc.

crucible n るつぼ

▶ The fusion can be made in an iron or nickel crucible; the iron one is soon spoiled, and even the nickel one does not last very long if the fusion is made over a free flame.

crude adj **1** 粗製の syn *natural, original, raw*

▶ The crude product of an organic reaction may contain a colored impurity.

2 おおざっぱな syn *gross, rough*

▶ The treatment is too crude to allow detailed predictions to be made, but it is evidently along the right lines.

crumble vi 砕ける syn *break up, disintegrate, fragment*

▶ Occasionally the pseudomorph may crumble to a dust, even with gentle handling.

crush vt 粉砕する syn *grind, pulverize*

▷ If lumps or large crystals of solid are slow in dissolving, the process can be hastened by crushing the material against the bottom of the flask with a flattened stirring rod.

crust n 地殻

▶ Oxygen is the most abundant element in the earth's crust, which consists primarily of Si, H, and metal oxides.

crustal adj 地殻の

▶ Sodium is the seventh most abundant element in crustal rocks and the fifth most abundant metal.

crusty adj 表面が硬くなった

▶ The efficiency of P_4O_{10} as a desiccant is greatly impaired by the formation of a crusty surface film of hydrolysis products unless it is finely dispersed on glass wool.

cryogenic adj 低温の

▶ Experimentally, a rectangular single crystal of n-type InSb is located in a powerful magnetic field and is cooled to cryogenic temperature by contact with liquid helium.

cryoscopic adj 凝固点降下の

▶ An appreciable dissociation of tetraarylhydrazines has been indicated by cryoscopic determinations of the apparent molecular weight in benzene.

cryptate n クリプテート

▶ An important idea in macrocyclic chemistry and also in the chemistry of cryptates is size-match selectivity.

crystal n 結晶

▶ Crystals normally have defects and impurities that profoundly affect their applications, as in semiconductors.

crystal class n 結晶族

▶ It is true that by careful growth of crystals additional faces may be developed that will give a more definitive indication of the crystal class.

crystal engineering n 結晶工学

▶ Our approach to the crystal engineering of coordination polymers was phrased in net-based terms rather than structure-based terms.

crystal face n 結晶面 syn *crystal plane*

▶ Specific crystal faces can be preferentially nucleated by the stabilization of particular transition states at the matrix surface.

crystal field n 結晶場

▶ If the crystal field is strong, then electrons are forced to pair in the lower t_{2g} set, and the configurations are known as low-spin.

crystal field stabilization energy n 結晶場安定化エネルギー

▶ The competition between Ni^{2+} and Al^{3+} shows that Al^{3+} has greater octahedral site preference energy than Ni^{2+} and well illustrates the presence of factors other than crystal field stabilization energies which strongly influence the problem.

crystal field theory n 結晶場理論

▶ The essence of the crystal field theory is that the five d orbitals, which are degenerate and equal in energy in the gaseous metal ion, become differentiated in the presence of the electrostatic field due to the ligands.

crystal form n 結晶形

▶ A crystal form includes all the faces of a crystal which are required by the symmetry if one face is present.

crystal growth n 結晶成長

▶ To start crystal growth, a seed crystal is placed in contact with the surface of a melt in a furnace with a temperature only slightly above the melting point of the sample.

crystal habit n 晶癖 ☞ habit

▶ Variation in crystal habit occurs without the slightest variation in angular relations of the faces and, therefore, without the slightest variation in the crystallographic symmetry.

crystal lattice n 結晶格子 ☞ lattice

▶ Having the most symmetrical structure, the para isomer fits better into a crystal lattice and has the highest melting point.

crystalline adj 結晶質の

▶ If a solid substance has the regular internal arrangement of atoms without smooth external faces, it is said to be crystalline or to have crystal structure.

crystalline polymer n 結晶性重合体

▶ The exclusion of reagents from highly crystalline polymers frequently is essential to their utility.

crystallinity n 結晶化度 syn *degree of crytallinity*

▶ A more or less random introduction of methyl groups along a polyethylene chain reduces the crystallinity sufficiently drastically to lead to a largely amorphous polymer.

crystallite n 微結晶

▶ The portion of a crystal whose constituent atoms, ions, or molecules form a perfect lattice, without strains or other imperfections, is called crystallite. Single crystals may be quite large, but crystallites are usually microscopic.

crystallization n 結晶化 syn *crystallizing*

▶ A moderate amount of branching in these polymers does not prevent them from crystallization but it does reduce their crystallinity.

crystallization temperature n 結晶化温度

▶ In annealing at higher than the original crystallization temperature, crystallinity does not increase monotonically but decreases at first and then increases, reaching a higher value only after an appreciable time has elapsed.

crystallize vi **1** 結晶化する syn *solidify*

▶ Polonium is unique in being the only element known to crystallize in the simple cubic form.

vt **2** 結晶化させる

▷ Many salts crystallizing in cubic lattices can undergo changes from one to another type of lattice.

crystallographic adj 結晶学的な

▶ Systematic analyses of crystallographic data for hydrogen bonds have revealed a range of geometries and have led to proposals for rules to rationalize or predict hydrogen-bonding patterns.

crystallographically adv 結晶学的に

▶ Crystal radii are ionic radii evaluated crystallographically as compatible with observed interionic distances in crystal.

crystallographic axis n 結晶軸

▶ In the hexagonal system, there are four crystallographic axes, the *c* axis and three *a* axes in the plane normal to *c*, which are customarily labeled a_1, a_2, and a_3.

crystal structure n 結晶構造

▶ Compared with the copper–gold system, that of iron–aluminum is somewhat more complex in that the two constituents have different crystal structures, iron being cubic body-centered at ordinary temperatures and aluminum cubic close-packed.

crystal structure analysis n 結晶構造解析

▶ The multitude of structures of inorganic compounds, along with the diversity of types of chemical bonds, makes crystal structure analysis an indispensable research tool.

cube n 立方体

▶ MgO is isometric in cubes, octahedrons, and grains with perfect 100 and imperfect 111 cleavages.

cubic adj **1** 立方体の

▶ The unit cell of a cubic crystal is a cube which when reproduced in parallel orientation would fill space to produce a cubic lattice.

2 三次の

▶ Mathematically, a cubic equation may have three real roots or one real and two complex roots.

cubic close-packed adj 立方最密充填の ☞ hexagonal close-packed

▶ The ions of one type might be packed in a simple cubic, hexagonal close-packed, or cubic close-packed arrangement, perhaps expanded to move the like-charged ions farther apart from one another.

cubic expansion coefficient n 体膨張係数 ☞ linear expansion coefficient

▶ The cubic expansion coefficient, the fractional increase in volume per degree at constant pressure, is approximately three times the linear coefficient in isotropic solids.

cubic system n 立方晶系 syn *isometric system*

▶ In the cubic system, the (100), (111), and (110) surfaces are close-packed, but all other surfaces are stepped.

cuboidal adj 立方体様の

▶ The Fe_3S_4 clusters retains the cuboidal structure with a vacant corner.

cue n 合図 syn *signal*

▶ The cues can be environmental signals such as temperature changes, nutrients in external environment, hormones, or other complex chemical signals exchanged by cells.

culminate vi 頂点に達する syn *peak*

▶ Their ambitious research program culminated in the synthesis of hemibrevetoxin B and the larger and more complex brevetoxins in the mid-1990s.

culmination n 極致 syn *peak, top*

▶ With respect to micelles, we note the early work of Harkins and collaborators and Philipoff and the culmination of this effort in the classic work of Luzzatti where the basic concepts of micelle formation were clarified.

cultivate vt 養う syn *advance, encourage, develope*

▶ If we fail to cultivate a fundamental interest in chemical processes, then it follows that our own resource base will be further depleted as funds are dissipated into peripheral disciplines.

culture n **1** 教養 syn *education*

▶ The justification for teaching chemistry in general education lies in the contribution it makes to general culture.

2 培養

▶ Penicillin is commercially produced by isolation from cultures of mutant strains of the mold *Penicillium chrysogenum.*

cumbersome adj 煩わしい syn *inconvenient, unwieldy*

▶ Many alcohols that are cumbersome to name by the IUPAC system may have structures that are more easily visualized when named by the carbinol system.

cumulate vt 一つにまとめる syn *accumulate, pile*

▷ Substances with cumulated carbonyl and carbon–carbon double bonds are called ketenes and have interesting and unusual properties.

cumulated double bond n 累積二重結合

▶ 1,2-Alkadienes and similar substance are said to have cumulated double bonds.

cumulative adj 累積した

▶ The cumulative effect of the lanthanide contraction makes the radii of the members of the third transition series very similar to those of the corresponding members of the second transition series.

cumulatively adv 累積して

▶ Lead exerts its poisonous action cumulatively, attacking, in particular, thiol sites on enzymes.

curd n 凝乳状物

▶ One gram of insoluble curd fibers of sodium arachidate could hold 82 g of ethanol without syneresis.

curdy adj 凝乳状の

▶ Silver nitrate produces no precipitation in a solution of free tartaric acid, but in solution of a neutral tartrate, a white, curdy precipitate is formed immediately.

cure vt **1** 治す syn *remedy*

▶ The proteins that cure transformed cells are called tumor suppressor proteins.

2 加硫する

▶ Natural rubber is cured by hot-molding or in open stream, at temperatures from 120 to 150 ℃ after addition of 3% sulfur, 1% organic accelerator, 3% zinc oxide, and fillers or reinforcing agents.

curiosity n 好奇心 syn *interest*

▶ Silicon polymers with alternating silicon and oxygen atoms in the molecular backbone were well known, but molecular species with catenated silicon chains were little more than laboratory curiosities.

curious adj **1** 知りたがる syn *fascinated*

▶ We have been curious about the chemical basis for iodine solvates.

2 珍しい syn *peculiar, strange, unusual*

▶ When aqueous ammonia or ammonium salts are electrolyzed at a low temperature using a mercury cathode, there is formed a curious type of amalgam, which, on warming puffs up and evolves gaseous nitrogen and hydrogen in the proportions corresponding to the radical NH_4.

curiously adv 珍しく syn *notably*

▶ The coordination number of the lanthanides in their solid compounds is curiously variable.

curl vi 丸くなる syn *loop, roll*

▶ The thinner kaolinite lamina begin to curl from the crystal edges.

curly bracket n 中括弧 ☞ bracket, parenthesis, square bracket

▶ When one plane is used to represent the whole form, it is enclosed in curly brackets, thus: {*hkl*}.

current n **1** 電流 ☞ electric current

▶ Any current passing from an electrode to an electrolyte causes a chemical reaction.

2 流れ syn *flow, stream*

▶ Substances may be dried at a high temperature in a current of air.

adj **3** 現在の syn *present, recent*

▶ One of more important current problems in polymer morphology is the elucidation of the precise nature of the structures that are found in practice.

current density n 電流密度

▶ Attempts to deposit Ni and Ni alloys from chloroaluminate melts found that the addition of benzene (20 wt%) produced a bright deposit at low current densities, but a black, finely divided powder was produced at higher current densities.

current efficiency n 電流効率

▶ The deposition of metals with large negative reduction potentials such as Cr and Zn is hindered by poor current efficiencies and hydrogen embrittlement of the substrate.

currently adv 今のところ

▶ Currently the most interesting approach is to prepare singly Ru-modified metalloproteins, in which the Ru serves as a second redox center attached to the surface of the protein.

cursory adj 粗略な syn *superficial*

▶ A cursory inspection of experimental data for isometric species indicates an inverse correlation between ΔH_f° and ΔH_v°.

curtail vt 削減する syn *diminish, reduce, short-*

en

▶ Only when the supply of oxygen is curtailed, when the supply itself is cut off, or when metabolic processes are so rapid that they reduce the local oxygen density in the body fluids, does the myoglobin unload its oxygen.

curvature n 湾曲 syn *bend*

▶ Difluorotoluene substitution in curved DNA sequences has been observed to affect bending only at certain locations, which suggests that local electrostatic interactions may be a chief cause of this curvature.

curve n **1** 曲線

▶ If one plots the rate of enzyme-catalyzed hydrolysis against the pH of the solution, one gets a bell-shaped curve.

vi **2** 曲がる syn *bend*

▶ In an electron microscope, the electrons are focused by a series of magnetic fields, i.e., the electron stream curves as it passes through the magnetic field.

vt **3** 曲げる syn *bend, twist*

▶ Certain sequences of DNA are inherently curved, leading to bending of the helix overall.

curved adj 湾曲した syn *bent*

▶ The curved arrows are not considered to have real mechanistic significance but are used primarily to show which atoms can be regarded as nucleophilic and which electrophilic.

cusped adj 先の尖った

▶ If the crystal has one very pronounced cleavage plane, the polar plot of the cleavage strength will have a sharply cusped, dimple-shaped minimum for the direction corresponding to the normal to the preferred cleavage plane.

custom n 習慣 syn *convention, practice*

▶ Formerly it was the almost universal custom to absorb carbon monoxide by means of a hydrochloric acid solution of CuCl.

customarily adv 習慣的に syn *generally, ordinarily, usually*

▶ Diazomethane is an intensely yellow gas, bp −23 ℃, which is customarily prepared and used in diethyl ether or dichloromethane solution.

customary adj 習慣的な syn *familiar, routine*

▶ Although the hydronium ion, H_3O^+, is present in water and confers acidic properties upon aqueous solutions, it is customary to use the symbol H^+ in place of H_3O^+ and to speak of hydrogen ion in place of hydronium.

customize vt 改造する syn *adapt, modify*

▷ Fluorescence spectroscopy with customized polarity-sensitive probes could be a powerful technique in estimating the local polarity of biomolecular cavities.

cut n **1** 切り口

▶ A schematic one-dimensional cut through the potential energy surface for reactions such as the recombination of methyl radicals is shown in Figure 1.

vt **2** 切断する syn *cut off*

▷ Uniaxial crystals have an ordinary refractive index which is the same when measured with a prism cut from the crystal in any orientation, since it is independent of the direction.

3 交差する syn *interupt*

▶ The solubility curve which is obtained when anhydrous sodium sulfate is present as the solid phase cuts the curve representing the solubility of the decahydrate at a temperature of about 32.4 ℃.

4 cut across 広く及ぶ

▶ Analytical chemistry cuts across several more or less classical or conventional areas of chemistry.

5 cut off 分離する syn *isolate*

▶ A black glass containing 9% nickel oxide cuts off almost all the visible radiation, while transmitting the 3650 Å wavelength freely.

cutting n 切り取ったもの

▶ Sodium residues and cuttings should be transferred at once to a bottle provided for the purpose and filled with kerosene.

cutting edge n 最前線

▶ The results of these cutting edge experiments are giving unprecedented information about the detailed dynamics of dissociation.

cuvette n **1** キュベット ☞ cell

▶ Either cylindrical or rectangular cuvettes may be used for circular dichroism.
2 キュベット
▶ The cuvette was placed on the Pt−Ir heating element of the hot stage with minimum exposure to the atmosphere.

cybotactic adj サイボタクチック
▶ When strong S_C-like fuctuations occur in the nematic, these are often known as cybotactic groups.

cybotaxis n サイボタキシス
▶ Neighboring molecules in a liquid tend to take up parallel or mutually oriented positions, particularly if the molecules are elongated or otherwise anisotropic; this phenomenon was given the name cybotaxis.

cyclability n 循環性
▶ Excellent cyclability can be obtained by replacing a small amount of the manganese on the octahedral sites by lithium, $Li_{1+\delta}Mn_{2-\delta}O_4$ ($\delta = 0.1$).

cycle n **1** サイクル syn *sequence, series, succession* ☞ thermodynamic cycle
▶ A composite reaction mechanism sometimes includes a cycle of reactions, such that certain reaction intermediates consumed in one step are regenerated in another.
vi **2** 循環する syn *recycle, return, rotate*
▶ After a short induction period, a steady state is established, and reaction cycles through the chain steps at a constant rate.

cyclic adj **1** 環式の
▶ Cyclic ketones almost always react more rapidly in addition processes than open-chain analogs.
2 巡回の syn *sequential*
▶ Functions, such as the internal energy, for which the cyclic integral is zero, are known as state functions and will be denoted by capital letters.

cyclic dimer n 環状二量体
▶ Carboxylic acids in the solid and liquid states mostly exist as cyclic dimers.

cyclic ether n 環状エーテル
▶ The angle strain in three- and four-membered cyclic ethers facilitates ring opening, whereas strainless five- and six-membered cyclic ethers are not attacked by Grignard reagents.

cyclic process n 循環過程
▶ If a system undergoes any series of changes such that it returns to its initial state, these changes constitute a cyclic process or cycle.

cyclic structure n 環状構造
▶ There are two isomeric forms of D-(+)-glucose because the cyclic structure has one more chiral center than Fischer's original open-chain structure.

cyclic voltammetry n サイクリックボルタンメトリー ☞ reduction potential
▶ Cyclic voltammetry can offer vital insight into the specific conditions necessary for clean demetallation.

cyclic voltammogram n サイクリックボルタンモグラム
▶ The appearance of a unique reversible process in the cyclic voltammogram indicates that the multiple ferrocenyl centers behave as independent, electronically isolated units.

cycling n サイクリング
▶ The average manganese valence above $Mn^{3.5+}$ seems to be the key factor responsible for improved capacity retention on cycling.

cyclization n 環化 ☞ ring closure
▶ Benzo-2-pyrone is commonly called coumarin and is obtained by the Perkin reaction between salicylaldehyde and acetic anhydride in the presence of triethylamine, as the result of spontaneous cyclization of the intermediate *o*-hydroxycinnamic acid.

cyclize vt 環化する
▶ On dehydration with sulfuric acid, *o*-benzoylbenzoic acid is cyclized to anthraquinone.

cycloaddition n 付加環化
▶ The light-catalyzed intermolecular cycloadditions between two identical alkenes are photodimerizations.

cyclodehydration n 脱水環化
▶ A simple, general procedure for the synthesis of five-membered oxygen-containing hetero-

cyclic systems involves the cyclodehydration of 1,4-diketones by acidic reagents, such as sulfuric acid, zinc chloride, etc.

cyclodextrin n シクロデキストリン

▶ Into its lipophilic interior a cyclodextrin typically takes as a guest, not an ion but a nonpolar organic molecule or the nonpolar end of an organic molecule.

cylinder n **1** 円柱

▶ Large cylinders of single-crystal silicon and other single crystals are produced by the Czochralski process.

2 ボンベ

▶ Gaseous carbon dioxide containing air but suitable for some purposes may be obtained from a cylinder of the liquid material.

cylindrical adj 円柱状の

▶ At the transition, *n*-alkane molecules become effectively cylindrical as a result of executing rotation-like motions about their long axes.

cylindrically adv 円柱状に

▶ The σ bond joining the carbon atoms is cylindrically symmetrical about a line joining the two carbon nuclei.

cytochrome n シトクロム

▶ The prosthetic group in all cytochromes is heme, which undergoes reversible Fe^{II}–Fe^{III} oxidation.

cytoplasm n 細胞質

▶ The intracellular contents between the nucleus and the plasma membrane is called cytoplasm.

cytosol n シトゾル

▶ The cytoplasm of eukaryotes consists of cytosol, a nutrient-rich, gel-like fluid, and many organelles.

D

damage n **1** 物の損傷 syn *destruction, harm, impairment*

▶ Damage to plants from pollution ranges from adverse effects on foliage to destruction of fine root systems.

vt **2** 損傷を与える syn *harm, impair*

▶ Any material or substance which in normal use can be damaging to the health and well-being of man is said to be hazardous.

damp adj **1** 湿気のある syn *moist* ☞ dry

▶ The induction period could be eliminated by the addition of a small drop of slightly damp solvent.

vt **2** 減じる syn *lessen, reduce, suppress*

▶ The parameter r_D determines how strongly the potential is damped from its pure Coulomb value.

damping n 制動

▶ A new mechanism in heavy-ion-induced reactions termed deeply inelastic collisions is characterized by damping of large amounts of collective nuclear energy through interactions with nucleonic modes of excitation.

danger n 危険 syn *hazard, risk*

▶ Esters of tertiary alcohols are prone to acid-catalyzed elimination, and the alkoxide method precludes danger from this reaction.

dangerous adj 危険な syn *hazardous, precarious*

▶ Potassium perchlorate and other perchlorates are oxidizing agents, somewhat less vigorous and less dangerous than the chlorates.

dangerously adv 危険なほどに syn *badly, severely*

▶ Diazomethane is highly toxic, dangerously explosive and cannot be stored without decomposition.

dangling bond n ダングリングボンド syn *unsaturated bond*

▶ The chemical inertness of alkali halide nanocrystals, based on their composition of closed-shell atomic ions, contrasts sharply to the reactivity of metal or semiconductor clusters which have unsaturated, or dangling, bonds at their surfaces.

dark n **1** 暗所 syn *blackness, darkness*

▶ Phosphorus seems a lively element, which glows in the dark and is essential for biological energy transfer and for heredity.

adj **2** 暗い syn *dim, shadowy*

▶ When viewed down an optic axis, anisotropic crystals appear to be isotropic, i.e., they are dark between crossed polarizers.

darken vi 薄黒くなる syn *blacken*

▶ Hypophosphoric acid is not reduced by zinc and dilute sulfuric acid and gives with silver nitrate a white precipitate which does not darken in the light.

dash n ダッシュ

▶ Dashes are inserted when the reagent causes no oxidation.

dashed line n 破線 syn *discontinuous line, broken line* なお，鎖線は dashed and dotted line ☞ dotted line, solid line

▶ There is a gradual, overall decrease in radius as the d shell is filled, as shown by the dashed line that passes through Ca^{2+}, Mn^{2+}, and Zn^{2+}.

dashpot n ダッシュポット

▶ The elastic elements are represented by springs, and the viscous by dashpots, the motion of which is retarded by a viscous liquid.

data n データ datum の複数形であるが，data を動詞の単数形と組み合わせる扱いも広く行われる．しかし，科学論文では複数形として扱うことが多い． syn *facts, record*

▶ The spectroscopic properties of this compound were consistent with the data available in the literature.

database n データベース

▶ The standard procedure is to use classical scattering theory with refractive index data as input data available from different databases.

date n **1** to date 現在まで syn *until now*

▶ To date, all applications of liquid crystal technology involve organic molecules, and in most instances these have evolved through the synthetic and physical studies of man-made materials.

vi **2 date back** さかのぼる syn *go back*

▶ The electroplating industry, which dates back well over 100 years, is based solely on aqueous solutions because of the high solubility of electrolytes and metal salts resulting in highly conducting solutions.

vt **3** 年代を定める

▶ Isotopic studies on meteorites have been extensively employed to date specific events in their history.

dated adj 時代遅れの syn *obsolete, old*

▶ Some of the examples used are now somewhat dated.

dating n 年代測定

▶ ESR dating fills a substantial gap between radiocarbon and potassium-argon dating schemes.

dative bond n 供与結合 syn *coordinate bond, dative bonding*

▶ An arrow can be used for the N→B bond in $H_3N \rightarrow BF_3$ to identify it as a dative bond, but usually the distinction is not made.

daughter n 派生したもの ☞ mother

▶ Studies of phase transitions which could be carried out as single-crystal—single-crystal transformations, as, for example, that of *p*-dichlorobenzene, have shown that the daughter grows in the mother crystal with an orientation which is random.

daylight n 昼光 syn *sunlight*

▶ Atmospheric chemistry that occurs in daylight derives reactions to favor products different from those formed at night.

dazzling adj まぶしい syn *bright, brilliant, splendid*

▶ Magnesium burns with dazzling brilliance in air to give MgO and Mg_3N_2.

d-d transition n d-d 遷移

▶ The spectrum of the enzyme containing a metal ion that shows d-d transitions provides information on the microsymmetry of the site of the metal.

deactivate vt 失活する ☞ activate

▶ If we are interested in studying the reactions of ground-state radicals, it is essential that the radical species are vibrationally deactivated.

deactivating group n 不活性化基

▶ A deactivating group directs meta simply because it deactivates the ortho and para positions even more than it does the meta.

deactivation n 失活 ☞ activation

▶ Electrophilic substitution of pyridine is hard to achieve, because of deactivation of the ring by the heteroatom.

deal n **1 a good deal** 相当な分量 syn *a lot*

▶ It is typical of van't Hoff's approach to problems that the derivation is by no means rigorous and that a good deal of intuition is involved.

2 a great deal 相当な分量 syn *a good deal, a lot*

▶ Sulfur diimidazole has been used a great deal for the construction of symmetrical trisulfides.

vi **3 deal with** 取り扱う syn *practice, take, treat*

▷ In dealing with rates, we compare the stability of the reactants with the stability of the transition state.

dealloy vt 脱合金化する

▷ Dealloying aluminum from an aluminum-nickel alloy leads to formation of the porous metal known as Raney nickel.

deaminate vt 脱アミノ化する

▶ The branched-chain amino acids are deaminated in muscle by a specific aminotransferase.

deamination n 脱アミノ化

▶ Deamination of cytosine yields uracil, which is thought to be why DNA has evolved to contain thymine, to enable the correction of this chemically inevitable corruption of the coded message.

dearomatization n 脱芳香化

▶ The design of a second generation of dearomatization agents, utilizing Re(I), W(0), and Mo(0), required a precise matching of electronic properties of these metals to the Os(II) system.

debate n **1** 論争 syn *argument, controversy, dis-*

cussion, dispute

▶ The mechanism by which this low oxidation state is stabilized for nickel, palladium, and platinum has been the subject of some debate.

vt **2 討論する** syn *argue, consider, discuss*

▶ The fact that scientists were willing to debate our hypothesis and spend time in the laboratory studying it was a promising sign that the topic was of substantial interest.

decade n 十年間

▶ It was recognized some decades ago that under certain conditions an atom of hydrogen is attracted by rather strong forces to two atoms, instead of only one, so that it may considered to be acting as a bond between them.

decant vt デカンテーションする

▶ The product can be decanted from the nickel catalyst.

decantation n デカンテーション

▶ The precipitated powder was washed by decantation, filtered, and dried in vacuum over P_2O_5.

decarboxylation n 脱炭酸

▶ Decarboxylation of *cis*-α-phenylcinnamic acid is effected by refluxing the acid in quinoline in the presence of a trace of copper chromite catalyst; both the basic properties and boiling point of quinoline make it a particularly favorable solvent.

decay n **1 減衰** syn *decline, fading*

▶ The decay of phosphorescence can be conveniently studied by means of spin resonance.

2 崩壊 syn *disintegration*

▶ The γ-rays that are used in Mössbauer spectroscopy are produced by decay of radioactive elements such as $^{57}_{29}Fe$ or $^{119}_{50}Sn$.

vi **3 衰退する** syn *deteriorate*

▶ Copper(II) chloride thus diffuses and allows the cell to decay through direct reaction of the electrode materials.

decay constant n 壊変定数 syn *disintegration constant*

▶ The decay of N_0 nuclides to give N nuclides after time t is given by $N = N_0\exp(-\gamma t)$, where γ is called the decay constant.

decay law n 減衰法則

▶ A different decay law will prevail if the electrons do not pass directly from trap to emitting state by a radiationless transition but are first excited to the conduction band.

decay time n 減衰時間

▶ By short we mean short compared with the overall decay time,

decelerate vt 減速する ☞ accelerate

▶ Upon application of a resonance excitation signal to the encap electrodes, the ions are accelerated and decelerated causing their trajectories of motion to enlarge.

decide vi **1 decide on 決定する** syn *fix, settle*

▶ When two phases are in equilibrium, which means that the pressure is not at our disposal if we have decided on a temperature.

vt **2 決定する** syn *conclude, determine*

▷ One of the principle factors deciding the magnitude of the velocity of advance is the density of kinks in the step.

decidedly adv 明らかに syn *definitely*

▶ The name ferrite is decidedly misleading as these is no acid anion in their structures.

decimal place n 小数点以下の桁数

▶ The index of refraction of solids is so nearly constant that the temperature range employed causes changes, in general, only in the fourth or fifth decimal place.

decimal point n 小数点

▶ The position of the decimal point is determined solely by the unit in which the quantity is expressed.

decinormal adj 十分の一規定の

▶ A large portion of the studies on the potentials of galvanic cells has been made using calomel electrodes containing normal or decinormal potassium chloride.

decision n 決定 syn *conclusion, determination*

▶ The prediction of the rate of a reaction proceeding by a stepwise mechanism involves a decision as to which step is rate-determining.

decisive adj 決定的な syn *critical, crucial, ultimate*

▶ The experimental verification of Einstein's theory of Brownian motion played a decisive part in establishing molecular reality.

decline n 1 衰微 syn *decrease, diminution, fall-off*

▶ Organolead antikock agents in gasolines are on the decline.

vi 2 低下する syn *decrease, diminish, fall*

▶ As the temperature is lowered, the atoms vibrate less and the resistance declines smoothly, until, if the material can become a superconductor it reaches a so-called critical temperature, T_c.

decolorization n 脱色 syn *decolorizing*

▶ Rapid decolorization of a bromine solution is characteristic of compounds containing the carbon–carbon double bond.

decolorize vt 脱色する

▶ Aromatic hydrocarbons with saturated side chains are distinguished from alkenes by their failure to decolorize bromine in carbon tetrachloride and by their failure to decolorize cold, dilute, neutral permanganate solutions.

decolorizing agent n 脱色剤

▶ Decolorizing agents refer to materials of highly absorbent character used to remove undesirable color and also bleaches involving a chemical reaction for removing color.

decompose vi 1 分解する syn *break down*

▶ Evaporation of solutions of lactose or maltose gives white solid products which are distinguishable because the temperature ranges at which they decompose differ by about 100°.

vt 2 分解する syn *degrade*

▶ Concentrated sulfuric acid decomposes all chlorates, setting free greenish-yellow chlorine dioxide gas, which colors the sulfuric acid yellow and explodes violently on warming.

decomposition n 分解 syn *breakdown, degradation*

▶ In simple decomposition, one substance breaks down into two simpler substances, e.g., water yields hydrogen and oxygen.

decomposition point n 分解点 syn *decomposition temperature*

▶ Acetylsalicylic acid decomposes when heated and does not possess a true, clearly defined mp. Decomposition points varying from 128 to 135 ℃ have been recorded.

decomposition temperature n 分解温度 syn *decomposition point*

▶ Decomposition temperatures will obviously be sensitive to impurities, such as oxygen, and will be influenced strongly by the presence of inhibitors, antioxidants, etc.

decontamination n 除染 syn *purification, refining*

▶ Separation of boron from borides is difficult, although partial decontamination can be achieved by flotation.

deconvolute vt デコンボリューションする

▶ If the generation of the transient species is not rapid, we have to deconvolute two different processes; the production of the transient from the photolysis pulse and consumption via reaction.

decorate vt 修飾する syn *furnish*

▶ The virus capsids were independently decorated with two different types of ligand to generate two populations of labeled virions.

decoration n 修飾 syn *attachment*

▶ Ferrocenecarboxylate was used for facile, covalent decoration of amine groups on the mosaic virus particle surface using standard coupling procedures.

decouple vt デカップルする ☞ couple

▷ The proton-decoupled spectrum tells us how many different carbons there are, and the proton-coupled spectrum tells us how many protons are attached to each of these carbons.

decoupling n デカップリング

▶ Proton dipolar broadening in ^{13}C spectra of solids could be removed by a high power version of the decoupling technique used in solution NMR spectroscopy.

decrease n 1 減少 syn *diminution, reduction*

▶ If the temperature of a solution is cycled up

and down while crystals are growing, it is found that there is a corresponding increase and decrease in thickness of the various layers.

vi **2 減少する** syn *decline, diminish*

▶ At elevated temperatures, the viscosity decreases by more than one order of magnitude.

vt **3 減少させる** syn *diminish, lessen, lower, reduce*

part ▶ Decreased viscosity can be obtained by decreasing the surface tension of the liquid, i.e., by increasing the free volume or by decreasing the ionic radius.

dedicated adj **専用の** syn *committed, devoted*

▶ Modern research infrared spectrometers incorporate dedicated computer capability for programmed operation, data collection, and data manipulation.

deduce vt **推測する** syn *conclude, derive, presume, surmise*

▶ Although the detailed structure cannot be deduced from the chemical formula, one can at least get an approximate idea of the type of silicate anion.

deduction n **推論** syn *conclusion, inference, reasoning*

▶ In spite of this limitation, we can still make deductions regarding the general nature of the forces involved.

deductive adj **演繹的な** syn *logical*

▶ Our knowledge of the structures of most organic molecules has been gained through a combination of experimental data and deductive reasoning.

deductively adv **演繹的に**

▶ Laws and relationships worked out deductively will be truer than merely empirical relationships induced from observations.

deem vt **…と思う** syn *consider, guess, take*

▶ The configuration at the allylic stereogenic center was deemed to be unimportant.

deep adj **1 深い**

▶ The bond angle provides no direct information about the degree of flexibility of the molecule; i.e., whether the potential minimum is shallow or deep.

2 深遠な syn *profound*

▶ For these purposes, a deeper understanding of the spectra is necessary.

3 濃い syn *intense*

▶ The absorption spectrum of the trapped electron gives rise to a deep coloration of the crystal, a beautiful deep purple in KCl.

deepen vi **濃くなる** syn *expand, intensify*

▶ The salts of $[IrX_6]^{2-}$ (X = F, Cl, Br) are comparatively stable, and their color deepens from red, through reddish-black, to bluish-black with increasing atomic weight of the halogen.

deepening n **濃くなること**

▶ The increasing dissociation of N_2O_4 can readily be followed by a deepening of the brown color due to NO_2 and an increase in paramagnetism.

deeply adv **深く** syn *intensely, profoundly, strongly*

▶ Zinc is deeply involved with proteins that bind to DNA and so help to control the expression of coded information.

deexcitation n **脱励起**

▶ Deexcitation of a molecule may occur either by emission of a photon or by collision.

default n **デフォルト** syn *defect, failure*

▶ Many metal-organic frameworks are known based on triangular, square, tetrahedral, and octahedral secondary building units, which lead to default structural nets.

defect n **1 欠陥** syn *inadequacy, shortcoming*

▶ The failure of the classical thermodynamic analyses to explain some complex transitions may be due to defects in the thermodynamic model.

2 格子欠陥 syn *imperfection, irregularity*

▶ Generally speaking, the slower the crystal growth process, the larger and more defect-free the crystals.

3 欠損 syn *error, mistake*

▶ The tiny defect in the hemoglobin molecules that results in sickle-cell anemia has been traced to a single gene, a segment of the DNA chain.

defective adj 欠陥のある syn *flawed, imperfect*

▶ The ledge defect might itself be defective: it might have kinks.

defect structure n 欠陥構造

▶ Inverse and disordered spinels are said to have a defect structure because all crystallographically identical sites within the unit cell are not occupied by the same cation.

defense n 防御 syn *protection, shield*

▶ Chemotherapy could complement but not substitute for the body's own defenses.

deficiency n 不足 syn *deficit, lack*

▶ Chromium(II) sulfide, CrS, is not an exactly stoichiometric compound, usually having a slight deficiency of chromium.

deficient adj 不足した syn *defective, insufficient, lacking*

▶ $K_2[Pt(CN)_4]\cdot 3H_2O$ is a very stable colorless solid, but by appropriate partial oxidation it is possible to obtain bronze-colored, cation-deficient $K_{1.75}[Pt(CN)_4]\cdot 1.5H_2O$.

deficit n 不足 syn *deficiency, loss*

▶ The electrode that is charged positively, i.e., that has a deficit of electrons, by the applied potential is called the anode.

definable adj 定義可能な

▶ Measurements of electronic transport on Li_xNiO_2 exhibit a small but definable activation energy for electronic transport associated with small polaron hopping in the mixed-valence $Ni^{4+/3+}$ state.

define vt **1** 定義する syn *describe, explain, interpret*

▶ The most widely accepted present-day definition is that due to Brønsted, who defined an acid as a species having a tendency to lose a proton.

2 限定する syn *fix, prescribe, specify*

▷ The procedure of incorporating the concentrations of pure solids and liquids into the equilibrium constants is equivalent to defining the activities of pure solids and liquids as unity.

definite adj 一定の syn *particular, specific*

▶ Solids can usually be arranged in a definite order according to their sorbing power.

definitely adv 確かに syn *certainly, surely*

▶ On the basis of these results, it was concluded that here there is definitely higher polymerization.

definition n **1** 定義 syn *description, explanation*

▶ The terms acid and base have been defined in a number of ways, each definition corresponding to a particular way of looking at the properties of acidity and basicity.

2 by definition 定義により syn *necessarily*

▶ Since a radical is by definition a molecule containing an odd electron, any reaction with a normal molecule, in which the electrons are all paired, must give another radical.

definitive adj 決定的な syn *conclusive, decisive*

▶ Although metals generally possess such properties as malleability, ductility, and high tensile strength, the definitive characteristic is their ability to conduct electricity that increases as the temperature decreases.

deflagrate vi 爆燃する

▶ Ammonium perchlorate deflagrates with a yellow flame when heated to 200 ℃.

deflagration n 爆燃

▶ Deflagration is usually initiated by contact with a flame or spark but may be caused by impact or friction.

deflect vt 偏向させる syn *change, deviate*

▶ Particles having permanent magnetic moments were deflected by the electromagnet.

deflection n 偏向 syn *shift*

▶ Knowing the velocity of cathode rays, Thomson could then determine the ratio of charge to mass of the particles by measuring the deflection of the beam produced by either the electrostatic field alone or the magnetic field alone.

deflocculant n 解こう剤 ☞ flocculant

▶ The remaining third of the sodium silicate produced is consumed in miscellaneous applications such as adhesives, cements, deflocculants, foundry applications, and vehicles in inorganic paints.

defluorinated adj 脱フッ素化した

▶ Defluorinated phosphate rock is the source of phosphate used in animal feeds and feed concentrations.

defoaming agent n 泡消し剤

▶ Defoaming agents are used to reduce foaming due to proteins, gases, or nitrogenous materials which may interfere with processing.

deform vi **1** 変形する

▶ As the polymers impinge on one another, they deform readily and form a dense gel structure.

vt **2** 変形させる syn *distort, misshape*

▶ The ions are deformed by the crystal field into similar polyhedra with linear dimensions d_1 and d_2.

deformable adj 変形可能な

▶ The hydride ion is readily deformable, and this constitutes a characteristic feature of its structural chemistry.

deformability n 変形性

▶ The polarizability of the anion will be related to its softness, that is, to the deformability of its electron cloud.

deformation n 変形 syn *distortion*

▶ The decrease in the energy cost of preparing the D_{2h} fragment is consistent with the view that the softer the metal atom is, the easier it is for any deformation of the structure to occur.

deformational adj 変形の

▶ Creep tests are generally carried out by applying a constant weight to a sample and measuring the deformational response as a function of time.

deformed adj 変形した syn *distorted*

▶ Softer atoms such as bromine are so deformed by the highly concentrated charge on the Al^{3+} ion that covalent overlap can occur.

defunct adj 消滅した syn *past*

▶ The now defunct Leblanc process for obtaining NaOH from NaCl signaled the beginning of large-scale chemical manufacture.

defy vt 受け付けない syn *baffle, resist, thwart*

▶ The products which have previously been examined were insoluble polymers which defied characterization

degas vt 脱気する

▶ All liquid samples were saturated with nitrogen and degassed on a vacuum line.

degassing n 脱気

▶ Solubility of oxygen in many organic solvents is about 10 times that in water and necessitates careful degassing if these solvents are to be used in the preparation and handling of oxygen-sensitive compounds.

degeneracy n 縮退

▶ The degeneracy of the t_{2g} levels may be removed by the Jahn–Teller effect.

degenerate adj **1** 縮退した

▶ The fifth electron in cyclopentadienyl radical can be placed in one of two degenerate and partially occupied bonding orbitals.

vi **2** 縮退する

▶ The methanol triplet converts to a 136-gauss doublet identified previously as CHO, and the ethanol quintet degenerates to a single line.

degradation n **1** 劣化 syn *deterioration*

▶ There are others with equally good or better optical properties which do suffer surface degradation.

2 分解 syn *decomposition*

▶ The structure of vitamin K_1 has been established by degradation and by synthesis.

degradative adj 分解の

▶ Fortunately, cholesterol is readily available, so that it was possible to use rather elaborate degradative sequences which would have been quite out of the question with some of the more difficultly obtainable natural products.

degrade vt 分解する syn *break, decompose*

▶ When microbes degrade petroleum products, polycyclic aromatic hydrocarbons are formed and eventually oxidized into carbon chains that can be broken down into carbon dioxide and water.

degrease vt 油を除去する

▷ A number of methods have been studied but by far the best adhesion is obtained by degreasing in a chlorinated solvent, followed by an aqueous pickle, rinse, dry, and then anodic etch in the ionic liquid prior to deposition.

degree n **1** 温度，角度の度

▶ Columns A and B give the maximum positive and negative deviations, to one hundred thousandth of a degree, of the individual observations from the mean of the thirty.

2 程度 syn *level, order, rank* ☞ kind

▶ Several approaches have been tried to overcome this problem with varying degrees of success.

3 a greater or lesser degree 多少とも

▶ The same is true of all monosubstituted cyclohexane derivatives to a greater or lesser degree.

4 to a degree 多少 syn *somewhat*

▶ Elemental sulfur did work to a degree, giving the desired product in 10% yield.

5 次数

▶ The degree of a point in a graph is the number of lines incident on that point.

degree of crystallinity n 結晶化度 syn *crystallinity*

▶ Differently prepared, or differently treated, samples of polymeric materials show different degrees of crystallinity.

degree of dissociation n 解離度

▶ Arrhenius calculated the degree of dissociation of 40 salts from Raoult's data and obtained values from 0.03 to 0.92.

degree of freedom n 自由度

▶ The number of degrees of freedom of a system is the number of independently variable factors taken from temperature, pressure, and composition of phases.

degree of polymerization n 重合度

▶ The pore network of polyurethane foam reflects the degree of polymerization and the microstructure of the polymer, both of which are controlled by the synthetic conditions employed.

degree of rotational freedom n 回転自由度

▶ If molecules are spheroids there are three degrees of rotational freedom, but the rotation about the axis of symmetry does not contribute to the specific heat, so that only two degrees of rotational freedom are to be counted.

dehydrate vi **1** 水分を失う

▶ Boric acid dehydrates progressively, forming a clear melt.

vt **2** 脱水する ☞ dewater

▶ Evaporation of aqueous solutions of LiOH under normal conditions produces the monohydrate, and this can be readily dehydrated by heating in an inert atmosphere or under reduced pressure.

dehydrating agent n 脱水剤 ☞ desiccant

▶ Because of its affinity for water, P_4O_{10} is widely used as a dehydrating agent.

dehydration n 脱水 ☞ desiccation

▶ In dehydration, the equilibria are shifted in favor of the alkene chiefly by the removal of the alkene from the reaction mixture by distillation.

dehydrogenate vt 脱水素する

▶ Lathosterol and 7-dehydrocholestrol are dehydrogenated by bromine to dienes and trienes that likewise remain in the mother liquor and are eliminated along with colored by-products.

dehydrogenation n 脱水素

▶ The characteristic biological function of the dehydrogenation enzymes involves conversion of the nicotinamide portion of NAD into dihydro structure.

dehydrohalogenation n 脱ハロゲン化水素

▶ The salts of aliphatic amines, which correspond to acids with K_a values of about 10^{-33}, are powerfully basic reagents and are particularly effective in causing dehydrohalogenation by the E2 mechanism.

deintercalate vt 挿入物を除く

▶ $LiMn_2O_4$ is a remarkable material. Li may be extracted yielding, when fully deintercated, a polymorph of MnO_2 which cannot be prepared by other means.

deintercalation n 挿入物の除去

▶ The structural changes that accompany

lithium deintercalation from $LiCoO_2$ will be considered in more detail.

deionization n 脱イオン化 ☞ demineralization, desalination

▶ A thoroughly mixed bed of strong-acid and strong-base resins is used for deionization of water.

delay n 遅延 syn *postponement*

▶ The delay in the emission of light is due to the time spent by the excited electron in an excited state within the luminescence center.

deleterious adj 有害な syn *detrimental, harmful*

▶ The abrupt change in unit-cell volume which accompanies the cubic-to-tetragonal phase transition in the spinel has deleterious effect on the cycling performance.

deletion n 欠落

▶ Deletions are chromosomal changes in which one or more genes or chromosomal segments are lost.

deliberate adj 計画的な syn *intentional, planned*

▶ It is useful to study an individual reaction to see how its rate is affected by deliberate changes in experimental conditions.

deliberately adv 故意に syn *intentionally, purposely*

▶ A substance having a distinctive, sometimes unpleasant odor is deliberately added to essentially odorless materials to provide warning of their presence.

delicate adj **1** 鋭敏な syn *critical, sensitive*

▶ Benzidine is oxidized by trivalent gold salts to form a blue dyestuff. This test is very delicate and has been obtained with 0.02 γ of gold.

2 微妙な syn *fine, subtle*

▶ The dipolar nature of the nitro group may result in the interactions in this region of the molecule being a delicate balance between attractive and repulsive forces.

delicately adv 微妙に

▶ These various explanations are not mutually exclusive but simply tend to emphasize differing aspects of an extremely complicated and delicately balanced situation.

delimit vt 限界を定める syn *define, specify*

▶ The nucleus is delimited from the cytoplasm by a membranous envelope called the nuclear membrane, which actually consists of two membranes forming a flattened sac.

delineate vt 定める syn *define, describe, specify*

▶ To delineate the role of laterally placed methyl groups upon mesomorphic character, a series of esters of the general formula **1** was prepared.

deliquescence n 潮解

▶ Sometimes the sodium salts have unsatisfactory properties, such as deliquescence, which make the potassium salts preferable for some uses, even though more expensive.

deliquescent adj 潮解性の

▶ Zinc chloride is extremely deliquescent, and it must, therefore, be introduced into the flask as rapidly as possible.

deliver vt 引き渡す syn *give, release*

▶ MnO_2 is more stable than CoO_2 in liquid solvents offering the possibility of obtaining the full capacity to deliver lithium compared with only half a lithium in the case of $LiCoO_2$.

delocalization n 非局在化

▶ Several complexes have been found to be anisotropic semiconductors with greater electron delocalization along the line of the metal–metal chain.

delocalization energy n 非局在化エネルギー

▶ The delocalization energy was defined as the calculated additional bonding energy which results from delocalization of electrons originally constrained to isolated double bonds.

delocalize vt 非局在化する

▶ If the excited electron in butadiene is delocalized over four carbon atoms rather than two, the excited state will be relatively stable.

delocalized electron n 非局在化電子

▶ The idea behind magnetic criteria is that delocalized π electrons produce a relatively high diamagnetic ring current under an external magnetic field.

delocalized orbital n 非局在化軌道

▶ The overlapping atomic orbitals in extended crystalline solids lead to delocalized orbitals that encompass the entire solid.

demagnetization n 消磁 ☞ magnetization

▶ During the processes of magnetization and demagnetization in an alternating magnetic field, energy is dissipated, usually as heat.

demagnetize vt 消磁する ☞ magnetize

▶ These small regions are all magnetized in the same direction, and it is very difficult to demagnetize them or change their magnetic orientation.

demand n **1** 需要 syn *desire, market*

▶ In spite of the great diversity of known colored structures, there still remains a demand for dyes with improved color properties.

vt **2** 必要とする syn *require*

▷ In the hydrolysis of alkyl halides, the theoretical rate sequences demanded solely by the operation of the inductive effect, viz. Me＞Et＞*i*-Pr＞*t*-Bu for the bimolecular (S_N2) and Me＜Et＜*i*-Pr＜*t*-Bu for the unimolecular (S_N1) mechanism, are experimentally verified.

demanding adj **1** 過酷な syn *difficult, hard*

▶ The thermal and chemical stability must be optimized so that the device can survive demanding operating conditions.

2 要求する syn *insistent*

▶ It is not surprising that the rate-determining step here is the one that involves the breaking of a bond, an energy-demanding process.

demarcation n 境界 syn *border line, boundary*

▶ The more readily ignition occurs, the more flammable the material; less easily ignited materials are said to be combustible, but the line of demarcation is often indefinite and depends on the state of subdivision of the material as well as on its chemical nature.

demineralization n 脱塩 ☞ deionization, desalination

▶ Sometimes strong-base exchangers are used after a cation exchanger for complete water demineralization.

demolish vt くつがえす syn *dispose, overturn*

▶ The classical concept of trajectory is demolished if we accept that the wave function is the basic feature of mechanics.

demonstrable adj 明白な syn *certain, evident, obvious*

▶ The methyl protons of the CH_3O groups in $(CH_3O)_2CHCH_3$ are too far from the others to give demonstrable spin-spin splitting.

demonstrably adv 明らかに syn *clearly*

▶ The thermodynamic temperature scale is demonstrably identical with the ideal gas temperature scale over the entire region.

demonstrate vt 示す syn *establish, exhibit, manifest, show*

▶ These transformations of cyclic sulfamidates demonstrate a new and straightforward approach to functionalized and enantiomerically pure lactams.

demonstration n 実証 syn *display, evidence, proof*

▶ The recent discovery of tetracyanoethylene and the demonstration of its versatile chemistry prompted us to explore the chemistry of other structures containing the highly electronegative cyano group.

demote vt 降格させる syn *downgrade*

▶ The contribution of electrostriction to apparent molal volumes is usually so large that it demotes the internal pressure effect of a solvent to a minor role.

demountable adj 取り外し可能な

▶ If path lengths of 0.1 mm or less are required, it is probably best to use demountable cuvettes where the sample is dropped onto a quartz disk or plate that is etched to a predefined depth and then another quartz disk carefully placed on top.

denaturant n 変性剤

▶ Proteins that can be readily removed from the membrane by using high salt or low denaturant concentrations are the extrinsic or peripheral membrane proteins.

denaturation n 変性

▶ Under the influence of heat, organic solvents, salts, etc., protein molecules undergo more or less irreversible changes, called denaturation, in which both the conformation of the chains and the degree of hydration are altered.

denature vt 変性する

▶ The extreme ease with which many proteins are denatured makes their study difficult.

denatured protein n 変性タンパク質

▶ Chymotrypsin attacks denatured proteins more rapidly than the natural proteins with their precisely folded chains.

dendrimer n デンドリマー

▶ Fullerene has been used as a spherical building block for the construction of globular dendrimers.

dendritic adj 樹枝状の

▶ With more rapid crystallization dendritic forms develop, first as saw-toothed outlines to (110) faces and ultimately as feathery aggregates of many individual platelets.

denitrification n 脱窒

▶ The loss of the labeled N may be by denitrification, leaching, possibly, ammonia volatilization, or a combination of these processes.

denominator n 分母 ☞ numerator

▶ The logarithm of a fraction is the logarithm of the numerator minus the logarithm of the denominator.

denote vt 意味する syn *indicate, mean*

▶ The prefix neo- is used to denote three methyl groups at the end of a chain.

dense adj 高密度の syn *compact*

▶ There is no fundamental chemical difference between the dense and open hybrid structures, though their properties and applications are often quite different.

densely adv 高密度に

▶ Inspection of a densely packed body-centered structure reveals that the coordination number in this structure is eight, with the nearest neighbors surrounding any given atom sitting at the corners of a cube.

densification n 高密度化

▶ We attribute the enhanced densification to an absence of aggregates of primary particles in the sols.

densify vt 密度を高める

▶ Many conventional hydrous oxides used as starting materials do not densify so well or so rapidly as the sol-gel materials.

density n 密度

▶ The density of polyethylene and other thermoplastic polymers is affected by the shape and spacing of the molecular chains.

density functional theory n 密度汎関数法

▶ Density functional theory is a form of quantum mechanics that uses the one-electron density function instead of the more usual wave function to describe a chemical system.

density-gradient centrifugation n 密度勾配遠心分離

▶ Density-gradient centrifugation can be used to separate the different organelles.

density of states n 状態密度

▶ The number of states in the interval $E + dE$ is known as the density of states.

dental adj 歯科の

▶ The constraints upon dental materials are so severe that lengthy procedures to ensure the safety and efficiency of new products must be undertaken.

dentate adj 配座の ☞ bidentate

▶ The larger the dentate character of a ligand the less labile its complexes.

dentin, dentine n 象牙質

▶ The bond region between the composite restorative material and dentin may be 50–300-μm thick, with gradients of intimately mixed components that have originated in either the substrate or the bonding agent, extending throughout the bond.

denude vt 表面侵食する syn *strip*

▷ As polyethylene crystals thicken on annealing, they become broken up into small islands of thickened crystal divided by denuded valleys or voids.

deodorant n 脱臭剤

▶ The cosmetic industry supplies a wide

variety of deodorants and antiperspirants, chiefly based on neutralization.

deoxygenate vt 酸素を除く

▷ Surface modification must be performed in an inert atmosphere with completely deoxygenated and dried reagents so as to prevent the formation of silicon oxides during the monomer formation.

deoxygenation n 脱酸素反応

▶ Having developed the deoxygenation rearrangement method for the preparation of 2-azabenzonorbornanes, our attention turned to developing an asymmetric access.

deoxyribonucleic acid, DNA n デオキシリボ核酸,

▶ DNA is the genetic material in most organisms. Its name is derived from the presence of the pentose sugar, deoxyribose, which is one of its major constituents.

deoxyribonucleoprotein n デオキシリボ核酸タンパク質

▶ Chemical analysis of the chromosomes has revealed them to be composed of giant molecules of deoxyribonucleoproteins, which are deoxyribonucleic acids bonded to proteins.

depart vi **1** 離れる syn *leave*

▶ The electron pair of the C−Br bond to be broken departs with the leaving bromide ion.

2 depart from それる syn *deviate*

▶ The curve departs from a straight line only by the entropy of mixing, which is definitely known, the same in liquid and solid.

departure n 離脱 syn *deviation*

▶ In practice, it is necessary to modify the well-known gas formula to allow for departures from ideal gas behavior.

depend vi 依存する syn *be dependent*

▷ Depending on the system and the conditions, there are a number of different relationships between the overall activation energy and the values for individual steps.

dependence n 依存 syn *reliance*

▶ Diamagnetism exhibits no dependence on temperature other than that due to the volume expansion of the material with temperature.

dependent adj 依存している

▶ Because the splitting of the triplet depends on the orientation of the magnetic field, the transitions are also orientation-dependent.

dephosphorylation n 脱リン酸

▶ Dephosphorylation of calcium-binding proteins and microtubule-stabilizing proteins may promote microtubule depolymerization and the movement of complementary chromatids to opposite poles.

depict vt **1** 叙述する syn *describe, express*

▶ We have depicted pinacol rearrangement as a two-step process with an actual carbonium ion as intermediate.

2 描く syn *draw, portray, sketch*

▷ Diagrams depicting structure in three dimensions are difficult to interpret, as well as being laborious to draw.

deplete vt 減少させる syn *consume, expend*

▷ When the metal nucleates at the electrode surface, the layer close to the electrode surface becomes depleted of metal complex and rich in counter anions.

depletion n 減衰 syn *loss*

▶ The reactions of nanocrystals of NaCl and NaBr with nitroxide pollutants provide one source of chlorine and bromine atoms for catalytic ozone depletion.

depolarization n 偏光解消 ☞ polarization

▶ If the incident beam is polarized with the vibration horizontal, the light scattered transversely may show a partial depolarization in which the horizontal component is more intense than the vertical component.

depolarization degree n 偏光解消度

▶ The depolarization degree of rotational Raman scattering is 3/4 for polarized or 6/7 for unpolarized incident light.

depolymerization n 解重合

▶ Highly polymeric $(NPCl_2)_\infty$ can be made by heating *cyclo*-$(NPCl_2)_3$ to 150–300°, though heating to 350° induces depolymerization.

depolymerize vt 解重合する

▶ Some methacrylates are efficiently polymerized and depolymerized in the solid state by UV irradiation.

deposit n **1** 析出物 syn *precipitate, sediment*

▶ If too strong a current was used, very often the deposits were spongy and did not adhere well to the electrodes.

vt **2** 析出させる syn *leave, set*

▶ Mixing alcoholic solutions of hydroquinone and quinone gives a brown-red solution, which then deposits a crystalline green-black 1:1 complex known as quinhydrone.

deposition n **1** 析出 syn *deposit*

▶ It is important to maintain the surface pressure within the monolayer at a constant value during deposition so that the density of molecules is stable and a uniform LB layer is produced.

2 供述 syn *statement*

▶ The guidelines for the deposition of crystallographic data must be followed.

depot n 貯蔵所 syn *storehouse*

▶ Like the calcium biominerals, biological iron oxides are used to strengthen soft tissues and to expand storage depots.

depress vt 降下させる syn *lower, reduce*

▶ If two substances are not identical, the second substance is a foreign substance with respect to the first and will, therefore, depress its melting point.

depressant adj 抑制作用のある syn *sedative*

▶ Ethanol is classified as a depressant drug.

depression n 降下 syn *decline*

▶ The discrepancy is made worse if we assume, as is probable for this anion on the evidence of its mobility and dielectric constant depression, a layer of partially immobilized water molecules around it.

depressurization n 減圧

▶ Because the restrictor is essentially open to atmospheric pressure on its free end, it allows for depressurization of the supercritical fluid with a concomitant release of the dissolved solute.

deprive vt **deprive…of** …から…を奪う syn *remove, withdraw*

▶ Trivial causes such as sintering were easily dismissed, and it was concluded that a strong metal-support interaction had somehow deprived these metals of one of their most characteristic properties.

deprotonate vt 脱プロトンする

▷ When pH = pK_a, equal quantities of the protonated and deprotonated species are present.

deprotonation n 脱プロトン化

▶ The most common entry to bis(carbene)s is through deprotonation of their bis(azolium) precursors.

depth n **1** 深さ syn *extent, measure*

▶ The quenching increases with the absorption coefficient of the crystal for the incident light, i.e., with a reduction in the penetration depth.

2 depth of focus 焦点の合う範囲

▶ Because of the depth of focus of scanning electron microscopy instruments, the resulting pictures have a definite three-dimensional quality.

3 in depth (a) 全く syn *profoundly*

▶ Liquids held between two solid surfaces or placed in capillaries behave as if the surface zone was rigidified in depth.

(b) 詳細に syn *in detail, throughghly*

▶ Although investigations of reactions believed to occur between molecular crystals and gases have been reported for many years, there have been few studies of such reactions in depth.

4 色の濃さ syn *brightness, intensity*

▶ If $TiCl_3$ is run into an acid solution of hydrogen peroxide, the latter is colored first yellow, then a deep orange, and as soon as the maximum depth of color is produced, it begins to fade upon the further addition of $TiCl_3$ until finally the solution becomes colorless.

depurination n 脱プリン

▶ While DNA is remarkably robust for a biopolymer, it is susceptible to degradation at low pH, in particular to depurination.

deregulation n 規制撤廃

▶ Mutations and deregulation of protein

kinases play causal roles in many human diseases.

derivation n 誘導 syn *origin, source*

▶ The derivation of the one-component phase rule was based on the argument that the chemical potentials of the phases in equilibrium must be equal.

derivative n 誘導体

▶ Thiols are derivatives of hydrogen sulfide in the same way that alcohols are derivatives of water.

derivatization n 誘導体化

▶ Chemical derivatization of the internal cysteines can be achieved using small thiol-reactive compounds.

derive vi **1 derive from** …に起因する syn *arise from, originate in, stem from*

▶ The rigidity of the porphyrin ring derives from the delocalization of the π electrons in the pyrrole rings.

vt **2 derive from** 誘導する syn *acquire, gain, get, obtain*

▷ Grignard reagents derived from aryl bromides are readily prepared and may be converted into organocadmium compounds by treatment with cadmium chloride.

desalination n 淡水化 ☞ deionization, demineralization

▶ Flash distillation appears to be the most effective method so far developed for seawater desalination.

desalt vt 塩分を除く

▶ Seawater can be desalted by electrodialysis on a large scale by placing it in the center chamber of a three-component container having two semipermeable membranes and a positive electrode in one end chamber and a negative electrode in the other.

descend vi 下る syn *fall*

part ▶ In descending order of effectiveness, the Hofmeister series for the cations is: lithium > sodium > potassium > ammonium > magnesium.

descent n 降下 syn *drop*

▶ During the descent, potential energy is converted back into kinetic energy, until we reach the level of the products.

describe vt 記述する syn *characterize, report, represent*

▶ The mirror plane describes symmetry with respect to a plane, the rotation axis describes symmetry with respect to a line, and the center of symmetry describes symmetry with respect to a point.

description n 描写 syn *characterization, explanation,*

▶ The VSEPR domain model gives a very approximate description of the electron distribution in a molecule that is based on the role of the Pauli exclusion principle in determining the electron density distribution.

descriptive adj 記述的な syn *detailed*

▶ Together with the closely related subjects of redox and coordination chemistry, acids and bases form the basis of descriptive inorganic chemistry.

deserve vt 値する syn *merit, warrant*

▶ The number of formally nonbonding electron pairs on each skeletal metal atom deserves comment.

deshield vt デシールドする

▶ Carbonyl carbon is both sp^2-hybridized and attached to electronegative oxygen and as a result is powerfully deshielded.

deshielding n デシールディング ☞ shielding

▶ The shifts in the position of NMR absorptions, arising from shielding and deshielding by electrons, are called chemical shifts. Shielding shifts the absorption upfield, and deshielding shifts the absorption downfield.

desiccant n 乾燥剤 syn *drier, drying agent*

▶ P_4O_{10} is used as a desiccant when a more powerful drying agent than $CaCl_2$ or other low-cost agent is needed.

desiccation n 乾燥 syn *drying*

▶ Anhydrous calcium chloride has a high water-absorption capacity but is not very rapid in its action; ample time must, therefore, be given for desiccation.

desiccator n デシケーター

▶ Many substances are hygroscopic and absorb moisture from the air, which can be removed by heating the substance or by simply allowing it to stand in a desiccator over calcium chloride, provided the substance itself undergoes no change by this treatment.

design n **1** 設計 syn *arrangement, intention, plan*

▶ The synthesis of these materials has evolved very rapidly over the last few years, such that several key elements of material design and mechanism are now well understood.

vt **2** 計画する syn *intend, plan*

▶ The present research was designed as an investigation of the bonding between alkali-metal atoms and networks of fused aromatic nuclei smaller than in graphite.

designate vt **1** 表す syn *specify*

▶ We cannot designate the strength of acid on an absolute basis.

2 名づける syn *call, name, term*

▶ A number of metals have several allotropic forms which are often designated by Greek letters, e.g., α-, γ-, and Δ-iron.

3 示す syn *denote, indicate*

▷ The iron crystal face designated (111) is about 430 times more active than the closest-packed (110) crystal face and 13 times more active than the simpler (100) face.

designation n **1** 名称 syn *name, term*

▶ The fluorine–fluorine bond has the same general shape as the hydrogen–hydrogen bond, being cylindrically symmetrical about a line joining the nuclei; it is given the designation of σ bond.

2 指定 syn *assignment*

▶ Compounds related to D-glyceraldehyde are given the designation D, and compounds related to L-glyeraldehyde are given the designation L.

designed adj 計画的な syn *intentional*

▶ The intensity of the laser light is very much greater than the fluorescence and in a poorly designed experiment would swamp the real signal.

desirability n 望ましさ

▶ Recent studies by Stewart on CH_4, CD_4, and CF_4 are especially pertinent and indicate the desirability of heat capacity, NMR, and neutron diffraction measurements at high pressures.

desirable adj 望ましい syn *worthwhile*

▶ To test the influence of inductive effects, it is desirable to prepare a series of compounds in which only inductive factors are varied and not steric or other factors.

desire n **1** 希望 syn *aim, intent*

▶ The desire to prepare charge-transfer systems with increased dimensionality of structural, and hence electric, properties has been prevalent in the work of many groups.

vt **2** 望む syn *demand, request, require*

▶ If very accurate *d* specings or intensities are desired, slower scanning speeds are used.

desired adj 目的の syn *desirable*

▶ All of these methods produced the desired product, and any of these methods were satisfactory for preparing complexes which were insoluble in water.

desorb vt 脱着する

▶ Pollutant will sorb onto and desorb from activated carbon particles until the rate of sorption equals the rate of desorption, and the system is in equilibrium.

desorption n 脱離 ☞ adsorption

▶ Desorption may be accomplished by heating, by reduction of pressure, by the presence of another more strongly adsorbed substance, or by a combination of these means.

desperately adv 不可避的に syn *extremely, intensely*

▶ Desperately needed are comparably sensitive experimental techniques that permit study of chemical behavior at the interface between a solid surface and a second medium at significant density.

despite prep にもかかわらず syn *even, irrespective, regardless, spite*

▶ The electron affinity is traditionally given a positive sign despite the negative enthalpy change in the reaction $X(g) + e^- \rightarrow X^-(g)$.

destabilization n 不安定化 ☞ stabilization

▶ The outcome of a destabilization in preparing the MX_2 monomer for its geometry in the dimer plus the stabilization upon dimerization is the dimerization energy.

destabilize vt 不安定にする ☞ stabilize

▶ The nitro group has an electron-withdrawing inductive effect; this tends to intensify the positive charge of the ring, destabilizes the carbocation, and thus causes a slower reaction.

destine vt 予定する syn *design, intend*

▶ Most of the HCl produced from salt or direct combination of the elements is destined for sale.

destined adj …するように定められて syn *designed, intended*

▶ Proteins destined to be retained on the cell surface have large stretches of hydrophobic amino acids that lodge in the hydrophobic lipid membrane, first in the vesicle and then in the plasma membrane.

destroy vt 破壊する syn *break down, demolish*

▶ Oxidation of naphthalene by oxygen in the presence of vanadium(V) oxide destroys one ring and yields phthalic anhydride.

destruction n 破壊 syn *breakdown*

▶ The catalytically induced destruction of hydrocarbon chains to smaller, more volatile fragment is called cracking.

destructive adj 破壊的な syn *damaging, devastating*

▶ Activated carbon is obtained by the destructive distillation of wood, nut shells, animal bones, or other carbonaceous material.

desulfonate vt 脱スルフォン化する☞ sulfonate

▶ To desulfonate we use dilute acid and often pass superheated steam through the reaction mixture; high concentration of water and removal of the relatively volatile hydrocarbon by steam distillation shift the equilibrium toward hydrocarbon.

desulfonation n 脱スルフォン化 ☞ sulfonation

▶ According to the principle of microscopic reversibility, the mechanism of desulfonation must be the exact reverse of the mechanism of sulfonation.

detachable adj 取り外せる syn *separable*

▶ The detachable proton is seen to be located on the oxygen next to the aluminum atom in the zeolite framework.

detachment n 脱離 syn *separation*

▶ The vast majority of experimental and theoretical studies of isolated multiply charged anions to date have focused on the ions' stability with respect to electron detachment.

detail n **1** 詳細 syn *particular*

▶ There is still some uncertainty as to detail.

2 in detail 詳細に syn *in depth, specifically, thoroughly*

▶ The thermal decomposition of some of the oxalates of the heavy metals has been studied in detail.

3 複数形で，詳細な情報 syn *particulars*

▶ The details of the atomic structure in the region of the conduction plane have been the subject of much crystallographic work but are still not well understood.

vt **4** 詳しく述べる syn *delineate, specify*

▶ It is more common to work at constant pressure, and temperature-composition diagrams are then required to detail the behavior of the liquid-vapor equilibrium.

detailed adj 詳細な syn *comprehensive, full, inclusive*

▶ In contrast to the study of ionic crystals, very few detailed investigations have been carried out on the thermal decomposition of organic substances in the solid state.

detain vt 引き止める syn *hold, keep*

▶ The species H_2^+ and H_3^+ need not detain us long for little chemistry is involved.

detect vt 検出する syn *discover, find, uncover*

▶ In general, small amounts of impurities are often difficult to detect by its melting or boiling point.

detectability n 検出能力

▶ The concentration of an impurity present at a level below analytical detectability may be computed from zone refining data if the effective distribution coefficient of the system is known.

detectable adj 検出可能な syn *discernible, observable, visible*

▶ Reaction may be proceeding so slowly that no detectable change will occur over a long period of time.

detection n 検出 syn *identification, recognition*

▶ Addition of bromine is extremely useful for detection of the carbon–carbon double bond.

detector n 検出器

▶ The resolution of a conventional IR spectrometer is usually limited by the decreasing amounts of energy reaching the detector as the slit width is reduced.

detergency n 洗浄力

▶ Foaming has no direct relation to detergency.

detergent n 洗剤

▶ Detergents are molecules in which a non-polar tail is joined to a polar head group.

deteriorate vi 劣化する syn *decline*

▶ Since the solution deteriorates on standing, the next step should be started at once.

deterioration n 劣化 syn *degradation*

▶ The intense irradiation of a fluorescent organic crystal or plastic solution with ionizing radiation causes a deterioration in its scintillation efficiency.

determinable adj 決定できる

▶ Frequently the determinable and meaningful combining ratio in the vapor is unrelated to any structural feature of the solid.

determinant n **1** 行列式 ☞ matrix

▶ From the properties of determinants, the wave function may be shown to obey the Pauli principle because the determinant vanished when two columns are identical and changes sign when two columns are interchanged.

2 決定因子 syn *factor*

▶ Some lymphocyte clones will have receptors with a high affinity for the determinant, binding strongly to it by the sum of multiple noncovalent forces.

determination n **1** 定量 syn *estimation*

▶ For the determination of very small amounts of titanium, it is advisable to use the colorimetric method proposed by Weller.

2 測定 syn *measurement*

▶ A number of proteins form well-defined crystals which, in principle at least, makes possible complete determination of their structures by X-ray diffraction.

determine vt **1** 定める syn *affect, influence, regulate*

▶ Ligands can, through their electronic or steric effect, their lipophilicity, or their chirality, help to determine the course of reaction.

2 決定する syn *define, fix*

▷ Direct nitration is obviously unsatisfactory when the orientation determined by substituent groups does not lead to the desired isomer.

detonate vi 爆発する syn *explode*

▶ Though kinetically stable in air, tetrasulfur tetranitride is endothermic ($\Delta H_f^\circ = +428$ kJ mol^{-1}) and may detonate when heated to its melting point or on shock.

detonation n 爆ごう syn *explosion*

▶ Detonation is a characteristic of high explosives, which vary considerably in their sensitivity to shock, nitroglycerin being one of the most dangerous in this respect.

detonator n 雷管

▶ $Pb(N_3)_2$ in particular is extensively used in detonators because of its reliability, especially in damp conditions.

detoxification n 解毒

▶ Nearly all foreign compounds undergo modification in vivo, which makes them less toxic and/or more water-soluble so that their metabolites are excreted in the urine. This process is called detoxification.

detoxification reaction n 解毒反応

▶ Molecular oxygen is used in a variety of oxidative detoxification reactions.

detoxify vt 解毒する syn *recover*

▶ Biological systems have to be detoxified from harmful products such as H_2O_2 and O_2^- which would cause ruin to their organic chemistry.

detrimental adj 有害な syn *deleterious, harmful, unfavorable*

▶ Reducing the catalyst loading to 5 mol% slowed the reaction but had no detrimental effect on the enantioselectivity.

deuterated adj 重水素化された

▶ Deuterated solvents were degassed using three freeze-pump-thaw cycles and were vacuum distilled from potassium or sodium.

deuteration n 重水素化

▶ A band which moves to about $1/\sqrt{2}$ of its originary frequency on deuteration is very probably associated with the O−H group.

develop vi **1** 発達する syn *expand, grow*

▶ It is reasonable that in thin crystals where less mechanical strain is produced by a given fraction of reaction the cracking does not develop.

vt **2** 生じさせる syn *appear, arise, emerge*

▶ Under the action of an applied mechanical stress, piezoelectric crystals polarize and develop electric charges on opposite crystal faces.

developer n 現像液

▶ Developers are used in photography to convert a latent image to a visible one by chemical reduction of a silver compound to metallic silver more rapidly in the portions exposed to light than in those not exposed.

development n **1** 展開 syn *evolution, expansion, extension*

▶ Other aspects of calixarene chemistry are in the early stages of development.

2 現像

▶ The latent subimage is able to initiate development directly when particularly vigorous developers are used.

deviate vi それる syn *deflect, turn*

▶ Small domains of other phases begin to precipitate as the composition deviates from 1 : 1.

deviation n **1** 逸脱 syn *variation*

▶ Deviations from ideal behavior are dealt with by using activities instead of concentrations.

2 偏差 syn *discrepancy*

▶ Except for one deviation of 0.20 kcal/mole, no other deviation is more than 0.08 kcal/mole, and the standard deviation for three degrees of freedom is 0.13 kcal/mole.

device n **1** 方策 syn *maneuver, strategy*

▶ The use of a large excess of one reactant is a common device of the organic chemist who wishes to limit reaction to only one of a number of reactive sites in the molecule of that reactant.

2 装置 syn *apparatus, instrument*

▶ The main use of semiconductors is in solid-state devices such as transistors, silicon chips, photocells, etc.

devious adj 遠回りの syn *indirect, zigzag*

▶ It was necessary to develop a number of more devious ways for determining Gibbs-energy changes from measurable thermal quantities.

devise vt 考案する syn *conceive, design, invent*

▷ The postulated mechanisms are essentially theories devised to explain the facts obtained by experiments.

devising n 工夫すること syn *invention*

▶ The devising of simple preparations for molecules isotopically substituted in different ways provides scope for considerable ingenuity.

devitrification n 失透

▶ For many glasses, crystallization or devitrification may occur at some temperature above the glass transition temperature and below the melting point.

devitrify vi 失透する

▶ The tendency to devitrify results from the unstable nature of glasses.

devoid adj **devoid of** …が欠けている syn *lacking, without*

▶ Antibonding orbital is usually devoid of

electrons..

devote vt …に当てる syn *commit*

▶ Considerable effort has been devoted to obtaining solids with high ionic conductivities to function as electrolytes in high-energy-density, solid-state batteries.

dewater vt 脱水する ☞ dehydrate

▶ Sludges and organic wastes are dewatered centrifugally.

dextran n デキストラン

▶ A commonly used gel matrix is dextran, a polymer of glucose, that is formed into very small beads for this purpose.

dextrorotatory adj 右旋性 ☞ levorotatory

▶ If the rotation of the plane of polarized light, and hence our rotation of the lens, is to the right, the substance is dextrorotatory; if the rotation is to the left, the substance is levorotatory.

diagnosis n 究明 syn *interpretation*

▶ One use for IR spectra is in structural diagnosis.

diagnostic n **1** 兆候

▶ This remarkable electric property was treated as a diagnostic for a new state of matter, the superconducting state.

adj **2** 診断の

▶ A circular dichroism spectrum of a protein solution is often used as a rapid diagnostic tool to assess the folded state of the protein.

diagonal n **1** 対角線 ☞ body diagonal, face diagonal

▶ Either the atom or bond values have the further advantage of allowing the extraction of all three diagonal elements of the susceptibility tensor rather than just one anisotropy.

adj **2** 斜めの syn *oblique*

▶ Boron shows many similarities to its neighbor, carbon, and its diagonal relative, silicon.

diagonally adv 斜めに

▶ Nitrogen and sulfur are diagonally related in the periodic table.

diagram n 図式 syn *figure, graphic, profile*

▶ A schematic free energy–temperature diagram does explain the observed behavior of the polymorphs and mesomorphs in this system.

diagrammatic adj 図式の syn *schematic*

▶ A diagrammatic derivation of this result is perhaps more revealing.

diagrammatically adv 図式で

▶ It is useful to plot diagrammatically four chief types of flow commonly observed.

dialysis n 透析

▶ The separation of small molecules from macromolecules in a solution by means of a semipermeable membrane such as parchment or collodion is called dialysis.

dialyzate n 透析物

▶ From the dialyzate was separated a thermostable substance of low molecular weight, which was called a coenzyme, since when it was added to the enzyme protein it restored the full catalytic properties.

diamagnetic adj 反磁性の

▶ In the majority of molecules and solids that we encounter, all of the electrons are paired, and the molecule or solid is said to be diamagnetic.

diamagnetic anisotropy n 反磁性異方性

▶ Determination of principal susceptibilities in the crystal is a useful tool in structural chemistry. This is particularly true for aromatic compounds, which often show striking diamagnetic anisotropy.

diamagnetic shielding n 反磁性遮へい

▶ The low-field shift is generally interpreted at least qualitatively, in terms of a decrease in diamagnetic shielding of the proton.

diamagnetic susceptibility n 反磁性磁化率

▶ Diamagnetic susceptibilities are approximately additive, provided that a constitutive correction is made for the presence of multiple bonds, ring closure, and certain other anomalous effects.

diameter n 直径 syn *width*

▶ Thermal motion of magnetite particles smaller than 100 Å in diameter is sufficient to prevent agglomeration due to magnetic interactions.

diametrically adv 直径に

▶ It is extremely improbable that the two electrons in the 1s orbital of helium will be diametrically opposite each other at all times.

diamond structure n ダイヤモンド型構造

▶ The diamond structure may be described as a ccp array of carbon atoms, with one set of tetrahedral sites occupied also by carbon atoms.

diaphragm n 隔膜

▶ The electrolytic cell contains anode and cathode compartments separated by a porous diaphragm or membrane to prevent mixing of the solutions.

diastereoisomeric adj ジアステレオ異性体の

▶ The concept of modification of zeolite supercages to create the chiral environment required to favor the formation of one of a pair of diastereoisomeric transition states has been pioneered in our recent research.

diastereoisomerization n ジアステレオ異性化

▶ Diastereoisomerization observed in solution proceeds probably with inversion on one of the three coordinate atoms.

diastereomer n ジアステレオマー syn *diastereoisomer*

▶ Using nitroethane, good stereoselectivity was obtained at the β-position, but the product was, again, a mixture of diasteromers at the γ-position.

diastereomeric adj ジアステレオマーの

▶ The most important general procedure for the resolution of a racemic modification involves its conversion into a pair of diastereomeric derivatives by reaction with an optically pure, optically active reagent, e.g., the formation of a pair of diastereomeric salts.

diastereomeric salt n ジアステレオマー塩

▶ Resolution of optically active acids through formation of diastereomeric salts requires adequate supplies of suitable optically active bases.

diastereospecificity n ジアステレオ選択性 syn *diastereoselectivity*

▶ Stereospecificity toward diastereomers is called diastereospecificity.

diastereotopic adj ジアステレオトピック

▶ Diastereotopic protons are nonequivalent, and no rotation will change this.

diatomic n 二原子

▶ Impact or other heating of smaller clusters generates the loss of a series of diatomics or small fragments.

diatomic molecule n 二原子分子

▶ For diatomic molecules, the bond energy is easily defined and frequently directly measurable.

diazo adj ジアゾの

▶ Diazo systems are approximately 10^{10} times less sensitive than silver halides and only sensitive to the UV-blue region of the spectrum.

diazo coupling n ジアゾカップリング

▶ A typical example of diazo coupling is afforded by formation of *p*-dimethylaminoazobenzene from benzenediazonium chloride and *N,N*-dimethylaniline.

diazotization n ジアゾ化

▶ Diazotization is usually carried out by mixing the amine and sodium nitrite in aqueous hydrochloric acid at 0 ℃.

diazotize vt ジアゾ化する

▷ Orange II is made by coupling diazotized sulfanilic acid with 2-naphthol in alkaline solution.

dibasic adj 二塩基性の syn *diprotic*

▶ Since phthalic acid is dibasic, treatment of 0.006 mole of it with 0.006 mole of $NaHCO_3$ gives a solution of sodium hydrogenphthalate.

dibasic acid n 二塩基酸

▶ When a dibasic acid dissolves in water, the two hydrogen atoms do not ionize to an equal extent.

dichromic adj 二色性の

▶ Both ruby and emerald are dichromic, but the effect is inconspicuous when the gems are viewed in normal light.

dichroism n 二色性 ☞ pleochromism

▶ Dichroism was first observed by Wollaston with potassium tetrachloropalladate(II), which shows red and green colors according to the direction of the light in the crystal.

dictate n **1** 指示 syn *direction, instruction, order*

▶ The orbitals are combined according to the dictates of group theory to form MOs for the cluster

vt **2** 規定する syn *determine, direct*

▶ Chemical regulation of redox and pH in localized biological environment may dictate the course of magnetite biomineralization.

die n **1** 鋳型 syn *mold*

▶ Laser beams are used in industry for cutting diamonds, which are used for wire-drawing dies.

vi **2** 死ぬ

▶ Cells that produce nonfunctional signal transduction proteins generally die.

dielectric n **1** 通常複数形で，誘電体

▶ Perpendicular to the chains the solids behave as transparent dielectrics or semiconductors.

adj **2** 絶縁体の

▶ Dielectric materials should be able to withstand high voltage without undergoing degradation and becoming electrically conducting.

dielectric breakdown n 絶縁破壊

▶ $FClO_3$ offers the highest resistance to dielectric breakdown for any gas and has been used as an insulator in high-voltage systems.

dielectric constant n 誘電率 syn *dielectric permittivity, permittivity*

▶ When the compound of interest is a liquid or solid material, measurements of the dielectric constant are generally made on solutions of the material in some inert nonpolar substance such as CCl_4 or benzene.

dielectric polarization n 誘電分極

▶ Fairbrother reported that iodine has an abnormally high dielectric polarization in the solvents dioxane, isobutylene, *p*-xylene, and benzene.

dielectric relaxation n 誘電緩和

▶ When the dielectric relaxation process is a decay of dipole orientation through molecular rotation, the observed or macroscopic relaxation time is longer than the molecular relaxation time, because it takes longer time for the material as a whole to reach equilibrium than it does for the individual molecules.

dienophile n ジエノフィル

▶ The nature of the dienophile seems to be the determining factor in 1,2- vs. 1,4-addition.

diet n 食餌 syn *food*

▶ Human diet requires proteins, carbohydrates, fats from both plants and animals, minerals from milk and meats, salt, vitamins from green vegetables and citrus fruits, and water.

dietary adj 食物の

▶ Ascorbic acid is a dietary factor which must be present in the diet of man to prevent scurvy.

differ vi **differ from** …と異なる syn *disagree, vary*

▶ The silicon halides differ markedly from those of carbon.

difference n **1** 相違 syn *discrepancy, dissimilarity, distinction*

▶ Infrared spectra result from differences in absorption by the bonds in the various functional groups.

2 差 syn *balance, remainder, rest*

▶ Differences in the field strengths at which signals are obtained for nuclei of the same kind but located in different molecular environments are called chemical shifts.

3 make a difference 重要である syn *be important*

▶ It makes a difference whether the ozone reacts with a neutral or with an acid solution.

different adj **1** 異なる syn *conflicting, dissimilar, unlike*

▶ It is well known that different molecules have different quenching efficiencies, more complex molecules being more efficient.

2 different from …と違う

▶ The spectrum of the dimer in EtOH is red-shifted with respect to the monomer, indicating that the geometry of the dimer formed in ethanol is different from that formed in water.

3 different to …と違う

▶ Spinels usually have cubic symmetry but some show tetragonal distortions in which one of the cell edges is of a different length to the other two.

differential n **1 微分**

▶ The accurate velocity is the limit of $\delta s/\delta t$ when $\delta t \to 0$; it is denoted by

$$u = \mathrm{d}s/\mathrm{d}t$$

where ds and dt are called differentials, or infinitesimals.

adj **2 微分の**

▶ The differential method has advantages, especially for work requiring precision.

3 差異のある

▶ Precise stoichiometry of borides is sometimes hard to achieve because of differential volatility or high activation energies.

differential heat n **微分熱**

▶ As the fraction of sites occupied by chemisorbed atoms or ions increases, lateral repulsion between the adsorbed species may be expected, and the differential heat will be thereby reduced.

differential heat of adsorption n **微分吸着熱**

▶ The variation of the differential heats of adsorption has not been widely applied to the study of specific surface.

differentially adv **特異的に**

▶ While differentially N,N-disubstituted amides generally prefer to exist as Z rotamers for steric reasons, it has been known for some time that *N*-alkyl acetanilides prefer to exist as the E rotamer, which places the aryl group and the amide oxygen in the trans conformation.

differential scanning calorimetry n **示差走査熱量測定**

▶ Differential scanning calorimetry determines the electric energy input rate necessary to establish zero temperature difference between a substance and a reference material against temperature as they both are subjected to a controlled temperature change.

differentiate vi **1 分化する**

▶ Under the proper culturing conditions, embryonic stem cells from humans and other species differentiate into a myriad of cells.

vt **2 区別する** syn *discriminate, distinguish, separate*

▶ Aldehydes are characterized and in particular are differentiated from ketones, through their ease of oxidation.

3 微分する

▷ If $s = at^2$, then ds/dt = $2at$ is found by differentiating s with respect to t.

differentiation n **1 識別** syn *distinction*

▶ Lacking a clear-cut qualitative distinction between these alternatives, a quantitative differentiation must be developed.

2 微分

▶ Since differentiation procedures always result in a loss of precision, ΔH° will be less precisely known than ln K_c.

3 分化

▶ In higher organisms, cells with specialized functions are derived from stem cells in a process called differentiation.

differently adv **異なって**

▶ Tertiary aromatic amines normally behave differently from aliphatic tertiary amines with nitrous acid in that they undergo C-nitrosation, pereferably in the para position.

difficult adj **困難な** syn *hard, problematic*

▶ Sulfanilic acid, being dipolar, is difficult to acetylate.

difficultly adv **難しく**

▶ The sulfides of the alkaline earths, particularly calcium sulfide, CaS, are difficultly soluble in water, but they are gradually changed by contact with water into soluble hydrosulfides.

difficulty n **1 困難** syn *problem, trouble*

▶ A practical difficulty with the emf method is that sometimes the electrodes do not operate reversibly.

2 with difficulty やっとのことで

▶ It is typical of aryl halides that they undergo nucleophilic substitution only with extreme difficulty.

diffract vt 回折する

▶ Colloidal particles, less than a wavelength of light in diameter, cannot reflect light but do diffract or scatter it.

diffraction n 回折 ☞ electron diffraction, X-ray diffraction

▶ Gold sols often have striking colors, ruby red, blue, green, and others, that are the result of diffraction of light by the gold sol particles with dimensions approaching the wavelengths of light.

diffractometry n 回折測定

▶ The other modern powder technique is diffractometry, which gives a series of peaks on a strip of chart paper.

diffuse adj **1** 広がった syn *scattered, spread*

▶ The electrons of the π bond of ethylene exist in a diffuse orbital above and below the plane of the molecule.

vi **2** 拡散する syn *disperse, scatter, spread*

▶ All of these factors will affect the rate at which metal ions diffuse to the electrode surface and the thermodynamics and kinetics of the metal reduction process.

3 普及する syn *distribute, penetrate*

▶ Polymer know-how has diffused throughout the industry, and polymer products were now firmly established in the economy.

diffuse double layer n 拡散二重層

▶ A diffuse double layer exists at interfaces, as a result of ionic distributions.

diffuseness n 散漫

▶ The diffuseness of the X-ray pattern from the atactic material indicates that in this form of the polymer there is little crystallinity.

diffuse reflectance n 拡散反射率 ☞ specular reflectance

▶ Diffuse reflectance spectra were plotted as the Kubelka-Munk function $f(R) = (1-R^2)/2R$ where R is the diffuse reflectance of the sample at a given wavelength relative to a MgO standard.

diffusible adj 拡散性の

▶ It is important to realize that the Nernst potential is due to the transfer of only an exceedingly small fraction of the diffusible ions, so that there is no detectable concentration change.

diffusion n 拡散 syn *dispersion, dissipation*

▶ Diffusion is typically many orders of magnitude slower in solids than in the liquid and gas phases; therefore, high temperatures are required for the reaction to proceed on a reasonable time scale.

diffusional adj 拡散の

▶ One envisages long-range diffusional processes involving the cations, with the oxygen ions being little disturbed.

diffusion coefficient n 拡散係数

▶ The diffusion coefficient depends on the ease with which the solute molecules can move.

diffusion control n 拡散律速

▶ If the rate we measure is almost exactly equal to the rate of diffusion, we speak of full microscopic diffusion control or of full encounter control.

diffusion-controlled reaction n 拡散律速反応

▶ In diffusion-controlled reactions, the reactive step is very rapid and the rate of reaction is controlled by the diffusive approach of the reactants.

diffusion layer n 拡散層

▶ An additional issue that has not been addressed is the structure of the diffusion layer during the electrolysis process.

diffusive adj 拡散性の syn *diffuse*

▶ The square-root-of-time dependence of the distance traveled is characteristic of diffusive motion.

diffusively adv 拡散によって

▶ In an inert solvent, the remaining iodine atoms have diffusively separated and so escaped geminate recombination.

digest vt **1** 温浸する

▶ A solution of sodium aluminate results when bauxite is digested under pressure with caustic soda.

2 消化する

▶ Starch is more water-soluble than cellulose, more easily hydrolyzed, and hence more readily digested.

digestible adj 消化できる

▶ The enzyme cellulase is widely used to make animal food more easily digestible.

digestion n **1** 温浸

▶ Digestions with hydrochloric acid, sulfuric acid, or caustic soda are all used to extract the mixtures of metal salts.

2 消化

▶ Digestion is responsible for the turnover of ca. 50 g of endogenous protein per day.

digestive adj 消化の

▶ Human digestive processes involve primarily the hydrolysis of complex carbohydrates to simple sugars, of proteins to a mixture of amino acids, and of fats to glycerol and higher fatty acids.

digestive enzyme n 消化酵素

▶ Chymotrypsin is a digestive enzyme,whose job is to promote hydrolysis of certain peptide links in proteins.

digit n 数字 syn *figure, number*

▶ The term digit denotes any one of the ten numerals including zero.

digital adj デジタル

▶ The two orientations of magnetization, ＋ and －, can be used to represent 0 and 1 in the binary digital system.

digitize vt デジタル化する

▷ The digitized data permit precise spectral subtraction so that the background spectrum can be virtually eliminated to reveal the spectral changes of interest.

dihedral angle n 二面角

▶ The angle of separation of any two specified hydrogens on the two carbon atoms is called the dihedral angle. In the staggered conformations of ethane the dihedral angle is equal to 60°, 180°, and 300°.

dilatancy n ダイラタンシー

▶ Dilatancy is a result of the indifference of the particles to each other and their consequent tendency under the influence of gravity to lie or pack rather closely together.

dilatometry n 膨張測定

▶ Dilatometry may be used to detect the density change on transformation.

dilemma n 窮地，ジレンマ syn *difficulty, trouble*

▶ The idea that energy is quantized resolves this dilemma; at ordinary temperatures the vibrations are not excited and not revealed in the specific heat capacity.

diluent n 希釈剤

▶ When the solvent is inert it is merely a diluent, and the neutralization product is obtained directly.

dilute adj **1** 希釈した

▶ In dilute enough solutions, a molecular interpretation needs, apparently, to take into account no other solute or solvent property.

vt **2** 薄める syn *weaken*

▶ The resulting clear solution was stirred at room temperature for 2 h to form a dark red solution which was then diluted by addition of saturated aqueous $NaHCO_3$ and water and extracted with dichloromethane.

dilute solution n 希薄溶液

▶ A dilute solution of ethanol can be concentrated by distillation to a constant boiling point mixture that contains 95.6 percent ethanol by weight.

dilution n 希釈

▶ Deviations from ideality disappear at infinite dilution in this Henry's law convention of standard states.

dimension n **1** 寸法 syn *size*

▶ Since the unit cells of most substances have dimensions that range from 2.5×10^{-8} to 10×10^{-8} cm, a special unit is used which is equal to 10^{-8} cm.

2 次元

▶ The cadmium chloride and cadmium iodide structures can be regarded as molecular arrangements in which each sheet constitutes a single molecule of infinite extent in two dimensions.

3 局面 syn *scope*

▶ Gas-liquid chromatography has added a new dimension for the analysis of volatile substances.

4 複数形で，規模 syn *span*

▶ In our concerted efforts to examine and elucidate the cognitive and contextual dimensions of the discipline of chemistry in an integrated way, methodological pluralism has been much in evidence.

dimensional adj **1** 寸法の

▶ The dimensional changes involved with magnetostriction are small.

2 次元の

▶ In the three-dimensional domain, it has been possible to study the atomically controlled formation of materials within templating structures consisting either of layered structures or of parallel aligned and ultrathin uniform channels.

dimensionality n 次元性

▶ A wide range of other inorganic families, especially phosphates, can form framework structures with varying dimensionalities.

dimensionally adv 寸法上

▶ Conductivities of 10 mS cm^{-1} still represent a difficulty for industrial plating applications particularly if dimensionally stable anodes are used for the electrodeposition of metals with large negative reduction potentials.

dimensionless adj 無次元の

▶ Where quantity involved is dimensionless, then clearly taking the logarithm leads to a dimensionless quantity.

dimer n 二量体

▶ The first step in this polymerization is formation of the dimer, which involves cyclopentadiene's acting as both diene and dienophile.

dimeric adj 二量体の

▶ Al_2Br_6 is a dimeric molecule which can be regarded as two $AlBr_4$ tetrahedra sharing a common edge.

dimerization n 二量化

▶ The heat of dimerization reported for heptoic acid suggests some major change in bond character from that prevailing in the lower paraffin acids.

dimerize vi 二量化する

▶ Monomeric RMgX can disproportionate to MgR_2 and MgX_2 by the Schlenk equilibrium or can dimerize to $RMgX_2MgR$.

diminish vi **1** 減少する syn *decline, decrease*

▶ The stability of hypohalite ions diminishes in the sequence $ClO^- > BrO^- > IO^-$.

vt **2** 減らす syn *decrease, lower, reduce*

▶ Sodium sulfate increases the viscosity of gelatin solutions, while chlorides and nitrates diminish it.

diminished adj 減少した syn *limited*

▶ Denaturation produces increased opalescence and has been shown to be accompanied by diminished hydration of the protein.

diminution n 減少 syn *decrease, decline*

▶ Since the act of solution of a gas in a liquid is necessarily accompanied by a diminution of volume, the effect of pressure will always be to increase the solubility of a gas in a liquid.

dimple n くぼみ syn *depression, pit*

▶ If a second close-packed layer is superimposed as tightly as possible on the first, the spheres in the second layer will rest in one or other of the two sets of dimples.

dinuclear compound n 複核化合物

▶ A large number of dinuclear compounds have been produced with a wide range of bridging groups or with no bridging groups at all for Cr^{II} and Mo^{II}.

diode n ダイオード

▶ A p-n junction can act as a diode for rectifying alternating current, the current passing more easily in one direction than the other.

dip vt 浸す syn *immerse, soak*

▶ In one arrangement, a polyacetylene film is dipped into a liquid electrolyte composed of $LiClO_4$ dissolved in propylene carbonate.

dipolar adj 双極子の

▶ One important class of molecules that has

received significant attention is dipolar molecules in which the polarization in one direction within the molecule is easier than that in the opposite direction.

dipolar ion n 双極イオン

▶ In dilute, weakly acidic or neutral solution, luminol exists largely as the dipolar ion and exhibits beautiful blue fluorescence.

dipolar structure n 双極構造

▶ Although alkylidenephosphoranes are frequently written as dipolar structures, they are better considered as hybrids of the contributing structures involving p_π-d_π bonding.

dipole n 双極子 ☞ permanent dipole

▶ The displacements of cations relative to its anionic neighbors give rise to dipoles and the high dielectric constants that are characteristic of ferroelectrics.

dipole-dipole interaction n 双極子-双極子相互作用

▶ A molecule accepts a certain amount of angle strain to relieve van der Waals or dipole-dipole interaction.

dipole moment n 双極子モーメント

▶ Since the C—O—C bond angle is not 180°, the dipole moments of the two C—O bonds do not cancel each other; consequently, ethers possess a small net dipole moment.

direct adj **1** 直接の syn *straight*

▶ Quinones of the more reactive, polycyclic, aromatic systems can usually be obtained by direct oxidation which is best carried out with chromium(VI) compounds under acidic conditions.

2 率直な syn *clear, plain, unambiguous, unmistakable*

▶ Although a considerable body of direct physical evidence was obtained for the nitronium ion, similar physical evidence for the simple halogen cations Cl^+, Br^+, and I^+ was not obtained.

vt **3** ある方向または目的物に向ける syn *aim, conduct, point*

▶ Our studies will be directed toward an understanding of macromolecule systems.

directed bond n 指向性結合

▶ Unlike the ionic bond, which is equally strong in all directions, the covalent bond is a directed bond.

direction n **1** 方向 syn *line, way*

▶ The SnS_6 octahedra in SnS are distorted along a [111] direction such that three sulfur atoms on one side of the tin are at ca. 2.64 Å, but the other three are repelled by the lone pair to a distance of ca. 3.31 Å.

2 通常複数形で，指示 syn *information, instruction*

▶ The following directions for determining magnesium will apply to a solution from which calcium has been removed as oxalate.

directional adj 方向性の

▶ Materials such as diamond and silicon carbide, which have very strong, directional, covalent bonds, can also be described as close-packed structures.

directionality n 方向性

▶ Electronic excitation of bulk crystals of alkali halides has the remarkable consequence that neutral halogen atoms are ejected from their surface, with high directionality.

directive adj 指向性の

▶ When the directive effect of one group opposes that of the other, it may be difficult to predict the major product.

directly adv 直接に

▶ Aryl halides are most commonly prepared by replacement of the nitrogen of a diazonium salt; this ultimately comes from a nitro group which was itself introduced directly into the ring.

dirt n 埃 syn *pollution*

▶ An appearance of complete wetting does not always mean that the dirt has been removed.

dirty adj 汚れた

▶ One pass may have the effect of reducing the impurity content only slightly, because a really dirty liquid phase cannot accumulate much more dirt.

disaccharide n 二糖

▶ To fully establish the structure of a disaccharide, we must know the identity of the component monosaccharides; the type of ring junction, furanose or pyranose, in each monosaccharide, as it exists in the disaccharide; the positions which link one monosaccharide with the other; and the anomeric configuration of this linkage.

disadvantage n 短所 syn *drawback, shortcoming*

▶ The general disadvantage in using photochemical methods in organic synthesis is the relatively low selectivity.

disaggregate vt 離解する

▷ The disaggregating action of urea was confirmed by changes in the absorption spectra of Methylene Blue in gelatin and in methylethyl cellulose.

disaggregation n 離解

▶ There must be some sort of disaggregation on passing from the viscous smectic to the more fluid nematic mesophase.

disagreeable adj 不快な syn *offensive, unpleasant*

▶ The volatile thiols, both aliphatic and aromatic, are like hydrogen sulfide in possessing characteristically disagreeable odors.

disagreement n 不一致 syn *difference, discrepancy*

▶ This disagreement among the reported thermodynamic values precludes a definite conclusion as to the dimerization driving force.

disappear vi 消滅する syn *fade, vanish*

▶ If the flame is observed through cobalt glass, the sodium flame disappears and the potassium flame appears pink.

disappearance n 消滅 syn *collapse, dissipation*

▶ The disappearance of tridymite and cristobalite with increasing pressure can be correlated with the lesser density of these phases relative to that of quartz.

disappointing adj 期待はずれの syn *inadequate, unsatisfacory*

▶ The materials prepared by high-temperature solid-state reaction at 850 ℃ gave disappointing cycling performance.

disappointingly adv 期待に反して

▶ Disappointingly, these conditions did not effect the desired rearrangement.

disassembly n 分解

▶ The abstraction of a hydride ion by NAD^+ does not immediately lead to the disassembly of the fatty acid.

disastrous adj 破滅的な syn *catastrophic, destructive*

▶ Carbon disulfide is very poisonous and can have disastrous effects on the nervous system and brain.

discard vt **1** 不要物を捨てる syn *abandon, dispose*

▶ The clear and colorless filtrate was discarded.

2 意見を捨てる syn *reject*

▶ If concerted four-center mechanisms for formation of methyl chloride and hydrogen chloride from chlorine and methane are discarded, the remaining possibilities are all stepwise mechanisms.

discern vt 認める syn *notice, perceive, realize*

▶ Certain trends in the stability of complexes can be clearly discerned.

discernible adj 識別できる syn *clear, perceptive, visible*

▶ After 3.5 hr, only diffuse powder rings were discernible.

discernibly adv 認識可能に

▶ Several of these lines are further discernibly split as the result of second-order spin–spin splitting.

discharge n **1** 放電 syn *electric discharge*

▶ Another technique is to increase the temperature suddenly, usually by the rapid discharge of a capacitor; this method is called the temperature-jump or T-jump method.

vt **2** 放電する

▶ The detailed mechanism as to how two H^+ ions are discharged and come together to form H_2 is not easily depicted.

3 脱色する syn *remove*

▶ The mixture was allowed to warm to room

temperature and stirred for at least 1 h, until all red coloration was discharged to yield a white or pale yellow suspension.

disciplinary adj 専門分野の

▶ The author delineates the various modes of production of the scientists of different disciplinary orientations.

discipline n 専門分野 syn *area, field, specialty, subject*

▶ Although physical chemistry is closely related to both inorganic and organic chemistry, it is considered a separate discipline.

disclose vt 明らかにする syn *expose, reveal, show, uncover*

▶ Since fused ionic compounds possess much higher conductivities, in general, than the corresponding solids, occurrence of melting is disclosed by a sudden rise in the conductivity-temperature curve.

discoidal adj 円盤状の syn *disk-shaped*

▶ Although $[NMe_4]^+$ and $[NH_4]^+$ are both spherical, the higher homologues have geometries which display substantial discoidal and cylindrical distortions.

discolored adj 変色した syn *dingy*

▶ Commercial *o*-phenylenediamine is usually badly discolored and give a poor result unless purified.

discomfort n 不快 discomfort と danger を一つの概念と見做して，動詞は単数形にしている． syn *inconvenience, nuisance*

▶ The discomfort and danger of smog is increased by the action of sunlight on the combustion products in the air.

discontinuity n 不連続 syn *break*

▶ The anomaly at 169 K corresponds to a pronounced discontinuity in the dielectric constant.

discontinuous adj 不連続な syn *discrete*

▶ A first-order transition has a discontinuous first derivative of the chemical potential.

discordance n 不一致 syn *disagreement, discrepancy*

▶ Although there is discordance between daily whole body energy intake and expenditure, our fuel reserves and, therefore, our weight remain remarkably constant over a year or more.

discordant adj 一致しない syn *contrary, different, opposite*

▶ These differences in sample form and possibly water content may account for the discordant results.

discotic liquid crystal n ディスコチック液晶

▶ Discotic liquid crystals are formed from molecules having more or less flat aromatic cores with usually six lateral substituents, normally alkoxy or ester groups, with at least five carbon atoms.

discount vt 低く評価する syn *disregard, ignore, overlook*

▶ At one time F_3ClO_2 was thought to exist in isomeric forms but the so-called violet form has now been discounted.

discouraging adj 思わしくない syn *disappointing*

▶ Inspection of the limited amount of experimental data is rather discouraging.

discover vt 発見する syn *find*

▶ It has been relatively recent in chemical chronology that compounds containing formal metal-carbon double and triple bonds were discovered.

discovery n 発見 syn *finding, recognition*

▶ The discovery of phosphate liquid crystals raised the possibility that other inorganic liquid crystals should exist and exhibit equally fascinating chemical and physical properties.

discredit vt 疑う syn *doubt, question*

▶ It is not the intent of the present work to discredit any of these explanations.

discrepancy n 不一致 syn *difference, disagreement, inconsistency*

▶ The source of these discrepancies lies in the use of the hard-sphere kinetic theory of gases in order to evaluate the frequency of collisions.

discrepant adj 矛盾した syn *contradictory, divergent, opposing*

▶ Vapor-pressure data for the crystalline aromatic hydrocarbons are notably discrepant and scanty.

discrete adj 個別的な syn *distinct, individual, separate*

▶ Metal clusters or polyhedra are defined as discrete molecules or ions in which the metal atoms describe a polyhedron.

discriminate vi **1 discriminate between** 二つを識別する syn *separate, tell*

▶ The biological oxidizing agent discriminates completely between two seemingly equivalent hydrogens in the ethanol molecule.

2 discriminate among 三つ以上を識別する

▶ The ability of a catalyst to discriminate among molecules on the basis of their shape is of great value in the cracking of straight-chain hydrocarbons.

vt **3** 識別する syn *differentiate, discern, distinguish*

▶ An intercalating pyrene pseudo-nucleotide was reported to discriminate DNA and RNA based on fluorescence intensity.

discrimination n 識別 syn *distinction*

▶ The enantiometric discrimination depends upon the relative sizes of the cavities and the sorbate molecules, with better selectivity being found when the size match is close.

discriminator n 識別法

▶ Medicinal chemists have found that the electrostatic potential surrounding a molecule is a much better discriminator of activity than electrostatic charge.

discuss vt 論じる syn *argue, debate*

▷ When discussing distillation both the gas and the liquid, compositions are of interest.

discussion n **1** 論議 syn *debate*

▶ Gold clusters have been found to have some rather unique structures and have caused a considerable amount of discussion.

2 under discussion 討議中 syn *in question*

▶ The apparent ionic radius will depend on the physical property under discussion and will differ for different properties.

disease n 病気 syn *illness, infection*

▶ There are a number of diseases that are caused by mutations of a single base pair in the promoters of important genes.

disentangle vt 解きほぐす syn *separate, simplify, solve*

▷ Bearing in mind this difficulty of disentangling the respective contributions of inductive and mesomeric effects of alkyl groups on the dipole moments, it is possible, with due caution, to detect probable hyperconjugative effects in the dipole moments of simple olefins, carbonyl derivatives, and nitriles.

disfavor vt 疎んじる syn *discredit*

▶ The need for very small bond angles in optimized C_{3v} conformation provokes the speculation that this dimer conformation should be disfavored for Be, Mg, and most of the Ca dimers, for the same reasons that a bent geometry is disfavored in their monomers.

disinfectant n 消毒薬

▶ *Pseudomonas aeruginosa* is a particularly problematic bacterium because it is resistant to many antibiotics and disinfectants commonly used to treat other microbes.

disintegrate vi **1** 崩壊する syn *decompose, shatter*

▶ With Lewis bases, such as $N(CH_3)_3$, the polymer $(AlH_3)_x$ disintegrates to give the tetrahedral, complete-octet compound $H_3AlN(CH_3)_3$.

vt **2** 崩壊させる syn *decompose, shatter*

part ▶ In small laminar fragments from disintegrated bulk polyethylene the molecular chains are perpendicular to the flat sheets.

disintegration n 崩壊 syn *collapse, decay*

▶ Low temperatures shift the equilibrium toward α-tin, and the lower density of this phase results in swelling and ultimately disintegration from the relatively large shifts in atomic position that are demanded by the phase transformation.

disk n **1** 円盤 disc とも綴られる.

▶ Data collected at a rotating glassy carbon disk electrode confirmed this $E_{1/2}$ value and verified that this was a one-electron reaction.

2 ディスク

▶ The disk format, wordprocessor format, file name(s), and the tile and authors of the article should be indicated on the disk.

dislocation n 転位 ☞ edge dislocation, screw dislocation

▶ As the dislocation moves and becomes associated with new half-planes, bonds on one side of the dislocation are broken and bonds on the other side are formed.

dislodge vt 除去する syn *clean*

▶ Martius Yellow is an acid, and although it is not dislodged from the adsorbent by either alkali or methanol alone, it can be eluted with methanolic alkali.

dismantle vt **1** 取り壊す syn *demolish*

▶ At the end of the experiment, the solid cell is dismantled, and the individual pellets are reweighed.

2 分解する syn *take apart*

▶ Fuel molecules are dismantled and their carbon atoms become oxidized progressively, finally producing CO_2.

disorder n 無秩序 syn *confusion*

▶ It is clear that the higher-temperature phases involve increasing degrees of disorder.

disordered adj 無秩序な syn *confused, indiscriminate*

▶ The nematic phase is the most disordered type of mesophase and is the one used in most display applications.

disordered phase n 無秩序相

▶ The electric resistance of the ordered phases Cu_3Au and CuAu is less than half that of the disordered phases of the same composition.

disordered state n 無秩序状態

▶ At the composition CuAu, the disordered state is cubic whereas the superstructure is tetragonal.

disparate adj 異なる syn *different, divergent, several*

▶ Difficulties in attaining high purities have frequently led to disparate values for some physical properties, while mechanical history has considerable effect on such properties as hardness.

disparity n 不釣合い syn *contrast, dissimilarity, diversity*

▶ Anionic conductivity is favored for salts with large cations, despite the disparity of size between cations and anions which still remains.

dispense vt **dispense with** …を不要にする syn *discard, throw away*

▶ Capillary chromatography dispenses with the granular materials normally packed into the column.

dispersal n 分散 syn *distribution, spread*

▶ With increased branching of the carbocation, dispersal of charge and the resulting stabilization is greater: dispersal over six hydrogens for the isopropyl cation and nine for the *tert*-butyl cation.

dispersant n 分散剤

▶ Before advent of synthetic detergents, $Na_4P_2O_7$ was much used as a dispersant for lime soap scum which formed in hard water.

disperse adj **1** コロイド粒子からなる

▶ In emulsions, the droplets are referred to as the disperse or dispersed phase and the second liquid as the dispersion medium.

vt **2** 分散させる syn *distribute, scatter, spread*

▶ An aliphatic amine is more basic than ammonia because the electron-releasing alkyl groups tend to disperse the positive charge of the substituted ammonium ion.

dispersed adj 分散している syn *scattered*

▶ The behavior of colloidal particles dispersed in an aqueous medium is greatly affected by the fact that the particles often carry an electric charge.

dispersed phase n 分散相

▶ Thus, colloids are systems in which the dispersed phase is in a state of subdivision in the dispersion medium such that at least one of its dimensions is in the range of one to a few thousand nanometers.

dispersible adj 分散可能な

▶ Thorium(IV) oxalate salts can be thermally decomposed to give water-dispersible oxides.

dispersion n **1** 分散 syn *spread*

▶ Solid-in-liquid colloidal dispersions can be precipitated by adding electrolytes which neutralize the electric charges on the particles.

2 物性の分散

▶ Dispersion is the term applied to the variation of a property with wavelength.

dispersion force n 分散力 syn *London force*

▶ Pictorially, the dispersion forces can be thought to arise from the synchronization of instantaneous dipoles in the interacting species.

dispersion medium n 分散媒

▶ The term disperse phase is used to refer to the particles that are present in the dispersion medium.

dispersive adj 分散的な

▶ Time-resolved EXAFS is also feasible through the use of the dispersive techniques in which the entire absorption spectrum is measured simultaneously.

displace vt 1 置換する syn *replace, supersede*

▶ Water acting as a base displaces the weaker base, chloride ion, from combination with the proton.

2 動かす syn *move, shift*

▶ A cation may be displaced from its normal lattice point to a distant interstitial position.

displacement n 1 置換 syn *substitution*

▶ Substitution of a octahedral complex by means of a displacement mechanism requires the intermediate formation of a seven-coordinated complex.

2 変位 syn *movement*

▶ The polarization by the displacement of atoms in the molecules requires about 10^{-14} to 10^{-16} sec.

display n 1 表示 syn *demonstration, presentation*

▶ For practical applications, such as optical and display devices and gas-liquid chromatography, nematic mesophases which exist at or near room temperature are desirable.

vt 2 見せる syn *exhibit, show*

▶ The lattice displays the three-dimensional periodicity of the crystalline materials and is independent of the detailed atomic content of the unit cell.

disposable adj 使い捨ての syn *discardable, unreturnable*

▶ Electrochemical biosensers have been extremely successful as disposable devices for the measurement of blood glucose for use by diabetics.

disposal n 1 処分 syn *discarding, throwing away*

▶ Complex anions are simple mixtures of two components so can be split to give the original components for recycling or disposal.

2 at a one's disposal 勝手に使えて syn *available*

▶ At this point, the total number of variables at our disposal is $P(C-1)+2$.

dispose vi 1 dispose of 処分する 受身可 syn *discard, throw away*

▶ Toxic fumes have to be disposed of.

vt 2 配列する syn *arrange, array*

▷ In the solid state, the typical E rotamer about the N−CO bond is observed, with the two anilide groups disposed anti.

disposition n 1 配置 syn *arrangement*

▶ All crystals display a regularity in the disposition of their faces.

2 性質 syn *character, nature*

▶ The extent and disposition of these deviations almost certainly result from the steric overcrowding in that region of the molecule.

disproportionate adj 1 不釣合いな syn *unequal, uneven*

▶ Often the water molecules serve to fill in the interstices and to bind together a structure which would otherwise be unstable because of disproportionate sizes of the cation and anion.

vi 2 不均化する

▶ When chlorine gas is passed into aqueous acid, a little of it disproportionates to form Cl^- and hypochlorous acid, HOCl.

disproportionation n 不均化 syn *dismutation*

▶ Another type of reaction by which a radical more complex than methyl can be destroyed is called disproportionation; two ethyl radicals can be transformed by a mutual hydrogenation and dehydrogenation into one molecule of ethane and one of ethylene.

disprove vt 誤っていることを示す syn *discredit, negate*

▶ Diazomethane was originally believed to possess the three-membered, diazirine ring structure, but this was disproved by electron-diffraction studies, which showed the linear structure to be correct.

dispute n 論争 syn *argument, debate, discussion*

▶ Although $[BH_2(NH_3)_2][BH_4]$ is commonly referred to as the diammonate of diborane, its molecular structure was in dispute for several years.

disqualify vt 失格させる syn *exclude, reject*

▶ The potentially short ground-state lifetimes can disqualify any physical observation of a relatively long time scale for the determination of point-group symmetry.

disregard vt 無視する syn *ignore, neglect*

▶ The separation between neighboring translational energy levels is so small that for the most purpose it is possible to disregard the quantization of translational energy even for atoms.

disrotatory adj 逆旋的な ☞ conrotatory

▶ The disrotatory rotation of butadiene molecule would force a plus lobe to interact with a minus lobe, and this necessarily would show repulsion leading to an antibonding orbital of the cyclobutene.

disrupt vt 構造を壊す syn *disturb, upset*

▷ Incorporation of a methyl branch on the carbon adjacent to the carboxylate carbon lowers the solid-to-liquid crystal transition temperature, apparently disrupting the interactions between chains.

disruption n 途絶 syn *interruption*

▶ Alpha particles cause damage in the plastic, probably because of disruption of the polymer linkage along the path of penetration.

disruptive adj 破壊的な syn *disrupting, uncontrollable*

▶ In contrast to this, fluorescent DNA base replacements can be considerably smaller and less disruptive to local structure and to interactions with other biomolecules.

dissect vt **1** 解体する syn *decompose*

▶ With suitable assumptions, the total dipole moment of a molecule can be dissected into component bond moments which are assumed to be relatively constant constitutive properties of the σ bonds of compounds.

2 詳細に調べる syn *analyze*

▶ Since most cellular processes are mediated by intermolecular protein–protein and protein–nucleic acid interactions, it is desirable to dissect the interactions between the products of these genes.

dissection n **1** 解体 syn *breakdown*

▶ Equation 1 represents the empirical dissection of the enthalpy of formation of a compound in the gas phase into a bond component and a steric component.

2 精密な吟味 syn *analysis*

▶ Dissection of factors influencing the strength of acids and bases without solvation effects is best carried out using proton affinity data.

dissimilar adj 似ていない syn *different, unrelated*

▶ The difference in temperature between the measuring junction and the reference junction of two wires made of dissimilar metals generates an electromotive force that is proportional to the temperature difference.

dissimilarity n 不同性 syn *difference, disparity, diversity*

▶ The dissimilarity of the two ends of an aldohexose molecule prevents the existence of meso compounds.

dissipate vt 散逸させる syn *disperse, scatter, spread*

▶ Some superheating of the liquid is almost inevitable, because the heat supplied is not all immediately dissipated by vaporization.

dissipation n 散逸 syn *dispersal, dispersion, scattering*

▶ In elastomeric materials, fatigue involves complete dissipation of their resilient energy by repeated cycles of low-order stresses.

dissociable adj 解離可能な

▶ The reaction of an aryl-substituted alkyl halide with molecular silver or with metallic mercury or zinc is frequently the most satisfactory method for the preparation of a dissociable ethane.

dissociate vi **1** 解離する syn *separate*

▶ The chemistry of these systems is interesting because many of these materials exhibit polymeric behavior in the solid state but dissociate upon dissolution to give the discrete bimetallic monomers.

vt **2** 解離させる syn *break up, separate*

▶ Hydrofluoric acid, unlike the other halogen acids, is only weakly dissociated in water.

dissociation n 解離 syn *breakup, separation*

▶ The formation of sulfonium salts from alkyl halides is reversible, and heating of the salt causes dissociation into its components.

dissociation energy n 解離エネルギー syn *dissociation enthalpy*

▶ Although all four C–H bonds in methane are the same, the four dissociation energies are not all the same, because there are adjustments in the electron distributions as each successive hydrogen atom is removed.

dissociation limit n 解離極限

▶ The vibrational levels of the molecule become closer and closer together as we approach the dissociation limit.

dissociation pressure n 解離圧

▶ Lithium hydride has a dissociation pressure of 27 mm at its melting point.

dissociative adj 解離的な

▶ Exciplex compounds are bound in an upper electronic state but dissociative in the ground state.

dissociative adsorption n 解離吸着

▶ Langmuir has also considered the dissociative adsorption for the case of each molecule occupying two sites.

dissolution n 溶解 syn *solution*

▶ The familiar dissolution of AgCl in aqueous ammonia is due to the formation of $[Ag(NH_3)_2]^+$.

dissolve vi **1** 溶解する

▶ When sulfuric acid dissolves in water, the acid H_2SO_4 gives up a proton to the base H_2O to form the new acid H_3O^+ and the new base HSO_4^-.

vt **2** 液体に溶かす

▷ To a solution of malononitrile (32.0 g, 0.48 mol) dissolved in hot water (500 cm^3) was added a solution of ninhydrin (32.0 g, 0.18 mol) dissolved in hot water (800 cm^3), with mechanical stirring.

3 固溶体を作る

▶ Calcium fluoride can dissolve small amounts of yttrium fluoride.

dissymmetric adj 不均斉な syn *chiral*

▶ A dissymmetric object cannot be superimposed on its mirror image.

dissymmetry n 不均斉

▶ Optical activity is caused by the dissymmetry of each individual molecule as a whole.

distance n 距離 syn *separation*

▶ The iron–carbon distances and, hence, the sizes of the interstitial sites are considerably larger in γ-Fe than in α-Fe.

distant adj 遠い syn *far, remote*

▶ In the Nb_2O_5 and Ta_2O_5 structures, there are two different types of metal atom. One has a distorted octahedral arrangement of six oxygens, while the second has two near oxygens and five more distant.

distil vi **1** 蒸留する

▶ Mixtures of the miscible liquids, carbon tetrachloride and toluene, distil at temperatures intermediate between the two boiling points, and the composition of the distillate changes progressively during the process.

vt **2** 蒸留する

▶ Dichloromethane was distilled from phosphorus pentaoxide, then from anhydrous potassium carbonate, and finally redistilled.

distillate n 留出物

▶ After about three-quarters of approximately 6 M hydrochloric acid have been distilled off, the next 10 to 15 percent of the distillate is collected as the standard acid.

distillation n 蒸留

▶ Reducing the pressure at constant temperature is one way of doing distillation.

distilled water n 蒸留水

▶ A washing bottle of distilled water is indispensable, and the classical design shown in Figure 1 is recommended over a variety of others.

distilling flask n 蒸留フラスコ

▶ If pure water is heated in the distilling flask with a small flame, the vapor pressure of the liquid increases until it becomes equal to the atmospheric pressure.

distinct adj **1** 別個の syn *different, separate*

▶ The ^{13}C spectrum of 3-methylheptane shows seven distinct peaks corresponding to the eight different carbon atoms.

2 明確な syn *clear, definite*

▶ Even with a fifth-normal solution the change of color is very distinct but less so with tenth-normal solutions.

distinction n **1** 区別 syn *difference, differentiation, discremination*

▶ There is, for many systems, no clear-cut distinction between physical and chemisorptions.

2 特色 syn *merit, quality, significance*

▶ Selenic acid has the distinction of being such a strongly oxidizing acid that it will dissolve gold and palladium.

distinctive adj 特色のある syn *characteristic, notable*

▶ Because of their distinctive electronic structures, the transition metals enter into the formation of a great many complex ions.

distinctly adv 明白に syn *clearly, notably, sharply*

▶ When potassium permanganate acts as an oxidizing agent in distinctly acid solution, the manganese is reduced from an oxidation state of +7 to +2; the manganese atom accepts five electrons from the substance oxidized.

distinguish vt 区別する syn *differentiate, discreminate*

▶ NMR, like any other technique, cannot distinguish between two environments unless they are unchanged over the period needed to measure them.

distinguishable adj 区別可能な syn *discernible, noticeable*

▶ Subunits are distinguishable from one another by their different rates of sedimentation in an ultracentrifuge cell.

distinguishing adj 際立って特徴的な syn *distinctive, salient*

▶ The distinguishing feature of fuel cells is that the reacting substances are continuously fed into the system, so that fuel cells do not have to be discarded when the chemicals are consumed.

distort vt ゆがめる syn *bend, deform*

▶ Occupation of high-energy orbitals is energetically unfavorable, and the system will distort, if able, to lower the energy of the HOMO and increase the HOMO–LUMO gap.

distortable adj ゆがみやすい

▶ Acids whose charge clouds are easily distorted interact most strongly with distortable bases, while acids that are not easily distorted interact most strongly with polar bases.

distortability n ゆがみやすさ

▶ Variations in the relative importance of polarity and distortability were proposed to explain the donor properties of ammonia and a series of amines.

distorted adj ゆがめられた syn *bent, deformed*

▶ $[XeF_8]^{2-}$ ion has been shown by X-ray crystallography to be a slightly distorted square antiprism.

distortion n ゆがみ syn *twist*

▶ In an applied field, there is always a polarization of the medium due to the polarization of the ions, that is to say, a distortion of their electronic structure.

distortion polarization n 変形分極 syn *electronic polarization*

▶ Distortion polarization, like diamagnetism, is exhibited by all substances and is relatively weak effect.

distribute vt 分布させる syn *disperse, scatter, spread*
▶ Germanium is widely distributed in the earth's crust but in such low concentration that it was discovered only about a century ago.
distribution n 分配 syn *allocation, partition*
▶ As the mobile phase moves past the stationary phase, repeated adsorption and desorption of the solute occurs at a rate determined chiefly by its ratio of distribution between the two phases.
distribution coefficient n 1 分配係数 syn *partition coefficient*
▶ The distribution coefficient is a ratio describing the solubility of the solute in the solvent to that in the original feed stream.
2 分布係数
▶ If a radioactive atom can be incorporated in the contaminant, its segregation may be followed readily even at low levels, and its distribution coefficient may be estimated.
distribution function n 分布関数
▶ Smectic A and C phases have long-range orientational order of the molecular long axes similar to, but with a distribution function sharper than, that for nematics.
distribution law n 分配の法則 syn *partition law*
▶ These particular velocities, most probable, average, and root-mean-square velocities, associated with the distribution law enter into various applications.
disturb vt 妨げる syn *affect, damage, upset*
▷ A variety of rare earth elements have been substituted for Y without greatly disturbing the superconducting characteristics of the material.
disturbance n 変動 syn *disorder, interference*
▶ Stirring, shaking, or other disturbances of the liquid phase during the crystallization process can affect the outcome.
disulfide bond n ジスルフィド結合 syn *disulfide linkage*
▶ Disulfide bonds hold the two peptide chains together and play an important role in determining the conformation and the physiological properties of insulin.
diverge vi 分かれる syn *divide, split, subdivide*
▶ In the lower oxidation state, the halides of vanadium, on the one hand, and niobium and tantalum, on the other, diverge still further.
divergence n 1 逸脱 syn *deviation, diversity, variation*
▶ Convergence and divergence can pose significant experimental and conceptual difficulties when trying to understand the physiological consequences of ligand binding to a particular receptor.
2 相違 syn *discrepancy*
▶ There is a wide divergence in the chemistry of the elements involved, as well as in the reducing agents and conditions used.
divergent adj 異なる syn *different, dissimilar, separate*
▶ The effect of grinding accounts for many divergent results obtained by different chemists who have analyzed the same original material.
diverse adj 多様な syn *different, various*
▶ The inorganic chemistry of boron is more diverse and complex than that of any other element in the periodic table.
diversified adj 多角化された syn *divided, mixed, spread*
▶ In highly diversified firms, a central research laboratory has a difficult time picking research areas of broad interest.
diversify vt 多角化する syn *change, divide, mix, spread*
▶ In the postwar era, chemical companies began to diversify their operations by moving into related technologies, product lines, or both.
diversity n 多様性 syn *extent, range, variety*
▶ The rich diversity of magnetic properties arising from unpaired electrons can also be demonstrated with paramagnetic and ferromagnetic materials.
divert vt 転用する syn *alter, change, switch*
▶ The additional pathway diverts glucose 6-

phosphate from the first step of glycolysis and reduces $NADP^+$ in two redox reactions.

divide vi **1** 分かれる syn *separate, split*

▶ The transition-state theory naturally divides into three parts: the potential-energy surfaces, the rate of passage of activated complexes over the barrier, and the statistical mechanics of the equilibrium between reactants and activated complex.

vt **2** 分割する syn *break up, separate, split*

▶ Chain reactions can be divided into three stages: initiation, the formation of radicals; propagation, the repeating cycle of reactants, usually two in number; and termination, the removal of radicals.

3 分離する syn *partition, segregate*

▷ By dividing out the concentration factors in the above formulas for rate, there will result the corresponding rate constants or specific reaction rates.

divided adj 分割された syn *separate*

▶ Nickel reacts with many nonmetals when heated, and the finely divided form is pyrophoric at room temperature.

divisible adj 分割可能な syn *separable*

▶ The carbon skeleton of 2-chloro-2,4,4-trimethylpentane can be seen to be divisible into two isobutene units.

division n **1** 分割 syn *partition, separation*

▶ The variation in the temperature of the atmosphere with altitude leads to its division into reasonably well-defined layers which contain significantly different chemical and kinetic problems.

2 境界 syn *border line, boundary*

▶ The division between basic oxides and acidic oxides falls in different periods at groups 12, 13, and 14.

3 割り算

▶ The thermodynamic equilibrium constant has the same numerical value as the practical equilibrium constant; it has simply been made dimensionless by division by the unit quantity.

divorce vt 通常受身で，分離する syn *separate*

▶ The Einstein equation $E = mc^2$ shows that energy cannot be completely divorced from mass, as the two are to some extent interconvertible.

do vi **1** する 同一の動詞の反復をさけた用法

▶ When methyl radicals combine to form ethane, they do so with zero activation energy.

vt **2** 行う

▶ Distillation can also be done at the vacuum of an oil or water pump with substantial reduction of boiling point.

aux **3** 肯定文を強調する用法

▶ An important elementary chemical process that alkali halide nanocrystals do undergo at moderate temperatures is adsorption of polar molecules like NH_3 or H_2O.

doctrine n 学説 syn *concept, principle, theory*

▶ Along with the doctrine of atomism, the electron theory of valence ranks as one of the most fundamental developments in the history of modern chemistry.

document vt 文書に記録する syn *record*

▶ The use of urea, thiourea, 4,4'-dinitrobiphenyl, hydroquinone, and various other materials for selective inclusion in a solid phase of various types of compounds is now widely recognized and documented.

documentary adj 文書の

▶ The documentary record is integrated, with each form of evidence contributing information to our total understanding.

documentation n 資料 syn *file, record, source*

▶ The documentation for any given drug is naturally dispersed among all of the individuals and institutions involved in these activities.

dodecahedral adj 十二面体の

▶ The dodecahedral structure has been observed in a number of octacoordinate chelates and ions; the classic example is the dodecahedral $Mo(CN)_8^{4-}$ ion.

doing n 行動すること syn *carrying out, execution, performance*

▶ One of the motivations for so doing is the possible applications of such networks in the area of chiral separations.

domain n 1 分野 syn *area, discipline, field*
▶ Apart from light-initiated processes, electrode surfaces with catalytic activity offer a new domain for chemical synthesis.
2 領域 syn *territory*
▶ These different functions generally occur via a physically independent domain of the protein.
domain wall n 磁壁 syn *magnetic domain wall*
▶ The technological applications of magnetic properties rely on the existence of domain walls and the ability to control the movement of the domain walls.
domestic adj 家庭の syn *household*
▶ Much more serious is the effect of volatile S compounds released into the atmosphere as a result of man's domestic and industrial activities.
domestically adv 家庭で
▶ The bleaching and sterilizing action of hypochlorites have long been used both industrially and domestically.
dominance n 優勢 syn *predominance, superiority*
▶ The dominance of ionic bonding has important consequences for the structures of alkali halide nanocrystals.
dominant adj 支配的な syn *influential, premominant, prevalent*
▶ Molecular shape is a dominant factor in determining the fundamental type of mesophase formed, as between smectic and columnar.
dominate vt 支配する syn *command, control, rule*
▶ The temperature dependence of conductivity in metals is dominated by the interaction of electrons with vibrations of the atomic lattice.
donate vt 供与する syn *give, provide, supply*
▷ Pyrrole possesses considerable aromatic character arising from the delocalization of four carbon π electrons and two paired electrons donated by the nitrogen heteroatom.
donation n 供与 syn *contribution*
▶ The binding of the organic substrate to the metal involves the donation of π electrons to orbitals of the metal atom.
donor n 供与体 ☞ acceptor
▶ For series of complexes involving different electron donors and the same acceptor, it has been found that there is a linear correlation between the ionization potential of the donor and the frequency of the charge-transfer band.
donor atom n 配位原子
▶ Where the ligand is composed of a number of atoms, the one which is directly attached to the acceptor is called the donor atom.
donor level n ドナー準位 ☞ acceptor level
▶ Electrons are excited from the donor levels into the conduction band to produce carriers in the conduction band or from the valence band into the acceptor levels to create holes in the valence band.
dopant n ドーパント
▶ A logical strategy is to choose a dopant that is about the same size as the atom to be replaced, so it readily fits into the structure.
dope vt 不純物を添加する
▶ Depending upon the desired electronic properties, silicon wafers are doped in a controlled fashion with electron-donating or -withdrawing impurities to render the intrinsic material more highly conducting.
doping n ドーピング
▶ Doping of La_2CuO_4 with the alkaline earth metals, Sr or Ba, suppresses the orthorhombic–tetragonal phase transition to below room temperature.
dormant adj 休眠の
▶ The nematode eggs can rest dormant in the soil for many years until the root of a nearby host plant releases a substance that will promote hatching.
dosage n 用量 syn *dose*
▶ Toxicity is objectively evaluated on the basis of test dosages made on experimental animals under controlled conditions.
dose n 用量 syn *amount, dosage, portion, quantity*
▶ Because mercury is toxic at very low doses, most current strategies for removing mercury from contaminated water supplies do not

remove enough mercury to meet acceptable standards.

dot n ドット syn *point, spot*

▶ In formulas of hydrates, the addition of the water molecules is conventionally indicated by a centered dot (e.g., $NaSO_4 \cdot 10H_2O$).

dotted line n 点線 鎖線は dashed and dotted line ☞ dashed line, solid line

▶ The dotted lines outline the unit cell in each structure.

double adj **1** 二重の syn *twofold*

▶ The photozyme plays a double role by providing hydrophobic reaction center as well as by acting as a photocatalyst.

vi **2** 二倍になる syn *increase*

▶ The superconducting transition temperature of the sulfur-nitrogen polymer does increase with pressure and has nearly doubled in value by ca. 9 kbar.

vt **3** 二倍にする syn *enlarge, increase*

▶ If the volume is halved, the density of molecules is doubled.

double bond n 二重結合 複合形容詞として使用

▶ The peptide bond is special in that it is planar with the attached atoms not readily able to rotate around the bond; in other words it has double-bond character.

double decomposition n 複分解 syn *metathesis reaction*

▶ The driving force for double decomposition usually depends on the fact that one of the products is a gas, a nonionic compound, or a solid that is only slightly soluble and, therefore, precipitates from solution.

double duty n **do double duty** 二重の役目を果たす

▶ Frequently the aromatic compound does double duty, serving as solvent as well as reactant.

double helix n 二重らせん

▶ At high temperatures the double helix comes apart, and the DNA becomes single-stranded. This process is called denaturation or melting.

double oxide n 複酸化物

▶ There are numerous double oxides $pMO_n \cdot qBi_2O_3$, e.g., $Bi_{12}GeO_{20}$.

double resonance n 二重共鳴

▶ By the method of double resonance, the spins of two sets of protons can be decoupled, and simpler spectrum obtained.

double-stranded adj 二本鎖の

▶ The surface of the helix shows alternating major and minor grooves that follow the twist of the double-stranded molecule along its entire length.

double-stranded DNA n 二本鎖 DNA

▶ Double-stranded DNA that is coiled in the direction of the double helix is described as positively supercoiled.

doublet n **1** 二重線 syn *pair*

▶ In alkali bromides and iodides, the first absorption peak is split into a doublet.

2 二重項

▶ The double peaks observed in the spectra of the alkali bromides and iodides are simply a consequence of the doublet nature of the ground state of the halogen atoms.

doubly adj 二重に

▶ During irradiation, Yb^{3+} ions transfer two photons of infrared radiation to nearby Eu^{3+} ions which are then raised into a doubly excited state and decay by the emission of visible light.

doubt n **1** 疑い syn *suspicion*

▶ There is little doubt that the solvent can affect considerably the magnitudes of the thermodynamic constants which describe the equilibrium of a complex with its components.

2 beyond doubt 疑う余地もなく syn *doubtless*

▶ Since CH_4, CCl_4, etc., have been shown beyond any possible doubt to have tetrahedral configurations, the simple orbital picture breaks down when applied to carbon.

3 in doubt 疑わしくて syn *doubtful, questionable*

▶ While the existence of definite hydrides of nickel and platinum is in doubt, the existence of definite palladium hydride phases is not.

4 no doubt 疑いなく syn *doubtless*

▶ There is no doubt that, in many systems, phosphorescence emission of the donor occurs along with the fluorescence from the excited charge-transfer state.

vt **5 疑う** syn *discredit, question, suspect*

▶ Apart from these extreme cases, significant discrepancies between experimental and calculated enthalpies of formation occur in three alkanes only, and in all of these there is reason to doubt the experimental values.

doubtful adj 疑わしい syn *questionable*

▶ In view of the low accuracy of the location of the hydrogen atom, it is doubtful if any definite conclusion can be drawn.

doubtless adv 確かに syn *presumably, probably, surely*

▶ When a strong acid, such as HCl, is dissolved in methanol, the proton is doubtless transferred to a CH_3OH molecule, to form a $CH_3OH_2^+$ ion, analogous to the H_3O^+ ion in water.

down adj **1 下方へ** syn *downward*

▶ In Figure 3 each triangle represents a trigonal prism in projection down *c*.

adv **2 下方へ** syn *downward*

▶ Oxoanion polymerization increases down group 6.

3 至るまで

▶ The side chain is oxidized down to the ring, only a carboxyl group remaining to indicate the position of the original side chain.

downfall n 没落 syn *breakdown, decay, destruction*

▶ The promise and possibly the downfall of chemical hydrogen storage schemes lie in the need to regenerate the fuel, thus approaching a sustainable solution to our transportation needs.

downfield adj **1 低磁場の** ☞ upfield

▶ An aromatic ring will show downfield absorption in both ^{13}C NMR and NMR.

adv **2 低磁場へ**

▶ Compared to other carboxylates, in $Mo_2[O_2C(CH_2)_6CH_3]_4$, σ_{xx} is shifted upfield by about 20 ppm, while σ_{zz} is shifted downfield by about the same amount.

downstream n 下流 ☞ upstream

▶ A signaling pathway which is initiated by signal transduction from one type of receptor can be divergent, branching to reach two different downstream targets.

downward adj 下方への syn *descending*

▶ A second downward passage through the air–water interface enables another layer to attach to the first and so on.

draft n 隙間風 syn *breeze, wind*

▶ To obtain so successful a result, one must use a very small flame protected from drafts and boil the liquid slowly and steadily enough to allow full heat equilibration between liquid and vapor in the column.

drag n **1 抗力** syn *control, power*

▶ The effect of the ionic atmosphere is to exert a drag on the movement of a given ion.

vt **2 引きずっていく** syn *draw, pull*

▶ Ions are attracted to solvent molecules mainly by ion–dipole forces; therefore, when they move, they drag solvent along with them.

drain vi **1 水が流れ出る** syn *drip, drop, release*

▶ A buret will not drain properly unless it is clean on the inside so that visible drops of liquid do not adhere to the sides when the liquid is withdrawn.

vt **2 排水させる** syn *draw off, tap*

▶ The wash liquor is drained out through the stopcock under nitrogen pressure..

drainage n 排水 syn *leak*

▶ Acid mine drainage and mobilization of radioactive mine tailings are subjects of continuing studies that should reduce adverse effects.

dramatic adj 目覚しい syn *specutacular*

▶ The chemical differences between zirconium and titanium, although not dramatic, are greater than those between yttrium and scandium.

dramatically adv 目覚しく

▶ Research into solid-state materials has grown dramatically in the past decade.

drastic adj 目立った syn *extreme, severe*

▶ Chemisorption may result in a rather drastic disruption of the bonding in an adsorbed molecule.

drastically adv 目立って

▶ While the polymers from ortho- and meta-substituted aniline monomers are essentially identical in chemical structure, their morphologies are drastically different.

draw vt **1** 図を描く syn *depict, portray, sketch*

▶ The difficulty lies in our inability to draw very explicit structures for the excited states of molecules with conventional bond diagrams.

2 引き出す syn *aquire, get, obtain*

▷ While for an exact prediction of chemical equilibria we need to know free energy rather than heat differences, the entropy terms in many of the reactions are usually sufficiently small to be neglected in drawing general qualitative conclusions.

3 引っ張る syn *drag, pull*

▶ The attractive Coulomb forces draw ions together from distances greater than the encounter distance.

4 流させる syn *get, take*

▷ Complete oxidation is easily achieved by drawing air over a nickel-silica catalyst for several minutes.

5 draw off 流し出す syn *pour, tap*

▶ The organic reaction product is distributed wholly or largely into the upper ether layer, whereas inorganic salts, acids, and bases pass into the water layer, which can be drawn off and discarded.

drawback n 欠点 syn *difficulty, disadvantage, shortcoming*

▶ Water does suffer from the drawback that it has a relatively narrow potential window.

drawing n **1** 図画 syn *design, representation*

▶ An alkane with n carbon atoms has the empirical formula C_nH_{2n+2} but may have one of several structural formulae, which are represented by graphlike drawings.

2 求引

▶ The drawing of negative charge to the fluorine end of the HOF molecule suggest that the O−H bond should be weakened and the O−F bond strengthened.

drier n 乾燥剤 syn *desiccant*

▶ The usual driers are salts of metals with a valence of two or greater and unsaturated organic acids.

drift n **1** ドリフト 電場による荷電体の移動

▶ In all kinds of electric conductivity at room temperature, it is well recognized that the motion of the charges carrying the current is merely a drift superimposed on a random motion that is already present in the absence of a field.

2 ドリフト 特性の徐徐な変化

▶ Environmental isolation of major system control components is required to avoid instrumental drift.

vi **3** 緩やかに移動する syn *migrate*

▶ When a field is applied, the positive ions have a tendency to drift in one direction, while the negative ions tend to drift in the opposite direction.

drift mobility n ドリフト移動度

▶ The drift mobility is the transmission velocity for a hole artificially introduced into the system.

drift velocity n ドリフト速度

▶ In a metal, the free electrons are executing a random motion; when a voltage is applied, a drift velocity is superimposed on this movement.

drill n **1** ドリル

▶ Tungsten carbide is extremely hard and is used for drill tips and abrasives.

vt **2** 穴をあける syn *bore, penetrate*

part ▶ Suspensions of bentonite clay in water display thixotropy, which is desirable in oil-well drilling fluids.

drip vi 滴る syn *drop*

▶ A slightly more involved situation occurs when the upper consolute temperature does not exist; in this case the distillate drips out of the still as a two-phase mixture.

drive n **1** 推進力 syn *effort, impetus*

▶ Initially the main drive to use ionic liquids

was the ability to obtain high concentrations of aluminum in a highly conducting aprotic medium for aluminum deposition.

vt **2** 駆り立てる syn *force, induce, oblige*

▶ The use of thin films of various inorganic materials in the electronics industry has recently driven interest in chemical vapor deposition.

driven adj 駆り立てられた syn *controlled*

▶ We see both an increase of dimensionality and multiple coordination changes during a temperature-driven quasi-topotactic dehydration reaction in the solid state.

driving force n 推進力 syn *drive, impulse*

▶ The driving force of a chemical reaction results from a tendency for the system to approach equilibrium.

drop n **1** 一滴 syn *droplet*

▶ Salicylic acid is acetylated readily on treatment with acetic anhydride in the presence of a few drops of concentrated sulfuric acid on a catalyst. The product is acetylsalicylic acid or aspirin.

2 drop by drop 一滴ずつ syn *dropwise*

▶ On adding silver nitrate solution, drop by drop to a neutral or alkaline solution of an alkali cyanide, a white precipitate is formed when the two liquids first come in contact with one another, but on stirring it redissolves owing to the formation of potassium dicyanoargentate(I).

3 降下 syn *descent, fall*

▶ Increasing the temperature led to a large drop in enantioselectivity.

vi **4** 下がる syn *decline, descend, fall*

▶ The influence of the surface on infrared spectra can no longer be neglected if the particle radius drops below 5–10 nm.

5 drop off 次第に少なくなる syn *decline*

▶ The inductive effect by a center of high electronegativity drops off rapidly with distance from that center.

vt **6** 液体をたらす syn *drip*

▷ Small volumes can be measured with capillary dropping tubes.

droplet n 小滴 syn *drop*

▶ An emulsion consists of droplets of one liquid dispersed in another liquid. The droplets are usually from 0.1 to 1 μm in diameter and, hence, are larger than the sol particles.

dropping mercury electrode n 滴下水銀電極

▶ If a steadily increasing potential is passed through a cell involving a dropping mercury electrode, the current that passes varies according to a pattern which depends on the nature of the ions present in the solution.

dropwise adv 一滴ずつ syn *drop by drop*

▶ The solution was stirred at −78 ℃ for 30 min, then dichlorophenylphosphine in tetrahydrofuran was added dropwise to the reaction vessel.

drug n 薬剤 syn *medicine*

▶ In many cases, it would appear that the binding of the drug to the receptor is of relatively low energy, certainly considerably less than that involved in normal covalent bonding.

drug delivery system n 薬物送達システム

▶ Silicone pressure-sensitive adhesives do not require extra ingredients because of their inherent chemical and thermal stability, and as a result they are ideal adhesive components in transdermal drug-delivery systems.

dry adj **1** 乾式の

▶ Many M^{III} fluorides (e.g., AlF_3) are quite insoluble when made by dry methods but hydrates (e.g., $AlF_3 \cdot 3.5H_2O$) produced from solutions dissolve readily in water.

2 乾いた syn *anhydrous, dehydrated*

▶ Sodium alkoxides are made by direct action of sodium metal on dry alcohols.

vi **3** 乾く

▶ Some types of printing inks dry by evaporation of a volatile solvent rather than by oxidation and polymerization of a drying oil or resin.

vt **4** 乾かす syn *dehydrate, desiccate*

▶ Solvents were dried from the appropriate

drying agent, distilled, degassed, and stored over a potassium mirror.

dry cell n 乾電池 syn *dry battery*

▶ Because MnO(OH) is so stable, MnO_2 can readily be reduced to it, a change that forms the basis of many alkaline dry cells.

dry ice n ドライアイス

▶ Bath temperatures in the range 0° to −30° can be maintained by adding pieces of dry ice as required to alcohol; the evolution of gaseous carbon dioxide keeps the bath stirred.

drying tube n 乾燥管

▶ A calcium chloride drying tube inserted in a cork that will fit either the flask or the top of the condenser is also made ready.

dual adj 二重性の syn *double*

▶ Several nitro-*p*-terphenyls exhibit well-defined dual charge-transfer properties; e.g., they complex with tetracyanoethylene, a typical electron acceptor, and with *N*,*N*,-dimethyl-*p*-toluidine, a representative electron donor.

duality n 二重性

▶ The duality of mechanism does not reflect exceptional behavior but is usual for electrophilic aromatic substitution.

ductile adj 延性のある

▶ Platinum is silvery-white and lustrous and is both malleable and ductile so that it is readily worked.

ductility n 延性

▶ In part, malleability and ductility depend on the numbers of close-packed planes and directions possessed by a structure.

due adj **1** 当然の syn *appropriate, correct, proper*

▶ Ozonizations must be carried out with due caution.

2 due to …に起因して 形容詞としては，be 動詞の後におく場合と，すぐ前の名詞または代名詞を修飾する場合に限るのが正しい． syn *ascribable, attributable*

▶ Of course the observed susceptibility of a radical will be the difference between the paramagnetic susceptibility due to the odd electron and the diamagnetic susceptibility due to the remaining paired electrons.

conj **3 due to** (prep) …の結果 syn *owing to*

▶ Due to isomorphous substitution in the clay structure of one atom with another of lower valence, the clay possesses a net negative charge.

dull adj 鈍い syn *dark*

▶ The most convenient laboratory preparation of ketene is to pass acetone vapor over a coil of resistance wire heated electrically to a dull red heat.

duly adv 正しく syn *appropriately, properly, suitably*

▶ Facts and experiments relating to any subject are never duly appreciated till they are made the foundation of a theory by which we are able to predict the results and foresee the consequences of certain other operations which were never before undertaken.

dumbbell n ダンベル

▶ Each 2p orbital is dumbbell-shaped.

dump n 集積場 syn *junkyard*

▶ The zone right at the end of the sample is the impurity dump: when the heater has passed its position, cooling occurs, and the dirty liquid simply cools to dirty solid, which can be discarded.

dumping n 廃棄物の投棄

▶ Dumping of wastes into lake and watercourses is strictly forbidden.

duplex adj 二重の

▶ DNA is a duplex molecule in which two polynuceotide chains are linked to each other through specific base pairing.

duplicate n **1 in duplicate** 正副二つ

▶ All volumetric titrations should be carried out in duplicate.

adj **2** 全く同じ syn *identical*

▶ Duplicate solubility determinations were made at each salt concentration.

vt **3** 複製する syn *copy, replicate*

▶ DNA molecules can duplicate themselves, that is, can bring about the synthesis of other DNA molecules identical with the originals.

duplication n 複製 syn *replication, reproduction*

▶ Duplications occur when one or more copies of a given gene are present on the same or different chromosomes.

durable adj 耐久性のある syn *persistent, stable*

▶ Molecular engineered layered structures are capable of serving as mechanically and chemically durable solid acid catalysts for effecting many proton-catalyzed reactions.

durability n 耐久性 syn *stability, strength*

▶ For the development of any catalyst the chemist seeks to maximize activity, selectivity, and durability.

duration n 持続時間 syn *length, period*

▶ Flash durations vary from a few microseconds to a few milliseconds.

during prep の間に syn *through, throughout*

▶ During and immediately after the exicitation of an electron, the crystal still preserves the equilibrium arrangement, owing to the fact that ions do not move during an optical transition.

dust n 粉じん syn *powder*

▶ Supersaturation may be achieved experimentally if the vapor is entirely dust-free.

dust explosion n 粉じん爆発

▶ Dust explosions are caused by reactive dusts such as sugar, soap, starch, coal, or even a petroleum mist.

dusty adj 埃っぽい

▶ Colloidal systems scatter light as is observed when a beam of sunlight passes through dusty air or through thin skimmed milk.

dye n **1** 染料

▶ Dyes are sometimes classified on the basis of their chief chromophores, e.g., $-NO$, nitroso dyes; $-NO_2$, nitro dyes; $-N=N-$, azo dyes; etc.

vt **2** 染める syn *stain, tint*

▶ An artificially thickened aluminum oxide film may be dyed to produce a variety of very bright surface colors.

dyeing n 染色

▶ Tin(IV) chloride, as used for a mordant in dyeing, is obtained as the solid salt or in a concentrated aqueous solution.

dye laser n 色素レーザー

▶ Experimentally the dye laser consists of a short-path-length cell, containing the dye solution, placed between two mirrors which constitute the resonant cavity.

dyestuff n 染料

▶ Just as there are dyestuffs which change color at a definite pH value of an aqueous solution and serve as indicators in alkalimetry and acidimetry, so there are certain organic substances which change color as a result of oxidation or reduction.

dynamic adj 動的な syn *dynamical*

▶ These ^{1}H NMR spectroscopic data together with a slightly broad phosphorus signal suggest dynamic behavior within the complex in the solution phase.

dynamic equilibrium n 動的平衡

▶ The complexes investigated all show evidence for dynamic equilibria in the solid state which involve reorientation of the naphthalene donor molecules in their molecular planes.

dynamics n **1** 単数扱いで，動力学

▶ The term dynamics relates to the actual forces, and these are conveniently dealt with in terms of potential-energy surfaces.

2 複数扱いで，適応様式

▶ In the last few years, protein dynamics have been revealed to play an important role in protein function.

dynamic surface tension n 動的表面張力

☞ static surface tension

▶ The dynamic surface tension corresponds to an instantaneous exposure of the interior of the solution before there has been time for diffusion, sorption, or any rearrangements to occur.

E

each pron **1 それぞれ** each of は名詞や代名詞の複数形の前に置く．これに続く動詞は単数形とする．否定の場合は，none of を使用する．

▶ There are five main AX structure types: rock salt, CsCl, NiAs, sphalerite, and wurtzite, each of which is found in a large number of compounds.

2 each other 互い

▶ Optically inactive ammonium sodium tartrate existed as a mixture of two different kinds of crystals, which were mirror images of each other.

adj **3 おのおの** 一集団のすべてのものに付いて記すとき，可算名詞の単数形の前に用いる．

▶ If a solution of a Cu(II) salt at a suitable concentration is treated with an excess of KI, there is liberated one atom of iodine for each atom of copper present.

adv **4 めいめいに**

▶ Nickel, palladium, and platinum form only one reasonably well-characterized oxide each, namely, NiO, PdO, and PtO_2.

early adj **1 初期の** syn *initial, original*

▶ It is worth considering some of the earlier theories that still have relevance today.

2 始めのほうの

▶ Our exploratory efforts have dealt with reduced halides of the early transition metals that lie along the metal-salt interface.

earn vt **得る** syn *get, make, receive*

▶ Energy that is earned by catabolism is spent on a wide variety of energy-consuming reactions.

earth n **the earth 地球**

▶ Vanadium has been estimated to comprise about 136 ppm of the earth's crustal rocks.

ease n **容易さ** syn *easiness, simplicity*

▶ From the ease with which triphenylmethyl can abstract hydrogen from other molecules, one could classify it as a powerful oxidizing agent.

easily adv **容易に** syn *smoothly, readily*

▶ Nitrosobenzene is too easily reduced to be formed by direct reduction of nitrobenzene.

easy adj **容易な** syn *straightforward*

▶ Bis(η-benzene)chromium forms a number of 1:1 molecular complexes with acceptors such as tetracyanoethylene; such behavior is in accord with its easy oxidation to salts of the $[(\eta\text{-}C_6H_6)_2Cr]^+$ ion.

eclipsed configuration n **重なり配座**

▶ In the boat conformer of cyclohexane, there are two pairs of carbon atoms which have eclipsed ethane conformations.

eclipsed form n **重なり形**

▶ In the eclipsed forms of ethane, the dihedral angle has values of 0°, 120°, 240°, and 360°.

eclipsing n **重なり**

▶ Each cyclopentane molecule assumes a puckered conformation which is the best compromise between distortion of bond angles and eclipsing of hydrogens.

ecological adj **環境上の** syn *environmental*

▶ During the last five years, there has been a growing awareness of the ecological problems associated with rapid population and industrial expansion.

economical adj **節約的な** syn *tight*

▶ The principal factor governing the crystal structures of most polymers is the need for economical packing of the molecular chains.

economically adv **経済的に**

▶ Alkyl bromide formation proceeds at a useful rate only in the presence of strong acid, which can be furnished by excess hydrogen bromide or, usually and more economically, by sulfuric acid.

economize vi **節約する** syn *save*

▶ To economize on adjustable parameters, I propose that the term E(C—C) be fixed by the experimental bond energy for diamond.

ecosystem n **生態系**

▶ Zeolites seem to be one of the few things we can add to the ecosystem without negative consequences.

edge n **1** 辺 syn *side*
▶ In ferroelectric, low-temperature KH_2PO_4, the hydrogens order themselves so that both are associated with the upper edge of each PO_4 tetrahedron.
2 端 syn *rim, side*
▶ The upper edge of the filter paper must always be below the rim of the funnel, as otherwise the filter cannot be washed satisfactorily.
3 刃
▶ The file is easily sharpened by light regrinding of the narrow edges.

edge dislocation n 刃状転位 ☞ screw dislocation
▶ If a positive and a negative edge dislocation meet on the same slip plane, they cancel each other out, leaving behind a strain-free area in the crystal.

edge-shared adj 辺を共有した
▶ Oxide ions in crystal lattices can be progressively removed by systematically replacing corner-shared octahedra with edge-shared octahedra.

edible adj 食用に適する syn *eatable*
▶ A major industrial application of catalyzed hydrogenation reactions is to the formation of edible fats from vegetable and animal oils.

effect n **1** 効果 effect on の形で多用される. syn *influence*
▶ The effect of substituents on the acidity of phenols is similar to their effect on carboxylic acids.
vt **2** 果たす syn *achieve, cause, execute, make*
▷ Since the separation effected in a fractionating column is dependent upon heat equilibration in multiple processes of vaporization and condensation, the efficiency of separation increases with increasing heat of vaporization of the liquids concerned.
3 変化をもたらす affect と意味の違いに注意せよ. syn *create, produce*
▶ Hydrogen bonding grossly effects the spectrum of any substance.

effective adj 有効な syn *capable, operative, useful*
▶ Evidently, light makes possible a very effective reaction path by which chlorine may react with methane.

effective charge n 有効電荷
▶ In making electrostatic calculations of the bonding energies of complexes, it is necessary to use the effective ionic charge instead of the usual charge.

effectively adv 事実上 syn *virtually*
▶ Water molecules are effectively tetrahedral since each H_2O molecule has two O−H bonds and two lone pairs which can be used in hydrogen bonding.

effectiveness n 効力 syn *usefulness*
▶ Microspheres, tiny particles that can be filled or coated with drugs, may be one way to improve drug effectiveness.

effector n エフェクター
▶ The binding of an effector molecule at a remote allosteric site can cause conformational changes at the active site which alters the selectivity and reactivity of the enzyme to its substrate.

efficiency n 効率 syn *effectiveness*
▶ The relative efficiency order for various terminal groups in promoting nematic thermal stability is well documented.

efficient adj 効率のよい syn *effective*
▶ Carbon tetrachloride is used as an efficient fire-extinguishing fluid for petroleum fires, although its tendency to phosgene formation makes it undesirable for confined areas.

efficiently adv 効率よく
▶ The structures of boron-rich borides are more efficiently dominated by inter-B bonding.

efflorescence n 風解
▶ Faraday demonstrated that the efflorescence of sodium carbonate decahydrate was facilitated when the crystal surface was scratched.

effluent n 流出液 ☞ influent
▶ The effluent from the column is mixed with ninhydrin solution, and the intensity of the blue color developed is measured with a photoelectric colorimeter and plotted as a function of

time at constant flow rates.

effort n 努力 syn *attempt, endeavor, venture*

▶ Great efforts are being made to obtain materials with higher T_c, with the goal of achieving superconductivity at room temperature.

effuse vt **流れ出させる**

▶ Hydrogen effuses through a porous plate four times as rapidly as oxygen.

effusion n **エフージョン**

▶ The Knudsen effusion method involves determination of the rate at which vapor escapes through a small orifice of known dimensions into an evacuated space.

e.g. abbr **たとえば** for example と読まれる．ラテン語 exempli gratia を略記したもので，ローマン体を用い，その前後にコンマを置く．

▶ In the ionic model, the valence band is made up of the top occupied anion orbitals, e.g., oxygen 2p in oxides, and chlorine 3p in chlorides, and the conduction band is composed principally of the lowest empty cation orbitals, e.g., 3s in compounds of sodium or magnesium.

eigenfunction n **固有関数**

▶ If the eigenfunction is real, its square multiplied by a small volume element d*x* d*y* d*z* is proportional to the probability that the electron is present in that element.

either pron **1 どちらか一方** either of に続く名詞は複数形，それに続く動詞は単数形を用いる．

▶ It is well known that the addition of hydrogen bromide to an olefinic compound may occur by either of two quite distinct types of mechanism, determined largely by the experimental conditions.

adj **2 二つのうちどちらかの** 可算名詞の単数形の前に置く．

▶ Microscopic observation of a noneutectic melt cooled to solidification reveals individual crystals of either solid dissolved in a uniform matrix of eutectic composition.

adv **3 いずれかを** either A or B を主語とする場合，動詞の数はより近い位置を占める B に一致させる．

▶ Either the intermolecular forces of attraction and repulsion or the potential energy is dealt with as a function of the distance between the molecules.

A と B は名詞に限らないが，文法上，対等の関係にあることを要する．

▶ Oxidation of almost all organic compounds involves either gaining oxygen atoms or losing hydrogen atoms.

三つのものを対象とした either A, B, or C も見受けられる．or を重ねて使用する場合については or を見よ．

▶ The structures and properties of many borides emphasize the inadequacy of describing bonding in inorganic compounds as either ionic, covalent, or metallic.

eject vt **放出する** syn *emit, expel*

▶ Auger emission occurs when electrons are ejected from the sample as an energy release mechanism.

elaborate adj **1 精巧な** syn *detailed, intricate, involved*

▶ It is difficult to distinguish between electronic and ionic conductivity without elaborate experimental technique.

vt **2 精巧に作り上げる** syn *develop*

▶ Metal-catalyzed carbonylations can elaborate diverse functional groups; the most commonly examined groups are alkyl halides, alkenes, and alkynes.

elapse vi **時が経過する** syn *go, pass*

▶ Henry's law holds true only for equilibrium condition; i.e., when enough time has elapsed so that the quantity of gas dissolved is no longer changing.

elastic adj **弾性の**

▶ The key to elastic behavior is to have a polymer that has either sufficiently weak forces between the chains or a sufficiently irregular structure to be very largely amorphous.

elastic collision n **弾性衝突**

▶ In an elastic collision, only the translational motion of the molecules is altered.

elastic modulus n **弾性率**

▶ In order to prevent undue extension of the cable in use, high elastic modulus materials must be incorporated in order to take the strain.

elastic scattering n 弾性散乱

▶ Infrared spectra of large particles exhibit in addition to characteristic absorption bands pronounced features due to elastic scattering of the light by the particles.

elasticity n 弾性

▶ Polymers showing a high degree of crystallinity do not show high elasticity, since there is not sufficient internal mobility to permit the conformational changes to occur.

elastomer n エラストマー

▶ An elastomer needs to have some crystalline regions to prevent plastic flow and, in addition, should have rather flexible chains.

elastomeric adj エラストマーの

▶ Ionic conductivity in polymer electrolytes is a property of amorphous, elastomeric phases.

electric charge n 電荷 ☞ charge

▶ Colloid particles in water carry an electric charge which can be acquired by various mechanisms, e.g., the ionization of surface acid or hydroxy groups and the differential solution or specific adsorption of ions.

electric conductivity n 電気伝導率 syn *electrical conductivity*

▶ Partially filled bands lead not only to good electric conductivity but also to good thermal conductivity.

electric discharge n 放電 ☞ discharge

▶ The volatile, colorless, crystalline KrF_2 is prepared by subjecting mixtures of krypton and fluorine to an electric discharge at low temperatures or by irradiating liquid krypton-fluorine mixture with ultraviolet light.

electric double layer n 電気二重層 syn *electrical double layer*

▶ The general conclusion to be drawn from the above analysis, namely, that an electric double layer is set up at the interface between Zn and ZnO is valid for any metal-oxide system.

electric energy n 電気エネルギー syn *electrical energy*

▶ Piezoelectric crystals have been used for many years as transducers for converting mechanical to electricl energy, and vice versa.

electric field n 電場

▶ Under influence of an applied electric field, the particle tracks are modified to give a net movement of the current carriers in the direction of the electric field.

electric insulator n 絶縁体 syn *electrical insulator*

▶ Dielectric materials are used principally in capacitors and electric insulators.

electric potential n 電位 syn *electrical potential*

▶ In a cell in which E is positive, the right-hand electrode has a higher electric potential than the left.

electric potential difference n 電位差

▶ If two electrolyte solutions of different concentrations are separated by a membrane, in general there will be an electric potential difference across the membrane.

electric resistance n 電気抵抗 syn *electrical resistance*

▶ It was observed from the electric resistance behavior that pentacene and hexacene undergo an irreversible transformation above 200 kbars, while smaller molecules such as anthracene and tetracene apparently do not.

electric resistivity n 電気抵抗率 syn *electrical resistivity*

▶ Electric resistivities were measured by the four-probe method, and for activation energies the temperature was varied from −20 to 50 ℃.

electric vector n 電気ベクトル

▶ The vibration direction is direction of the electric vector in the terminology of electromagnetic radiation.

electride n エレクトライド

▶ A compound composed of an alkali-metal cation and an electron in which the electron functions as a chemical element in salt formation is called an electride.

electrify vt 帯電させる

▶ The smokes produced by an arc are highly electrified.

electrocatalysis n 電極触媒反応

▶ Electrocatalysis includes the advantages of catalytic rate control, including specificity, and adds the opportunity to inject or extract electric energy.

electrochemical adj 電気化学的な

▶ Electrochemical techniques are the basis of a wide variety of calculations and measurements in chemistry.

electrochemical gradient n 電気化学勾配

▶ The electrochemical gradient is composed of both a proton chemical potential difference and a membrane potential and is called the proton motive force.

electrochemical equivalent n 電気化学等量

▶ Experiments with electrolysis enabled Faraday to formulate his idea of electrochemical equivalents, that is, a relationship between the amount of electricity and the quantity of substance decomposed.

electrochemically adv 電気化学的に

▶ Both oxygen and hydrogen peroxide are difficult species to measure electrochemically because of their poor electrode kinetics.

electrochemical reaction n 電気化学反応

▶ The Tafel law describes how the rate of an electrochemical reaction is affected by the potential difference at the electrode.

electrochemical reduction n 電気化学的還元

▶ Polarography is not a technique that can be readily adapted to automated, in situ analysis, and it is almost exclusively limited to electrochemical reductions.

electrocrystallization n 結晶電析

▶ Before electrocrystallization is initiated the solvent and anionic derivative are placed in the cathode compartment of the cell, while the solvent, anionic derivative, and organic donor are loaded in the anode compartment.

electrode n 電極 ☞ anode, cathode

▶ Electrodes are essential components of both batteries and electrolytic cells; in batteries the negative plate is the anode and the positive plate the cathode, whereas in electrolytic cells the reverse is the case.

electrodeposition n 電析

▶ An important advantage of electrodeposition is its ability to coat complex shapes having small and irregular cavities with exact thickness control.

electrode potential n 電極電位

▶ One of the most important practical applications of electrode potentials is to titrations.

electrode process n 電極過程

▶ The choice of the metal used for the electrodes can also affect the various electrode processes, and this increases the separation factor of deuterium between the gaseous and liquid phases still further.

electrode reaction n 電極反応

▶ The electrode reactions proceed because the electrons produced at one electrode can travel through the external circuit and be used up at the other electrode.

electrodialysis n 電気透析

▶ Electrodialysis is well-suited for demineralization of waters which contains 5g/L or less of dissolved solids.

electrokinetic adj 電気運動学的な

▶ The term electrokinetic is used to describe those effects in which either an electric potential brings about movement or movement produces an electric potential.

electrokinetic potential n 界面動電位

▶ The potential difference between the bulk of the solution and the position of closest approach of the ions to the surface is known as the electrokinetic potential, or ζ potential.

electroless adj 無電解の

▶ It is realistic that ionic liquids could have applications in such fields as electroless deposition where the chemistry of aqueous solutions limits the systems that can be studied.

electroless plating n 無電解めっき

▶ Hydrated sodium hypophosphite has been increasingly used for the electroless plating of Ni onto both metals and nonmetals.

electroluminescence n エレクトロルミネセンス

▶ Electroluminescence is a solid-state phenomenon involving p- and n-type semiconductors and is observed in many crystalline substances.

electrolysis n 電気分解

▶ Electrolysis of benzoic acid in pyridine gives some 4-phenylpyridine together with other products.

electrolyte n 電解質 ☞ strong electrolyte, weak electrolyte

▶ It is, in many cases, difficult or impossible to find electrodes that are reversible to both ion constituents of a binary electrolyte.

electrolytic adj 電解の

▶ Alkaline MnO_2 cells use electrolytic MnO_2 and zinc powder in the KOH electrolyte.

electrolytically adv 電解によって

▶ It is possible to deposit zinc electrolytically with a current of 0.2–0.3 ampere from a neutral solution to which sodium acetate and a few drops of acetic acid have been added.

electrolytic conductivity n 電解伝導度

▶ Electrolytic conductivity in a true solid electrolyte is an inherent property of the crystal and is directly associated with the crystal structure.

electrolytic oxidation n 電解酸化

▶ The electric conductivity of graphite is found to increase as a result of the uptake of hydrogensulfate ions during electrolytic oxidation in sulfuric acid.

electrolytic reduction n 電解還元

▶ Another technique for obtaining low oxidation state is by electrolytic reduction using cyclic voltammetry.

electrolytic refining n 電解精錬

▶ In electrolytic refining of copper, the anode mud contains the relatively inert metals platinum, silver, and gold and is usually treated for the recovery of these metals and other rare elements.

electrolytic solution n 電解液

▶ In studying thermodynamic properties of electrolytic solutions, both the standard state potential and the activity coefficients of the solute are of fundamental importance.

electrolyze vt 電解する

▶ When a solution of NaC_2H_5 in $Zn(C_2H_5)_2$ is electrolyzed, a mixture of ethane and ethylene is liberated at the anode.

electromagnetic wave n 電磁波 syn *electromagnetic radiation*

▶ Absorption of electromagnetic wave by molecules can occur not only by electronic excitation of the type described for atoms but also by changes in the vibrational and rotational energies.

electromotive adj 起電力の

▶ In general, a metal higher in the electromotive series tends to displace one lower down.

electromotive force, emf n 起電力

▶ If the resultant emf is positive, the reaction will be expected to proceed spontaneously in the direction written, but if the emf is negative, the reverse reaction would be favored.

electron-accepting adj 電子受容性の syn *electron-withdrawing*

▶ The reaction tolerated incorporation of both electron-donating and electron-accepting substituents on the phenyl ring of this substrate.

electron acceptor n 電子受容体 ☞ electron donor

▶ Plants can use NO_3^- as the electron acceptor instead of CO_2.

electron affinity n 電子親和力

▶ Electron affinities of individual molecules have been deduced from measurements of the potential of polarographic addition of electrons to the molecules in solution at the surface of a mercury electrode.

electron beam n 電子線

▶ The application of Auger spectroscopy to surface chemistry involves the use of an electron beam to excite electrons near the surfaces of solids, resulting in the emission of electrons.

electron bombardment n 電子衝撃

▶ Metals can be evaporated from a target in vacuum by ion bombardment, electron bombardment, or heating.

electron cloud n 電子雲 ☞ charge cloud

▶ At extremely short interatomic distances, the inner electron clouds of the interacting atoms begin to overlap, and Pauli repulsion becomes extremely large.

electron compound n 電子化合物

▶ The term electron compound denotes in an alloy system an intermediate phase where crystal structure is determined by the establishment of a certain electron-to-atom ratio.

electron configuration n 電子配置

▶ Using electron configuration as the criterion, we ordinarily recognize four general types of elements; the inert gas elements, the representative or main group elements, the transition elements, and the inner transition elements.

electron deficiency n 電子欠損

▶ The fact that trimethoxyborane and triaminoborane have little or no tendency to complex with amines is taken as evidence supporting $p_\pi - p_\pi$ bonding, since this would reduce electron deficiency at boron and discourage coordination.

electron-deficient adj 電子不足の ☞ electron-rich

▶ Having more valence atomic orbitals than valence electrons, electron-deficient compounds present ambiguities in the use of the paired electron bond.

electron delocalization n 電子非局在化

▶ The difference between the observed magnetic susceptibility and that calculated from local group values should provide a quantitative measure of electron delocalization and hence aromaticity

electron density n 電子密度

▶ Because of this delicate balance between electron density and intermolecular forces, the van der Waals radius of light atoms is not an absolute property but depends on the bonding state in which it is held in a molecule.

electron density map n 電子密度図

▶ Accurate electron density maps not only tell us the positions of the nuclei but also provide information as to the density of the electron charge around the nuclei.

electron diffraction n 電子回折

▶ Most electron diffraction work has been confined to gases.

electron distribution n 電子分布

▶ The electron distribution in one atom will affect and be affected by the electron distributions in the surrounding atoms.

electron-donating group n 電子供与性基

syn *electron-releasing group*

▶ By chemical insertion of electron-donating or electron-withdrawing groups, molecules can become one-electron donors or one-electron acceptors.

electron donation n 電子供与

▶ Optical basicity represents the extent of electron donation by a ligand in a coordinate bond and should be related in some way to the electronegativity difference between central atom and ligand.

electron donor n 電子供与体 複合形容詞として使用. ☞ electron acceptor

▶ Certain molecules contain an effective electron-donor site, while in another part of the same molecule a potential electron-acceptor site exists.

electron donor-acceptor complex n 電子供与体-受容体錯体 ☞ charge-transfer complex

▶ A description of electron donor-acceptor complexes has become the cornerstone for virtually all subsequent work in this field.

electronegative adj 電気陰性の ☞ electropositive

▶ The most electronegative elements are those located in the upper right-hand corner of the periodic table.

electronegativity n 電気陰性度

▶ The electronegativity of an atom is a

measure of the net attractive force experienced by an outermost electron interacting with the nucleus.

electron-electron repulsion n 電子間反発 syn *electron repulsion*

▶ The ionization energy and electron affinity of F differ because of the extra electron−electron repulsion energy involved in forming F^- as compared to F^+.

electron emission n 電子放射

▶ There could be no electrochemical reaction path in classical mechanics, because, according to it, the rate of electron emission to a solution at room temperature is negligible.

electron energy loss spectroscopy n 電子エネルギー損失分光

▶ The structure is further defined by determining the vibrational frequencies of the adsorbate on the surface using IR spectroscopy and electron energy loss spectroscopy.

electroneutrality n 電気的中性

▶ In mica phlogopite, one quarter of the Si atoms in talc are effectively replaced by Al, and extra K^+ ions are added to preserve electroneutrality.

electroneutrality principle n 電気的中性の原理

▶ A further consequence of the electroneutrality principle is that the highest oxidation states of the metals are stabilized by strongly electronegative, non-π-accepting ligands.

electron exchange n 電子交換

▶ At still smaller separations, overlap of electron orbitals may occur and electron-exchange forces assume significance.

electron gun n 電子銃

▶ The wavelength of electrons is related to velocity, which is governed by the voltage through which they are accelerated in an electron gun.

electronic adj **1** 電子の

▶ The importance of electronic and steric balance is apparent in complexes such as $Ni(PPh_2Et)_2Br_2$, which can be isolated as a green tetrahedral complex and also as a brown square planar complex.

2 電子工学の

▶ Silicon tetrachloride is available in extremely high purity as a result of demands for electronic grade material by the semiconductor industry.

electronic absorption spectrum n 電子吸収スペクトル

▶ Electronic absorption spectra in the range 200-1400 nm were acquired on a Cary 2300 spectrophotometer.

electronically adv **1** 電子的に

▶ In extended crystalline solids, a vast number of atoms interact electronically with one another.

2 電子工学的に

▶ Modern instrumentation allows the differentiation to be obtained electronically.

electronic conductivity n 電子伝導率 ☞ ionic conductivity, electrolytic conductivity

▶ The tendency of oxides and sulfides to exhibit electronic conductivity is connected with their ability to depart from stoichiometry.

electronic conductor n 電子導体 ☞ ionic conductor

▶ In metallic or electronic conductors, the current is carried by a flow of electrons, the atomic nuclei remaining stationary.

electronic configuration n 電子配置

▶ Transitions between one electronic configuration and another are responsible for most of lines in the emission and absorption spectra of atoms.

electronic energy level n 電子エネルギー準位

▶ The absorption and emission spectra of the hydrogen atom, which arise from transitions between the quantized electronic energy levels of the atom, also have counterparts in doped semiconductors.

electronic polarization n 電子分極 syn *distortion polarization*

▶ The electrons in the sample molecules will be displaced with respect to the nuclei toward the positive pole of the applied electric field. This effect is known as the electronic polariza-

tion.

electronic repulsion n 電子反発

▶ In advanced MO methods, explicit consideration is given to electronic repulsions.

electronics n エレクトロニクス

▶ Molecular electronics is a term that covers both the use of molecular materials in electronics and electronics at the molecular level.

electronic spectrum n 電子スペクトル

▶ The appearance of the vibrational structure of an electronic spectrum can be accounted for in terms of the Franck–Condon principle.

electronic state n 電子状態

▶ Density functional theory describes the electronic states of atoms, molecules, and materials in terms of the three-dimensional electronic density of the system, which is a great simplification over wave function theory.

electronic structure n 電子構造

▶ The basic concepts of band theory provide the key to understanding why certain organic materials behave like metals; the conduction properties depend on the electronic structure of the energy levels in these solids.

electronic transition n 電子遷移

▶ The blue color of oxygen in the liquid and solid phases is due to electronic transitions by which molecules in the triplet ground state are excited to the singlet states.

electron impact n 電子衝撃

▶ Ethyl radicals in the molecular beam are ionized by electron impact and then mass selected by a quadrupole so that only mass 29 is detected by the electron multiplier.

electron microscope n 電子顕微鏡

▶ The amorphous calcium phosphate phase appears spherical in the electron microscope (diameter ca. 300–1000 Å), unlike the needle-like crystals of hydroxyapatite.

electron microscopic adj 電子顕微鏡による

▶ The particles in silica sols are spherical, and electron microscopic examination of gels shows that these too are based on roughly spherical particles, generally about 100 Å in diameter which are themselves composed of still smaller particles, about 15 Å in diameter.

electron microscopy n 電子顕微鏡法

▶ With the technique of direct lattice imaging, electron microscopy has found great application for structural studies of crystallographic shear phases.

electron-pair acceptor n 電子対受容体

▶ On the Lewis theory, H^+ is an electron-pair acceptor and OH^- an electron-pair donor.

electron-pair donor n 電子対供与体

▶ AsF_6 and SbF_6 are extremely powerful fluorinating and oxidizing agents, and they also have a strong tendency to form complexes with electron-pair donors.

electron repulsion n 電子間反発 syn *electron-electron repulsion*

▶ In the HMO method, we neglected to incorporate electron-repulsion effects not only between the two electrons in one orbital but between electrons in different orbitals.

electron-rich adj 電子過剰の ☞ electron-deficient

▶ The binding in these complexes results from attractive forces between electron-rich and electron-poor substances.

electron scattering n 電子散乱

▶ Important characteristics of the Cooper-pair electrons are that their spins are paired, and the combined momentum of pairs is not affected by electron scattering.

electron spectroscopy n 電子分光法

▶ Electron spectroscopy techniques measure the kinetic energy of electrons that are emitted from matter as a consequence of bombarding it with ionizing radiation or high-energy particles.

electron spectroscopy for chemical analysis n X線光電子分光法 ☞ X-ray photoelectron spectropcopy

▶ The electrons that are produced in ESCA cannot escape from solids unless they are ejected within 2 to 5 nm of the surface.

electron spin n 電子スピン

▶ The magnetic properties of unpaired electrons are regarded as arising from two causes, electron spin and electron orbital motion.

electron spin resonance n 電子スピン共鳴 syn *electron paramagnetic resonance*

▶ The most instructive ESR spectra are obtained from single crystals where anisotropic and isotropic interactions may be separated.

electron transfer n 電子移動

▶ Oxygen is much more electronegative than hydrogen, although not enough to effect complete electron transfer.

electron-transport chain n 電子伝達鎖

▶ A group of porphyrin-containing enzymes called cytochromes are part of the electron-transport or respiratory chain in all cells capable of aerobic metabolism.

electron-withdrawing adj 電子求引性の syn *electron-accepting*

▶ The more strongly electron-releasing or electron-withdrawing the substituents, the more important the polar factor.

electro-optic adj 電気光学的な syn *electro-optical*

▶ Electro-optic devices that use sublimated layers of various fluorescent dyes have been demonstrated with high efficiencies, high brightness outputs, and low dc drive voltages.

electro-optical adj 電気光学的な syn *electro-optic*

▶ Successful applications of liquid crystals have been developed in the area of electro-optical displays.

electroosmosis n 電気浸透

▶ If a membrane separates two identical liquids or solutions and a potential is applied across the membrane, there results a flow of liquid through the pores of the membrane. This phenomenon is known as electroosmosis.

electrophile n 求電子試薬 syn *electrophilic reagent*

▶ In nucleophilic substitution, the substrate is acting as an electrophile, and in electophilic addition, the substrate is acting as nucleophile.

electrophilic adj 求電子の ☞ nucleophilic

▶ Electrophilic attack by NO^+ occurs at the position of highest electron availability in primary and secondary amines.

electrophilic addition n 求電子付加 ☞ nucleophilic addition

▶ Electrophilic substitution, like electrophilic addition, is a stepwise process involving an intermediate carbocation.

electrophilic aromatic substitution n 求電子芳香族置換

▶ Most electrophilic aromatic substitutions entail an addition intermediate called a σ complex.

electrophilic substitution n 求電子置換 syn *electrophilic replacement*

▶ Substitution of more than one hydroxyl group on an aromatic ring tends to make the ring particularly susceptible to electrophillic substitution, especially when the hydroxyls are meta to one another.

electrophilic substitution reaction n 求電子置換反応

▶ Unlike most other electrophilic substitution reactions, sulfonation shows a moderate isotope effect: ordinary hydrogen is displaced from an aromatic ring about twice as fast as deuterium.

electrophilicity n 求電子性 ☞ nucleophilicity

▶ Electrophilicity and Lewis acidity do not necessarily parallel one another, for each is markedly dependent upon the reacting center on the substrate.

electrophoresis n 電気泳動

▶ Electrophoresis is important in the study of proteins because the molecules of such materials act like colloidal particles.

electrophoretic effect n 電気泳動効果

▶ The moving ionic atmosphere drags solvent molecules with it, and as a result there is an additional retardation; this is known as the electrophoretic effect.

electrophoretic mobility n 電気泳動移動度

▶ The velocity of movement of a solid cylindrical particle through a liquid under the influence of an applied field of unit potential gradient is called the electrophoretic mobility.

electrophotography n 電子写真

▶ Surface charges on organic photoconductive insulators are important in certain forms of xerography, which is a type of electrophotography.

electroplate vt 電気めっきする

▶ By working at high current density and low temperatures, we can electroplate from aqueous solution metals that lie above hydrogen in the activity series.

electroplating n 電気めっき

▶ The most common application of electrolytes is in electroplating of metals in which dissolved metal salts are the electrolytes.

electropolishing n 電解研磨

▶ Electropolishing is actually the reverse of electroplating.

electroporation n エレクトロポレーション

▶ In electroporation, an instrument called an electroporator produces a brief electric shock that introduces DNA into bacterial cells without killing most of the cells.

electropositive adj 電気陽性の ☞ electronegative

▶ Magnesium is more electropositive than the amphoteric Be and reacts more readily with most of the nonmetals.

electrorefining n 電解精錬 syn *electrolytic refining*

▶ Hydrogen overvoltage makes possible, for instance, zinc and nickel plating and the electrorefining of iron.

electrospray ionization n エレクトロスプレーイオン化法

▶ Electrospray ionization is a generic technique for transferring solution-phase multiply charged anions into the gas phase for detailed characterization.

electrostatic adj 静電気的な

▶ The electrostatic forces between these positive ions and the negatively charged layers make mica considerably harder than kaolinite and talc.

electrostatically adv 静電気的に

▶ The inner and outer surfaces of the virus protein cage provide electrostatically dissimilar environments.

electrostatic effect n 静電効果

▶ A small average movement of electrons will cause a very large electrostatic effect at the short distances which correspond to atomic diameters.

electrostatic force n 静電気力

▶ A great deal of energy is necessary to overcome the powerful electrostatic forces holding together an ionic lattice.

electrostatic interaction n 静電気的相互作用

▶ The design of cavities acting receptors for other anions is feasible; the major requirement is to provide suitable electrostatic interactions and the correct array of hydrogen bonds for the anion, while hindering as much as possible the hydration of the hydrogen bond donor sites.

electrostatic repulsion n 静電気的反発

▶ The S_N2 mechanism is impeded by electrostatic repulsion between the nucleophile and the π cloud in the halide.

electrostriction n 静電収縮

▶ The increase of internal pressure by dissolved electrolytes is achieved through electrostriction, namely, a volume-reducing process which involves polarization and attraction of solvent molecules around the ionic species.

elegant adj 手際のよい syn *clever, ingenious*

▶ This elegant explanation was substantiated experimentally by showing that the conductivity of an ionic crystal in the low-temperature region could be varied appreciably by the incorporation of altervalent ions.

elegantly adv 手際よく

▶ The distinction between Al and Si atoms in minerals can be made simply and elegantly by calculating experimental cation valences.

element n **1** 元素

▶ All elements heavier than lead are unstable and radioactive.

2 行列の要素

▶ If all the elements on one side of the principal diagonal are zero, the value of a determinant is equaled to that of the principal term.

3 成分 syn *component, constituent*

▶ We have decomposed the elements of the molecular magnetic susceptibility tensor into additive group contributions.

4 複数形で，初歩 syn *essentials, fundamentals*

▶ A genuine analysis of a problem in chemical physics or physical chemistry would demand far more than a cursory account of some elements of physics grafted onto chemistry or some elements of chemistry grafted onto physics.

elemental adj **1 基本的な** syn *basic, fundamental*

▶ Exchange forces between electrons represented empirically by the Weiss field provide the coupling between the elemental magnets which are fundamental to ferromagnetism and antiferromagnetism.

2 本源的な syn *original, primordial*

▶ It was hoped that elemental levels of trace and minor elements would be determined primarily by dietary factors.

3 元素の syn *elementary*

▶ In some cases, two different compounds from the same elements are prepared by varying only the elemental proportions.

elementary adj **1 元素の** syn *elemental*

▶ Of the alkali metals, only lithium will react with elementary nitrogen as N_2.

2 初歩の syn *basic, fundamental*

▶ There are other very large areas where inorganic elements are essential and where much elementary chemistry is not understood.

elementary reaction n **素反応**

▶ An elementary reaction is a reaction that occurs in a single step, with no experimentally detectable reaction intermediates.

elementary step n **素過程** syn *elementary process*

▶ The analysis of a composite mechanism involves explaining the overall behavior in terms of the kinetics of the elementary steps.

element of symmetry n **対称要素**

▶ The simplest elements of symmetry are axis of symmetry, center of symmetry, and plane of symmetry.

elevated adj **高い** syn *high*

▶ Pure metallic borides are best prepared by heating powdered boron with the powdered metal in an inert atmosphere at an elevated temperature.

elevation n **上昇** syn *enhancement, promotion*

▶ If consumed, dinitrophenol will cause a rapid elevation in fuel oxidation and oxygen consumption.

elicit vt **誘い出す** syn *bring out, evoke*

▶ Each new foreign cell that invades the body elicits a new supply of specific antibodies, which are stored in the blood as γ-globulins.

eliminate vt **除く** syn *exclude, omit, reject, remove*

▶ Great attention has to be paid to the geometry of the reactive aggregate that is built on the clay surface, as reactions may be enhanced or eliminated by its exact architecture.

elimination n **1 脱離**

▶ Most E2 eliminations tend to yield the most stable alkene, which is usually the most highly substituted alkene.

2 除去 syn *exclusion, removal*

▶ Fluorescent nanoparticles could be favorably used for tumor elimination.

elimination reaction n **脱離反応**

▶ The theory of elimination reactions developed in a way remarkably similar to the way the theory of nucleophilic substitution developed.

ellipsoidal adj **楕円体の**

▶ If the core is nonspherical, the simplest assumption that we can make about its shape is that it is ellipsoidal, either prolate or oblate.

elliptically adv **楕円形に**

▶ Since the velocity of the extraordinary ray varies elliptically with angle to the optic axis, so also, inversely, does its index of refraction.

elongated adj 引き伸ばされた ☞ flattened

▶ The majority of dyes of the cyanine type have flat elongated molecules with the charged groups at the ends of the long axis.

elongation n 伸長 ☞ flattening

▶ Magnesium metal is quite remarkable in that it can be stretched out to an elongation up to several times its original length.

else adv そのほかに syn *besides, in addition*

▶ This method is suitable only for the determination of manganese in solutions of pure manganese salts containing nothing else except alkali and ammonium salts.

elsewhere adv どこかよそに syn *somewhere else*

▶ The glove-box system, electrochemical cell, glassy carbon working electrode, and thermostatted furnace have been described elsewhere.

elucidate vt 明らかにする syn *clarify*

▶ One of the most complex terpenoid biosynthesis which has been elucidated is the synthesis of cholesterol from the thirty-carbon-atom acyclic triterpenoid, squalene.

elucidation n 解明 syn *explanation, interpretation*

▶ Stewart and Warren were concerned with the measurement and elucidation of diffraction peaks that were characteristic of spacings much greater than those attributable to atomic nearest neighbors.

elude vt …からすり抜ける syn *escape*

▶ The P−N bond is one of the most intriguing in chemistry, and many of its more subtle aspects still elude a detailed and satisfactory description.

eluent n 溶離液 syn *eluant*

▶ The eluent liquid is collected in numbered fractions, each of which is evaporated for examination of the residue.

elute vt 溶離する

▶ In an investigation of an unknown mixture, one elutes first with petroleum ether, then with petroleum ether–benzene mixtures varying from 4:1 to 1:4, then with benzene, benzene–ether mixtures, ether, and ether–methanol mixtures.

elution n 溶離

▶ Solvents employed for elution, listed in the order of increasing eluent power, are; petroleum ether, benzene, ether, methanol.

emanate vi 発する syn *merge, originate, radiate*

▶ In these diagrams, as each B has four valence orbitals four lines emanate from each open circle.

embark vi **embark upon** …に乗り出す syn *initiate, launch, start*

▶ We embarked upon a study of 2,5-dibenzylidenecyclopentanone for four reasons.

embarrassment n 当惑 syn *dilenma, problem, trouble*

▶ Zeise's salt and a few closely related complexes remained as chemical curiosities and a considerable theoretical embarrassment for over 100 years.

embed vt 埋め込まれる syn *inbed*

▶ The hydrocarbon tail embeds itself into the oily phase, and the polar head sticks into the water.

embody vt **1** 体現する syn *exemplify, symbolize*

▶ The law of mass action embodies one of the most important principles utilized in analytical chemistry.

2 取り入れる syn *include, incorporate*

▷ In spite of detail embodied in our final equation, the agreement with experiment is not favorable.

embrace vt 含む syn *cover, include*

▶ The absorbing chromophore embraces the electrons of the unsaturated substituent as well as those of the ring.

embrittle vi 脆化する

▷ Bosch empirically discovered the embrittling effect of hydrogen on steel, which was vividly demonstrated by the alarming tendency of the reactors to blow up at frequent intervals.

embrittlement n 脆化

▶ In metals, the primary cause of embrittlement is exposure to hydrogen, though other factors such as corrosion also are involved.

embryo n 萌芽 syn *beginning, origin*

▶ The formation of embryos involves an increase in free energy, since the entropy loss associated with the transfer of a free molecule to a small cluster is not fully compensated for by the attractive energy between the molecule and the cluster.

embryonic adj **1** 萌芽的な syn *developing, immature, undeveloped*

▶ The membrane surface acts as a charged template for ion localization and subsequent formation of the embryonic crystallite.

2 胚の

▶ Expression of a particular gene may be lethal to a transgenic organism in its embryonic form.

emerge vi **1** 明らかになる syn *appear, be revealed, turn out*

▶ The concept of bond orientational order has emerged in recent years from theoretical considerations of two-dimensional systems.

2 出現する syn *arise, come out, result*

▶ Although many compounds emerge from preparative procedures in crystal form, these crystals are rarely suitable for structure determination.

emergence n 出現 syn *appearance*

▶ The photodimerizations have been shown to occur initially at sites of emergence of nonbasal dislocations.

emergent adj 出現する

▶ Line defects can be detected by reflected light microscopy because in the region of an emergent dislocation at the crystal surface, the crystal structure is in a stressed condition.

eminently adv 抜きん出て syn *exceedingly, extremely, very*

▶ The Gouy method is eminently satisfactory for metals, alloys, and other materials capable of being fabricated in the form of long uniform cylinders providing the materials are not ferromagnetic.

emission n **1** 発光 syn *radiation*

▶ In general, organic compounds exhibit both a short-lived emission and a long-lived emission.

2 複数形で，排出物 syn *exhaust*

▶ Air pollution by SO_2 gas emissions has been the subject of increasing legislation and control.

emission spectroscopy n 発光分光法

▶ In emission spectroscopy, the spectra of elements are in the form of lines of distinctive color; those of molecules are groups of lines called bands.

emission spectrum n 発光スペクトル

▶ Emission spectra are usually mirror images of the corresponding absorption spectra when plotted on a frequency scale.

emissive adj 発光の

▶ Electroluminescence offers great promise for fabricating low operating voltage, large area emissive panels and displays.

emit vt 放射する syn *emanate, radiate*

▷ For black and white screens, mixtures of the blue-emitting $ZnS:Ag^+$ and yellow-emitting $(Zn, Cd)S:Ag^+$ are used.

emitter n **1** エミッター

▶ In $YVO_4:Eu^{3+}$, the vanadate group absorbs energy in the cathode ray tube but the emitter is Eu^{3+}.

2 放射体

▶ A nuclide which is neutron-rich is normally a β^- emitter, one which is neutron-deficient may be a β^+ emitter.

emphasis n 強調 emphasis and importance を一つの概念と見做して，動詞は単数形にしている． syn *attention, significance*

▶ Considerable emphasis and importance has been placed upon the determination of magnetic properties of transition-metal complexes.

emphasize vt 強調する syn *call attention, stress*

▶ These observations emphasize need for a better theoretical understanding of the nature and mechanism of radiationless transitions within molecules.

empirical adj 経験的な syn *experimental, prac-*

tical, pragmatic

▶ The Trouton rule for estimating the enthalpy of vaporization of a liquid is one of the oldest and best known of all empirical thermodynamic rules.

empirical expression n 経験式

▶ A number of empirical expressions have been suggested to account for the variation of surface tension with temperature.

empirical formula n 実験式

▶ Many with the empirical formula ABC_3 have the cubic perovskite ($CaTiO_3$) structure. Examples are $CaZrO_3$, $LaAlO_3$, and $KMgF_3$.

empirically adv 経験的に

▶ Although much is known about the theory of crystal growth, the crystallization of compounds is still based on methods established empirically through changes of solvent, temperature, and ionic strength.

empiricism n 経験主義

▶ Since so many factors are involved, the development of ion-selective electrodes is necessarily done on the basis of a good deal of empiricism.

employ vt 利用する syn *apply, use, utilize*

▶ Most laboratory studies employ specific surface of single crystals, for the simple reason that the different planes show different behavior.

employment n 利用 syn *application, use, utilization*

▶ The employment of ferrites in electromagnetic devices is associated with their very square hysteresis loops, their high-speed response, and their high flux output.

empty adj 空の syn *unfilled, vacant*

▶ The orbital picture of the benzyl cation is similar to that of the benzyl free radical except that the p orbital that overlaps the π cloud is an empty one.

emulsification n 乳化

▶ Emulsification can be completely spontaneous in strict absence of mechanical stirring and even downward against the direction of gravity for a light oil emulsifying in heavier water.

emulsify vt 乳化する

▷ Soaps exert cleaning action by emulsifying the oily components of soils.

emulsifying agent n 乳化剤

▶ It is frequently of great advantage to have the emulsifying agent formed in the interface by chemical reaction, as in the formation of soap from aqueous alkali and oil containing fatty acid.

emulsion n エマルション

▶ All emulsions consist of a continuous phase and a disperse phase: in an oil-in-water emulsion, water is the continuous phase and oil the disperse phase; in a water-in-oil emulsion, oil is the continuous phase and water droplets are the disperse phase.

enable vt ＋目的語＋ to do something の形で，…することを可能にする syn *allow, approve, permit*

▶ Careful control of temperature, pressure, and reaction time enables the yield of the various intermediate boranes to be optimized.

enamel n **1** ほうろう syn *glaze*

▶ Porcelain enamel is composed of various blends of low-sodium frit, clay, feldspar, and other silicates, ground in a ball mill, and sprayed onto a metal surface to which it bonds firmly after firing, giving a glass-like fire-polished surface.

2 歯のエナメル質

▶ Because lactic acid dissolves enamel and dentin in teeth, it leads to cavities and tooth decay.

enantiomer n 鏡像異性体 syn *optical isomer*

▶ Enantiomers are usually separated by forming a compound with one enantiomer of a second optically active compound and utilizing differences in physical properties between the products.

enantiomeric adj 鏡像異性体の

▶ Absolute configuration and enantiomeric purity were retained in all cases.

enantiomerically adv 鏡像異性体として

▶ When the dithiane oxide is used in the enantiomerically enriched form, the active sites are able to discriminate between the enantiomers of butan-2-ol.

enantiopure adj エナンチオピュア

▶ Despite the mixture of products obtained, the shorter reaction times and good yields of enantiopure products represent significant contributions to this area.

enantioselection n エナンチオ選択

▶ The mechanism by which enantioselection occurs has become a matter of intense debate.

enatioselective adj エナンチオ選択性の

▶ When rate enhancement is observed, then unmodified active sites are not a significant source of reaction and do not interfere with enantioselective reactions.

enantioselectivity n エナンチオ選択性

▶ The attainment of high enantioselectivities requires the absence of unmodified active sites.

enantiospecificity n エナンチオ特異性

▶ Enantiospecificity is the rule for the countless reactions taking place in the chiral medium provided by the optically active enzymes of living organisms.

enantiotopic adj エナンチオトピック

▶ The word enantiotopic means in mirror-image places.

enantiotropic adj 互変的な ☞ monotropic

▶ Enantiotropic forms are those having a definite transition temperature, below which one form and above which the other form is stable, and at the transition temperature both forms are in equilibrium.

enantiotropy n 互変 ☞ monotropy

▶ There is no absolutely safe generalization relating enantiotropy and monotropy to the properties of the polymorphs, except location of the transition temperature.

encapsulate vt カプセルに包む

▶ When *cis*-$[Mn(bpy)_2]^{2+}$ complexes are encapsulated within the micropores of zeolite Y and X, they can act as effective alkene epoxidation catalysts.

encapsulation n カプセル化

▶ In organic reactions, encapsulation of the cation by lipophilic ligand may enhance the nucleophilic activity of the anion.

enclose vt 取り囲む syn *confine, surround*

▶ Cells are enclosed by a membrane about 7000-pm thick and composed of double layers of protein separated by lipids.

encode vt 暗号化する syn *code*

part ▶ Nucleosides and nucleotides play a much broader role in biology than simply forming a double helix and encoding the copying of that helix.

encompass vt **1** 包含する syn *embrace, include*

▷ Numerous, commercially important, organic reactions encompassing hydration of olefins and the dehydration of alkanols could be smoothly effected by the appropriate clay catalyst under mild conditions.

2 取り囲む syn *enclose, surround*

▷ For certain applications, low loss materials are required, one essential for which is that the area encompassed by the hysteresis loop should be as small as possible.

encounter n **1** 出会い syn *meeting*

▶ Four separate situations may be visualized in which encounters are either in the gas phase or in solution and either between noninteracting molecules or between molecules which attract or repel each other strongly.

vt **2** 出会う syn *come upon, meet*

▶ Cis-trans isomerism is very frequently encountered.

encourage vt **1** 助長する syn *aid, advance, promote*

▶ A uniform size in the compacted powder encourages nonporous microstructures to form on densification.

2 励ます syn *inspire, stimulate*

▶ We were encouraged in this aim by reports that N-alkylated oligo(*p*-benzamides) and oligo (*m*-benzamides) adopt a well-defined, helical secondary structure.

encouraging adj 励みになる syn *favorable, promising*

▶ The results are extremely encouraging and confirm both the value of scanning tunneling microscopy and potential for understanding surface processes and structures by using a multidisciplinary approach.

end n **1** 物事の結末 syn *conclusion, termination*

▶ The following procedure leads to the same end, but the results are not quite so reliable.

2 末端 syn *extreme, extremity*

▶ The positive end of the halogen dipole attacks the aromatic compound, while the negative end is complexed with the catalyst.

3 目的 syn *aim, object, purpose*

▶ The calculation of the potential energy surface is not necessarily an end in itself.

adj **4** 最後の syn *final, last*

▶ Relative rate methods with gas chromatographic analysis of the end reaction mixture provide an alternative analysis technique.

vt **5** 終える syn *cease, stop, terminate*

▶ After heating was ended and whilst still warm, the red organic layer was decanted to distillation flask under a nitrogen atmosphere.

endeavor n 努力 syn *attempt, effort*

▶ The chemical study of materials in general and the paths to high-performance materials in particular owe a good deal of the processes of cold, logical endeavor.

endergonic adj 吸エルゴンの ☞ exergonic

▶ Processes which require energy input to drive them have a positive ΔG and are termed endergonic.

end group n 末端基 syn *terminal group*

▶ A methylation that yields 0.25% of tetra-*O*-methyl-D-glucose shows that for every end group there are about 400 chain units.

ending n 語尾 syn *suffix*

▶ The ending -ate is a characteristic ending for the name of anions of oxoacids and their derivatives.

endless adj 無限の syn *infinite, limitless, unlimited*

▶ The structure of the iodine–dioxane complex involves endless chains of dioxane and iodine molecules in which both halogen atoms within a molecule participate equally in intermolecular bonding with separate donor molecules.

end member n 固溶体の末端メンバー

▶ In general, two solids are most likely to be completely miscible in one another if the substituting atoms or ions are similar in chemistry and size and if the end members of the series have the same structure.

endogenous adj 内因性の ☞ exogenous

▶ If a transgene randomly inserts into an endogenous gene, or its control region, the normal function of the endogenous gene can be disrupted.

endogenously adv 固有に

▶ The liver synthesizes cholesterol and repackages both the diet-derived and endogenously synthesized forms into lipoproteins for export to the peripheral tissues.

end-on adj エンドオンの

▶ In end-on views of the models, the eclipsed conformation is seen to have the hydrogens on the forward carbon directly in front of those on the back carbon.

endotherm n 吸熱 ☞ exotherm

▶ The DTA curve of **1c** measured at a heating rate of 5°/min showed two well-separated exotherms followed by the final endotherm due to melting of **3**.

endothermic adj 吸熱の ☞ exothermic

▶ In some cases, energy in the form of heat, light, or electricity is required to initiate the change; then the reaction is called endothermic.

endothermically adv 吸熱して ☞ exothermally

▶ Many sodium salts dissolve endothermically in water because the heat liberated by hydration of the ions is inadequate to break down the lattice.

endothermic reaction n 吸熱反応 ☞ exothermic reaction

▶ Other things being equal, an exothermic reaction will proceed more readily on a thermodynamic basis than will an endothermic reaction.

endow vt 付与する syn *supply*

▶ A screw dislocation would endow a singular surface with a perpetual step against which molecules could be continuously organized into the lattice.

end point n 終点

▶ With phenolphthalein, the end point is reached when 1 mole of sodium carbonate has reacted with 1 mole of hydrochloric acid.

end product n 最終生成物

▶ The end products of elemental combustion are oxides; of organic compounds are carbon dioxide and water.

energetic adj **1** エネルギーの

▶ Dislocations are good candidates as nucleation centers from both an energetic and structural point of view.

2 強力な syn *powerful, strong*

▶ The electrons that are produced in ESCA are not very energetic and are rapidly absorbed by solid matter.

energetically adv エネルギー的に

▶ The most energetically stable combination has the fewest nodes between atoms, and the most unstable combination has the most nodes between atoms.

energetics n エネルギー論

▶ Their results are unable to account for the most salient characteristic of perhalate energetics, namely, the fact that both perbromates and periodates are very much less stable to reduction than are perchlorates.

energize vt エネルギーを与える syn *activate, excite*

▶ There is negligible reaction unless H_2 is energized photolytically or by a high-current arc.

energy n エネルギー

▶ An impurity with an excited state of energy lower than that of the host will quench the fluorescence of the host.

energy band n エネルギーバンド ☞ conduction band, valence band

▶ The occupancy of these energy bands is critically important: when the energy gap between the valence band and the conduction band is large, the material is an insulator.

energy barrier n エネルギー障壁

▶ The geometric structure of a molecule depends upon the magnitude of the energy barrier which prevents conversion into any of its geometric isomers.

energy conversion n エネルギー変換

▶ If a circuit is prepared by making low resistance contacts with metals, one to the n-type material and one to the p-type material, a variety of energy conversion schemes can be realized once a voltage source or light source is introduced into the circuit.

energy gap n エネルギーギャップ ☞ band gap, forbidden band

▶ Which of the two positions gives more stable Cr^{2+} ion depends on the energy gap between the higher and lower levels, and this in turn depends on the ligands.

energy level n エネルギー準位

▶ For a pair of molecules, the interaction leads to a splitting of the single-molecule energy level into a pair of levels.

energy loss n エネルギー損失

▶ An additional source of energy loss in an alternating magnetic field is associated with electric currents called eddy currents that are induced in the material.

energy metabolism n エネルギー代謝

▶ A deeper understanding of the regulation of energy metabolism can be gained experimentally by examining the outcome when parts of the system are inhibited or activated by added chemical species.

energy transfer n エネルギー移動

▶ Energy transfer is of considerable importance in crystal spectroscopy.

enforce vt 強いる syn *demand, impose, require*

▶ High-fidelity DNA polymerases tightly surround the incipient base pair to enforce the size and shape of the incoming nucleobase.

engage vi engage in …に携わる syn *embark, enter, participate*

▶ Hydrogen bonds are important because the protein chain contains a sequence of groups that can engage in hydrogen bonding.

engaged adj 従事して syn *busy, interested*

▶ A few basic points outlined here are of particular relevance to chemists engaged in the design and study of new materials.

engineer vt **1** 設計する syn *devise, invent, plan*

▶ Less dramatic, but no less surprising, is the fact that organic polymers may now be engineered so as to be freely conducting.

vt **2** 工作する syn *arrange, manipulate, set up*

▶ Plants can be engineered to produce a wide range of pharmaceutical proteins in a broad array of crop species and tissues.

engulf vt 巻き込む syn *envelop, plunge*

▶ Calcium ions can be engulfed by cryptands and crowns of a suitable size.

enhance vt 高める syn *amplify, elevate, improve, increase*

▷ Enhancing interest in fluorescence has been the realization that this property can arise in a wide variety of molecules and structures.

enhancement n 増大 syn *elevation, improvement*

▶ It is tempting to link the observation of the rate enhancement effect with the enantioselection that is achieved with this catalyst system.

enhancer n エンハンサー

▶ Enhancers greatly influence the frequency at which the gene is transcribed,

enjoy vt 経験する syn *have*

▶ The rule has enjoyed greater success than the attempts to explain how it works.

enlarge vt 大きくする syn *expand, lengthen*

▶ The complete miscibility of the lower alcohols with water falls off as the alkyl group is enlarged.

enlargement n 拡大 syn *expansion, swelling*

▶ The rock powder is heated in a hard glass tube, closed at one end, containing one or more enlargements to provide for the condensation of the water.

enolate n エノラート

▶ A range of variously substituted 1, 2- and 1,3-cyclic sulfamidates react efficiently with S-substituted enolates to give α-thiosubstituted lactams.

enolate anion n エノラートアニオン

▶ Acetone forms an enolate anion easily but is relatively poor as an acceptor.

enolization n エノール化

▶ 5-Nitro-2, 3-dihydro-1,4-phthalazinedione is soluble in alkali by virtue of enolization.

enormous adj 莫大な syn *huge, immense*

▶ The areas involved in a porous body are enormous; for example, 2.6 million square meters in one kilogram of charcoal.

enormously adv 非常に syn *exceedingly, vastly*

▶ Enzymes are proteins that act as enormously effective catalysts for biological reaction.

enough pron **1** 十分な量 syn *plenty*

▶ In softening water by use of calcium hydroxide or sodium carbonate, enough of the substance is used to cause magnesium ion to be precipitated as magnesium hydroxide and iron as iron(II) hydroxide or iron(III) hydroxide.

adj **2** 十分な syn *adequate, sufficient*

▶ Enough ammonium salt should be present to prevent the precipitation of $Mg(OH)_2$ upon adding aqueous ammonia.

adv **3** 修飾する形容詞，副詞の後に置いて十分に syn *sufficiently*

▶ Other homologous series of compounds with *n*-alkyl groups that have been studied thoroughly enough show similar regularities in melting points and enthalpies of fusion but always with significant differences from the behavior of the *n*-alkanes.

enrich vt 濃縮する syn *refine, upgrade*

part ▶ Use of ^{15}N-enriched reagents shows that all the N from HNO_2 goes quantitatively to the internal N of N_2O.

enrichment n 濃縮 syn *refinement*

▶ Enrichment of boron, chlorine, and sulfur isotopes has been achieved by a chemical reaction accompanied by irradiation from a laser beam.

ensemble n **1** 集合 syn *aggregate, collection*

▶ Diffusion in the particle as well as evaporation and recondensation of the particle ensemble is likely to contribute to this dynamical behavior.

2 アンサンブル

▶ Since the ensemble is an isolated system, the entropy of the most probable state of the ensemble must be a maximum.

ensure vt 確実にする syn *assure, confirm, quarantee, secure*

▷ The ionic liquids with optimum conductivity and viscosity are generally highly fluorinated ensuring good shielding of charge from the cation.

entangle vt もつれる syn *tangle*

▶ In a jelly the unit is the primary particle, but these are sufficiently linked so that the whole liquid is entangled in the loose framework formed by aggregation.

enter vi **1 enter into** …に携わる syn *engage, participate*

▶ A large number of polynitro compounds enter into addition compounds which, in at least superficial aspects, resemble quinhydrones.

vt **2** 入り込む syn *penetrate*

▶ A variety of atoms can enter a silicon crystal and introduce energy levels within the band gap in accord with their ionization properties.

enthalpy n エンタルピー

▶ We shall assign zero enthalpy to each element at 298.15 K when it is in the standard state of the most stable form of the element at this temperature.

enthalpy of combustion n 燃焼エンタルピー syn *heat of combustion*

▶ Enthalpies of formation of organic compounds are generally deduced by well-known thermochemical methods from their enthalpies of combustion.

enthalpy of formation n 生成エンタルピー syn *heat of formation*

▶ The tabulated values of enthalpies of formation are enthalpies of reactions written in such a way that the stoichiometric coefficient of the regarded product is 1 and that the reactants are elements in their standard states.

enthalpy of hydration n 水和エンタルピー

▶ The ability of difluorine to oxidize water is a result, in part, of the low F—F bond strength, together with the fairly high electron affinity of the fluorine atom; but it is reinforced by the high favorable enthalpy of hydration for the small fluoride ion.

enthalpy of sublimation n 昇華エンタルピー syn *heat of sublimation*

▶ The enthalpies of sublimation, vaporization, and fusion are related by

$$\Delta H_s = \Delta H_v + \Delta H_m$$

which is an exact relationship if all three quantities are evaluated at the same temperature.

enthalpy of vaporization n 蒸発エンタルピー syn *heat of vaporization*

▶ As the enthalpy of vaporization is an important factor in determining the standard electrode potential of a metal, the metals of the highest boiling point will tend to be noble metals.

enthusiasm n 熱意 syn *interest*

▶ Opportunities to study semiconductor surfaces have attracted considerable enthusiasm and effort.

enthusiastic adj 熱烈な syn *eager, keen*

▶ Polyethyene was discovered by two enthusiastic, applied scientists who felt it was worth subjecting ethylene to very high pressure at the requisite temperature to stimulate polymerization.

entire adj 全体の syn *whole*

▶ Good quality crystals should show a sharp extinction; i.e., on rotating the crystal relative to the polars, extinction should occur simultaneously throughout the entire crystal.

entirely adv 完全に syn *completely*

▶ A natural DNA is replaced by an entirely different structure having fluorescent emission.

entity n 存在 syn *existence, object*

▶ An important recent development has been

the spectroscopic detection of transition species, which are molecular entities having configurations intermediate between those of the reactants and products.

entrance n 入り口 syn *access*

▶ In a mesoporous system, initial adsorption is likely to occur at the exterior surface or entrances to pores.

entrap vt 捕捉する syn *trap*

▷ Assembly of the reagents within the zeolite supercage leads to entrapped complexes that are too large to diffuse out of the zeolite host.

entrapment n 捕捉

▶ Entrapment is limited by the fraction of dispersion volume and to particles whose diameter is smaller than the inner volume of vesicles.

entrench vt 確立する syn *establish*

▶ Despite the fact that the Haber–Bosch process requires high temperature and pressure, it is efficient and well entrenched, and it can produce large volumes of product in short time periods.

entropic adj エントロピー的な

▶ The contraction of an elastomer is an entropic process, being controlled almost entirely by the entropy increase.

entropically adv エントロピー的に

▶ Self-assembly is usually entropically driven in an aqueous system, where association of modules is accompanied by exclusion of ordered water molecules.

entropy n エントロピー

▶ For a gaseous chemical reaction, there is an increase of entropy in the system if there is an increase in the number of molecules.

entropy of activation n 活性化エントロピー

▶ Since electrostriction leads to entropy loss, the result is that a reaction between like charges has a negative entropy of activation and a low preexponential factor.

entropy of fusion n 融解エントロピー

▶ In addition to simple positional defects, it seems possible that the overall entropy of fusion of metals might include extensive cooperative positional disorder as part of the mechanism of melting.

entropy of transition n 転移エントロピー

▶ From the foregoing considerations, it was concluded that the entropy of transition of typical tetrahedral molecules is about 21 J K^{-1} mol^{-1} or of the order of *R*ln 12.

entropy of vaporization n 蒸発エントロピー

▶ The entropies of vaporization of most non-hydrogen-bonded compounds have values in the neighborhood of 88 J K^{-1} mol^{-1}.

entry n 立ち入り syn *admission*

▶ The plasma membrane is a selective filter that controls the entry of nutrients and other molecules needed for cellular processes.

enumerate vt 列挙する syn *list, name, specify*

▷ One of the major barriers to using microwave techniques in materials synthesis is the paucity of data enumerating the dielectric permittivity of solids as a function of temperature.

enumeration n 列挙 syn *account, specification*

▶ The appearance of high-speed computers in recent years has considerably facilitated several types of isomer enumeration.

enunciation n 発表 syn *expression*

▶ Subsequent to the enunciation of the theory of growth of imperfect crystals, many workers began investigating the faces of various crystals for spiral growth features with step heights of molecular dimensions.

envelop vt 包む syn *cover, enclose, wrap*

▷ The protective colloid operates by enveloping the original colloid with a film of the stable colloid and exposing its stabilizing agent.

envelope n 覆い syn *cover*

▶ The boundary is the envelope, real or imaginary, across which the interactions between the system and the surroundings take place.

environment n 環境 syn *atmosphere, circum-*

stance, surrounding

▶ Water is a major constituent of our bodies and of the environment in which we live.

environmental adj 環境の syn *green*

▶ Extreme cold, pressure from living at great depths, high salinity, and other environmental constraints are hardly a barrier because aquatic organisms have adapted to their difficult environments.

environmental chemistry n 環境化学 ☞ green chemistry

▶ In the last few decades, we have become increasingly aware of the issues environmental chemistry encompasses and of its remarkable power to treat those issues.

environmentally adv 環境的に

▶ The fact that the two vibrational modes show different shifts indicates an unusually large and environmentally sensitive shift of the excited-state vibrational mode responsible for the splitting.

environmental monitoring n 環境モニタリング

▶ Measurements of the pH of aqueous solutions are vital in a broad diversity of area ranging from environmental monitoring to clinical chemistry.

environmental pollution n 環境汚染

▶ Environmental pollution is an important problem that is receiving a lot of attention.

envisage vt 想像する syn *conceive, imagine*

▶ For the synthesis of porous materials, networks are often envisaged where rigid organic molecules and metal atoms or clusters replace bonds and atoms in classical inorganic structures.

enzymatic adj 酵素による ☞ enzymic

▶ Enzymatic hydrolysis of cellulose leads to the disaccharide cellobiose.

enzymatically adv 酵素的に

▶ In the pancreas, β-cells synthesize both insulin chains as one polypeptide that is secreted, then enzymatically cut, and folded to join the two subunits.

enzyme n 酵素

▶ Enzymes exist in an ensemble of conformations that exhibit thermal motion on the picosecond to millisecond time scales.

enzyme activity n 酵素活性

▶ Evidence shows that certain other amino acid residues are also vital to enzyme activity.

enzyme electrode n 酵素電極 ☞ amperometric enzyme electrode

▶ The main applications of enzyme electrodes have been in the clinical analysis of biological fluids.

enzymic adj 酵素的な ☞ enzymatic

▶ A very large group of proteins display enzymic activity, which is to say that they have the ability to catalyze specific organic or even inorganic reactions.

enzyme-substrate complex n 酵素-基質複合体

▶ The destabilization of the enzyme-substrate complex can be imagined to be due to distortion of bond angles and lengths from their previously more stable configuration.

epimer n エピマー

▶ Galactose and glucose are epimers of each other, differing in absolute configuration only around the carbon atom in position 4.

epimeric adj エピマーの

▶ From the equilibrium mixture between an aldonic acid and its epimer formed by the treatment with pyridine, the epimeric aldonic acid is separated and reduced to the epimeric aldose.

epimerization n エピマー化

▶ The epimerization takes place, while the monosaccharide is conjugated to uridine diphosphate (UDP); thus, galactose is substituted for glucose on UDP-glucose.

epitactic adj エピタクチックな

▶ Epitactic reactions refer to the two-dimensional orientational relations that exist when a new phase forms on the surface of another.

epitaxial adj エピタキシャルな

▶ This kind of growth, wherein the atoms of the deposited layer are in registry with those beneath on the substrate, is called epitaxial growth.

epitaxially adv エピタキシャルに

▶ Paraffin wax crystallizes epitaxially from solution onto the surface of strips of cold-drawn polyethylene and polyethylene sebacate.

epitaxy n エピタキシー

▶ For epitaxy to occur, the force pattern of the substrate must match the force pattern of the adhering crystals in the contact plane within fairly close limits.

epitope n エピトープ

▶ To detect a particular protein, antibodies are employed which bind specific regions, known as epitopes, on a protein.

epoxidation n エポキシ化

▶ Epoxidation of the cyclic allylic ethers was an attractive option that ultimately proved to be successful.

epoxidize vt エポキシ化する

▶ The encapsulated *cis*-$[Mn(bpy)_2]^{2+}$ catalysts are able to epoxidize a broad range of alkenes, and high epoxide selectivities are observed with 1-hexene, cyclohexene, 1-dodecene, and cyclododecene.

equal adj **1** 等しい syn *identical, same*

▶ Either procedure can be used with equal success.

2 other things being equal ほかのことは同じとして ☞ other things being the same

▶ Other things being equal, boiling point rises with increasing molecular size.

vt **3** 匹敵する syn *correspond, match*

▶ *s*-Tetrachloroethane is an excellent solvent for many substances which dissolve with difficulty in benzene, glacial acetic acid, etc., although it does not equal nitrobenzene.

equality n 同等 syn *uniformity*

▶ Quite appreciable deviations from equality in the vertical columns in Table 1 occur even for related substances.

equalize vt 等しくする syn *compensate, offset*

▶ The addition of MnO_2 to the molten glass produces red-brown Mn^{III} which equalizes the absorption across the visible spectrum so giving a colorless, i.e., grey glass.

equally adv 同様に syn *alike*

▶ Isolation of large complex ions is facilitated by attempting to precipitate them as salts of equally large counter ions.

equate vt 数を等しくする

▶ Because the change in energy is independent of the path, the heat of formation of MX(c) from the elements can be equated to the algebraic sum of the other thermochemical quantities with the appropriate thermodynamic signs.

equating n 等式化

▶ The treatment of the equilibrium aspects of adsorption proceeds via an examination of the adsorption and desorption processes and the equating of their rates at equilibrium.

equation n 方程式

▶ This important equation enables the chemical potential of a component of a liquid mixture to be expressed in terms of the amount of the species present.

equation of state n 状態方程式

▶ The equation of state does not include all the experimental information which we must have about a system or substance.

equatorial adj エクアトリアル位の ☞ axial

▶ There is a single NMR signal for all twelve protons in a cyclohaxene molecule, since their average environments are identical: half equatorial and half axial.

equatorial bond n エクアトリアル結合

▶ The bonds holding the hydrogens that are in the same plane of a cyclohexane ring lie in a belt about the equator of the ring and are called equatorial bonds.

equidense adj 等密度の

▶ For a given molecular size and shape, several almost equidense and, hence, almost equienergetic arrangements are often possible.

equidistant adj 等距離の

▶ In the case of the sodium chloride structure, the nearest neighbors of a cation are six equidistant anions, and vice versa.

equilateral triangle n 正三角形

▶ The $[B_{12}H_{12}]^-$ ion is a regular icosahedron of atoms, each of the twenty faces being an equilateral triangle.

equilibrate vt 平衡させる

part ▶ A ketone and its corresponding enol are called tautomers, a name given to equilibrating isomers which differ only in the position of a mobile proton.

equilibration n 平衡化

▶ The equilibration of the mixture of *cis*- and *trans*-stilbenes into *trans*-stilbene was achieved using iodine and sunlight and proved most convenient.

equilibrium n 平衡

▶ The mixture was stirred an additional 15 minutes to allow the system to come equilibrium, and the mixture was then filtered.

equilibrium constant n 平衡定数

▶ The variation of an equilibrium constant with the temperature provides a useful way of obtaining enthalpy changes for chemical reactions.

equilibrium diagram n 平衡状態図 syn *phase diagram*

▶ Two types of curve arise, depending on whether the curve of partial miscibility in the solid state intersects an equilibrium diagram of type I or type III.

equilibrium geometry n 平衡構造

▶ A much more troublesome difficulty arises because it is not possible to obtain molecules in their equilibrium geometry for experimental study.

equilibrium state n 平衡状態

▶ Thermodynamics is directly concerned only with equilibrium states, in which the state functions have constant values throughout the system.

equimolar adj 等モルの

▶ Equimolar amounts of two compounds to be compared are mixed together and allowed to react with a limited amount of a particular reagent.

equip vt 備え付ける syn *furnish, provide*

▷ A solution of 95 g of sodium nitrite in 375 mL of water was placed in a 1-L three-neck flask, equipped with a mechanical stirrer, a thermometer, and a dropping funnel with stem extending to the bottom of the flask and cooled to 0°.

equipment n 装置 不可算名詞である. syn *apparatus*

▶ With suitable equipment, we can easily demonstrate that the electric conductivity of the high-temperature form of Cu_2HgI_4 is much larger than that of the low-temperature form.

equivalence n 等価

▶ The equivalence of conformations which differ only by a rotation about a single bond is often described by the statement that the rotation is free, or that free rotation exists, about such a bond.

equivalent n **1** 当量

▶ Since 8 g of oxygen combines with 1.008 g of hydrogen, the latter is considered equivalent to 8 g of oxygen.

2 相当するもの syn *copy, replica*

▶ Ultramarine blue, which is the synthetic equivalent of the mineral lapis lazuli, contains radical anions, S_3^- and S_2^-.

adj **3** 当量の

▶ Phenylglyoxylic acid is formed in good yield by oxidation of mandelic acid with an equivalent amount of permanganate in alkaline solution at a low temperature.

4 同等の syn *alike, comparable, similar*

▶ To be chemically equivalent, carbons must be stereochemically equivalent: in an achiral medium diastereotopic carbons will give different signals, and enantiotopic carbons will not.

equivalent ionic conductivity n 当量イオン伝導率

▶ At infinite dilution, it is profitable to write

$$\Lambda_0 = \lambda^+_0 + \lambda^-_0$$

where λ^+_0 and λ^-_0 are the equivalent ionic conductivities at infinite dilution.

equivalent point n 当量点

▶ The pH of the equivalent point is the same as the pH of an aqueous solution of the salt.

equivocal adj あいまいな syn *ambiguous, un-*

certain, vague

▶ More equivocal but often adequate approaches include differences in gross crystal structure, melting point comparisons, and infrared and ^{1}H NMR analysis of the hydroxyl group.

erasable adj 消すことができる

▶ Phototropic substances could be used as the basis of a photochemical memory with an erasable image.

Erlenmeyer flask n 三角フラスコ

▶ To aqueous silver nitrate solutions of varying silver ion concentration contained in glass-stoppered Erlenmeyer flasks were added 0.1–1.0-g samples of aromatic hydrocarbon.

erode vt 損なう syn *deplete, diminish, reduce*

▶ The addition of large amounts of water, however, did not result in further enhancement of the product enantioselectivity but instead substantially eroded the yield.

erosion n **1** 侵食 syn *abrasion, corrosion, wear*

▶ If some kind of erosion process can be used to remove surface atoms layer by layer, Auger electron spectroscopy can provide a depth profile of the composition of a solid at and below its surface.

2 エロージョン syn *loss, reduction*

▶ Both polar protic and polar aprotic solvents led to almost complete erosion of the enantioselectivity, suggesting that these solvents disrupt interactions necessary for stereocontrol..

errata n 正誤表 syn *misprint*

▶ Page 1, line 3, for Saunder's read Saunders'.

▶ Page 2, line 7 from foot, after N, 20.98 add %.

▶ Page 3, last line, prefix reference number[6].

▶ Page 4, last line but one, The symbol C_{6h} should be C_{6v}.

▶ Page 5, eq 1, in the last term for $1/R$ read R.

▶ Page 6, eqs 2 and 3, transpose t_1 and t_2.

erratically adv 不規則に syn *randomly*

▶ Latent heats of sublimation vary erratically between 281 kJ for manganese and 515 kJ for vanadium, with the exception of that for zinc which has the much lower value of 125 kJ.

erroneous adj 誤った syn *incorrect, mistaken, wrong*

▶ One contribution factor to the erroneous predictions is that the ions are not hard spheres, as assumed in the radius-ratio model but have sizes that vary with the coordination number.

error n **1** 誤り syn *fault, inaccuracy, mistake*

▶ The agreement between the two sets of values was within the experimental error of the Cary spectrophotometer.

2 in error 誤った syn *incorrect, wrong*

▶ The boiling point of 135–137/13 mm recorded in *Organic Syntheses* would appear to be in error.

escalate vi 拡大する syn *rise*

▶ As the size of devices on silicon wafer shrink, the surface characteristics play increasingly crucial roles in the proper functioning of the device since the ratio of surface atoms to bulk escalates.

escape n **1** 排出 syn *release*

▶ Anhydrous NaH_2PO_4, when heated to 170° under conditions which allow the escape of water vapor, forms the diphosphate $Na_2H_2P_2O_7$.

2 脱出 syn *departure*

▶ In hexane, only an electron which travels more than 29 nm from its parent ion will have a high probability of escape from the attractive force.

vi **3** 逃げる syn *break out*

▶ The hydrogen atom escapes with a surface oxygen, with the formation of water.

especial adj 特別な syn *particular, special*

▶ The present paper describes some results which are of especial interest in that lattice energy is found to be simple function of the gaseous ion radii.

especially adv 特に syn *particularly*

▶ The compound is not especially soluble, and so for large scale reactions it was added as a fine suspension in THF.

essence n **1** 本質 syn *cornerstone, principle, significance*

▶ These arguments are very qualitative but

give the essence of some of the more important dynamic solvent effects.

2 in essence 本質的に syn *basically, essentially*

▶ In essence, the Pauli exclusion principle states that there can be only one electron in each possible energy state of the atom.

essential n **1** 複数形で，本質的要素 syn *body, essence, heart*

▶ The surroundings may affect the rates of reaction, but the essentials of the reaction remain the same.

adj **2** 本来の syn *intrinsic*

▶ Only in rare instances can all the essential faces be grown.

3 不可欠の syn *important, indispensable, necessary*

▶ Certain elements are essential in that they are absolutely necessary for life processes.

essential amino acid n 必須アミノ酸 ☞ nonessential amino acids

▶ Two of the nonessential amino acids, tyrosine and cysteine, are derived from essential amino acids and may be considered to be breakdown products of them.

essential element n 必須元素

▶ The phenomenon of an essential element becoming toxic at higher than normal concentration is not rare.

essential fatty acid n 必須脂肪酸

▶ Linoleic, linolenic, and arachidonic acids are called essential fatty acids by biochemists because such acids are necessary nutrients that are not synthesized in the animal body.

essentiality n 不可欠性 syn *essentialness*

▶ Recent curiosity as to the biological significance of a wide range of trace elements detectable in animal tissues in relatively constant concentrations has stimulated an extensive series of studies to investigate their essentiality.

essentially adv 本質的に syn *largely, mainly, primarily*

▶ The scanning tunneling microscope is essentially a surface tool and can provide electronic and other information in addition to an image of the physical structure of the surface.

essential oil n 精油

▶ Essential oils are derived from the leaves, stem, flower, or twigs of plants and usually carry the odor or flavor of the plant.

establish vt 確立する syn *entrench, secure*

▶ Equilibrium between hexaphenylethane and triphenylmethyl radicals is rapidly established so that oxygen readily converts the ethane into the relatively stable triphenylmetyl peroxide.

established adj 確立した syn *prevalent, standard*

▶ All solvents were purified by established methods starting with commercial materials.

establishment n 確立 syn *realization*

▶ The establishment of scales of Lewis basicity for other molten salt systems is also shown to be feasible.

esterification n エステル化

▶ Esterification may be driven to completion by removing the ester and/or water as they are formed.

esterify vt エステル化する

▶ In triacylglycerols, the three hydroxyl groups of glycerol are each esterified, invariably with three different fatty acids.

estimate n **1** 概算 syn *approximation, evaluation*

▶ Estimates of core energy have not so far been made with much accuracy for any very realistic model of an actual crystalline substance.

vt **2** 見積もる syn *evaluate, gauge, guess*

▶ One way to estimate the relative strengths of the $M_2 \cdots O$ interactions is to compare the structural parameters shown for a closely related series of compounds.

estimation n 評価 syn *approximation, estimate, gauge, guess*

▶ Detailed discussion of both qualitative and quantitative estimation of stabilities will be reserved for Chapter 3.

etc. abbr **その他** et cetera を略記したもので，ローマン体を用いる．前に置くコンマの使用法は and の場合と同様で，A etc.や A, B, etc. とする．文の最後にあるとき以外は，後にコン

マを付ける. syn *and others, and so forth*

▶ For oxo acids having more than one OH group, we can determine values of pK_1, pK_2, etc.

etch vi **1** エッチングをする

▶ Grain boundaries etch away preferentially, and low-angle grain boundaries are seen as rows of dislocation etch pits.

vt **2** エッチングを施す

▶ The aluminum plates were initially etched to assist adhesion, and then the silica support formed in situ from sodium silicate.

etching n エッチング

▶ Much of the early work on dislocations was done by metallurgists using crystals of LiF and the technique of etching followed by examination with optical microscopy.

etch pit n エッチピット ☞ pit

▶ Dislocation densities may be determined by counting the number of etch pits per unit area.

ethanolic adj エタノールの ☞ alcoholic.

▶ [$RhCl(PPh_3)_3$] is readily obtained by refluxing ethanolic $RhCl_3 \cdot 3H_2O$ with an excess of PPh_3.

ethereal adj エーテルの

▶ Instead of removing the ether and dissolving the residue in ligroin, it is simpler to add the ligroin to the ethereal solution, distil, and so displace the ether by the less volatile crystallization solvent.

ethylenic adj エチレンの

▶ The abnormal addition of hydrogen bromide is possible with most olefins in which the ethylenic double bond is not conjugated with a carbonyl group.

eutectic n **1** 共晶 syn *eutectic mixture*

▶ Electrolysis of CaH_2 dissolved in a LiCl−KCl eutectic at 360 ℃ does give hydrogen at the anode in an amount that accords with Faraday's laws.

adj **2** 共融の

▶ A mixture of eutectic composition melts sharply at the eutectic temperature to form a liquid of the same composition.

eutectic mixture n 共融混合物 syn *eutectic*

▶ Eutectic mixtures have a number of important applications, and their study is essential to an understanding of the formation and properties of alloys.

eutrophication n 富栄養化

▶ Eutrophication refers to the unintentional enrichment or fertilization of either fresh- or saltwater by chemical elements or compounds present in various types of industrial wastes.

eutrophying agent n 富栄養化剤

▶ Phosphates and nitrogenous compounds in detergent and chemical-processing wastes are particularly effective eutrophying agents.

evacuatable adj 排気可能な

▶ Infrared spectra were measured with an evacuatable Nicolet model 8000 FT IR spectrometer equipped with a TGS and a cooled MCT/InSb infrared detector.

evacuate vt 排気する syn *empty, exhaust*

▶ The X-ray tube must be evacuated to prevent absorption of the electrons by gas molecules, and the target must be water-cooled, otherwise the electron beam would burn a hole in it.

evacuation n 排気 syn *exhaust*

▶ Under ordinary experimental conditions, surfaces are covered by a layer of adsorbed molecules, which are removed only by prolonged evacuation.

evaluate vt 評価する syn *approximate, estimate, gauge*

▶ Varying successively the initial concentration of each component will give the order with respect to each component, following which the rate constant k may be evaluated approximately from any one run.

evaluation n 評価 syn *approximation, estimate, estimation*

▶ The evaluation of magnetic susceptibility tensor has been a problem of theoretical and experimental interest for many years.

evaporate vi **1** 蒸発する syn *vaporize*

▷ Solvent evaporating from the surface condenses on the walls and lid, from which it drains back into the outer chamber.

vt **2** 蒸発させる syn *vaporize*

▶ If the solution is evaporated to dryness, the residue will be found to be a solid of low and wide melting range, since it contains considerable by-products.

evaporating dish n 蒸発皿

▶ The precipitate is allowed to settle, the excess of liquid decanted off, and the suspension transferred to a evaporating dish which is heated on a boiling-water bath without stirring until red tin(II) oxide begins to be formed near the sides of the dish.

evaporation n 蒸発 syn *vaporization*

▶ By evaporation in a vacuum, the aqueous solution of chloric acid can be concentrated until its density is about 1.28, when it corresponds to $HClO_3 \cdot 7H_2O$ with 40.1 percent $HClO_3$.

evaporation to dryness 蒸発乾固

▶ It is sometimes desirable to remove perchloric acid from a solution to be analyzed for the alkali metals. This is best accomplished by evaporation to dryness and igniting with ammonium salt.

蒸着

▶ Following evaporation of a p-type or an n-type dopant onto a pure silicon substrate, the sample is heated to allow the dopant to diffuse into the silicon.

evaporative adj 蒸発による

▶ Temperatures below 4 K are reached by evaporative cooling of the liquid helium.

evaporite n 蒸発残留物

▶ In the desert regions of northern Chile $NaNO_3$ occurs with other evaporites such as NaCl, Na_2SO_4, and KNO_3.

even adv **1** …でさえ syn *still*

▶ Acetyl derivatives of primary and secondary amines are usually solids suitable for melting point characterization and are readily prepared by reaction with acetic anhydride, even in the presence of water.

2 even if たとえ…としても syn *though*

▶ Ions of comparable size can replace one another even if the charges differ.

3 比較級を強めていっそう syn *still, yet*

▶ An even more exciting aspect is that the M_3C_{60} phase becomes superconducting on lowering the temperature.

evenly adv **1** 平等に syn *equally*

▶ Halogen withdraws electrons through its inductive effect and releases electrons through its resonance effect. For halogen, the two effects are more evenly balanced, and we observe the operation of both.

2 均等に syn *uniformly*

▶ If the initiator molecules are evenly distributed, then they should be consumed during the formation of nanofibers.

even number n 偶数

▶ Fats are the source of straight-chain acids of even carbon number ranging from six to eighteen carbons.

event n **in any event** いずれにしても syn *anyway, in any case*

▶ Complications of this sort are fortunately rather rare and, in any event, can usually be easily detected, when they do occur.

even though conj …であるのに syn *despite, in spite of*

▶ The silicates, Mg_2SiO_4 and Zn_2SiO_4 can each dissolve about 20 percent of the other one in solid solution formation, even though their crystal structures are quite different.

eventual adj 最終の syn *final, last, ultimate*

▶ The eventual equilibrium dimer geometries are the outcome of optimizing all interactions.

eventually adv ついには syn *finally, ultimately*

▶ Eventually, the reaction chain is terminated by steps that consume but do not form free radicals: combination or disproportionation of two free radicals.

ever adv **1** いつも syn *at any time*

▶ For any atom Y, only one substance of formula CH_3Y has ever been found.

2 if ever たとえ…にしても

▶ Absolutely free rotation seldom if ever occurs in nature.

every adj **1** すべての 可算名詞の単数形の前に置いて，一集団の個々全部を意味する．every の代わりに all を用いるときには名詞は複数形にする．個々の意味では each を用いる．

▶ If reaction occurred at every collision between given reacting molecules, the rate would be much greater than is usually observed.

2 every so often 時々 syn *often*

▶ As in halogenation of alkanes, every so often a reactive particle combines with another one or is captured by the wall of the reaction vessel, and a chain is terminated.

3 十分な

▶ This would be a grossly impractical synthesis, since there is every reason to anticipate that separation of the desired isomer would be very difficult.

4 …ごと

▶ Hemoglobin binds one H^+ for every two dioxygen molecules released.

5 every other 一つおきの syn *alternate*

▶ With $CdCl_2$, the chloride ions are close-packed, and only every other layer of octahedral holes is filled with the smaller Cd^{2+} ions.

everyday adj ありふれた syn *common, familiar, usual*

▶ The color rendering index is a measure of the degree to which the appearance of everyday colored objects, when illuminated by a light source, matches their appearance in daylight.

everyday life adj 日常生活

▶ High-tech materials and advanced devices already play a prominent role in everyday life.

everything pron 何でもみな syn *all, whole*

▶ Underlying everything is the need for a better understanding of the detailed mechanism of radiationless transitions in complex molecules both in fluid and in rigid media.

everywhere adv どこでも syn *throughout*

▶ In any chemical system, the electronegativity, χ, must be constant everywhere at equilibrium, and a good approximation is $\chi=(I+A)/2$.

evidence n **1** 証拠 不可算名詞である． syn *ground, proof*

▶ Much evidence has accumulated about the efficiencies of energy transfers between molecules.

vt **2** 明示する syn *demonstrate, display, show*

▶ Strain within the metallocyclopentane ring across all crystal structure is evidenced by closing down of both the $N-C_{methylene}-C_{aryl}$ and $Pd-C_{aryl}-C_{methylene}$ angles with concurrent opening of the external $Pd-C_{aryl}-C_{aryl}$ angle which average 109.0, 112.3, and 129.9°, respectively.

evident adj **1** 外的証拠が存在して，明白な syn *clear, obvious, plain*

▶ It has become evident that these are modified by the structure of the rest of the molecule.

2 はっきり見える syn *discernible, perceptible, recognizable*

▶ The most evident property of trivalent organophosphorus compounds is their ability to utilize the unshared electron pair on phosphorus to form bonds to other atoms.

evidently adv 明らかに syn *clearly, obviously, plainly*

▶ Copper occurs in nature as Cu_2O, known as cuprite, whose color, evidently depending on particle size, may be yellow, orange, red, or dark brown.

evil adj 不快な syn *bad, offensive*

▶ Methyl isocyanide is a volatile liquid with an evil odor, far more pungent than that of methyl cyanide.

evoke vt 記憶を呼び起こす syn *elicit, recall*

▶ The term molten salts evokes an image of high-temperature, viscous and highly corrosive media.

evolution n **1** 漸進的発展 syn *advance, development, progress*

▶ Throughout the evolution of this chemistry, cyclic voltammetry has provided critical information about the structure and reactivity of dihapto-coordinated aromatics.

2 進化

▶ Evolution tends to proceed by modifying structures and functions that are already present in an organism rather than producing new ones.

3 発生 syn *formation, production*

▶ Metallic sodium dissolves, without evolution of hydrogen, in air-free solutions of aromatic ketones in dry ether, giving dark blue products.

evolutionarily adv **進化的に**

▶ We hypothesized that the enzyme has evolved a slightly too-large active site to allow for evolutionarily beneficial mutations.

evolutionary adj **進化の**

▶ The iron(II) compounds within biological systems may reflect an evolutionary history from a primitive reducing atmosphere on earth.

evolve vi **1 発展する** syn *develop*

▶ The rules should evolve as new structures become available, but we have found that the most important relationships are usually seen from the first 10 or so structures in a series.

2 進化する

▶ Historically it is thought that we evolved with the need to store metabolic fuel for times of famine so that we are inclined to overeat in times of plenty and store this fuel as triglycerides.

vt **3 放出する** syn *result*

▶ One hundred milliliters of the commercial hydrogen peroxide will evolve 1000 mL of oxygen, i.e.,10 times its own volume.

exact adj **正確な** syn *accurate, correct, precise*

▶ Collision frequency and, hence, rate depends in a very exact way upon concentration.

exacting adj **骨の折れる** syn *difficult, hard, tough*

▶ Tanning by any method is a time-consuming and exacting process, requiring careful control of pH, temperature, humidity, and concentration factors.

exactly adv **正確に** syn *accurately, correctly, precisely*

▶ We observe discontinuous changes in the IR spectra that occur exactly at the DSC transition temperatures.

exaltation n **エキサルテーション**

▶ The diamagnetic susceptibility exaltation of a compound is the difference between the susceptibility exhibited by a compound and that predicted for the identical but not cyclically delocalized structural counterpart.

examination n **調査** syn *analysis, investigation, study*

▶ Quantitative examinations of the dependence of the conductivity on pressure have been made in some instances.

examine vt **調査する** syn *investigate, study*

▶ The behavior of these crystal forms on heating at 10 ℃/min was examined with the aid of a Du Pont Model 900 thermal analyzer equipped with a differential scanning calorimeter cell.

example n **1 実例** syn *case, instance, sample*

▶ The derivatives of molybdenum(II) halide systems were one of the first examples of a cluster system to be identified by crystallographic techniques.

2 for example たとえば syn *for instance, to illustrate*

▶ The ionic bond results from transfer of electrons, as, for example, in the formation of sodium chloride.

exceed vt **超える** syn *surpass*

▶ For the first time, T_c exceeded the boiling point of nitrogen, and sample cooling costs, which are an important consideration for technological applications, became relatively inexpensive.

exceedingly adv **極めて** syn *considerably, extremely, very*

▶ An exceedingly unstable mercury hydride (HgH_2) is also formed from HgI_2 and $LiAlH_4$ in an ether solvent at −135 ℃, but it decomposes at slightly higher temperatures.

excel vi **卓越する** syn *dominate, surpass*

▶ Most prominent schools of chemistry are grounded in experimental work and excel in devising and refining new apparatus for chemical research.

excellent adj **優れた** syn *exceptional, outstanding*

▶ The *p*-phenylene esters of the para-substituted benzoic acids provide an excellent system for the study of structural effects on mesophase stability because they are so versatile.

except prep **1** …のほかは but よりも強い除外を意味する. syn *excepting, excluding, exclusive of*

▶ Oxygen's electronegativity is highest of all the elements except fluorine.

2 except for …を除いて syn *but for, except*

▶ In general, the vibrational-entropy contribution is small but, except for wide vibrational spacings as in CO, not negligible.

exception n **1** 例外 syn *anomaly, departure, irregularity*

▶ Ethylene oxide, the simplest cyclic ether, is an outstanding exception to the generalization that most ethers are resistant to cleavage.

2 with…exceptions …を除いて

▶ With certain exceptions, we can assume that the order of an elementary reaction indicates the number of molecules that enter into the reaction.

3 with the exception of (prep) …を除いては syn *except*

▶ It will be noted that, with the exception of the fluorides, the melting points of the silicon, germanium, and tin halides are extremely close to one another.

4 without exception 例外なく syn *always, entirely*

▶ Amino acid residues on the outside of soluble proteins are, almost without exception, polar.

exceptional adj 例外的な syn *especial, special, unusual*

▶ Potassium perchlorate is exceptional for its low solubility, 0.75 g in 100 g of water at 0 ℃.

exceptionally adv 並外れて syn *extremely, very*

▶ C_{60} would exhibit an exceptionally high electron affinity and could accept up to 12 electrons under appropriate circumstances.

excess n **1** 過剰 syn *overabundance, surplus*

▶ In order to complete the reaction and obtain a good yield of product, it is necessary to use a large excess of water.

2 in excess of 以上の syn *above, over*

▶ Ethylene may be polymerized with peroxide catalysts under high pressure at temperatures in excess of 100°.

adj **3** 余分の syn *excessive*

▶ To this mixture an excess portion of selenium was added, and the solution was heated under reflux for 12–14 h.

excessive adj 過大な syn *excess, superfluous*

▶ An excessive quantity of decolorizing agent must be avoided, since it may also adsorb some of the compound which is being purified.

excessively adv 過度に syn *exceedingly, overly*

▶ The calculations of *K*, which are excessively tedious, were carried out for a selection of points considered as representative of the data.

exchange n **1** 交換 syn *interchange*

▶ The exchange of Al^{3+} for Si^{4+} has created an excess of one unit of negative charge per substitution.

vt **2** 交換する syn *interchange*

▶ Zeolites have been used to remove water from organic solvents, to exchange sodium ions for calcium or magnesium ions in water softeners, and to prepare synthnetic gasoline from such feedstocks as methanol.

exchangeable adj 交換可能な

▶ The relative quantities of exchangeable atoms or ions present in the medium used for synthesis may or may not preserved in the resulting solid.

exchange integral n 交換積分

▶ The magnitude of exchange integral represents the energy associated with exchanging electrons between the nuclei.

exchange interaction n 交換相互作用

▶ The exchange interaction narrows the line by a rapid flipping of spins.

exchanger n イオン交換体 ☞ ion exchanger

▶ Strong-acid exchangers easily remove all cations in solution. They have highly reactive sites such as the sulfonic group ($-SO_3H$), phosphonic group ($-PO_3H_2$), or hydroxyl group ($-OH$).

excimer laser n エキシマーレーザー
▶ The excimer laser radiation is produced by passing a pulsed electric discharge through a mixture of helium, a noble gas, and either F_2 or HCl.

exciplex n エキシプレックス
▶ The exciplexes are probably best described as weak complexes of a slightly perturbed excited state similar to ground-state donor-acceptor complexes.

excise vt 切除する syn *cut, eliminate*
▶ A number of DNA repair enzymes are known to bind and excise bases involved in damaged or mismatched pairs in DNA.

excision repair n 切除修復
▶ Of interst is the mechanism by which these base excision repair enzymes identify such pairs as damaged.

excitation n 励起
▶ These Rydberg lines arise from the excitation of a π electron to orbitals, which are sufficiently large that they resemble orbitals of a one-electron atom.

excitation energy n 励起エネルギー
▶ An electronically excited molecule may emit a quantum of energy by fluorescence or may transfer its excitation energy by collision with other molecules.

excitation spectrum n 励起スペクトル
▶ The S_1 state can emit to a number of vibrational states in the S_0 level making the fluorescence signal an approximate mirror image of the excitation or absorption spectrum.

excite vt **1** 励起する syn *energize*
▶ The electrons of a carbon-carbon double bond of an alkene are excited to higher energy states by light of wavelength 2000 to 1000 Å.
2 引き起こす syn *cause, generate*
▶ Luminescence can be excited by irradiation with γ-ray, X-ray, electrons, α-particles, and energetic particles generally.
3 そそる syn *arouse, provoke, stimulate*
▶ The nature of bonding in CO has excited much attention because of the unusual coordination number and oxidation state of carbon.

excited state n 励起状態
▶ Excited states are obtained by promoting an electron from one of these occupied orbitals to an orbital which is vacant in the ground state.

exciting adj 興奮させる syn *impressive, interesting*
▶ Of all fields of chemistry, the study of the nucleic acids is perhaps the most exciting, for these compounds are the substance of heredity.

exciton n 励起子
▶ There are certain additional excited levels, exciton levels, in which electron-hole pairs move throughout the crystal.

exclude vt 除外する syn *eliminate, preclude, reject*
▶ The latter mechanism can be excluded on the basis of symmetry arguments and with reference to a schematic potential-energy surface.

exclusion n 排除 syn *removal*
▶ All experiments were carried out in standard Schlenk glassware or in glove box with strict exclusion of air and moisture.

exclusive adj **1** 排他的な syn *incompatible*
▶ The crystal field theory is not an exclusive property of the electrostatic method but can be incorporated into the molecular orbital theory.
2 exclusive of (prep) …を除外して syn *apart from, except for*
▶ All chromosomes exclusive of the sex chromosomes are called autosomes.

exclusively adv もっぱら syn *entirely, only*
▶ If chlorine is bubbled into boiling toluene that is exposed to ultraviolet light, substitution occurs almost exclusively in the side chain.

excrete vt 排泄する syn *discharge*
▶ The treatment for toxic metals, such as lead, is to inject a chelating agent to form a soluble complex that can be excreted.

excursion n 偏位運動 syn *run*
▶ On each excursion in and out alternate layers are deposited upon the solid, and a builup film of a known number of layers can be placed upon the solid.

execute vt 実施する syn *accomplish, carry out,*

perform

▷ Normal alkane molecules become effectively cylindrical as a result of executing rotation-like motions about their long axes.

execution n 実行 syn *performance, realization*

▶ For the successful execution of any radical chain reaction it is necessary to be able to generate the initial radical on a substrate in a site-specific manner.

exemplify vt 例示する syn *illustrate, represent*

▷ The unique affinity between aluminum and silicon, exemplified by the ability of $[AlO_4]^{5-}$ to replace $[SiO_4]^{4-}$ in aluminosilicate minerals, is also evident in aqueous solution chemistry.

exercise vt 働かせる syn *exert*

▷ In hydrogen fluoride, a number of organic compounds containing oxygen, nitrogen, or sulfur atoms capable of exercising a donor function can take up protons and yield conducting solutions.

exergonic adj 発エルゴンの ☞ endergonic

▶ Spontaneous processes are termed exergonic and have a negative value of ΔG.

exert vt 及ぼす syn *exercise*

▶ The nitro groups in the nitroanilnes exert a more powerful electric effect than in the nitrobenzoic acids.

exfoliate vt 剥離させる

▶ For many applications, it is necessary to completely exfoliate the graphite into discrete graphene sheets.

exhaust n **1** 排気 syn *emission*

▶ The environmental problems of NO_x from automobile exhaust fumes and from photochemical smog are well known in all industrial countries.

vt **2** 余すところなく述べる syn *treat thoroughly*

▶ The applications of this technique to the many problems encountered in surface chemistry have by no means been exhausted.

exhaust gas n 排気ガス

▶ As the exhaust gases rapidly cool, the radicals are quickly consumed and the high temperature equilibrium concentration of NO is frozen into the mixture.

exhaustive adj 徹底的な syn *complete, thorough*

▶ The following discussion is to be considered illustrative rather than exhaustive.

exhaustively adv 徹底的に

▶ A yellow precipitate forms rapidly, which can be isolated by filtration and exhaustively washed with diethyl ether to yield the product.

exhibit vt 示す syn *display, show*

▷ Metal-metal interactions are a key feature of the structures of solids exhibiting superconductivity.

exist vi 存在する syn *be found, be present, happen, occur*

▶ Empirical relationships exist between ionization potentials of donors and the energy of the intermolecular charge-transfer absorption for the corresponding complexes with a given acceptor.

existence n 存在 syn *actuality, fact, presence*

▶ The existence of several high-temperature phases in these longer *n*-alkanes indicates that more than one mechanism is involved in the disordering that occurs before melting.

exocyclic adj 環外の

▶ A further effect of the intramolecular overcrowding of the substitutents on the picryl ring is to be seen in the values for the exocyclic angles at C(1).

exogenous adj 外因性の ☞ endogenous

▶ Unlike other exogenous pathogens, viruses are unable to replicate by themselves and use host replication machinery.

exon n エクソン ☞ intron

▶ Introns are excised or spliced out of the primary transcript to leave segments that are to be expressed, called exons, in the mRNA,

exoskeleton n 外骨格

▶ The exoskeletons of insects and crustaceans contain large amounts of the aminopolysaccharide.

exotherm n 発熱 ☞ endotherm

▶ Devitrification appears as an exotherm and is followed by endotherm at a higher temperature that corresponds to the melting of the same crystals.

exothermally adv 発熱して ☞ endothermally

▶ Aliphatic nitriles are prepared by treatment of alkyl halides with sodium cyanide in a solvent that will dissolve both reactants; in dimethyl sulfoxide, reaction occurs rapidly and exothermically at room temperature.

exothermic adj 発熱の ☞ endothermic

▶ When energy is given off as result of rupture of chemical bonds, the change is said to be exothermic.

exothermic reaction n 発熱反応 ☞ endothermic reaction

▶ Uranium metal, in the form of small lumps, was dissolved in dioxygen-free hydrobromic acid, resulting in an exothermic reaction.

exotic adj 珍しい syn *singular, unfamiliar, unique, unusual*

▶ Although multiply charged anions can reasonably be regarded as exotic, metastable species in the gas phase, they are common components of polar solutions and ionic solids.

expand vi **1** 膨張する syn *swell*

▷ Bismuth is unusual in expanding on solidifying from the melt, a property which it holds uniquely with Ga and Ge among the elements.

vt **2** 拡大する syn *enlarge, extend*

▶ The ability of phosphorus to expand its octet and to act as both donor and acceptor provides scope for a wide variety of covalent chemistry.

expandable adj 膨張しやすい

▶ The expandable nature of montmorillonite and hectorite makes them particularly suitable for colloidal studies.

expanded adj 拡大された syn *enlarged*

▶ X-ray diffraction experiments and elemental analysis established the alloy to be $AuAl_2$, with the large Au atoms in fcc positions and the smaller Al atoms in expanded tetrahedral holes.

expanded film n 膨張膜

▶ On less compression or upon warming, the two-dimensional liquid may approximately double in area, producing the expanded liquid film.

expansion n **1** 膨張 syn *swelling*

▶ The unit cell undergoes a small contraction or expansion as the composition varies across a solid solution series.

2 拡大 syn *enlargement*

▶ Another more obvious manifestation of octet expansion is the existence of compounds such as the pentahalides PX_5 in which phosphorus forms five σ bonds.

3 展開式

▶ The expansion of the determinant is the algebraic sum of the products formed by taking one element from each column and row, each with the proper sign prefixed according to the rule of inversions.

expansion coefficient n 膨張率 ☞ cubic —, linear—

▶ The relations between the expansion coefficients of molecular crystals and the structure of the constituent molecules exist on two levels of sophistication.

expansive adj 広大な syn *comprehensive, extensive*

▶ As a result of the expansive literature on heterocycle chemistry, a myriad of information is known on how to modify the basic structural and electronic components of N-heterocyclic carbenes.

expect vt 期待する syn *assume, presume, suppose*

▶ As with the Grignard reagents, aryllithiums react as one might expect by analogy with alkyllithiums.

expectation n 期待 syn *assumption, supposition*

▶ Bonds tend to be shortened, relative to the expectations for nonpolar bonds, in proportion to the electronegativity difference of the component atoms.

expectation value n 期待値

▶ The mean value of an observable is equal to the expectation value of its corresponding operator.

expel vt 排除する syn *dislodge, eject, throw out*

▶ At low magnetic field strengths, the magnetic field is completely expelled from the superconductor's interior.

expend vt 費やす syn *consume, deplete, spend*

▶ Any pure stoichiometric crystal in thermal equilibrium above 0 K will necessarily contain Frenkel and Schottky defects, since although energy is expended in forming them, their production leads to an increase in entropy.

expenditure n 消費量 syn *expense*

▶ Generally speaking, to bring nonbonded atoms to near-bonding distances requires a very large expenditure of energy.

expense n **1** 費やすこと syn *impairment, loss*

▶ It should be emphasized that this improvement is attained at very little expense in terms of complexity.

2 at the expense of …を犠牲にして

▶ This strain is relieved by an increase of the C(4)-C(3)-Cl(1) angle at the expense of the C(2)-C(3)-Cl(1) angle.

expensive adj 高価な syn *costly, valuable*

▶ Ninhydrin is an expensive reagent widely used for identification of amino acids by paper-strip chromatography.

experience n **1** 経験 syn *common sense, knowledge*

▶ It is common experience that proton transfer between electronegative atoms such as oxygen or nitrogen is very fast whereas that involving carbon is usually quite slow.

vt **2** 経験する syn *encounter, feel, meet, sense*

▷ Difficulty experienced in starting a previously unexplored Grignard reaction may be due to inadequate purification of the halide rather than to its lack of reactivity.

experiment n **1** 実験 syn *investigation, trial*

▶ In all experiments, potassium was used as reducing agent, tetrahydrofuran as solvent.

vi **2 experiment with** …を実験する syn *examine, investigate, test, try*

▶ Kopp experimented with a large number of organic compounds and found that it is very difficult to avoid impurities and superheating.

experimental adj 実験的な syn *empirical*

▶ The experimental dipole moment is obtained as a scalar having no direction.

experimental animal n 実験動物

▶ Since control animals were chosen to resemble the other experimental animals as closely as possible, they could be viewed as differing in only one variable, namely, that they were untreated.

experimental error n 実験誤差

▶ The difference in the values for isopropyl and *tert*-butyl compounds is within the experimental error.

experimentally adv 実験的に

▶ Experimentally, pressure, volume, and temperature of a gas cannot be arbitrarily chosen to describe the state of a fixed amount of gas.

experimental value n 実験値

▶ The experimental values for the number of water molecules associated with an ion in solution tend to depend on the method used to evaluate them.

experimentation n 実験 syn *experiment, research*

▶ Despite extensive and diligent experimentation, no sign of BH_3 was ever encountered by Stock.

expert n 専門家 syn *professional*

▶ The invention must not be obvious to an expert in the art.

explain vt 説明する syn *define, interpret*

▶ The existence of AlF_6^{3-}, but only $AlCl_4^-$ and not $AlCl_6^{3-}$, can be explained and is a property of the relation between the size of the ligand and the size of the central ion.

explanation n 説明 syn *interpretation*

▶ It should be noted that the HSAB principle is not an explanation or a theory, but a simple rule of thumb which enables the user to predict qualitatively the relative stability of acid-base adducts.

explicable adj 説明可能な

▶ Most of the chemical properties of molecular systems are explicable in terms of the distribution of electron density within the molecules.

explicit adj はっきりした syn *distinct, exact, unambiguous, well-defined*

▶ It is not possible to draw a very explicit structure for the excited state with conventional bond diagrams because the excited electron is not in a normal bonding orbital.

explicitly adv 明確に

▶ For ions such as $Fe(H_2O)_6^{2+}$ and $Fe(H_2O)_6^{3+}$, the total energy is given explicitly as a function of the distance, the equilibrium value of which was taken as 2.21 Å for iron(II) and 2.05 Å for iron(III).

explode vi **1** 爆発する syn *burst*

▶ Sodium peroxide explodes with powdered aluminum or charcoal, reacts with sulfur with incandescence, and ignites many organic liquids.

2 急増する

▷ Finally in this brief section on the exploding field of coordination polymers, we should mention that there is growing body of work on systems that contain more than one metal or more than one ligand type, or both.

vt **3** 爆発させる

▷ The determination of oxygen by combustion may be effected by exploding it with hydrogen or by conducting a mixture of the two gases through a glowing platinum capillary.

exploit vt 利用する syn *manipulate, use*

▶ The dramatic combustion of powdered magnesium in air is exploited by manufacturers of flares and fireworks.

exploitation n 利用

▶ Although the electro-optical properties of liquid crystals have been known for a long time, their exploitation in practical devices has to wait for developments in materials technology.

exploration n 探究 syn *investigation, research, study*

▶ The reports of rare-earth-metal iodide clusters containing dicarbon prompted us to make a systematic exploration of interstitials in what appeared to be only isolated examples of zirconium clusters.

exploratory adj 探究の syn *experimental, tentative, trial*

▶ Exploratory absorption measurements of the iodine solutions were made using quartz cells ranging from 1 mm to 10 cm in length.

explore vt 探究する syn *examine, investigate, search, study*

▶ Over the past two decades, we have explored the organic chemistry of dihapto-coordinated aromatic ligands.

explosibility n 爆発性

▶ Ether is avoided in industrial processes because of its fire hazard, high solubility in water, losses in solvent recovery incident to its volatility, and oxidation of ether on long exposure to air to a peroxide, which in a dry state presents the hazard of explosibility.

explosion n **1** 爆発 syn *burst*

▶ Although present in only low concentrations, these peroxides are very dangerous, since they can cause violent explosions during the distillations that normally follow extractions with ether.

2 急激な増加 syn *expansion, increase*

▶ In the field of reaction mechanisms, there has been an explosion of research on chemical intermediates in the last two decades, and silicon chemistry is no exception to this.

explosion limit n 爆発限界

▶ Since the temperature rise can be controlled by the addition of inert gas, explosion limits dependent on temperature and pressure can be studied.

explosive n **1** 火薬類 syn *gunpowder*

▶ 1, 3, 5-Trinitrobenzene has excellent properties as an explosive but is difficult to prepare by direct nitration of benzene.

adj **2** 爆発性の

▶ Liquid diazomethane is an explosive compound, and explosions may also occur in the gaseous state if the substance is dry and undiluted.

3 急激な

▶ Liquid crystals have experienced an explosive growth in the last twenty years.

explosively adv 爆発的に

▶ Beryllium dialkyls are explosively hydrolyzed by water.

exponential adj 指数的な

▶ In many phosphors, marked divergence from a simple exponential law is observed.

exponential function n 指数関数

▶ The decay of the polarization is assumed to be an exponential function of time.

exponentially adv 指数的に

▶ The mole fraction of solute present in the saturated solution increases exponentially with temperature.

expose vt **1 expose to** …にさらす syn *bring into contact with*

▷ Doping may be achieved simply be exposing polyacetylene to gaseous or liquid dopant.

2 現わす syn *reveal, show, uncover*

▶ The surge of activity in LB film research over the past 15 years has exposed many potential applications.

exposed adj **1** 露出した syn *naked, open*

▶ It is assumed that only collisions with the exposed surface can lead to the sticking of a molecule to the surface.

2 exposed to …にさらされた syn *subject*

▶ Potassium bromide exposed to bromine vapor shows a series of absorption bands in the ultraviolet.

exposure n 露出 syn *disclosure, unveiling*

▶ Films of C_{60} changed from yellow to magenta on exposure to alkali-metal vapor, and the conductivity increased by several orders of magnitude and then decreased.

express vt **1** 表す syn *denote, represent*

▶ Raoult's law can be expressed by the equation

$$P = P_0 x$$

in which P is the partial pressure of the solvent above the solution, P_0 is the vapor pressure of the pure solvent, and x is the mole fraction of solvent in the solution.

2 発現させる

▶ These genes may only be expressed by cells at certain times, in response to specific cues from inside or outside of the cell, to make proteins as needed.

expressible adj 表現可能な

▶ Thermodynamics was a method of explanation grounded in direct, precise measurement and expressible with mathematical rigor.

expression n **1** 表現 syn *indication, representation*

▶ The expression order of magnitude usually applies to extremely large or extremely small units.

2 式

▶ The standard Gibbs' energy increment of a chemical reaction may be related to the equilibrium constant, K, for the reaction by the expression

$$\Delta G^\circ = -RT \ln K$$

3 発現

▶ Testosterone stimulates expression of genes involved in muscle growth and hair growth because these genes contain androgen-response elements.

expulsion n 放出 syn *removal*

▶ Expulsions of neutral molecules from ions are observed in the mass spectrum of phthalic anhydride. The molecular ion expels CO_2, followed by CO and then by C_2H_2.

ex situ adv 本来の場所でない

▶ Additionally, the polymerization needs to be quenched as soon as possible to avoid the formation of new ex situ polyaniline.

extend vi **1** わたる syn *range, spread*

▶ The feature which enables one to distinguish a carboxylic acid from all the other carbonyl compounds is a broad absorption band which extends from 3300 cm^{-1} to 2500 cm^{-1}.

vt **2** 拡張する syn *broaden, widen*

▶ The contaminant depresses the melting point and extends the melting range.

extended adj **1** 延長した syn *outstretched, spread out*

▶ In a typical bulk polymer, the average length of a fully extended molecule is about 1 μ, whereas the thickness of individual crystallites, measured in the directions of the chain axes, are considerably less.

2 長い syn *long*

▶ With the elevated temperatures and extended times used in such solid-state reactions, the products are usually the thermodynamically stable phases; compounds that are stable only at relatively low temperatures may be inaccessible by this synthetic approach.

3 広範囲な syn *general*

▶ Mixed halides PX_2Y and PX_2Y_3 form an extremely useful extended series with which to follow the effect of progressive substitution on various properties.

extended X-ray absorption fine structure, EXAFS n X線吸収広域微細構造

▶ The method of EXAFS is chemically specific because the X-ray absorption edges of different atoms fall at different energies.

extension n 1 伸張 syn *expansion, stretching*

▶ The amount adsorbed is obtained from the increased length of the spring previously calibrated for extension with known weights.

2 拡大 syn *development*

▶ The Heitler-London treatment of the hydrogen molecule and its extensions to other molecules are often called valence-bond theory.

extensive adj 1 広範囲な syn *big, considerable, substantial*

▶ Nitrogen forms an extensive series of oxides: N_2O, NO, N_2O_3, NO_2, N_2O_4, N_2O_5, and NO_3.

2 示量性の ☞ intensive

▶ A physical property is classified as extensive if its value depends on the amount of substance in the sample. Examples include volume, internal energy, and mass.

extensively adv 広範囲に syn *thoroughly, widely*

▶ Aluminum occurs very extensively in nature, principally in the form of silicates, of which the feldspars and micas with their decomposition products are important examples.

extensive property n 示量性 ☞ intensive property

▶ The ionic contributions to the extensive properties can be related to the dielectric constant of the medium and to its first and second derivatives with respect to temperature and pressure.

extent n 1 程度 syn *magnitude*

▶ Zinc is such a strong reducing agent that it easily reduces nitric acid, the extent of the reduction depending upon the concentration of the acid.

2 to some extent ある程度まで syn *partially, somewhat*

▶ Cross-linked polymers have, at least to some extent, a three-dimensional array of covalent bonds.

3 大きさ syn *dimension, size*

▶ The spatial extent of molecules is related to the electron density distribution on their outside.

extent of reaction n 反応進行度

▶ The extent of reaction is the change in the amount of any reactant divided by its stoichiometric coefficient.

exterior n 1 外側 syn *outside, surface*

▶ The pure paraffins exhibit a wide range of surface tensions, although they all can expose only paraffin chains and their exteriors are assumed to consist entirely of methyl groups.

adj 2 外側の syn *external, outer*

▶ The most conspicuous structural features of the product cluster are μ_6-S atom and the large exterior Fe−(μ_6-S)−Fe angle, features present in the native P^N cluster.

external adj 1 外からの syn *exterior, outer*

▶ The Daniell cells can be made to behave in a reversible fashion, by balancing their potentials by an external potential so that no current flows.

2 無関係な syn *extraneous, foreign*

▶ The equilibrium between the neutral amino acid and its zwitterion does not require external acid or base.

externally adv 外部に

▶ There are three systems whose externally exposed groups are identical: pure oleic acid, a close-packed film of oleic acid on water, and a solution of sodium oleate.

external pressure n 外圧 ☞ internal pressure

▶ When the sum of the two partial pressures equals, the external pressure boiling occurs.

extinction n 吸光

▶ By rotating the sample and stage, the extinction directions can be seen if the crystals are anisotropic; from the nature of extinction, information on the quality of the crystals may be obtained.

extinction coefficient n 吸光係数 syn *absorptivity*

▶ The difference in extinction coefficient between the two circularly polarized components of the light can be measured and is known as the circular dichroism.

extinguish vt 火や光を消す syn *quench*

▶ When the concentration of naphthacene in anthracene is 0.1 part per million, the anthracene fluorescence is nearly extinguished; at 0.3 ppm it is completely extinguished.

extra adj 余分の syn *additional*

▶ Adamantane has an extra methylene bridge linked to the six-membered ring that stabilizes a cagelike structure.

extracellular adj 細胞外の

▶ Organisms have had to evolve highly specific and very effective mechanisms in order to control the balance between intra- and extracellular concentrations of inorganic ions.

extracellular fluid n 細胞外液

▶ The predominance of calcium biominerals over other group 2 metals can be explained by the low solubility products of carbonates, phosphates, pyrophosphates, sulfates, and oxalates and the relatively high levels of Ca in extracellular fluids.

extract n **1** 抽出液 syn *concentrate, essense*

▶ Diethyl ether is so highly volatile that it is easily removed from an extract at a temperature so low that even highly sensitive compounds are not liable to decompose.

vt **2** 抜き出す syn *remove, withdraw*

▶ This deficiency of electrons is caused by the tendency of the reaction in the cell to extract electrons from that electrode.

3 取り出す syn *derive, draw, obtain*

▶ It is easier to extract quantitative information about the relative transition moment directions from the stretched-film spectra.

4 抽出する

▶ If a complex can be extracted into an organic solvent, it may consist in the solid of neutral molecules held together by van der Waals' forces.

extractable adj 抽出可能な

▶ Oxine forms a large number of extractable complexes but is, therefore, of less use for selective analyses.

extraction n **1** 抽出

▶ Extraction to achieve separation can be treated empirically and does not require a knowledge of how the partition coefficient changes with concentration.

2 採取 syn *separation*

▶ Aluminum combines so powerfully with oxygen that the powdered metal can be used for the extraction of other metals, such as chromium and manganese, whose oxides are too stable to be reduced with coke.

extractive adj 抽出の

▶ The relatively low critical temperature and pressure of CO_2, coupled with its wide availability and low cost, toxicity, and reactivity have made it the substance of choice for a variety of extractive processes.

extractor n 抽出器

▶ In a Soxhlet extractor, solvent vapor rises in the tube on the right, and condensed solvent drops onto the solid, leaches out soluble material, and carries it to the boiling flask, where nonvolatile extracted material accumulates.

extraneous adj 本質的でない syn *peripheral, superfluous*

▶ Control over the concentration of extraneous ions such as Mg^{2+} can be responsible for polymorph selectivity in carbonate and phosphate mineralization.

extraordinarily adv 非常に syn *exceeding-*

ly, extremely

▶ An extraordinarily wide variety of reactions of organic compounds are known to occur under the influence of visible and ultraviolet light.

extraordinary adj 異常な syn *remarkable, uncommon, unusual*

▶ While we may pay lip service to the extraordinary catalytic activity of enzymes based on C, H, N, and O alone, we must observe that they achieve their activity with rather poor attacking chemistry.

extraordinary ray n 異常光線 ☞ ordinary ray

▶ In general, the vibration direction of the polarized extraordinary ray is in the plane defined by the propagation direction and the optical axis.

extrapolate vt 補外する ☞ interpolate

▷ A limiting value of the polarization of the second component may be obtained by determining its value for a number of low mole fraction, n_2, and extrapolating to $n_2 = 0$.

extrapolation n 補外 ☞ interpolation

▶ If the sample is dissolved in a nonpolar solvent, interaction between the solute molecules may be eliminated by extrapolation of a series of measurements taken at different concentrations to infinite dilution.

extreme n **1** 極端 syn *bound, limit*

▶ The distinction between coordinated and noncoordinated perchlorate is sometimes hard to make, and there is an almost continuous gradation between the two extremes.

adj **2** 極端な syn *severe, strict*

▶ Extreme purity is crucial in the solid state, especially when crystalline specimens are being dealt with.

extremely adv 極端に syn *exceedingly, very*

▶ Apart from their very high melting point and extreme hardness, all of these interstitial carbides, nitrides, and borides are extremely inert chemically.

extremity n 末端 syn *end, termination*

▶ An electret is a more or less permanently electrified body having electric charges of opposite sign at its extremities.

extrinsic adj 外因性の ☞ intrinsic

▶ At temperatures below the intrinsic range, the electric properties are controlled by impurities, and here we speak of impurity conductivity or extrinsic conductivity.

extrinsic semiconductor n 不純物半導体 syn *impurity semiconductor*

▶ Silicon may be converted into an extrinsic semiconductor by doping with an element from either group 13 or group 15 of the periodic table.

extrudable adj 押し出し可能な

▶ Molten glass is extrudable into extremely fine filaments.

extrude vt 押し出す syn *protude*

▶ Plastics can be extruded as sheets or pipes, painted on surfaces, or molded to form countless objects.

extrusion n 押し出し加工

▶ Extrusion involves rheological principles of some complexity, critical factors being viscosity, temperature, flow rate, and die design.

exudation n **1** 浸出 syn *leak*

▶ Exudation of a portion of the liquid from jelly or gel is called syneresis.

n **2** 浸出物

▶ Natural gums occur as exudations from various trees and shrubs in tropical areas and differ from natural resins in both chemical composition and solubility properties.

exude vt 発散する syn *emit, leak*

▶ Wheat and oats, adapted to grow on alkaline soils, have evolved the ability to exude various polyamino-acid chelating agents through the root tips to solubilize the iron so that it may be absorbed.

F

fabricate vt 組み立てて製造する syn *assemble, form, set up*

▷ Attention has increasingly turned to the possibility of fabricating photovoltaic solar cells based upon molecular or polymer light absorbers.

fabrication n 製作 syn *assembly, production*

▶ Fabrication of thin films has been accomplished by a variety of techniques, most of which involve deposition on a substrate having lattice constants closely matched with that of the substance.

face n **1** 面 syn *plane, side*

▶ On the right-hand face of the diagram, there is a curve for the diatomic molecule $H^{\beta}-H^{\gamma}$, the distance r_1 now being large.

2 面 ☞ crystal face

▶ Crystals vary not only in their angular relations and symmetry but also in their size, the relative development of their faces, and the number and kind of faces or forms that are present.

3 face to face 直面して syn *confronting, facing*

▶ Interpreting the electrical, optical, and mechanical alterations produced by irradiation brings us face to face with both the transient species and such permanent effects as cross-linking and chain scission.

4 on the face of it 一見したところでは syn *apparently, seemingly*

▶ On the face of it, the earth's atmosphere looks singularly uninteresting from a kinetic viewpoint, consisting primarily of two very stable gases, nitrogen and oxygen.

vt **5** 直面する syn *confront, encounter*

▶ With virtually no precedent to go on, we were faced with two fundamental problems.

face-centered cubic adj 面心立方の

▶ Face-centered cubic metals have four close-packed planes and six close-packed directions which are suitable for slip.

face diagonal n 面対角線 ☞ body diagonal

▶ The planes of atoms can shift relative to one another in each of two directions parallel to the face diagonal and shear in each of two directions parallel to the cell edges.

face-diagonal adj 面対角の

▶ There are six equivalent face-diagonal planes in the CsCl structure.

face value n 額面の値

▶ In the first column, the weights are named by their face values.

facile adj 容易な syn *smooth*

▶ The reason for facile substitution reactions of halogen in the α-position is at least partly related to the electron-attracting characteristics of the neighboring carboxyl function.

facilitate vt 容易にする syn *advance, assist, promote*

▶ To facilitate comparisons between chemical shifts measured at different frequencies, shifts in cps are often divided by the oscillator frequency and reported as ppm（parts per million）.

fact n **1** 事実 syn *event, happening*

▶ The low basicity of aromatic amines is due to the fact that the amine is stabilized by resonance to a greater extent than is the ion.

2 in fact 実際は syn *actually*

▶ Molybdenum has in fact a higher melting point than niobium, although a lower heat of atomization.

factor n **1** 要因 syn *cause*

▶ Factors which can influence micelle size include temperature, pressure, ionic strength, charge, hydrocarbon chain length, the nature of the head group, and the type of counter ion.

2 因子 syn *component, constituent*

▶ Several water-soluble factors appear to be quite important in animal and plant metabolism.

3 係数 syn *coefficient*

▶ Acid-base catalysis does not contribute to rate enhancement by a factor greater than ca. 100.

4 因数

▶ In elementary algebra, a^m is defined as the product of *m* factors each equal to *a*, where the index *m* is a positive integer and *a* is any positive or negative, integral or fractional, quantity.

factual adj 事実に関する syn *actual, real, true*

▶ Such older publications are worth referring to for the wealth of factual information that is now largely forgotten.

fade n **1** 次第に低下すること syn *decline*

▶ Probably all these factors play a role in the capacity fade of the spinel material.

vi **2** 次第に消えてゆく syn *pale*

▶ If phenolphthalein is used as the indicator, the red color fades gradually.

fading n 退色

▶ The participation of oxygen and water vapor in the light fading of dyes has been known for many years.

fail vi できない syn *be unsuccesful*

▶ Glucose fails to give a hydrogensulfite addition compound and, although it will react amines, the products are not the expected Schiff's bases.

failing n 複数形で, 欠点 syn *defect, fault, flaw, shortcoming*

▶ Its failings call our attention to forces in solids other than purely electrostatic ones acting on billiard-ball-like ions.

failure n **1** 不成功 syn *failing*

▶ The failure to separate optical isomers is, of course, not proof that they do not exist.

2 破損 syn *breakdown*

▶ Failure in the metals occurs through a ductile process.

faint adj かすかな syn *feeble, pale, weak*

▶ Anhydrous N_2H_4 is a fuming, colorless liquid with a faint ammoniacal odor first detectable at a concentration of 70–80 ppm.

faintly adv かすかに

▶ Oxalic acid precipitates white, granular tin(II) oxalate from neutral or faintly acid solutions of tin(II) salts.

fair adj まずまずの syn *average, not bad*

▶ High-molecular-weight acids are advantageous since they are likely to give crystalline esters, and these may usually be separated with fair ease by fractional crystallization.

fairly adv かなり syn *moderately, rather*

▶ A fairly simple electrostatic picture of complexes assuming point charges and dipoles could account for many of their properties.

fall n **1** 低下 syn *drop*

▶ Although several factors which contribute to the fall in heat with coverage have received precise expression, the problem as applied to real cases is still largely unsolved.

vi **2** 下がる syn *decrease, diminish, lower*

▷ The N–I temperatures alternate typically, the points fitting two falling curves, the upper for even and the lower for odd members of carbon atoms in the *n*-alkyl chain.

3 ある範囲に入る

▶ The P–O interatomic distance in these compounds generally falls in the range 154–158 pm, the small value being consistent with considerable double-bond character.

4 fall back on よりどころにする syn *depend, employ, rely, use*

▶ Many theoreticians have fallen back on simple predictability of observables as a criterion of success of a theory.

5 fall off 減じる syn *decline, decrease, diminish*

▶ Inductive effects of substituents fall off rapidly with distance.

6 fall into 属する

▶ The oxides of the alkali metals fall into three classes: the normal oxides, M_2O, containing the O^{2-} ion, the peroxide, M_2O_2, containing the O_2^{2-} ion, and the superoxides, MO_2, containing O_2^- ion.

7 fall on 光が…に当たる

▶ Hertz found that when suitable radiation falls on a metal surface, electrons are emitted from the surface.

fallacy n 誤った推論 syn *error, illusion, mistake*

▶ A puzzling fallacy of the late 20th century is

that salts are completely ionized in solution.

fall-off n 減衰 syn *decline, drop*

▶ By careful choice of dimensions of the sector, most of the steep fall-off can be eliminated, and the ripples in the molecular scattering curve then become more pronounced.

fallout n 降下物

▶ In order to protect life from ^{90}Sr, one of the most dangerous fallout isotopes from nuclear fission, it is necessary to prevent or control such absorption, while ion is still in the hydrated surface.

false adj 誤った syn *erroneous, incorrect, wrong*

▶ The assumption that the interaction between ligands and a given metal ion becomes less ionic and more covalent as the optical electronegativity of the ligand decreases may hold for the halides, but as general principle it is false.

familiar adj **1** よく知られている syn *common, well-known*

▶ It is of interest to note that many of the familiar oxo acids are difficult or impossible to isolate as pure substances.

2 familiar with …に通じている syn *aware*

▶ It is important that research workers are familiar with these general limitations of MO theory and also the particular limitations of approximate methods.

familiarity n 熟知 syn *knowledge, understanding*

▶ Because of our familiarity with many ionic crystals, we are naturally prone to draw some rather general conclusions concerning their properties.

familiarly adv 通例 syn *well*

▶ The class of reaction is familiarly known as Friedel–Crafts alkylation.

family n **1** 周期表の族

▶ Owing to the small size and extreme electronegativity of fluorine, the structural chemistry of this halogen is very different from that of the other members of the family.

2 族 syn *group*

▶ The trend in band gaps of this family of isoelectronic, isostructural solids is reflected in their colors, which can be predicted based on the color of transmitted light.

3 系統群 syn *class, type*

▶ Carbohydrate-processing enzymes are classified by primary sequence similarity into families.

famous adj **1** 有名な syn *well-known*

▶ The most famous example of the systems which show both lower and upper consolute temperatures is the system nicotine–water.

2 すばらしい syn *notable, prominent*

▶ To say that potential-energy surfaces are famous is not to imply that they are known.

far adv **1** 距離に関して，はるかに syn *considerable, much*

▶ The proton of an aldehyde group, $-CHO$, absorbs far downfield, at δ 9–10.

2 程度に関して，はるかに syn *considerable, much*

▶ The allotropy of sulfur is far more extensive and complex than for any other element.

3 as far as する限り

▶ This tetrahedral arrangement is the one that permits the orbitals to be as far apart as possible.

4 by far 断然 最上級，比較級を強調する syn *incomparably*

▶ Both indium and thallium can form compounds in the（III）and（I）states; but for thallium the（I）state is by far the more stable.

5 far apart 遠く離れて syn *distant, far*

▶ In an extremely dilute solution, the cations and anions are so far apart that they have insignificant interactions.

6 far from （a）…から遠くに

▶ The actual conditions under which such reactions take place are usually far from being those of equilibrium.

（b）どころか

▶ The problems involved are far from simple, there being even in a small protein, such as insulin, on the order of 700 atoms to be located.

7 in so far as （conj）…する限りにおいて

▶ The presence or absence of asymmetric atoms is important only in so far as it determines

the symmetry or dissymmetry of the molecules.
8 so far これまで syn *thus far, until now*
▶ So far a relationship between structural preferences in the gas phase and in the solid phases of the group 2 dihalides has not been examined systematically.
far-reaching adj 遠くまで及ぶ syn *unlimited, wide*
▶ With appropriate modifications, Dalton's theory was of far-reaching significance for chemistry.
farther adj **1** もっと先の syn *further*
▶ Because proline has the nitrogen of its amino group incorporated into a ring, it cannot form a hydrogen bond with a carbonyl group farther along the chain.
adv **2** もっと先に syn *far, further*
▶ In all directions outward from the center, the light path through the crystal has made an angle with the optic axis which is small near the center of the figure and larger farther out.
fascinate vt 魅惑する syn *attract*
▶ Laddered polyether natural products of marine origin continue to fascinate and inspire those engaged in target-directed synthesis and the development of new synthetic methods.
fascinated adj 魅せられた syn *interested*
▶ Over one hundred years later, chemists remain fascinated with the challenge of nitrogen fixation.
fascinating adj 魅力的な syn *interesting*
▶ Alkali halides are not the only interesting nanometer structure but they are unique in providing fascinating insights into ionic-bonding structures.
fascination n 魅力 syn *attraction*
▶ Since then, interest in the field has spread widely because of the intrinsic fascination of a material which exhibits facile diffusion while remaining a solid.
fashion n 方法 syn *manner, mode, way*
▶ Due to selective precipitation materials prepared in this fashion are often no more homogeneous than hand-ground mixtures.
fashionable adj 流行の syn *current, modern*
▶ 'Shake the bottle' may be a fashionable instruction for certain types of salad dressing, but it should not be necessary in a well-formulated consumer product.
fast adj **1** 速い syn *rapid, swift*
▶ Oxygen is very electrophilic and reacts fastest with the *tert*-butyl radical, which is the most nucleophilic radical.
2 固定した syn *firm, fixed*
▶ One of the oldest known methods of producing wash-fast colors is with the aid of metal hydroxides to form a link between the fabric and the dye.
fast reaction n 高速反応
▶ High flow rates are required both to promote mixing and, for fast reactions, to ensure that the reaction is not complete in the initial portion of the flow tube.
fat n 脂肪
▶ The even numbers are a natural consequence of the way fats are synthesized in biological systems.
fatal dose n 致死量
▶ White phosphorus is highly toxic, and ingestion, inhalation, or even contact with skin must be avoided; the fatal dose when taken internally is about 50 mg.
fate n 結末 syn *end, future, outcome*
▶ Radiation-induced conductivity measurements offer one approach to determining the fate of the electrons and ions.
fatigue n 疲労 syn *tiredness, wearness*
▶ The accumulation of lactate is often cited as a cause of fatigue during intense muscle contraction, but the effect is actually due to a low pH in muscles.
fault n 欠陥 syn *defect, shortcoming, weakness*
▶ It is easy for a theoretician to find faults in this simplistic prescription.
favor n **1 in favor of** (a) …に有利に syn *in support of*
▶ Excess acid is used to shift the equilibrium in favor of a high concentration of hydrobromic acid.
(b) …のほうを選んで syn *in behalf of*

▶ The greater proportion of the work on powdered and sintered materials has been performed in simple apparatus where the limitations have been neglected in favor of ease of operation.

vt **2 助ける** syn *assist, encourage, facilitate, promote*

▶ The fact that the stable isopoly anions are compact rather than extended aggregates indicates that polymerization is favored by charge reduction and the highly favorable entropy change accompanying elimination of water molecules with the formation of compact structures.

3 有利である syn *prefer*

▶ Amperometric approaches have generally been favored, although this is not always possible.

favorable adj **有利な** syn *appropriate, beneficial, suitable*

▶ The low molecular weight of water makes this a favorable liquid for two-phase distillation of organic compounds.

favorably adv **有利に** syn *advantageously, positively*

▶ Domain wall migration may occur such that favorably oriented domains grow in size at the expense of unfavorably oriented ones.

favored adj **有望な** syn *popular, preferred*

▶ The favored oxidation states of gold are well illustrated, as is often the case, by the halides.

favorite adj **人気のある** syn *favored, ideal*

▶ Octyl alcohol is a favorite defoaming agent for aqueous systems.

F center n **F 中心** ☞ color center

▶ It was first suggested by de Boer that the F centers are electrons trapped at vacant anion sites, a hypothesis which seems to have found general acceptance.

feasibility n **可能性** syn *practicability*

▶ The feasibility of isolating the product by distillation depends upon its boiling point and the boiling points of contaminants.

feasible adj **実行可能な** syn *achievable, attainable, realizable*

▶ In order for heterolytic cleavage of the carbon–oxygen bond to be feasible, the electronegativity of the oxygen must be enhanced by conversion into a positive species.

feature n **1 特徴** syn *character, characteristic, property*

▶ A general feature of many of the salts is a framework comprising ions of one type, often the relatively large anions.

vt **2 特徴をなす** syn *involve, participate*

▶ All complexes with π-donor ligands feature electron donation from filled π-ligand orbitals to the metal and donation from the metal into empty π*-ligand orbitals.

featureless adj **特徴のない** syn *flat*

▶ In considering the structure of mesophases, we will regard our molecules as either featureless rods or discs.

feeble adj **微弱な** syn *weak*

▶ The transition-metal oxides deviate from stoichiometry either grossly, e.g., TiO, or to a small degree, e.g., NiO, and they range from metallic conductors, TiO, to very feeble semiconductors, NiO.

feed n **1 供給材料** syn *supply*

▶ Ball mills can be adapted to continuous operation in which the feed enters at one end and is discharged at the other.

vt **2 食物などを与える** syn *provide*

▷ One experimental problem is the difficulty of feeding suitably labeled precursors to plants.

feedback n **フィードバック**

▶ Negative feedback checks the runaway acceleration of an intermediate concentration and differentiates the oscillatory reaction from an explosion, where the feedback is always positive.

feedstock n **原料油**

▶ After some time, the hydrodesulfurization activity of the catalyst drops, as a result of poisoning of the surface by deposits of carbonaceous material and by metals such as vanadium and nickel that are extracted from the feedstock.

feel vt **感じる** syn *note, perceive, see, sense*

▶ Optically active reagents do feel the difference between mirror-image environments.

ferment n **1** 興奮 syn *agitation, excitation*

▶ There is great intellectual ferment now in inorganic chemistry, much of it at the interfaces with sister disciplines.

vt **2** 発酵させる

▶ The particular beverage obtained depends upon what is fermented, how it is fermented, and what is done after fermentation.

fermentation n 発酵

▶ Fermentation of sugars by yeast, the oldest synthetic chemical process used by man, is still of enormous importance for the preparation of ethyl alcohol and certain other alcohols.

Fermi level n フェルミ準位

▶ As donors are added to a neutral semiconductor, enhancing the concentration of conduction-band electrons and the likelihood that electrons will be found occupying the higher energy levels of the conduction band, the Fermi level will shift toward the conduction-band edge.

Fermi resonance n フェルミ共鳴

▶ A special case is Fermi resonance which involves intensity transfer between coupled vibrational levels.

fern-like adj シダ状の

▶ The α form crystallized from the melt in fern-like structures.

ferrimangetic adj フェリ磁性の ☞ antiferromagnetic, ferromagnetic

▶ If the alignment of the spins is antiparallel but with unequal numbers in the two orientations, a net magnetic moment results and the behavior is ferrimangetic.

ferrite n フェライト

▶ Ceramic ferrites are made by pressing powdered ingredients into a sheet, then sintering or firing.

ferroelectric n **1** 通常複数形で，強誘電体 ☞ antiferroelectric

▶ Ferroelectrics exhibit a saturation polarization at high field strength and a remnant polarization, which is the value retained as the applied voltage is reduced to zero after saturation.

adj **2** 強誘電性の ☞ antiferroelectric

▶ Ferroelectric KH_2PO_4 and antiferroelectric $NH_4H_2PO_4$ are both built of isolated PO_4 tetrahedra that are linked together by K^+ and NH_4^+ ions and hydrogen bonds.

ferroelectricity n 強誘電性

▶ There are strong analogies between these magnetic properties and corresponding electric properties such as ferroelectricity.

ferromagnetic adj 強磁性の ☞ antiferromagnetic, ferrimagnetic

▶ Ferromagnetic materials can form the basis for digital recording media such as computer disks.

ferromagnetism n 強磁性 ☞ antiferromagnetism

▶ When the temperature dependence of ferromagnetism is considered, it is found that at a particular temperature known as the Curie point the magnetization falls to zero.

fertile adj 実りの多い syn *fruitful, productive*

▶ Polymorphism is a fertile field of research for the physicist or chemist interested in the sold state.

fertilizer n 化学肥料

▶ By using ^{15}N to label the fertilizer, the amount of nitrogen in the harvested crop that is derived from the applied fertilizer can be measured.

few pron **1** 複数扱いをして，少数 syn *not many*

▶ Few other neutral ligands besides water form a $[TiL_6]^{3+}$ complex. Urea is one of these few.

adj **2** 少ないことを強調して少数の syn *not many*

▶ Elastomers and thermoplastics typically have long polymer chains with few, if any, chemical bonds acting as cross-links between the chains.

比較級は数の比較に用い，名詞の複数形と組み合わせる．

▶ The coordination number of Ni^{II} rarely exceeds six, and its principal stereochemistries are octahedral and square planar with rather fewer examples of trigonal bipyramidal. square pyramidal, and tetrahedral.

3 a few 少数の syn *several*

▶ A few other reagents, such as thiols, thioacids, and hydrogensulfites, can also add abnormally to olefins in the presence of oxidizing agents.

fiber n **1** 繊維 syn *fibril, filament, thread*

▶ Often the X-ray study is made on a fiberlike material in which the macromolecules are more or less ordered along the fiber axis.

2 光学繊維 ☞ optical fiber

▶ Information is transmitted through the fiber by light from a laser or light-emitting diode.

fiber-reinforced adj 繊維強化の

▶ Although high strength ceramics can be made reproducibly with ideal sinterable powders, research is concentrating on fiber-reinforced matrices.

fibril n 原繊維，フィブリル syn *fiber*

▶ When fibrils can grow freely in three dimensions, the sheaf of fibrils formed at each nucleus fans out to become a globular spherulite.

fibrous adj 繊維状の syn *filamentous*

▶ A very low-density form of fibrous silica has been made by the disproportionation of crystalline SiO.

fibrous protein n 繊維状タンパク質

▶ Protein are divided into two broad classes: fibrous proteins, which are insoluble in water, and globular proteins, which are soluble in water or aqueous solutions of acids, bases, or salts.

field n **1** 分野 syn *area, discipline*

▶ The field of metal-containing liquid crystals is now growing quite fast.

2 電気力，磁気力などの場 ☞ electric field, magnetic field

▶ The polarization of one atom or ion by another is equivalent to allowing the electrons of the first ion to move into the field of the second.

field effect n フィールド効果

▶ The field effect operates through space or, in solutions, through the solvent or the low-dielectric cavity provided by organic solutes.

field effect transistor n 電界効果型トランジスター

▶ Field effect transistor devices made from crystalline thin films of a thiophene hexamer with an organic insulator have recently reported with mobilities of 5×10^{-1} $cm^2 V s^{-1}$.

field emission n 電界放射

▶ The preferential coverage of certain planes during chemisorption is observed by the field emission technique.

field ion microscope n 電界イオン顕微鏡

▶ The field ion microscope was the first instrument capable of providing direct images of individual atoms on a solid surface.

figure n **1** 数字 syn *number, numeral*

▶ Two major results emerge from the figures in Table 2.

2 図形 syn *diagram, drawing, illustration, picture*

▶ For convenience, aliphatic rings are often represented by simple geometric figures: a triangle for cyclopropane, a square for cyclobutane, a pentagon for cyclopentane, a hexagon for cyclohexane, and so on.

vi **3** 重要な位置を占める syn *feature, participate*

▶ Holes derived from cubic, octahedral, and tetrahedral arrangements of packed anions figure prominently in describing extended structures.

vt **4 figure out** 理解する syn *interpret, understand*

▶ To figure out what the most stable conformation of a particular molecule should be, one ideally should consider all possible combinations of bond angles, angles of rotation, and even bond lengths and see which combination results in the lowest energy content.

filament n フィラメント syn *fiber, thread*

▶ The filament, usually made of tungsten wire, is heated to produce a beam of electrons which is accelerated by a high voltage on to the target.

filamentous adj 糸状の syn *threadlike*

▶ In animals acetyl-CoA carboxylase is a

filamentous polymer of molecular weight, 4–8 MDa, made from 230-kDa monomers.

file n **1** ファイル syn *data, documentation*

▶ The files should be saved in the native format of the word processing program used.

2 やすり

▶ A triangular file is unsatisfactory because after only brief use it becomes worn and produces too wide a scratch.

vt **3** やすりをかける

part ▶ The scratch is best made by pressing one edge of the file firmly against the tube and slightly rotating the tube away from the body, while filing in the same direction.

fill vt **1** 満たす syn *fill up*

▷ The electrons in the partially filled conduction band easily absorb thermal energy and thereby increase their kinetic energy.

2 fill up いっぱいに満たす syn *fill*

▶ As lithium is inserted into the intercalation host forming a continuous range of solid solutions, the electrons fill up the band, while the mutual repulsions between Li^+ ions rises.

filled band n 充満帯 syn *valence band*

▶ The current is carried by both the electron in the conduction band and the positive hole in the filled band.

filled circle n 黒丸 ☞ hatch, open, shaded

▶ Use of the experimental radii proposed by Blandamer and Symons would raise all of the filled circles in Figure 1.

filler n 充填剤

▶ Fillers have neither reinforcing nor coloring properties, and the term should not applied to materials that do, i. e., reinforcing agents or pigments.

filling n 充填

▶ The filling of the d-electron level has some impact on case of formation of high-coordination structures.

film n **1** 膜 syn *layer, membrane*

▶ A long fatty acid or alcohol on water produces a film whose thickness is the length of one molecule.

2 写真フィルム

▶ Apart from considerations of the extra time involved, the amount of background radiation detected by the film increases with exposure time and, consequently, weak lines may be lost altogether in the background.

filter n **1** 沪過器

▶ The remainder of the crystals on the filter was washed with 40 mL of methyl cellosolve by reslurrying and refiltering.

2 沪光板

▶ Filters are preferable to a monochrometer since they waste far less of the incident light.

vt **3** 沪過する syn *separate*

▷ X-ray powder patterns were obtained using a General Electric XRD-3 diffractometer, equipped with Cu Kα radiation filtered through nickel foil (λ=1.5418 Å).

4 filter off 沪別する

▶ The insoluble phthalyl hydrazide is filtered off, leaving the amine hydrochloride in solution from which the amine may be liberated and isolated in the appropriate manner.

5 filter out 選別して取り除く syn *eliminate*

▶ In the spectrum of X-rays emitted by copper, the Kα line is the most intense, and it is desired to filter out all the other wavelengths, leaving the Kα line for diffraction experiments.

filtering flask n 吸引瓶 ☞ suction flask

▶ A filtering flask is not connected to the water pump directly but through a second vessel that serves as a reservoir for equalization of pressure and as a trap for mother liquor that may be carried over in a foam from a filtering flask.

filtrate n 沪液

▶ The combined filtrates were diluted with 50 mL of heptane and washed three times with 40-mL portions of dilute HCl.

filtration n 沪過 syn *separation*

▶ The oil was further purified by addition of 100 mL of *n*-heptane, the insoluble white solids that precipitated were removed by filtration via Celite plug.

final adj 最終の syn *last, ultimate*

▶ Hexachloroethane and hexachloro-1, 3-butadiene are often the final products of exhaustive chlorination of many higher aliphatic compounds.

finality n 決定的なこと

▶ Finality has certainly not been reached.

finally adv 最後に syn *eventually, lastly, ultimately*

▶ The carbonyl carbon is the one that finally bears the the −OH group in the product: here the number of hydrogens defines the alcohol as primary, secondary, or tertiary.

find vt **1** 見出す syn *reveal, uncover*

▶ It is much more frequently possible to find electrodes which are reversible to one of the constituents.

2 手に入れる syn *acquire, gain, obtain*

▶ To find the molecular formula, we must determine the molecular weight; today, almost certainly by mass spectrometry.

finding n 発見 syn *discovery*

▶ These experiments led to the unexpected finding that even aromatic molecules such as benzene would bind to this powerful π base.

fine adj **1** 精製した syn *refined*

▶ Fully purified pinacolone has a fine camphor-like odor.

2 細かい syn *minute*

▶ The acid-catalyzed xerogels have extremely fine microstructural features, and the low electron-density contrast in the xerogel suggests that the pores are extremely small and evenly spread.

3 微妙な syn *delicate, subtle*

▶ After this coarse approach, a fine positioning mechanism, such as a differential screw or a piezoelectric walker, allows the tip to be moved toward the surface with greater precision.

adv **4** 細かく syn *finely*

▶ Materials of eutectic composition are generally fine grained and uniformly dispersed, as revealed by high magnification microscopy.

finely adv 細かく

▶ Hydrogenation of a nitro compound to an amine takes place smoothly when a solution of the nitro compound in alcohol is shaken with finely divided nickel or platinum under hydrogen gas.

fineness n 細かさ

▶ If the material is homogeneous, it is merely a matter of grinding a portion until it is of suitable fineness.

fine splitting n 微細分裂

▶ Such intimate details can be obtained even for quite large molecules, and it is this aspect which makes the fine splittings of NMR spectra of great value in molecular structure studies.

fine structure n 微細構造

▶ The experimental observations that led to the discovery of the spin of the electron were mainly those of the fine structure of the spectral lines.

fine-tune vt 微調整する

▶ One of the unique attributes of ionic liquids is the ability to fine-tune their physical and chemical properties by selection of the appropriate cations and anions

fingerprint n 指紋

▶ X-ray diffraction has been in use in two main areas, for the fingerprint characterization of crystalline materials and for determination of their structure.

finish n 仕上げ syn *polish, surface*

▶ Visually the deposit formed without acetonitrile is black and powdery whereas the one with a brightener has a mirror finish.

finish line n ゴール地点

▶ It is not imperative that the chromatogram be extended to the point where the solvent has exactly reached the finish line.

finite adj 有限の syn *limited*

▶ If conditions were found where both reactions proceed at a finite rate, equilibrium would ultimately be established and then the rates of the reactions in each direction would be the same.

fire n **1** 火薬 syn *combustibles*

▶ There is a whole history of the use of fire and flame throwers in warfare, dating back to Greek fire in the seventh century.

vt **2** 焼く syn *ignite*

▷ Firing the resulting pellet at high temperature leads to a reaction between the pore walls and the coating.

firing n 焼成

▶ Our solution-reaction synthesis involves the reaction between Li_2CO_3 and $Mn(CH_3CO_2)_2$ in aqueous solution followed by firing at the exceptionally low temperature of 200 ℃.

firing temperature n 焼成温度

▶ Alternative synthesis routes to the spinel have been reported which possess the common factor that they employ low-firing temperatures, and this significantly improves the capacity retention on cycling.

firm adj 確固とした syn *decisive, definite*

▶ Firm evidence that heat is related to motion and is, therefore, a form of energy came only toward the end of the eighteen century.

firmly adv しっかりと syn *tightly*

▶ Atoms within a covalent molecule are bound firmly together by covalent bonds.

firmness n 堅固 syn *stability, strength*

▶ Firmness of binding and chemical inertness are not the only factors which control siderophile tendency.

first n **1** 一番目 syn *beginning, earliest*

▶ The most stable hydride of P is phosphine (phosphane), PH_3. It is the first of a homologous series P_nH_{n+2} ($n = 1-6$) the members of which rapidly diminish in thermal stability, though P_2H_4 and P_3H_5 have been isolated pure.

2 at first 最初は syn *in the beginning, initially*

▶ What at first appears to be complicated reaction is actually a sequence of simple steps involving familiar, fundamental types of reactions.

3 first of all まず第一に syn *above all, in the first place*

▶ First of all, studies of crystal structures show that magnesium tends to form six-coordinate structures, while zinc forms four-, five-, and six-coordinate structures with approximately equal ease.

adj **4** 第一の syn *basic, primary*

▶ As a first approximation, a diatomic molecule can be considered to be two point masses with a fixed value of the interatomic distance.

adv **5** まず syn *firstly*

▶ In order to understand gases, one could first propose some hypotheses about the nature of gases.

6 初めて syn *initially*

▶ This relationship was first given in 1889 by Nernst and is known as the Nernst equation.

7 まず第一に syn *to begin with*

▶ First, each crystalline solid has a unique characteristic X-ray powder pattern, which may be use as a fingerprint for its identification if the pattern for the known material has been recorded.

first derivative n 一次導関数

▶ ESR spectra are usually presented as the first derivative of the absorption, rather than as absorption itself.

first-order reaction n 一次反応

▶ Both S_N1 and S_N2 solvolyses appear to be first-order reactions.

first principles n 第一原理

▶ Obviously, trying to calculate even a simple NMR spectrum from first principles is difficult.

Fischer projection n フィッシャー投影

▶ The convention of the Fischer projections is such that the east and west bonds of the asymmetric carbon are considered to extend out of the plane of the paper and the north and south bonds extend behind the plane of the paper.

fission n **1** 結合の分裂 syn *rupture*

▶ The photolysis of tetramethyltetrazene, $(CH_3)_2N-N=N-N(CH_3)_2$, in a hydrocarbon glass at liquid-nitrogen temperatures appears to proceed with the fission of an N−N single bond, forming the radicals $(CH_3)_2N$ and $(CH_3)_2N_3$.

2 核分裂 ☞ nuclear fission

▶ The fission of a heavy nucleus into two lighter ones releases a large amount of energy, together with other neutrons which, if conditions are right, may themselves produce fission.

fissionable adj 核分裂性の

▶ The fissionable ^{235}U nucleus constitutes only

0.72 % of natural uranium and must be separated from the bulk ^{238}U.

fission product n 核分裂生成物

▶ Rare-earth elements occur as fission products of uranium and plutonium and are the only source of promethium.

fit n **1** 適合性

▶ The theory of paramagnetic susceptibilities is in such good condition that not only must a fit of experimental data illustrate the correct qualitative behavior, but quantitative agreement of experiment and theory must be found.

vi **2** ぴったりはまる syn *conform*

▶ The more symmetrical a compound, the better it fits into a crystal lattice; hence, the higher the melting point and the lower the solubility.

3 適合する syn *agree*

▶ The two equations fit about equally well, frequently agreeing with each other better than with the experimental data; so that here there is no basis for a choice between the two.

4 fit in はめ込む syn *get, take*

▶ The carbon atoms do not dissolve well in ferrite, because the carbon atom is too big to fit in the interstices.

5 fit into …にうまくおさまる

▶ Both naphthalene and azulene satisfy the criteria for aromaticity, since each contains 10 ring carbons each of which has a p orbital suitably oriented for π-bond formation, and in both cases there are exactly 10 electrons to be fitted into the π orbitals.

vt **6** 適合する syn *befit, match, suit*

▶ Such a rate law does fit the observed decomposition data of ammonia on a platinum surface.

7 備える syn *equip, furnish, provide*

▷ The reaction is carried out in a 200-mL round bottom flask fitted with a cork holding a thermometer with the bulb extending to the bottom of the flask.

fitting adj 適切な syn *proper, suitable*

▶ The name chalcogen is fitting, owing to the tendency of the chalcogens to occur in nature as ores with Cu.

five-membered ring n 五員環

▶ Conjugated five-membered rings are always associated with enhanced electron affinity as a result of the well-known stability of the cyclopentadienyl anion.

fix vt 決める syn *determine, specify*

▷ The electronegativity for the ligand would be the most important property in fixing the amount of π bonding.

fixation n 固定 ☞ nitrogen fixation

▶ Carbohydrates are formed in green plants as the result of photosynthesis, which is the chemical combination or fixation of carbon dioxide and water by utilization of energy gained through absorption of visible light.

fixed adj 決まった syn *definite, established*

▶ The synthesis, separation, and characterization of these isomers generally follow no set of fixed rules.

flake n 薄片 syn *scale*

▶ Many metals in powder or flake form will ignite and burn rapidly, whereas most are noncombustible as bulk solids.

flaky adj 薄片状の

▶ Antimony is very brittle and forms bluish-white, flaky, lustrous crystals of high electric resistivity.

flame n **1** 炎 syn *fire*

▶ Flame temperatures of about 2800° can be obtained by combustion of acetylene with pure oxygen.

vt **2** 炎に当てる syn *burn*

▷ Flame drying was accomplished by alternately flaming the reaction vessel under vacuum and filling with argon once ambient temperature was reached.

flameproof adj 燃えない

▶ Urea phosphate, $NH_2CONH_2 \cdot H_3PO_4$, has been used to flameproof cotton fabrics.

flame retardant n 難燃剤

▶ Ammonium phosphates are much used as flame retardant for cellulosic materials.

flame spectrum n 炎光スペクトル

▶ Thallium was named after the characteristic

bright green line in its flame spectrum.

flammability n 引火性

▶ Diethyl ether is highly volatile, and the flammability of its vapor makes explosions and fires unless proper precautions are observed.

flammable adj 可燃性の syn *combustible, inflammable*

▶ Cellulose acetate is less flammable than cellulose nitrate and has replaced the nitrate in many of its applications.

flash n フラッシュ syn *spark*

▶ A flash of light of extremely short duration and not particularly high intensity can induce the rapid reaction between hydrogen and chlorine, the process being complete in an instant.

flash chromatography n フラッシュクロマトグラフィー

▶ The crude material was purified by flash chromatography, followed by recrystallization from hot ethanol to afford the title compound as white prisms.

flash lamp n 閃光ランプ

▶ Light intensities several thousand times higher than those of continuous sources can be produced for short periods of time with the aid of flash lamps.

flash photolysis n 閃光光分解

▶ Another advantage of flash photolysis over flow methods is that because the reactant species are generated and monitored in the center of the reaction cell there are no possible complications from wall-catalyzed reactions.

flash point n 引火点

▶ Substances differ greatly in their combustibility, that is, in their ignition points（solids and gases）or their flash points（liquids）.

flash pyrolysis n フラッシュ熱分解

▶ Flash pyrolysis of certain solid wastes yields synthetic fuel oil and other useful products.

flat n **1** 平板

▶ Infrared spectra of crystalline films, obtained from the melt between cesium iodide flats, were measured with Perkin-Elmer spectrometers, models 21 and 421.

adj **2** 平らな syn *even, plane, smooth*

▶ The part of the molecule immediately surrounding carbonyl carbon is flat; oxygen, carbonyl carbon, and the two atoms directly attached to carbonyl carbon lie in a plane.

adv **3** 平らに syn *precisely, wholly*

▶ The isolated images of phthalocyanine molecule lying flat on single-crystal Cu have been successfully compared with a Hückel orbital calculation, with the charge-density contours evaluated at 2 Å above the molecular plane.

flatten vt 平らにする syn *smooth*

▶ Very small drops of any liquid are almost exactly spherical; larger ones are flattened by their weights.

flattening n 扁平化 ☞ elongation

▶ For many years, K_2CuF_4 was thought to be the only example of a Cu^{2+} compound in which the tetragonal Jahn–Teller distortion was a flattening rather than an elongation.

flavor n **1** 味 syn *taste*

▶ The sense of feel as related to flavor encompasses only the effect of chemical action on the mouth membranes such as heat from pepper, coolness from peppermint, and the like.

vt **2** 味を付ける

▷ The flavoring vanillin occurs naturally as glucovanillin in the vanilla bean, although it is also obtained commercially by oxidation of eugenol, which itself is a constituent of several essential oils.

flavoring agent n 着香料

▶ Some sulfur compounds have been identified as natural favoring agents in foodstuffs, e. g., 3,6-dimethyl-1,2,4,5-tetrathiane is a volatile constituent of cooked mutton.

flaw n **1** 欠陥 syn *defect, fault, imperfection*

▶ Optical fibers and diode light sources can help to identify structural flaws that might be developing in a material.

vt **2** ひびを入れる syn *break, crack*

▷ The crystals should be brought to room temperature gradually over a period of a number of hours to avoid flawing of the crystals due to thermal stress.

3 損なう syn *damage*

▶ The second preparation is flawed by the fact that the chlorination is not specific for the tertiary hydrogen.

flaw-free adj **欠陥のない** syn *perfect*

▶ Silica glass fiber, provided that its surface is protected as noted previously, can be made so sufficiently perfect and flaw-free that kilometer lengths can withstand an extension of 1 percent or more.

fleeting adj **つかの間の** syn *brief, momentary, temporary, transient*

▶ Benzyl and allyl free radicals are extremely reactive, unstable particles, whose fleeting existence has been proposed simply because it is the best way to account for certain experimental observations.

fleetingly adv **つかの間に** syn *briefly, temporarily, transiently*

▶ Although chromium(IV) compounds exist only fleetingly in solution, we have probably all used chromium(IV), as the oxide CrO_2, which is the ferromagnetic material present in many audiotapes.

flexibility n **たわみ性** syn *flexibleness*

▶ The ions from β-diketone have greater flexibility as far as adjusting to preferred donor atom separations for a given metal atom.

flexible adj **1 しなやかな** syn *ductile, elastic*

▶ Diethylenetriamine being flexible is stereochemically relatively undemanding.

2 融通のきく syn *amenable, manageable*

▶ The stereochemistry of Mg and the heavier alkaline-earth metals is more flexible than that of Be.

3 容易な syn *easy, facile*

▶ Synthetic strategies for the construction of fused polyether natural products involving ring-closing metathesis offer flexible, efficient and rapid access to fragments of the type found in fused laddered ether marine natural products.

flip vi **1 ひっくり返る** syn *spin, turn, twist*

▶ When the molecules cannot flip from one position to another in a field, the contribution of the orientation of molecular dipoles to the dielectric constant is absent.

vt **2 ひっくり返す** syn *spin, turn, twist*

▶ There is the more serious requirement that every other ring be flipped upside down before the new structure can pack in its preferred state.

flipping n **向きの変化**

▶ The change from the yellow form to the white form of dimethyl 3,6-dichloro-2,5-dihydroxyterephthalate involves not only a change in hydrogen bonding but also the flipping of every other aromatic ring 180°.

float vi **1 浮かぶ** syn *drift, sail*

▶ If the area per molecule is large, a gaseous monolayer is usually formed, in which most of the molecules float freely and independently on the surface, with their chains spending most of the time on the interface, moving at random with no long-range order or organization.

vt **2 浮かべる**

▷ When a lens of molten fatty acid floating on hot water is cooled and solidified, the upper surface is like paraffin and not wettable by water, whereas the lower surface remains freely wettable.

flocculant n **凝集剤** ☞ coagulant

▶ Flocculants are used in water purification, liquid waste treatment, and other special application. Inorganic flocculants are lime, alum, and iron(III) chloride; polyelectrolytes are examples of organic flocculants.

flocculate vi **1 凝集する** ☞ coagulate

▶ The nature of the net charge on the particles is clearly important in questions of the tendency of colloidal particles to come together, or flocculate.

vt **2 凝集させる**

▶ Cationic reagents are used to flocculate and collect minerals that are not flocculated by oleic acid or soaps.

flocculation n **凝集** ☞ coagulation

▶ The addition of salts of strongly sorbed positive ions will serve to neutralize the negative charge on the clay and bring about flocculation.

floor n 底 syn *base, bottom*

▶ There may be large deposits of methane hydrates beneath the ocean floor.

flotation n **1** 浮揚

▶ The density was measured by flotation in a mixture of methylene dibromide and carbon tetrachloride.

2 浮選

▶ PbS is first concentrated from low-grade ores by froth flotation then roasted in a limited supply of air to give PbO.

flourish vi 隆盛である syn *blossom, develop, grow*

▶ The possibility that the allyl group can act as an η^3 ligand was recognized independently by several groups in 1960, and since then the field has flourished.

flow n 流れ syn **1** *stream*

▶ At infinite dilution, the current can be attributed to the independent flow of positive and negative ions.

vi **2** 流れる syn *stream*

▶ If two objects are placed in contact with one another, thermal energy may flow from one object to the other one.

flowing adj 流れる

▶ The clearing point is detected by the transformation of the viscous column to a free-flowing transparent liquid.

flow system n 流通系 syn *flow reactor*

▶ Flow systems are useful in studies of very rapid reactions; a convenient technique is the stopped-flow method, in which a rapid flow is stopped suddenly and an analysis is made of the change of concentration with time.

fluctuate vi 変動する syn *change, shift, vary*

▶ Brownian movement will cause the particle to fluctuate from this position.

fluctuation n 変動 syn *change, variation*

▶ Temperature control is provided by a well-stirred oil bath which will control temperature fluctuations within 0.1 ℃.

fluctuation-dissipation theorem n 揺動散逸定理

▶ Parr and Chattaraj's proof of the principle of maximum hardness is based on a combination of statistical mechanics and the fluctuation-dissipation theorem.

fluid n **1** 流体 syn *gas, liquid, solution, vapor*

▶ The term fluid applies not only to liquids but to gases and to finely divided solids.

adj **2** 流動性の syn *flexible, formless, mobile*

▶ Liquid crystals are fluid like liquids but also have partial translational and/or orientational order, reminiscent of solids.

fluidity n 流動度

▶ The fluidity $\phi = 1/\eta$ is almost a linear function of temperature, especially at higher temperatures; the $\phi - t$ lines are parallel in homologous series, and at the boiling points the fluidities are nearly equal for nonassociated liquids.

fluidization n 流動化

▶ In fluidization, a finely divided solid is caused to behave like a fluid by suspending it in a moving gas or liquid.

fluidized adj 流動する

▶ The fluidized catalysts, e.g., alumina-silica gel, is brought into intimate contact with the suspending liquid or gas mix, usually a petroleum fraction.

fluoresce vi 蛍光を発する

▶ When doped with Sb^{3+} apatite fluoresces blue, and with Mn^{2+} it fluoresces orange-yellow, the two together give a broad emission spectrum that approximates to white light.

fluorescence n 蛍光

▶ Fluorescence effectively ceases as soon as the excitation source is removed.

fluorescence label n 蛍光標識

▶ It is quite common to tag synthetic oligonucleotides with fluorescence labels added at either end of the strand.

fluorescence microscopy n 蛍光顕微鏡法

▶ Fluorescence microscopy was developed as a powerful technique to visualize the location of specific molecules in a cell.

fluorescence spectroscopy n 蛍光分光

法

▶ Intramolecular photoinduced electron transfer was monitored by decrease of intensity in steady-state fluorescence spectroscopy.

fluorescence spectrum n 蛍光スペクトル

▶ The fluorescence spectrum always arises from a level lower than that reached by absorption and is in consequence displaced toward longer wavelengths compared to the absorption spectrum.

fluorescence yield n 蛍光収率

▶ For 9,10-dibromoanthracene the fluorescence yield fell to about one-half at 3050 Å and then stayed constant throughout the second transition before falling again.

fluorescent adj 蛍光性の

▶ Fluorescent proteins, originally isolated from jellyfish and corals, are now commonly used as reporters and are available in a range of colors for simultaneous detection of multiple signals.

fluorescent dye n 蛍光染料

▶ Fluorescent dyes are used for labeling molecules in biochemical research.

fluoridation n フッ化物添加

▶ The chemicals most commonly used for city fluoridation programs are fluosilicic acid, sodium silicofluoride, and sodium fluoride.

fluorinate vt フッ素化する

▷ The solubilities of some fluorinated metal chelates in supercritical CO_2 are 2–3 orders of magnitude higher than the nonfluorinated analogues.

fluorinating agent n フッ素化剤

▶ AgF_2 is thermally stable but is a vigorous fluorinating agent used especially to fluorinate hydrocarbons.

fluorination n フッ素化

▶ The nitrosyl halides can be made by direct halogenation of NO with X_2, though fluorination of NO with AgF_2 has also been used.

fluorite structure n 蛍石型構造 ☞ antifluorite structure

▶ In the fluorite structure, the calcium ions are arranged at the corners and face centers of a cubic unit cell, and the fluoride ions are at the centers of the eight cubelets in which the cell may be divided.

fluorocarbon n フルオロカーボン

▶ The extraordinary solvent properties of fluorocarbons have come to light only in 1940s.

fluorophore n 発蛍光団

▶ Certain fluorophors display considerable sensitivity to the presence of hydrogen-bonding solvents.

flush vt **1** 水で流す syn *rinse, wash*

▶ There are people who believe the chemical industry flushed mercury into the streams in full knowledge of the hazards involved.

2 どっと流す syn *purge*

▶ If hydrogen is flushed on to the nickel at room temperature, a large loss of magnetization occurs almost instantly, but some of this is recovered over a period of a few minutes.

flux n **1** 融剤

▶ A flux promotes the fusing of minerals or metals or prevents the formation of oxides.

2 流束

▶ The term flux is used to denote the quantity that crosses a unit area of a given surface in a unit of time.

foam n **1** 泡 syn *bubble, froth*

▶ The interchangeable terms foam and froth are restricted to systems in which gas is the enclosed or discontinuous phase.

2 泡状の物質

▶ Plastics can be made into flexible and rigid foams by use of a blowing agent.

vi **3** 泡立つ syn *froth*

▶ Pure liquids do not foam; the bubbles break as soon as they reach the surface.

vt **4** 泡立てる syn *bubble*

▷ The principal use of foamed metals is in absorption of shock impact without elastic rebound.

focal-conic texture n フォーカルコニック組織

▶ The typical textures obtained with smectic A liquid crystals are focal-conic textures.

focus n **1 焦点**
▶ If the fragment has a higher refractive index than the liquid, it tends to bring parallel light to a focus above the fragment.
2 興味の的 syn *center*
▶ In terms of organic ligands, much of the recent focus has been on connectivity through oxygen atoms of carboxylic acid groups.
vi **3 集中する** syn *concentrate*
▶ Increasing research activity has focused on understanding the role of diffusion in mesoporous systems.
vt **4 集中させる** syn *center, concentrate*
▶ We focus attention on the use of surfactant vesicles and their polymerized counterparts as colloidal catalysts and drug carriers.
5 焦点を合わせる
▶ The liquid is illuminated by strong light, usually from a mercury lamp, and the scattered light is focused by a lens into a spectrograph, in which the light is separated into a spectrum which is recorded on a photographic plate.

fog n **1 霧** syn *mist, smog*
▶ Synthetic fogs can be produced on a laboratory scale by ultrasonic vibrations.
vt **2 曇らせる** syn *obscure*
▶ A crystal of silver bromide which has been sensitized by exposure to hydrogen sulfide is fogged with the formation of a surface latent image both by exposure to light and by treatment with bromine at a very low vapor pressure in the dark.

foil n **1 箔** syn *film, sheet*
▶ For copper radiation, a sheet of nickel foil is very effective in filtering out all the wavelength other than the Kα line.
vt **2 妨げる** syn *hamper, offset, thwart*
▶ All attempts to isolate the element either by chemical reactions or by electrolysis were foiled by the extreme reactivity of the free fluorine.

fold n **1 折り目** syn *pleat*
▶ The folds in the hydrocarbon chains would probably take place by rotations of 120° into gauche positions around three C—C bonds.
vi **2 折り重なる** syn *gather*
▶ As electrons are removed from the neutral six- and eight-membered rings there is a strong tendency for them to fold up into more closely packed cluster structures.
vt **3 折りたたむ** syn *bend, pleat*
▶ Molecules of globular proteins are folded into compact units that often approach spheroidal shapes.
suff **4 倍** nine までは綴り，10 からは数字とハイフンを用いる．
▶ The rate of flow of various liquids through a nitrocellulose membrane was proportional to the pressure and had the same temperature coefficient as the viscosity, where the viscosity of the different liquids varied by 200-fold.

folded structure n **折りたたみ構造**
▶ Although many synthetic polyamino acids have no well-defined conformation and appear to exist in solution as nearly random coils, most biological polypeptides adopt a well-defined folded structure.

folding n **1 折りたたみ**
▶ The folding of a protein chain into a compact globular conformation removes nonpolar groups from contact with water.
adj **2 折りたたみの**
▶ Some proteins, such as the keratins of hair and feathers, are fibrous, and they are organized into sheetlike shapes with regular repeating folding pattern.

follow vi **1 随伴する** syn *accompany*
▶ If the transfer of an atom is rate-determining, then the usual adiabatic assumption can be made that the electrons can easily follow the motions of the nuclei.
2 it を主語として，**…の結果として起こる** syn *result*
▶ From the relationship between ΔG and the cell emf, it follows that the more positive the oxidation potential of a half reaction, the greater will be its thermodynamic tendency to take place.
3 as follows …は次のとおり 名詞の複数形や二つ以上の名前，もの，語が関与する場合も，

as follow とはしない.

▶ The arguments of van't Hoff are briefly as follows.

4 follow on 生じる

▶ SI electrical units follow on from the MKSA (meter, kilogram, second, ampere) system of units which has been used by electrical engineers for many years.

vt **5** …に続く syn *succeed*

▷ In many solid-state syntheses, several cycles of heating for many hours, each followed by regrinding at room temperature, and a final heating may be needed to form homogeneous sample.

6 従う syn *conform, obey, observe*

▶ As in almost all reactions of organic halides, reactivity toward alcoholic silver nitrate follows the sequence RI > RBr > RCl.

7 たどる syn *pursue, track*

▷ Measurement of the resistance offers a convenient and accurate means of following the course of the reaction.

following adj **1** 以下の syn *subsequent*

▶ On the basis of these data, it is possible to make the following broad generalizations.

prep **2** …について syn *subsequent to*

▶ Following a survey of the reactions of several solid aromatic hydrocarbons with bromine, a quantitative study was performed of the decomposition of a perylene-bromine complex in the solid state.

food n 食品 syn *diet*

▶ With the exception of sodium chloride and water, all foods are derived from plants, either by direct consumption or by ingestion of animal tissue or animal products which are derived metabolically from vegetable sources.

food additive n 食品添加物

▶ A food additive is a substance or mixture other than a basic foodstuff that is present in food as a result of any aspect of production, processing, storage, or packaging

for prep **1** …に対する

▶ The proper solvent for mercury is nitric acid.

2 …の間 syn *during, over*

▶ La[Co(CN)$_6$] can be converted to $LaCoO_3$ by heating in air for a short time.

3 …の場合には

▶ Proteins move to the anode for sufficiently basic solutions, to the cathode for sufficiently acidic solutions, and show no electrophoretic effect at the isoelectric point.

4 動詞と結合して，…を求めて syn *looking for, seeking*

▶ In intermolecular competition, a mixture of labeled and unlabeled reactants compete for a limited amount of reagent.

5 …の理由で syn *because of, for the sake of*

▶ The fugacity is the pressure adjusted for lack of ideality.

6 …に対して 割合を表して.

▶ Treatment of these coagula with solutions of other electrolytes results in ion exchange, equivalent for equivalent.

7 …と…を比べてみて 前後に同一名詞を置く.

▶ Weight for weight, aluminum has about twice the electric conductivity of copper and so is much used for cables.

8 …が…する 不定詞の主語関係を示す.

▶ For a range of simple substitutional solid solutions to form, there are certain requirements that must be met.

conj **9** だから syn *as, because, since*

▶ Thus, the distribution constant K_D of *cis*-2-pentence between carbon tetrachloride and 1N potassium nitrate solution is regarded as a better measure of the coordination reaction, for it seems reasonable that the activity coefficients of the olefins in dilute solutions of carbon tetrachloride are alike at similar concentrations.

forbid vt 禁止する syn *prohibit*

▶ Transition-metal cations are usually colored. As many of the colors are weak the corresponding excitations must be forbidden.

forbidden band n 禁止帯 ☞ band gap, energy gap

▶ In terms of the band theory, the model involves the postulate of a discrete and localized impurity level spatially located in the vicinity of the impurity and situated on the energy scale in the forbidden band of the parent lattice slightly below the conduction band.

forbidden transition n 禁制遷移 ☞ allowed transition

▶ Phosphorescence is attributed to a forbidden transition from the lowest triplet state of the molecule to the ground state.

force n **1** 力 syn *power*

▶ Stress is measured in units of force per unit area; strain is the extent of the deformation.

vt **2** 強いてさせる syn *constrain, enforce*

▶ In $BaTiO_3$, the large cation forces the structure to expand with the result that, if the crystal is subjected to pressure or an electric field, the titanium ions can move within their cavities.

force constant n 力の定数

▶ These bending force constants have a key advantage over the bond angle data in this regard.

forced adj 強制の syn *artificial, unnatural*

▶ Forced cooling of the sample causes crystallization, and upon heating, the crystal-to-liquid crystal transition can again be observed.

forcefully adv 力強く syn *mightily*

▶ Slight pressure is exerted outward with the thumbs and at the same time the tube is forcefully pulled apart.

forecast n **1** 予測 syn *anticipation, prediction*

▶ It is obviously impossible to make accurate forecast of future developments.

vt **2** 予測する syn *anticipate, predict*

▶ The concept of homology, when used to forecast the properties of unknown members of series, works most satisfactorily for the higher molecular weight members.

forefront n 先頭 syn *front*

▶ Chemistry and chemists will be at the forefront of such research.

foregoing adj **1** 前述の syn *preceding, previous*

▶ In practice, however, the actual application of this method is by no means as simple as the foregoing discussion would make it appear to be.

2 名詞的に，前記のもの

▶ It should be evident from the foregoing that many complications may enter into the prediction of reaction rates.

foreign adj 異質の syn *extraneous, extrinsic, unrelated*

▶ The adsorption of a product, or a foreign substance, can compete for the catalyst surface and thereby inhibit the reaction.

foremost adj 主要な syn *best, main, superior*

▶ It is doubtful whether 50 years ago any scientist believed that organic conductors would ever be a foremost area of research.

foresee vt 予知する syn *envisage, predict*

▶ As the sophistication of synthetic organic chemistry advances, striking innovation in chemical catalysis can be foreseen.

foreseeable adj あまり遠くない syn *predictable*

▶ Developments in computational chemistry in the foreseeable future will allow inexpensive calculation of enthalpies of formation for most organic substances which are comparable in accuracy with the best available experimental methods.

forever adv 永久に syn *continuously*

▶ The sequence of reactions must stop at some time, since the polymerization cannot go on forever.

forget vt 忘れる syn *disregard, ignore*

▶ It should not be forgotten that primary and secondary amines are also able to act as acids.

form n **1** 形 syn *conformation, configuration*

▶ There is considerable evidence which shows that the equatorial form of methylcyclohexane predominates over the axial form in the equilibrium mixture.

2 結晶形 ☞ crystal form

▶ Crystals in general may exhibit one or more of a large number of possible forms including the cube, tetrahedron, octahedron, rhombohedron, and other polyhedra.

3 形態 syn *structure*

▶ There are several solid forms of phosphorus, all containing single bonds.

4 形式 syn *condition, state*

▶ Many sulfonic acids have considerable commercial importance as detergents in the form of their sodium salts.

vi **5 形をなす** syn *set up*

▶ In the Czochralski process, a crystal forms as a melt passes through a temperature gradient from a hot zone to a cooler zone, and impurities remain in the melt.

vt **6 形成する** syn *constitute, make up*

▶ Halides frequently form hydrates which differ in properties from the anhydrous materials.

7 作り出す syn *make, produce*

▶ In plant nutrition, the essential requirements are carbon dioxide and water, from which the plant forms carbohydrates by photosynthesis.

formal adj **形式上の** syn *customary*

▶ In formal sense, H_2O_2, with its oxygen atoms in the oxidation state −1, can be prepared by oxidizing O^{2-} or reducing molecular O_2.

formal charge n **形式電荷**

▶ The sum of the formal charges in a molecule is zero; the sum of the formal charges for an ion is the same as the charge on the ion.

formalism n **形式論**

▶ Localized three-center bond formalism can readily be used to rationalize the structure and bonding in most of the non-*closo*-boranes.

formalize vt **形式化する**

▶ InTe, TlS, and TlSe have a structure which can be formalized as $M^{I}[M^{III}X_2]$.

formally adv **形式的に**

▶ Formally, addition is the opposite of elimination; just as elimination generates a multiple bond, so addition destroys it.

format n **方式** syn *form, style*

▶ Most popular wordprocessor file formats are acceptable, though we cannot guarantee the usability of all formats.

formation n **形成** syn *generation, production*

▶ Metallic calcium decomposes water with evolution of hydrogen and formation of the hydroxide.

former adj **代名詞的に用い，前者**

▶ Both X-rays and secondary electrons are emitted by the sample; the former are used for chemical analysis, and the latter are used to build up an image of the sample surface which is displayed on a screen.

三つ以上について述べるときは，first と last を用いる．

formerly adv **以前に** syn *previously*

▶ Such semiaxes are entirely based upon experiment and are free from certain assumptions made formerly.

formidable adj **1 非常に優れた** syn *important, impressive, strong*

▶ On the basis of formidable experimental evidence, it is believed that complex formation between like and/or unlike molecules takes place in many systems.

2 難しい syn *difficult, overwhelming*

▶ Formidable problems must be overcome in developing catalytic colloidal semiconductors capable of photoreducing water efficiently.

formula n **1 方策** 複数形は formulas または formulae syn *prescription, procedure*

▶ When most commercial materials were found too sensitive to irradiation, subsequent attention was turned to developing new formulas in search for higher stability.

2 公式 syn *prescription, rule*

▶ From the results of heat capacity and latent enthalpy measurements, other useful thermodynamic properties may be calculated by standard formulas.

3 化学式 ☞ chemical formula

▶ A formula like $Si_{0.283}Ge_{0.717}$ means that at any site where a silicon or germanium atom could be found in the crystal structure, there is a 28.3% chance that the atom is silicon and a 71.7% probability that the atom is germanium.

formulate vt **1 作り上げる** syn *conceive, devise*

▶ On the basis of this evidence, we formulate the following mechanism.

2 式で表す syn *denote*

▶ The allyl cation can be formulated as two equivalent structures.

3 調合する syn *compose, form*

▷ It is common practice to broaden the temperature range of desired phases by formulating eutectic mixtures.

formulation n **1** 明確な記述 syn *conception*

▶ The atomic-orbital formulation immediately suggests two equivalent Kekulé structures.

2 式 syn *composition*

▶ Substitution of fluoride for hydroxide is tunable over the complete range of formulations $Ca_5(PO_4)_3(OH)_xF_{1-x}$ $(0 \leqq x \leqq 1)$.

3 調合製品 syn *assembly*

▶ Phosphate-silicate formulations, once widely used as builders, have been restricted for environmental reasons.

forthcoming adj まもなく現れる syn *coming*

▶ These questions will be considered in forthcoming papers of this series.

fortuitous adj 偶然の syn *accidental, incidental*

▶ This agreement must be regarded as fortuitous in view of the very serious approximations involved in the London equation.

fortuitously adv 偶然に syn *arbitrarily*

▶ Electrophilic substitution of $B_{10}H_{14}$ follows, perhaps fortuitously, the sequence of electron densities in the ground-state molecule.

fortunate adj 幸運な syn *fortuitous*

▶ We were fortunate to obtain a crystalline material from an orange solution of selenium in oleum which proved to be the hydrogendisulfate of Se_4^{2+}.

forward adj 前方へ syn *advance*

▶ The advantage of relaxation techniques is that they allow both forward and reverse rate coefficients to be measured.

forward reaction n 正反応 ☞ reverse reaction

▶ The same catalyst, hydrogen ion, that catalyzes the forward reaction, esterification, necessarily catalyzes the reverse reaction, hydrolysis.

forward scattering n 前方散乱 ☞ backward scattering

▶ The term forward scattering means that, following reaction between atom M and molecule RX, the product MX emerges in roughly direction in which M was traveling before the collision.

fossil fuel n 化石燃料

▶ Acidic substances and their precursors are formed when fossil fuels are burned to generate power and provide transportation.

foster vt 促進する syn *promote, stimulate*

▶ The rapid growth in the areas of both chemical biology and molecular imaging has fostered the search for new synthons beyond those traditionally used in the pharmaceutical chemistry.

foundation n 基礎 syn *basis, fundamental*

▶ Medium effects on the structure, conformation, and reactivity of small and large molecules have been at the foundation of physical chemistry.

four-center reaction n 四中心反応

▶ Such trans additions can hardly involve simple four-center reactions between one molecule of alkene and one molecule of an addend X-Y.

fourfold axis n 四回軸

▶ If you have a fourfold axis, it causes repetition every 90°.

Fourier transform infrared spectroscopy, FTIR n フーリエ変換赤外分光法

▶ Broad band FTIR investigations are particularly useful for the characterization of aerosols.

four-probe adj 四端子法による ☞ potential probe

▶ Four-probe DC conductivity measurements were conducted on 15 different crystals of perylene・$4I_2$.

fourth adj 四番目の

▶ The bonding of the carbonium center is considered to involve three, two-electron covalent bonds and a fourth two-electron-three-center bond.

fourth power n 四乗べき 複合形容詞と

して使用

▶ The fourth-power dependence of the scattering on the wavelength should be noticed.

fraction n **1** 小部分 個々の集合体と見做すときは，動詞は複数形とする． syn *bit, piece, portion, segment*

▶ Only a very small fraction of the molecules populate the higher vibrational levels at ordinary temperatures.

2 分数 ☞ denominator, numerator

▶ When the units along different crystallographic axes are different, we can still describe positions in terms of fractions of these units without knowing their actual length.

3 フラクション syn *section*

▶ The third major fraction of biological membranes consists of carbohydrate, which is always covalently bound either to the lipid or to the protein components.

fractional adj 分数の syn *partial*

▶ The intrinsic viscosity represents the fractional change in the viscosity of a solution per unit concentration of polymer at infinite dilution.

fractional crystallization n 分別結晶

▶ It is possible by procedure of fractional crystallization described below to isolate both isomers in pure condition.

fractional distillation n 分別蒸留

▶ Fractional distillation repeats the boiling and condensation cycle several times.

fractionally adv **1** わずかに syn *partly, somewhat*

▶ Catalysis is the loosening of the chemical bonds of two or more reactants by another substance, in such a way that a fractionally small percentage of the latter can greatly accelerate the rate of the reaction, while remaining unconsumed.

2 精留して

▶ Cyclopentadiene was fractionally distilled from its dimer through a ten-theoretical-plate column and collected in a receiver cooled by Dry Ice.

fractionate vt 精留する

▶ Carbon tetrachloride was dried ($CaCl_2$) and fractionated through a 1-m column packed with glass helices.

fractionating column n 精留塔

▶ A fractionating column effects in a single operation what amounts to several separate simple distillations.

fractionation n **1** 分別

▶ In distillation, fractionation is done by means of a tower or column in which rising vapor and descending liquid are brought into contact.

n **2** 分画

▶ The biochemical roles of subcellular organelles could not be studied properly until they had been separated by fractionation of the cells.

fracture n **1** 破砕 syn *breakage, cleavage, rupture*

▶ Bismuth is normally regarded as a hard, brittle metal, yet when it has been zone-refined it forms rods which can be bent without fracture.

vi **2** 割れる syn *break, cleave, crack, rupture*

▶ Porous oxides appear to fracture at grain boundaries.

fracturing n 破砕 syn *breakdown*

▶ When a mechanical stress is applied to an ionic solid in certain directions, like-charged ions are forced into proximity with one another, and the repulsive electrostatic forces lead to fracturing.

fragile adj 壊れやすい syn *brittle*

▶ Charcoal is too fragile to be used for widening tubes of small diameter and for this use carbon rods are preferable.

fragility n 壊れやすさ syn *weakness*

▶ The disadvantages of the torsion type of balance are fragility and the difficulty of construction which requires special tools, techniques, and conditions

fragment n **1** フラグメント syn *part, piece, portion*

▶ Proof of structure of a new compound is best accomplished by degradation: cleavage by ozone, periodate, or permanganate, followed by identification of the fragments formed.

vi **2 分解する** syn *break, disintegrate, shatter, split*

▶ Graphite has been used, but it fragments following electrolysis at high overpotentials leaving a black powdered residue at the base of the cell.

vt **3 分解する** syn *break, disintegrate, shatter, split*

▷ Fragmented nuclear membranes in many cells are stored as stacks of endoplasmic reticulum.

fragmentary adj **断片的な** syn *incomplete, scattered*

▶ There is fragmentary information on the precipitation of an insoluble polonium(IV) carbonate, iodate, phosphate, and vanadate.

fragmentation n **フラグメンテーション** syn *separation*

▶ Fragmentation of NH from a pyridine derivative seems to be a drastic change, but loss of CH_3 is reasonable.

fragment ion n **フラグメントイオン**

▶ There are two important factors which determine the intensities of fragment ions in the mass spectrum: the stability of the ion and the energy relationships of the bonds broken and formed in the reactions leading to the ion.

fragrance n **香気** syn *aroma, odor, perfume, smell*

▶ Most essential oils have a pleasant odor and are the basis of perfumes and fragrances.

frailty n **1 もろさ** syn *fragility, weakness*

▶ A major problem with charge-transfer complexes or organic superconductors is the extreme frailty of the crystals of these materials.

2 短所 syn *defect, flaw, imperfetion*

▶ The best method is rigorous mathematical solution, without recourse to the frailties of chemical intuition.

framework n **1 骨格** syn *structure*

▶ The longest continuous chain of carbon atoms is taken as the framework on which the various alkyl groups are considered to be substituted.

2 枠組み syn *context, outline, scenario*

▶ The approach is also a very general one, so general that one can, within its framework, seek answers to questions.

fraught adj **fraught with …を伴って** syn *accompanied with*

▶ The calculation of transition moment directions by semiempirical procedures is fraught with uncertainty.

free adj **1 自由な** syn *able, allowed*

▶ In a liquid composed of molecules with a dipole moment, the dielectric constant is relatively high because the molecular dipoles are free to orient in the field.

2 遊離した

▶ Semicarbazide is not very stable in the free form and is used as the crystalline hydrochloride.

3 自由に移動する syn *independent, uncontrolled*

▶ This difficulty showed that there was something fundamentally wrong with the simple free electron picture of a metal, and that has proved to be assumption of the Maxwell–Boltzmann statistics for the electrons.

4 free from 免れて syn *exempt from*

▶ The first step in growth from solution is the preparation of a small single crystal of the desired substance as free from flaws as possible.

5 free of …のない syn *without*

▶ It is hard to overemphasize the hazards met in using diethyl ether, even when it is free of peroxides.

vt **6 除去する**

▶ Alkanes, which are insoluble in sulfuric acid, can be freed from alkene impurities by washing with sulfuric acid.

free acid n **遊離酸**

▶ Attempts to prepare thiosulfuric acid by acidification of stable thiosulfates are invariably thwarted by the ready decomposition of the free acid in the presence of water.

freedom n **自由** syn *independence*

▶ A molecule in the gas phase or in solution

has more freedom of motion than one that is attached to a surface.

free electron n 自由電子

▶ Metals have a resistivity that increases slowly with temperature, owing to increased scattering of the free electrons by crystal atoms with temperature.

free energy n 自由エネルギー ☞ Gibbs energy

▶ Free energy may be separated into its enthalpy and entropy components.

freely adv 自由に syn *readily*

▶ Alkali ions are able to move freely within the conduction planes but cannot penetrate the dense, spinel blocks.

free radical n 遊離基, フリーラジカル

▶ In each step, the consumption of a free radical is accompanied by the formation of a new, bigger free radical.

free rotation n 自由回転

▶ In cyclic compounds, free rotation about the σ bonds comprising the ring is obviously not possible as long as the ring remains intact.

freestanding adj 支えなしで立っている

▶ Where the host material can be dissolved or removed chemically from the inclusion material, then freestanding one-dimensional structures will result.

free valence n 自由原子価

▶ Other free neutral radicals react instantly with triphenylmethyl, and this type of addition process has been used as a diagnostic test for a free valence.

freeze vi **1** 凝固する syn *solidify*

▶ Some change of order occurs when a substance freezes or boils, and so it will be accompanied by a change of entropy.

vt **2** 凍結する syn *fix, pin, stop*

▶ At lower temperatures, the disorder which happens to prevail is frozen, and changes in temperature below this region have no effect on the number of defects.

freezing n 凍結

▶ Freezing normally proceeds from the outside, and the air is accumulated in the retreating liquid phase.

freezing point depression n 凝固点降下

▶ Molecular weights of high-boiling liquids and slightly volatile solids are generally determined by measurements of freezing point depressions or boiling point elevations of solutions in suitable solvents.

frequency n **1** 振動数

▶ Solid indigo absorbs at very low frequencies since in it each molecule is surrounded by highly polar dye molecules and is, therefore, effectively embedded in a medium of high dielectric constant.

2 頻度

▶ If the gaseous molecules interact with each other at a distance, then this will influence their frequency of collision.

frequency distribution n 頻度分布

▶ A frequency distribution in increments of five is shown in Figure 1.

frequency factor n 頻度因子

▶ To a good approximation, most reaction rates vary with temperature according to the Arrhenius equation

$$k = A \exp(-E_a/RT)$$

where A is a constant often called the frequency factor, and E_a is known as the activation energy.

frequently adv しばしば syn *continuously, often, regularly*

▶ Indeed, it is frequently useful and instructive to consider types of reactions rather than types of compounds.

fresh adj **1** 塩分のない syn *clean, pure*

▶ Less than 2.7% of the total surface water is fresh.

2 新しい syn *new, renewed*

▶ Ethylene dibromide continually produces a fresh magnesium surface, which is sufficiently active to be able to react with the aryl chloride.

freshly adv 新たに syn *newly*

▶ The freshly precipitated aluminum hydroxide is readily soluble in dilute acids; but after standing some time in a salt solution or after long boiling, it becomes more difficultly soluble.

freshwater adj 淡水の 名詞は fresh water
▶ Freshwater pollution typically occurs closer to populated areas and poses a serious threat to human health by contaminating sources of drinking water.

friction n 摩擦
▶ A single monolayer of oily substance on glass or porcelain is enough to decrease static friction to low values.

frictional adj 摩擦の
▶ The viscosity of a fluid is a measure of the frictional resistance it offers to an applied shearing force.

fridge n 冷蔵庫 syn *refrigerator*
▶ The deep red filtrate was maintained under a nitrogen atmosphere and stored in the fridge.

fringe n フリンジ ☞ interference fringe
▶ When strongly divergent plane-polarized light passes through slices of a uniaxial or biaxial crystal cut in different ways relative to the optic axes, colored fringes of various forms, overlaid by dark crosses or brushes, appear.

frit n **1** フリット
▶ Finely powdered glass may be called a frit. The term frit is also used for finely ground minerals, mixed with fluxes and coloring agents which turn into a glass or enamel on heating.
2 溶結したガラス粉
▶ The reference electrode compartment was separated from the bulk melt with a fine porosity frit.

fritting n フリット化
▶ The process of sintering or fritting comprises the agglomeration of a powdered solid into a more or less compact mass by heating at a temperature below the melting point.

from prep **1** 出発点，起点を表して，…から
▶ If manganese from any oxidation state is converted into manganese at some other state, the work performed is theoretically the same whether the oxidation takes place all at once or in several stages.
2 原因，動機，理由を述べて，…から
▶ Although titanium is very reactive, it has a good resistance to corrosion from a thin protective oxide coating on its surface.
3 起源，由来を表して，…から
▶ Naphthols can be made from the naphthylamines by direct hydrolysis under acidic conditions.
4 区別，相違を表して，…から
▶ The iodoform test distinguishes a methyl ketone from all aldehydes but acetaldehyde.
5 変化，推移を表して，…から…へ
▶ The metabolic fate of blood glucose varies from tissue to tissue.
6 選択を表して，…の中から
▶ The crude material is purified by crystallization from water, in which it is very soluble.

front n **1** 前面 syn *forefront*
▶ The spread of a reaction front from a nucleation site throughout the crystal may be highly anisotropic.
adj **2** 正面の syn *foremost*
▶ A front and side view of the boiling point apparatus is shown in Figures 1a and 1b, respectively.

frontier n 最先端 syn *front line*
▶ Another frontier full of promise involves the development of homogeneous stereoselective catalysts.

frontier orbital n フロンティア軌道
▶ Of all the easily calculated molecular electronic properties, frontier molecular orbitals have been perhaps the most successful for rationalizing organic reactivity.

front-side n 前面
▶ An S_N2 reaction between *cis*-3-methylcyclopentyl chloride and hydroxide ion would give the cis alcohol by front-side attack but the trans alcohol by back-side attack.

frosted adj つや消しの
▶ Even good platinum ware is likely to become frosted by strong ignition.

froth n **1** 泡 syn *bubble, foam*
▶ A frothing agent is used to stabilize the bubbles in the form of a froth which can be easily separated from the body of the liquid.
vi **2** 泡立つ syn *bubble, foam*
▶ A 10% solution of ethylene glycol in water

froths when it is bubbled, but the froth is destroyed or does not form on beating.

frothing n 泡立ち syn *bubble, foam*

▶ The Claisen flask with two necks has the advantage of minimizing the chance of liquid being carried over into the distillate by frothing or spattering.

frozen adj 凍結の syn *immovable*

▶ Frozen free radicals may be generated by intense γ radiation penetrating a sample at very low temperatures.

fruitful adj 実りの多い syn *fertile, productive, prolific*

▶ One of the most fruitful approaches to an understanding of the adsorbed state is through studies of the reactions that the molecules of this state undergo.

fruition n 実現 syn *achievement, realization*

▶ Whether these applications come to fruition will depend on whether some formidable problems in the realm of materials science can be overcome.

frustrated adj じゃまをされた syn *dissatisfied, unsuccessful*

▶ The undecahedral species may be considered as frustrated nonacoordinate structures in that close approach of the ninth halogen is inhibited by the steric hindrance of the other eight halogen atoms.

frustration n 挫折 syn *disappointment*

▶ We have ourselves experienced the frustration of not being able to reproduce an experimental result that was undoubtedly obtained earlier.

fuel n **1** 燃料

▶ The simplest type of fuel cell uses hydrogen and oxygen as fuel.

2 エネルギー源としての食糧 syn *nutriment*

▶ The acetyl coenzyme A that is fuel for respiration comes not only from carbohydrates but also from the breakdown of amino acids and fats.

vt **3** 刺激する syn *encourage, provoke, stimulate*

▶ Specific interest in chalcogenides has been fueled by the discovery of their superconducting properties, their application as battery materials, and their utility as hydroprocessing catalysts.

fuel cell n 燃料電池

▶ Fuel cells differ from batteries in that electricity is produced from chemical fuels fed to them as needed, so that their operating life is theoretically unlimited.

fugacity n フガシティー

▶ The ratio of the fugacity of a substance in any state to the fugacity in the standard state is known as the activity and is given the symbol *a*.

fugitive adj つかの間の syn *fleeting, transient*

▶ There is evidence for fugitive species such as nitrosyl azide N_3NO and nitryl azide N_3NO_2.

fulfill vt 果たす syn *accomplish, achieve, complete*

▶ The promise of ESCA as a local structure probe, applicable to virtually all elements of the periodic table, is as yet only partly fulfilled.

full n **1 in full** 詳しく syn *completely, thoroughly*

▶ We now report the synthesis of a variety of thiepines in full as well as reactions that were attempted with these compounds.

adj **2** 最大限の syn *complete, exhaustive, extensive*

▶ The unit cell is defined as the smallest repeating unit which shows the full symmetry of the crystal structure.

3 満ちた syn *filled, saturated*

▶ For Na the 3s band is exactly half-full, because each contributing atom has a $3s^1$ or half-full electronic configuration.

full-color adj フルカラーの

▶ To make full-color light-emitting diode displays, we need devices with output wavelengths throughout the visible spectrum.

fullerene n フラーレン

▶ The formation of giant fullerenes with molecular formulas up to C_{400} had previously been observed in the laser vaporization not only of graphite but also of other carbon-rich materials like polycyclic aromatic hydrocarbons and higher oxides of carbon.

full-fledged adj 十分に発達した syn *mature*

▶ The full-fledged positive charge of the $-N(CH_3)_2^+$ group has a powerful attraction for electrons.

fully adv 完全に syn *completely, entirely, thoroughly*

▶ The fully developed negative charge makes $-O^-$ much more powerfully electron-releasing than $-OH$.

fume n **1** 煙霧 syn *smoke, vapor*

▶ Certain tertiary alcohols are so prone to dehydration that they can be distilled only if precautions are taken to protect the system from the acid fumes present in the ordinary laboratory.

vi **2** 煙を発する syn *smoke*

▶ Titanium(IV) chloride fumes strongly when exposed to moist air, forming a dense and persistent white cloud..

fuming adj 煙を出す

▶ $SeOF_2$ and $SeOCl_2$ are colorless, fuming, volatile liquids.

fuming sulfuric acid n 発煙硫酸 syn *oleum*

▶ Solutions of SO_3 in H_2SO_4 are called fuming sulfuric acid or oleum.

fumigant n くん蒸剤

▶ Fumigants are used chiefly in enclosed or limited areas and also are applied locally to soils, grains, fruits, and garments.

function n **1** 官能基 syn *functional group*

▶ Lithium aluminum hydride reduces not only aldehydes and ketones but other types of carbonyl functions as well.

2 関数

▶ Since the absorption spectra of solutions of Rhodamine is a function of concentration, aggregation of Rhodamine 6G has been postulated in both aqueous solution and organic solvents.

vi **3** 機能する syn *act, operate, perform*

▶ Solid solutions of *p*-terphenyl in polystyrene functioned efficiently as scintillators.

functional n **1** 汎関数

▶ A functional is a recipe for turning a function into a number, just as a function is a recipe for turning a variable into a number.

adj **2** 関数の

▶ The pressure p of a gas at constant temperature is a single-valued function of the volume v, the exact form of the functional dependence being given by Boyle's law: $p = f(v) = k/v$.

3 機能の syn *operating, working*

▶ The capability of making polyaniline nanofibers individually at desired positions is needed to fabricate single nanofiber-based functional devices.

functional group n 官能基 syn *function*

▶ IR and Raman spectra are much used for the straightforward identification of specific functional groups, especially in organic molecules.

functionality n 機能性

▶ Nitroalkanes can be easily prepared, and the nitro group is a versatile functionality that can readily modified, enabling access to a range of products.

functionalizable adj 官能基化される

▶ Calixarenes provide rigid, functionalizable, ion-selective filters around which ionophore or channel frameworks can be constructed.

functionalization n 官能基化

▶ In addition to allowing rapid chain extension, cross metathesis allows selective functionalization of the side chain to be performed without affecting the enol ether.

funcltionalize vt 機能化する

▷ Metal-organic framework structures with the same net can often be made by functionalizing the organic linker units, and this allows surface chemistry to be tailored.

functionalized adj 機能性の

▶ Functionalized resins may be prepared by introducing the desired functional group(s) into suitable commercial resins or directly by copolymerization.

functionally adv 機能上

▶ Functionally, this is very important because there is no bulk precipitation of iron oxide, and iron storage and transport can be closely regulated by the levels of ferritin biomineralization within the organism.

functioning n 機能 syn *action, operation*

▶ The sequence of amino acids in the peptide chain is of critical importance in the biological functioning of the protein.

function of state n 状態関数

▶ Both of the quantities ΔU and ΔH are functions of state only.

fundamental n 1 複数形で, 基礎 syn *basics, elements, essentials, principles*

▶ With a knowledge of the fundamentals of wave mechanics, we may now apply this very powerful tool to gain an understanding of electronic structure of atoms.

adj 2 基礎の syn *basic, essential, main, primary, principal*

▶ Fundamental research has no predetermined goal or purpose.

fundamental frequency n 基音振動数

▶ Overtone absorptions come at just twice the frequency of the fundamental frequency.

fundamentally adv 根本的に syn *largely, principally, ultimately*

▶ The breaking of a covalent bond can take place in two fundamentally different ways, depending upon what happens to the two electrons making up the bonding pair.

fungicidal adj 殺菌性の

▶ Some alkylmercury(II) halides, such as ethylmercury(II) chloride, have fungicidal properties and are used to preserve seeds.

fungicide n 殺菌剤

▶ Copper sulfate is widely used as a fungicide.

fungus n 菌類 複数形は fungi

▶ Fungi are important sources of antibiotics and drugs that lower blood cholesterol.

funnel n 漏斗 ☞ separatory funnel

▶ When the stoppered funnel is shaken to distribute the components between the immiscible solvents ether and water, pressure always develops through volatilization of ether from the heat of the hands, and liberation of a gas will increase the pressure.

furnace n 炉

▶ Most sodium silicate is produced via the furnace route, in which sodium carbonate and sand react at ca. 1500 ℃ to give sodium silicate glass of various compositions.

furnish vt 与える syn *afford, provide, supply*

▶ An instance of the acid-catalyzed replacement of a basic bidentate ligand is furnished by studies on the $Fe(bipy)_3^{2+}$ complex.

further adj 1 さらにまた syn *additional, another, extra, other*

▶ The heteropoly anions contain a further cationic element in addition to that present in the isopoly anions.

2 もっと程度の進んだ syn *farther*

▶ Further investigations of this effect are in progress.

adv 3 もっと先に syn *farther*

▶ Reaction does not stop at the monoalkylphosphine stage but proceeds further, to give varying amounts of dialkyl- and trialkylphosphines.

4 さらに syn *also, besides, furthermore, too*

▶ Once an oxide of the alkali metals is produced, it reacts further with water with considerable evolution of heat.

vt 5 促進する syn *advance, favor, promote*

▶ These conditions are tragic for the patients and their families, and yet they have furthered the understanding of normal carbohydrate metabolism.

furthermore adv そのうえ syn *besides, moreover*

▶ The Goldschmidt values are more directly related to experiment and, furthermore, show the irregularities in the radii of the first transition series which are predicted because of the crystal field of the neighboring anions in the solid state.

fuse vt 1 溶融する syn *melt*

▷ Thiocyanates and selenocyanates can be made by fusing the corresponding cyanide with S or Se.

2 縮合させる syn *combine, unite*

▷ Azulene is isomeric with naphthalene and has a five- and a seven-membered ring fused through adjacent carbons.

fused ring n 縮合環 syn *condensed ring*

▶ A number of molecules which contain fused

aromatic rings are known to be powerful carcinogens.

fused salt bath n 溶融塩浴

▶ The container for the fused salt bath must be of metal, for the melt expands sufficiently on solidification to crack glass or porcelain.

fusible adj 可融性の

▶ A bath of a fusible metal alloy allows better temperature control: Wood's metal (mp 71 ℃): bismuth, 4; lead, 2; cadmium, 1; tin, 1; Rose's metal (mp 94 ℃): bismuth, 9; lead, 1; tin, 1.

fusion n **1** 融解

▶ Silicides are usually prepared by direct fusion of the elements, but coreduction of SiO_2 and a metal oxide with C or Al is sometimes used.

2 縮合 syn *union*

▶ Quinoline and isoquinoline are the two analogs of naphthalene which can result from the fusion of a benzene ring to a pyridine ring.

3 融合

▶ Through fusion with lysosomes, the remnants are digested, liberating fatty acids and cholesterol ester into the cytoplasm.

4 核融合 ☞ nuclear fusion

▶ An endothermic nuclear reaction yielding large amounts of energy in which the nuclei of light atoms unite to form helium is called fusion.

future n **1** 将来 syn *time to come*

▶ With recent developments in crystal packing theory and availability of increasingly powerful computer programs for the calculation of potential energy minima in organic crystals, one may predict the possibility of complete void design in the foreseeable future.

adj **2** 将来の syn *coming, subsequent*

▶ The question of the refraction of various electron groups involving nitrogen will form the subject of a future communication.

G

gain n **1** 増加 syn *increase, increment*

▶ The gain in energy, associated with formation of nitrogen by decomposition of a diazonium ion, is not sufficient to make production of aryl cations occur readily at less than 100 ℃.

2 取得 syn *acquisition*

▶ Many organic reactions are known in which a reactant must first to be converted to a salt, either by loss or by gain of a proton.

vi **3** 増す syn *increase*

▶ The contribution of the randomness of the system gets more important as the reaction temperature is raised because $T\Delta S$ gains in importance.

vt **4** 手に入れる syn *acquire, get, obtain*

part ▶ Reduction reactions usually involve gaining hydrogen and, in many cases, losing a hetroatom.

galvanic cell n ガルバニ電池

▶ Galvanic cells are of interest in themselves as sources of electric energy, as arrangements for storing electric energy, and as devices for furnishing definite reproducible values of electromotive force.

gap n **1** ギャップ syn *opening, space* ☞ band gap, energy gap

▶ A miscibility gap in the liquid state is not to be expected in a binary system with continuous solid solution.

2 相違 syn *difference, discrepancy, divergence, inconsistency*

▶ We attempted to identify the structural and chemical trends that are beginning to emerge from the literature and to pinpoint some of the areas where there are important gaps and opportunities.

gas n 気体

▶ The thionyl chloride method is particularly convenient since the inorganic products are gases, which are easily removed.

gas chromatograpy n ガスクロマトグラフィー

▶ Molecular sieves can separate gases based on molecular size, as in gas chromatography.

gaseous adj 気体の

▶ Hydrogen bromide, like hydrogen chloride, is gaseous, and its aqueous solution is an even stronger acid.

gaseous reaction n 気体反応 syn *gas-phase reaction*

▶ Chemisorption is of essential importance in the catalysis of gaseous reactions and is much more specific than physical types of sorption.

gasify vt ガス化する

▶ Peat can be gasified for production of methanol after mechanical dewatering.

gasification n ガス化

▶ Gasification of coal produces carbon monoxide and hydrogen, from which ammonia and other products can be made.

gas-permeable adj 通気性の

▶ The production of ammonium ion is monitored as ammonia, with a glass electrode which is covered with a thin film of electrolyte and a gas-permeable membrane on which the enzyme is coated.

gas phase n 気相

▶ The ionization energy of an element is the minimum energy required to remove an electron from one of its atoms in the gas phase.

gas-phase reaction n 気相反応 syn *gaseous reaction*

▶ High-temperature gas-phase reactions that have half-lives between 10^{-3} and 10^{-6} s are studied conveniently by means of shock tubes.

gate n ゲート

▶ In a conventional field effect transistor, the current flowing between the source and the drain is influenced by a voltage applied to the gate.

gather vt 集める syn *accumulate, assemble, collect*

part ▶ To complement the information gathered from the NMR data in the solid state, X-ray crystal structure determinations were carried out for 1-(phenylazo)-2-naphthol at 213 and 300 K.

gating n ゲートの開閉

▶ Voltage gating of some K^+ channels occurs when arginine residues react to transmembrane potential by moving through the phospholipid membrane to mechanically close the channel.

gauche adj ゴーシュの

▶ NH_2OH can exist as cis and trans configurational isomers and in numerous intermediate gauche conformations.

gauge n **1** 計器 syn *measure*

▶ For measurements of pressure below 1 mm, standard ionization gauges of various types, thermocouple gauges, and thermister or other resistance-type gauges have been used.

2 寸法 syn *dimension, magnitude, size*

▶ In materials technology, the term gauge is often synonymous with thickness, especially in the metals, rubber, and plastics fields.

vt **3** 評価する syn *determine, evaluate, measure*

▶ In general, solvent-solute interaction is a much more specific effect than can be gauged from a simple series of dielectric constants.

gauze n 金網

▶ The electrodes must always reach to the bottom of the beaker, and the top of the gauze electrode should be nearly covered by solution.

gel n ゲル

▶ A gel is an interconnected, rigid network with pores of submicrometer dimensions and polymeric chains whose average length is greater than micrometer.

gelatin n ゼラチン

▶ Gelatin is strongly hydrophilic, absorbing up to 10 times its weight of water and forming reversible gels of high strength and viscosity.

gelatinize vi **1** ゼラチン状になる

▶ $Na_5Fe(SiO_3)_4$ is hexagonal in prismatic crystals which are slowly attacked by water and gelatinize with HCl.

vt **2** ゼラチン化する

▶ Addition of 1 N potassium chloride gelatinizes silicic acid in one day, but in the presence of a few drops of alkali the gelatinization is instantaneous.

gelatinous adj ゼラチン状の

▶ Ammonia produces a gelatinous precipitate of aluminum hydroxide, which, as a colloid, is somewhat soluble in water but insoluble in the presence of ammonium salts.

gelation n ゲル化

▶ At gelation, the viscosity increases sharply, and a solid object in the shape of the mold results.

gel filtration n ゲル沪過

▶ Gel filtration relies on diffusion of protein molecules into pores of a gel matrix in a column.

geminal adj ジェミナル

▶ The enthalpies of formation of alkanes and other hydrocarbons can be calculated empirically by means of an additive bond energy scheme which included a term for each geminal H···H interaction in methylene and methyl groups.

geminate recombination n 対再結合

▶ In solution, the separation of the iodine atoms is hindered by the surrounding solvent molecules, and there is now a nonzero probability that they will reform the iodine molecule, undergoing a process termed geminate recombination.

gene n 遺伝子

▶ Genes occur on the chromosomes of every living cell, where they are arranged in a linear order.

general n **1 in general** 一般に syn *chiefly, mainly, ordinarily*

▶ Perhaps one of the most ingenious applications of Mössbauer spectroscopy in general and of quadrupole interactions in particular comes from the field of xenon chemistry.

adj **2** 普遍的な syn *common, popular*

▶ Despite the greater complexity, the same general principles may be applied.

generality n 一般性 syn *generalization*

▶ The method is of considerable generality and is limited only by the ease with which the particular diastereomers are separable and the ease of regenerating the pure enantiomer without concurrent racemization.

generalization n 一般化 syn *generality*

▶ The more stable the free radical, the more easily it is formed. This is an extremely useful generalization.

generalize vt 一般化する

▶ These definitions can be generalized, but they are sufficient for the present discussion.

generalized adj 一般化した syn *broad, general*

part ▶ According to Brönsted's generalized definition of acids and bases, an acid is any substance which can donate protons, a base any substance which can accept them.

generally adv **1** 一般に 口語では多くの場合 syn *in general, usually*

▶ A less generally recognized property of very acidic media is their ability to stabilize monoatomic cations of transition metals in unusually low oxidation states.

2 generally speaking 一般的にいって syn *usually*

▶ Generally speaking, covalent interactions can lead to a change in coordination number or to a departure from the regularity which would be predicted on the simple ionic model.

generate vt **1** 生じる syn *create, make, produce*

▶ Overlap of two carbon sp^3 orbitals, containing one electron each, generates a carbon–carbon bond.

2 招く syn *cause, initiate, inspire*

▶ Materials with unusual properties may generate applications that did not exist before.

generation n **1** 生成 syn *production*

▶ The strongly electron-withdrawing nature of the nitro group facilitates generation of the nitronate anion under mild conditions.

2 世代 syn *age, contemporaries*

▶ The flow of genetic information from an individual cell to its next generation requires the duplication of its genetic blueprint, the genome.

generator n 発生装置

▶ Carbon monoxide is prepared by dropping 85–90% formic acid into concentrated sulfuric acid contained in a 1-L distilling flask as the generator.

generous adj 寛大な syn *ample*

▶ The author is indebted to NRC for the generous support of this work.

genetic adj 遺伝の syn *hereditary*

▶ Genetic mechanism is the same in all organisms, ranging from the lowest forms of life, both plant and animal, to man.

genetic code n 遺伝暗号

▶ The basis is the genetic code, which describes how various combinations of nucleotides can be interpreted as individual amino acids.

genetic control n 遺伝的支配

▶ The substances responsible for genetic control in plants and animals are present in and originate from the chromosomes of cell nuclei.

genetic information n 遺伝情報

▶ The genetic information inherent in DNA appears to depend on the arrangement of the bases along the phosphate–carbohydrate backbone.

genome n ゲノム

▶ Every cell of any organism that is able to reproduce itself contains the full genetic instruction set that is known collectively as its genome: double-stranded DNA makes up this coded set.

genre n ジャンル syn *class, kind, sort, type*

▶ If materials of this genre can integrate into bone, they may be able to deliver bioeffectors and drugs, from the hydrophobic region of the mesolamellar calcium dodecylphosphate phase, to stimulate bone growth and combat disease.

gentle adj 穏やかな syn *mild, moderate*

▶ Uranium is almost always precipitated by means of ammonia as ammonium uranate and changed to U_3O_8 by gentle ignition in a platinum crucible with free access of air.

gently adv 穏やかに

▶ To a gently boiling solution of 3.0 g of TCNQ in 350 mL of acetonitrile was added a hot solution of 1.0 g of H_2TCNQ in 50 mL of acetonitrile.

genuine adj 本物の syn *authentic, original, real, true*

▶ Few genuine pyrroles occur in nature, but certain compounds formally derived from the pyrrole system are of very great biochemical importance.

geochemical adj 地球化学の

▶ The primary principle of geochemical classification of the elements must obviously be their tendency to form ionic, covalent, or metallic bonds.

geomagnetic adj 地球磁場の

▶ The magnetic properties of mixed-valence phases are utilized by several types of bacteria as means of navigation in the ambient geomagnetic field.

geometric adj 幾何学的な syn *geometrical*

▶ The most important geometric characteristic of hydrogen bonds is that the distance between the proton and the acceptor atom is shorter than the sum of their van der Waals radii.

geometrical configuration n 幾何学的配置

▶ The acyclic $SSNSS^-$ anion is planar with cis-trans configuration, though a different geometrical configuration occurs in the $[AsPH_4]^+$ salt.

geometrically adv 幾何学的に

▶ Hydrogen atoms were positioned geometrically after each cycle of refinement.

geometric isomer n 幾何異性体 syn *geometrical isomer*

▶ Stereoisomers are also referred to as geometric isomers, and the specific arrangement in space is referred to as configuration.

geometric isomerism n 幾何異性 syn *geometrical isomerism*

▶ Geometric isomerism is found wherever free rotation about a bond is not possible.

geometry n 幾何学的形

▶ The lines in structural diagrams merely define the geometry of the clusters of boron atoms.

germicidal adj 殺菌の

▶ The germicidal action of ordinary soaps lies almost entirely in the bacteria being carried away in the rinsing.

germination n 発芽

▶ Ethylene has been found to function as a hormone by enhancing fruit ripening, leafdrop, and germination as well as growth of root and seedling.

get vi **1** 場所に到達する syn *arrive, reach*

▶ The attractive force is rapidly overcome by repulsive forces when the atoms get very close to one another.

2 状態に到達する syn *become*

▶ The boiling points and melting points rise because the intermolecular forces increase as the molecules get larger.

3 get out of 抜け出す syn *depart, leave*

▶ The combination of azeotropic formation and partial miscibility is quite common, because both properties indicate a tendency for the two kinds of molecules to get out of each other's way.

vt **4** 得る syn *acquire, obtain, secure*

▷ One way of getting wider nematic ranges is to mix two miscible mesomorphic components with approximately the same mesophase-isotropic transition points.

giant molecule n 巨大分子 syn *macromolecule*

▶ Metals, ionic, and covalent crystals are simply giant molecules and are chemical systems which have electronegativity and hardness.

Gibbs energy n ギブズエネルギー ☞ free energy

▶ Modern treatments of equilibrium are based on the idea that equilibrium is a state of minimum Gibbs energy and that forward and reverse processes are occurring at equal rates.

gigantic adj 巨大な syn *big, huge, large*

▶ Only two cross-links per polymer chain are required to connect together all the polymer molecules in a given sample to produce one gigantic molecule.

give vt **1** 与える syn *confer, provide, turn out*

▶ From many criteria, the order of π bonding

for common ligands is usually given as: CS > CO ≒ PF_3 > N_2 > PCl_3 > C_2H_4 > PR_3 ≒ AsR_3 > R_2S > CH_3CN > pyr > NH_3 > R_2O.

2 もたらす syn *cause, induce*

▷ Solutions A and B were mixed together with rinsing, giving a clear, homogeneous solution of two metal ions in the exact stoichiometric ratio desired.

3 give off 放つ syn *release*

▶ Solid potassium chlorate gives off oxygen when it is heated, and so it is used in fireworks.

4 give up 渡す syn *pass, submit*

▷ Manganese dioxide is an oxidizing agent, giving up oxygen rather readily to other substances.

given adj **1 特定の** syn *noted, specified, stated*

▶ For a given crystal structure, there is usually a preferred plane or set of planes on which dislocations can occur and also preferred directions for dislocation motion.

2 接続詞的に用いて，…**を仮定すれば** syn *assumed*

▶ Porous hybrid frameworks would be generally inferior to conventional zeolitic materials for applications in heterogeneous catalysis, given their relative instability, their lack of strong acidity, and their relatively costly synthesis.

glacial adj 氷のような syn *freezing, icy*

▶ The term glacial is applied to a number of acids, e.g., acetic and phosphoric, which have a freezing point slightly below room temperature when in a highly pure state.

glance n **1 一見** syn *look*

▶ A cursory glance at any bioinorganic chemistry textbook will reveal many processes, vital to health, that rely on the movement of inorganic species across cell membranes.

2 at first glance 一見したところでは

▶ At first glance, this mechanism may appear complex.

glass n ガラス

▶ Systematic studies of phosphorescence were first made by Lewis and Kasha using mixtures of organic solvents which form clear rigid glasses at liquid nitrogen temperatures.

glass electrode n ガラス電極

▶ The glass electrode comprises a thin glass membrane which is filled with an aqueous solution of hydrochloric acid.

glass transition n ガラス転移

▶ The glass transition represents the upper temperature limit at which the glass can really be used.

glass transition temperature n ガラス転移温度

▶ The experimental glass transition temperature occurs when the relaxation time attains a constant value on the time scale of the experiment.

glassware n ガラス器具

▶ Glassware and porcelain are now made of better quality and can be used in chemical work without much contamination.

glass wool n ガラスウール

▶ The reaction vessel is packed with glass wool so that the wall surface is greatly increased.

glassy adj ガラス状の

▶ At 90° crystals of **1** were transformed to an orange glassy material which showed no X-ray diffraction pattern and appeared isotropic when viewed through crossed polarizing filters.

glassy crystals n ガラス性結晶

▶ Orientational disorder in plastic crystals may be quenched to yield glassy crystals which exhibit glasslike transitions.

glaze n うわぐすり syn *enamel*

▶ A glaze may refer to a vitreous coating on pottery or enamelware, the mixed dry powders of the batch to be used for coating or a water suspension of these materials.

glide n **1 すべり** syn *slide, slip*

▶ Glide of dislocations is the normal process of plastic deformation of crystals, and the only one taking place at low temperature.

vi **2 すべる** syn *slide, slip*

▶ Frequently the dislocations glide most easily when they lie parallel to particular crystallographic planes, close-packed planes in general.

glisten vi きらめく syn *shinning*

▷ Quinhydrone, in the form of dark green crystals having a glistening metallic luster, is conveniently prepared by the partial oxidation of hydroquinone with a solution of iron alum.

global adj **1** 包括的な syn *comprehensive, general*

▶ In a delocalized system, we can expect electron-electron repulsions to be less and energy spacing to be reduced. Thus, the global softness of such systems is increased.

2 世界的な syn *international, worldwide*

▶ The global challenge for chemists working on conducting CT salts is the preparation of new molecular systems which meet the stringent requirements, both at the intra- and intermolecular level, for high conductivity or superconductivity.

globular adj 球状の syn *spherical*

▶ Crystals of globular molecules usually undergo transformations in the solid state in which the molecules become disordered in orientation; in some cases they may even rotate freely.

globular protein n 球状タンパク質

▶ The characteristic feature of the globular proteins is that lipophilic parts are turned inward, toward each other, and away from water.

globule n 小滴 syn *drop, droplet, sphere*

▶ The ester condensation does not proceed satisfactorily if the sodium used is in the form of chunks or slices; hence, the first operation is preparation of powered sodium, a mixture of fine particles and globules.

globulin n グロブリン

▶ Globulins are coagulated by heat, insoluble in water, but soluble in dilute solutions of salts, strong acids, and strong alkalis.

gloss vt **gloss over** 言い紛らわす syn *discount, disregard, overlook*

▶ A crucial difficulty was glossed over in our earlier treatment by assuming that all partition functions were of equal magnitude.

glove box n グローブボックス

▶ Some operations can be carried out in evacuated systems or glove boxes, but it is desirable to maintain clean room facilities for convenient handling of pure materials.

glow n 輝き syn *brilliance, brightness*

▶ In luminescence, the localized electron traps provided by impurity atoms substituted in the host lattice are of paramount importance in regulating the glow characteristics of phosphors.

glow discharge n グロー放電

▶ I_2O_5 results from the direct oxidation of I_2 with oxygen in a glow discharge.

glowing adj 白熱(赤熱)している syn *incandescent, luminous*

▶ The so-called color scale of temperature depends on the color of the light emitted by a glowing body.

glue n **1** にかわ syn *adhesive, paste*

▶ Most familiar glues are those derived by boiling animal hides, tendons, or bones, which are high in collagen.

vt **2** 接着剤で付ける

▶ Highly polished as well as original crystal surfaces can be strongly glued together by pure chemicals.

glycan n グリカン syn *polysaccharide*

▶ The incorporation of pure, chemically synthesized glycans at specific sites in proteins to produce homogeneous glycoproteins could potentially afford active glycoproteins.

glycoconjugate n 複合多糖 syn *complex carbohydrate*

▶ Carbohydrates linked to other biomolecule classes are termed glycoconjugates.

glycogen n グリコーゲン

▶ The amount of glycogen stored is very small; the liver normally stores ca. 100 g of glucose as glycogen, while the muscles, in total, store ca. 250 g of glycogen.

glycolysis n 解糖

▶ The end product of glycolysis, lactic acid, collects in the muscle, and the muscle feels tired.

glycoprotein n 糖タンパク質

▶ The multiple glycoforms of glycoproteins

makes the study of their biological function extremely difficult.

glycosidic bond n グリコシド結合

▶ A disaccharide consists of two monosaccharides linked by a single glycosidic bond.

go vi **1 向かう** syn *move*

▷ The comparisons of frequency shift data will be influenced by the reference state since there are infrared shifts observed in going from the gas to the liquid phase.

2 進む syn *proceed*

▶ Most, if not all, other substitution reactions of gaseous paraffin hydrocarbons are considered to go by radical-chain mechanism.

3 go back さかのぼる syn *begin with, originate*

▶ The calculation of the maximum translational lattice frequency from elastic constants goes back to Debye's classical paper.

4 go into (a) …に入る

▶ In MFe_2O_4, the large divalent ions go preferentially into octahedral sites, and Fe^{3+} ion is distributed over both tetrahedral and octahedral sites.

(b) **費やされる** syn *pursue, take up, undertake*

▶ A great deal of effort has gone into designing crystal structures and synthesizing new compounds with high lithiumion conductivities.

5 go on 続けて…する syn *continue, last, proceed*

▶ It goes on to discuss the choice of complex to be elaborated in order to generate a liquid-crystalline material and then to discuss the properties which may be introduced.

6 go to great pain 骨身を惜しまず…する syn *spare no pains*

▶ Van't Hoff went to great pains to show that the thermodynamic equations were consistent with the experimental results.

goal n **目的** syn *aim, object, purpose*

▶ A simple, rational empirical model for the estimation of the enthalpies of vaporization ΔH_v° of organic liquid seems to be a worthwhile goal.

good adj **1 よい** syn *fair, satisfactory*

比較級・最上級は better, best

▶ Moderate to good yields and good enantioselectivities were obtained in all cases.

2 完全な syn *excellent, fine*

▶ Good single crystals of Tl_xWO_3 have been made by vapor-phase reaction.

3 的確な syn *appropriate, right, suitable*

▶ The van der Waals equation gives a reasonably good interpretation of the behavior of gases, but a number of refinements to it have to made for more exact agreement with experiment.

goods n **品物** syn *material*

▶ Biotechnology is application of scientific and engineering principles to the processing of any organic or inorganic substance by biological agents to provide goods and services.

good solvent n **良溶媒** ☞ poor solvent

▶ In a good solvent, it is expected that the polymer chains will be solvated and will open up, while in a poor solvent they will tend to remain coiled up.

govern vt **決定する** syn *control, determine, guide, regulate*

▷ The factors governing the course of these reactions are not fully understood, but steric effects play some role.

gradation n **徐々に変化すること** syn *step*

▶ The series of straight-chain alkanes shows a remarkably smooth gradation of physical properties.

grade n **等級** syn *category, level, rank*

▶ Final purification for semiconductor-grade germanium is effected by zone refining.

graded adj **勾配を付けた**

▶ A standard graded index optical fiber has a core diameter of 50–62.5 μm and a cladding diameter of 125 μm.

gradient n **勾配** syn *slope* ☞ concentration—, potential—, temperature—

▶ The amount of substance ds diffusing through a unit area in time dt is proportional to the gradient of concentration c with linear distance x, thus.

$$\frac{ds}{dt} = D\frac{dc}{dx}$$

gradual adj 徐々の syn *gentle, moderate, slow*
▶ Phase transitions were determined by observation of powdered samples on gradual heating in a melting point apparatus which had been calibrated against pure compounds of known melting points.

gradually adv 徐々に syn *evenly, slowly*
▶ The electron density does not decrease abruptly to zero at a certain distance from the nucleus but decreases only gradually with increasing radius.

graduation n 目盛 syn *scale*
▶ In the graduation of thermometers, the melting point of ice and the boiling point of water at 760 mm pressure are used as fixed points.

graft vt グラフトする syn *implant*
▶ Trimethylsilyl groups are often grafted onto performed catalysts to modify the surface adsorption properties.

grafting n グラフトすること
▶ Grafting of organic functionality onto silica is a standard method of preparing hybrid materials but has the drawbacks of poor control over loading.

graft copolymer n グラフト共重合体
▶ In graft copolymers, a branch of one kind is grafted to a chain of another kind.

grain n 粒子 syn *granule, fragment, particle*
▶ In steels with pearlite texture, the hardness depends on the size, amount, and distribution of the very hard Fe_3C grains.

grain boundary n 結晶粒界
▶ TiO_x migration appears to occur via grain boundaries, since it is far too rapid to be accounted for by diffusion of titanium and oxygen through the metal.

grain size n 粒径
▶ In both cases, codeposition starts at potentials positive of the pure aluminum redox potential, and as the Al content increases the grain size of the alloy decreases.

grained adj …な粒の 他の形容詞とともに複合語を作る.
▶ Alabaster is a compact, massive, fine-grained form of $CaSO_4 \cdot 2H_2O$ resembling marble.

gram n グラム 文頭では，数値，単位ともに綴る.
▶ One gram of sucrose dissolves in 0.5 mL of water at 25℃ and in 0.2 mL at the boiling point, but the substance has such marked crystallizing power that in spite of the high solubility it can be obtained in beautiful, large crystals.

grant vt **take for granted** (a) 当たり前のことと思う syn *accept, expect*
▶ Crystallization is a process taken for granted by most practicing chemists.
(b) よく考えもしないで正しいと思う syn *assume, presume, reckon, suppose*
▶ The purity of a commercial sample should never be taken for granted.

granular adj 粒状の syn *granulated, particulate*
▶ When mandelic acid that has been crystallized from benzene is allowed to stand in contact with the mother liquor for several days, the needles gradually change into granular, sugar-like crystals of a molecular compound containing one molecule each of mandelic acid and benzene.

granulate vt 粒状にする syn *grind, mill, pulverize*
part ▶ $(NH_4)_2HPO_4$ in granulated or liquid form consumes more phosphate rock than any other single end-product.

granule n 細粒 syn *grain*
▶ Starch occurs as granules whose size and shape are characteristic of the plant from which the starch is obtained.

graph n グラフ syn *chart*
▶ In some chemical applications, we allow points to represent chemical atoms so that lines represent chemical bonds and graphs, therefore, correspond to molecules.

graphene n グラフェン
▶ Post-treatment of the exfoliated graphite oxide with hydroquinone yielded high-quality graphene sheets.

graphic n **1** 図表 syn *diagram*
▶ Graphics and photographs should not be

embedded in the text of the manuscript.
adj **2 図表による** syn *pictorial, schematic*
▶ Phase diagrams summarize in graphic form the ranges of temperature and composition over which phases or mixtures of phases are stable under conditions of thermodynamic equilibrium.
graphically adv **図表を用いて**
▶ If a series of numerical values of *y* corresponding with values of *x* is given, an approximate value of the integral can be found graphically.
graphitization n **黒鉛化**
▶ The structural and molecular changes established by fundamental researches on graphitization constitute an important section of polymer and solid-state chemistry.
graphitize vt **黒鉛化する**
▷ The high-temperature carbonization of coal yields metallurgical coke, a poorly graphitized form of carbon.
graph theory n **グラフ理論**
▶ Graph theory has been found to a useful tool in quantitative structure–activity relationship and quantitative structure–property relationship research.
grasp vt **1 しっかりとつかむ** syn *hold*
▶ After the scratch has been made, the tube is grasped with the scratch away from the body and thumbs pressed together at the near side of the tube just opposite the scratch.
2 理解する syn *appreciate, understand*
▶ The properties of determinants are easily grasped by considering determinants of the second and third orders.
grateful adj **感謝して** syn *appreciative, thankful*
▶ I am particularly grateful to my collaborators for stimulating discussions.
gratifying adj **満足な** syn *rewarding, satisfying*
▶ In boric oxide glass or sulfate glass, it is gratifying to observe that the two optical basicity values are similar.
gratifyingly adv **満足できる**
▶ Although the experimental critical compression factor is less than 3/8, the discrepancy is gratifyingly small.
gravimetric adj **重量による測定の**
▶ Oxides will be preferable to chalcogenides in order to achieve a high gravimetric energy density.
gravimetric analysis n **重量分析**
▶ Gravimetric analysis involves precipitation of a compound which can be weighed and analyzed after drying.
grayish adj **灰色がかった**
▶ Green or grayish green nickel oxide, NiO, is obtained when the carbonate, nitrate, or hydroxide of nickel is heated.
grease n **1 グリース** syn *lubricant, oil*
▶ Greases range in consistency from thin liquids to solid blocks and in color from transparent to black.
vt **2 グリースを塗る**
▶ The cover of the desiccator must not be greased, for grease is attacked by iodine vapors, forming hydriodic acid, which might cause contamination.
great adj **1 程度，質に関して，大きい** syn *extensive, huge, immense, large*
▶ Anthracene is more reactive than phenanthrene and has a greater tendency to add the 9, 10-positions than to substitute.
2 多量の syn *huge, large, tremendous*
▶ A great amount of study has been conducted on the optical properties of crystals.
3 大変な syn *considerable, extreme, marked*
▶ A subject of great importance in homogeneous catalysis is the bonding of various, neutral molecules to transition metal atoms.
4 十分な syn *close*
▶ Phosgene is highly toxic and should be handled with great caution.
greatly adv **大いに** syn *exceedingly, highly, profoundly*
▶ The strength of titanium can be greatly increased by adding small amounts of tin or aluminum, and such alloys have the highest ratio of strength to weight of any engineering

metal.

green chemistry n グリーンケミストリー

☞ environmental chemistry

▶ The TiO_2-supported Au-Pd alloy nanocrystals give significantly enhanced activity for alcohol oxidation using a green chemistry approach with O_2 under mild solvent-free conditions.

green fluorescent protein n 緑色蛍光タンパク質

▶ Green fluorescent protein was originally isolated from the light-emitting organ of the jellyfish *Aequorea Victoria.*

greenhouse effect n 温室効果

▶ Gradual rise in average global temperature due to absorption of infrared radiation by increasing amounts of carbon dioxide in the air is called greenhouse effect.

greenish adj 緑色を帯びた

▶ Glass always contains iron at least in trace amounts, and this imparts a greenish color.

grid n グリッド

▶ These problems are solved by quickly diluting the reaction extract to the optimal concentration for transmission electron microscopy experiments and depositing it immediately onto sample grids placed on a piece of filter paper.

Grignard reagent n グリニャール試薬

▶ The scope of Grignard reagents in syntheses has been greatly extended by a recently developed method for preparing very reactive Mg.

grind vt 細かく砕く syn *crush, granulate, pulverize*

▷ Crystalline quinhydrones, prepared by grinding together a solid quinone and hydroquinone, have been found to undergo a thermal self-oxidation-reduction on heating or even on standing at room temperature.

grinding n 粉砕

▶ Physical shock, especially grinding, is usually an effective way to convert a metastable crystal form to the stable form.

groove n 溝

▶ DNA contains two different grooves on opposite sides of the base pairs. Although both grooves are of similar depth, the major groove is considerably wider than the minor groove.

gross adj 全般的な syn *overall, whole*

▶ Compared to this gross effect, the fact that selenate and perbromate are somewhat stronger oxidants than tellurate and periodate, respectively, appears to be of only secondary importance.

grossly adv 大いに

▶ It is important to realize that the addition of several methyl groups to the molecule can grossly affect the molecular shape and, as a direct consequence, the intermolecular interactions in the crystalline state.

ground n **1** 根拠 syn *base, basis, foundation*

▶ This is not acceptable on physical grounds, as it is not consistent with the small or negative apparent molar volumes of the smaller cations.

2 論議の立場 syn *argument, justification, rationale*

▶ The order of ligands in the spectrochemical series cannot be explained on purely electrostatic grounds.

3 on the ground の理由で

▶ The sublimation energy of Hf is uncertain and has been assumed equal to that of Zr on the ground that the sublimation energies of Nb and Ta in the next group are virtually the same.

adj **4** 基底の

▶ The equilibrium ionic arrangement found by X-ray analysis is that for the ground electronic state.

vt **5** 立脚する syn *base, derive*

▶ Though catalyst development in this field is grounded in discoveries some 50 years old, contemporary advances have made this field a burgeoning one.

ground adj **1** 粉にした

▶ A finely ground alloy of nickel and aluminum was treated with strong sodium hydroxide solution which dissolved the aluminum with evolution of hydrogen and left finely divided nickel, usually pyrophoric.

2 すり合わせた

▶ The smooth finish on ground joints is achieved by performing the final grinding operation with fine powder, with water as the vehicle.

ground configuration n 基底配置

▶ In an octahedral field, the ground configuration t_{2g}^2 permits three degenerate arrangements and so corresponds to a T ground term.

ground state n 基底状態

▶ Resonance stabilizes the excited state more than it stabilizes the ground state and, thus, reduces the energy difference between them.

ground water n 地下水

▶ Ground water in limestone regions may contain a large amount of calcium ion and hydrogencarbonate ion.

group n **1** 族 ☞ family

▶ The word group should not be applied to a number of elements of similar properties that are not actual group in the periodic table; the proper term for these is series.

2 基 ☞ functional group

▶ The activation or deactivation caused by some groups is extremely powerful: aniline is roughly one million times as reactive as benzene, and nitrobenzene is roughly one-millionth as reactive as benzene.

3 集団 group of に続く名詞は複数形，動詞は集合を強調するときは単数形，個々を強調するときは複数形を用いる． syn *assemblage, assembly*

▶ A cubic crystal in which the atom or group of atoms at the center of the cubic cell is indistinguishable from that at the corners has a body-centered cubic cell.

4 群論における群

▶ A set of operations which transform a symmetrical figure into itself are said to form group.

vt **5** 分類する syn *classify*

▶ It has been customary to group the binary hydrides of the elements into various classes according to the presumed nature of their bonding: ionic, metallic, covalent, polymeric, and intermediate or borderline.

grouping n **1** 分類 syn *category*

▶ The variety of silicates that occur in nature is almost limitless, but it is possible to place them in four general groupings depending on whether the silicate complex is finite or whether it combines to form infinite chains, sheets, or three-dimensional framework structures.

2 配置 syn *arrangement, organization*

▶ Azo compounds possess the −N=N− grouping.

group theory n 群論

▶ The systematic study of symmetry is known as group theory, which leads to useful rules for designating the different types of molecular wave functions.

grow vi **1** 発展する syn *develop, expand*

▶ This area of research is growing and new, interesting materials and observations are continually being reported.

2 次第になる syn *become, get*

▶ Our immune systems tend to grow weaker in the advanced years of our lives.

vt **3** 育てる syn *cultivate, produce*

▷ If a cubic crystal growing from solution happens to have more atoms added to it along one direction than another, it may grow long in that direction.

growing adj 増大する syn *increasing*

▶ These is growing interest in enantiotopic syntheses, which yield directly optically active amino acids.

grown adj 成長した syn *prepared*

▶ Electron diffraction is potentially a very useful adjunct to X-ray diffraction in cases where the polymer under study is available in the form of solution-grown lamellar crystals.

growth n **1** 成長 syn *development, enlargement*

▶ These factors control the nucleation and growth mechanisms and affect the material morphology.

2 発展 syn *advance, progress*

▶ The growth of knowledge and chemical understanding has imposed greater problems in higher education.

growth regulator n 成長調節物質

▶ Growth regulators are compounds that in small concentrations regulate the physiology of plants and animals.

guarantee n **1** 保証 syn *assurance*

▶ The many three-component equations have little to recommend them; they involve additional computational labor without any guarantee of increased accuracy.

vt **2** 確約する syn *assure, ensure, promise*

▶ The presence of the twelve five-membered rings in the fullerenes guarantees that these molecules will all possess six low-lying energy levels.

guard vi **1** 警戒する syn *control, mind*

▶ It is necessary to guard against making too broad generalizations based on limited observations.

vt **2** 防ぐ syn *protect*

▶ Contamination by the crucible material must be guarded against.

guess n **1** 推測 syn *postulate, speculation*

▶ By using these steps, it is often possible to make a good guess at the symmetry and unit cell of a crystalline substance.

vt **2** 推測する syn *postulate, speculate*

▷ All these factors must be assessed in guessing the most likely mechanism for a given system.

guest n ゲスト ☞ host

▶ Metal ions may be incorporated as guests within an existing organic assemblies or may, by virtue of their unique coordination requirements, determine the architecture and structure of the mesophase.

guest molecule n ゲスト分子

▶ In the study of the intercalation of organic molecules by smectic clays, for example, useful deductions could be made on the disposition of the guest molecule within the interlamellar spaces.

guidance n 手引き syn *direction, instruction*

▶ Construction of tables of properties of starting material, reagents, products, and byproducts provides guidance in regulation of temperature and in separation and purification of the product.

guide n **1** 基準 syn *criterion, standard*

▶ Since the lowest energy absorption band in the electronic spectrum is related to $(I - A)$, we can also use the spectra of molecules as a guide to reactivity.

vt **2** 手引きをする syn *direct*

▶ A direct application of thermodynamics to solutions would be difficult if we did not have the ideal gas equations to guide our way.

guideline n 指針 syn *instruction*

▶ In considering the types of polyhedral linkage, Pauling's third rule for the structures of complex ionic crystals provides a very useful guideline.

guiding adj 指針となる syn *influential*

▶ The guiding principle for the quantitative discussion is to look for the temperature at which one phase has the same chemical potential as the solvent in the solution.

H

habit n **1** 習慣 syn *manner, practice*

▶ The habit should be acquired of rejecting at each stage of the work all figures which have no influence on the final result.

2 晶癖 ☞ crystal habit

▶ A second habit consisted of thin hexagonal plates shown by precession photographs to have the same unit cell dimensions and symmetry as the more common crystals.

half n **1** 半分

▶ According to bond energies, cleavage of one half of a carbon–carbon double bond requires 63 kcal, while cleavage of a carbon–carbon single bond requires 83 kcal.

pron **2** 半分のもの half of に続く名詞，代名詞が単数形であれば動詞は単数形を，複数形であれば動詞は複数形を用いる．

▶ Half of the molecular orbitals will be bonding, i.e., more stable than the original atomic orbitals, and half will be antibonding, or less stable.

adj **3** 半分の half および次の half of は限定詞を伴う名詞の前に用い，half の前に a は付けない．

▶ The total attractive energy is given by half the sum of the cationic and anionic energies, since simply adding the two would take each interaction into account twice.

half cell n 半電池

▶ A platinum electrode covered with finely divided platinum is effective for many half cells.

half-filled adj 半分満たされた

▶ The water molecule can be constructed by allowing overlap between the two half-filled sp^3 orbitals of the oxygen atom and the 1s orbitals of two hydrogen atoms.

half-integral adj 半整数の

▶ Doubly charged molecular ions are often apparent as low-intensity peaks at half-integral mass values.

half-life n 半減期

▶ If the half-life is less than 10^{-3} s, the reaction will be largely completed by the time that it takes for mixing to be achieved.

half-wave potential n 半波電位

▶ The half-wave potential, $E_{1/2}$, in polarography is the potential at which the current is one-half its value at the plateau for a species when current is plotted against potential.

halfway adj 中間の syn *intermediate, middle*

▶ For a given reaction, the half-life $t_{1/2}$ of a particular reactant is the time required for its concentration to reach a value that is halfway between its initial and final values.

Hall coefficient n ホール係数

▶ Information regarding the mobility of electrons and holes can be derived from measurements of conductivity and Hall coefficients.

hallmark n 特徴 syn *characteristic, feature*

▶ This intimate association of inorganic and organic phases is a hallmark of biomineralization.

hallucinogen n 幻覚剤

▶ Other derivatives of the decarboxylated form of tryptophan produce powerful psychic effects and are classified in the group of drugs known as hallucinogens.

haloform reaction n ハロホルム反応

▶ The course of the haloform reaction depends on the formation of the enolate ion of the methyl ketone.

halogen n ハロゲン

▶ The halogens other than fluorine form stable compounds corresponding to nearly all values of the oxidation number from −1 to +7.

halogenate vt ハロゲン化する

▷ Ring-halogenated alkenylbenzenes must be prepared by generation of the double bond after halogen is already present on the ring.

halogenation n ハロゲン化

▶ With rare exceptions, halogenation of alkanes is not suitable for the laboratory preparation of alkyl halides.

halt n **1** 停止 syn *termination*

▶ If the sample has the eutectic composition, the cooling proceeds at an approximately constant rate until eutectic temperature is reached, then there is a long eutectic halt, while the whole of the sample solidifies.

vt **2 停止させる** syn *cease, end, stop, terminate*

▶ Whenever unoxidized aluminum is exposed to air or water, the oxide forms immediately and halts the reaction between the aluminum and its chemical environment.

halve vt 半減する

▶ If air is kept in contact with water at standard atmosphere pressure, each kg of water dissolves 0.017 g of oxygen at 20 ℃; if this pressure is halved the water dissolves only 0.0085 g of oxygen.

Hamiltonian n ハミルトニアン

▶ To describe the state of any quantum mechanical system, one need simply write down the Hamiltonian and solve the Schrödinger equation for the wave function.

hammer vt ハンマーで打つ syn *mold*

▷ On hammering a metal, an enormous number of dislocations are generated which in a polycrystalline material are in a large number of orientations.

hamper vt 妨げる syn *hinder, interfere, prevent*

▶ Fundamental studies of multiply charged anions were hampered for many years by the lack of a broadly applicable technique for generating the gas-phase species.

hand n **1 in hand** 手元に

▶ With these results in hand, we were interested in the effects of substitution at both the azabicyclo[2.2.1] framework and the aromatic ring.

2 on the one hand 一方では

▶ On the one hand, are those cases in which the bond may be considered profitably as a single σ bond.

3 on the one hand…on the other (hand) 一方では…また他方では

▶ Cage structures range from clathrate compounds on the one hand to metal-metal clusters and boranes on the other.

4 out of hand (a) 手に負えなくなって syn *uncontrolled*

▶ Consequently the mixture is cooled during the early stages to prevent the reaction from getting out of hand, and later heat is applied to force the diluted components to react.

(b) **即座に**

▶ For the preparation of an alkyl aryl ether, there are two combinations to be considered; here, one combination can usually rejected out of hand.

handedness n 掌性 syn *chirality*

▶ A chiral catalyst is not always sufficient to promote a preference for a particular handedness in products.

handful n 少数 handful of に続く名詞は複数形，動詞も複数形を用いる． syn *few*

▶ A handful of reports highlighting the superior selenating abilities of **1** have appeared, though there is still a need to extend this family of compounds not least to develop more soluble analogues.

handicap n 不利な条件 syn *limitation, restraint, restriction*

▶ For classical techniques, a major handicap was the time taken for mixing to occur.

handle vt 取り扱う syn *deal with, treat*

▶ XeF_6 cannot be handled in glass or quartz because of a stepwise reaction which finally produces the dangerous XeO_3.

handling n 取り扱い syn *operation, treatment*

▶ The great reactivity of the halogen fluorides makes it difficult to prevent contamination in handling.

hang vi 垂れ下がる syn *suspend*

▶ When a thin filament hangs under its own weight, the weight tends to elongate the filament.

haploid n 一倍体

▶ A genome is the set of chromosomes that corresponds to the haploid set of a species.

haploid genome n 一倍体ゲノム

▶ The human haploid genome contains some 3×10^9 DNA base pairs corresponding to a linear length of DNA of about 1 m divided in 23 chromosomes.

happen vi 偶然生じる syn *occur, take place*
▶ In extreme cases, two substances may happen to have the same unit cell parameters and, therefore, the same *d* spacings, but since different elements are probably present in the two, their intensities are quite different.

happening n 出来事 syn *event, occurrence, phenomenon*
▶ It is important, if one is to obtain a definite picture of the happenings on a surface, to have some way of estimating the surface area.

happily adv 幸いにも syn *fortunately*
▶ Phosphorus forms an enormous number of oxides, sulfides, and oxosulfides, but happily all the molecular structures seem to be built on the same model.

hard adj **1** 硬い syn *rigid, stiff*
▶ One substance is less hard than a second substance when it is scratched by the second substance.
2 原子，分子，イオンが硬い ☞ hard acid, hard base, hardness
▶ Compounds containing hard anions such as fluoride, chloride, sulfate, and nitrate rarely have variable stoichiometry.
3 難しい syn *difficult, laborious*
▶ Hydrogens attached to doubly bonded carbons are harder to abstract than ordinary primary hydrogens.
4 極度の syn *strict*
▶ Measurements of the semiconductivity of the oxide in a hard vacuum are made over a range of temperature for various compositions.

hard acid n 硬い酸 ☞ soft acid
▶ The term hard and soft acids and bases refer to the ease with which the electron orbitals can be disturbed or distorted.

hard base n 硬い塩基 ☞ soft base,
▶ It is found empirically that the most stable complexes are those of hard acids with hard bases or of soft acids with soft bases.

harden vi **1** 硬くなる syn *set, stiffen, solidify*
▷ The key ingredient in rapid-hardening Portland cement is Ca_3SiO_5.
vt **2** 硬くする syn *reinforce, strengthen*
▶ Iridium has been used to harden platinum, but the alloy is less resistant to the action of reagents than pure platinum.

hardening n 硬化 ☞ precipitation−, work−
▶ Hardening processes in alloys often involve precipitation; e.g., the formation of cementite precipitates in the hardening of steels.

hardening agent n 硬化剤
▶ Precipitates, as found in steel and almost all other alloys, are used as hardening agents because they pin dislocations and strengthen metals.

hardly adv とても…でない syn *barely, scarcely, seldom*
▶ The situation is hardly clear-cut; we can write the hydrated proton as H^+, H_3O^+, $H_9O_4{}^+$, or $H_{11}O_5{}^+$.

hardness n **1** 硬度 syn *Mohs' hardness*
▶ A crystal which scratches and also is scratched by one of the minerals of this scale has the same hardness as that mineral.
2 水の硬度
▶ The sum of the temporary and permanent hardness of water represents the total hardness.
3 磁性の硬度
▶ The hardness of magnets can be increased if ways can be found of either pinning or reducing the ease of motion of the domain walls.
4 原子，分子，イオンの硬度 ☞ hard
▶ With use of density functional theory as a basis, the hardness of a chemical system has been rigorously defined.

hard sphere n 剛体球 複合形容詞として使用
▶ The decreased coordination number in the Ca structures as X gets larger has been explained in terms of hard-sphere repulsion limiting the number of halides that can congregate around a metal center.

hard-sphere potential n 剛体球ポテンシャル
▶ A hard-sphere potential is a model intermolecular potential in which there is no

interaction between two particles until they are at a specified distance, when the potential energy climbs abruptly to infinity.

hardware n 機器 syn *devices, equipment*

▶ The hardware needed for the cryoscopic measurements is quite simple, consisting essentially of a flask equipped with stirrer and thermometer.

harm n **1** 害 syn *damage*

▶ In the formation of Cu_2I_2 from a solution of a Cu(II) salt, a large excess of potassium iodide does no harm.

vt **2** 害する syn *damage, injure*

▶ The precipitate of sulfur formed during the reduction of the iron(III) salt by H_2S does not harm if the above directions are followed closely.

harmful adj 有害な syn *dangerous, destructive, detrimental*

▶ The binding of an antibody to a bacterial toxin may block the attachment of the toxin to its cell receptor and thus neutralizes its harmful effects.

harmless adj 無害な syn *benign, mild, safe*

▶ Some metals that are harmless in solid or bulk form are quite toxic as fume, powder, or dust.

harmonic adj 調和の ☞ anharmonic

▶ When a point moves uniformly in a circular path, its projection on any fixed diameter of the circle describes a simple harmonic motion in that line.

harmonically adv 調和して

▶ The position of the oscillating particle varies harmonically with a frequency ω.

harmonic oscillator n 調和振動子

▶ A particle vibrating under the influence of a restoring force that obeys Hooke's law is called a harmonic oscillator.

harmony n **in harmony with** …と調和して syn *in agreement with*

▶ They are entirely in harmony with the hypothesis that the vapor is almost exclusively composed of monomeric, dimeric, and trimeric species.

harvest n **1** 収穫 syn *crop, output*

▶ By sampling the soil at harvest, it is possible to see how much of the labeled N remains in the soil unused and whether it is present as organic N or as inorganic N which may be lost by leaching.

vt **2** 取り入れる syn *obtain, receive*

▶ Transition-metal ions exhibit variable redox behavior and may harvest or emit photons, in addition to carrying a number of unpaired electrons at a given center.

hasten vt 促進する syn *accelerate, advance*

▶ Drying can be hastened by drawing air through the apparatus at the suction of the water pump.

hatch vt 陰影を付ける ☞ shaded

▷ The hatched circles represent the positions of the methyl groups in the propylenediamine complex.

have vt **1** 所有の意味でもっている syn *keep, maintain, possess*

▶ Many alkaloids have isoquinoline and reduced isoquinoiline ring systems.

2 属性としてもっている syn *bear, possess*

▶ Van der Waals forces have a very short range; they act only between the portions of different molecules that are in close contact.

3 必要とする syn *oblige, require*

▷ To avoid having to separate mixtures, synthetic routes have been devised that can give rise to primary, secondary or tertiary amines exclusively.

aux **4** 動詞の過去分詞と結合して完了形を作る.

▷ Having measured equilibrium constants over a temperature range, the value of ΔH° can be obtained by using eqs 4–6.

hazard n 偶発性の危険 syn *danger, risk*

▶ Because of the hazards of using hydrogen and the cost of preparing it from other fuels, attention is being given to the development of fuel cells using other materials.

hazardous adj 危険な syn *dangerous, precarious, tricky*

▶ The ability to replace environmentally

hazardous chemicals, together with the ability to fine tune the composition and morphology of alloys, is clearly advantageous to already mature market products.

head vi 向く syn *point, steer*

▶ The two chains head in opposite directions; that is, the deoxyribose units are oriented in opposite ways, so that the sequence is C-3, C-5 in one chain and C-5, C-3 in the other.

heading n 見出し syn *category*

▶ Under this heading, we shall briefly describe three types of carbonyl derivative not referred to previously.

head-to-head adj 頭-頭の

▶ A study of solid state and fluid solution photodimerization of three dichloroanthracenes shows that the head-to-head photodimers are predominant in crystals whereas the head-to-tail photodimers are favored in fluid solutions.

head-to-tail adj 頭-尾の

▶ A high degree of regularity is imposed by the procedures used in polymerization, and a head-to-tail arrangement is overwhelmingly abundant.

heart n 核心 syn *center, core, focus*

▶ The catalyzed chemical reaction has become the heart of the modern chemical process.

heat n **1** 熱 syn *warmth*

▶ If heat is absorbed in the reaction and the products contain more energy than the reactants, ΔE, if the process is at constant volume, and ΔH, if the process is at constant pressure, are positive.

vt **2** 熱する syn *boil, warm*

▷ Anhydrous Al salts cannot be prepared by heating the corresponding hydrate for reasons closely related to the amphoterism and hydrolysis of such compounds.

heat bath n 熱浴

▶ The solvent molecules can act as a heat bath and pass energy on to the reactants to enable them to overcome the activation barrier or remove energy in an association reaction.

heat capacity n 熱容量

▶ The experimental heat capacities of gases are measured values extrapolated to low pressures to remove any nonideal contribution.

heater n ヒーター

▶ High temperatures（300–1300 K）can be routinely attained by a wide range of conventional resistive heaters.

heat exchanger n 熱交換器

▶ Heat exchangers are used to produce steam from the heat developed in nuclear reactors for power generation.

heating n 加熱

▶ The power of a CO_2 laser can be used to provide instantaneous heating.

heating rate n 加熱速度

▶ At a heating rate of 0.3 K/min, the difference between the sample temperature and that of the heating block is constant.

heat interchange n 熱交換

▶ Fresh vapor from the flask forces its way up through the descending condensate with attendant heat interchange, since the vapor is hotter than the liquid.

heat of adsorption n 吸着熱 heat は不可算名詞であるが, heat of adsorption, heat of combustion, heat of sublimation などは加算名詞である.

▶ Heats of adsorptions are generally listed without sign.

heat of atomization n 原子化熱

▶ From the heat of atomization of methane, we can calculate E(C—H) simply by dividing by four the number of C—H bonds in the molecule.

heat of combustion n 燃焼熱 syn *enthalpy of combustion*

▶ The strain in ring compounds can be evaluated quantitatively by comparison of the heats of combustion per CH_2 group.

heat of formation n 生成熱 syn *enthalpy of formation*

▶ The heats of formation of halides per halogen atom tend to fall numerically in passing across the periodic table from metals to nonmetals.

heat of fusion n 融解熱

▶ Generally a low melting point and a low heat of fusion will favor increased solubility.

heat of liquefaction n 液化熱

▶ Physical adsorption is interpreted as a reversible process in which the main binding forces are of van der Waals type, and the adsorption heat corresponds roughly to the heat of liquefaction or sublimation of the adsorbate.

heat of solution n 溶解熱

▶ The heat of solution can be positive or negative because the lattice energy and the heats of solvation of the ions are comparable.

heat of sublimation n 昇華熱 syn *enthalpy of sublimation*

▶ Diamond and graphite have a high melting point and almost identical heats of sublimation.

heat of vaporization n 蒸発熱

▶ The high heat of vaporization, the high surface tension, and a boiling point that is high for a substance of so low molecular weight are all consequences of a high degree of association of molecules in liquid water.

heat pump n ヒートポンプ

▶ The heat pump works on exactly the same principle as the refrigerator, but the purpose is to bring about heating instead of cooling.

heat theorem n 熱定理

▶ In view of the Nernst heat theorem, it is convenient to adopt the convention of assigning a value of zero to the entropy of every crystalline substance at the absolute zero.

heat transfer n 熱伝達

▶ At high temperatures, the large coefficient of heat transfer by radiation increases the difficulty of adiabatic operation.

heat treatment n 熱処理 複合形容詞として使用 syn *thermal treatment*

▶ After each heat-treatment period which lasted for 2 hr, the ESR spectrum of each sample was recorded.

heavily adv **1** 大いに syn *strongly, very much*

▶ Organic and inorganic chemists frequently rely heavily on a variety of modern theoretical ideas in the area of structure and reactivity.

2 密に syn *to a great extent*

▶ The larger rubidium ion is somewhat less heavily hydrated than the potassium ion and so has smaller hydrated radius, which results in a higher ionic mobility.

heavy adj **1** 重い ☞ heavy metal

▶ The heavier alkali metals form a wide variety of hydrated carbonates, hydrogencarbonates, sesquicarbonates, and mixed-metal combinations of these.

2 密度が大きい syn *compact, dense*

▶ The chlorinated hydrocarbon solvents are heavier than water rather than lighter; hence, after equilibration of the aqueous and nonaqueous phases, the heavier lower layer is drawn off into a second separatory funnel for washing, and the upper aqueous layer is discarded.

3 大量の syn *thick*

▶ Cladding is generally 5–20% of total thickness but may be heavier depending on the properties desired.

4 線などが太い ☞ light

▶ Heavy circles represent atoms lying in the plane of the paper, light circles atoms lying in planes 2 Å above and below that of the paper.

heavy chain n 重鎖 複合形容詞として使用

▶ A striking observation is that there is a specific metal-binding oxidation site in the heavy-chain but not in the light-chain polypeptide subunit.

heavy hydrogen n 重水素

▶ Substitution of one heavy hydrogen on the methyl group of an ethyl group (XCH_2CH_2D) produces a triplet resonance.

heavy metal n 重金属 ☞ light metal

▶ Most solutions of salts of heavy metals contain colloidal particles produced by hydrolysis.

heavy water n 重水

▶ Heavy water is manufactured by the electrolytic enrichment of normal water.

height n 高さ syn *elevation, level*

▶ Peaks are of equal height, indicating equal

numbers of each.

heighten vt 高める syn *enhance, intensify*

part ▶ There is a renewed and heightened interest today for two reasons.

helical adj らせん状の syn *coiled, spiral*

▶ Much progress has been made, and, in particular, the idea of a helical macromolecule chain has been very fruitful.

helical structure n らせん構造

▶ Hydrogen bonding between other pairs of bases would not allow them to fit into the double helical structure.

helix n らせん 複数形は helices または helixes syn *coil, spiral*

▶ The helices in any one crystal of α-quartz can be either right-handed or left-handed so that individual crystals have nonsuperimposable mirror images and readily separated by hand.

Helmholtz layer n ヘルムホルツ層

▶ Adding small ions such as Li^+ to an ionic liquid will decrease the Helmholtz layer thickness considerably and should make metal ion reduction easier.

help n **1 be of help** 役に立つ syn *be of assistance, serve*

▶ Along with the other shortcomings, the simple theory is of no help in predicting and explaining magnetic, spectral, and kinetic properties of complexes.

vt **2** 助ける syn *aid, assist, support*

▷ Attempts have been made to use similarity as a helping guideline in the complex area of toxicology, i.e., if one molecule is toxic, will a similar one have the same effect..

helpful adj 有益な syn *beneficial, profitable, useful*

▶ Rearrangement of the alkyl group of an alcohol is very common in dehydration, particularly in the presence of sulfuric acid, which is very helpful in carbonium ion formation.

heme n ヘム

▶ The active center of the cytochromes is the heme group. It consists of porphyrin ring chelated to an iron atom.

hemisphere n 半球

▶ If the meniscus is small, it may be considered to be a hemisphere.

hemoglobin n ヘモグロビン

▶ In the common form of hemoglobin, there are four heme sites.

hence adv それゆえに syn *accordingly, consequently, therefore, thus*

▶ Pure alkanes show no ultraviolet absorption above 2000 Å and, hence, are often excellent solvents for the determination of the ultraviolet spectra of other substances.

herald vt 告知する syn *mean, show*

▶ The use of ionic liquids heralds not only the ability to electrodeposit metals that have hitherto been impossible to reduce in aqueous solutions but also the capability to engineer the redox chemistry and control metal nucleation characteristics.

herbicide n 除草剤

▶ The synthesis of numerous indoleacetic acid analogs led to the first herbicide, 2,4-dicloro-phenoxyacetic acid, or 2,4-D.

here adv **1** ここで

▶ Polarized microscopy, X-ray diffraction, and ^{13}C NMR studies reported here indicate that $Mo_2(O_2CR)_4$ systems behave like M = Cu and Rh systems.

2 注意を引くために文頭に用いて，ここに

▶ Here is another example of a relatively unreactive reagent displaying selectivity.

3 here and there あちこちに syn *around*

▶ Nearly all crystals actually have various kinds of imperfections scattered here and there in their structure which spoil their unending perfection.

hereditary adj 遺伝的な syn *genetic*

▶ The hereditary information is stored as the sequence of bases along the polynucleotide chain.

heredity n 遺伝

▶ Nucleic acids control heredity on the molecular level.

herring-bone adj 矢はず模様の

▶ IBr forms black crystals in which the IBr molecules pack in a herring-bone pattern similar to that in I_2.

hertz, Hz n ヘルツ（周波数の単位）

▶ The NMR coupling constant, *J*, between the two central protons was found to be 7.3 Hz.

hesitate vi ちゅうちょする syn *hold back*

▶ Einstein did not hesitate to apply his calculations to molecules of sucrose in water, relating the viscosity and diffusion to the Avogadro number, which he thus deduced as being equal to 6.56×10^{23}.

heteroatom n ヘテロ原子

▶ The kind of heteroatom present is indicated by the prefix, *oxa, thia, aza* for oxygen, sulfur, or nitrogen, respectively.

heterocycle n 複素環

▶ Organic ligands for metals are quite often constructed from flat aromatic heterocycles; thus, this structural feature is shared in common with DNA bases.

heterocyclic adj 複素環式の ☞ homocyclic

▶ The use of heterocyclic enones provided products in moderate yields and good enantioselectivities.

heterocyclic compound n 複素環式化合物

▶ Most aromatic and heterocyclic compounds can be divided into rings which share six electrons per ring.

heterogeneity n 不均一性 syn *diversity, variety* ☞ homogeneity

▶ The heterogeneity of a piece of granite is obvious to the eye.

heterogeneous adj 不均一な syn *dissimilar, diverse, various*

▶ A reaction that occurs partly or entirely at the interface between phases is heterogeneous.

heterogeneous catalysis n 不均一系触媒作用

▶ Heterogeneous catalysis is usually studies by passing the gaseous reagent, or reagents, through a tube having a section containing the catalyst.

heterogeneous catalyst n 不均一系触媒 ☞ homogeneous catalyst

▶ The heterogeneous catalyst should be removed from the reaction mixture via filtration and reused in consecutive reactions to ensure that its catalytic activity is retained.

heterogeneous equilibrium n 不均一系平衡

▶ If the system is in heterogeneous equilibrium, the chemical potential of each component is the same in all phases containing it.

heterogeneously adv 不均一に ☞ homogeneously

▶ These results provide the first example of a gas-phase enantioselective reaction heterogeneously catalyzed by a zeolite.

heterogenize vt 不均一化する ☞ homogenize

▷ Heterogenized metal catalysts can be easy to store and handle and simple to recover at the end of a reaction by filtration or decantation.

heterologous expression n 異種発現

▶ The main advantage of heterologous expression is that it allows production of virus-like particles that would be unlikely to assemble and accumulate in the natural host cells.

heterolysis n ヘテロリシス

▶ Acid is needed to convert the alcohol into the protonated alcohol, which can then undergo heterolysis to lose the weakly basic water molecule.

heterolytic adj 不均一な ☞ homolytic

▶ In addition to the normal homolytic dissociation of N_2O_4 into $2NO_2$, the molecule sometimes reacts as if by heterolytic fission: $N_2O_4 \leftrightharpoons NO^+ + NO_3^-$.

heterolytic cleavage n 不均一開裂

▶ Heterolytic cleavage has been defined as bond breakage in which one of the fragments retains both electrons.

heteronuclear adj 異核の ☞ homonuclear

▶ We can describe the heteronuclear system by placing a charge $-q$ on the more electronegative atom and a charge $+q$ on the less electronegative atom.

heteronuclear diatomic n 異核二原子分子

▶ When the vibrational spectrum of a gaseous heteronuclear diatomic is analyzed under high resolution, each line is found to consist of a large number of closely spaced components.

heteropolyacid n ヘテロポリ酸

▶ Tungstates enter into many condensed complexes, either alone or with other oxoanions, forming heteropolyacids and salts.

heteropolyanion n ヘテロポリ陰イオン ☞ isopolyanion

▶ In these heteropolyanions, the heteroatoms are situated inside cavities or baskets formed by MO_6 octahedra of the parent M atoms and are bonded to oxygen atoms of the adjacent MO_6 octahedra.

heterotopic adj ヘテロトピック ☞ homotopic

▶ Together, enantiotopic and diastereotopic ligands are known as heterotopic ligands, that is, ligands in different places.

hexagon n 六角形

▶ Hexagon-based close packing is ubiquitous in nature.

hexagonal adj 六方晶の

▶ An ideal hexagonal unit cell has a c/a ratio of 1.632.

hexagonal close-packed adj 六方最密の ☞ cubic close-packed

▶ The two most dense three-dimensional packing arrangements of atoms for the metallic elements are hexagonal close-packed and cubic close-packed.

hexagonal system n 六方晶系

▶ All crystals which belong to the hexagonal system have three equal axes making angles 60° in one plane and one unequal axis normal to this plane.

hierarchical adj 階層的

▶ The hierarchical constructions may exhibit unique properties that are not found in the individual components.

hierarchy n 階層 syn *graded series*

▶ A feature of self-assembly is hierarchy, where primary building blocks associate into more complex secondary structures that are integrated into the next size level in the hierarchy.

high adj **1** 数値が大きい syn *elevated*

▶ In metal clusters, high coordination numbers have been observed for metal ions of high d^x configuration and low spin.

2 含有量が高い syn *rich*

▶ The success of the salt bridge can be attributed to the high concentration of KCl at the solution junction.

3 高度の syn *superior*

▶ The vertical resolution of the scanning tunneling microscope is extremely high, typically 0.1Å.

4 強烈な syn *strong, violent*

▶ A number of aromatic polynitro compounds have important uses as high explosives.

high-energy phosphate n 高エネルギーリン酸塩

▶ An instantly available store of high-energy phosphate is provided by creatine phosphate.

high-frequency adj 高周波の

▶ The high-frequency resistance is effectively that of the bulk material, whereas at zero and low frequencies the effect of surface or contact areas predominates.

highlight n **1** 最も興味ある出来事

▶ A striking highlight of the literature of the past decade on inclusion chemistry has been the careful design and synthesis of new host materials.

vt **2** 強調する syn *emphasize, stress*

▶ Electric phenomena highlight well the intriguing often surprising properties of inorganic materials.

highly adv 非常に syn *extremely, quite, very*

▶ Phenylglyoxylic acid is highly soluble in water but can be isolated easily as the phenylhydrazone.

high-performance liquid chromatography n 高速液体クロマトグラフィー

▶ High-performance liquid chromatography is an exciting field with new developments continually affecting many disciplines.

high polymer n 高重合体

▶ The term elastomer originally referred to

synthetic thermosetting high polymers having the ability to be stretched to at least twice their original length and to retract very rapidly to approximately their original length when released.

high pressure n 高圧

▶ High pressure has proved to be a powerful tool in the study of electronic phenomena in solids.

high-purity adj 高純度の

▶ Silicon is now invariably made by the reduction of quartzite or sand with high-purity coke in an electric arc furnace.

high-resolution spectroscopy n 高分解能分光学

▶ Tunable lasers are likely to revolutionize high-resolution spectroscopy, especially in the IR.

high-spin adj 高スピンの ☞ low-spin

▶ If the crystal field is weak, then the maximum possible number of electrons remain unpaired and the configurations are known as spin-free or high-spin.

high-temperature phase n 高温相 ☞ low-temperature phase

▶ An understanding of the features of the shape-memory cycle lies in the relationship between the structures of the high- and low-temperature phases.

hill n 小山 syn *elevation*

▶ The catalyst does not affect the net energy change of the overall reaction; it simply lowers the energy hill between the reactants and products.

hinder vt 妨げる syn *hamper, impede, interfere*

▶ Complexity can hinder the use of NMR spectra for qualitative analysis and structure proofs.

hindered rotation n 束縛回転 syn *restricted rotation*

▶ The particular kind of diastereomers that owe their existence to hindered rotation about double bonds are called geometric isomers.

hindsight n 後からの考え syn *retrospect*

▶ With the knowledge of hindsight, we can see that this extraordinary omission arose primarily because the classical coordination chemists did not explore the use of temperature as a variable during synthesis.

hinge vi **hinge on** で定まる syn *depend*

▶ The production of ClO_2 obviously hinges on the redox properties of oxochlorine species.

hint n **1** ヒント syn *implication, suggestion*

▶ The molecular signals that initiate the separation of chromatids are unknown, but there are hints as to what is occurring.

vi **2 hint at** …をにおわせる 受身も可 syn *imply, suggest*

▶ The variable oxidation states and the marked ability to form coordination compounds with a wide variety of ligands are barely hinted at in group 3.

hinterland n 奥地 syn *interior*

▶ Fascination with spectroscopy has left the kinetics and thermodynamics of inorganic complexes in biology in an obscure hinterland.

histological adj 組織学的な

▶ There are over 250 different histological types of cells in the human body.

history n 経歴 syn *background, experience, past*

▶ In the low-temperature range, the conductivity is in general much less reproducible and is particularly dependent upon the thermal history of the specimen and its degree of purity.

hitherto adv 今までのところ syn *formerly, previously*

▶ Frequently, it is desirable to estimate the enthalpy of a chemical reaction involving a hitherto unsynthesized compound, that is, a substance for which no enthalpy data are available.

hold vi **1** 適用できる syn *apply, function, operate*

▶ Henry's law holds fairly accurately for slightly soluble gases, but marked deviations are found with the very soluble gases, e. g., ammonia, hydrogen chloride, etc., in water.

vt **2** 保持する syn *keep, maintain*

▶ In hydrated copper sulfate, one of the H_2O

molecules is held much more tenaciously than the other four.

vt **3** …を有する syn *have*

▶ Catalysis in ionic liquids is an exciting and burgeoning area of research which holds considerable potential for industrial application.

hold back 引き止める syn *reserve*

▶ Filtration through ordinary filter paper holds back all but the smallest particles visible in the microscope.

holder n 支持器 syn *support*

▶ The electrode holder consists of two brass rods insulated from one another by means of an intervening layer of mica.

hold-up n 停滞液またはガス

▶ The hold-up in the metal sponge-packed column is so great that if a chaser solvent is not used in the following procedure the yield will be only about one third that reported.

hole n **1** 正孔 syn *positive hole*

▶ Experiments show that only 0.057 eV is needed to ionize the holes provided by aluminum atom dopants in silicon.

2 空隙 syn *interstice, void*

▶ Between two close-packed layers there are interstitial sites or holes with octahedral and tetrahedral symmetry.

3 空孔 syn *cavity*

▶ The hole in the crown ether can be larger than the cation and still bind it.

hollow n **1** くぼみ syn *cavity, hole, pit*

▶ The next close-packed layer is placed such that the spheres rest in the hollows of the first layer.

adj **2** 中空の syn *empty, vacant*

▶ Each particle of tobacco mosaic virus consists of 2130 identical protein subunits arranged in a helical motif around a single stranded RNA molecule to produce a hollow tube of 18 nm by 300 nm with a 4-nm wide central channel.

homing adj 帰巣性の

▶ Tiny particles of Fe_3O_4 are present in many homing animals such as whales, pigeons, salmon, and bees.

homoallylic adj ホモアリルの

▶ This precedent led us to consider the application of nitrogen-directed homoallylic radical rearrangement to the synthesis of isoquinuclidine-containing alkaloids.

homocyclic adj 同素環の ☞ heterocyclic

▶ More frequent success has attended the synthesis of stable homocyclic polysilane derivatives.

homogenate n ホモジェネート

▶ The homogenate is layered onto a discontinuous or continuous concentration gradient of sucrose solution.

homogeneity n 均一性 syn *uniformity* ☞ heterogeneity

▶ Colloidal particles contain 10^3–10^9 molecules, but alkoxides interact in solution at the molecular level, producing mixed metal–oxygen linkages, which results in high chemical homogeneity and low reaction temperatures.

homogeneous adj **1** 均一な syn *uniform* ☞ heterogeneous

▶ A solution is homogeneous when it is thoroughly mixed, since it is impossible to distinguish any difference between different portions of the solution.

2 同質の syn *similar,*

▶ Alkali metals form a homogeneous group of extremely reactive elements which illustrate well the similarities and trends to be expected from the periodic classification.

homogeneous catalyst n 均一系触媒 ☞ heterogeneous catalyst

▶ Homogeneous catalysts are soluble and active in a liquid reaction medium.

homogeneous hydrogenation n 均一系水素化

▶ In homogeneous hydrogenation, the complex metal ion breaks the hydrogen–hydrogen bond and transfers the hydrogens, one at a time, to the double bond.

homogeneously adv 均一に ☞ heterogeneously

▶ Sulfur was generated homogeneously in acidified thiosulfate solution until the critical supersaturation was attained.

homogenous reaction n 均一系反応

▶ There are fewer truly homogeneous reactions than superficial observation would suggest; a solid-state change may seem to be of this type but on further investigation prove to be more complicated.

homogenization n 均質化

▶ Care is taken not to destroy the organelles by excessive homogenization.

homogenize vt 均質にする ☞ heterogenize

▷ So-called homogenized milk is not truly homogeneous; it is a mixture in which the fat particles have been mechanically reduced to a size that permits uniform dispersion and consequent stability.

homoleptic adj ホモレプチック（錯体の配位子がすべて同じ）

▶ Design features intended to avoid premature precursor chemistry included homoleptic ligand sets to avoid ligand scrambling.

homolog n 同族体 syn *homologue*

▶ The members of a homologous series are called homologs.

homologation n ホモログ化

▶ The homologation of benzene rings produces a surface, and ultimately graphite, but with the inevitable dangling bonds due to the unsatisfied carbon valencies at the periphery of the sheet.

homologous adj **1** 同族の

▶ The acid fluoride of perbromic acid, BrO_3F, is much more reactive than the homologous ClO_3F.

2 相同の syn *analogous, similar*

▶ Although the amino acid sequences of myoglobin and hemoglobin are homologous and they adopt the same globin fold, there are important differences in patches where the polypeptide chains in hemoglobin make contact with one another.

homologous series n 有機，無機化合物を問わず同族列

▶ In the different members of a homologous series of phases, such as the reduced WO_3 phases, the integrity and composition of each member is given by the separation distance of regularly repeating crystallographic shear planes.

homology n 相同関係

▶ Unknown structures can be modeled based on their homology with a known structure.

homolytic adj 均一な ☞ hetrolytic

▶ The photochemical and thermal decomposition of ClO_2 both begin by homolytic scission of a Cl–O bond.

homomorphic adj 同形の

▶ Involvement of the out-of-plane d_{xz} and d_{yz} orbitals on the phosphorus with the singly occupied p_z orbital on N gives rise to the possibility of heteromorphic (N−P) pseudo-aromatic $p_\pi-d_\pi$ bonding (with d_{xz}) or homomorphic (N−N)$p_\pi-p_\pi$ bonding (through d_{yz}).

homonuclear adj 等核の ☞ heteronuclear

▶ The bond in a diatomic molecule can be truly covalent only if the molecule is homonuclear.

homonuclear diatomic molecule n 等核二原子分子

▶ All homonuclear diatomic molecules having nuclides with nonzero spin are expected to show nuclear spin isomers.

homopolar adj 等極の

▶ For homopolar diatomic acceptors such as I_2, Br_2 or Cl_2, the infrared forbidden stretching vibrations of the free molecules become infrared active and decrease in frequency on interaction with an electron donor.

homopolymer n ホモポリマー

▶ A polymer made up of identical repeating units is called a homopolymer.

homotopic n ホモトピック ☞ heterotopic

▶ Sets of ligands in identical stereochemical environments are called homotopic, that is, ligands in the same place.

homotropic adj ホモトロピックな

▶ If an effector of an enzyme is also a substrate, it is called a homotropic effector; while if it is a nonsubstrate, it is called heterotropic.

honeycomb n 蜂の巣状のもの

▶ The honeycomb structure of zeolites is open, with channels which have free diameters in the range 4–7 Å.

hood n フード

▶ If the hood is directly connected with the chimney, it often happens that on a windy day a considerable amount of dust falls into the hood.

hop vi 飛び回る syn *bound, jump*

▶ In hopping semiconductors, the electrons are localized on individual atoms but are able to hop to adjacent atoms provided they gain sufficient energy.

hopeless adj 見込みがない syn *impossible*

▶ Treatments of the solid state have been limited to simplified, ideal models, and complete quantitative treatment of organic solids is hopeless at present.

hopping semiconductor n ホッピング半導体

▶ Semiconductors for which band theory is not applicable are termed hopping semiconductors.

horizon n **1** 視野 syn *range, scope*

▶ The use of these techniques has allowed us to measure the rates of even the fastest known reactions and has widened our kinetic horizons enormously.

2 on the horizon 起こりかけて syn *forthcoming*

▶ New detectors of greater sensitivity and selectivity are on the horizon.

horizontal adj 水平な syn *flat*

▶ It is generally recognized that vertical similarity of the elements in the periodic table is the rule with the exception of the rare earths and elements of groups 8, 9, and 10, for which horizontal similarity seems to apply.

hormone n ホルモン

▶ Some proteins act as hormones regulating various metabolic processes.

host n **1** ホスト ☞ guest

▶ An interesting recent development is the use of iron(III) sulfates as hosts for lithium intercalation.

2 a host of 多数 syn *many*

▶ Given the knowledge that a particular reaction will proceed at a suitable rate, a host of practical considerations are necessary for satisfactory operation.

adj **3** ホストの

▶ From ESR spectra, one can obtain information on the paramagnetic ion and its immediate environment in the host structure.

host organism n 宿主生物

▶ Viruses present in their outer shell carbohydrate structures of the host organism.

hot adj **1** 熱い syn *heated*

▶ Hot concentrated sulfuric acid attacks all borides with evolution of sulfur dioxide.

2 最新の syn *new, recent*

▶ Studying viral genomes is another hot area of research.

3 動作が激しい syn *excited*

▶ Time of flight and infrared emission measurements have shown that the CO_2 is both translationally and vibrationally hot as it leaves the surface.

hot plate n ホットプレート

▶ The solution has a temperature gradient from bottom to top, being hottest on the bottom in contact with the hot plate.

hot stage n 融点測定装置のホットステージ

▶ Attempts to detect the suspected intermediate picryl imidate **2** were made by heating a powdered sample of **1** to 118° on a hot stage.

household n **1** 世帯 syn *house*

▶ The discovery of a room-temperature superconductor has the potential to bring superconductive devices into every household.

adj **2** 家庭用の syn *domestic*

▶ A solution of sodium hypochlorite made by electrolysis of sodium chloride solution is a popular household sterilizing and bleaching agent.

housing n 箱 syn *case, container*

▶ The brass housing containing the sample in its mounting was immersed in a thermostat bath which regulated to ±0.05° over the range

−40 to 90°.

hover vi 空中に停止する syn *drift, float, hang, suspend*

▶ If a superconducting pellet is cooled below its critical temperature and a light, strong magnet is placed above the pellet, it hovers over the pellet.

how adv **1** どんなふうに

▶ An important aspect of chemical kinetics is concerned with how rates depend on temperature.

2 どれほど

▶ For each carbon, the multiplicity of the signal depends upon how many protons are attached to it.

however adv しかしながら 導入の語として文頭で使用した場合には，次にコンマを要する．二つの節の間に置くときは，前はセミコロンで，後はコンマで区切る．その他の位置では，前後ともコンマで区切る． syn *nevertheless, nonetheless, notwithstanding, regardless, in spite of*

▶ Rate studies never really prove a mechanism; they may, however, provide conclusive evidence that a mechanism is incorrect.

Hückel molecular orbital method, HMO method n ヒュッケル分子軌道法

▶ In the HMO method, we established a series of one-electron wave functions but then proceeded to put two electrons into each.

huge adj 大きな syn *enormous, great, large, tremendous*

▶ The origin of this huge difference in ligand-exchange rate constants lies in the large ligand field stabilization energy of the d^3 Cr^{3+} ion.

human adj 人間の

▶ The effects of synthetic compounds on the human or animal organism, effects that were at best unpredictable and in many cases useless or harmful, could be systematically studied.

humidity n 湿度 ☞ relative humidity

▶ Cobalt(II) chloride changes color as the humidity of the environment changes.

hump n こぶ syn *bulge, bump*

▶ One can conveniently describe the ultraviolet spectrum in terms of the position of the top of the hump, $\lambda_{max,}$ and the intensity of that absorption, ε_{max}.

humped adj こぶのある

▶ The variation of crystal field stabilization energy with the number of d electrons is a double-humped curve.

hundred n **1** 100 複数形で数百 syn *many*

▶ There are hundreds of amino acids linked in a particular manner in a typical protein.

adj **2** 100 の

▶ Many colloidal solutions of this low concentration exhibits an increase above the viscosity of the solvent many hundred times greater than the value predicted by Einstein's formula.

hundredth n 100 分の 1

▶ The trinitrotoluene polymorph is probably as unstable as will ever be observed, since it always transforms to the stable form, in a few hundredths of a second after it forms.

hunting ground n 探し場所

▶ Gold would provide the hunting ground for many leading edge catalysis studies in recent years.

hybrid n **1** 混成

▶ According to the resonance theory, a carboxylate ion is a hybrid of two structures which, being of equal stability, contribute equally.

adj **2** 混成の ☞ hybrid orbital

▶ Acetylene is an organic compound which is often formulated with sp hybrid bonds.

3 ハイブリッドの syn *mixed*

▶ Hybrid mesoporous silica-perfluorosulfonic acids have recently been reported as strong Brønsted solid acid catalysts.

hybridization n **1** 混成

▶ In beryllium, uncoupling of the $2s^2$ electrons gives the configuration $1s^2 2s^1 2p^1$, and if this followed by hybridization two equivalent sp orbitals are produced, the bonds from which are oppositely directed along a straight line.

2 ハイブリッド形成

▶ DNA hybridization assays are aimed at detecting the presence of unique nucleic acid sequences in the genome of organisms or in DNA molecules of interest.

hybridize vi **1 混成する**
▶ We can account for the actual geometry of water by postulating that the various orbitals of oxygen can mix, or hybridize.
2 雑種を生じる
▶ Plants from different species generally do not hybridize, so a genetic trait cannot be isolated and refined unless it already exists in a plant strain.
vt **3 混成する**
▷ An sp^3-hybridized carbon atom has a tetrahedral geometry, with each of the four outer shell electrons occupying one of the sp^3 orbitals.

hybrid material n ハイブリッド材料 ☞ composite material
▶ Hybrid inorganic-organic framework materials contain both inorganic and organic moieties as integral parts of a network with infinite bonding connectivity in at least one dimension.

hybrid orbital n 混成軌道
▶ More electronegative substituents prefer hybrid orbitals having less s character, and more electropositive substituents prefer hybrid orbitals having more s character.

hydrate n 水和物 ☞ solvate
▶ Any statement as to the vapor pressure of a hydrate has a definite meaning only when the second solid phase produced by the dissociation is given.

hydrated adj 水和した syn *hydrous*
▶ When a solution of Co(II) salt is treated with sodium hydroxide, a blue precipitate of hydrated CoO is formed that changes to pink on standing or heating.

hydration n 水和 ☞ solvation
▶ In the large cavities of the framework silicates, the lack of specific strong interactions enables the degree of hydration to vary continuously over very wide ranges.

hydration energy n 水和エネルギー
▶ It is not possible to allocate the single-ion hydration energies with any certainty.

hydration isomerism n 水和異性
▶ Hydration isomerism occurs when water may be inside or outside the coordination sphere.

hydration number n 水和数
▶ The hydration numbers for the diamagnetic cations Be^{II}, Al^{III}, and Ga^{III} have been directly measured by ^{17}O NMR as 4, 6, and 6, respectively.

hydration shell n 水和殻 syn *hydration sphere*
▶ A rationalization of the effective radii is that one should consider the ion and its first hydration shell as an entity to go into the Born equation.

hydraulic adj 油圧の
▶ The linear silicon polymers remain fluid to low temperatures and are very stable thermally, which makes them useful as hydraulic fluids and lubricants.

hydride n 水素化物
▶ We tend to think of hydrogen chiefly in its proton character. Actually, its hydride character has considerably more reality.

hydridic adj 水素化物性の ☞ protic
▶ The solid-state structure of neat H_3NBH_3 exhibits short BH⋯HN intermolecular contacts; the hydridic hydrogen atoms on boron are 2.02 Å away from the protic hydrogen atoms on nitrogen of an adjacent molecule, a distance less than the van der Waals distance of 2.4 Å.

hydroboration n ヒドロホウ素化
▶ Hydroboration involves addition of BH_3 to the double bond, with hydrogen becoming attached to one doubly bonded carbon, and boron to the other.

hydrocarbon n 炭化水素
▶ Chemisorption of hydrocarbons may result in formation of chemisorbed hydrogen atoms and hydrocarbon fragments.

hydrodesulfurization n 水素化脱硫
▶ Hydrodesulfurization is a process for removing sulfur from petroleum feedstocks.

hydrodynamic adj 流体力学的
▶ Another hydrodynamic property of solutions of macromolecules which is dependent on the molecular weight is the viscosity of a

solution of the macromolecule material.

hydroformylation n ヒドロホルミル化

▶ The first example of homogeneous transition-metal catalysis in an ionic liquid is the platinum-catalyzed hydroformylation of ethene in tetraethylammonium trichlorostannate.

hydrogel n ヒドロゲル

▶ The properties of the hydrogel are influenced both by the structure of the polymer network and by the water, which acts just like a conventional plasticizer.

hydrogenate vt 水素化する

▶ It is usual to hydrogenate sodium metal in the presence of a surface-active agent in kerosene in order to overcome difficulties due to the surface film of hydride, which can lead to imcomplete conversion.

hydrogenation n 水素化

▶ Hydrogenation may occur either as direct addition of hydrogen to the double bonds of unsaturated molecules, resulting in a saturated product, or as reaction of hydrogen with the molecular fragments given by rupture of the bonds of organic compounds.

hydrogenation agent n 水素化剤

▶ Hydrogen appears to migrate more selectively than expected if the hydrogenation agent were a free H atom.

hydrogenation catalyst n 水素化触媒

▶ In the presence of hydrogenation catalysts, unsaturated compounds undergo not only hydrogenation but also isomerization: shift of double bonds or stereochemical transformations.

hydrogen bond n 水素結合 syn *hydrogen bonding*

▶ Typical hydrogen-bond distances emphasize that the hydrogen bond is too long for much covalent character to be expected.

hydrogen bonding n 水素結合 複合形容詞として使用 syn *hydrogen bond*

▶ Hydrogen-bonding-type interactions of phenol with donors are not completely electrostatic in nature.

hydrogen electrode n 水素電極

▶ The hydrogen electrode is assumed to be ideally reversible and of so small surface that its adsorption may be neglected.

hydrogen ion concentration n 水素イオン濃度

▶ The conductivity of a solution of an acid provides an approximate measure of the hydrogen ion concentration in the solution.

hydrogen ion exponent, pH n 水素イオン指数

▶ It is quite common to express the acidity of a solution in terms of the so-called hydrogen ion exponent, pH, which is the value log（$1/c$）when c is the concentration of H^+ in moles per liter.

hydrolysis n 加水分解 ☞ solvolysis

▶ If halide ion is present during hydrolysis of benzenediazonium or *p*-nitrobenzenediazonium ion, there is obtained not only the phenol but also the aryl halide; the higher the halide concentration, the greater the proportion of aryl halide obtained.

hydrolytic adj 加水分解の

▶ This distortion leads to a loss of resonance energy in the bond and enhances its susceptibility to hydrolytic attack.

hydrolytic enzyme n 加水分解酵素 syn *hydrolase*

▶ Man cannot utilize cellulose as food because he lacks the necessary hydrolytic enzymes.

hydrolytically adv 加水分解に

▶ Above room temperature potassium ozonide KO_3 decomposes to the superoxide KO_2, and the compound is also hydrolytically unstable.

hydrolyzable adj 加水分解する

▶ $OSCl_2$ reacts rigorously with water and is particularly valuable for drying or dehydrating readily hydrolyzable inorganic halides.

hydrolyze vi **1** 加水分解する ☞ solvolyze

▶ Dilute soap solutions hydrolyze to form acid soaps.

vt **2** 加水分解する

▶ The nitrile group can be hydrolyzed to a carboxylic acid in either strong acid or strong base.

hydrophilic adj 親水性の

▶ Activated alumina is very hydrophilic and is often used for the drying or dehydration of gases and liquids.

hydrophobic adj 疎水性の ☞ lipophilic

▶ Important work on varied nucleobase shapes has given useful evidence of the importance of close hydophobic packing.

hydrophobic colloid n 疎水コロイド

▶ The flocculation of hydrophobic colloids can be made to occur by adding an electrolyte, particularly one with positive ions of high charge.

hydrophobic group n 疎水基

▶ Small aggregates, usually dimers or trimers, are often formed in dilute aqueous solutions of molecules containing hydrophobic groups.

hydrosol n ヒドロゾル

▶ By making the brown solution of nickel sulfide slightly acid with acetic acid and by boiling, the hydrosol is coagulated and can be removed by filtration.

hydrosphere n 水圏

▶ Iodine is considerably less abundant than the lighter halogens both in the earth's crust and in the hydrosphere.

hydrostatic pressure adj 静水圧

▶ Valuable information about reactions in solution is provided by studies of the influence of hydrostatic pressure on their rates.

hydrothermal reaction adj 水熱反応

▶ One potential problem is that chiral organics may not survive hydrothermal reaction conditions enantiomerically intact.

hydroxo n ヒドロキソ（金属への配位子としての OH^-）

▶ The accurate determination of pK_a values is rendered very difficult by the pronounced tendency of almost all hydroxo complexes to undergo olation.

hydroxylation n ヒドロキシ化

▶ In humans, cytochrome P-450 catalyzes the hydroxylation of drugs, steroids, pesticides, and other foreign substances.

hydroxylic adj 水酸基の

▶ The hydrocarbon by-product is easily eliminated since it is much more soluble in hydrocarbon solvents than the hydroxylic major product.

hygroscopic adj 吸湿性の

▶ Concentrated sulfuric acid is very hygroscopic and is used for drying gases, etc.

hyperbolic adj 双曲線の

▶ When the fractional saturation is plotted against the partial pressure of O_2, the curve is not hyperbolic but sigmoidal.

hyperchromic effect n 濃色効果

▶ The hyperchromic effect of DNA denaturation results from the unstacking of the base pairs in the helix.

hyperconjugation n 超共役

▶ The concept known as hyperconjugation received support not only from experiments deliberately designed to detect the effect but also from its mathematical derivation as the molecular orbital method gained power and versatility.

hyperfine adj **1** 超純粋な

▶ Hyperfine Si is one of the purest materials ever made on an industrial scale.

adj **2** 超微細の

▶ Only the magnetic nuclei produce hyperfine lines; consequently the role of nonmagnetic carbon and oxygen atoms must be inferred from their assumed association with the hydrogen atoms.

hyperfine structure n 超微細構造

▶ The hyperfine structure of the ESR spectra of the monopositive and mononegative ions arises from the interaction between the magnetic moment of the unpaired electron and the magnetic moments of the protons in the aromatic system.

hypothesis n 仮定 syn *assumption, postulate, supposition*

▶ If a hypothesis continues to agree with the results of experiment, the hypothesis is called a theory or law.

hypothesize vt 仮定する syn *presume, speculate, suppose*

▶ We hypothesized that steric effects, rather than these hydrogen bonds, were main arbiters of DNA replication fidelity.

hypothetical adj 仮想の syn *assumed, supposed*

▶ Two conditions must be met in establishing the standard state, namely, that the activity coefficient and the mean molality both equal unity, and the only way for this to occur is to postulate a hypothetical solution.

hypsochromic shift n 浅色移動 ☞ bathochromic shift

▶ Whereas alkylation always give rise to a bathochromic shift of the bands in the benzenoid spectra, the azulene band undergoes either a bathochromic or a hypsochromic shift, depending on the position of substitution.

hysteresis n ヒステレシス

▶ Hysteresis ranging from a few degrees to several hundred degrees are commonplace.

hysteresis loop n 履歴曲線

▶ A critical requirement for magnetically based information storage components is that they should be soft with low eddy current losses and a certain type of hysteresis loop, either square or rectangular.

I

ice n **1 氷**

▶ Ice is unusual in having a density less than that of the liquid phase with which it is in equilibrium.

vt **2 氷で冷やす** syn *chill, cool, freeze*

▶ The flask is held over the ice bath in readiness to ice it if the reaction becomes too vigorous.

ice-cold adj 氷のように冷たい

▶ The black needles were collected and washed on the funnel with ice-cold acetonitrile and then with dichloromethane; yield 3.16 g.

icosahedral adj 二十面体の

▶ There is an icosahedral arrangement of six water and six perchlorate oxygen atoms about the barium ion in $Ba(ClO_4)_2 \cdot 3H_2O$ with Ba−O distances ranging from 2.96–3.18 Å.

idea n **1 着想，アイディア** syn *suggestion, thought*

▶ New ideas must be tested, and alternative explanations must be considered.

2 概念 syn *conception, image, understanding*

▶ A large number of nucleophilic substitutions are listed below to give an idea of the versatility of alkyl halides.

ideal n **1 理想** syn *perfection*

▶ This ideal is often difficult to attain, and actual location of the atoms in space by X-ray or electron diffraction techniques may be required to settle difficult structural questions.

adj **2 理想的な** syn *complete, excellent, perfect*

▶ It can be recognized that all gases tend to become ideal as the pressure approaches zero.

ideal dilute solution n 理想希薄溶液

▶ Systems obeying Henry's law are less ideal than those obeying Raoult's law and are referred to as ideal dilute solutions.

idealistic adj 理想上の syn *hypothetical, theoretical*

▶ The modern procedure is based on the recognition that our interpretation of binary temperature-composition diagram is idealistic.

ideality n 理想性

▶ Activity coefficients are usually omitted, although even in dilute solutions the departure from ideality is probably significant.

idealization n 理想化 syn *perfection*

▶ The curves shown in Figure 1 are idealizations that may be or may not be achieved in practice.

idealize vt 理想化する

▶ The positions of the hydrogen atoms were idealized and refined with riding thermal parameters.

ideally adv 理想的に syn *in principle, theoretically*

▶ In ideally ionic crystal structures, the coordination numbers of the ions are determined largely by electrostatic considerations.

ideal solution n 理想溶液

▶ Ideal solutions must obey Raoult's law of vapor pressures exactly throughout the complete range of concentrations over which they can be considered as ideal.

identical adj 詳細にわたって，同じ syn *indistinguishable, interchangeable, same*

▶ The best way to measure the relative reactivities of different compounds toward the same reagent is by the method of competition, since this permits an exact quantitative comparison under identical reaction conditions.

identically adv 全く同じ syn *alike, nearly*

▶ A precipitate of identically the same chemical composition can be formed from any soluble barium salt by using a solution of any soluble sulfate.

identifiable adj 確認できる syn *discernible*

▶ Many naturally occurring substances are hydroaromatic; conversion into identifiable aromatic compounds gives important information about their structures.

identification n 確認 syn *characterization, designation, verification*

▶ The spectroscopic identification of intermediates and kinetic studies are of importance.

identify vt 確認する syn *recognize, specify*

▶ Ordinary paper chromatography can separate and identify 40 sugars.

identity n **1** 独自性 syn *individuality, uniqueness*

▶ It is characteristic of the complex that it retains its identity, more or less, even in solution, though partial dissociation may occur.

2 同一性 syn *agreement, indistinguishability*

▶ If a substance under study is to be compared with a compound that is known but not available for mixed melting point determination, probable identity can be established by showing that several derivatives of the substance in question all correspond in melting point with known derivatives.

i.e. abbr すなわち that is の意味のラテン語 id est を略記したもので，ローマン体を用い，その前後にコンマを置く．

▶ Bonds or groups within a molecule sometimes vibrate with a frequency, i.e., have an energy-level pattern with a spacing that is little affected by the rest of the molecule.

if conj **1** もしも…ならば syn *supposing*

▶ The reaction is quantitative if the correct conditions are maintained. 誤解の恐れのないときは，主語や動詞を省略することができる．

▶ If the sugar residue is glucose, the derivative is a glucoside; if frustose, a fructoside; if galactose, a galactoside.

2 たとえ…としても

▶ The relative lack of eight-coordinate structures, CsCl, CsBr, and CsI being the only known alkali metal examples, is commonly found, if hard to explain.

事実に反することを述べるときには過去時制を用い，主語が単数形でも動詞は was ではなく were を使用する．その場合，主節には助動詞 should, would, could, might を使用する．

▶ If the configuration of the Cu^{2+} ion were $d^1_{z}2d^2_{x^2-y^2}$, we should expect to find two short and four long bonds.

3 if any もしあるなら

▶ Many papers that deal with measurements made on organic single crystals report little, if any, information on how the crystals were grown.

4 if not …でないとしても syn *otherwise*

▶ The formation of the postulated tetramer would be considerably more difficult, if not impossible, in this instance.

ignite vi **1** 発火する syn *burn, fire, inflame*

▶ The 18-electron carbonyl, $Re_2(CO)_{10}$, is much less stable than the technetium one and ignites spontaneously in air.

vt **2** 点火する syn *light*

▶ Once ignited, the reaction between Ni and Al is sufficiently exothermic that it is self-sustaining.

3 強く熱する syn *heat*

▷ Small amounts of zinc sulfate may be changed into oxide by igniting over the blast lamp.

ignition n 燃焼 syn *fire, flame*

▶ All acetates are decomposed on ignition with the evolution of combustible vapors and gases, leaving behind either the carbonate, oxide, or the metal itself.

ignorance n 不案内 syn *inexperience, unfamiliarity*

▶ The use of the term receptor to denote the site of action of a compound is in itself an indication of our ignorance concerning the molecular action of the drug.

ignore vt 無視する syn *disregard, overlook*

▶ The Walden rule is normally only valid for ions at infinite dilution where ion–ion interactions can be ignored which is clearly not the case in ionic liquids.

ill adj 悪い syn *dangerous, detrimental, harmful*

▶ Only a small portion of the sun's spectrum reaches the surface of the earth, and parts of the UV portion that are largely screened can cause various ill effects to living systems.

ill-defined adj 不明確な syn *indefinite, indistinct, unspecific*

▶ These magnetic materials were ill-defined with compositions varying according to the starting material used.

illuminate vt **1** 照射する syn *light up, shed light on*
▶ When a mixture of hydrogen and chlorine is illuminated with blue light, reaction immediately begins.
2 解明する syn *clarify, explain, reveal*
▷ The extensive and unifying correlation which attends the consistent application of the hypothesis and its success in illuminating obscure and unexplained experimental data provide the most conclusive evidence for its reality.
illuminating adj 啓蒙的な syn *informative, instructive*
▶ Particularly illuminating is a comparison of two well-known rubbery polymers poly(*cis*-isoprene) and poly(isobutene), which crystallize below room temperature.
illumination n 照射 syn *lightening*
▶ Illumination of the metal surface by any radiation with wavelength shorter than the photoelectric threshold causes the emission of photoelectrons.
illustrate vt **1** 例証する syn *demonstrate, exemplify*
▶ The similarities and the differences that exist between carbon and silicon are illustrated by comparison of the chemistry of organosilicon compounds with that of analogously constituted carbon compounds.
2 図解する syn *explain, represent, show*
▷ Halogen-molecule bridges have also been observed in the bromine-acetone complex which contains endless chains involving the basic structure illustrated in Figure 1.
illustration n 実例 syn *demonstration, example*
▶ The reactivity of $AuAl_2$ provides an excellent illustration of the amphoteric character of aluminum: both acids and bases dissolve the aluminum from the intermetallic compound and thus permit the gold to be recovered essentially quantitatively.
illustrative adj 例となる syn *representative*
▶ The work cited below is illustrative rather than exhaustive.
image n **1** 画像 syn *figure, picture*
▶ For semiconductors, the constant current image includes spectroscopic information about electron densities of states.
vt **2** 推察する syn *imagine, presume, suspect*
▶ Whenever a crystallographic shear plane is imaged, irregularity in the fringe spacing occurs.
3 画像化する
▷ Imaging individual oxygen atoms on the GaAs(110) surface, with the Ga and As atoms resolved, it was shown that the oxygen atoms bond at an interchain bridging position.
imaginable adj 想像できる syn *plausible, possible, tenable*
▶ Bringing together the four components Zn^{2+}, Cu^{+}, CN^{-}, and $NMe_4{}^{+}$ in aqueous solution under the simplest imaginable conditions led to the spontaneous assembly of crystalline material with precisely the structure intended.
imaginary adj **1** 架空の syn *fictitious*
▶ Ordinarily, thermodynamics deals with a finite system, isolated from its neighbors by an imaginary closed surface.
2 虚数の
▶ The square root of a negative quantity is imaginary.
imagination n **1** 想像 syn *ingenuity, insight*
▶ The apparent nonexistence of perbromates could not but challenge the imagination of chemical theorists.
2 空想 syn *thinking, thought*
▶ A thermodynamically reversible process can only occur in our imagination.
imaginative adj 想像力に富む syn *ingenious, inventive*
▶ The novel properties of nanoparticles are being explored in many imaginative ways.
imaginatively adv 想像的に
▶ Chemical laws are intuitively discovered, and chemical compounds are imaginatively constructed.
imagine vt 想像する syn *conceive, consider, image, think of*

▶ Proteins can supply all the diversity of ligand donors and stereochemical controls within cavities which could be imagined.

imaging n イメージング

▶ For successful atomic resolution imaging, precautions must be taken against vibration and thermal gradients.

imbalance n 不均衡 syn *difference, disparity, inequality*

▶ Even in atoms in molecules which have no permanent dipole, instantaneous dipoles will arise as a result of momentary imbalances in electron distribution.

imitate vt 模造する syn *copy, duplicate, mimic*

▶ Synthetic organic catalysts have been developed which imitate the action of enzymes such as chymotrypsin.

imitation n 模倣 syn *copying, mimicry*

▶ The term biomimetic chemistry means imitation or mimicry of natural organic processes in living systems.

immaterial adj 重要でない syn *insignificant, trivial, unimportant*

▶ Whether the structure of the phosphoryl group is written as P=O or P^+-O^- is immaterial provided we recognize its hybrid character.

immeasurable adj 計ることができない syn *measureless*

▶ Even a seemingly immeasurable variation in stoichiometry can cause a materials to be highly colored.

immeasurably adv 計れないほど syn *exceedingly, far, well*

▶ The transport number for the silver ions is unity, while the transport number for the halide ions is immeasurably small.

immediate adj 即座の syn *instantaneous*

▶ The optical method provides us with an immediate visual check both on the reciprocal space and on real space distributions and their relative scale and orientation.

immediately adv **1** すぐ近くに syn *closely, intimately*

▶ Solvents were dried, purified under N_2 by using standard methods, and were distilled immediately before use.

2 即座に syn *instantaneously, instantly, promptly*

▶ Since proton transfers are often very fast reactions, acid-base equilibria are established immediately.

immense adj 莫大な syn *enormous, huge, tremendous, vast*

▶ Immense numbers of molecules of different kinds are found in the complicated structures of plants and animals.

immensely adv 非常に syn *highly, vastly*

▶ Only one monocarboxylic acid anhydride is encountered very often; however, this one, acetic anhydride, is immensely important.

immerse vt 浸す syn *dip, plunge, sink, soak*

▷ A seed crystal immersed in a saturated solution will neither grow nor dissolve.

immersion n 浸漬 syn *plunge*

▶ Mixtures of TlBr and TlI are recommended for immersion media for high index of refraction determinations.

immiscibility n 混和しないこと

▶ Many glasses exhibit a liquid immiscibility texture on a submicrometer scale, and this can be seen by electron microscopy.

immiscible adj 混和しない ☞ miscible

▶ In the distribution of a substance between two immiscible solvents, the ratio of the concentrations remains constant for a given molecular species.

immobilization n 固定化 syn *fix, freeze*

▶ Immobilization of biomolecules and the assembly of biomolecules in defined arrays is a desired requirement of nanotechnolgy

immobilize vt 固定する syn *fix, freeze, pin*

▶ Rubidium atoms in $RbAg_4I_5$ are immobilized in sites that have a distorted octahedral environment of I^- ions.

immoral adj 不道徳な syn *bad, evil, wrong*

▶ The invention must be useful for a purpose not immoral and not injurious to the public welfare.

immune adj **1** 免疫の

▶ Even a high state of immunity to a particular

pathogen may not necessarily provide protection, because a sufficiently large dose of bacterial toxin or virus might overwhelm the immune defenses.

2 影響されない syn *protected, unaffected*

▶ The absence of any other oxidation state of comparable stability for nickel implies that compounds of Ni^{II} are largely immune to normal redox reactions.

immune system n 免疫系

▶ Cells and molecules of the immune system are capable of recognizing and destroying specific foreign substances.

immunity n 免疫

▶ Immunity is a physiological condition, developed after contact with an immunogen, that specifically aids the body in combating pathogenic organisms of which the immunogen is a part.

immunization n 免疫化

▶ Immunization is the process of activating the immune system.

immunize vt 免疫にする

▷ The blood globulins are used in immunizing against specific diseases and in medical research.

impact n **1** 影響 syn *effect, influence*

▶ Overall, the trend observed here may not be altogether surprising when the steric impact of the ligand is considered.

vt **2** 強い影響を与える syn *affect, change, modify*

▷ The major factor impacting on ATP generation in muscles of unfit subjects is a suboptimal blood supply.

impair vt 損なう syn *damage, harm, injure, weaken*

▷ The composition of the electrolyte, KF/HF, can vary within fairly wide limits without impairing the operation conditions or efficiency.

impaired adj 損なわれた syn *defective*

▶ To understand how this highly selective process occurs, how it can become impaired, and how impairment may be treated, model compounds are useful tools.

impairment n 損傷 syn *damage, harm*

▶ Pyrolysis of polyacrylonitrile makes it electrically conductive without impairment of its structure.

impart vt 与える syn *afford, confer, give, grant*

▶ The heated sodium atoms impart an orange-yellow color to the flame.

impedance n インピーダンス

▶ To derive quantitative data, it is necessary to employ impedance measurements over the low-frequency dispersion range.

impede vt 妨げる syn *hamper, hinder, inhibit, obstruct*

▶ The progress of chemistry might well have been impeded instead of enhanced by thermochemistry had it been necessary to measure experimentally the heat of reaction of every chemical system studied.

impediment n 妨害 syn *barrier, block, restraint, restriction*

▶ The weakness of the O−H···Cl hydrogen bond ensures that there is very little impediment to complete ionic dissociation.

impenetrable adj 入り込めない syn *dense, impermeable, inaccessible, thick*

▶ The primary purpose of the present investigation is the calculation of the volume occupied by a molecule, i.e., impenetrable for other molecules with thermal energies at ordinary temperatures.

imperative adj 必須の syn *essential, mandatory, necessary, obligatory*

▶ It is imperative that systematic studies are carried out into the thermodynamics of metal ion reduction in ionic liquids with nominally noncoordinating anions.

imperfect adj 不完全な syn *defective, deficient, incomplete*

▶ Most real crystals are imperfect, investigations of the mechanical properties of solids show that no macroscopic specimen ever exhibits the theoretical strength of a perfect crystal.

imperfect gas n 不完全気体

▶ In an imperfect gas near the critical point, the density fluctuations are abnormally great, being visible in the phenomenon of opalescence, the irregular scattering of light, giving the material a milky appearance.

imperfection n **1** 欠陥 syn *defect, error, fault, flaw*

▶ Imperfections normally exist in rather high concentrations in all real materials and may be classified as point defects, dislocations, or line defects, and plane defects.

2 不完全 syn *inadequacy*

▶ The gaseous imperfections are assayed on the assumption that there is no higher polymerization.

imperfectly adv 不完全に

▶ A recent neutron diffraction study of the hexagonal UO_3 suggests that this form is an imperfectly crystalline form of rhombohedral UO_3.

impermeable adj 不浸透性の syn *tight*

▶ Few plastics are impermeable to the slow diffusion of oxygen or water vapor.

impetus n 刺激 syn *drive, impulse, stimulas*

▶ Great impetus was given to this field by Cohen and Schmidt with generalization that in many such photochemical reactions the nature of the product may be explained by the crystal structure of the starting material.

impinge vi **impinge on** 衝突する syn *infringe on*

▶ Jets of the two solutions impinge on one another and give very rapid mixing.

impingement n 衝突

▶ When hydrolysis is more rapid, polymers are larger and more highly cross-linked; on impingement the polymers will not deform as readily.

implant n インプラント

▶ The body responds to an alumina implant by surrounding it with a fibrous capsule which can be several micrometers thick.

implantable adj 移植可能な

▶ For implantable devices, the toxicity of the materials, immunogenicity, thrombogenicity, and resistance to biodegradation are extremely important.

implantation n 移植 syn *graft, transplant*

▶ Alumina does not cause adverse reactions on implantation and is generally classified as being inert, or nearly inert.

implicate vt …と関係がある syn *associate, include, involve*

▶ The bioorganometallic chemistry and cellular studies of molybdocene dichloride implicate interaction with cellular thiols as the key reaction related to biological activity.

implication n **1** 複数形で，影響 syn *conclusion, effect, outcome, result*

▶ We have to consider the differences in the magnitude of the deformation energies and the possible implications for the geometric preferences in the dimer.

2 意味合い syn *hint, suggestion*

▶ Although reference has been made here to covalent bonds, the implication is simply that the bonds are largely covalent.

implicit adj 暗黙の syn *implied, indirect, tacit*

▶ It is current practice to use a single Kekulé structure to represent benzene with the implicit understanding that the C−C bonds are all equivalent.

implicitly adv 暗黙裡に

▶ A more elaborate calculation, while implicitly taking these effects into consideration, may successfully reproduce the experimental enthalpy of formation without recognizing any anomaly.

implied adj 暗黙の syn *implicit, potential, tacit*

▶ The implied reversal of the polarity of B−H and C−H bonds is an important factor in discussing hydroboration and other reactions.

imply vt 暗示する syn *hint, suggest*

▶ Deviations from ideal stoichiometry in ionic crystals imply changes in the charge numbers of some of the ions, usually the cations.

importance n **1** 重要性 syn *consequence, significance, value*

▶ The importance of layered $LiMnO_2$ may be that it opens a new avenue of exploration that will lead to optimized materials.

2 of importance 重要な syn *important, serious, significant*

▶ Because of the low values of the electronegativity for most of these metals, π bonding, from the metal to the ligand, is usually of greatest importance.

important adj 重要な syn *critical, significant, vital*

▶ In order to avoid the formation of crystal defects and dislocations at the interface, it is important that lattice dimensions of the layer are the same as the substrate on which it is epitaxially deposited.

importantly adv 重要なことには syn *highly, notably, significantly*

▶ More importantly, colloidal dispersions of nanofibers in a nonsolvent can be made to retain the nanofibrillar morphology, thus enabling further processing.

impose vt 課する syn *demand, force, require*

▶ The rapidity of electron transfer imposes strict geometrical requirements.

imposition n …を及ぼすこと syn *application*

▶ The imposition of pressure upon nominally high-oxidation-state transition-metal compounds, such as those of iron(III) salts, leads to the production of low-oxidation-state analogues, iron(II) salts.

impossible adj 不可能な syn *hopeless, impracticable, inconceivable*

▶ Formation of a high-molecular-weight polymer requires a reaction that proceeds in very high yields, and purification of the product by distillation, crystallization, etc., is difficult, if not impossible.

impossibly adv 極端に

▶ Completely undescriptive names should be avoided, although clearly this is not always practical since the best descriptive names may be impossibly unwieldy.

impoverish vt 貧しくする syn *weaken*

▶ Concern has been expressed that increasing levels of N_2O following denitrification may eventually impoverish the ozone layer in the stratosphere.

impracticable adj 実行不可能な syn *impossible, unattainable*

▶ Since fusion is impracticable because of their high melting points, Mo and W in the form of powders are converted to the massive state by compression and sintering under H_2 at high temperatures.

impractical adj **1** 考え，計画が実行不可能な syn *ineffective, useless*

▶ While it is impractical at the moment to carry our general theoretical studies much further, the second virial coefficient has been calculated for several of these models.

2 実際的でない syn *unrealistic*

▶ It is impractical that ionic liquids will compete with all aqueous solutions in large markets where the current technology works effectively.

imprecise adj 不正確な syn *inaccurate, indefinite, wrong*

▶ The term organometallic is somewhat vague since definitions of organo and metallic are themselves necessarily imprecise.

imprecisely adv 不正確に syn *ambiguously, unclearly, vaguely*

▶ Since these two natural radioactive decay series produce different isotopes, the relative atomic mass of lead can be stated only very imprecisely.

impregnate vt 浸み込ませる syn *penetrate, permeate, saturate, soak*

▶ Bones and shells consist of organic macromolecules (collagen and chitin) that are impregnated with inorganic crystals (calcium phosphate and calcium carbonate).

impregnation n 含浸

▶ Activation of charcoal consists in heating in air, steam, or carbon dioxide, with or without previous impregnation with materials such as zinc salts, to burn away more than half of the carbon and leave an ultraporous, skeletonized structure.

impress vt 強く認識させる syn *affect, in-*

fluence, strike

▶ In examining the external appearance, or habit, of a crystal, we are immediately impressed by the series of naturally occurring plane faces which usually appear to have some kind of regularity of arrangement.

impression n 感じ syn *impact, notion, sensation, sense*

▶ Some impression of the variation in reported values may be seen by comparisons of the several values on the biphenyl compounds.

impressive adj 印象的な syn *exciting, powerful, striking*

▶ Even more impressive is the recent observation that the rate of exchange in the system $Co(phen)_3^{3+}$–$Co(phen)_3^{2+}$ is some 10^5 times faster than for $Co(en)_3^{3+}$–$Co(en)_3^{2+}$.

impressively adv 印象的に

▶ These species play a role in an impressively large number of biological processes.

imprison vt 閉じ込める syn *confine, detain, trap*

▶ In clathrate compounds, the guest molecules are imprisoned in discrete closed cavities or cages.

improbable adj ありそうにもない syn *doubtful, questionable, unlikely*

▶ Collisions between four molecules are so improbable that it is unlikely that there are any elementary reactions having a molecularity of 4 or more.

improper adj 誤った syn *erroneous, imprecise, wrong*

▶ Improper and overuse of antibiotics in humans and farm animals has led to dramatic increases in antibiotic-resistant bacteria.

improve vt 改良する syn *enhance, reform, upgrade*

▶ The accuracy with which molecular geometry can be determined has greatly improved, and significance is now placed on quite small differences.

improved adj 改良された syn *better*

▶ One report shows that nanocrystalline aluminum has improved hardness over bulk aluminum.

improvement n **1** 進歩 ☞ advance, progress

▶ This procedure certainly introduces some improvement, but the evaluation of the steric factor cannot be done in an entirely satisfactory way.

2 向上 syn *enhancement*

▶ The improvement in reaction rate was realized owing to the homogenization of reagents and catalyst under supercritical conditions.

impulse n 推進力 syn *impetus, incentive*

▶ A reversible change cannot actually occur, since an impulse which moves a system out of an equilibrium state is necessarily finite, although it may be very small.

impure adj 不純な syn *contaminated*

▶ Some high-boiling substances decompose at the boiling point, particularly if impure, but can be freed from contaminating substances by steam distillation at a lower temperature at which they are stable.

impurity n 不純物 syn *contaminant, foreign matter*

▶ More typical sugars are obtainable in crystalline form only with difficulty, particularly in the presence of a trace of impurity.

impurity effect n 不純物効果

▶ Interpretation of the data appears to be complicated by impurity effects.

in prep **1** …の中で

▶ The test reagent is made by chilling 1 mL of acetic anhydride in a 10-mL Erlenmeyer flask in an ice bath and adding one drop of concentrated sulfuric acid.

2 …に関して

▶ Proton magnetic resonance studies of solid adamantine reveal an abrupt diminution in the second moment at 143 K, which is interpreted as involving a rotational transition with an activation energy of about 21 kJ mol^{-1}.

3 …の状態で

▶ Hypochlorous acid and most of its salts are known only in aqueous solution.

4 方法，…で

▶ If the changes in pressure are plotted as a function of time in the usual way, it is found that the order of the reaction is close to 2.

5 形状，…で

▶ $Pb(UO_2)_2P_2O_8 \cdot 8.4H_2O$ is pseudotetragonal in eight-sided basal plates or less regular flakes.

6 …において

▶ A particularly important practical use of the hydroquinone–quinone oxidation–reduction system is in photography.

7 …において　副詞の役割を果たす前置詞句に用いられる．

▶ In conducting a reaction, the required amount of magnesium is placed in the flask, and the apparatus is assembled and swept with nitrogen.

8 in and out　出たり入ったり

▶ Cells maintain their internal environment in a relatively stable state, continuously pumping ions in and out, turning over proteins, etc.

inability n 無力 syn *incapability*

▶ In spite of their modest polarity, alkyl halides are insoluble in water, probably because of their inability to form hydrogen bonds.

inaccessible adj 到達しがたい syn *unattainable, unavailable*

▶ Alkynes, like alkenes, only absorb ultraviolet radiation very strongly at wavelengths in the relatively inaccessible region below 2000 Å.

inaccuracy n 不正確 syn *error*

▶ X-ray analyses cannot compete with spectroscopic methods in accuracy, partly due to the inaccuracies in the observed intensities.

inaccurate adj 不正確な syn *imprecise, incorrect, wrong*

▶ The disadvantage is the rather inaccurate size distributions obtained by this method.

inactivate vt 失活する

▶ Glycogen synthase is inactivated when specific serine residues are phosphorylated.

inactive adj 不活性な syn *inert, passive*

▶ While these complexes slowly effected dehydrogenation of H_3NBH_3, the reaction was accompanied by precipitation of catalytically inactive black solids.

inactivity n **1** 不活性 syn *inertness, passivity*

▶ Optical inactivity requires that one molecule of a compound act as the mirror image of another.

2 無活動 syn *rest*

▶ During periods of physical inactivity, creatine phosphate is formed from creatine and ATP simply by reversal of the creatine kinase reaction.

inadequacy n 不適切 syn *dearth, defect, lack*

▶ The substantial difference between the two Pt–P distances and the wide deviation of the angles of Pt from 90° emphasize the inadequacy of describing the bonding of such complicated species in terms of simple localized bonding theory.

inadequate adj 不適切な syn *imperfect, insufficient, unsatisfactory*

▶ Over any extended temperature range, it is now well known that the two-parameter Arrhenius equation is inadequate and that the three-parameter equation must be used in its place.

inadequately adv 不適切に syn *badly, poorly, unsatisfactorily*

▶ Aside from structure, the physical properties most studied, though still inadequately, are magnetism and electronic spectra.

inadvertent adj 偶然の syn *accidental, unintentional*

▶ Some of trace gases in the atmosphere absorb and retain solar energy and may eventually cause inadvertent climate change with catastrophic consequence.

inadvertently adv 偶然に syn *accidentally, by chance*

▶ The first stable, carbon free radical to be reported was the triphenylmethyl radical prepared inadvertently by Gomberg in 1900.

inapplicable adj 応用できない syn *improper, inappropriate*

▶ Away from this region, the crystal field predictions are sometimes right and sometimes wrong, and often inapplicable.

inappreciable adj わずかな syn *negligible, sparse*

▶ If we propose to replace the actual ion, as far as its effects on the space outside it are concerned, by a sphere of equivalent electrostatic radius, we must clearly take a radius outside which the electron density is inappreciable.

inappreciably adv わずかに

▶ Calcium oxalate is inappreciably soluble in water and dilute acetic acid containing dissolved ammonium oxalate but readily soluble in mineral acids.

inappropriate adj 不適当な syn *improper, incompatible, unfit*

▶ The double layer models used for aqueous solutions are inappropriate in ionic liquids.

inborn adj 先天性の syn *inherited, native*

▶ Phenylketonuria is due to an inborn error of phenylalanine metabolism.

incalculable adj 予知できない syn *immeasurable, uncountable*

▶ The economic consequences of electrochemical reactions at surfaces are almost incalculable.

incandescence n 白熱 syn *glow, illumination*

▶ Phosphorus reacts spontaneously with O_2 and the halogens at room temperature, the mixtures rapidly reaching incandescence.

incandescent adj 白熱光を発する syn *glowing, luminous*

▶ Like other high-melting solids, lime becomes incandescent when heated to near its mp.

incapable adj できなくて syn *unable, unfit*

▶ Molecular orbital theory in even its extended forms is quite incapable of predicting geometries for higher-coordination polyhedra.

incidence n **1** 入射

▶ If the molecular beam of metal is directed on to the specimen at an oblique incidence, the metal adheres to the sides of the projection that is accessible to the metal atoms.

2 発生 syn *occurrence*

▶ Sufficient work has been carried out to confirm the incidence of both S_E1 and S_E2 mechanisms.

incident adj 入射する

▶ In neutron diffraction, an incident beam of neutrons is scattered by the atoms of a crystal.

incidental adj 偶然の syn *accidental, fortuitous*

▶ The intense integration of inorganic and organic elements is not incidental but essential for life.

incident light n 入射光

▶ Index of refraction of a substance varies with the wavelength of the incident light, temperature, and pressure.

incidentally adv 偶発的に syn *accidentally, by chance*

▶ Incidentally, the previous calculation that the activation energy of the second step of the thermal reaction was 17.6 kcal is in excellent agreement with the observed overall activation energy of the photochemical reaction, 17.6 kcal.

incineration n 灰化

▶ Disposal of solid and liquid organic waste materials by burning at temperatures from 1200 to 1500 ℃ is called incineration.

incipient adj 発端の syn *initial*

▶ In the hydrolysis of an ester, the initial step is the attack of the strong nucleophile, OH^-, on the incipient carbonium ion of the ester.

inclination n 傾き syn *angle, slope*

▶ All crystals may be classified as belonging to six different crystal systems, distinguished by the relative lengths and angular inclinations of the assumed crystallographic axes.

incline vt **1** 気にさせる syn *lean, tend*

▶ This observation inclines us to view with considerable skepticism the postulated tetramer.

2 傾ける syn *tilt*

▶ The planes of the benzene rings in the two crystallographically independent molecules are inclined at an angle of 12°4' to each other.

inclined adj 傾斜した syn *tilting*

▶ Barium sulfide is easily roasted back to sulfate by heating the inclined crucible with a small flame at the back so that there is ready access of air into the crucible.

include vt 含む syn *embody, incorporate*

▶ Other oxidizing agents which may be used include, for example, iodine and arsenic pentaoxide.

including prep …を含めて 補足的説明を加える非制限的用法の場合はコンマで区切る. syn *inclusive, through*

▶ All molecules, including those without a permanent dipole, attract each other.

inclusion n **1** 算入 syn *incorporation, involvement*

▶ The inclusion of the ionic terms in the wave function is actually a quantitative application of the resonance concept.

2 介在物

▶ Any substance completely enclosed by another is known as an inclusion, no matter what its nature or origin may be. Accordingly, inclusions may be gaseous, liquid, or solid.

inclusion compound n 包接化合物 syn *clathrate compound*

▶ Numerous X-ray studies of α-cyclodextrin with various guests reveal that both cage-type and channel-type crystalline inclusion compounds are formed.

inclusive adj すべてを含んだ syn *broad, comprehensive, general, wide*

▶ All metals of the first transition series from titanium to nickel inclusive are now known to from neutral bis(π-cyclopentadienyl) complexes, and those of vanadium, chromium, cobalt, and nickel are isomorphous with ferrocene.

incoherent scattering n 非干渉性散乱

▶ The measured cross sections of the neutrons in a transmission experiment arise from the neutron absorption, from inelastic scattering caused by the lattice vibrations of the solid, and from incoherent elastic scattering.

incoming adj 入射の syn *incident*

▶ Exposing crystalline solids to X-rays yields diffraction patterns because the atoms in the crystal scatter the incoming radiation, and interference occurs among the many resulting waves..

incomparably adv 比較にならないほど syn *exceedingly, far, highly*

▶ Electron beams have a wavelength incomparably smaller than that of light.

incompatible adj 相容れない syn *conflicting, contradictory, opposed*

▶ This bonds, of course, can only exist at an interface, since metal atoms and cations are mutually incompatible within a solid lattice.

incomplete adj 不完全な syn *defective, deficient, imperfect*

▶ Other examples, in which only incomplete series of alkyl substituents were examined, showed similar trends.

incompressible adj 非圧縮性の

▶ Instead of being incompressible, ions are probably quite elastic, by virtue of flexibility in the outer sphere of influence of an ion while the inner core remains unchanged.

inconceivable adj 想像も及ばない syn *incredible, unthinkable*

▶ One of the delights of exploratory synthesis is the discovery of new compounds, reactions, structures, etc., which are not only unexpected but also inconceivable on the basis of past experience and intuition.

incongruent adj 調和しない syn *discordant, inconsistent*

▶ The formation of most of the new phases from metal and melt involves very incongruent processes since the compositions of metal-saturated MX_3 melts usually correspond to only a small amount of reduction.

incongruent melting n 分解溶融

▶ In contrast to congruent melting, incongruent melting occurs for each of the compounds in the Au–K system shown in Figure 1.

incongruently adv 不調和に ☞ congruently

▶ The published MCl–$CrCl_2$ phase diagrams show that the K_2CrCl_4 and Rb_2CrCl_4 phases are congruently melting, while Cs_2 $CrCl_4$ melts incongruently.

inconsistency n 矛盾 syn *confusion, differ-*

ence, discrepancy, paradox

▶ Just as in the cases with entropies, the convention of assigning zero heat capacity to aqueous hydrogen ion can lead to thermodynamic inconsistencies.

inconsistent adj 矛盾する syn *contradictory, discordant, improper*

▶ Although the amino acids are commonly shown as containing an amino group and a carboxylic group, certain properties are inconsistent with this structure.

inconspicuous adj 目立たない syn *insignificant, modest*

▶ The variation of the heat of dimerization with increasing size of the hydrocarbon substituent of the lower alkanoic acids is relatively inconspicuous.

inconvenience n 不都合 syn *disadvantage, discomfort, nuisance, trouble*

▶ An inconvenience of homogenous hydrogenation has been difficulty of separating the catalyst from the product once reaction is over.

inconvenient adj 不便な syn *awkward, cumbersome, troublesome*

▶ The electrolytic oxidation of bromate was found to be rather inconvenient, and a more practical synthetic method was sought.

incorporate vt 取り入れる syn *combine, embody, include*

▶ The concentrations of pure solids and pure liquids are always incorporated into the equilibrium constant.

incorporation n 編入 syn *inclusion*

▶ It was clear that incorporation of this reaction sequence into a general strategy for polyether construction would provide additional synthetic diversity and flexibility.

incorrect adj 間違った syn *erroneous, false, wrong*

▶ It is often stated that the voltage of a cell is determined by the difference between the Fermi levels of the two electrodes; this is incorrect.

incorrectly adv 間違って syn *badly*

▶ We imagined, incorrectly as it turns out, that the achiral thiepine would be easy to synthesize and serve as a useful model for reactions of its chiral counterpart.

increase n **1** 増加 syn *expansion, extension, growth*

▶ The lowering of surface tension of water is parallel to the increase in molecular weight of a fatty acid within the homologous series.

vi **2** 増大する syn *develop, expand, grow*

▶ The solvent power increases with increasing boiling point; for example, ethanol dissolves about twice as much of a given solute as does methanol.

3 増大させる syn *enlarge, expand, extend*

▶ One way to increase the conductivity could be to add a small inorganic cation such as Li^+ that could have increased mobility compared to the large organic cation.

increasing adj ますます増える syn *continuing, progressive*

▶ Increasing attention is being devoted to the heat capacity of organic solids at temperatures below the boiling point of helium.

increasingly adv ますます syn *progressively, successfully*

▶ Increasingly accurate determinations of interatomic distances have shown that these may differ by amounts much greater than the experimental error without any obvious chemical explanation.

increment n 増分 syn *addition, gain, increase*

▶ The uniformity in packing density is responsible for the approximate atom increment additivity of the zero-point volume of many nonionic compounds.

incremental adj 増分の

▶ Each incremental increase in gas pressure produces a larger increase in the amount of gas adsorbed, up to the limit of a pressure equal to the vapor pressure of the material being adsorbed, at which pressure the adsorption isotherm ascends vertically as condensation occurs.

incubate vt 保温する

▶ A tissue slice is incubated at 37 ℃ with

glucose 6-phosphate in a suitable buffer solution.

incubation n 保温

▶ In all cases ignition is preceded by a period of incubation.

incur vt 招く syn *arouse, invite, provoke*

▷ The section of *cyclo*-S_{20} molecule has the smallest dihedral angles thereby incurring increased repulsion between adjacent nonbonding lone pairs of electrons.

incursion n 侵入 syn *attack*

▶ There has been some tendency to overemphasize hyperconjugation and to neglect the possible incursion of other factors which might be significant in any interpretation of the experimental observations.

indebted adj 負うところがある syn *obligated*

▶ We are indebted to London for the foundations of a satisfactory theory of the interaction between nonpolar molecules.

indebtedness n 恩義 syn *obligation*

▶ The author wishes to indicate his indebtedness to several of his colleagues who have read parts of the manuscript and made valuable comments.

indeed adv **1** 実際に syn *really*

▶ Indeed, a number of theoretical studies suggest that anionic gold species could be important in the overall reaction scheme.

2 本当に syn *certainly, definitely, surely, undoubtedly*

▶ In many of these compounds, the M−O distances are very short indeed.

indefinite adj **1** 不明確な syn *imprecise, uncertain*

▶ The concept of chain length is of great importance but somewhat indefinite.

2 無限の syn *immeasurable, limitless, unlimited*

▶ Each SiO_4 tetrahedron can share two oxygens to form single chains of indefinite length.

indefinitely adv 無期限に

▶ Superconductivity is associated with zero resistance to the flow of electric current, implying that the current could, in principle, flow indefinitely.

indefiniteness n 不明確さ syn *ambiguity*

▶ The calculations provide a convenient opportunity to estimate the uncertainties introduced into the trimer determination by the indefiniteness of the equation of state of carbon tetrachloride.

indentation n くぼみ syn *depression*

▶ The hardness number for a given sample is usually determined by very empirical methods, such as the scratch test or the indentation caused by dropping a weight on the sample.

indented adj でこぼこのある syn *rough, uneven*

▶ Another factor to be borne in mind is that excessive shrinkage occurs on cooling, causing a deeply indented meniscus to form.

independent adj **1** 独立の syn *individualistic, separate*

▶ The independent optical data can be provided in the form of refractive index.

2 無関係な syn *unrelated*

▶ There are certain resemblances among metal ions that can be discussed in terms of oxidation state but which are relatively independent of electron configuration.

independently adv **1** 別々に syn *separately*

▶ Because benzene and water are immiscible, the two liquids independently exert pressures against the common external pressure.

2 依存しないで syn *apart from, freely*

▶ On withdrawing the vapor phase, the pressure remains constant so long as any of the dissociating compound is present, independently of the degree of the decomposition.

independent vatiable n 独立変数

▶ The independent variables in a one-component system are limited to temperature and pressure because the composition is fixed.

inderminancy n 不確定性

▶ For an electron confined to a region the size of an atom, the indeterminancy in momentum is about 1×10^{-24} kg m s^{-1}, but as the mass of the electron is 0.9×10^{-30} kg this corresponds to an indeterminancy in the velocity of as much as 1×10^{6} m s^{-1}.

indeterminate adj 不確定の syn *ambiguous, uncertain, vague*
▶ The total number of organic compounds is indeterminate, but some 6,000,000 have been identified and named.

inderminately adv 不確定に
▶ In determining relative strengths of the stronger acids such as HCl or H_2SO_4 in an aqueous system, the equilibrium constants are all found to be indeterminately large.

index n 指数 複数形は indices
▶ Some researchers consider that the relatively flat low-index planes are more suitable for adsorption of the cinchona modifier.

indicate vt 1 表す syn *designate, specify*
▶ A_i indicates an atom A at an interstitial site, V_B a vacancy at a B site.
2 示す syn *evidence, manifest, reveal, show*
▷ Many disaccharides have reducing properties indicating that one of the sugar residues has the easily opened hemiacetal function.
3 望ましい syn *recommend, require*
▶ Platinum is indicated for the electrodes since it is an inert material and will take no chemical part in the reaction.

indication n 1 複数形で，兆候 syn *evidence, data*
▶ All indications are that an alkyl carbonium ion once dissociated from its accompanying anion is planar.
2 表示 syn *hint, implication, suggestion*
▶ Absorption at a frequency that is characteristic of a particular group can then be taken as an indication of the presence of that group in the compound being studied.

indicative adj indicative of …を示して syn *indicating, signifying, suggesting*
▶ A sharp melting point is usually indicative of the high purity of a substance.

indicator n 1 指示薬 ☞ acid-base indicator, adsorption indicator
▶ The indicators used in acidimetry and alkalimetry are dyestuffs which are of one color in acid solutions and another color in basic solutions.
2 指標 syn *clue, key, sign*
▶ These rules are to be used as indicators of hydrogen-bond preferences of a particular functional group, whether in the solid state or in solution, in the absence of other competing forces.

indifferent adj 無関係な syn *tolerable*
▶ Nonionic detergents have the practical advantage of being indifferent to hard water and to salts, acids, and bases.

indigestion n 消化不良
▶ Aluminum seems to do no harm to healthy tissue, and the hydroxide is often used to treat indigestion by lowering acidity of the stomach.

indirect adj 間接的な syn *secondary, subordinate, subsidiary*
▶ The evidence supporting the belief that simple carbonyl compounds of the types just mentioned are in equilibrium with small amounts of their respective enols is somewhat indirect, but it seems to be entirely conclusive.

indirectly adv 間接的に
▶ Not all organic halides react satisfactorily with lithium, but often the desired lithium compound can be prepared indirectly by an exchange reaction with a readily available lithium compound such as phenyllithium.

indiscernible adj 識別できない syn *indistinguishable*
▶ The C≡C absorption is sometimes of such low intensity that it may be indiscernible.

indiscriminate adj 無差別な syn *careless*
▶ Carbon tetrachloride is very commonly employed as a cleaning solvent, although its considerable toxicity entails some hazard in indiscriminate use.

indiscriminately adv 無差別に syn *randomly*
▶ Alkanes are comparatively unreactive, and the reactions they do undergo take place more or less indiscriminately over the molecule to yield complex mixtures.

indispensable adj 不可欠な syn *essential*
▶ Electron microscopy is now and has been for the past decade an indispensable tool for the materials scientist and solid-state chemist.

indistinct adj 不明瞭な syn *obscure, vague*
▶ The electrons surrounding the atomic nucleus are not confined to a hard shell but rather have a smoothly varying distribution, meaning that the edge of an atom is indistinct.

indistinguishability n 区別が付かないこと syn *identity, sameness*
▶ The factor $1/N!$ appears because of the indistinguishability of the molecules.

indistinguishable adj 区別が付かない syn *alike, identical, undifferentiated*
▶ The exchange energy has no counterpart in classical mechanics and arises in the quantum-mechanical treatment as a result of the fact that the electrons, being indistinguishable, can be interchanges.

indistinguishably adv 区別なく
▶ Delocalization of the electrons indistinguishably over all six centers, as in benzene, is expected to correspond to a more stable electron distribution than any in which the electrons are considered to be localized in pairs between adjacent carbons.

individual n **1** 個人 syn *person*
▶ Most work in physics before World War II was done by individuals working alone or perhaps with a few others.
2 個々のもの syn *one*
▶ To apply statistics to any problem, we must have a great many individuals whose average properties we are interested in.
adj **3** 個々の syn *distinct, particular*
▶ The synthetic control on the scale of individual atoms and molecules will play a crucial role in the creation of new materials and devices.

individualistic adj 個性の強い syn *independent, individual, unique*
▶ Complexity is helpful because it makes the NMR spectra more individualistic and better suited as fingerprints for characterization of organic molecules.

individuality n 個性 syn *distinctiveness, identity, originality*
▶ The atoms retain at least some of their individuality when combined in a molecule.

individually adv 個々に syn *one by one, separately*
▶ To study nanocrystals individually and in detail, special techniques are employed for their formation, separation, and detection.

induce vt 引き起こす syn *cause, generate, produce*
▶ Experimental evidence verifies that extraction of small quantities of lithium from $LiCoO_2$ does indeed induce metallic behavior.

induced dipole n 誘起双極子 ☞ permanent dipole
▶ The symmetry of coordination in the crystal will tend to reduce the magnitude of the induced dipole.

induced heterogeneity n 誘起不均一性
▶ A second matter to consider is to what extent the kinetic behavior is affected by interactions between adsorbed molecules, that is, by induced heterogeneity.

induced magnetic field n 誘起磁場
▶ In circulating, electrons generate secondary magnetic fields: induced magnetic fields.

induced polarization n 誘起分極
▶ With ferroelectrics, the simple linear relation between the induced polarization and applied voltage does not hold.

induction n 誘導 syn *initiation*
▶ Conceivably, induction of the chlorination of methane by near-ultraviolet light could occur through activation of either methane or chlorine or both.

induction period n 誘導期
▶ When the rhodium precatalyst was added to a toluene solution of amine-borane, black particles appeared after an induction period.

inductive adj 誘起の
▶ Where effects are essentially inductive, as for all meta- and some para-substituents, σ values seem to be true constants, leading to good linear relations for many different reactions and equilibria.

inductive effect n 誘起効果
▶ The inductive effect depends upon the

intrinsic tendency of a substituent to release or withdraw electrons, acting either through the molecular chain or through space.

inductively adv 誘起的に

▶ Carbon-carbon cleavage, a key step in the haloform mechanism, is made possible by the presence of three electronegative halogen atoms which inductively stabilize the developing carbanion.

industrial adj 工業の syn *technical*

▶ The partial oxidation of propene to acrolein is the start of important industrial processes.

ineffective adj 効果的でない syn *inadequate, useless, vain*

▶ It is of interest that, whereas silica gel alone is relatively ineffective in separating mixtures of closely related aromatic hydrocarbons, silica gel containing 5% 2,4,7-trinitrofluorenone provides a clean separation.

ineffectively adv 効果的でなく

▶ The effective nuclear charge increases steadily since electrons added to the valence shell shield each other very ineffectively.

ineffectual adj 無力な syn *inoperative, unproductive, useless*

▶ Chromatography becomes more difficult to apply as molecular size grows and becomes ineffectual in separating macromolecules and colloidal particles in the range 0.01 to 1 micron in diameter.

inefficient adj 非能率な syn *wasteful*

▶ Most of the modern methods of generating electricity are inefficient.

inelastic adj **1** 伸縮性のない syn *inflexible, rigid*

▶ Gutta percha is stiff, hard, and inelastic when cold.

2 非弾性の（運動エネルギーの損失がある）

▶ Almost every encounter between a particle of an aerosol and a surface or another particle is inelastic and results in permanent adhesion.

inelastic collision n 非弾性衝突

▶ In an inelastic collision, the internal state of one or both molecules is changed but the molecules themselves remain the same.

inequality n 不等式

▶ This inequality shows that these spontaneous processes must lead to an increase in entropy of the universe.

inert adj **1** 不活性な syn *inactive, unreactive*

▶ The electrode material can, in principle, be any inert metal.

2 不活発な syn *unresponsive*

▶ An alkyl substituent is spectroscopically rather inert.

inert atmosphere n 不活性雰囲気

▶ Under an inert atmosphere, the phase changes are reversible; thus, the phase behavior may be termed enantiotropic.

inert complex n 反応不活性錯体

▶ Metal ions of the d^3, d^6, and d^8 systems are among the more inert complexes.

inertness n 不活性 syn *inactivity*

▶ In contrast to the relative inertness of solid Si to gaseous and liquid reagents, molten Si is an extremely reactive material.

inert pair effect n 不活性電子対効果

▶ Lead is a soft, dense metal with toxic compounds in which it often exhibits the inert pair effect by retaining its $6s^2$ electrons.

inert solvent n 不活性溶媒

▶ A solution of benzophenone in an inert solvent like dry ether or benzene becomes intensely blue if it is treated with metallic sodium or with sodium amalgam.

inevitable adj 必然的な syn *necessary, unavoidable*

▶ A low interfacial tension, a high viscosity, or the inevitable electric charges are each insufficient factors to stabilize an emulsion.

inevitably adv 必然的に syn *necessarily*

▶ The reason that enthalpies of adsorption must be negative is that the adsorption process inevitably involves a decrease in entropy.

inexpensive adj 費用のかからない syn *cheap, economical, reasonable*

▶ Anhydrous sodium sulfate is a neutral drying agent, is inexpensive, and has a high water-absorption capacity.

infect vt 感染させる syn *affect, attack*

▶ Some antiviral drugs block viruses from binding to the surface of cells and infecting cells; others block viral replication after the virus has infected body cells.

infected adj 感染した syn *diseased*

▶ Plant viral particles can be obtained in gram scales from 1 kg of infected leaf material within 2 to 4 weeks.

infection n 感染 syn *disease*

▶ Humane and scientific considerations gave the problem of bacterial infections enormous significance for medicine.

infectious disease adj 感染症

▶ There are the antibiotics which, after isolation, purification, determination of structure, and manufacture, have been widely used in medicine to control infectious diseases.

infer vt 推論する syn *conclude, deduce, derive, surmise*

▶ The presence of liquid among reacting solids can be inferred from the phase diagram for the system, if it is available.

inference n **1** 推論 syn *assumption, conclusion, deduction*

▶ The significance of this inference should be treated with some caution.

2 by inference 推論して

▶ Nonbonding electron pairs which are by inference stereochemically active will be considered quasi-ligands.

infinite adj 無限の syn *endless, limitless, unlimited, vast*

▶ Fibrous sulfur consists of infinite chains of S atoms arranged in parallel helices whose axes are arranged on a close-packed net 463 pm apart.

infinite dilution n 無限希釈

▶ The dissociation is a reversible reaction, and all electrolytes may be considered to be completely ionized at infinite dilution.

infinitely adv 無限に syn *extremely, vastly*

▶ Nonideal behavior is to be expected for all but infinitely dilute electrolyte solutions.

infinitesimal adj 無限小の syn *little, minute, slight, tiny*

▶ The direction of change can be reversed by an infinitesimal change in the external pressure.

infinitesimally adv 無限小に

▶ Metallic behavior is associated with partly filled bands in which it is possible for a large number of electrons to move easily into infinitesimally higher energy states within the band.

infinity n 無限大 syn *endlessness*

▶ The reciprocal of infinity gives us zero as the Miller index for that intercept.

inflame vi 燃え盛る syn *fire, ignite*

▶ Rb_9O_2 inflames with H_2O and melts incongruently at 40.2° to give Rb_2O and Rb.

inflammable adj 可燃性の syn *combustible, flammable*

▶ Aluminum tetrahydroborate is spontaneously inflammable unless the air is completely dry.

inflammation n 炎症 syn *infection, irritation*

▶ Because of its very low solubility in water, uric acid precipitates in the joints, and the sharp, needlelike crystals cause inflammation.

influence n **1** 影響 syn *effect, impact*

▶ A nitro group usually has a rather strong influence on the properties and reactions of other substituents on an aromatic ring.

2 under the influence of …の影響を受けて

▶ Aromatic aldehydes under the influence of strong aqueous or alcoholic alkali undergo simultaneous oxidation and reduction yielding the alcohol and corresponding carboxylate salt.

vt **3** 影響を及ぼす syn *affect, change, impress, modify*

▶ The activity coefficients of strong electrolytes varies at different concentrations and is influenced greatly by the presence of other ions.

influent adj 流れ込む ☞ effluent

▶ In most real process, the effluent solution from an ion-exchange column is not in equilibrium with the influent solution.

influential adj 大きな影響を及ぼす syn *classical, important, leading*

▶ Heck's cobalt-carbonyl-mediated heterocycle carbonylation is among the most influential early publications in this field.

influx n 流入

▶ The macroporous structure of the material would facilitate the influx of cells and blood vessels to the site of the bone implant.

information n 情報 syn *data, fact, knowledge*

▶ In many cases, optical fibers are replacing metal wire, owing to the much greater density of information that can be carried.

informative adj 有益な syn *illuminating, instructive, revealing*

▶ One of the techniques used to provide less accurate, but informative, surveys of thermal properties of materials is differential thermal analysis.

infrared n 赤外線

▶ The hydrogen-oxygen bond of a hydroxyl group gives a characteristic absorption in the infrared and, as we might expect, this absorption is considerably influenced by hydrogen bonding.

infrared active adj 赤外活性の ☞ infrared inactive

▶ Any molecular vibration that results in a change of dipole moment is infrared active.

infrared inactive adj 赤外不活性の ☞ infrared active

▶ Centrosymmetric vibrational modes are IR inactive.

infrared radiation n 赤外線

▶ Any vibrationally excited HCl molecules produced may be detected by observing the infrared radiation which they emit.

infrared spectroscopy n 赤外分光法

▶ Infrared spectroscopy is sensitive to the presence of various nonplanar conformers and to some types of intermolecular order.

infrared spectrum n 赤外スペクトル

▶ Of all the properties of an organic compound, the one that gives the most information about the compound's structure is its infrared spectrum.

infrequently adv **1** まれに syn *rarely, seldom*

▶ A similar scheme is possible for anion substitution but is not considered further because anion substitution occurs rather infrequently in solid solutions.

2 not infrequently しばしば

▶ Not infrequently, a material which seems to consist of a single pure substance cannot be adequately described by means of any individual structure.

infusible adj 不融性の ☞ fusible

▶ Zinc oxide is a white infusible powder, which becomes yellow when heated but turns white again on cooling.

ingenious adj 巧妙な syn *clever, inventive, skillful*

▶ They made an ingenious experiment to decide between the two alternatives.

ingenuity n 巧妙さ syn *art, skill*

▶ Much ingenuity has been expended in designing appropriate syntheses, but no new principles emerge.

ingest vt 摂取する syn *consume, take, use*

▶ If we ingest less nitrogen than we require for normal growth and tissue repair, then we use the nitrogen that is stored in muscle proteins.

ingestion n 摂取

▶ All inorganic compounds of mercury are highly toxic by ingestion, inhalation, and skin absorption.

ingredient n 成分 syn *component, constituent, element*

▶ The trinitro derivatives of *m*-*t*-butyltoluene and 1,3-dimethyl-5-*t*-butybenzene possess musklike odors and have been used as ingredients of cheap perfumes and soaps.

inhalation n 吸入

▶ Nitrobenzene should be handled with care since it has a considerable toxicity both by inhalation and by absorption through the skin.

inhale vt 吸い込む syn *breathe in, suck*

▶ Nitrobenzene is toxic if gotten on the skin or if the vapor is inhaled.

inherent adj 固有の syn *basic, essential, intrinsic*

▶ The approximations and assumptions inherent in the model were least likely to apply to the lightest members of the series.

inherently adv 本質的に syn *normally*

▶ A system that has a large number of accessible states is inherently more likely to exist than one with a more limited accessibility.

inherit vt 遺伝する syn *acquire, succeed to*

▷ Sucrase deficiency comes about either by profound damage to the small intestine or more usually from an inherited defect.

inhibit vt 抑制する syn *discourage, impede, obstruct, supress*

▶ High concentrations of Na^+ cations in a test solution interfere with pH measurements since they will inhibit the H_3O^+-Na^+ exchange reaction.

inhibited adj 抑制された syn *restrained*

▶ There are light-induced, radical-trap inhibited reactions of bromine and hydrogen bromide with alkenes.

inhibition n 阻害 syn *blockage, constraint, interference, prevention*

▶ The slowing down of a reaction upon addition of a constituent to the reaction mixture is called inhibition.

inhibitor n 阻害剤

▶ An inhibitor is a substance that diminishes the rate of a chemical reaction. In contrast to a catalyst, an inhibitor may be and frequently is consumed during the course of reaction.

inhibitory adj 阻害の syn *preventive*

▶ Binding of ADP or AMP allosterically activates phosphofructokinase, but high concentrations of ATP are inhibitory.

inhomogeneity n 不均等性 ☞ homogeneity

▶ If the rate at which a macroscopic inhomogeneity in a system relaxes toward uniformity is small compared to the rate of establishment of equilibrium between the molecular degrees of freedom, local values of thermodynamic function can be defined.

initial adj 初期の syn *incipient*

▶ Initial experiments have been promising, but much development is still needed.

initially adv 初めに syn *originally*

▶ We may consider the addition of progressively more and more copper to initially pure gold.

initial velocity n 初速度

▶ The initial velocity is used because enzyme degradation may occur during the reaction or inhibition by reaction product may arise, thus yielding results that can be difficult to interpret.

initiate vt 始める syn *begin, commence, give rise to*

▶ Acidification of aqueous solutions of the yellow, tetrahedral chromate ion, CrO_4^{2-}, initiates a series of labile equilibria involving the formation of the orange-red dichromate ion, $Cr_2O_7^{2-}$.

initiation n 開始 syn *beginning, onset, start*

▶ The initiation step involves rupture of the weakest C–C bond.

initiator n 開始剤

▶ Polymerization requires the presence of a small amount of an initiator. Among the commonest of these initiators are peroxides, which function by breaking down to form a free radical.

inject vt **1** 注入する syn *drive, force, introduce*

▶ The monomer solution was first heated to boiling, and then the initiator solution was injected into the flask using a syringe.

2 導入する syn *bring in, introduce*

▶ The dangers of space-charge formation and electrode effects inject uncertainties into many experiments of recombination process.

injection n **1** 注射 syn *shot*

▶ Passive immunity may be acquired either artificially by an injection of antibodies from other organism or naturally by passage of some antibodies from a mother to her child across the placenta.

2 注入

▶ Much of the injection of CO_2 into the atmosphere must cease during the next few decades.

injure vt 損傷する syn *damage, harm, impair*

▶ Long contact with hot carbon injures

platinum, forming some carbide.

injurious adj 有害な syn *bad, detrimental, harmful*

▶ Metallic mercury vaporizes in air and is injurious to humans.

inlet n 注入口 ☞ outlet

▶ The lithium wire is collected under heavy paraffin oil and cut into uniform 12-cm lengths which are placed in a glass tube having a stopcock at the bottom and a nitrogen inlet near the rubber stopper.

inner adj 内部の syn *central, interior*

▶ The effective nuclear charge is somewhat less than the true nuclear charge because of screening by inner electrons and can be estimated empirically.

inner-core electron n 内殻電子 syn *inner-shell electron*

▶ In a metal such as aluminum, the inner-core electrons are localized in discrete atomic orbitals on the individual aluminum atoms.

innermost adj 最も内側の ☞ outermost

▶ There are specific examples, such as $Cr(H_2O)_6^{3+}$, in which the innermost layer is quite firmly bonded, so that the water molecules exchange with other solvent molecules only very slowly.

inner salt n 分子内塩

▶ Aminosulfonic acids exist as inner salts; they are soluble in alkali but not in acid.

inner shell n 内殻 複合形容詞として使用 ☞ outer shell

▶ For inner-shell transitions, much larger energies are involved and fall in the X-ray region.

inner-shell ionization potential n 内殻イオン化ポテンシャル

▶ While it is largely true that inner-shell ionization potentials are unaffected by the nature of the molecule, it is not wholly true, and small but significant shifts can be detected which depend on the local environment of the atoms.

inner sphere n 内圏 ☞ outer sphere

▶ The species bonded to the metal ion in the coordination sphere are said to be in the inner sphere.

inner-sphere mechanism n 内圏機構 ☞ outer-sphere mechanism

▶ Many oxidation–reduction reactions have been shown to occur by a ligand-bridging or inner-sphere mechanism in which substitution of the coordination shell of one of the metal ions occurs.

inner transition element n 内部遷移元素

▶ Those elements falling in the d block are classified as transition elements, and those falling in the f block, inner transition elements.

innocuous adj 無害の syn *harmless, safe*

▶ Residual solutions containing alkali cyanides should be rendered innocuous by the addition of an excess of sodium hypochlorite before being washed down the main drain of the laboratory with a liberal supply of water.

innovation n 新機軸 syn *alteration*

▶ Rapid development of our understanding of genetics and molecular biology has led to exciting new innovations and applications in biotechnology.

innovative adj 革新的な syn *imaginative, revolutionary*

▶ Innovative methods for recapture and recycle of valuable substances, such as metals that would otherwise contribute to water pollution, are needed.

innumerable adj 無数の syn *immeasurable, limitless, many, unlimited*

▶ The electric charges produce the innumerable colloidal electrolytes, many of whose properties are shared in lesser degree by all charged colloids.

inoculate vt 接種する syn *inject*

▶ If no crystals separate, the solution is inoculated with a trace of seed.

inoculation n 接種 syn *injection, vaccination*

▶ Changes have been induced by pricking the crystal with a pin or by inoculation with reaction product.

in-plane adj 面内の ☞ out-of-plane

▶ Carbon–hydrogen bending in alkenes and aromatic rings is both in-plane and out-of-plane, and of these the latter kind is more useful for spectroscopic analysis.

input n 入力

▶ For rotation to occur, the necessary energy input would be roughly equal to the difference in energy between a double and a single carbon–carbon bond.

inquire vt 問う syn *examine, investigate, study*

▶ Our desire to probe these connections and to inquire whether the extended solids of the group 2 dihalides remember the gas-phase structures has motivated a comprehensive examination of structural preferences in the monomers, dimers, and solids.

inquiry n 調査 syn *exploration, investigation, research*

▶ The thermodynamic development can be made without any inquiry into the molecular reason for the behavior of the system.

inscribe vt 内接させる ☞ circumscribe

▶ The perovskite structure features corner-shared TiO_6 octahedra, with Ca^{2+} ions at the corners of cubic unit cells in which the octahedra are inscribed, leading to the formula $CaTiO_3$.

inscribed circle n 内接円

▶ In recent years, benzene has often been represented as a hexagon with an inscribed circle.

insect n 昆虫

▶ Irradiation is effective in inhibiting sprouting and preventing insect infestation of stored grains.

insecticide n 殺虫剤

▶ DDT is a hard insecticide, one which is stable to environmental destructive forces and exists for months to years, not only in the soil and water systems but in plant and animal life as well.

insensitive adj 鈍感で syn *insensible*

▶ The $S_2 \rightarrow S_1$ radiationless process is relatively insensitive to medium.

insensitivity n 鈍感 syn *tolerance*

▶ The insensitivity to external reagents is the best evidence that the nature of these reactions is primarily one of dissociation.

inseparable adj 切り離すことができない syn *close, indistinguishable*

▶ The words sulfur and environmental pollution have now become inseparable, at least in the mind of the public.

分離できない

▶ Pyridine **5** was inseparable from the starting enone but was generated in excellent conversion and moderate enantioselectivity.

insert vt 挿入する syn *intercalate, introduce, place in, put in*

▶ Intercalation or insertion solids may be defined as hosts into which atoms may be inserted or removed without a major disruption of the structure.

insertion n 挿入 syn *addition, insert*

▶ The incorporation of carbonyl functionality into organic molecules through the transition-metal-mediated insertion of carbon monoxide is among the most important and synthetically useful catalytic transformations.

insertion reaction n 挿入反応

▶ Carbon disulfide is rather more reactive than CO_2 in forming complexes and in undergoing insertion reactions.

inset n 挿入図

▶ The inset in Figure 1 shows a detailed view of agglomerated and coagulated primary particles with a primary particle size of about 100–300 nm.

inside n **1** 内部 syn *interior*

▶ The inside of a cell is crowded with molecules, and these are in continuous vigorous motion that is driven by thermal energy.

adj **2** 内部の syn *interior, internal*

▶ The simplest conjugated cyclic polyolefin that could have a strainless planar ring containing trans double bonds, except for interference between the inside hydrogens, is [10]annulene.

prep **3** …の内部に syn *within*

▶ If the extraordinary ray is the slow ray,

relative to the ordinary ray, then the ellipsoid will lie inside the sphere, and the crystal will be said to be optically positive.

insight n 洞察 syn *imagination, intuition, understanding*

▶ Observation of catalysts under in situ conditions has provided insights into the nature of the species present on the catalyst surface and their relative concentrations.

insignificant adj ささいな syn *minor, negligible, trivial, unimportant*

▶ Breaking the π bond requires about 70 kcal of energy; at room temperature an insignificant proportion of collisions possess this necessary energy and, hence, the rate of this interconversion is extremely small.

in situ adv 本来の位置で syn *in proper position, set up*

▶ In situ electrochemical ESR techniques have been developed in which the electrode may be located completely within the ESR cavity, allowing short-lived radicals to be observed.

insofar as conj …する限りにおいて syn *in so far as*

▶ Ionic bonding is nondirectional insofar as it is purely electrostatic.

insolubilize vt 不溶化する ☞ solubilize

▶ One process for removing phosphates involves addition of a metal ion source to waste effluents so as to insolubilize dissolved phosphates, after which the particles are agglomerated by anionic polymers.

insolubility n 不溶性 ☞ solubility

▶ Its total insolubility limited the investigation to IR and electronic spectra and made any significant characterization impossible.

insoluble adj 不溶性の ☞ soluble

▶ The solubility products of magnesium hydroxide, carbonate, and phosphate are so small that these substances may be regarded as insoluble.

inspect vt 検査する syn *check, examine, investigate, study*

▷ The strength of the metal–metal bonding can be examined by inspecting a series of MO_2 oxides.

inspection n 検査 syn *check, examination*

▶ Inspection of a ball-and-stick model of an ethyl derivative in the staggered configuration shows that one of the CH_3 hydrogens should not have exactly the same chemical shifts as the other two.

inspire vt 触発する syn *activate, prompt, stimulate*

▶ The ease of fabricating high-quality photonic crystals from porous silicon and its biocompatibility have inspired the conception of various biosensing schemes using this material.

inspired adj 見事な syn *imaginative*

▶ Schrödinger's equation is an inspired postulate that permits the calculation of the wave function.

instability n 不安定性 syn *fluctuation*

▶ A combination of aggressive chemical reactivity and/or thermal instability prevents the determination of physical properties in some instances.

install vt 取り付ける syn *establish, introduce, place, position*

▷ The rate of effusion may be determined by measuring the loss in weight of sample installed in a small cell.

instance n 実例 syn *case, event, example*

▶ Pyridine, in its rare instances of electrophilic substitution, affords 3-substituted pyridines.

instant n 瞬間 syn *moment*

▶ The magnetic field that a secondary proton feels at a particular instant is slightly increased or slightly decreased by the spin of the neighboring tertiary proton.

instantaneous adj 瞬間的な syn *immediate*

▶ The rates of formation of a charge-transfer complex and decomposition into the components are so high that the reactions appear to be instantaneous by normal techniques.

instantaneously adv 直ちに syn *immediately, promptly, rapidly*

▶ In some cases, such as hydrogen with nickel, the hydrogen is instantaneously chemisorbed as single atoms at all temperatures from that of liquid air up.

instantly adv 直ちに syn *instantaneously, quickly*

▶ The quality of crystals cannot be instantly recognized in situ; examination with a microscope is necessary.

instead adv **1** その代わりに syn *rather*

▶ To take an example from group 2, $CaCO_3 \cdot 6H_2O$ does not crystallize with the expected hexaaquocalcium (II) cation surrounded by several carbonate anions; instead there are discrete $Ca^{2+}CO_3^{2-}$ ion pairs surrounded by water molecules, and the calcium has eight oxygen neighbors, six from water molecules and two from the carbonate.

2 instead of (prep) …の代わりに syn *for*

▶ We should find a state function, instead of path-dependent quantities such as heat or work.

instruction n 指令 syn *direction, guideline, information*

▶ A gene is a sequence of nucleotides that provides cells with the instructions for the synthesis of a specific protein or particular type of RNA.

instructive adj 有益な syn *informative, helpful, revealing*

▶ Although thermodynamic arguments can be developed without any reference to the existence and behavior of atoms and molecules, it is, nevertheless, instructive to interpret the arguments in terms of molecular structure.

instrument n 器具 syn *apparatus, device, tool*

▶ Important in the development of fluorescence as a tool has been the development of a wide array of instruments that take advantage of fluorescence.

instrumental adj **1** 助けになって syn *contributory, helpful, useful*

▶ Lysosomes are instrumental in intracellular digestion and the digestion of material from outside the cell.

2 器具の

▶ The instrumental problem was solved by Dr. K., who designed a Langmuir balance capable of detecting film pressures as small as 0.005 dyn/cm on the floating barrier.

instrumentation n 器具類

▶ There is actually very little instrumentation involved in carrying out an extraction with a supercritical fluid.

insufficient adj **1** 不適当な syn *inadequate, unsatisfactory*

▶ Thus, a model based on purely ionic bonds holding C^{4+} and $4Cl^-$, for example, is insufficient in these cases, and important covalent bonding must exist.

2 不十分な syn *meager, scant, scanty, scarce*

▶ When methane and ethylene are exploded with a quantity of oxygen insufficient for complete combustion, the carbon burns to carbon monoxide, and hydrogen is set free.

insufficiently adv 不十分に syn *badly*

▶ All of these preparative methods provide an anion which is insufficiently basic to react with the cation.

insulate vt **1** 隔離する syn *isolate, segregate, separate*

▶ Separation of functional groups by two or more carbons of a saturated hydrocarbon chain usually serves to insulate them from pronounced interaction.

2 断熱する

▶ In the type of calorimeter shown in Figure 2, the bomb is surrounded by a water jacket which is insulated as much as possible from the surroundings.

insulator n **1** 絶縁体 ☞ electric insulator, thermal insulator

▶ Insulators differ from semiconductors only in the magnitude of their conductivity, which is usually several orders of magnitude lower.

2 絶縁体

▶ Enhancers and silencers are separated from promoters by DNA sequences called insulators that prevent inappropriate activation of other genes in the same region.

insulin n インスリン

▶ The lack of insulin results in an elevated

blood glucose concentration that can cause a number of health problems.

insurance n 備え syn *assurance, guarantee*

▶ The amount of sodium bromide is arbitrarily set at 1.2 times the theory as an insurance measure.

insure vt 確実にする syn *ensure, warrant*

▷ The semipermeable nature is essential to insuring that a separation takes place.

intact adj 損なわれていない syn *complete, entire, perfect, whole*

▶ Several important reactions of alcohols involve only the oxygen–hydrogen bond and leave the carbon–oxygen bond intact.

intake n 摂取

▶ Intake of even moderate amounts of ethanol together with barbiturates or similar drugs is extremely dangerous and may even be fatal.

integral n **1** 積分

▶ Some of the integrals in the Schrödinger equation are replaced by empirical parameters from reference compounds or other information chemists already have at hand.

adj **2** 不可欠な syn *essential, fundamental*

▶ In the past few years, a wide variety of polymers containing metal atoms as an integral part of the polymeric structure have been prepared.

3 積分の

▶ The heat of solution tabulated in various reference manuals is the integral molar enthalpy of solution at infinite dilution.

4 整数の

▶ Charge-transfer complexes usually involve simple integral ratios of the components.

integrate vt **1** 融合する syn *combine, consolidate, unite*

▶ Numerous stringent criteria have to be satisfied for a biomaterial to be acceptable as a bone implant, including the ability to integrate into bone and not cause any side effects.

2 積分する

▷ The relative heights of the principal steps of the integrated spectrum show the three groups of lines at 9.8, 2.4, and 1.0 ppm to arise from one, two, and three hydrogens, respectively.

integrated adj 組織化された syn *organized, systematic*

▶ The hallmark of an integrated chemical, physical, or biological system is the assembly of components into a particular architecture that performs a certain function.

integrated circuit n 集積回路

▶ Microlithography is an essential step in producing electronic integrated circuits.

integration n **1** 積分

▶ This constant of integration may be evaluated using the boundary condition that $x = 0$ when $t = 0$.

2 取り込み syn *inclusion, incorporation*

▶ Integration of cholesterol into a lipid bilayer renders it less leaky to small solutes and ions.

integrity n 完全な状態 syn *completeness*

▶ Raman spectroscopy of the precursor (cetyltrimethylammonium)$_4Ge_4S_{10}$ proved that the cluster integrity remained intact.

intellectual adj 知的な syn *mental*

▶ The particular method of mentally building molecules that we are learning to use is artificial: it is a purely intellectual process involving imaginary overlap of imaginary orbitals.

intellectually adv 知的に

▶ A very effective and intellectually satisfying way of going beyond information connected with a particular occasion is to create a model.

intelligent adj 気のきいた syn *bright, clever, smart*

▶ To understand the reason for Markownikoff's rule, it is desirable to discuss some of the principles that are important to intelligent prediction of the course of an organic reaction.

intelligible adj 理解できる syn *clear, plain, understandable*

▶ Their seemingly unusual stoichiometries only became intelligible after structural elucidation by X-ray crystallography.

intend vt 意図する syn *aim, contemplate, design, plan, propose*

▶ In all cases, the ligands behaved almost

exactly as was intended; the unpredictability arose in the behavior of the metal center.

intended adj 所期の syn *intentional*

▶ The entire, fully intended reaction sequence was designed.

intense adj **1** 強烈な syn *deep, profound, strong*

▶ The intense colors exhibited by sodium tungsten bronzes are one of their most characteristic features.

2 熱烈な syn *enthusiatic*

▶ The host-guest relationship is of intense interest to the organic chemist.

intensely adv **1** 強烈に syn *deeply, profoundly*

▶ Acid solutions of titanium sulfate are colored intensely yellow when treated with hydrogen peroxide.

2 懸命に syn *actively*

▶ Heterogeneous catalysis by acidic zeolites is one of the most intensely investigated topics of chemistry.

intensification n 増大 syn *enlargement*

▶ Gibson and Loeffler were the first to report on the intensification of the visible spectrum of a mixture of an electron donor and an electron acceptor when the system is compressed.

intensify vi **1** 強くなる syn *increase*

▶ The effect intensifies in the order Pt > Pd > Ni because of the greater spatial extent of the orbitals with larger *n* value.

vt **2** 強める syn *strengthen*

▶ Just as electron release by alkyl groups disperses the positive charge and stabilizes a carbocation, so electron withdrawal by halogens intensifies the positive charge and destabilizes the carbocation.

intensity n **1** 集中 syn *concentration, focus*

▶ Not all issues merit the same intensity of study.

2 強度 syn *strength*

▶ Intensities are taken as peak heights, unless very accurate work is being done.

intensity borrowing n 強度借用

▶ A subject of considerable interest in spectroscopy is intensity borrowing between electronic and vibrational states.

intensive adj 集中的な syn *concentrated, intensified*

▶ The nonmetal-to-metal transition in condensed systems has been the subject of intensive experimental and theoretical investigations.

intensively adv 集中的に syn *thoroughly*

▶ β-Lactams have been rather intensively investigated following the discovery that the important antibiotic penicillin G, produced by fermentation with *Pencillium notatum*, possesses a β-lactam ring.

intensive property n 示強性 ☞ extensive property

▶ Properties which are characteristic of the individual phases of the system and are independent of the amounts of the phases are known as intensive properties.

intensive variable n 示強性変数

▶ Volume and heat capacity are typical examples of extensive variables, whereas temperature, pressure, viscosity, concentration, and molar heat capacity are examples of intensive variables.

intent n 意図 syn *aim, intention, plan, purpose*

▶ The intent is to show that absolute electronegativity and hardness can also be used in a novel and rational way.

intention n 意図 syn *aim, intent, plan, purpose*

▶ The intention was that Cu(I) would take on a tetrahedrally disposed set of four nitrile donors from separate tetranitrile component [$C(C_6H_4CN)_4$] which would in turn bind four Cu(I) centers at the corners of a tetrahedron.

intentional adj 意図的な syn *deliberate, planned, purposeful*

▶ We need to differentiate between what we may term intentional and unintentional seeding.

intentionally adv 意図的に syn *deliberately, purposefully*

▶ A common method to stabilize porous silicon from degradation is to intentionally grow an oxide layer on the surface to slow further oxidation and avoid rapid changes in the refractive index during sensing.

interact vi 相互に作用する syn *cooperate*

▶ Both aliphatic and aromatic Grignard reagents interact with molecular oxygen in ether solution with formation of the hydroxyl derivatives.

interaction n 相互作用 syn *cooperation, synergy*

▶ Basicity involves interaction with a proton; nucleophilic power and leaving ability involve interactions with carbon.

interaction energy n 相互作用エネルギー

▶ Even in HCl, the calculations indicate that the orientation effect constitutes less than 20% of the total interaction energy at room temperature.

interatomic adj 原子間の

▶ When atoms are brought together to form a crystal, the interatomic interactions spread out the degenerate atomic energy levels into bands.

interatomic distance n 原子間距離

▶ Within each helix, the interatomic distance S−S is 206.6 pm, the bond angle S−S−S is 106.0°, and the dihedral angle S−S−S−S is 85.3°.

intercalate vt 挿入する syn *insert*

▶ Many substances can be intercalated between the layers of graphite, but one of the longest known and best studied is potassium.

intercalation n インターカレーション

▶ A range of composition has been reported, but the intercalation isotherm of the iron(III) chloride on graphite at 350 ℃ shows two compositions, $C_{12}FeCl_3$ and C_7FeCl_3.

intercalation compound n 層間化合物

▶ Lithium is rarely in the form of Li atoms in the intercalation compounds of interest here but instead exists as Li^+ ions along with their charge-compensating electrons.

intercellular adj 細胞間の

▶ The extracellular concentrations of Na^+ and K^+ are 140 mM and 5 mM, respectively, but the required intercellular concentrations are 5 mM and 150 mM in humans.

intercept n 1 切片 syn *cut*

▶ A plot of ln *v* against ln *c* will give a straight line if the reaction is of simple order; the slope is the order *n*, and the intercept is ln *k*.

vt 2 捕捉する syn *catch, trap*

▶ The intermediate bromonium ion was not intercepted by the ether oxygen during formation of the intermediate bromohydrin.

interchange n 1 交換 syn *exchange*

▶ In an interchange reaction between fluorides and metal iodides in aqueous solution, soft metal ions prefer I^- and hard metal ions prefer F^-.

vt 2 交換する syn *exchange*

▷ If the number of interchanged atoms is large and especially if it increases significantly as a function of temperature, this takes us into the realm of order-disorder phenomena.

interchangeable adj 交換可能な syn *equal, equivalent, identical*

▶ In the rock salt and zincblende structures, the cation and anion positions are interchangeable, and it is immaterial whether the origin coincides with an anion or a cation.

interchangeably adv 交換可能で

▶ The terms melting point and freezing point are often used interchangeably, depending on whether the substance is being heated or cooled.

interconnect vt 相互に連結させる syn *link, unite*

▶ These chains are interconnected so as to produce a structure with one-dimensional channels.

interconnected adj 1 相互に連結した syn *linked, tied*

▶ Zeolite structures are characterized by the presence of tunnels or systems of interconnected cavities.

2 相互に関連した syn *related*

▶ Internal pressure, electrostriction, and apparent molar volumes are interconnected phenomena.

interconversion n 相互変換

▶ The interconversion of two conformers of a

compound requires only rotations about single bonds.

interconvert vt 相互変換する

▶ The simple boranes rapidly interconvert with each other at very moderate temperatures.

interconvertibility n 相互変換性

▶ The interconvertibility of FeO, Fe_3O_4, and γ-Fe_2O_3 arises because of their structural similarity.

interconvertible adj 相互変換のできる

▶ If the activation energy is small, the different geometrical arrangements are so readily interconvertible that the corresponding substances cannot be separated.

intercrystallization n 相互晶出

▶ $KAl(SO_4)_2 \cdot 12H_2O$ intercrystallizes with many other alums, and such intercrystallizations seem to be submicroscopic lamellar intergrowths.

interdependence n 相互依存 syn *link, relation*

▶ Because of the complexity and interdependence of the many parameters that govern platelet activation and coagulation, few, if any, satisfactory thromboresistant materials have been reported.

interdependent adj 相互依存の syn *related*

▶ The bonding in Zeise's salt is considered to arise from two interdependent components.

interdiffusion n 相互拡散

▶ Because there is a lattice mismatch of ca. 14 percent between CdTe and GaAs, interdiffusion between layers of GaAs and CdTe becomes a problem at high temperatures.

interdisciplinary adj 学際的な

▶ Materials science is an interdisciplinary field, and its challenges are increasingly being met by teams whose members have backgrounds that span the scientific and engineering disciplines.

interelectronic adj 電子間の

▶ A number of the small fluorine atoms can often fit around a central atom, attracting negative charge to themselves, and so reducing interelectronic repulsion.

interest n **1** 関心 syn *attention, concern, curiosity, notice*

▶ Thermochemical data are not always available for systems of interest especially when these interests include the synthesis of new compounds.

vt **2** 関心をもたせる syn *attract, fascinate*

▶ We shall be interested to see how this competition is affected by such factors as the structure of the halide and the particular nucleophilic reagent used.

interesting adj 興味を起こさせる syn *attractive, fascinating*

▶ In brief, molecular polyhedra of high coordination number have scope, interesting chemistry, and significant research challenge to those interested in synthesis and structural chemistry.

interestingly adv 興味深いことに

▶ Interestingly, a significant part of the hydrolysis comprises hydroxide ion attack of the chelate structure at a ligand position.

interface n **1** 界面 syn *surface*

▶ When an aqueous solution of a diamine and an organic solution of a diacid chloride are brought together, they react to form a thin film of a polyamide at the interface between the aqueous and organic phases.

2 接点 syn *boundary*

▶ A tempting question has been to probe the interface or boundary between neutral and ionic crystals.

interface reaction n 界面反応 ☞ heterogeneous reaction

▶ The evidence so far available points to the surface reaction proceeded by a different mechanism from the interface reaction.

interfacial adj 界面の

▶ In a heterogeneous material, charge may accumulate at the interfaces between phases, giving rise to a so-called interfacial polarization.

interfacial angle adj 面角

▶ All the corresponding interfacial angles on all crystals of the same species are constant;

that is, they all have the same value.

interfacial energy n 界面エネルギー syn *surface energy*

▶ Both these factors influence the interfacial energy such that there may be an ensemble of nucleation profiles that are crystallographically specific and dependent on the nature of the substrate.

interfacially adv 界面で

▶ The interfacially polymerized polyaniline tends to disperse into the body of the aqueous solution after the reaction.

interfacial polymerization n 界面重合

▶ It turns out that both interfacial polymerization and rapidly mixing reactions are very robust synthesis that can be performed under a wide range of conditions.

interfacial tension n 界面張力

▶ The interfacial tension between two liquids is often reduced practically to zero by the addition of a surface active material.

interfere vi **1** 妨げる syn *hamper, hinder, impede, obstruct*

▶ The presence of the reaction product from the ethylmagnesium bromide does not interfere with isolation of the desired product if this is of significantly higher molecular weight.

2 干渉する syn *approach, touch*

▶ Inspection of models shows that a strainless structure can only be achieved with two or more of the double bonds in trans configurations and then with a large enough ring that the inside hydrogens do not interfere with one another.

interference n **1** 干渉 syn *intervention*

▶ The broadening of the diffraction lines results from interference effects that lead to only incomplete constructive and destructive interference.

2 妨害 syn *barrier, block, impediment*

▶ Evidently the steric interference between the carboxyl and phenyl groups in *trans*-α-phenylcinnamic acid is greater than that between the two phenyl groups in *cis*-α-phenylcinnamic acid.

interference fringe n 干渉縞

▶ Thicknesses above 1μ can be measured under the polarizing microscope by counting interference fringes between crossed polaroids in highly convergent light.

interfering adj 干渉する syn *intrusive*

▶ After the separation from interfering elements has been accomplished, it is best to determine the titanium volumetrically or colorimetrically.

interferometric adj 干渉測定の

▶ The concept of interferometric biosensing takes advantage of the difference in the phase of light reflected at the top surface and base of a thin film.

interferon n インターフェロン

▶ Interferons are signal proteins, responsible for inducing the synthesis of proteins that inhibit the replication of viruses.

interhalogen compound n ハロゲン間化合物

▶ The halogens combine exothermically with each other to form interhalogen compounds.

interionic adj イオン間の

▶ The powerful interionic forces in solid sodium chloride are overcome only at a very high temperature.

interionic attraction n イオン間引力 ☞ electrostatic force

▶ Strong interionic electrostatic attractions cause lithium hydride to have high melting and boiling points like those of sodium chloride, lithium fluoride, etc.

interior n **1** 内部 syn *inside*

▶ To determine the course of a reaction, it is necessary to make quantum-mechanical calculations corresponding to a number of points in the interior of the diagram.

adj **2** 内部の syn *inner, inside, internal*

▶ By altering the charge of the interior surface of the virus cage from cationic to anionic, the resulting particles favored the encapsulation of cationic species.

interlamellar adj 層間の syn *interlayer*

▶ It is possible to replace the Na^+, Ca^{2+}, and

other alkali-metal or alkaline-earth-metal ions originally present in the interlamellar spaces separating the negatively charged aluminosilicate sheets by interlamellar, hydrated protons by the simple expedient of ion exchange with dilute mineral acid.

interlink vt 相互に連結する

▷ Solutions of Grignard reagents can contain a variety of chemical species interlinked by mobile equilibria.

interlock vi からみあう syn *fit together, match*

▶ The ability of the ligands on successive molecules to interlock may be a prerequisite to the formation of the Ni(dmg)$_2$-type structure.

interlocking n からみあい

▶ Bulky side groups are particularly unfavorable since, by interlocking, they may provide serious barriers to the motion of one chain relative to another.

intermediate n **1** 中間体 syn *intermediate product*

▶ The term intermediate implies that a species has a detectable lifetime.

adj **2** 中間の syn *halfway, middle*

▶ The negative charge on the HOF fluorine is intermediate between that on the fluorines in OF_2 and that on the fluorine in HF.

intermediate complex n 中間複合物

▶ In halogenation experiments, the kinetics suggests that an intermediate complex is formed between the hydrocarbon and a bromine molecule and that a second molecule of the halogen is necessary to break the halogen–halogen link.

intermediate product n 中間生成物 syn *intermediate*

▶ The degradation of the amino acid apparently occurs through an intermediate product of condensation of the acid with ninhydrin.

intermetallic compound n 金属間化合物

▶ Coordination numbers of nine or more are found in AB_2 and AB_3 salts or polymeric lattices and in intermetallic compounds.

intermolecular adj 分子間の

▶ While the critical volume would be the simple measure related to intermolecular distance, it is unsatisfactory from the empirical viewpoint.

intermolecular force n 分子間力 ☞ van der Waals force

▶ All these particles have in common that they are held together by weak intermolecular forces which are in general weaker than typical chemical bonds by about two orders of magnitude.

intermolecular interaction n 分子間相互作用

▶ Hydrogen bonds are not the only type of intermolecular interactions that are useful for directing molecular self-assembly.

intermolecularly adv 分子間に

▶ The rearrangement of β- to α-*p*-nitrophenol involves a negligible change in molecular volume, no change in crystal symmetry, no alteration in the order of attachment of hydrogen bonds, and a seemingly relatively minor alteration in the arrangement in the chains of intermolecularly hydrogen-bonded nitrophenol molecules.

internal adj 内部の syn *inner, inside, interior*

▶ The external regularities exhibited by crystals naturally lead us to expect a similar regularity in the internal arrangement.

internal conversion n 内部変換

▶ Internal conversion between excited singlet states is extremely efficient and results in the lowest excited singlet being the only state which fluoresces.

internal degree of freedom n 内部自由度

▶ An isolated molecule has internal degrees of freedom and a given amount of energy.

internal energy n 内部エネルギー

▶ Excitation of a molecule, with consequent increase of internal energy, may occur either by absorption of a photon or by collision with another molecule.

internally adv 内部で

▶ These solutions contain a solvent-dependent rapidly established equilibrium mixture of a yellow compound doubly hydrogen-bonded internally to the ester carbonyl group, a white species doubly hydrogen-bonded internally to chlorine, and a hybrid species with an internal hydrogen bond to oxygen at one end and chlorine at the other.

internal oxidation-reduction n 内部酸化還元

▶ Chromium(II) nitrate has not been prepared because of internal oxidation-reduction.

internal pressure n 内圧力 ☞ external pressure

▶ In an ideal gas the internal pressure is zero, and in real gases the term is usually small compared to the external pressure.

internal rotation n 分子内回転

▶ The extent of the freedom of rotation is commonly expressed as the height of the potential energy barrier to internal rotation or equivalent conformation changes.

international adj 国際的な syn *global, universal, worldwide*

▶ The overwhelming need for lightweight and compact sources of portable electricity has resulted in a massive international effort into the development of radically new rechargeable batteries.

internuclear distance n 核間距離 syn *internuclear separation*

▶ If a beam of electrons is accelerated through about 40 kV and passed through a sample of gas, diffraction effects will occur since the wavelength of the electrons is of the order of magnitude of internuclear distances in molecules.

interpenetrate vt 互いに貫入する

▶ The spaces within a single network were so large that they were occupied by a second identical network that interpenetrated the first and vice versa.

interpenetration n 1 相互貫入

▶ Interpenetration of electron clouds of molecules in contact is resisted by the Pauli exclusion principle; it is assisted by the attractive forces between the molecules.

2 浸透

▶ Interpenetration can be attractive for certain gas storage applications, since it can increase the available surface area per unit volume.

interplanar spacing n 面間隔 syn *lattice spacing*

▶ In any case where molecular shapes and interplanar spacings of starting material and product suggest minimal disruption of the crystal, the products separate as microcrystallites with no net orientation with respect to the parent crystal.

interplay n 相互作用 syn *connection, link*

▶ The modes of thermal decomposition of halates and their complex oxidation-reduction chemistry reflect the interplay of both thermodynamic and kinetic factors.

interpolate vt 補間法を行う ☞ extrapolate

▶ We take the values for Ca^{2+}, Mn^{2+}, and Zn^{2+} as reference points and interpolate values for other dipositive ions in the absence of crystal field effects.

interpolation n 補間

▶ For Ni^{2+} the total heat of hydration is 2180 kJ, and the value obtained by interpolation from those for Mn^{2+} and Zn^{2+} is 2060 kJ.

interpret vt 解釈する syn *clarify, elucidate, explain*

▶ The NMR spectra of alkanes are reasonably characteristic but difficult to interpret, because the chemical shifts between the various kinds of protons are rather small.

interpretation n 解釈 syn *clarification, elucidation, explanation*

▶ Many types of organic compounds exhibit characteristic mass spectral behavior, a knowledge of which is useful in the interpretation of their spectra.

interpretive adj 解釈上の syn *explanatory*

▶ As far as simple-chain alkanes are concerned, measurements were made many years ago by Stewart and co-workers, some interpretive work on the Stewart data was undertaken by Warren.

interrelation n 相互関係 syn *connection*

▶ The brief outline of the principal structural features of the known mesophases of rod- and disc-shaped molecules shows that a good descriptive understanding of the structures and of their interrelations now exists.

interrelationship n 相互関係 syn *connection*

▶ We illustrate how we have applied coordination chemistry as a useful guide to the syntheses of new solid-state materials and to the interrelationships among them.

interrupt vt 中断する syn *cease, end, halt, stop, terminate*

▶ In thin sections of crystals, cleavage is seen as one or more sets of more or less narrow, straight and parallel cracks, which are continuous or interrupted, according to the quality of the cleavage.

interruption n 中断 syn *interval, rest*

▶ In photochemical reactions, a periodic interruption of the beam of light can lead to interesting kinetic consequences.

intersect vi 交差する syn *cross*

▶ Two planes intersect in a straight line.

intersection n 交点 syn *joint, junction*

▶ In the two-phase region, the composition of the solid solution is given by the intersection of the tie line at the temperature of interest.

interstellar adj 星間の

▶ A major uncertainty in studying reactions in the interstellar medium is the degree to which the heavy elements have been depleted in the gas by freezing onto the cold grains or by incorporation into the refractory grain material in the atmospheres of cool stars.

interstellar molecule n 星間分子

▶ Organic compounds found in some meteorites might provide clues to those compounds that were precursors to life, to interstellar molecules, and to cometary material.

interstice n 隙間 ☞ hole, interstitial position, interstitial site, void

▶ Perovskite may be regarded as a framework structure constructed from corner-sharing octahedra (TiO_6) with Sr^{2+} ions placed in twelve-coordinate interstices.

interstitial n **1** 格子間原子 ☞ interstitial atom

▶ Because the product of the concentrations of cation vacancies and Ag^+ interstitials is constant, the interstitial Ag^+ concentration must decrease with increasing Cd^{2+} concentration

adj **2** 格子間の

▶ This can only mean that the oxygen excess is being accommodated by virtue of vacancies in the iron lattice, rather than as interstitial oxygen ions.

interstitial atom n 格子間原子

▶ Another interesting aspect of octahedral clusters is that several of them contain a central interstitial atom.

interstitial carbide n 侵入型炭化物

▶ Interstitial carbides are infusible, extremely hard, refractory materials that retain many of the characteristic properties of metals.

interstitial compound n 侵入型化合物

▶ As the nuclear charge increases and the atomic radius decreases across the first transition series, we find fewer interstitial compounds.

interstitial site n 割り込み位置 syn *interstitial position*

▶ In any close-packed structure, the number of octahedral interstitial sites is the same as the number of packing sites and the number of tetrahedral interstitial sites is twice as great.

interstitial solid solution n 侵入型固溶体

▶ In interstitial solid solutions, the introduced species occupies a site that is normally empty in the crystal structure.

intersystem crossing n 項間交差

▶ Where the radiationless transition occurs between states of different multiplicity, Kasha has chosen the term intersystem crossing.

interval n **1** 時間の間隔 syn *period*

▶ If the velocity is not constant, an average value over a small interval is found by dividing the small distance δs traveled in that instant by the very small time interval δt: $v=\delta s/\delta t$.

2 温度の間隔

▶ As far as temperature intervals are concerned, the degree Celsius is the same as the kelvin.

3 at intervals ここかしこに

▶ If the silicon atoms in $Al_2(OH)_4Si_2O_5$ are replaced at intervals by aluminum, then the layers take on a negative charge and unipositive cations must be taken into the lattice, holding it together.

intervalence n **混合原子価** ☞ mixed-valence

▶ Intervalence charge-transfer transitions between isolated clusters of Fe^{2+} and Fe^{3+} ions are responsible for the blue coloration of aquamarine.

intervalence band n **混合原子価吸収帯**

▶ It has not yet proved possible to irradiate within the contour of the intervalence band of the Creutz-Taube ion owing to the lack of exciting lines in this low-wavenumber region.

intervening adj **介在する** syn *intermediate*

▶ In general, the magnitude of the spin-spin splitting effect of one proton on another proton depends on the number and kind of intervening chemical bonds and on the spatial relations between the groups.

intervention n **介入** syn *interference*

▶ This intervention of extraneous nucleophilic agents in the reaction mixture is evidence against a one-step mechanism.

intimate adj **1 詳細な** syn *detailed, thorough*

▶ There is evidence that gives us a rather intimate view of just how the rearrangement step takes place.

2 密接な syn *close, thorough*

▶ The superconducting compound, $YBa_2Cu_3O_{7-x}$, is readily prepared by heating an intimate mixture of yttrium oxide, barium peroxide, and cupper (II) oxide at approximately 930 ℃ for 10-12 hours.

intimately adv **密接に** syn *well*

▶ This condition would be that the compounds are formed by the mechanism of direct union of the elements themselves, while intimately mixed in a very finely divided active state.

into prep **1** …の中に

▶ Silver and mercury (II) cations are precipitated by the introduction of hydrogen sulfide into solutions which are 0.3 M in acid.

2 …になる

▶ Acetone on treatment with metallic reducing agents is converted to a considerable extent into the product of bimolecular reduction, pinacol.

in toto adv **全部で** syn *completely, entirely, wholly*

▶ We must make every effort to enlarge our interests and responsibilities in belief that inorganic chemistry in toto will be strengthened by its engagement in new areas of scientific enquiry.

intracellular adj **細胞内の**

▶ When the diet is rich in cholesterol, intracellular cholesterol concentrations increase in the liver and the biosynthesis of cholesterol is suppressed.

intracolumnar adj **カラム内の**

▶ Intracolumnar stacking distances shorter than those in the solid state indicated that their hexagonal mesophases resulted from a crankshaft or north-south-east-west arrangement.

intracrystalline adj **結晶内の**

▶ The higher melting point and lower solubility of a para isomer are only special examples of the general effect of molecular symmetry on intracrystalline forces.

intradimer adj **二量体内の**

▶ The effect of compression on the intradimer interactions dominates over the van der Waals intermolecular interactions.

intralayer adj **層内の**

▶ The intralayer C-C distance in graphite is twice the covalent radius of aromatic carbon, and the interlayer C-C distance is twice the van der Waals radius of carbon.

intramolecular adj **分子内の**

▶ Many of the substances which can form intramolecular hydrogen bonds turn out to be volatile with steam.

intramolecularly adv 分子内で

▶ An X-ray study of the thermally induced nitrite–nitro isomerization and the photochemically induced nitrite–nitro isomerization of Co(III) complexes has shown that both occur intramolecularly by rotation of the NO_2 group in its own plane.

intramolecular vibration n 分子内振動

▶ In deriving the theory of corresponding states, we assumed that the intramolecular vibrations were unaffected by the presence of adjacent molecules.

intravesicular adj 小胞内の

▶ For example, the intravesicular precipitation of Fe(II), Fe(III), and Fe(II)/Fe(III) solutions by base addition gave discrete particles of spherulitic magnetite, goehite, and ferrihydrite, respectively, while the corresponding products in bulk solution were gel-like ferrihydrite, needle-shaped lepidocrocite, and coarse-grained magnetite.

intricate adj 複雑な syn *complex, complicated, involved*

▶ In the present case, the problem is more intricate, but every step is guided by a clear appreciation of the underlying physical ideas.

intriguing adj 興味をそそる syn *interesting*

▶ The nature of the interaction between the lipase and the triglycerides in the micelles is intriguing because the active site of the enzyme must bind to a hydrophobic substrate, while the enzyme as a whole is immersed in an aqueous environment.

intrinsic adj 真性の syn *genuine, proper, real, true*

▶ Except at low temperatures, a highly purified semiconductor often exhibits intrinsic conductivity, as distinguished from the impurity conductivity of less pure specimens.

intrinsically adv 本質的に syn *substantially*

▶ Theories of liquids are intrinsically more complicated than those of gases.

intrinsic membrane protein n 膜内在性タンパク質

▶ Intrinsic or integral membrane proteins can be removed only by treating the membranes with detergents or with organic solvents.

intrinsic protein n 内在タンパク質

▶ Some intrinsic proteins require a fluid environment for their activity, while others operate in a fixed, immobile condition.

intrinsic viscosity n 固有粘度

▶ Most directly related to the nature of the individual solute molecules is the intrinsic viscosity, which has the effect of macromolecule intermolecular interaction removed by the extrapolation to infinite dilution.

intrinsic volume n 固有体積

▶ In fact, the partial molar volumes of very small ions are negative implying that the electrostriction term dominates the intrinsic volume for these ions.

introduce vt **1** 導入する syn *inject, insert, put in*

▶ Many different cations may be introduced into the spinel structure, and several different charge combinations are possible.

2 提出する syn *bring in, present*

▷ Ground-state geometries are most simply rationalized by using nonbonding repulsion model first introduced by Sidgwick and Powell, later extended by Gillespie and Nyholm.

introduction n **1** 挿入 syn *entry*

▶ Introduction of eq 1 into eq 2 gives rise to eq 3.

2 導入 syn *initiation*

▶ The introduction by Sony in 1990 of the world's first commercially successful rechargeable lithium battery represented a revolution in the power source industry.

intron n イントロン

▶ Genes are interrupted by stretches of DNA that do not contain protein coding information, called introns.

intuition n 直感 syn *common sense, insight*

▶ Comparison of the parameters determined for the various donors is in agreement with chemical intuition and experiment.

intuitive adj 直感的な syn *intuitional*

▶ The functionalization of liquid crystals, which is just beginning, represents an intuitive attempt by chemists to exploit molecular self-assembly and organization.

intuitively adv 直感的に

▶ The use of a reduced dielectric constant is intuitively more plausible than that of effective radii.

invade vt 侵入する syn *occupy*

▷ Viruses function by invading the cells of the host and by supplying a genetic pattern that destroys the normal functions of the cells and sets the cellular enzymes to work synthesizing more virus particles.

invader n 侵入者 syn *intruder, outsider*

▶ Antibodies are proteins that form part of the immune response in mammals in the presence of foreign invaders.

invaluable adj 非常に有益な syn *valuable*

▶ Depression of melting point or nondepression is invaluable in the identification of unknowns.

invariably adv 常に syn *universally*

▶ Crystals are invariably imperfect because the presence of defects up to a certain concentration leads to a reduction of free energy.

invariant adj 不変の syn *fixed*

▶ When three phases are present, there is freedom to choose neither pressure and temperature. The system is invariant; there are no degrees of freedom.

invent vt 創案する syn *conceive, create, devise, originate*

▶ Two approaches, interfacial polymerization and rapidly mixed reactions, have been invented to make pure nanofibers by suppressing the secondary growth of polyaniline, which would otherwise produce irregularly shaped particles.

invention n 発明 syn *conception, creation*

▶ The chief requirement of an invention is that it be unobvious to a person having ordinary skill in the art to which the claim pertains and knowing everything that has gone before.

inventive adj 独創的な syn *clever, imaginative, ingenious*

▶ The inventor is the one who contributes the inventive concept in workable detail not the one who demonstrates it or tests it.

inverse n **1** 逆数

▶ The softness is simply the inverse of the hardness.

adj **2** 逆数の syn *reverse*

▶ The inverse sixth power attractive term arises from the interaction of instantaneous dipole moments.

3 in inverse proportion to …と逆比例して

▶ The relative amounts of the phases are in inverse proportion to the distance of the phase lines from the overall composition.

inversely adv 逆比例して

▶ The mean square velocity is inversely proportional to the molecular mass.

inverse spinel n 逆スピネル ☞ normal spinel

▶ In an inverse spinel such as $NiFe_2O_4$, all the A^{2+} and half of the B^{3+} cations are in octahedral holes, and the other half of the B^{3+} cations are in tetrahedral holes.

inversion n **1** 順序の反転

▶ There will be of different magnitudes for hard and soft ions; hence, inversions in the binding order of the halide ions can occur with changes in solvent.

2 配置の反転

▶ In $MnFe_2O_4$, the overall magnetic moment is insensitive to the degree of inversion since both cations, Mn^{2+} and Fe^{3+}, are d^5.

3 逆位

▶ Inversions occur when a breakage in one of the chromosomes occurs, and the segment rotates 180° before it rejoins.

inversion of configuration n 立体配座の反転

▶ Only two examples have been reported where substitution in an optically active complex yields an optically active product with inversion of configuration.

invert vt 反転する syn *reverse, turn*

▶ Displacement of the chloride of active *s*-butyl chloride by chloride ion inverts configuration at the atom undergoing substitution.

invertebrate n 無脊椎動物

▶ Of considerable biochemical interest is the astonishing ability of certain invertebrates to accumulate vanadium in their blood.

investigate vt 研究する syn *analyze, examine, explore, study*

▶ Fluorides of Kr, Xe, and Rn have now been discovered. The Xe compounds have been most thoroughly investigated.

investigation n 研究 syn *examination, exploration, search, study*

▶ By far the most stable square complexes are those of platinum(II), and as a result the syntheses and reactions of these compounds have been the subject of extensive investigations.

investigator n 研究者

▶ The various colors of iodine in different solvents have attracted the attention of investigators over half a century.

invisible adj 目に見えない syn *undetectable*

▶ The propylamine Schiff base melts to a smectic phase with a very narrow temperature range, which is readily observed via DSC but is almost invisible microscopically.

invite vt 招く syn *attract*

▶ Excited-state quenching phenomena have invited a great range of descriptions.

in vitro adv 試験管内で ☞ in vivo

▶ A considerable body of literature from 1910s and 1920s pointed to the antibacterial effects in vitro of azo comounds.

in vivo adj **1** 生体内の ☞ in vitro

▶ The in vivo significance of the aminotransferases is not always understood.

adv **2** 生体内で

▶ In vivo, glycogen is stored as cytoplasmic granules that are associated with a lot of water, this lowers the effective energy yield from a glycogen particle to ca. 6 kJ g^{-1}.

invoke vt **1** 引き合いに出す syn *advance*

▶ There is no need to invoke hyperconjugation in molecules like propylene and propyne, and the changes in C–C bond length are due simply to changes in hybridization of carbon.

2 引き起こす syn *elicit, evoke, inspire*

▷ The literature of chemical physics contains many observations of phenomena that can be explained by elegant theories invoking new concepts.

involatile adj 不揮発性の syn *nonvolatile*

▶ Organosodium and organopotassium compounds are involatile and do not readily dissolve in nonpolar solvents.

involve vt 関係する syn *comprise, contain, include*

▷ Except for autoreactions, in which a molecule or crystal may undergo chemical change as in thermal decomposition, without involving other chemical substances, all reactions of a compound involve the close approach of other substances.

involved adj 複雑な syn *complex, complicated, intricate*

▶ The details of the methods are actually quite involved, but, even so, agreement between theory and experiment is often not good.

involvement n 掛り合い syn *concern, connection, implication*

▶ Copper(III) is generally regarded as uncommon, being very easily reduced, but it has recently received attention because of its apparent involvement in some biological processes.

inward adj 内に向かう syn *interior*

▶ At the surface of a liquid, there is practically no force attracting the surface molecules away from the liquid, and there is, therefore, a net inward attraction on the surface molecules.

iodinate vt ヨウ素化する

▶ IBr almost invariably brominates rather than iodinates aromatic compounds.

iodination n ヨウ素化

▶ One way of achieving direct iodination is to convert molecular iodine to some more active species with an oxidizing agent such as nitric acid.

iodometric titration n ヨウ素還元滴定

▶ The quantitative oxidation of $S_2O_3^{2-}$ by I_2 to form tetrathionate and iodide is the basis for the iodometric titrations in volumetric analysis.

ion association n イオン会合

▶ With salts having ions of different valencies, such as Na_2SO_4, ion association will lead to the formation of species such as $Na^+SO_4^{2-}$.

ion bombardment n イオン衝撃

▶ By plasma standards, microwave-induced plasmas are generally considered cold but can contain enough heat to cause exposed objects to reach temperatures in excess of 1000 ℃ via ion bombardment.

ion channel n イオンチャンネル

▶ While the design and synthesis of novel ionophores is an interesting challenge, models of ion channel activity are of more immediate relevance.

ion exchange n イオン交換

▶ The ion exchange is restricted to near the surface of the glass membrane where it is in contact with the internal or test solutions and where there exists a hydrated zone of the glass.

ion-exchange vt イオン交換する

▶ Part or all of the Na^+ ions in β-aluminas may be ion-exchanged for a variety of other monovalent ions, even though most of these ion-exchanged materials are not thermodynamically stable.

ion-exchange chromatography n イオン交換クロマトグラフィー

▶ Ion-exchange chromatography relies on the electrostatic interaction between a charged protein and a stationary ion-exchange resin particle carrying a charge of the opposite sign.

ion exchanger n イオン交換体 ☞ exchanger

▶ All ion exchangers, whether natural or synthetic, have fixed ionic groups that are balanced by counter ions to maintain electroneutrality.

ion-exchange resin n イオン交換樹脂

▶ The glass functions as an ion-exchange resin, and an equilibrium is established between Na^+ cations in the surface of the glass matrix and hydrogen ions in solution.

ionic adj イオン性の

▶ Molten sodium chloride is an ionic liquid,but a solution of sodium chloride in water is an ionic solution.

ionic activity coefficient n イオン活量係数

▶ The standard state of an electrolytic solute is that of a hypothetical solution at a mean ionic molality of unity where the reference state is so chosen that the mean ionic activity coefficient approaches unity as the concentration of the solute approaches zero.

ionically adv イオンによって

▶ With metal electrodes and ionically conducting electrolytes, interfacial polarization occurs associated with charge build-up at the interface.

ionic atmosphere n イオン雰囲気

▶ An important consequence of the existence of the ionic atmosphere is that the conductivity should depend on the frequency if an alternating potential is applied to the solution.

ionic bond n イオン結合 syn *ionic bonding*

▶ The ionic bond is concerned only with electrostatic attractions and repulsions due to electric charges, permanent electric dipoles, and induced dipoles.

ionic charge n イオン電荷

▶ For reactions between ions the electrostatic effects are very important, and the preexponential factors of such reactions depend on the ionic charges.

ionic conduction n イオン伝導

▶ Ionic conduction often involves the motion of the cations and can occur by a process in which a cation moves from an occupied site into a vacancy.

ionic conductivity n イオン伝導率 ☞ electronic conductivity

▶ It is relatively straightforward to measure the total ionic conductivity of a polymer electrolyte by using standard small-signal ac or four-electrode dc techniques, which enable the bulk resistance to be distinguished from any impedances associated with the electrode–electrolyte interfaces.

ionic conductor n イオン伝導体 ☞ electronic conductor

▶ In electrolytic or ionic conductors, the current is carried by ions, as in solutions of acids, bases, and salts and in many fused compounds.

ionic crystal n イオン結晶

▶ The prime bonding force in ionic crystals is the nearest neighbor cation–anion attractive force, and this force is maximized at a small cation–anion separation.

ionicity n イオン性 syn *ionic character*

▶ The term ionicity has been used to indicate the degree of ionic character in bonds.

ionic lattice n イオン格子

▶ Only in the study of ionic lattices, have we retained the concept of relating properties to the individual ions comprising the crystal.

ionic liquid n イオン液体

▶ Since the discovery of air-sensitive chloroaluminate ionic liquids, subsequent research concentrated on their use in electrochemical applications.

ionic mobility n イオン移動度

▶ The average velocity with which an ion moves toward an electrode under the influence of a potential of 1 volt applied across 1-cm cell is known as the ionic mobility.

ionic product n イオン積

▶ The ionic product of water can be written as

$$K_w = [H_3O^+][OH^-]\ mol^2\ L^{-2}$$

ionic radius n イオン半径 ☞ atomic radius, covalent radius, van der Waals radius

▶ The ionic radius of O^{2-} is assigned the standard value of 140 pm, and all other ionic radii are derived from this.

ionic strength n イオン強度

▶ In all cases, sufficient potassium nitrate was contained in the aqueous silver nitrate solution to maintain an ionic strength of unity.

ionizable adj イオン化しうる

▶ If the sample is heated in phosphorus vapor to achieve uniform incorporation, ionizable electrons from these donor phosphorus atoms can combine with the valence band holes, shifting the Fermi level toward the conduction band.

ionization n イオン化

▶ Ionization is most effective in water because its high dielectric constant lowers the ionic bonding forces in the solute molecules enough to cause separation of their constituent atoms.

ionization constant n イオン化定数

▶ The constant K_a, characteristic of the acid, is called its acid constant or ionization constant.

ionization energy n イオン化エネルギー syn *ionization potential*

▶ The ionization energy is the minimum energy required to remove an electron from a free neutral gaseous atom in its lowest energy or ground state.

ionization equilibrium n 電離平衡

▶ Ionization equilibria of very weak acids must be measured in solvents which are less acidic than water, such as ammonia, ethers, or, in some cases, even hydrocarbons.

ionization isomerism n イオン化異性

▶ Ionization isomerism involves interchange of a ligand anion with an associated anion outside the complex.

ionization potential n イオン化ポテンシャル syn *ionization energy*

▶ The sharp lines in the ethylene spectrum are the first members of a Rydberg progression, which extrapolates to an ionization potential of 10.45 eV.

ionize vi **1** イオン化する

▶ In aqueous solution, H_2SO_4 ionizes as a strong dibasic acid, and both acid and normal salts are known.

vt **2** イオン化する

▶ The small amount of energy needed to ionize these dopants means that at room temperature the amount of thermal energy available is sufficient to ionize a substantial fraction of them.

ionizing power n イオン化能

▶ Reaction by S_N1 is favored by solvents of high ionizing power, that is, by polar protic solvents that help to pull the leaving group out of the molecule.

ionizing radiation n 電離放射線

▶ The term ionizing radiation is restricted to electromagnetic radiation at least as energetic as X-rays and to charged particles of similar energies.

ionizing solvent n イオン化溶媒

▶ The most efficient ionizing solvents will be those that are most effective in solvating both anions and cations.

ionomer n イオノマー

▶ An ionomer resin is a copolymer of ethylene and a vinyl monomer with an acid group, such as methacrylic acid.

ionophore n イオノホア（イオンを運ぶもの）

▶ Crystallization is induced by trapping inorganic phosphate within the vesicles and then loading this inner space with Ca^{2+} ion by using ionophore molecules sited in the membrane.

ionophoric adj イオノホアの

▶ In an attempt to imitate the ionophoric properties of the polyether antibiotics, lipophillic crown ethers bearing carboxylic acid function have been synthesized.

ion pair n イオン対　複合形容詞として使用.

▶ In metal-ammonia solutions, ion-pair formation appears to account for the effect of cation-electron interaction.

ion-selective electrode n イオン選択性電極

▶ The production of ammonium ion is monitored with an ammonium ion-selective electrode, onto which the enzyme is coated.

ion-selective membrane n イオン選択性膜

▶ In an ion-selective field effect transistor, an ion-selective membrane acts as the gate instead of metal, and the electrical potential it exerts on the source-to-drain current depends on the extent to which substrate ions have entered the membrane.

ion source n イオン源

▶ The pressure of the sample in the ion source of a mass spectrometer is usually about 10^{-5} mm, and, under these conditions, buildup of fragments to give significant peaks with m/e greater than M^+ is rare.

ion transport n イオン輸送

▶ Transmembrane ion transport can be studied directly through either a lipid bilayer or a cell membrane.

iridescence n 虹色 syn *opalescence*

▶ When the thickness of the SnO_2 film is similar to the wavelength of visible light, then thin-film interference effects occur and the glass acquires an attractive iridescence.

irradiate vt …を照らす syn *radiate, shed*

▶ If a mixture of dry hydrogen and chlorine is irradiated by visible light, the reaction, $H_2 + Cl_2 \rightarrow 2HCl$, occurs with explosive violence.

irradiation n 照射

▶ Polyethylene is degraded by 1850-Å irradiation but is crosslinked by 2537-Å irradiation.

irreducible representation n 既約表現

▶ An irreducible representation giving a unique set of characters is said to belong to a particular symmetry species.

irregular adj 不規則な syn *changeable, erratic, uncertain, variable*

▶ Although the nature of the change was similar in many crystals, the time scale was characteristically irregular.

irregularity n 不規則 syn *abnormality, inconsistency, peculiarity, twist*

▶ Any irregularity in a crystal, such as a foreign atom in solid solution, a vacant lattice point, or a surface crack, will lead to a modification of the crystal field.

irregularly adv 不規則に syn *randomly, unsystematically*

▶ Examination using scanning and transmission electron microscopy shows that the as-synthesized polyaniline powders are mostly irregularly shaped particles.

irreproducible adj 再現性のない

▶ These results were fairly irreproducible, with reaction times varying between 48 and 84 h at 45 ℃.

irrespective adj **irrespective of** 関わりなく syn *independent of*

▶ Bands arise irrespective of whether the bonding in the extended solid is principally metallic, covalent, or ionic.

irreversible adj 不可逆の

▶ Any finite expansion that occurs in a finite time is irreversible.

irreversible process n 不可逆的過程

▶ Irreversible processes are those for which the forces acting on the boundary are not balanced.

irreversible reaction n 不可逆反応

▶ Dehydrohalogenation is an irreversible reaction, so that orientation is determined by the relative rates of competing reactions.

irreversible wave n 非可逆波

▶ On a platinum electrode, two irreversible waves were noted with cathodic peak potentials at −1.01 and −1.38 V.

irrigation n 灌漑 syn *flushing*

▶ Controlled-release fertilizers have advantages of more uniform supply of nutrient, lower labor costs, and reduced leaching losses in areas of irrigation and high rainfall.

irritating adj 刺激性の syn *painful*

▶ Ozone is a powerful oxidizing agent, is very irritating to the eyes, and causes damage to the respiratory system.

irrotationally adv 非回転的に

▶ Near cations a number of water molecules are irrotationally bound so that orientation of their permanent dipoles cannot make its normal large contribution to the dielectric constant.

island n 島に似たもの

▶ On the Si(100) surface, the growth of Ag islands has been observed by STM.

isobar n 等圧式

▶ The congruent melting curve of a binary compound is continuous through the maximum, and the lower part of the diagram may be considered as two simple eutectic-type isobars set side by side.

isobaric adj 等圧の

▶ Gravimetric measurements are typically performed under isobaric conditions and can provide accurate adsorption-desorption kinetics.

isoelectric focusing electrophoresis n 等電点電気泳動

▶ Isoelectric focusing electrophoresis separates a mixture of proteins based on their different isoelectric points.

isoelectric point n 等電点

▶ The hydrogen ion concentration of the solution in which a particular amino acid does not migrate under the influence of an electric field is called the isoelectric point of that amino acid.

isoelectronic adj 等電子の

▶ Although H_3NBH_3 is isoelectronic with ethane, the structures, bonding, and reactivity of the two compounds are very different.

isoionic point n 等イオン点

▶ At some intermediate pH, the isoionic point, the protein does not migrate in an electric field.

isolability n 単離の可能性

▶ Interconvertibility of stereoisomers is of great practical significance because it limits their isolability.

isolable adj 単離可能な

▶ The distinction between conformational isomers and configurational isomers is that the latter are presumed stable and isolable, while the former are not.

isolate vt 単離する syn *segregate, separate*

▶ From this filtrate, dark red prismatic crystals were isolated by filtration and identified as **1**.

isolated adj **1** 孤立した syn *separate, unrelated*

▶ A system is said to be isolated if neither matter nor heat is permitted to exchange across the boundary.

2 単離した syn *separated*

▶ An isolated sample of pure NH_4Cl contains a

single component, provided that the only phases present are solid NH_4Cl and a gas, even if this gas consists entirely of separate molecules of NH_3 and HCl.

isolated system n 孤立系

▶ An isolated system is one which cannot exchange matter or energy with surroundings.

isolation n 単離 syn *segregation*

▶ The isolation and structural characterization only in the case $NiBr_2$ derivative suggests that the nature of the halide group may have a stabilizing influence for this particular conformation.

isomer n 異性体

▶ The nitrations mentioned give mixtures of ortho and para isomers, but there are usually easy to separate by distillation or crystallization.

isomeric adj 異性体の

▶ Depending upon which hydrogen atom is replaced, any of a number of isomeric products can be formed from a single alkane.

isomerism n 異性

▶ The field of isomerism is subdivided into structural isomerism and stereoisomerism.

isomerization n 異性化

▶ The treatment of sugars with alkali can cause extensive isomerization and even decomposition of the chain.

isomerize vi 異性化する

▶ Azulene is less stable than naphthalene and isomerizes quantitatively on heating above 350 ℃ in the absence of air.

isometric adj 等軸の syn *cubic*

▶ The isometric phase of $KAl(SO_4)_2 \cdot 12H_2O$, called potassium alum, forms octahedral or cubic crystals without cleavage.

isometric crytal n 等軸晶 syn *cubic crystal*

▶ In isometric crystals, the amount of absorption bears no relation to the crystal form.

isometric system n 等軸晶系 syn *cubic system*

▶ Each form of the isometric system completely encloses space and can occur alone.

isomorphous adj 同形の

▶ Dolomite is an isomorphous mixture of calcium and magnesium carbonates.

isomorphous replacement n 同形置換

▶ The replacement of one atom by another without structural change is often observed in inorganic and mineralogical structures and is called isomorphous replacement.

isopolyanion n イソポリ陰イオン ☞ heteropolyanion

▶ As fragments of oxide structures, the isopolyanions contain cubic close-packed arrangements of oxide ions with metal ions in octahedral holes.

isosbestic point n 等吸収点

▶ A wavelength at which the absorption coefficients of two substances are equal is known as the isosbestic point.

isostructural adj 等構造の

▶ Just because two phases are isostructural, it does not follow that they will form solid solutions with each other.

isotactic adj アイソタクチック

▶ Polymerization of styrene with Ziegler catalysts produces isotactic polystyrene, in that all of the phenyl groups are located on one side of the chain or the other.

isotactic polymer n アイソタクチックポリマー

▶ Isotactic polymers show the most crystallinity, while atactic and syndiotactic polymers are largely amorphous.

isotherm n **1** 気体の等温式

▶ At high temperatures, the real and ideal isotherms do not differ very much.

2 吸着等温式 ☞ adsorption isotherm

▶ The various isotherms of the Langmuir type are based on the simplest of assumption; all sites on the surface are assumed to be the same, and there are no interactions between adsorbed molecules.

isothermal adj 等温の

▶ Isothermal systems are of most significance because temperature can then be considered an independent variable.

isotope n 同位体，同位元素

▶ Different isotopes have, by definition, different masses, and because of this their chemical properties are not identical.

isotope effect n 同位体効果

▶ The vapor pressure isotope effect is typically one to two orders of magnitude smaller than chemical equilibrium or kinetic isotope effects.

isotopic adj 同位体の

▶ For ordinary purposes, natural isotopic mixtures known as pure substances are to be treated as single substances.

isotopic abundance n 同位体存在度

▶ It is clear that substantial progress has been made in the last two decades to interpret the mysterious variety of isotopic abundances which comprise the elements used by chemists.

isotopically adv 同位体について

▶ It has been shown by the use of N_2H_4 isotopically enriched in ^{15}N that both the N atoms of each molecules of N_2 originated in the same molecule of N_2H_4.

isotropic adj 等方性の ☞ anisotropic

▶ When the surface representing a particular property of a crystal is spherical, the crystal is said to be isotropic with respect to that property.

isotropically adv 等方的に ☞ anisotropically

▶ The first stage is to determine isotropic temperature factors, which assumes that the atoms are vibrating isotropically.

isotropic liquid n 等方性液体

▶ The Schiff bases formed from propylamine to at least nonylamine pass through viscous fluid anisotropic phases on heating, before clearing to isotropic liquids.

isozyme n イソ酵素

▶ Isozymes are multiple forms of an enzyme that catalyze the same chemical reaction but differ in net charge or catalytic efficiency, or both.

issue n **1** 問題点 syn *point, problem, subject, topic*

▶ A further issue that has been largely overlooked in ionic liquids is the importance of the concentration of metal ions.

2 at issue 問題となっている syn *in dispute, uncertain*

▶ Explanations have to be rational and enable us to understand the events at issue.

3 雑誌の号 syn *number*

▶ Corrections may be published in the forthcoming issue.

it pron 不定詞句，動名詞句，that 節などを代表する形式主語

▶ For many applications, it is necessary to disrupt this layered packing to introduce small guests between the individual sheets.

italicize vt イタリック体で印刷する

▶ If there is possible ambiguity as to which proton or carbon, the assignment refers to the appropriate atom is italicized.

item n 品目 syn *matter, thing*

▶ Use of polyvinyl chloride in thinner items such as films and package coatings is permissible.

iteration n 反復 syn *repetition*

▶ Both the equilibrium constant and molar absorptivity of the dimer are treated as variables and their values determined by iteration.

iterative adj 反復の syn *periodic, recurrent, repetitive*

▶ The first approach that we devised for the rapid iterative preparation of fused cyclic ethers involved sequential ring-closing metathesis of an acyclic enol ether.

itself pron **1** それ自身

▶ In glycolysis, D-glucose is itself broken down into three-carbon compounds.

2 by itself それだけで

▶ By itself, a single characterization test seldom proves that an unknown is one particular kind of compound.

J

jacket n 覆い syn *cover*

▶ In the adiabatic calorimeter, an adiabatic process is achieved by surrounding the inner water jacket by an outer water jacket which by means of a heating coil is maintained at the same temperature as the inner jacket.

jelly n ゼリー

▶ To study the mechanical properties of the jelly, the jelly should be motionless and formed loose from the walls, and a whole lump of it should be investigated.

jet n 噴出口 syn *outlet*

▶ The flask is placed beneath the buret, around the jet of which cotton wool is wrapped to serve as a loose stopper to exclude air from the flask.

job n 仕事 syn *operation, task*

▶ The separation of a racemic modification into enantiomers is a special kind of job and requires a special kind of approach.

join n **1** 接合線 ☞ tie line

▶ Compositions between CaO and SiO_2 can be regarded as forming a binary join in the ternary system Ca−Si−O.

vt **2** 接続する syn *connect, link, unite*

▷ A class of minerals known as the amphiboles contain double chains of SiO_4 tetrahedra joined to form rings of six tetrahedra.

3 join up 連携する syn *associate, combine, link, unite*

▶ The octahedra are then joined up, often with some distortion, by sharing edges or corners, or by almost any combination of these.

joint n **1** 継ぎ目 syn *junction*

▶ The tubes are removed a short distance above the flame, and the ends are pressed lightly and evenly together on the same axis and then pulled slightly to reduce the thick ring of glass at the joint.

2 継ぎ手

▶ The receiver is attached by a ground-glass joint to the side tube.

adj **3** 共同の syn *combined, cooperative, mutual*

▶ Through the joint efforts of biochemists, photochemists, and synthetic organic chemists, progress in unraveling the details of these photochemical processes has been accelerated.

judge vt 判断する syn *conclude, decide, find*

▷ Judging by the success of the nearest-neighbor approximation in other fields, it seems reasonable to assume a quasi-crystalline model in which each molecule possesses a coordination shell of z nearest neighbors beyond which orientational correlation does not extend.

judicious adj 思慮分別のある syn *logical, rational, reasonable*

▶ Photoelectron spectroscopy allows the direct measurement, and with judicious effort the assignment, of all important valence ionization potentials of a molecule.

jump n **1** ジャンプ syn *hop*

▶ Approximately 1 in 10^5 of tetracene impurity in anthracene will reduce the anthracene fluorescence by half; hence, the excitation makes about 10^5 jumps from one anthracene to another in the crystal before radiating.

vi **2** 飛び移る syn *hop, spring*

▶ The current carriers undoubtedly jump from site to site even when there is no electric field applied to the material.

junction n 接合箇所 syn *joint, linking, meeting*

▶ Electric potential differences occur at the junction of ionic solutions.

just adv **1** ほんの syn *at best, merely, only*

▶ The diagram is just an approximate representation of the valence orbitals of the group 2 dihalides.

2 ちょうど syn *exactly, precisely*

▶ Nitrides solvolyze in liquid ammonia to produce basic solutions of amides, just as oxides in water produce basic solutions of hydroxides.

justifiable adj もっともな syn *rational, rea-*

sonable

▶ Simplification of eq 2 to 3 is quite justifiable, in all practical cases, since the concentration of solvent generally is in such great excess.

justification n 正当化する根拠 syn *explanation, ground, reason*

▶ The principal justification for a chemist carrying out structural studies is the potential importance of this information to an understanding of chemical reactions and chemical reactivity.

justify vt 正当化する syn *explain, rationalize, support*

▶ The behavior of the metals in the series $6s^26p^{0-3}$ demonstrates that the stabilization to be gained by forming three M−Cl bonds in MCl_3 is not enough to justify disruption of the $6s^2$ pair.

K

keep vi **1** とどまる syn *remain, stay*
▶ A pure permanganate solution will keep indefinitely, provided it is kept free from dust and reducing vapors.
2 維持する syn *hold, maintain, preserve, retain*
▷ In experiments, the dependence of *y* on each independent variable is found by keeping all the others constant.
3 keep away 近づけない syn *exclude from*
▶ 2-Methyl-1,4-naphthoquinone must be kept away from light, which converts it into a pale yellow, sparingly soluble polymer.
4 keep out 中に入れない syn *exclude from*
▶ Water vapor can be kept out by use of calcium chloride tubes.

keto-enol tautomerism n ケト-エノール互変異性
▶ Although the keto-enol tautomerism in a given solvent will be independent of the basic catalyst, it will be affected by a change of solvent.

ketone body n ケトン体
▶ While the production of ketone bodies is associated with starvation, it is possible that ketone body formation occurs much earlier.

key n **1** 手がかり syn *clue, explanation, guide, indication*
▶ The movement of dislocations is the key to plastic deformation.
adj **2** 重要な syn *critical, crucial, essential, important, main*
▶ Infrared spectroscopy has played a key role in the study of hydrogen bonding.

kill vt 殺す syn *destroy, eliminate*
▶ Bioremediation is ineffective when the polluted environment contains high concentrations of very toxic substances such as heavy metals and chlorine-rich organic molecules because these compounds typically kill microbes.

kind n **1** 種類 syn *class, sort, species, type, variety*
▶ The experimental methods for determining molecular weights of macromolecular substances will measure some kind of an average value.
2 a kind of …の一種 syn *alike, close, similar*
▶ Peptides of one kind of amino acid are often highly insoluble partly because of strong intermolecular forces.
3 本質 syn *character, manner, nature*
▶ There are also marked differences between mercury and cadmium but often in degree rather than in kind.

kinetic adj 速度論的な
▶ Kinetic absorption spectroscopy is an ideal method of following fast reactions.

kinetically adv 速度論的に
▶ Diamond is kinetically stable, although thermodynamically metastable.

kinetic control n 速度支配 ☞ thermodynamic control
▶ Biological systems have evolved strategies to offset these thermodynamic limitations by establishing kinetic control of nucleation and crystal growth.

kinetic energy n 運動エネルギー ☞ potential energy
▶ X-rays are produced when electrons with sufficient kinetic energy strike the atoms of any element.

kinetic isotope effect n 動的同位体効果
▶ The determination of the presence or absence of a kinetic isotope effect is another way to establish the mode of elimination.

kinetics n 速度論
▶ The essential feature of all methods of studying the kinetics of a reaction is to determine the time dependence of concentrations of reactants or products.

kingdom n 界 syn *area, field*
▶ Yeast are single-celled eukaryotic microbes that belong to a kingdom of organisms called fungi.

kink n **1** 曲がり syn *angle, bend, knee*
▶ The results both on the Hall constant and on

the conductivity of various specimens of copper(I) oxide show certain kinks when log σ is plotted against $1/T$.

2 キンク（表面の欠陥）

▶ By the Frank mechanism, molecules add to kinks in the exposed step of the screw dislocation.

kit n **一式** syn *collection, set*

▶ The objective of developing a universal strategy for the efficient synthesis of fused polyethers containing any combination of ring sizes based on a small but highly efficient synthetic tool kit remains as our ultimate goal.

knee n **急激な変化** ☞ kink

▶ Graphs of log σ as a function of reciprocal temperature invariably possessed a knee dividing the graphs into two distinct parts.

knock n **1 ノック** syn *blow*

▶ Knock is caused by spontaneous oxidation reactions in the cylinder head resulting in loss of power and characteristic ignition noise; antiknock additives virtually eliminate it.

vt **2 失わせる** syn *pull apart*

▶ In a partially evacuated tube through which an electric discharge is passing, the collision of high-velocity electrons with atoms or molecules may knock one or more electrons from an atom or molecule, leaving it with a positive electric charge.

knocking n **ノッキング**

▶ Experiments with pure compounds have shown that hydrocarbons of differing structures differ widely in knocking tendency.

know vt **知る** syn *identify, recognize*

▶ Phosphorus compounds are known with three, four, and five covalent bonds to phosphorus.

knowledge n **1 知識** syn *data, information*

▶ Knowledge of the wavelength λ responsible for a particular spot and the angle θ of reflection makes possible the calculation of the distance d between atomic planes.

2 見聞 syn *awareness, understanding*

▶ It is a matter of common knowledge that at the melting point, a solid and a liquid can exist in equilibrium with each other in any proportions, as can a liquid and vapor at the boiling point.

3 to the best of one's knowledge …の知っている限りでは,

▶ To the best of our knowledge, this is the first example of morphological regeneration of a porous material.

known n **1 既知物質**

▶ The best known of these metalloenzymes is nitrogenase which has been isolated in an active form from about twenty different bacteria and in a pure form from a handful of these.

adj **2 既知の** syn *familiar, well-known*

▶ All samples were synthesized according to known procedures.

L

label n **1** 標識 syn *identification, mark, tag*
▶ The label, which permits visualization of the probe, often emits a radioactive signal, is fluorescent, or is an enzyme that catalyzes the formation of a colored product.
2 表示 syn *designation, name*
▶ The axis labels for graphs must be pure numbers, such as the quotient of the symbol for the physical quantity and the symbol for the unit used.
vt **3** 明示する syn *identify*
▶ First of all, the initial oxidation labels the D-glucose unit that contains the free aldehyde group.
4 ラベルをつけて，分類する syn *categorize, classify*
▶ The energies and wave functions have been labeled with the quantum number *n*.
5 標識をつける
▶ Large proteins need to be labeled with NMR-active ^{15}N or ^{13}C atoms.
vt **6** 呼ぶ syn *call, designate, name*
▶ Researchers are free to label the systems they work with as they see fit.

labeled compound n 標識化合物
▶ Evidence, chiefly from kinetics and experiments with isotopically labeled compounds, indicates that even this seemingly different reaction follows the familiar pattern for carbonyl compounds.

labile adj 置換活性な syn *changeable, unstable, variable*
▶ Those complexes that react completely within about one minute at 25 ℃ should be considered labile and those that take longer should be considered inert.

lability n 不安定性
▶ The kinetic lability of the $Mo_2 \cdots O$ intermolecular interactions is fast on the NMR time scale in the mesophase since there is rapid rotation about the Mo−Mo axes.

labilization n 置換活性化
▶ The trans effect may be defined as the labilization of ligands trans to certain other ligands.

laboratory n **1** 実験室
▶ Although almost all naturally occurring iron is in an oxidation state of either +2 or +3, compounds of iron(0) can be made in the laboratory.
adj **2** 実験室用の
▶ The reaction between a silicon halide and an organometallic reagent provides a versatile laboratory synthesis of organosilicon compounds.

laborious adj 困難な syn *difficult, hard, tough*
▶ We have preferred a treatment which, though more laborious, permits a more straightforward comparison of the coefficients of the acid species with a wider variety of model compounds.

laboriously adv 苦心して syn *earnestly*
▶ In 1848, Pasteur laboriously separated a quantity of the ammonium sodium salt of racemic tartaric acid into two piles of mirror-image crystals.

lack n **1** 不足 syn *deficiency, shortage*
▶ The reason is the lack of experimental information, for instance on the particle number, size, and shape when optical data are derived directly from particle measurements.
vt **2** …がない syn *be deficient in, need, require*
▶ Although macroporous bioceramics show excellent biocompatibility, they are inherently brittle and lack the mechanical strength or toughness needed for high stress use.

lacking adj 不足して syn *deficient*
▶ Ideal FeO appears not to exist, the stable form always lacking some iron, the composition being $Fe_{0.95}O$.

ladder n **1** はしご
▶ The silicate anion of xonotlite is an infinite double chain or ladder with a crosslink or rung at every third tetrahedron in each chain.
vt **2** はしごをかける

▷ Much of the early pioneering work on new synthetic methodology for the assembly of laddered polyethers was performed by Nicolaou and co-workers.

lag vi **1 lag behind …に遅れる** syn *delay, fall behind*

▶ The temperature of sample and reference should be the same until some thermal event occurs in the sample, in which case the sample temperature either lags behind or leads the reference temperature.

vt **2 保温材で覆う** syn *insulate, wrap*

▷ Temperature variation was effected by means of a small furnace wound on a Pyrex tube lagged with asbestos.

lamella n ラメラ 複数形は lamellae syn *flake, scale*

▶ Single crystals of polyethylene are lamellae about 100 Å thick, with the molecules oriented normal to their planes and folded sufficiently regularly to give remarkably smooth upper and lower crystal faces.

lamellar compound n **成層化合物**

▶ Removal of the intercalated substance from a number of the lamellar compounds leaves a small amount of material which is very tenaciously held in a so-called residue compound.

laminar adj **層状の** syn *lamellar, layered*

▶ Flow circular dichroism may suffer additional scattering artifacts if the flow is not laminar.

landmark n **画期的な出来事** syn *monument, turning point*

▶ The interpretation of the electronic spectrum of aqueous Ti^{III} was an early landmark in the development of crystal field theory.

Langmuir-Blodgett film n **ラングミュア・ブロジェット膜** syn *built-up multilayer*

▶ The attachment of long hydrophobic aliphatic chains to the active molecules enables Langmuir-Blodgett films to be deposited on a suitable substrate.

language n **1 言語** syn *style, words*

▶ The hereditary message is written in a language of only four letters, A, G, T, C.

2 専門用語 syn *terminology*

▶ The analysis of these interactions is most conveniently made through the language of perturbation theory; here, as everywhere, the role of symmetry is paramount.

lanthanide n **ランタニド** syn *lanthanoid, rare-earth element*

▶ Since most lanthanides behave very like lanthanum in most of their reactions, it is not surprising that they occur together in nature and that it is very difficult to separate one from another.

lanthanide contraction n **ランタニド収縮**

▶ The lanthanide contraction occurs because, although each increase in nuclear charge is exactly balanced by a simultaneous increase in electronic charge, the directional characteristics of the 4f orbitals cause the $4f^n$ electrons to shield themselves and other electrons from the nuclear charge only imperfectly.

lapse n **短い時間** syn *interval*

▶ The time lapses involved in flash photolysis are approximately 1/100,000 second.

large adj **1 形などが大きい** syn *big, great*

▶ These macromolecules are sufficiently large so that individual molecules have colloidal dimensions.

2 数・量などが大きい syn *huge, massive*

▶ The small molecules elute only after a larger volume of buffer has been passed through.

3 範囲・規模などが大きい syn *extensive, sizable, wide*

▶ Chemical shifts are much larger in ^{13}C NMR than in proton NMR.

largely adv **主として** syn *as a rule, chiefly, mainly*

▶ The size of artificial crystals depends so largely upon a very exact control of the rate of growth.

large-scale adj **大規模の**

▶ Fluorspar CaF_2 is the only large-scale source of fluorine.

laser n **レーザー**

▶ Lasers have the advantage of short pulse

duration, high repetition rate, a narrow, precisely defined wavelength range, and a well-defined spatial profile.

laser action n レーザー作用

▶ A population inversion, in which the population of an excited state exceeds that of a lower state, is needed to obtain laser action.

laser dye n レーザー色素

▶ Coumarins, the largest class of laser dyes for the blue-green region, are highly sensitive.

laser-induced fluorescence n レーザー誘起蛍光

▶ Laser-induced fluorescence offers greater sensitivity and applicability, especially for molecular radicals, at the cost of greater experimental complexity.

laser magnetic resonance n レーザー磁気共鳴

▶ Laser magnetic resonance has been applied to the detection of radicals in discharge flow systems.

last adj **1** 最後の syn *final*

▶ Manganese is the last member of the first transition series that gives an accessible d^0 compound.

2 すぐ前の syn *latest, most recent, newest*

▶ The interest in the growth of sizable single crystals of organic compounds has increased rapidly in the last ten years.

vi **3** 持続する syn *continue, persist, remain*

▷ For loadings lasting a very short time, nothing at all happens, since the sluggish dashpot must first start to move before the system can experience an elongation.

lastly n 最後に syn *finally*

▶ Lastly, it is critical that we learn from experience.

late adv 遅くに syn *lately*

▶ A number of experiments done late in the nineteenth century and early in this gave results totally at variance with the predictions of classical physics.

latent image n 潜像

▶ A stable latent image consists of three or more atoms of silver or gold formed during exposure.

later adv 後ほど syn *subsequently*

▶ The phenomena of coacervation will be dealt with later.

lateral adj 側方の ☞ longitudinal

▶ For a range of lateral substituents of varying size, polarizability, and polarity, the N−I transition temperatures fall in proportion to the size of the substituent.

lateral diffusion n 側方拡散

▶ It is usual for membrane vesicles to fuse with each other in a process that also involves the lateral diffusion of membrane constituents

laterally adv 側方に syn *sideways*

▶ With the aim of understanding the role of laterally placed methyl groups upon mesomorphic properties, nine compounds were prepared corresponding to all independent combinations of methyl and hydrogen in positions A, B, C, and D.

latter adj 代名詞的に用いて後者

▶ No advantage is obtained by dissolving the desired quantity of sublimed iodine in a definite volume of solution, for the latter cannot be kept very long unchanged.

lattice n **1** 結晶格子 ☞ crystal lattice

▶ The CsCl crystal is body-centered at ordinary temperatures but undergoes a transition at 445 ℃ to the simple NaCl type of lattice, the heat of transition being 1.34 kcal.

2 格子 syn *mesh, network*

▶ The plates in the lead storage battery are lattices made of a lead alloy, the pores of one plate being filled with spongy metallic lead and those of the other with lead dioxide.

lattice constant n 格子定数

▶ These can often be differentiated from other products on the basis of powder patterns and lattice constants.

lattice defect n 格子欠陥 syn *crystal defect, lattice imperfection*

▶ The reactions of metals with electronegative elements, e. g., oxygen, sulfur, and halogens, were among the first to be interpreted in terms of the migration of lattice defects.

lattice energy n 格子エネルギー syn *crystal energy*

▶ The experimental value of lattice energy usually exceeds appreciably the theoretical value for salts such as HgS, HgSe, and PbO_2, for which polarization is important.

lattice plane n 格子面

▶ Those lattice planes that are least tightly bound to adjacent planes will be preferred as cleavage planes.

lattice point n 格子点

▶ The body-centered cubic cell has a lattice point at its center as well as at its corners.

lattice spacing n 面間隔 ☞ spacing

▶ Above 0 K, any steps on the（100）face will have a large number of kinks in them; there will be roughly one kink per ten lattice spacings at temperatures of about 300 K.

lattice vibration n 格子振動

▶ The effect of lattice vibrations will be to broaden the absorption lines into bands.

lattice water n 格子水 ☞ water of crystallization

▶ Water molecules which occupy definite lattice positions but are not associated directly with either anion or cation are called lattice water.

launch vt 事業を始める syn *establish, found, open, start*

▶ Perkin's discovery of a mauve dye made from aniline launched the synthetic dye industry, which soon had the capability to produce a rainbow of colors in innumerable shades.

law n 法則 syn *principle, postulate, theory*

▶ Both the first law and the second law place limits on how much useful energy or work can be obtained from a given source.

lay vt **1** 置く syn *deposit, place, position,*

▶ Completion of the first monolayer brings the solubility to a minimum and restores it as soon as the second monolayer is laid on with the polar groups outward.

2 lay down 定める syn *dictate*

▶ Form I should be the form most stable at room temperature. No rigid convention can be laid down for the use of the higher numerals.

layer n **1** 層 syn *film, sheet*

▶ As the crystal grows, the supply of material may happen to be greater on one side than another, so that more layers will be added there.

vt **2** 層にする syn *cover, spread*

▷ X-ray diffraction quality crystals of the stable complexes could be obtained by layering a dichloromethane solution with diethyl ether.

layered adj 層状の syn *lamellar, laminar*

▶ One study on imidazolium salts suggested that a layered model of alternating anions and cations was appropriate to describe the structure.

layer lattice n 層状格子 syn *layer structure*

▶ Layer lattices, when occur in oxides, indicate a predominantly covalent character in the metal−oxygen bonds.

layer spacing n 層間隔

▶ The layer spacing in smectic A phases is normally close to but somewhat shorter than the molecular length.

layer structure n 層状構造 syn *layer lattice*

▶ Crystalline H_3PO_4 has a hydrogen-bonded layer structure in which each $PO(OH)_3$ molecule is linked to six others by hydrogen bonds which are of two lengths, 253 and 284 pm.

leach vt 浸出する syn *drain, filter, percolate*

▶ The bulk of the organic N does not usually leach, but it is known that biomass N in organisms and colloidal organic particles can be mobile.

leaching n 浸出

▶ Selective leaching to form macroporous materials has been applied to engineer fuel cells.

lead n **1** 優勢 syn *advanced position, leading position*

▶ When it comes to quantitative calculations, the electrostatic plus crystal field theory is well in the lead.

2 手本 syn *direction, guidance, model, standard*

▶ Regarding the original purpose of this investigation, i.e., the synthesis of liquid crystals transparent in the ultraviolet region, our results, while not immediately successful, seem to provide a good lead.

3 導線 syn *cable, wire*

▶ The bundle of leads makes several loops in the vacuum space to lengthen the conduction path.

adj **4 先導する** syn *first, foremost, leading*

▶ Since the lead compound in cholesterol synthesis is acetyl-CoA and since mitochondria produce this, almost any cell with mitochondria can synthesize cholesterol.

vi **5 首位に立つ** syn *exceed, surpass*

▶ Stearic acid leads all other fatty acids in industrial use, primarily as a dispersing agent and accelerator activator in rubber products and in soaps.

6 帰着する syn *bring about, result in*

▶ Microwave measurements on ozone lead to a bond angle of 116.8 ± 0.5° and an interatomic distance of 127.8 pm between the central O and each of the two terminal O atoms.

vt **7 導く** syn *cause, induce*

▶ The relative ease with which both s electrons are lost from atoms of the alkaline earth metals leads to compounds in which only +2 oxidation state is found.

leading adj **主要な** syn *chief, foremost, important, prime*

▶ The two benzene rings of thiepine are twisted so that we see one leading edge is slightly above the other.

lead storage battery n **鉛蓄電池**

▶ The electrolyte in the lead storage battery is a mixture of water and sulfuric acid with density about 1.290 g cm^{-3} in the charged cell.

leak n **漏れ口** syn *crack, hole, opening*

▶ If there is no leak around the barrier, the sweeping effectively removes all insoluble films, leaving a clean mobile liquid surface.

leaky adj **漏れやすい**

▶ For all agricultural systems, the nitrogen cycle is leaky, with losses occurring both to the atmosphere (NH_3, NO, N_2O) and to ground and surface waters (NO_3^-, NO_2^-, NH_4^+).

lean vi **lean toward** **…におもむく傾向がある** syn *favor*

▶ Alcohols undergo necleophilic substitution by both S_N2 and S_N1 mechanisms, but alcohols lean more toward the unimolecular mechanism.

leap n **1 躍進** syn *increase, growth, rise*

▶ The past decade or two have seen a considerable leap in the level of our understanding of the electronic structures of small molecules.

vi **2 飛び上がる** syn *jump*

▶ If only one in a million silicon atoms is replaced with a phosphorus atom, then at temperatures at which most of the donors are ionized, the value of *n* leaps to ca. 10^{17} cm^{-3}, roughly a ten-million-fold increase in charge carrier density.

learn vt **知る** syn *discover, find out, investigate*

▶ The properties of any kind of matter are most easily and clearly learned and understood when they are correlated with its structure.

least pron **1 at least** **少なくとも** syn *at the minimum*

▶ The unit cell, by definition, must contain at least one formula unit, whether it be an atom, ion pair, molecule, etc.

2 to say the least **控えめにいっても** syn *as a matter of fact, indeed*

▶ The subject is developing at an astonishing rate where many of the ideas are controversial to say the least.

adv **3 最低に** syn *minimum*

▶ It is only necessary to add some comments on BeH_2, which is the most difficult of these hydrides to prepare and the least stable.

least squares n **最小二乗法** 複合形容詞として使用

▶ Inclusion of positional and anisotropic thermal parameters for the bromine atoms in the least-squares refinement, along with positional and isotropic thermal parameters for the other non-hydrogen atoms, gave a final *R* factor on all the observed data of 0.15.

leave vi **1** 去る syn *depart*
part ▶ Before leaving the methyl acetate example, a mechanism should be suggested to show how the hydrolysis is base-catalyzed as well as acid-catalyzed.
vt **2** 残す syn *allow to remain, give, yield*
▶ In contrast to the B−Cl bond, the N−H bond in borazines undergoes few reactions that leave the ring intact.
3 leave aside 考慮に入れない
part ▶ Leaving aside these anomalies, the mean discrepancy is less than 1.1 kJ mol^{-1}.
4 leave behind 後に残す
▶ This equilibrium mixture of α and β forms may be extracted with acetone to give the β form in solution and leave behind the insoluble α form.

leaving group n 離脱基
▶ Among the halogens, iodine is the best leaving group and fluorine the poorest (I > Br > Cl > F).

ledge n 棚（表面の欠陥） syn *step*
▶ The surface defect formed by a screw dislocation is a ledge, possibly with some kinks.

left n 左 ☞ right
▶ Atomic radii progressively decrease from left to right across the periodic table.

left-hand adj 左手の ☞ right-hand
▶ By convention, a potential difference corresponding to an external flow of electrons from the left-hand electrode to the right-hand electrode is said to be a positive potential difference.

left-handed adj 左回りの ☞ right-handed
▶ Solid quartz loses its optical activity when melted. This is because the optical activity can result from a crystalline arrangement of atoms in a right- or left-handed spiral.

legend n 説明文 syn *caption, title*
▶ Equations, tables, and legends should be numbered consecutively and separately throughout the paper.

legitimate adj 合理的な syn *acceptable, justifiable, logical, reasonable*
▶ In the long-range limit, it is legitimate to speak of the H atom and the H^+ ion as two distinct entities of H_2^+.

leisure n **with leisure** ゆっくりと syn *at one's convenience, when convenient*
▶ The flow method is an example of a steady-state method in that, at each point in the observation tube, a constant set of concentrations is maintained which can be measured with reasonable leisure.

lend vt **lend itself to** に役に立つ，適している syn *be applicable, be suitable, fit, suit*
▶ Most of our knowledge of silicate structures comes from the study of naturally occurring silicate minerals, which are crystalline and lend themselves to investigation by X-ray crystallographic methods.

length n **1** 距離的な長さ syn *dimension, period*
▶ If one assumes that the 18 C−C lengths in the three benzene rings of the molecule of **1** are chemically equivalent, then the average C−C length is 1.383 Å with a root-mean-square deviation of 0.009 Å.
2 at length (a) ついに syn *eventually, finally, ultimately*
▶ Such crystals with regions of solid solution often remained unchanged for some days when cooled to room temperature until at length separation of the product was indicated by development of an opaque region.
(b) 詳細に syn *completely, in depth, thoroughly*
▶ The ionic-covalent transition in the halides of group 15 has already been discussed at length.

lengthen vi **1** 長くなる syn *enlarge, expand, extend, grow*
▶ As the chains lengthen, the excess volume so generated cannot be accommodated in the aggregate.
vt **2** のばす syn *extend, increase, prolong*
▶ When a small quantity of tetraethyllead is added to the hydrocarbon-oxygen charge, the induction period is greatly lengthened.

lengthening n 長くなること syn *extension*
▶ In all these complexes, the lengthening of the X−X bond from that in the free halogen

molecule is notable.

lengthy adj 長い syn *long, prolonged*

▶ These topics would require lengthy treatment.

Lennard-Jones potential n レナード・ジョーンズポテンシャル

▶ The Lennard-Jones potential is a model intermolecular potential energy that expresses the potential energy of two molecules as a function of their separation in terms of two parameters.

lens n レンズ ☞ ocular lens, objective lens

▶ With powdered substances immersed in liquids, nearly all the fragments are thinner on the edge than in the center and, therefore, act as imperfect lenses to refract the light, provided they differ in index from the liquid.

lenticular adj レンズ状の

▶ When bubbles are generated in a jelly, they have the shape of lenticular capsules, even when the jelly has appeared to be optically isotropic.

less adj **1** 量や程度に関してより少ない little の比較級で，名詞の単数形と組み合わせる．

▶ The K_b of aniline is 10^{-10}, which is 10^6 less than that of cyclohexylamine.

adv **2** 量や程度に関してより少なく little の比較級

▶ The alkali-metal molybdenum bronzes are analogous to, but less well-known than, those of tungsten.

lessen vt 減らす syn *decrease, diminish*

▷ Oxalate ligands are small and their charge more dispersed, thus lessening the Coulombic repulsion still further.

lesser adj より少ない syn *minor, smaller*

▶ In general, copolymerization shows, to a greater or lesser extent, a tendency to alternation of monomer units.

lesson n 教訓 syn *advice, message, warning*

▶ Some important lessons are to be learnt from this episode.

let vt させる syn *allow to, permit to*

▷ A highly effective method of purifying a solid substance consists in dissolving it in a suitable solvent at the boiling point, filtering the hot solution by gravity to remove any suspended insoluble particles, and letting crystallization proceed.

lethal adj 致命的な syn *deadly*

▶ Poisonous gases vary in toxicity from nerve gases, which are lethal, to tear gases, which cause only temporary disability.

lethal dose n 致死量 syn *fatal dose*

▶ The lethal dose of sarin for man may be less than 1 mg.

leuco form n ロイコ形

▶ That alkaline solutions are required for solubilization of the leuco form of most vat dyes restricts the use of such dyes to fabrics such as cotton and rayon which are reasonably stable under alkaline conditions.

level n **1** 程度 syn *degree, grade, rank, stage*

▶ Milk is a complex system exhibiting several levels of dispersion, from the molecular through colloidal and into microscopic size range.

2 準位 syn *rank, stage*

▶ The formula for calculating the population of the available levels is known as the Boltzmann distribution.

vt **3** 水平にする syn *flatten, smooth*

▶ A solvent which is more basic than water, e. g., liquid ammonia, will level the strength of a larger group of acids.

leveling effect n 水平化効果

▶ The leveling effect is one phenomenon that must be taken into account when comparing relative acid and base strengths.

leveling off n 水平になること syn *plateau*

▶ The initial rise is taken as corresponding to the strong tendency of the surface to bind the gas molecules, and the leveling off can be attributed to the saturation of these forces.

levitate vt 空中に浮揚する

▶ The height at which the magnet is levitated reflects the tendency to minimize the total energy of the system.

levitation n 浮揚

▶ Levitation is readily induced in a superconducting material, associated with the diamagnetism.

levorotatory adj 左旋性の ☞ dextrotatory

▶ The fact that a substance is dextro- or levorotatory is conveniently specified explicitly in its name by the prefix (＋) or (－), respectively.

Lewis acid n ルイス酸

▶ Supported Lewis acids often possess both Lewis and Brønsted acid sites, and the relative importance of these sites depends upon the application.

Lewis base n ルイス塩基

▶ A coordination compound is formed when a Lewis base is attached to a Lewis acid by means of a lone pair of electrons.

liable adj しやすくて 望ましくないことに用いる. syn *likely, prone*

▶ If heated above 100 ℃, RuO_4 decomposes explosively to RuO_2 and is liable to do the same at room temperature if brought into contact with oxidizable organic solvents such as ethanol.

liberate vt 遊離させる syn *deliver, release, set free*

▶ Like water, alcohols are acidic enough to react with active metals to liberate hydrogen gas.

liberation n 遊離 syn *release*

▶ The liberation of an equivalent amount of hydrogen ion in the form of acid has been verified for arsenic trisulfide and gold sols by Freundlich.

libration n 秤動 ☞ swing

▶ The lattice heat capacity of hydrogen-bonded crystals distinctly smaller than that of ordinary van der Waals crystals because of the high libration frequency of molecules tied to each other by hydrogen bonds.

lid n 容器などの蓋 syn *cap, cover*

▶ The lid of the crystallizer is now removed and is replaced by filter paper, and growth by evaporation begins.

lie vi **1** …に位置する syn *can be found, be stituated*

▶ It is found that the atomic polarization usually lies between 5 percent and 15 percent of the electronic polarization.

2 lie in …にある lay は lie の過去形 syn *be, belong*

▶ Apart from salvarsan, the most striking successes of chemotherapy lay in the treatment of tropical diseases.

lieu n **in lieu of** (prep) …の代わりに syn *in place of*

▶ In lieu of preparing seed crystals from solution, they may be prepared by subliming some of it onto a slightly cooled surface.

life n 寿命 syn *duration, existence, lifetime*

▶ Phosphorescence of long life is never observed except under conditions where the molecule is constrained as in rigid media, in highly viscous liquids, or in the adsorbed state.

life cycle n ライフサイクル

▶ Factors are naturally occurring plant growth regulators produced in minute quantities, show high activity in a species-specific manner, and have a role in the maintenance of the plant's life cycle.

life-span n 寿命 syn *life*

▶ Despite its short life-span, radon has been forced into compound formation, and tracer experiments has indicated the existence of RnF_2 and RnI^+.

lifetime n 寿命 syn *life*

▶ A very short lifetime of an ion pair is better treated as a collision between the ions rather than as molecule formation.

lift vt **1** 引き上げる syn *elevate, raise*

▶ A molecule in the triplet state may be lifted back up to the excited singlet state S_1 either by thermal energy or perhaps by absorption of a quantum of infrared radiation.

2 解除する syn *end, terminate*

▶ This coupling lifts the degeneracy of the uncoupled molecular energy levels.

ligand n 配位子

▶ Ligands are most conveniently classified according to the number of donor atoms which they contain and are known as uni-, bi-, ter-, quadri-, quinqi-, and sexadentate accordingly as the number is 1, 2, 3, 4, 5, or 6.

ligand field n 配位子場 複合形容詞として使用.

▶ In so far as the relative energies of the absorption bands are independent of the ligand-field parameter *Dq*, they can be used to evaluate the differences in electron–electron repulsion between the two excited states.

ligand-field stabilization n 配位子場安定化

▶ Ligand-field stabilization greatly favors planar complexes for Cu^{II}, Ni^{II}, and Co^{II}, but $CuCl_4^{2-}$, $NiCl_4^{2-}$, and a few other complexes of these metals are tetrahedral.

ligand-field theory n 配位子場理論

▶ In ligand-field theory, the orbitals of the ligands are combined with those of the metal in the manner of molecular orbital theory; many of the conclusions and predictions of crystal field theory are not thereby affected substantially.

ligation n 配位子化

▶ In the ligation reaction, one metal ion, usually Mg^{2+}, forms a bridge between the carbanion oxygen of 5′-ribose residue and a phosphate oxygen atom.

light n **1** 光 syn *radiation*

▶ When a beam of light passes through a medium, a certain amount of the light is scattered and can be detected by making observations perpendicular to the incident beam.

2 in the light of …を考慮して syn *considering, in view of*

▶ From an experimental point of view, it is convenient to view the chemical bond in the light of four idealized types: ionic, covalent, metallic, and van der Waals.

adj **3** 明るい syn *bright, brilliant, luminous, shining*

▶ The emergent beam has a component that is vibrating parallel to the vibration direction of the analyzer; hence, anisotropic substances appear light.

4 軽い syn *lightweight*

▶ The lighter molecules move, on the average, much faster than the heavy ones.

5 軽微な syn *easy, moderate, tolerable, undemanding*

▶ In the initial stages of light exercise, the demand for ATP is met by the full oxidation of glucose through the Krebs cycle.

6 淡い syn *faint, slight*

▶ The crude light cream-colored solid can be nitrated directly if it is thoroughly dry.

vt **7** 火を付ける syn *fire, ignite*

▶ Thin strips of zirconium burn in the air when lighted with a match.

light-emitting diode n 発光ダイオード,

▶ When a voltage is applied across the light-emitting diode, nothing will happen unless the energy is sufficient to excite an electron from its localized bond.

light-induced adj 光誘起の

▶ Light-induced, free radical chlorination or bromination of alkylbenzenes with molecular chlorine or bromine gives substitution on the side chain rather than on the ring.

lighting n 照明 syn *illumination*

▶ Lighting phosphors exploit the efficient production of radiation at 254 nm by a low pressure mercury-vapor discharge.

lightly adv 軽く

▶ The ionic conductivity of lightly compacted powders of silver bromide in the low-temperature region is largely due to the motion of silver ions over the surfaces of the crystalline particles.

light metal n 軽金属 ☞ heavy metal

▶ There is a general trend for highly directional covalent bonds to be associated with lighter elements.

light scattering n 光散乱

▶ The unambiguous result from the combined techniques of absorbance and fluorescence probe solubilization, light scattering, and NMR is that semifluorinated hydrocarbon/fluorocarbon mixtures do not form binary molecular solutions at high solute concentration.

light-sensitive adj 感光性の

▶ Pale yellow light-sensitive crystals were prepared by evaporation at room temperature from a solution of phenothiazine in acetone and absolute alcohol.

light sensitivity n 感光性

▶ UV absorbers are added to unsaturated substances (plastics, rubbers, etc.) to decrease light sensitivity and consequent discoloring and degradation.

lightweight adj 軽量の syn *light*

▶ This problem can be relieved, and the capacity increased by using interpenetrating networks or by using lightweight hybrid materials with smaller cavities.

like n **1** 似たもの syn *same kind, similar type*

▶ As with other solutes in ionic liquids, the general rule of like dissolving like is applicable.

adj **2** 同様な syn *equal, identical, similar*

▶ The frequency of collision between the two ions of like sign will be reduced on account of the electrostatic repulsions between the like charges.

vi **3** 気に入る syn *prefer, want*

▶ We cannot distribute the molecules in any way we like among the energy levels; we are restricted to distributions that are consistent with the total energy of the system.

prep **4** たとえば…のような syn *for example, namely, such as*

▶ Alkali metals like rubidium and cesium behave like transition metals under pressure.

5 …に似た syn *similar to*

▶ Silicon has a structure like diamond, but, since the bonding is weaker, it is more volatile.

6 同様に syn *in the same way as, similarly to*

▶ Selenium, like sulfur, exists in several allotropic forms.

suff **7** 名詞に付けて，…のようなを意味する形容詞を造る.

▶ All polymers except the highly crosslinked or almost totally crystalline ones exhibit a relaxation process accompanying the gradual change from glass-like to rubber-like or liquid-like behavior of the amorphous parts.

likelihood n 可能性 syn *probability, strong possibility*

▶ The likelihood of two materials having the same cell parameters and *d* spacings decreases considerably with decreasing crystal symmetry.

likely adj **1** もっともらしい syn *conceivable, probable, reasonable*

▶ It seems likely that irradiation produces excitons which are trapped at imperfections.

2 likely to do …しそうで syn *liable to do*

▶ A particular kind of polarity which is likely to result in the formation of addition compounds is that which may produce hydrogen bonds.

liken vt なぞらえる syn *compare, equate, match*

▶ The packing of the S_8 molecules within the crystal has been likened to a crankshaft arrangement extending in two different directions and leads to a structure which is very complex.

likewise adv 同様に syn *in the same manner, similarly*

▶ Likewise 4,4'-dinitro-*p*-terphenyl has been shown to have characteristic properties of an electron donor and an electron acceptor as demonstrated by its various charge-transfer complexes.

limb n 腕 syn *branch*

▶ The platinum electrodes are placed near the top of the two limbs of a U-tube.

limit n **1** 制限 syn *limitation, restriction*

▶ While there is a limit to how much glycogen can be stored in a cell, there is no limit to the size of triglyceride stores.

2 複数形で，範囲 syn *district, region, zone*

▶ The pH of the suspension was varied by addition of varying amounts of sodium hydrogencarbonate, and it was found that red tin(II) oxide could only be obtained between the pH limits 4.9–6.2.

3 within limits 適度に syn *moderately*

▶ Within limits, ions can expand or contract as the situation demands.

4 within the limits of …の範囲内で

▶ Within the limits of the experimental error,

all of the points fall on the line which has the slope required by the theory.

vt **5 限定する** syn *define, determine, fix*

▶ Recognition of the importance of hydrogen bonding in protein structures limits, to a very great extent, the structures that need to be considered.

6 制限する syn *confine, delimit, restrict*

▶ Hydrogen sulfide burns to water and sulfur dioxide in abundant air, but to water and sulfur if the oxygen supply is limited, thus somewhat resembling arsine.

limitation n 制約（不利な条件） syn *limit, restriction, restrain*

▶ A limitation to use of nitrobenzene is that at the boiling point it has a pronounced oxidizing action.

limited adj 限られた syn *meager, narrow, restricted*

▶ Direct oxidation of carbon in a limited supply of oxygen or air yields CO.

limiting adj 極限の syn *rigorous*

▶ The various bond models represent grossly oversimplified limiting cases.

limitless adj 無限の syn *extensive, unlimited, vast*

▶ In spite of the relatively lower stability of organic crystals, especially to temperature variations, the almost limitless variety of compositions and structures available make them suitable for many practical applications.

line n **1 線** ☞ dashed line, dotted line, parallel line, solid line, wriggly line

▶ The double vertical line indicates a salt bridge separating the two solutions.

2 境界線 syn *border line, boundary*

▶ Fillers are similar to extenders and diluents in their cost-reducing function; exact lines of distinction between these terms are difficult to draw.

3 方向 syn *course, direction, strategy*

▶ Work along this line is in progress.

4 in line with …と一致する syn *in accord with*

▶ In line with the rule of like dissolves like, each nonpolar end of a soap molecule seeks a nonpolar environment.

5 情報 syn *data, information*

▶ There are many lines of evidence supporting this, and it would take a full monograph to describe them all in detail.

vi **6 line up** 並ぶ syn *arrange, assemble, organize*

▶ During mitotic metaphase, the chromosomes line up in the center of the spindle.

vt **7 一列に並べる** syn *align, array, straighten*

▶ Providing a molecule has a sufficiently large dipole moment, it may be lined up in a strong electric field to such an extent that the solution shows optical dichroism.

linear adj 直線的な syn *straight*

▶ It is interesting to note that there is not a linear relationship between ^{13}C chemical shift and the carbonyl stretching frequency in the infrared.

linear combination n 線形結合

▶ The donor-acceptor bond can be described by a linear combination of covalent（charge-transfer）and electrostatic wave functions.

linear combination of atomic orbitals n 原子軌道の線形結合

▶ The linear combination of atomic orbital method is far easier to use in actual numerical calculations than the valence-bond method.

linear dependence n 線形従属

▶ The fact that the catalyst does enter into reaction may be shown by measuring the rate at a variety of catalyst concentrations; generally a linear dependence is found.

linear dichroism n 直線二色性

▶ Linear dichroism is the difference in absorption of light linearly polarized parallel and perpendicular to an orientation axis.

linear expansion coefficient n 線膨張係数 ☞ cubic expansion coefficient

▶ The linear expansion coefficient is defined as the fractional increase in length per degree at constant pressure.

linearization n 直線化

▶ The dimerization process is highly exothermic, and the dimerization energies are quite large compared to the computed monomer linearization energies.

linearize vt 直線化する

▷ The relationship between the structural variation in the monomer and the solid crystal structures has been examined in further detail by reference to the linearized bending force constant of the metal dihalide monomers.

linearly adv 直線的に

▶ The heat capacities increased linearly with the molar mass of the ionic liquid.

linear molecule n 直線形分子

▶ Linear molecules have a C_∞ axis, and the rotation group is C_∞.

linear polymer n 線状重合体

▶ These linear polymers are characterized by covalently bound skeletons that extend throughout the length of the molecule.

lined adj 裏の付いた

▶ A fluorescent lamp consists of a glass tube lined on the inside with a coating of phosphor material and filled with a mixture of mercury vapor and argon.

line defect n 線欠陥 syn *dislocation*

▶ A dislocation is a line defect; it runs somewhat like a string through a crystal.

line shape n 線形

▶ The factors which affect the width and shape of an ESR line are broadly similar to those which are of importance for NMR line shapes.

line width n 線幅

▶ The ESR spectrum of the electrolytically produced solution of the cation radical showed only a single line with a peak-to-peak line width of 2.7 G and a *g* value of 2.0064.

lining n ライニング syn *backing*

▶ Sodium β-alumina forms in the refractory lining of furnaces by reaction of soda from the melt with alumina in the refractory bricks.

link n **1** 因果関係 syn *connection, interdependence, relation, relationship*

▶ The change of 11-*cis*-retinal to 11-*trans*-retinal is the beginning of the visual process; it is the link between the impingement of light and the series of chemical reactions that generates the nerve impulses.

2 結合 syn *bond*

▶ In unsymmetric hydrogen bonds, more electron density is concentrated in the shorter link.

vi **3** 連結する syn *connect, join, unite*

▶ CdI_6 octahedra link up at their edges to form infinite sheets.

vt **4** 関連付ける syn *associate, connect, relate*

▶ The activity of some receptors is strongly linked to the extent of fluidity of the membrane around them.

linkage n 結合 syn *bond, connection*

▶ The linkage of tetrahedra at all four corners generates SiO_2, silica, which exists in at least three crystalline forms: quartz, cristobalite, and tridymite.

linkage isomer n 結合異性体

▶ Differences in infrared spectra arising from the differences in bonding are often used to distinguish between linkage isomers.

linkage isomerism n 結合異性

▶ Several instances of linkage isomerism have been established in compounds of the type *trans*-$[M(PR_3)_2(SCN)_2]$.

linker n リンカー

▶ The use of covalent linkers has proved successful, with examples indicating the more stable binding of the enzyme to the support.

linking n **1** 連結 syn *connection, junction*

▶ Successive linking of oxygen atoms of one, two, three, or four vertices of the tetrahedron generates a wide variety of discrete ions, rings, chains, sheets, and three-dimensional structures.

adj **2** 連結する

▶ The rigidity of the central core structure is frequently achieved by using aromatic rings connected either directly or through a linking unit.

lipase n リパーゼ

▶ Lipase hydrolyzes the ester bond between the fatty acid residues and the glycerol backbone of triglycerides.

lipid n 脂質

▶ Lipids are defined as water-insoluble compounds extracted from living organisms by weakly polar or nonpolar solvents.

lipid bilayer n 脂質二分子膜

▶ Mitochondria are bounded by two lipid bilayers, the inner one being highly folded.

lipophilic adj 親油性の ☞ hydrophilic

▶ An alcohol is a composite of an alkane and water and contains a lipophilic, alkane-like group and a hydrophilic, water-like hydroxyl group.

lipophilicity n 親油性

▶ The lipophilicity of dialkylimidazolium salts can be increased by increasing the chain length of the alkyl groups.

lipoprotein n リポタンパク質

▶ The triglyceride is transported to the periphery in lipoproteins, and from these it enters adipose and other tissues.

liposome n リポソーム

▶ Liposomes are small-diameter hollow microspheres made of lipid molecules, similar to fat molecules present in cell membranes.

liquefaction n 液化

▶ The propane–butane fraction is separated from the more volatile components by liquefaction.

liquefy vt 液化する

▶ Methane is colorless and, when liquefied, is less dense than water.

liquid n 液体 syn *fluid, solution*

▶ Over the past few years, many measurements have been made of the solvation forces in a variety of different liquids, including inert nonpolar liquids, polar liquids, hydrogen-bonding liquids, and aqueous electrolyte solutions.

liquid ammonia n 液体アンモニア

▶ Many important synthetic organic reactions are carried out in liquid ammonia; this is a good solvent for many organic compounds having a range of polarities and also for the metals, lithium, potassium, sodium, and calcium.

liquid chromatogram n 液体クロマトグラム

▶ Liquid chromatograms may be used for the isolation of pure samples weighing many grams, and they may be used for the concentration of small amounts of impurities..

liquid chromatography n 液体クロマトグラフィー

▶ Liquid chromatography is an analytical method based on separation of the components of a mixture in solution by selective adsorption.

liquid crystal n 液晶 複合形容詞として使用

▶ Liquid-crystal phases may be classified on the basis of the shape of molecules which give rise to the phase, the method of inducing the mesophase, and whether we are dealing with low- or high-molar-mass systems.

liquid-crystalline adj 液晶性の

▶ At higher concentrations, above about 40 percent surfactant, a series of liquid-crystalline phases can form which, as their name implies, have properties intermediate between those of solids and liquids.

liquid crystallinity n 液晶性

▶ The phenomenon of liquid crystallinity was discovered almost a hundred years ago for cholesteryl benzoate.

liquid junction n 液絡

▶ The use of a salt bridge can best be justified by the empirical results that emf's so obtained are generally in satisfactory agreement with results from cells without liquid junctions.

liquid membrane n 液体膜

▶ The liquid membrane is composed of a solution of the cyclic peptide valinomycin dissolved in diphenylether.

liquid-phase reaction n 液相反応

▶ Flow systems are not limited to the study of gas-phase reactions but, with certain alterations, can be applied to the study of liquid-phase reactions.

liquid–solid chromatography n 液–固クロマトグラフィー

▶ Liquid–solid chromatography was originally developed for the separation of colored substances, hence, the name chromatography which stems from the Greek word chroma meaning color.

liquidus n 液相線

▶ The liquidus and solidus are continuous through the thermal maximum or minimum and do not show a discontinuity such as is observed for peritectics and eutectics.

list vt 表にする syn *tabulate*

▷ The compounds listed in this table are by no means exhaustive.

literally adv 文字通り syn *actually, in fact, really, truly*

▶ Dye research consisted of the synthesis and evaluation of literally thousands of compounds.

literature n 文献 syn *facts, information*

▶ Literature values are often wildly discordant, and care should be taken in interpreting the data.

lithiate vt リチオ化する

▶ All the bromine-containing compounds were lithiated with four equivalents of *tert*-butyl lithium and lithiated as expected.

lithiation n リチオ化

▶ The degree of lithiation was detected in each case by quenching with methyl iodide and analyzing the degree and position of methylation.

lithium battery n リチウム電池

▶ In general, lithium batteries offer improvements over other batteries in life, capacity, low-temperature performance, and cost.

lithosphere n 岩石圏，リソスフェア

▶ Oxygen comprises 23% of the atmosphere by weight, 46% of the lithosphere, and more than 85% of the hydrosphere.

little pron **1** ほんの少し 量，程度を表す．動詞は単数形を用いる． syn *bit*

▶ In spite of a steady increase of interest in organic reactions in the solid state, relatively little is known about the mechanistic details of thermally induced intramolecular rearrangement of molecular crystals.

adj **2** 程度を表して，ほとんどない syn *insignificant, negligible*

▶ The size of the genome has little correlation with the complexity of the organism.

living adj 生きている syn *live*

▶ The role of the Ca^{2+} ion in living systems extends far beyond making structural components such as bones and teeth.

living cell n 生細胞

▶ In every living cell, there are found nucleoproteins: substances made up of proteins combined with natural polymers of another kind, the nucleic acids.

living thing n 生き物

▶ The world of living things is made up of a great variety of chemical substances, many of which fall into the category of macromolecules.

load n **1** 負荷 syn *pressure, weight*

▶ When a load is applied to any elastic body so that the body is deformed or strained, the resulting stress is proportional to the strain.

vt **2** 詰める syn *fill, pack, pile, stack*

▶ Organic compounds often are available as hairly, crystalline powders, which are not easily loaded into zone-refining tubes.

loading n **1** 負荷

▶ With loadings of longer duration the dashpot begins to move.

2 装填

▶ These reactions typically require several hours, gentle heating, and modest catalyst loading to attain single dehydrogenation of substrate.

lobe n 丸い突出部

▶ Each 2p orbital consists of two lobes with the atomic nucleus lying between them.

local adj 局所的な syn *particular, specific*

▶ Local interactions within molecules in a particle and between neighboring molecules can lead to characteristic features in infrared particle spectra.

localize vt 局所に制限する syn *fix, locate, position*

▶ When a positive hole is localized on any halogen ion, the latter becomes a halogen atom.

localized adj 局在的な ☞ delocalized

▶ Localized corrosion processes are inhibited by stirring because this dissipates the local environment.

localized electron pair n 局在化電子対

▶ The concept of a localized electron pair implies that there exists a region of real space in which these is a high probability of finding two electrons of opposite spin and for which there is a correspondingly small probability of exchange of these electrons with the electrons in other regions.

localized orbital n 局在化軌道

▶ Localized orbitals or groups of orbitals may interact with each other directly, through space, or indirectly, through other bonds in the molecule.

locally adv 局所的に

▶ Pauling's second rule merely implies that charges in the crystal are neutralized locally.

locate vt **1** 位置する syn *fix, place, position, situate*

▶ The two substituents may be located so that the directive influence of one reinforces that of the other.

2 探し出す syn *discover, find*

▶ The task is to locate and assure the preservation of just those record.

located adj 位置して

▶ A hydroxyl ion moves from the support to Au^{3+} species located at the periphery of the particle creating an anion vacancy.

location n 位置 syn *position, site, situation*

▶ The location of the odd electron is intimately involved with the stabilization of free radicals by substituent groups.

lock vt **lock in** 閉じ込めて動けなくする syn *fix, retain*

▶ Upon crystallization, chirality emerges since the conformations are locked in.

lock-and-key adj 鍵と鍵穴の

▶ The construction kit consists of complementary organics and inorganics that spontaneously assemble through lock-and-key intermolecular interactions.

locked adj 組み合わされた syn *fixed*

▶ In an antiferromagnetic solid, adjacent atoms have their net electron spins locked together in an antiparallel array.

locus n 遺伝子座 複数形は loci

▶ Crucially, the transgene is bordered by regions that are homologous to the endogenous locus to facilitate site-specific recombination.

lodge vi 止まる syn *reside, stay, stop*

▶ The glass funnel has the advantage that the inside, where residues often lodge, is visible.

logarithm n 対数

▶ pH is defined as the logarithm of the reciprocal of the hydrogen ion concentration of a solution.

logarithmic adj 対数的な

▶ From the nature of entropy and probability, it can be deduced that the relationship is logarithmic.

logarithmically adv 対数的に

▶ In accordance with the Nernst equation the emf varies logarithmically with the hydrogen-ion concentration and, therefore, varies linearly with the pH.

logical adj 論理的な syn *plausible, reasonable, sound, valid*

▶ It is logical and convenient to divide organic compounds into families with similar properties in the same way that the periodic classification of elements divides the elements into groups or families having related properties..

logically adv 論理的に syn *clearly, consequently, naturally*

▶ Both ortho, para orientation by activating groups and meta orientation by deactivating groups follow logically from the structure of the intermediate carbocation.

London force n ロンドン力 syn *dispersion force*

▶ London forces are extremely short range in action and the weakest of all attractive forces of interest to the chemist.

lone pair n 孤立電子対 syn *unbonding electron pair, unshared electron pair*

▶ All B−N distances in borazine are equal, indicating some degree of delocalization of the nitrogen lone pairs instead of alternating single and double bonds.

long adj **1** 距離が長い syn *elongated, extended, extensive*

▶ Although cellophane is not a crystal, it contains long molecules oriented in the fabrication of the sheet and is optically anisotropic.

2 時間，過程などが長い syn *lengthy, prolonged*

▶ In the pharmaceutical industry, the rates of solid-state reactions are important because drugs should survive storage for very long times without undergoing chemical change.

3 長さで syn *in length*

▶ Consistent with partial double-bond character, the C(2)－C(3) bond in 1,3-butadiene is 1.48 Å long, as compared with 1.53 Å for a pure single bond.

adv **4** 長い間 syn *for a long time*

▶ Macroscopic manifestations of molecular chirality have long been of great interest to chemists.

5 as long as (conj) …する限りは syn *providing*

▶ As long as chemists depict molecular structures in terms of bonds, they will be attracted to the notion that molecular properties can be expressed as additive functions of bond properties.

6 no longer もはや…でない syn *not any more*

▶ If the gas is not ideal, this expression no longer applies.

longevity n 寿命 syn *permanence, standing*

▶ Several examples of radical longevity are known.

longitudinal adj 縦方向の ☞ lateral

▶ When a thin filament hangs under its own weight, the surface tension tends to cause longitudinal shrinking.

long-lived adj 長寿命の ☞ short-lived

▶ Direct observation of stable, long-lived carbocations, generally in highly acidic systems, only became possible in the last decade.

long-range adj 長距離の ☞ short-range

▶ We shall deal primarily with the long-range transfer of energy characterized by dipole-dipole coupling.

long-range order n 長距離規則度 ☞ short-range order

▶ Short-range order cannot be as readily detected by X-ray methods as long-range order.

long run n **in the long run** ついには syn *eventually, finally*

▶ In the long run, we must obtain alternative sources of energy.

long-standing adj 長年の syn *existing for a long time*

▶ The preparation of crystalline electrides has been a long-standing goal of our research.

long-term adj 長期の syn *durable, permanent*

▶ Reaction miniaturization is of particular interest to the pharmaceutical industry where long-term objectives include the desire to perform multiple functions such as synthesis, screening, detection, and biological evaluation on a single integrated device.

look n **1** 一見 syn *glance*

▶ A closer look at the chemical world shows that there are particles in the size range of about 100 to 10,000 Å that are of great interest and importance.

vi **2** 見える syn *appear, seem*

▶ The dotted line shows how the diagram might have looked if the solid compound had survived to a real or congruent melting point.

3 look at 考察する syn *regard, study*

▷ Looking first at the aliphatic acids, we see that the electron-withdrawing halogens strengthen acids.

4 look on みなす syn *consider, regard*

▶ Determinations of the intrinsic viscosity for different molecular-weight fractions of the same polymer lead to the expression, which is best looked on as being empirical

$$[\eta] = KM^a$$

where K and a are empirical constants which depend on the solvent, the polymer, and the temperature.

lookalike n うり二つ syn *exact match, perfect match*

▶ The need for a numerical measure of just how good a lookalike any molecule may be demands a quantitative similarity index.

loop n **1** 曲線 syn *coil, ring* ☞ hysteresis loop

▶ If the curve forms a closed loop or several such loops, each loop traced out clockwise encloses a positive area, and if traced out counterclockwise it encloses a negative area.

vi **2** 輪になる syn *coil, turn, wind*

▶ In one model for the condensed chromosome, the solenoid loops onto the scaffold 18 times, forming a disclike structure.

looping n 輪になること

▶ Enhancers and promoters come into proximity by looping of the intervening DNA.

loose adj **1** あいまいな syn *imprecise, inaccurate, vague*

▶ Rare metals is a loose term for the less common metallic elements.

2 締りのない syn *bulky, voluminous*

▶ With dry, loose powders, it is generally open to doubt whether the resistance being measured is significant unless special precautions are taken.

loosely adv おおまかに syn *broadly, freely, roughly*

▶ The loosely held π electrons are particularly available to a reagent that is seeking electrons.

loosen vt 緩める syn *relax, relieve*

▶ The acetylation of the lysine side chain neutralizes the positive charge of the lysine and will loosen the association between the histone and DNA.

loosening n ほぐすこと syn *freeing, releasing*

▶ Since polymorphic changes require a loosening and rearrangement of lattice units, it is to be expected that lattice mobility is high at and near the transition temperature.

lose vt 失う syn *give up, yield*

▶ Fluorescence is the short-lived re-emission of absorbed radiation at a longer wavelength, the excited molecule having first lost some or all of its excess vibrational energy through collisions.

loss n **1** 喪失 syn *depletion, diminution, reduction*

▶ What is disturbing about the phenomenon of disappearing or elusive polymorphs is the apparent loss of control over the process.

2 損失 syn *wastage, waste*

▶ The capacity loss arises because the particles contract on deintercalation resulting in a loss of mutual contact.

lot n **a lot** たくさん syn *many, much*

▶ It is possible to obtain quite a lot of useful information from a cursory examination of powdered samples using the polarizing microscope.

low adj **1** 程度，高さが低い syn *feeble, weak*

▶ Elemental tin has low toxicity, but most of its compounds are toxic.

2 数量が低い syn *scant, scanty, short, sparse*

▶ Glazes must be low in sodium and are usually mixtures of silicates and flint, lead compounds, boric acid, calcium carbonate, etc.

low-density polyethylene n 低密度ポリエチレン

▶ Low-density polyethylenes have highly branched and widely spaced chains, whereas high-density polyethyenes have comparatively straight and closely aligned chains.

low-dimensional solid n 低次元固体

▶ Much of the enormously increased interest in low-dimensional solids has arisen from the synthesis of the so-called molecular or synthetic metals..

low-energy electron diffraction n 低速電子回折

▶ Low-energy electron diffraction is the most commonly employed technique for characterizing both the structure of the surface and the arrangement of the adsorbate molecules on it.

lower adj **1** より低い syn *farther down*

▶ The cost of producing plant material with recombinant proteins is often significantly lower than producing recombinant proteins in bacteria.

2 同族列について低級の

▶ Certain of the lower carboxylic acids are known to form dimers which are not entirely dissociated even in the gaseous state and which show little, if any, tendency to come together into still larger aggregates.

vt **3 低くする** syn *decrease, diminish, drop, reduce*

▷ Intuitively, we might suppose that lowering the dietary consumption of cholesterol would automatically result in a decreased concentration of cholesterol in blood plasma.

lowering n 低下 syn *decrease*

▶ When q_1 and q_2 are of opposite sign the energy of interaction is negative, indicating a lowering of energy of the system in bringing the charges together.

low-lying adj 低い位置にある

▶ By preferentially filling the low-lying levels of the d electrons can stabilize the system compared to the case of random filling of the d orbitals.

low-molecular-weight adj 低分子量の

▶ The diffusion of low-molecular-weight reagents in high polymers is governed by the solubility and the mobility of the small molecule in polymer phase.

low-spin adj 低スピンの ☞ high-spin

▶ Cytochrome P-450, containing low-spin Fe^{III}, is converted to high-spin on binding a substrate.

low-temperature phase n 低温相 ☞ high-temperature phase

▶ The low-temperature phase of NH_4Cl has the CsCl space lattice for N and Cl.

lubricate vt 円滑に運ぶようにする syn *facilitate, smooth*

▶ The diluent molecules become part of the mobile species and lubricate the ion movement.

lumen n 細胞内腔

▶ The polypeptide chain passes through the membrane and into the lumen.

luminescence n ルミネセンス

▶ The luminescence is usually excited by optical means, but irradiation with electrons, positive ions, or X-ray is also effective.

luminescent adj 発光性の

▶ Although limited to the relatively few luminescent molecular structures, optical spectroscopy readily provides direct access to the excited molecules.

luminescent center n 発光中心

▶ The foreign ion and its immediate neighborhood where it interacts with the host lattice are referred to as a luminescent center.

luminosity n 光度

▶ A continuum light source such as a xenon arc or a tungsten−halogen lamp illuminates the entrance slit of a high luminosity grating spectrometer.

luminous adj 輝く ☞ bright, brilliant, shining

▶ After the blowing operation has been completed, the glass should be annealed to relieve strains by rotating the piece in a luminous flame until it is coated evenly with a layer of soot.

lump n **1** 不定形の塊 syn *mass, piece*

▶ If a lump of solid is left in contact with a solvent, it will dissolve until the solvent has become saturated.

2 in a lump ひとまとめで syn *collectively*

▶ Quantum mechanically the excitation energy must remain in a lump and cannot be spread out.

luster n 光沢 syn *glow*

▶ As x in the series Na_xWO_3 rises and electrons begin to occupy the 5d band of tungsten, the materials become electrically conducting and exhibit metallic luster, hence the name bronze.

lustrous adj 光沢のある syn *shiny*

▶ Cleavage is perfect when it is obtained with great ease and yields smooth, lustrous surfaces, as in mica, calcite, etc.

lying adj 横たわっている

▶ Dislocations lying at the interface between two different phases may be of particular interest in connection with chemical reactions in the solid state.

lyophilic adj 親液性の ☞ hydrophilic

▶ If a material readily goes into colloidal suspension in a liquid, it is called lyophilic.

lyophilic colloid n 親液コロイド

▶ Lyophilic colloids are stabilized by the formation of an adsorbed layer of molecules of the dispersing medium about the suspended particles.

lyophilic sol n 親液ゾル

▶ Under certain circumstances, it is possible to cause a lyophilic sol to coagulate and yield a semirigid jellylike mass that includes the whole of the liquid present in the sol.

lyophobic adj 疎液性の ☞ hydrophobic

▶ A lyophobic material exists in the colloidal state but with a tendency to repel liquids.

lyophobic colloid n 疎液コロイド ☞ hydrophobic colloid

▶ Lyophobic colloids are generally stabilized by the adsorption of ions and coagulate when the charge is neutralized.

lyophobic sol n 疎液ゾル ☞ hydrophobic sol

▶ The addition of electrolytes to lyophobic sols frequently causes coagulation and precipitation.

lyotropic liquid crystal n リオトロピック液晶 ☞ thermotropic liquid crystal

▶ Lyotropic liquid crystals are formed from compounds with amphiphilic properties and solvents.

lyotropic series n 離液系列

▶ The lyotropic series of ions is of general significance in a large number of fields in which colloids do not occur at all, as well as in such behavior of colloids as salting out, swelling, gelation, and many other properties of colloidal systems.

lysis n 溶解

▶ Necrosis is believed to be due to osmotic lysis or enzymatic attacks and is associated with increased cellular volume.

lysosome n リソソーム

▶ Lysosomes are involved in digesting a whole range of biological material, exemplified by the destruction of a whole bacterium with all its different types of macromolecules.

M

machinery n 機構 syn *mechanism, movement*
▶ Although humans store most of their energy as fatty acids within triglycerides, metabolic machinery to convert these into carbohydrates is not present.
macrocycle n 大員環
▶ Metal ions that are too large for the cavity of the macrocycle are simply coordinated out of the plane of the donor atoms in conformers.
macrocyclic adj 大環状の
▶ In these out-of-plane positions the size of the macrocyclic cavity is unimportant, and stability is controlled by the same geometrical factors that control stability in open-chain ligands.
macrocyclic ligand n 大環状配位子
▶ Tripodal ligands typically possess only a single relevant binding conformation, whereas macrocyclic ligands are more conformationally flexible.
macromolecular adj 高分子の
▶ Macromolecular substances are always inhomogeneous in the molecular sense, the molecules having a range of sizes.
macromolecule n 巨大分子 ☞ polymer
▶ Solutions of certain macromolecules, such as starch and proteins, exhibit colloidal behavior, and although they may involve single molecules, they are conveniently classified as colloidal systems.
macrophage n マクロファージ
▶ A macrophage is a cell type that is involved in engulfing foreign material such as bacteria and damaged host cells.
macropore n マクロ細孔
▶ A direct approach to the determination of pore distribution for macropores involves the use of liquids.
macroporosity n マクロ細孔性
▶ Volatile components can be mixed with an hydroxyapatite powder which will create macroporosity during firing of the mixture.
macroporous adj マクロポーラス
▶ Macroporous resins are generally more rigid than microporous resins and have the advantage that reactions using them are less sensitive to the choice of reaction solvent.
macroscopic adj 巨視的な ☞ microscopic
▶ A dispersion is uniform only on a macroscopic scale, for close inspection shows that it consists of grains or droplets of one component in a matrix of the other.
macroscopically adv 巨視的に
▶ Macroscopically, the effects of hydrogen bonding are seen indirectly in the greatly increased melting and boiling points of such species as NH_3, H_2O, and HF.
magical adj 魔法のような syn *extraordinary, fascinating*
▶ ATP is called by biochemist as an energy-rich molecule, but there is nothing magical about this.
magic angle spinning n マジックアングルスピニング
▶ In the magic angle spinning technique, the sample is rotated at a high velocity at a critical angle of 54.74° to the applied magnetic field.
magnetic adj 磁性の ☞ antiferromagnetic, ferrimagnetic, ferromagnetic
▶ Magnetic oxides, especially ferrites such as $MgFe_2O_4$, are modern-day materials with uses in transformer cores, magnetic recording and information storage devices, etc.
magnetically adv 磁気的に
▶ Magnetically hard materials are not easily demagnetized and, therefore, find uses as permanent magnets.
magnetic circular dichroism spectrum n 磁気円二色性スペクトル
▶ A magnetic circular dichroism spectrum is measured by first placing the sample in a magnetic field that is aligned parallel with the direction of propagation of light and then measuring the circular dichroism spectrum.
magnetic dipole n 磁気双極子

▶ A magnetic dipole placed in a magnetic field experiences a force on each pole that tends to turn the dipole about its center; i.e., the magnet experiences a torque.

magnetic domain n 磁区

▶ The crystals must be aligned in chains and have dimensions compatible with that of a single magnetic domain, if they are to function efficiently as biomagnetic compasses.

magnetic field n 磁場

▶ As the magnetic field strength increases, more and more magnetic field lines enter $YBa_2Cu_3O_{7-x}$ until, at sufficiently high magnetic field strengths, the material reverts to being a nonsuperconducting solid.

magnetic moment n 磁気モーメント

▶ In the absence of a magnetic field, an average of half of the protons of an organic molecule will have magnetic moments oriented in one direction and half in the opposite direction.

magnetic properties n 磁性

▶ Magnetic properties depend on the presence of unpaired electrons, especially in d or f orbitals.

magnetic rotation n 磁気旋光

▶ The magnetic rotation is an additive and constitutive property like the molar volume and molar refraction.

magnetic scattering n 磁気散乱 ☞ neutron scattering

▶ The magnetic scattering is due to the interaction of the magnetic moments of the neutrons with the permanent magnetic moments of the paramagnetic atoms.

magnetic susceptibility n 磁化率 ☞ susceptibility

▶ In the solid state, pentaphenylcyclopentadienyl forms dark red crystals; a cryoscopic determination of its molecular weight in benzene has agreed well with the monomeric formula, and its magnetic susceptibility shows the presence of one unpaired electron per molecule of monomer.

magnetization n 磁化

▶ Ferromagnetic materials have a preferred or easy direction of magnetization; in iron, this is parallel to the axes of the cubic unit cell.

magnetize vt 磁化する

▶ The tape is magnetized as it passes through the magnetic field of the recording head.

magnification n 倍率 syn *enlargement, expansion*

▶ An optical microscope of good quality has a magnification of about 10^3 and is useful for particles greater than 2000 Å in linear dimensions.

magnify vt 拡大する syn *enlarge, expand*

▶ The difficulties are magnified in the two-electron case, where in a typical closed-shell calculation one of the interacting levels is occupied but the other one is empty.

magnifying glass n 拡大鏡 syn *lens*

▶ The salt grains in table salt show the cubic form nicely under a magnifying glass.

magnitude n **1** 大きさ syn *extent, size*

▶ Much valuable information can be obtained from the magnitudes of nuclear spin-spin coupling constants.

2 重要さ syn *consequence, importance, significance*

▶ The manipulation of materials during and after purification is a problem of some magnitude.

main n **1 in the main** 概して syn *essentially, generally, mainly, mostly*

▶ In the main, the chemistry of group 3 elements concerns the formation of a predominantly ionic +3 oxidation state arising from the loss of all three valence electrons.

adj **2** 主要な syn *chief, important, prime, principal*

▶ One of the main differences between ionic liquids and aqueous solutions is the comparatively high viscosity of the former.

main-group element n 主族元素 複合形容詞として使用

▶ It seems more satisfactory to attempt to describe the main-group-element electron-precise clusters in terms of localized bonding and nonbonding electron pairs.

mainly adv 主として syn *chiefly, predominant-*

ly, principally

▶ Direct combination of alkali metals with oxygen yields Li_2O from lithium, mainly Na_2O_2 from sodium, and mainly the orange superoxides from K, Rb and Cs.

mainstream adj 主流の syn *common, popular*

▶ Gold had for many years considered to be too noble to be of mainstream chemical interest.

maintain vt 維持する syn *continue, keep, persist, preserve, retain*

▶ Ordinary sulfur melts at 388 K to a pale yellow liquid of low viscosity; it consists of S_8 rings, a structure which is maintained up to 433 K.

maintenance n 維持 syn *continuation, persistence*

▶ Globular proteins serve a variety of functions related to the maintenance and regulation of the life process.

major adj より大きい syn *chief, larger, main*

▶ The controlled growth of crystalline materials is of major importance both in fundamental and applied chemistry.

majority n **1** 大部分 一つの集合とみなす場合には，動詞は単数形，個々のものを強調する場合には，複数形を用いる． syn *bulk, more than half, mass, the greater number, the greater part*

▶ Almost all organic compounds can be prepared in the crystalline state, and a majority are crystalline under ambient conditions.

adj **2** 大多数の

▶ For example, in MgO, the intrinsic thermally produced majority disorder consists of equal numbers of cation and anion vacancies known as Schottky defects.

make vt **1** 造る syn *create, produce*

▶ Styrene is made from ethylbenzene by abstracting a molecule of hydrogen from the ethyl group to form a vinyl group.

2 行う syn *accomplish, do, execute, perform*

▶ Measurements were made on a Cary model 14 recording spectrophotometer thermostatted at 22°.

3 させる 受動態のあとでは，to＋不定詞を伴う． syn *cause, induce, provoke*

▶ The requirement for high concentrations was made difficult to accomplish by the limited solubility of the rod compounds, especially as the rod length was increased.

4 make up (a) 構成する syn *be composed of, comprise, constitute, form*

▶ The vapor of phosphorus pentaoxide is made up of dicrete P_4O_{10} molecules, and these also form the structural unit in the common form of the solid.

(b) 作成する syn *prepare*

▶ It is worth noting that a 1 M aqueous solution cannot be made up accurately by dissolving one mole of solute in 1 liter of water, because the volume of the solution is in general different from that of solvent.

make-up n 配置 syn *arrangement, configuration, constitution*

▶ To be effective these synthetic materials must duplicate the stereochemical make-up of the natural pheromones.

making n 製造 syn *manufacture, preparation, production*

▶ Usually the isolation and purification of a product take much more time and effort than the actual making of it.

malfunction vi 機能しない syn *fail*

▶ The presence of covalently attached glucose residues generally causes a protein to malfunction.

malleability n 展性 syn *flexibility*

▶ The malleability and ductility of metals can be traced to the ease with which the sea of electrons adjusts to permit different bonding arrangements.

malleable adj 打ち延ばしのできる syn *flexible, plastic*

▶ Face-centered cubic metals are malleable and ductile whether in the form of single crystals or polycrystalline pieces.

malnutrition n 栄養失調

▶ A diet based on one source of protein, e.g., corn, can lead to malnutrition.

manageable adj 扱いやすい syn *amenable, controllable, tractable*

▶ It is useful first to cleave the protein into smaller, more manageable segments.

mandate vt 指図する syn *dictate*

▶ The need for an unconventional material is mandated by the fact that most of the elements of interest react with silica, other ceramics, Au, and Pt to form highly stable compounds.

mandatory adj 必須の syn *essential, obligatory, requisite*

▶ The presence of a Lewis acid is mandatory for carbonylation of epoxides to β-lactones in cobalt-carbonyl-based systems.

maneuver n 方法 syn *plan, scheme, strategy, tactic*

▶ The obvious advantage of a concentrated attack employing the various forms of spectroscopy, coupled with electric conductivity measurements, deserves mention as a maneuver to circumvent weakness of the individual methods.

manganese nodule n マンガン団塊

▶ Colloidal particles of the oxides of manganese, iron, and other metals are continuously being washed into the sea where they agglomerate and are eventually compacted into the manganese nodules.

manifest adj **1** 明白な syn *apparent, evident, unmistakable, unquestionable*

▶ Isomers are compounds with the same chemical composition but different structures, and the possibility of their occurrence in coordination compounds is manifest.

vt **2** 示す syn *disclose, exhibit, reveal, show*

▶ The polarity of the carbonyl group is manifested in many of the other properties of aldehydes and ketones.

3 manifest oneself 現れる syn *demonstrate*

▶ This is the so-called inert pair effect and manifests itself structurally by a distortion of the metal ion coordination environment.

manifestation n 現れ syn *appearance, display, example, indication*

▶ Surface tension is a manifestation of the attraction between neighboring molecules.

manifold n **1** 多様性 二つの名詞句を合せて一つの概念とみなして，動詞は単数形を用いる． syn *diversity, variety*

▶ There is ample evidence that this manifold of orbitals and their electron occupancy is of tremendous importance in controlling angular geometry, relative bond lengths, ligand site preferences, reactivity, and many other facets of molecular structure.

2 多岐管

▶ At the user level, the apparatus required to generate and apply microwave-induced plasmas is not complex essentially comprising a microwave cavity and gas manifold.

adj **3** 多方面にわたる syn *diverse, varied, various*

▶ Causes for failure may be manifold, but prominent among them is the existence of a very stable crystalline phase for one of the components.

manipulate vt 操作する syn *control, handle, operate*

▶ Progress in the development of analytical instrumentation has enabled scientists to characterize and manipulate matter with ever-greater spatial resolution.

manipulation n 操作 syn *control, handling, operation*

▶ All manipulations were carried out in an atmosphere of dry argon using standard Schlenk tube techniques or under vacuum in a sealed all-glass apparatus.

manipulative adj 取り扱いの syn *calculating*

▶ So many pitfalls, both manipulative and theoretical, are encountered in the use of fusion curves that extreme care must be taken in obtaining and interpreting the thermal data.

man-made adj 人為の syn *artificial, created, synthetic*

▶ The photoreactions generated in the photozyme can greatly aid the natural biodegradation processes of man-made pollutants.

manner n 仕方 syn *method, mode, procedure, way*

▶ The conductivity of an ionic liquid is strongly dependent upon temperature, and in an analogous manner to viscosity it is found to change in an Arrhenius manner.

manometric adj 圧力測定による

▶ The most useful methods for following gaseous reactions are manometric.

mantle n マントル syn *cover*

▶ Heating in a mantle prevents some of the loss of heat by radiation, and the temperature is obtained nearly as high as that in a crucible over a blast lamp.

manual n **1** 手引き syn *directions, guide, instructions*

▶ The methods of determining boiling points at atmospheric pressure are described in all the practical manuals.

adj **2** 手動の

▶ Automated procedures are much more efficient than manual and effect notable cost savings provided that the machinery is reliable.

manually adv 手動で

▶ A barrier, which may be moved either manually or by a screw-drive arrangement, is used to vary the surface area of the larger section of the Langmuir trough.

many pron **1** 多数のもの many of には the で始まる名詞の複数形が続く syn *abundance, hundreds, lots, plenty, thousands*

▶ Many of the important heat treatments of steels involve phase transformations across the austenite–ferrite phase boundary.

adj **2** 多くの syn *numerous*

▶ Very many different proteins can be isolated from the whole variety of living matter.

3 as many…as… と同数の

▶ If as many independent moments of inertia are obtained as the unknowns sought, then solution of the appropriate set of simultaneous equations will lead to the desired structure.

4 many more ずっと多く 複数形の名詞を伴う.

▶ Many more examples of displacement reactions of exactly the same type could be given.

map n **1** 図 syn *chart, projection, trace*

▶ With modern high-quality X-ray diffraction work, it is possible to obtain fairly accurate maps of the distribution of electron density throughout ionic crystals.

vt **2** 図に描く syn *delineate, draw, sketch*

▶ The course of a reaction is conveniently mapped by means of a potential-energy surface, in which energy is plotted against suitable parameters which describe the reaction system.

3 調査する syn *record, trace*

▶ When ionizing radiation passes through the cloud chamber, the ions formed in its path act as nuclei for condensation, and the trajectory is mapped out as a streak of condensed water.

4 区画する syn *mark out*

▶ A slurry of ice and water is a two-phase system even though it is difficult to map the boundary between the two phases.

margin n 差 syn *allowance, extent, range*

▶ Since the conductivity of oxide semiconductors is a function of purity, method of preparation, and after-treatment, the specific conductivities quoted by various authors may vary by a considerable margin.

marginally adv 少しばかり syn *slightly*

▶ Attempts at increasing oxygen stoichiometry under high oxygen pressures appear marginally successful in the sense that neither the oxygen content nor T_c can be raised substantially.

marine adj 海産の syn *oceanic*

▶ Close to three fourths of all marine organisms can release light through a process known as bioluminescence.

mark vt 示す syn *designate, identify, indicate, signify, specify*

▶ The transition temperature marks the disappearance of long-range order over many interatomic distances.

marked adj 著しい syn *prominent, remarkable, significant*

▶ An alkaline solution of luminol displays particularly marked chemiluminescence when oxidized with a combination of hydrogen peroxide and potassium hexacyanoferrate(III).

markedly adv 著しく syn *especially, notably, particularly*

▶ X-ray powder patterns are markedly affected by the size of the crystallite producing the pattern; diffraction lines are broadened because minute crystals lack resolving power.

marker n マーカー syn *label, tag*

▶ The great potential for structural diversity of carbohydrates makes these molecules very useful as cell-recognition markers.

market n **on the market** 市場に出ている syn *available*

▶ The effectiveness of carbohydrate-based antibacterial vaccines is evidenced by the presence on the market of a variety of antibacterial vaccines.

martensite n マルテンサイト

▶ Martensite is essentially a distorted form of the bcc ferrite structure, distorted because it contains large amounts of carbon trapped during the transformation.

martensitic transformation n マルテンサイト転移

▶ In any martensitic transformation, the high-temperature phase is called austenite, and the low-temperature phase is called martensite, regardless of the crystal structure of the phases.

marvel n 驚異 syn *curiosity, wonder*

▶ In the inner mitochondrial membrane, there reside many copies of a molecular complex that is a marvel of nanoengineering.

mask vt 隠す syn *conceal, cover, obscure, screen*

▶ The magnetic contribution to the specific heat is often masked by the lattice contribution at high temperatures.

mass n **1** 質量 単位は kg, weight の単位は N

▶ At 1 ℃ min^{-1}, the mass loss from polyamino-borane corresponded to 1 equiv H_2, whereas heating at 10 ℃ min^{-1} gave an additional mass loss of 12%.

2 塊 syn *chunk, lump*

▶ When large crystals of silver halides are reduced with developers, the silver halide around the development centers is transformed into compact crystalline masses of silver.

3 量 syn *quantity*

▶ There is a considerable mass of physical data, in widely differing fields, which is in harmony with the concept of hyperconjugation and which can be consistently explained in terms of its operation.

massive adj **1** 塊状の syn *bulky*

▶ Silicon in the massive, crystalline form is relatively unreactive except at high temperatures.

2 非常な syn *enormous, tremendous*

▶ The massive complexity of a molecular electronic system will require appropriate architectures.

massively adv 非常に

▶ The structure of sodium-β-alumina can accommodate supernumerary Na ions, and the compound, even in the form of single crystals, is massively defective, having typically 20–30% more Na than indicated by the idealized formula.

mass spectrometry n 質量分析法

▶ Mass spectrometry is accurate to better than 0.01%; it requires only picomoles of sample, and it can be used to determine masses of molecules ranging in size from single amino acids to proteins larger than 100 kDa.

mass spectrum n 質量スペクトル

▶ A mass spectrum is a record of the damage done to molecules when they are bombarded in the gas phase by an electron beam in an instrument called a mass spectrometer.

master vt 熟達する syn *grasp, learn, understand*

▶ Photosynthesis in green plants is an excellent example of how solar technology can be mastered by living organisms.

mastery n 熟達 syn *command, control*

▶ For a mastery of solid-state chemistry, it is obviously of prime importance to have an understanding of and an ability to control the nucleation process.

mat n マット
▶ Crystals of the charge-transfer complexes have been grown in highly oriented mats by slowly evaporating ether solutions prepared from equimolar quantities of resublimed TCNE and the reagent grade donor compounds.

match n **1** 釣り合うもの syn *parallel*
▶ The masses of the individual peptides in the group are compared by using a computer program with a database of predicted fragmentation patterns for various proteins. A positive match identifies the protein.
vt **2** 匹敵する syn *equal, resemble*
▶ Carbon and hydrogen are well matched with respect to electron-attracting power.
3 調和する syn *accord, agree, fit, suit*
▶ The angular geometries of simple main group molecules are well matched by the predictions of the theoretical tools.

matched adj 釣り合った syn *equal*
▶ Differential absorption spectra were obtained in matched 1- or 5-cm cells on a Beckman model DK-2A spectrophotometer.

matching n **1** 整合 syn *accordance, agreement, coincidence*
▶ Another approach, involving the spreading of Langmuir monolayers on the surface of supersaturated solutions, has shown that ion binding, lattice matching, and stereochemical recognition are important factors responsible for oriented nucleation.
adj **2** 調和する syn *comparable, corresponding*
▶ In the isoelectronic series of compounds successive replacement of CO by NO is compensated by a matching decrease in atomic number of the metal center.

material n 物質 syn *matter, substance*
▶ Cadmium iodide is a layered material in both its crystal structure and properties.

material balance n 物質収支
▶ The term material balance used by chemical engineers denotes a precise list of all the substances to be introduced into a reaction and all those that will leave it in a given time. the two sums being equal.

materials chemistry n 材料化学
▶ Hierarchy has been introduced into materials chemistry, and purely synthetic integrated chemical systems that are designed to achieve a particular function are becoming a reality.

mathematical adj 数学的な
▶ Mathematical functions and expressions that contain the logarithm of a quantity are commonplace in the physical sciences.

mathematically adv 数学的に
▶ The observed diffracted X-ray beams may be combined mathematically, as a Fourier transform, to give crystal structure in the form of an electron density map.

matrix n **1** 行列 複数形は matrices
▶ A matrix is distinguished from a determinant by enclosing it in square brackets instead of parallel lines.
2 基質
▶ Although the organic matrix clearly plays an important biomechanical role, it is also active at the molecular level in controlling crystal nucleation.
3 マトリックス（細胞）
▶ The inner space of the mitochondrion is called the matrix.

matrix-isolated adj マトリックス分離した
▶ ESR results on matrix-isolated CH_3 suggest that it is, or is very close to, being planar.

matrix isolation n マトリックス分離 複合形容詞として使用
▶ A few simple aliphatic compounds that generate free radicals photochemically have been studied by the matrix-isolation method.

matter n **1** 物質 syn *material, substance*
▶ In considering the liquid-crystal state, it is important to realize that it represents a discrete state of matter, lying between the solid and liquid states.
2 事柄 syn *issue, question, subject*
▶ It is quite another matter to see how this symmetry may apply to an actual crystal.
3 a matter of …の問題
▶ The entire question of oxidation state is an arbitrary one, and the assignment of appropri-

ate oxidation states is often merely a matter of convenience.

4 as a matter of fact 事実 syn *actually, indeed, really*

▶ The simple argument we have given so far does not give a very adequate interpretation of the entropy of melting, and as a matter of fact no very complete theory is available.

5 no matter how いかにあっても syn *irrespective of*

▶ As an azeotrope, it gives a vapor of the same composition and, hence, cannot be further concentrated by distillation no matter how efficient the fractionating column used.

6 no matter what 何であっても syn *regardless of*

▶ No matter what the reaction conditions or nucleophile size, methyl and primary alkyl halides almost always substitute by an S_N2 mechanism, whereas tertiary halides react exclusively by the alternate pathway, the S_N1 mechanism.

maturation n 熟成 syn *development, maturity, progress*

▶ Maturation of organic solar cells from laboratory concept into disruptive technology requires substantial advances in power conversion efficiency.

mature adj **1** 成熟した syn *fully developed, fully grown*

▶ While photochemistry of gases and liquids is a mature research field and photochemistry on semiconductors is relatively well-studied, photochemistry at metal surfaces is a newly emerging area.

vi **2** 成熟する syn *develop, grow up*

▶ The field is maturing fast and awaits the input of many who would not even have considered it worthwhile or interesting.

maturity n 完全な発達 syn *full development, full growth*

▶ Commercial instruments have become available, underlining the maturity of the technology.

mauve adj 藤色の

▶ An ammoniacal solution of nickel cyanide produces a pale mauve clathrate when shaken vigorously with benzene.

maximize vt 最大限度にする syn *broaden, enlarge, expand, increase*

▶ In order to reduce interference between substituents in adjacent molecules and to maximize the possibilities for intermolecular hydrogen bonding with such a molecular configuration, half of the molecules in a stack must rotate almost 180° about the Cl−Cl intramolecular axis.

maximum n **1** 最大値 syn *extreme, most, peak, top, upper limit*

▶ Values of internal pressure of water rise with increasing temperature until reaching a maximum in the region of 150 ℃.

adj **2** 最高の syn *extreme, greatest, highest, most*

▶ For maximum catalytic effect, nickel is usually prepared in a finely divided state.

may aux できる canよりやや控えめな表現に用いられる. ☞ might

▶ Initial rates may be determined by getting the slope of the pressure-time curve at zero time.

meager adj 乏しい syn *insufficient, poor, scanty, sparse*

▶ Until recent years, knowledge of the magnetic properties of organic solids has been relatively meager and slowly acquired in comparison to our knowledge of the magnetism of inorganic solids.

meagerness n 乏しさ syn *lack, need*

▶ Any meagerness here is in documentation, not the stability of a radical.

mean n **1** 平均値 syn *average, middle*

▶ Since there is no reason for one direction to be favored over the others, the mean of the u_x^2 values will be the same as the mean of the u_y^2 values and the mean of the u_z^2 values.

adj **2** 平均の syn *average, center, middle*

▶ The refractive index for the yellow light of sodium is commonly used as an approximate mean value for all parts of white light.

3 no mean feat 至難のわざ

▶ The task facing these enzymes with respect to maintaining efficient and highly specific catalysis is no mean feat.

vt **4** 意味する syn *denote, imply, refer to, represent, signify*

▶ The large band gap of simple compounds means that the energy required to produce electronic carriers is very high, and any electric conductivity shown by these solids is generally caused by the motion of ions rather than electrons.

mean activity n 平均活量

▶ In view of the absence of any other arrangement of the activities, it is convenient to introduce a mean activity of the ions.

meaning n 意味 syn *content, sense,*

▶ All symbols have their usual meaning.

meaningful adj 有意義な syn *important, relevant, serious, significant*

▶ A meaningful discussion of chemical bonding can be carried out in terms of Hartree–Fock one-electron density distributions and their dependent properties.

meaningfully adv 有意義に syn *importantly, notably, significantly*

▶ There are insufficient data to comment meaningfully about stereochemistry in the higher-coordinate structures.

meaningless adj 無意味な syn *ineffective, unimportant, vain*

▶ It is meaningless to speak of the molecularity if the mechanism is a composite one.

mean lifetime n 平均寿命

▶ The reciprocal of the unimolecular constant, k, is the mean lifetime, τ.

means n **1** 手段 複数形も同じく means. syn *approach, manner, method, procedure, process, way*

▶ The combination of IR spectroscopy in the carbonyl region and ^{13}C NMR provides a valuable means of determining the structures of organometallic compounds.

2 by means of (prep) …によって syn *employing, through, using, utilizing*

▶ By means of these complexes, good separation of isomers such as the xylenes, cymenes, methylnaphthalenes, and others has been realized.

3 by no means 決して…でない syn *absolutely not, not at all, not conceivably*

▶ Metal alkyls are by no means the only compounds which, on pyrolysis, give rise to free radicals of short life.

mean square velocity n 二乗平均速度

▶ The average over the squares of the velocities is called the mean square velocity.

measurable adj **1** 測定できる

▶ Few hydrocarbons are sufficiently acidic to be measurable in the usual ways.

2 ある程度の syn *moderate, perceptible*

▶ Complex ions, particularly those containing water and ammine ligands, are frequently acids of measurable strength.

measure n **1** 測定 syn *assessment, estimation, evaluation, measurement*

▶ In differential scanning calorimetry, the equipment is designed to allow a quantitative measure of the enthalpy changes that occur in a sample as a function of either temperature or time.

2 程度 syn *extent, magnitude*

▶ All colloidal systems possess some measure of stability, otherwise they would not exist.

3 in large measure 大いに syn *largely*

▶ The energy change accompanying a particular reaction in large measure determines the position of equilibrium or extent of reaction.

4 尺度 syn *gauge, scale*

▶ Temperature is a convenient measure of the amount of energy which is available .

5 複数形で，処置 syn *approach, means, method, procedure*

▶ Potassium cyanide is dangerous if safety measures are not observed.

vi **6 measure up** 必要とされるだけの資格がある syn *fulfill, match, meet*

▶ The editors reserve the right to reject papers which do not measure up to the above standards.

vt **7** 測定する syn *determine, estimate, evaluate,*

gauge, rate

▶ We have to measure the potential difference at constant composition and, therefore, without permitting any current to flow.

measurement n 測定 syn *assessment, determination, estimation*

▶ The results of a series of such measurements are shown in the figure.

mechanical adj 力学的な

▶ Mechanical stress can cause dislocations to move through the grain until they are trapped or annihilated.

mechanically adv 機械的に

▶ In some runs, free TCNQ could be mechanically separated from the crystals of complex, while in others microscopic examination showed that the surfaces of some of the crystals were plated with a yellow layer.

mechanism n 機構 syn *means, process, way*

▶ The mechanism of a reaction should give a detailed stereochemical picture of each step as it occurs.

mechanistic adj 機構的な syn *mechanical*

▶ The temperature-jump technique is now well established in the mechanistic field and has enabled a great range of very fast reactions to be investigated.

mechanistically adv 機構的に

▶ The vigorous hydrolysis of primary nitroalkanes by refluxing with strong mineral acids, though mechanistically obscure, is commercially applicable.

mediate vt 仲介する syn *adjust, moderate, regulate, settle*

▷ Unlike metals, ionic compounds are thought to be conductive due to ion mobility mediated by crystal defects.

mediator n 媒介者

▶ The redox mediator can be chosen so that the sensor has a fast reaction with the reduced enzyme, good stability, and fast electrode kinetics.

medical adj 医学的な syn *medicinal, therapeutic*

▶ Availability of the active compound in purified form often gave rise to new medical investigations.

medicinal adj 医薬の syn *therapeutic*

▶ Steroids, many of which are of great biological and medicinal importance, are in a class of rearranged and degraded triterpenoids characterized by the tetracyclic system.

medicinal chemistry n 医薬品化学

▶ The design of synthetic molecules that are capable of selectively perturbing the function of individual biomolecules is at the heart of chemical biology and medicinal chemistry.

medicinally adv 医薬上

▶ Medicinally, metal chelates are used against Gram-positive bacteria, fungi, viruses, etc.

medicine n 医薬 syn *drug, pharmaceutical*

▶ Essential oils are widely used as perfumes, food flavors, medicines, and solvents.

medium n **1** 媒質 複数形は media syn *atmosphere, environment*

▶ Media of higher basicity may be obtained by using solutions of alkali-metal alkoxides in the corresponding alcohols.

adj **2** 中程度の syn *average, middle, ordinary*

▶ The application of mass spectrometry to organic molecules involves bombardment with a beam of medium-energy electrons in high vacuum and analysis of the charged particles and fragments so produced.

medium ring n 中員環 複合形容詞として使用

▶ In an effort to construct the synthetically challenging medium-ring cyclic ethers found in many fused polyether natural products, we turned our attention to an investigation of the more reactive allylic ethers as cyclization precursors.

meet vi **1** 遭遇する syn *come across, encounter, see*

▶ The liquid surface may meet the solid at a finite angle of contact θ, or the liquid may spread on the solid, in which $\theta = 0$.

2 meet with 出会う syn *encounter, experience, undergo*

▶ X-ray diffraction has met with considerable

success in unraveling the structures of the macromolecules in the solid state.

vt **3 満たす** syn *fulfill, observe, satisfy*

▶ ESR spectra of solids often show broad absorption peaks; therefore, certain conditions must be met in order to get sharp peaks from which useful information may be obtained.

meiosis n **減数分裂**

▶ Prior to cell division by either mitosis or meiosis, DNA must be replicated in the cell.

Meissner effect n **マイスナー効果**

▶ Since no other known phenomena produce diamagnetism of a comparable magnitude, the Meissner effect is taken as sufficient evidence for superconductivity.

melt n **1 溶融物**

▶ Below the glass transition temperature, the uncrystallized melt becomes a metastable glass.

vi **2 融解する** syn *fuse, liquefy*

▷ Substances melting as high as 300℃ have been distilled with success at the pressure of an ordinary oil pump.

3 融合する syn *disappear, merge*

▶ When examined on an atomic scale, the concepts of particle and wave melt together, particles take on characteristics of waves, and waves the characteristics of particles.

meltable adj **融解できる** syn *fusible*

▶ Meltable safety devices, heat regulators, and sprinklers use alloys of In with Bi, Cd, Pb, and Sn.

melting n **1 融解** syn *fusion*

▶ Melting is the change from the highly ordered arrangement of particles in the crystal lattice to the more random arrangement that characterizes a liquid.

adj **2 融解の**

▶ 2,4-Dinitrophenylhydrazones are often preferable for characterization because they are higher melting, less soluble, and more stable than the corresponding phenylhydrazones.

melting point n **融点** ☞ melting temperature

▶ The closer the two melting points, the more easily the unstable form can be obtained, and its melting point can usually be obtained easily before a solid-solid transformation occurs.

melting temperature n **1 融解温度** ☞ melting point

▶ When a polymer containing crystallites is heated, the crystallites ultimately melt, and this temperature is usually called the melting temperature and is symbolized as T_m.

2 変性温度

▶ The melting temperature T_m is the temperature at which half of the DNA has been denatured.

member n **構成員**

▶ Tetracyanoethylene is the first member of a class of compounds called cyanocarbons.

membered adj **…員の**

▶ Both $Te_2Se_4^{2+}$ and $Te_3Se_3^{2+}$ cations have very similar structures consisting of a three-membered ring to which is attached a three-atom chain forming a five-membered ring.

membrane n **膜** syn *film, sheet*

▶ Complications arise with ions that are too large to diffuse through the membrane, and the diffusible ions then reach a special type of equilibrium, known as the Donnan equilibrium.

membrane electrode n **膜電極**

▶ The importance of membrane electrodes is that some of them are highly selective to particular ions.

membrane potential n **膜電位**

▶ The membrane potentials can be measured and exploited by locating reference electrodes such as calomel electrodes in each of the solution phases, using the following arrangement: internal reference electrode | internal solution (constant composition) | membrane | external solution | external reference electrode.

membrane protein n **膜タンパク質**

▶ Membrane proteins and peptides are peripherally associated with or embedded within the phospholipid bilayer that surrounds the cell and its organelles.

membranous adj **膜状の**

▶ Mitochondria are membranous organelles of

great importance in the energy metabolism of the cell.

memory n 記憶 syn *retention*

▶ Of the many alloys that exhibit the shape-memory effect, the memory metal comprising nickel and titanium atoms is one of the most accessible and dramatic.

meniscus n メニスカス

▶ When the liquid and the material of the capillary repel each other just beneath the meniscus, the pressure must be greater than atmospheric, and in order to equalize hydrostatic pressures at the same depth throughout the fluid the surface of the liquid falls, which gives capillary depression.

mental adj **1** 知能の syn *abstract, conceptual, intellectual, rational*

▶ Electron density maps show that our mental picture of atoms as spheres is essentially correct, at least on a time average.

2 精神の syn *psychological*

▶ An indole derivative commonly known as seronin, which is actually 5-hydroxytryptamine, is of interest because of its apparent connection with mental processes.

mention n **1** 言及 syn *note, reference*

▶ Certain general points which emerge from the data in this table call for explicit mention.

vt **2** …に言及する syn *cite, indicate, note, quote, refer to*

▶ It is worth mentioning at this point that the triclinic phase appears to be metastable at room temperature for $Mo_2[O_2C(CH_2)_6CH_3]_4$.

mercuration n 水銀化

▶ The mercuration of butadiene rubbers is reported to yield electric conductors.

mercury electrode n 水銀電極

▶ The mercury electrode is assumed to be completely polarizable; that is, no ions from solution can be discharged at it.

mercury-in-glass thermometer n 水銀温度計

▶ Standard mercury-in-glass thermometers may be used closer to ambient conditions.

mere adj 単なる syn *just, nothing but, only*

▶ Solutions of intrinsically stable colloids are readily prepared by mere contact with the appropriate solvent.

merely adv 単に syn *basically, essentially, only, purely, simply*

▶ Selenium forms a number of sulfides, SeS_7, Se_2S_6, Se_4S_4, merely by replacing one or more sulfur atoms in the S_8 ring.

merge vi 溶け合う syn *blend, coalesce, combine, join, mix, unite*

▶ Chemistry is not an isolated discipline, for it merges into physics and biology.

merit n **1** 長所 syn *advantage*

▶ In order to assess fully the merits of a particular precursor, an increased understanding of the mechanism of growth in metal organic chemical vapor deposition is required.

vt **2** 値する syn *be worthy of, deserve, warrant*

▶ Several other features of our data merit comment.

mesh adj **1** メッシュの

▶ The raw materials are ground to pass 200-mesh sieves.

vt **2** かみ合う syn *entangle, involve*

▶ We propose that this structure provides a model for the packing of longer *n*-alkyl carboxylates in that the alkyl chains are meshed in two different ways.

meso compound n メソ化合物

▶ A meso compound is one whose molecules are superimposable on their mirror images, even though they contain chiral centers.

mesogen n メソゲン

▶ Replacement of a terminal ring hydrogen in a mesogen by any substituent, which does not destroy the linearity of the molecule or broaden it, enhances T_{NI} values.

mesomorphic adj 中間相の

▶ Phase diagrams have been constructed of binary mixtures of a mesomorphic metal complex and a mesomorphic organic compound in order to establish the identity of certain phases.

mesophase n 中間相

▶ A mesophase is a phase intermediate in

character between a solid and a liquid.

mesopore n メソ細孔 ☞ micropore

▶ Mesopores are normally defined as 20–500-Å diameter; micropores as those smaller than 20-Å diameter.

mesoporous adj メソ細孔の ☞ microporous

▶ A vast number of new mesoporous materials and their applications have been reported, with extended ranges of accessible pore sizes and increasingly complex and intricate functional groups.

mesoscale adj メソスケールの

▶ It is naive to assume that reactants will diffuse through mesoscale channels within the catalyst as single molecules.

message n メッセージ syn *implication, meaning*

▶ One kind of RNA carries a message to the ribosome, where protein synthesis actually takes place.

messenger RNA n メッセンジャー RNA

▶ Messenger RNA is then translated into a sequence of amino acids by ribosomes and related molecular machinery to give a polypeptide of definite length.

messy adj 面倒な syn *bad, dirty, rough*

▶ Thermal decomposition of a higher halide is often the best route to lower halides, although this method can be rather messy, giving an impure product.

metabolic adj 代謝の

▶ Sugar is an important source of metabolic energy in foods, and its formation in plants is an essential factor in the life process.

metabolically adv 代謝によって

▶ Fructose and galactose are metabolically transformed and enter the early stages of the glycolytic pathway.

metabolic pathway n 代謝経路

▶ The simplest form of regulation of a metabolic pathway is the inhibition of an enzyme that catalyzes the formation of a precursor compound by the product of the pathway.

metabolism n 代謝

▶ Metabolism decomposes organic matter and releases energy by converting nonliving material into cell constituents.

metabolite n 代謝生成物

▶ The chemicals involved in metabolism are called metabolites.

metabolize vt 代謝する

▶ Only one of the chiral forms of a dissymmetric substance such as a sugar can be metabolized by a given organism.

metal n **1** 金属

▶ Aluminum is a white metal, very light and strong in comparison with its weight.

adj **2** 金属の

▶ The higher borides are more likely to show variation in composition by metal vacancies.

metal carbonyl n 金属カルボニル

▶ Metal carbonyls are remarkable for the way in which, almost without exception, they conform to the effective atomic number rule.

metal chelate n 金属キレート

▶ Metal chelates are found in biological systems, e.g., the iron-binding porphyrin group of hemoglobin and the magnesium-binding chlorophyll of plants.

metal cluster n 金属クラスター

▶ Metal clusters or polyhedra are explicable only on consideration of the structures as a whole.

metal complex n 金属錯体

▶ The reactants become ligands in a new metal complex, often taking the place of some of the old ligands.

metal hydride n 金属水素化物

▶ The closest M–M approach in metal hydrides is often less than for the metal itself: this should occasion no surprise since this is a common feature of many compounds in which there is substantial separation of charge.

metal ion buffering n 金属イオン緩衝

▶ Calcium concentration is controlled either by a precipitation or by a complex-formation equilibrium to give metal ion buffering.

metallic adj **1** 金属の

▶ A strip of copper placed in a solution of a zinc salt does not cause metallic zinc to deposit.

2 金属状の

▶ Partial oxidation of Rb at low temperatures gives Rb_6O which decomposes above −7.3 ℃ to give copper-colored metallic crystals of Rb_9O_2.

metallic bond n 金属結合 syn *metallic bonding*

▶ Metallic structures and bond are characterized by delocalized valence electrons, and these are responsible for properties such as the high electric conductivity of metals.

metallic glass n ガラス合金

▶ Many of metallic glasses are very hard, have high yield points, and yet show considerable ductility when bent.

metallic luster n 金属光沢

▶ Silicon is a semiconductor with a distinct shiny, blue-gray metallic luster.

metallization n 金属で覆うこと

▶ Metallization of the virus fibers with silver led to anisotropically conductive arrays of wires of length up to multiple centimeters.

metallize vt 金属被覆する

▷ Ozone is best prepared by flowing O_2 at 1 atm and 25° through concentric metallized glass tubes to which low-frequency power at 50–500 Hz and 10–20 kV is applied to maintain a silent electric discharge.

metallocene n メタロセン

▶ The sandwich structure of the metallocenes gave the first clue that organic ligands could be bonded to metals via their π systems.

metalloenzyme n 金属酵素

▶ Metalloenzymes not involving oxidation-reduction often serve as Lewis acids and/or they can alter the acidity or basicity of ligands.

metalloid n メタロイド syn *semimetal*

▶ Germanium is a metalloid with a similar electric resistivity to Si at room temperature but with a substantially smaller band gap.

metallomesogen n メタロメソゲン

▶ Since the inorganic backbone of the $M_2(O_2CR)_4$ polymers, the $M_2\cdots O$ interactions, controls access to the mesophase, we investigated the effects of variations in the properties of R, the organic substituents, on the phase behavior of the metallomesogens of chromium and molybdenum.

metalloprotein n 金属タンパク質

▶ Electron transfers between metal centers in metalloproteins more than 100 nm apart are no longer regarded as exceptional.

metallurgical adj 冶金の

▶ The metallurgical microscope is suitable for looking at the surfaces of materials, especially opaque ones.

metal-metal bond n 金属-金属結合

▶ For the influence of metal-metal bonds, a slight contraction of the metal-metal distance in the rutile structure would have been expected.

metal organic chemical vapor deposition n 有機金属気相成長法

▶ The adducts formed in situ during metal organic chemical vapor deposition process can be readily prepared without isolation of the hazardous alkyls of which they are composed.

metal organic vapor phase epitaxy n 有機金属気相エピタキシー

▶ In metal organic vapor phase epitaxy (MOVPE), highly purified volatile compounds of the elements to be incorporated into the semiconductor are passed over the heated substrate where they decompose to give the thin film.

metasilicate n メタケイ酸塩 ☞ orthosilicate

▶ Chain metasilicates formed by corner-sharing of SiO_4 tetrahedra are particularly prevalent in nature.

metastability n 準安定性

▶ The existence of the repulsive Coulomb barrier can confer the metastability of a multiply charged anion even when system is exothermic with respect to electron detachment or fragmentation.

metastable adj 準安定な

▶ Since the metastable form will always be more soluble and more volatile and will have the lower melting point, it follows that supercooling is necessary in order to crystallize the metastable form from the vapor, melt, or solution.

metathesis n メタセシス

▶ Evidently metathesis is merely a matter of solubility not a class of chemical reactions.

metathetical reaction n メタセシス反応 syn *double decomposition*

▶ Tetrabutylammonium hexafluorophosphate was obtained from metathetical reaction of tetrabutylammonium iodide in ethanol with 65% hexafluorophosphoric acid, recrystallized three times from 95% ethanol, and dried under vaccum at 100 ℃.

methanolic adj メタノール性の ☞ alcoholic, ethanolic

▶ Cyanil can be easily prepared from commercially available bromanil and methanolic NaCN.

method n 方法 syn *approach, means, procedure, route, way*

▶ Methods normally satisfactory for the growth of inorganic and metallic crystals often must be greatly modified before they are suitable for the growth of organic crystals.

methodical adj 組織的な syn *ordered, organized, systematic*

▶ By a systematic and methodical repetition of the process of fractional crystallization, a practically complete separation of the components can be effected.

methodology n 方法論 syn *method, practice, procedure, scheme*

▶ Core conversion is one of several synthetic methodologies directed toward the attainment of weak-field clusters.

methylate vt メチル化する

▶ Bacteria typically methylate adenine residues, which protect genomic DNA from restriction by restriction endonucleases and tag the parent strand during mismatch repair.

methylation n メチル化

▶ Methylation labels every free −OH group.

metric adj 計量的な

▶ A graph is a topological concept rather than a geometrical concept; hence metric lengths, angles, and three-dimensional spatial configurations have no meaning.

micellar adj ミセルの

▶ Early investigations of reactivities in micelles were prompted by the expected analogies between micellar and enzymatic catalysts.

micelle n ミセル

▶ Micelles are transient species which form and break up very quickly, but an average spherical micelle might contain 30–100 surfactant molecules.

micro adj 極小の syn *diminutive, minute, tiny*

▶ Chromatography may be used to separate micro amounts of impurities.

microbalance n 微量てんびん

▶ A study using quartz crystal microbalance with probe beam deflection analysis could quantify the depletion of the diffusion layer of metal and the built up of anions.

microbe n 微生物 syn *microorganism*

▶ Microbes are necessary for normal body functions and for many natural processes in the environment.

microbial adj 微生物の

▶ Fermentation is an important microbial process that produces many food products and beverages.

microchemistry n 微量化学

▶ Specifically, microchemistry refers to carrying out various chemical operations on samples ranging from 0.1 to 10 mg; this often involves use of a microscope.

microcrystal n 微結晶 syn *crystallite*

▶ The characteristic curves present the statistical behavior of the whole assembly of microcrystals in the emulsion.

microcrystalline adj 微結晶性の

▶ Marble is a microcrystalline form of calcium carbonate, and limestone is a rock composed mainly of this substance.

microcrystallite n 微結晶

▶ The formation of further product is accompanied by the separation of the product as microcrystallites randomly oriented.

microdroplet n 微液滴

▶ By compartmentalizing reactions in aqueous microdroplets of water-in-oil emulsions, reaction volumes can be reduced by factors of up to 10^9 compared to conventional microtiter-plate-based systems.

microemulsion n マイクロエマルション

▶ Microemulsions are but a small part of a wide range of self-assembling systems which are now under study.

microenvironment n 微環境

▶ Fluorescent tags must be highly sensitive to small changes of the microenvironment caused by the binding of small molecules.

micronization n 微粉化

▶ For the micronization of thermally labile drugs, supercritical CO_2 is a particularly suitable solvent because it has low critical data.

micronize vt 微粉にする

▶ Supercritical CO_2 has low critical data and is well suited to micronize sensitive thermally labile drugs.

microorganism n 微生物 syn *microbe*

▶ Microorganisms in either their natural state or genetically modified forms have served as useful tools in a variety of fascinating ways.

micropore n ミクロ細孔 ☞ mesopore

▶ The apparent diffusion coefficient dropped as the reaction proceeded, as progressively less space becomes available within the micropore for the incoming reactant.

microporous adj 微孔性の ☞ mesoporous

▶ Whereas microporous materials are normally limited to gas phase reactions, mesoporous supports are particularly suited to liquid phase applications as they have pores of a size that allows diffusion of liquid to the active centers within the regular internal channels of the material.

microreactor n ミクロ反応器

▶ The microemulsion droplet can be effectively considered as a microreactor, because reactions will take place in the very limited size domain provided by the droplet.

microscope n 顕微鏡 ☞ electron−, optical−, polarizing−

▶ The reflected light microscope is similar to the transmission one except that the source and objective lens are on the same side of the sample.

microscopic adj **1** 顕微鏡による syn *microscopical*

▶ There seems to have been no detailed microscopic observation of oriented single crystals of α-resorcinol as they underwent polymorphic transformation.

2 微小な syn *diminutive, small, tiny*

▶ The microscopic particles of powder serve as nuclei on which crystallization occurs.

3 微視的 ☞ macroscopic

▶ A system at equilibrium remains in the same macroscopic state, even though its microscopic state is changing rapidly.

microscopically adv **1** 顕微鏡で

▶ The rate of growth of the CBr_4 crystals was measured micoscopically during growth and was in the order of 1 mm/hr for a 1 ℃ temperature difference.

2 顕微鏡でなければ見えないほどに

▶ Crystals of some substances are nearly always small microscopically, while those of other substances are frequently of large size.

microscopic reversibility n 微視的可逆性

▶ By microscopic reversibility, a given reaction between specific states has the same probability as its reverse.

microscopy n 顕微鏡法 不加算名詞である.

▶ The purity of a sample may be checked rapidly by microscopy provided the impurities form a separate crystalline or amorphous phase.

microsphere n ミクロスフェア

▶ Polymeric or proteinaceous microspheres are used to introduce drugs to special locations in the body.

microstructural adj 微構造の

▶ Microstructural defects in metals provide

the mechanistic basis for these properties.

microstructure n 微構造

▶ The physical behavior of a polymer depends not only on its size and its chemical composition but also on the details of its microstructure.

microstructured adj 微構造化された

▶ The channels of the silica framework impart a microstructured environment.

microwave n **1** マイクロ波

▶ A significant limitation of solid-state microwave synthesis is that many technologically important classes of solids do not couple significantly with microwaves and are essentially microwave transparent.

adj **2** マイクロ波の

▶ Microwave heating occurs principally via coupling between the electric component of microwave electromagnetic radiation and the reactants or reaction medium.

microwave absorption n マイクロ波吸収

▶ Microwave absorptions can be characterized as essentially pure rotational spectra.

microwave-induced plasma n マイクロ波誘導プラズマ

▶ A microwave-induced plasma is essentially a partially ionized gas at nonequilibrium where the electron temperature is significantly higher than that of the ions.

microwave spectrum n マイクロ波スペクトル

▶ It is possible to obtain rotational moments of inertia from microwave spectra and from these the bond angles and bond distances for simple molecules.

mid- pref 中間部分の

▶ In the mid-infrared region, classical scattering theory can only be applied to particles with sizes above about 10 nm.

middle n **1** 中央 syn *center*

▶ A pure semiconductor for which $n = p$ will have its Fermi level in about the middle of the band gap.

adj **2** 真ん中の syn *central, halfway, mean*

▶ The middle unit has the effective stoichiometry $YCuO_2$ and is characterized by four missing edge-center oxygens.

might aux もしかしたら…であろう may よりも程度の弱い可能性を表す.

▶ The covalent radius of gallium appears to be slightly smaller than that of aluminum in contrast to what might have been expected.

migrate vi 移動する syn *go, move*

▶ This catalyst is also shape-selective, permitting only molecules of certain sizes to migrate through the channels.

migrating group n 移動基

▶ With the allyl aryl ethers, the migrating group ordinarily goes to the ortho position of the aromatic ring if there is at least one such position that is not already occupied.

migration n 移動 syn *displacement, movement, transfer*

▶ The migration of Na^+ ions in NaCl is difficult and is associated with a considerable activation energy barrier.

migrative adj 移動性の

▶ By migrative orientation we mean the alignment due to translational motion of nonspherical particles through a viscous or porous medium.

migratory aptitude n 移動しやすさ

▶ It is generally true in 1,2-shifts that aryl groups have greater migratory aptitudes than alkyl groups.

mild adj 穏やかな syn *modest*

▶ Sodium tetrahydroborate, being mild reagent, is much more selective.

mildly adv 穏やかに syn *modestly*

▶ Finely divided mildly abrasive minerals such as calcite are use in many hard surface cleaners.

milky adj 白濁した syn *cloudy, white*

▶ A homogeneous solution might distil as a milky, two-phase mixture.

mill vt 粉砕する syn *crush, grind, powder, pulverize, triturate*

▶ It is difficult to obtain chemical homogeneity at the atomic level when mixtures are milled, especially if one component is present as a minor phase.

Miller index n ミラー指数

▶ A Miller index of 2 indicates that the plane cuts the relevant axis at half the cell edge.

million n 複数形で，無数

▶ Red blood cells contain millions of hemoglobin molecules, each of which consists of four polypeptide chains.

millionfold adj 百万倍の

▶ A change from one solvent to another can bring about a millionfold change in reaction rate.

mimetic adj 模倣の syn *simulated*

▶ Biological-membrane-inspired, membrane mimetic chemistry has become an important discipline.

mimic vt まねる 過去分詞は mimicked と綴る． syn *copy, imitate, simulate*

▶ Changing the density of a supercritical fluid allows the fluid's solvating power to mimic a wide variety of liquid solvents.

mimicry n 模倣 syn *imitation*

▶ Design and mimicry in truly inorganic models is stretched beyond its limit by the presence of folded proteins.

mind n **keep in mind** 記憶する syn *recall, remember, retain*

▶ It must be kept in mind that the phase rule applies only to systems in states of reversible heterogeneous equilibrium and not at all to systems in course of transformation, regardless of the slowness of this transformation.

mineral n **1** 鉱物

▶ Minerals can be synthesized to achieve purity greater than that found in natural products.

adj **2** 無機の

▶ In plants, nutrients include numerous mineral elements as well as nitrogen, carbon dioxide, and water.

mineral acid n 鉱酸

▶ Like ammonia, amines are converted into their salts by aqueous mineral acids and are liberated from their salts by aqueous hydroxides.

mineralization n 鉱化

▶ Mineralization occurs mainly in the autumn and spring when the soil is moist and warm.

mineralize vt 鉱化する

▶ Interestingly, some yeasts mineralize nanometer-size intracellular CdS particles within short chelating peptides of general structure, $(\gamma\text{-Glu-Cys})_n\text{Gly}$.

mineral oil n 鉱油

▶ Boric acid glass is hygroscopic and slowly turns cloudy, but coating with mineral oil retards this process.

miniaturization n 小型化

▶ The heat generated from conventional metallic materials limits the degree of miniaturization that is possible.

minimal adj 最小の syn *least, minimum, nominal, smallest*

▶ The red morpholine salt was collected, dried, and recrystallized from a minimal amount of methanol to give 12.5 g of the purified product.

minimize vt 最小にする syn *diminish, lessen, reduce*

▶ In very dilute solutions of alcohols in nonpolar solvents, hydrogen bonding is minimized.

minimum n **1** 最小値 syn *least, lowest*

▶ Between adjacent anions and cations, the electron density passes through a broad minimum.

adj **2** 最小の syn *least, lowest, minimal, nominal, reduced*

▶ Atoms or ions of the crystal may be found at metastable positions in the interstices of the lattice, between the normal sites of minimum potential energy.

minor adj **1** 比較的重要でない syn *insignificant, trivial, unimportant*

▶ Physical adsorption is probably of minor importance in heterogeneous catalysis.

2 より少ない syn *lesser, secondary, smaller, subsidiary, subordinate*

▶ The rates of production of minor products can often be related to the rates of the individual steps in the mechanism.

minority carrier n 少数担体

▶ Under forward bias, the electrons reaching the p-type side and the holes reaching the n-type side are minority carriers in these regions.

minus sign n マイナス記号 syn *negative sign*

▶ The Miller index for an intercept on the negative part of any axis is distinguished by a minus sign directly over it.

minute n 1 分

▶ In partially polymerized vesicles, acid and base transfer occurs in the minute to hour time-scale.

adj 2 微小な syn *little, small, tiny*

▶ The energy differences between allowed orientations of the nuclear spins are very minute.

minutely adv 精密に syn *exactly, precisely, rigorously*

▶ Chemical shift depends on the fact that the spacings of the nuclear energy levels depend minutely on the chemical environment of the nucleus.

mirror n 1 鏡 syn *glass*

▶ At one end of the ruby crystal rod is a mirror for reflecting the light pulse back through the rod.

vt 2 反映する syn *reflect, represent, reproduce*

▶ This spectral evolution with decreasing size nicely mirrors increasing surface fraction.

mirror image n 鏡像

▶ At low concentrations the fluorescence of pyrene solutions is violet, shows fine structure, and is a mirror image of the first absorption band.

mirror plane n 鏡映面

▶ A mirror plane exists when two halves of a molecule or complex ion can be interconverted by carrying out the imaginary process of reflection across the mirror plane.

miscibility n 混和性

▶ Miscibility of two mesophases is proof of their identity, although nonmiscibility provides no information.

miscible adj 混和性の ☞ immiscible

▶ The lower alcohols, where the hydrophilic −OH group constitutes a large part of the molecule, are miscible with water.

misfortune n 不運 syn *bad luck*

▶ The expression for H_3 was, of necessity, even less accurate than that for H_2, but chemists were accustomed to this sort of misfortune.

mislead vt 誤解させる syn *misguide*

▶ Alcohols can be distinguished from alkenes by the fact that alcohols give a negative test with bromine in carbon tetrachloride and a negative Baeyer test so long as we are not misled by impurities.

misleading adj 誤解させる syn *ambiguous, confused, erroneous*

▶ It has become apparent that use of this simple model can frequently be misleading, and it is necessary to make use of the more complete and more complex ligand field treatment.

misleadingly adv 紛らわしく

▶ Another example entails the use of phosphoric acid but in a form that is misleadingly called solid phosphoric acid.

mismatch n 不適当な組み合わせ syn *misalliance*

▶ Within the domains the structure is relatively perfect, but at the interface between domains these is a structural mismatch.

misorientation n 誤った配向

▶ The dislocation model for grain boundaries is useful for angular misorientations of up to several degrees.

misplace vt 置き違える syn *displace, lose*

▶ Apart from the fact that atoms are vibrating, a number of atoms are inevitably misplaced in a real crystal.

misreading n 誤読 syn *misunderstanding*

▶ Some antibiotics, by altering the ribosome, cause misreading of the code and, with this, the production of defective proteins and death to the organism.

misrepresentation n 誤って伝えられること syn *falsification, misstatement*

▶ In spite of this obvious misrepresentation,

the space-filling polyhedron approach has the advantage that shows the topology or connectivity of a framework structure.

missing adj 欠けている syn *absent, short*

▶ Aliphatic absorption is strongest at higher frequency and is essentially missing below 900 cm^{-1}.

mist n 霧 syn *fog, haze, smog*

▶ Mists or fogs composed of atomized particles of oil are used for insecticidal purposes in orchards.

mistake n 誤り syn *error, fault*

▶ It is a mistake to use more decolorizing carbon than that actually needed, for an excess may adsorb some of the sample and cause losses.

mistakenly adv 誤って syn *incorrectly*

▶ In 1895, Ramsay identified helium as the gas previously found occluded in uranium minerals and mistakenly reported as nitrogen.

misunderstanding n 誤解 syn *confusion, misreading*

▶ Perhaps the cause of the misunderstanding lies in the method by which we are accustomed to observe the preparation of the silver—ammonia complex.

misunderstood adj 誤解された syn *misconceived*

▶ The relationship between the cell voltage and the lithium chemical potentials in the electrodes can be derived in another way which serves to highlight a frequently misunderstood fact.

mitochondrial adj ミトコンドリアの

▶ The energy derived from the oxidation of carbohydrates is coupled to the synthesis of ATP via a series of redox reactions, the mitochondrial electron-transport chain.

mitochondrion n ミトコンドリア 複数形は mitochondria

▶ All eukaryotic cells contain membrane-bound organelles called mitochondria, where enzymes for electron and proton transport and ATP synthesis reside.

mitosis n 有糸分裂

▶ Cytokinesis may begin in late telophase of mitosis or it may be delayed until mitosis is complete.

mix n **1 混合物** syn *alloy, blend, combination, mixture*

▶ Uranium hexafluoride is a mix of $^{238}UF_6$ and $^{235}UF_6$ in a ratio of 140:1.

vi **2 混ざる** syn *coalesce, combine, merge*

▶ Two liquids do not completely mix if one liquid has a much greater cohesion than the other.

vt **3 混ぜる** syn *blend, combine, unite*

▶ There is no difficulty, using optical or other techniques, in following the course of a very rapid reaction, but for hydrodynamic reasons it is impossible to mix two solutions in less than about 10^{-3} s.

4 結びつける syn *associate with, join with*

▶ The group vibrations are in reality appreciably mixed with other modes both in the ligand itself and in the complex as a whole.

mixed adj **1 混合の** syn *contaminated, hybrid, impure*

▶ It can be only mentioned here that isopoly and hetropoly anions also give rise to highly colored mixed oxidation state species: the tungsten bronzes and the heteropoly blues.

2 mixed with …と混ざった syn *associated, involved*

▶ Monazite is a phosphate of the rare earths and occurs in so-called monazite sand mixed with quartz, rutile, zircon, etc.

mixed crystal n 混晶

▶ Theoretical studies show that interactions between host and guest in mixed crystals are comparable in magnitude with second-order effects in pure crystals.

mixed oxide n 混合酸化物

▶ Mixed oxides of Fe^{IV} such as $M^{I}_4FeO_4$ and $M^{II}_2FeO_4$ can be prepared by heating Fe_2O_3 with appropriate oxide or hydroxide in oxygen.

mixed-valence adj 混合原子価の ☞ intervalence

▶ The main features of interest in the rich and varied chemistry of mixed-valence complexes are the relationships between gross physical properties, molecular structure, and the extent of valence-electron delocalization.

mixed-valence compound n 混合原子価化合物

▶ Perhaps the best known of mixed-valence compounds is Prussian blue, potassium iron (III) hexacyanoferrate(II).

mixing n **1** 混合 syn *blending, mixture*

▶ The lack of long-range order in liquids permits mixing together, without difficulty, molecules of very different sizes and shapes.

2 混成 syn *hybridization*

▶ The presence of low-lying unoccupied MOs capable of mixing with the ground state accounts for the polarizability of soft atoms.

mixture n **1** 混合物 syn *blend, composite*

▶ Isobutylene with sulfuric acid dimerizes to an 80 : 20 mixture of 2,4,4-trimethyl-1-pentene and 2,4,4-trimethyl-2-pentene.

2 結合 syn *association, combination*

▶ Storage is a mixture of energized rates of uptake and thermodynamic equilibria.

mobile adj **1** 流動性の syn *fluid, formless*

▶ Anhydrous HSO_3F is a colorless, dense, mobile liquid which fumes in moist air.

2 可動性の syn *movable, transportable*

▶ Because electrons are mobile electrically charged particles, they are easily displaced in a solid by applied electric fields or by mechanical deformation.

mobile phase n 移動相 ☞ stationary phase

▶ Chromatography involves the flow of a mobile (gas or liquid) phase over a stationary phase, which may be a solid or a liquid.

mobility n **1** 移動度

▶ In metals, the number of mobile electrons is large and essentially constant, but their mobility gradually decreases with rising temperature due to electron–phonon collisions.

2 可動性 syn *motion, movability*

▶ That some polymers are highly elastic is related to the fact that they consist of long molecular chains arranged in such a fashion that there is some internal mobility.

mobilization n 動態化

▶ The near depletion of liver glycogen and the mobilization of fatty acids as alternative fuel to glucose occur after about 1 day of starvation.

mobilize vt 動態化する syn *activate*

▶ If muscle uses fatty acid oxidation to supply ATP, then less glycogen is mobilized.

mode n **1** 方式 syn *approach, manner, method, way*

▶ The experimental data may be subjected to two independent but complementary modes of analysis, which should yield concordant results.

2 機能形態 syn *condition, state, status*

▶ The exact mode of action of inhibitors is still uncertain in most cases.

model n **1** 模型 syn *representation*

▶ Within this model for the columnar structure, the 4.7-Å spacing seen in the X-ray diffraction pattern is consistent with the expected average layer-to-layer separation along the columns.

adj **2** モデルの syn *representative*

▶ Preliminary studies concerning enol ether ring-closing metathesis were performed using model systems in order to demonstrate that the reaction could be used to prepare fused systems possessing six- or seven-membered rings.

vt **3** モデル化する syn *form, make, produce*

▶ Infrared spectra of large molecular aggregates can be modeled either by classical scattering theory or by quantum chemical calculations.

modeling n モデル化

▶ Modeling plays a key function in the understanding of the spectroscopic features of these complex systems with their many degrees of freedom.

moderate adj **1** 並みの syn *average, medium, modest, ordinary*

▶ Alkyl fluorides may be prepared in moderate yield by interaction of an alkyl bromide with anhydrous potassium fluoride in the presence of dry ethylene glycol as a solvent for the inorganic fluoride.

vt **2** やわらげる syn *reduce, relax, relieve*

▶ PCl_5 reacts violently with water to give HCl and H_3PO_4, but in equimolar amounts the reaction can be moderated to give $POCl_3$.

moderately adv 適度に syn *comparatively, rather, somewhat*

▶ In order to have significant activity, the catalyst should be extensively covered by the adsorbate, and this requires moderately strong adsorption.

moderation n 減速 syn *diminution*

▶ The original Skraup procedure employed nitrobenzene as the oxidant, in which case the reaction is frequently very vigorous and requires moderation by the addition of iron(II) sulfate.

moderator n 減速材

▶ If ordinary water is used as moderator or coolant, a concentration of 2-3% ^{235}U is necessary to compensate for the inevitable absorption of neutrons by the protons of the water.

modern adj 現代の syn *contemporary, current, latest, new, recent*

▶ The principal objective of modern electrochemistry is to study the behavior and reactions of ions in a variety of environments.

modest adj 控えめな syn *moderate, reasonable*

▶ Reactions of aldehydes proceeded in, at best, modest enantioselectivities and yields.

modestly adv 控えめに syn *barely, sparsely*

▶ In nonclassical carbonium ions, such as the norbornyl ion, the photoelectron spectrum indicates only two modestly electron positive carbon atoms.

modification n **1** 変更 syn *alteration, change, revision*

▶ Innumerable modifications of the Ostwald viscometer have been described, all the same in principle.

2 変態 syn *alterative, variant, variation*

▶ A metastable red modification of SnO is obtained by heating the white hydrous oxide.

modifier n 調節剤

▶ The solubility of both polar and nonpolar solids in a supercritical fluid may be enhanced through the use of a modifier.

modify vt 変更する syn *adjust, change, reform, revise*

▶ The alkyl groups attached to the double bonded carbons modify the reactions of the double bond; the double bond modifies the reactions of the alkyl groups.

modular adj 組み立てユニットの

▶ The basic architecture of such a bis(carbene) has several modular features, including the linker adjoining the carbene moieties, pendant substitutents on the nitrogen atoms, as well as the heteroatoms.

modularity n モジュール方式

▶ Modularity is a design concept in which readily interchangeable parts are used to rapidly create new products.

modulate vt 調節する syn *adjust, regulate, transform*

▶ A number of research groups are actively studying the chemistry of synthetic analogues of zinc enzymes as part of concerted effort to establish how and why the chemistry of zinc is modulated by its coordination environment.

modulation n 調節 syn *regulation, variation*

▶ The modulation of the conductivity of the polymer by redox reactions can be utilized to fabricate microelectrochemical transistors.

modulus n モジュラス 複数形は moduli

▶ The adiabatic moduli changed very little with temperature until this reached one-third of the absolute mp, when there was a sudden increase.

modulus of rigidity n 剛性率

▶ The modulus of rigidity of a wire may be determined by means of the torsion pendulum.

Mohs hardness n モース硬さ ☞ hardness

▶ The Mohs hardness is based on the concept of scratch hardness and is used chiefly for minerals.

moiety n 部分

▶ Attachment of the boron moiety of the reagent takes place more readily to the less crowded carbon of the double bond.

moist adj 蒸気や霧で湿った syn *damp*
▶ Since lithium chloride is a very hygroscopic salt, it is necessary to weigh it out of contact with moist air.
moisten vt 湿らせる syn *damp, wet*
▶ Lead tetraacetate as supplied by manufacturers is moistened with glacial acetic acid to prevent hydrolytic decomposition.
moisture n 湿気 syn *damp, wet*
▶ All xenon fluorides must be protected from moisture to avoid formation of xenon trioxide.
moisture content n 含水量
▶ Grinding can easily reduce the moisture content of a sample of gypsum from 20 to 5 percent.
molality n 質量モル濃度
▶ The usual units of molality are mol/kg, for which the symbol m is frequently used; to avoid confusion with the symbol for meter, it is best to use italics.
molar conductivity n モル伝導率
▶ In all cases the molar conductivity diminishes as the concentration is raised, and two patterns of behavior can be distinguished.
molar enthalpy of solution n モル溶解エンタルピー
▶ The temperature dependence of solubility should be related to the differential molar enthalpy of solution at saturation.
molar extinction coefficient n モル吸光係数
▶ The molar extinction coefficient ε is a measure of the absorption efficiency at the wavelength λ_{max}.
molar heat capacity n モル熱容量
▶ The vibrational modes are not making their expected contributions to the molar heat capacity; in other words, the assumption of equipartition of energy is not generally valid.
molarity, M n モル濃度
▶ The solution contains 20.2 percent of hydrogen chloride, and its concentration is thus a little less than 6 M.
molar ratio n モル比
▶ The yield of polyaniline from interfacial polymerization is comparable to that by conventional synthesis with the same monomer-to-oxidant molar ratio.
molar refraction n モル屈折 ☞ atomic refraction
▶ Molar refractions for the sodium D line in benzene solution have been determined for several substituted benzenes containing both electron-releasing and electron-withdrawing groups.
molar volume n モル体積 syn *molecular volume*
▶ As the molar volume of diamond is much smaller than that of graphite, it follows that diamond can be made from graphite by application of a suitably high pressure, provided that the temperature is also sufficiently high to permit movement of the atoms.
mold n **1** 鋳型 syn *die, template*
▶ The terms die and mold are virtually synonymous in the sense of a negative cavity into which a molten metal or plastic is introduced under pressure, the former being used in reference to metals and the latter for plastics, rubber, etc.
2 かび syn *fungus*
▶ The *Penicillium chrysogenum* mold generates penicillin G, which is still the largest volume penicillin.
vt **3** 形成する syn *form, make, shape*
▶ Celluloid was the first chemical product that could be molded by heat and pressure into a number of different shapes.
moldable adj 成型可能な syn *ductile, flexible, malleable, soft*
▶ Gutta percha softens at 60 ℃ to moldable condition and melts at 100 ℃.
molding n 成型 syn *casting*
▶ Plastics may be shaped by either compression molding or injection molding.
molecular adj **1** 分子の
▶ Entropy changes in chemical reactions can be understood on a molecular basis.
2 分子状の
▶ The highest-valence metal oxides, M_2O_7 and

MO_4, may be regarded as simple volatile molecular oxides.

molecular association n 分子会合

▶ The deviation from ideality in such cases may be understood as a measure of the extent of molecular association.

molecular beam n 分子線

▶ A molecular beam of metal travels in straight lines in high vacuum.

molecular beam epitaxy n 分子線エピタキシー

▶ A second technologically important method of semiconductor growth that can afford better control is molecular beam epitaxy.

molecular complex n 分子錯体

▶ Naphthalene forms a molecular complex with picric acid, and biphenyl does not; since the complex is much less soluble than either component, naphthalene can be isolated easily from the mixture in the form of this derivative.

molecular compound n **1** 分子化合物

▶ Boranes are colorless, diamagnetic, molecular compounds of moderate to low thermal stability.

2 分子間化合物 syn *addition compound, molecular complex*

▶ The molecular compounds with amines were decomposed by digestion with 5M hydrochloric acid.

molecular crystal n 分子結晶

▶ Molecular crystals typically have their constituent molecules tightly packed and have little freedom of motion of the sort needed for reaction.

molecular diameter n 分子直径

▶ Around the cation, the dielectric constant rises from 5 at the surface of the ion to its bulk value of 78 at a distance of one or two molecular diameters.

molecular diffusion n 分子拡散

▶ An important aspect of monitoring single molecule on a flat substrate is related to problems of molecular diffusion.

molecular distillation n 分子蒸留

▶ A molecular distillation is distinguished by the fact that the distance from the surface of the liquid being vaporized to the condenser is less than the mean free path of the vapor at the operating pressure and temperature.

molecular dynamics n 分子動力学

▶ In the method of molecular dynamics, the classical equations of motion of *N* interacting particles are solved by step-by-step numerical integration.

molecular electronics n 分子エレクトロニクス

▶ The replacement of inorganic semiconductors and metals used in the electronic industry by organic materials is termed molecular electronics.

molecular ion n 分子イオン

▶ If one electron is removed from the parent molecule, there is produced M^+, the molecular ion, whose m/e value is the molecular weight of the compound.

molecularity n 反応分子数

▶ The molecularity of an elementary reaction is the number of reactant particles that are involved in each individual chemical event.

molecular mechanics n 分子力学

▶ The molecular mechanics calculation models steric effects in molecules by use of simple equations to represent contributions to steric strain, such as bond-length and bond-angle deformation, torsional strain, and van der Waals interactions between nonbonded atoms.

molecular model n 分子模型

▶ Crown ethers were so named because their molecular models resemble a crown.

molecular orbital n 分子軌道関数

▶ In the molecular orbital theory of diatomic molecules, molecular orbitals can form when an atomic orbital from one atom overlaps with an atomic orbital from the second atom to give a lower-energy bonding molecular orbital and a higher-energy antibonding molecular orbital.

molecular orbital method n 分子軌道法

▶ The successful application of MO methods to carbonium ion equilibria prompts extension to reactions in which carbonium ions are intermediates.

molecular polarizability n 分子分極率

▶ To obtain a calculation of molecular weight from a measured turbidity, it is necessary to have a value of the molecular polarizability.

molecular rearrangement n 分子転位

▶ Certainly, a more detailed understanding of such transformations will be of great help in studies of molecular rearrangement in the solid state.

molecular recognition n 分子認識

▶ Phenomena involving molecular recognition in complexation have long been of great interest in organic chemistry.

molecular rotation n 分子回転

▶ Crystals in which there is no molecular rotation at room temperature are relatively hard and brittle, whereas those in which the molecules rotate at room temperature are soft and waxy.

molecular sieve n 分子ふるい

▶ Molecular sieves of synthetic zeolite preferentially adsorb nitrogen from air at ambient temperature, giving 95% oxygen and 5% argon.

molecular spectroscopy n 分子分光学

▶ In molecular spectroscopy, a beam of electromagnetic radiation is passed through a material.

molecular spectrum n 分子スペクトル

▶ Molecular spectra in the near infrared region correspond to changes in vibrational energy levels accompanied by changes in rotational energy levels.

molecular structure n 分子構造

▶ The science of organic chemistry rests on a simple premise: that chemical behavior is determined by molecular structure.

molecular symmetry n 分子対称

▶ Even though the vector sum of the individual bond moments may not give an accurate quantitative estimate of the total dipole moment, the latter, being a vector quantity, must depend on the molecular symmetry.

molecular vibration n 分子振動

▶ Molecular vibrations cause the crystal field splitting to oscillate continuously and so broaden the absorption bands.

molecular volume n 分子容 syn *molar volume*

▶ The molecular volume could be matched to some properties of simple gases, liquids, and solids.

molecular weight n 分子量 ☞ number-average−, weight-average−

▶ Polymerization reactions lead to products in which there is a range of molecular weights. Measurements of molecular weight will, therefore, lead to some kind of an average value.

molecular weight distribution n 分子量分布

▶ The molecular weight distribution in a sample of a polymer has a significant effect on its physical properties.

mole fraction n モル分率

▶ Stated differently, the mole fraction solubility of a substance in an ideal solution is the same in all solvents.

molten adj 溶融した syn *fused, liquid*

▶ Scandium metal, which is obtained by electrolysis of the molten chloride, is moderately electropositive.

molten salt n 溶融塩 syn *fused salt*

▶ One of the most interesting aspects of molten salt chemistry is the readiness with which metals dissolve.

moment n **1** 瞬間 syn *instant, minute, second*

▶ A wave equation cannot tell us exactly where an electron is at any particular moment or how fast it is moving.

2 モーメント ☞ dipole−

▶ Lone pairs of electrons may have moments equally as large as the bond moments.

momentary adj 瞬間的な syn *brief, fleeting, temporary*

▶ Although the momentary dipoles and induced dipoles are constantly changing, the net result is attraction between the two molecules.

moment of inertia n 慣性モーメント

▶ The reason that we do not count the rotation about the axis of the linear molecule is that the mass is almost completely concentrated in the atomic nuclei, so that the moment of inertia corresponding to these rotations is very small.

momentum n 運動量 複数形は momenta

▶ Knowing the position and momentum of the particle at some initial time gives the trajectory at all later times.

monitor vt 絶えず監視する syn *check, follow, observe, scan*

▷ The kinetics of the decay of the F-center absorption in KBr can be measured on a spectrometer by monitoring the absorption over a period of time.

monoatomic adj 単原子の

▶ Only monoatomic substances which have no internal degrees of freedom can reasonably be expected to obey the theory of corresponding states in the solid phase.

monobasic acid n 一塩基酸 syn *monoprotic acid*

▶ Hypophosphorus acid forms white crystals mp 26.5° and is a monobasic acid pK 1.1.

monochromatic adj 単色の

▶ It is possible to filter out all but one main peak called the Kα radiation, and X-ray diffraction experiments are almost always carried out with this highly monochromatic Kα radiation.

monochromatic light n 単色光

▶ Light of a narrow band of wavelengths, if sufficiently narrow, is usually called monochromatic light.

monochromatize vt 単色化する

▶ In a typical experiment, a beam of electrons from a hot filament is monochromatized by means of a carefully stabilized accelerating potential.

monoclinic adj 単斜晶系の

▶ The common mica, muscovite, is a monoclinic crystal with point-group symmetry $2/m$ and cleavage parallel to {001}.

monodentate ligand n 単座配位子 syn *unidentate ligand*

▶ The CN^- ion can act either as a monodentate or didentate ligand.

monodispersed adj 単分散の syn *monodisperse*

▶ The light scattered from a monodispersed sulfur sol can exhibit different colors when viewed at different angles to the incident beam.

monograph n モノグラフ syn *paper, treatise*

▶ The object of this monograph is to give a coherent review of the phenomenon now generally known as hyperconjugation.

monolayer n 単分子層 syn *monomolecular layer*

▶ Spreading a solution of a surfactant or lipid in a volatile solvent results in monolayer formation at the air-water interface.

monomer n 単量体，モノマー ☞ dimer, trimer, tetramer, polymer

▶ When polymerization occurs in a mixture of monomers, there will be some competition between the different kinds of monomers to add to the growing chain and produce a copolymer.

monomeric adj 単量体の ☞ polymeric

▶ Nitric oxide is a colorless, monomeric, paramagnetic gas with a low mp and bp.

monomer unit n 単量体単位 syn *repeating unit*

▶ The process of addition of monomer units to the growing chain can be interrupted in different ways.

monomolecular adj 単分子の syn *unimolecular*

▶ An aqueous solution of glyoxal contains monomolecular species and reacts weakly to acid.

monomolecular reaction n 単分子反応 syn *unimolecular reaction*

▶ Experimentally, the methyl isocyanide isomerization has been the most intensively studied monomolecular reaction involving a small molecule.

mononuclear adj 単核の

▶ Rather than bonds, we choose as our fundamental parts mononuclear fragments of the system with boundaries defined in real space.

mononuclidic adj 単核種の

▶ The atomic weights of only the 20 mononuclidic elements can be regarded as constants of nature.

monotonically adv 単調に

▶ The elution volume from a Sephadex column is a monotonically decreasing function of molecular weight.

monotonous adj 単調な syn *monotonic, ordinary, routine, uneventful*

▶ Any success at the application of tabulated heat of solution data for simple covalent substances dissolved in weakly interacting solvents must be attributed to the monotonous consistency of the sign of integral molar enthalpy of solution at infinite dilution.

monotropic adj 単変の ☞ enantiotropic

▶ A monotropic form must be obtainable only from a metastable vapor or from a metastable liquid.

monotropy n 単変 ☞ enantiotropy

▶ There is a relation known as monotropy, in which the second form has or appears to have no region of stability anywhere on the *P*–*T* diagram, so that it is always metastable with respect to the stable form.

monovalent adj 一価の syn *univalent*

▶ A wide variety of monovalent cations enter the tungsten bronze structure; similar series occur with MoO_3 in the molybdenum bronzes.

Monte Carlo method n モンテカルロ法

▶ In the Monte Carlo method, one generates a sequence of particle configurations in such a way that the probability of a particular configuration of potential energy V appearing in the sequence is proportional to $\exp(-V/kT)$.

month n 月

▶ Thick prismatic crystals often showed cracking after the beginning of development of the yellow color which occurred in a period of a number of months at room temperature.

mordant n 媒染剤

▶ The most important mordants are trivalent chromium complexes.

mordant dye n 媒染染料

▶ A mordant dye is a dye requiring use of a mordant to be effective.

more pron **1** より多くの量 syn *a greater quantity*

▶ When the dissolved salt reduces the internal pressure of the aqueous solution, more of the nonelectrolyte is able to dissolve (salting in).

adv **2** もっと syn *further*

▶ Etching with an acid removes material from grain boundaries more rapidly because of the higher energy of these sites.

3 more and more ますます

▶ In the development of chemical analysis, more and more emphasis is placed upon the detection and identification of very small quantities of material.

4 more or less 多少 syn *nearly, rather, relatively, somewhat*

▶ The more or less covalent CuO is without acid character, and the similar PbO is weakly basic.

5 more so 前の形容詞や副詞を受けて，なおさらそうだ

▶ Thiophene is less reactive than pyrrole and furan but more so than benzene.

6 more than… …して余りある

▶ Although lithium would be expected to be the more electron-attracting than hydrogen because of its higher nuclear charge, this is more than offset by the greater atomic radius and the screening effect of the two inner-shell electrons of lithium.

7 more than one 一つより多い 複数を意味するが，これに続く名詞や動詞は単数形を用いる．

▶ Most substances can exist in more than one phase or state of aggregation.

8 no more than わずかに syn *at best, merely, only, simply*

▶ The amount should be no more than enough to cover the bottom of the tube and form a layer about 0.5-mm thick.

moreover adv さらに syn *besides, further, furthermore, in addition*

▶ Moreover, a solution of ammonium in liquid ammonia is formed at the cathode when a solution of ammonium chloride in that solvent is electrolyzed.

morphological adj 形態的な

▶ Unlike inorganic materials, many polymeric materials are known to have well-defined basic morphological units at the nanometer scale.

morphology n 形態 形態学の意味では不加算名詞である.

▶ Needle or plate-like morphologies of hydroxyapatite crystallize at regular intervals along the collagen fibers with the long *c* axes oriented parallel to the fibrils.

mortar n 乳鉢 ☞ pestle

▶ When the sample is ground in a steel mortar or in a steel ball mill, it will be contaminated with a little iron.

mosaic adj モザイクの

▶ One type of imperfection in so-called single crystals is the presence of a domain or mosaic texture.

Mössbauer spectrum n メスバウアースペクトル

▶ Further insight into the subtlety of the effects which give rise to electric field gradients at the nucleus of atoms is obtained by considering the Mössbauer spectra of many unsymmetrically substituted organotin compounds.

most pron **1 the most** 最大限度 単数形の扱いをする. syn *(upper) limit, maximum*

▶ A difference of 15 percent in the radii of the metal atoms that replace each other is the most that can be tolerated if a substantial range of substitutional solid solutions is to form.

2 たいてい 複数形の扱いをする. syn *majority*

▶ Although certain derivatives of quinoline can be made from quinoline itself by substitution, most are prepared from benzene derivatives by ring closure.

ただし, most of が量を表すときには動詞は単数形とする.

▶ Liquid zinc is insoluble in liquid lead, and the solubility of silver in liquid zinc is about 3000 times as great as in liquid lead. Hence, most of the silver dissolves in the zinc.

3 at most せいぜい syn *just, merely*

▶ Like alkanes, alkenes are at most only weakly polar.

adj **4** たいていの syn *nearly all*

▶ For most dyes on most fabrics, the gross light-fading rate is increased by the presence of moisture and, especially, of oxygen.

adv **5** 最も syn *guite, very*

▶ Oximes, semicarbazones, and dinitrophenylhydrazones are most often used to characterize aldehydes and ketones.

mother adj 本源の syn *native, natural* ☞ daughter

▶ The relative orientation of the mother and daughter phases is random, and it has been suggested that the transformation possibly goes through an intermediate amorphous state.

mother liquor n 母液

▶ Colored impurities, like colorless impurities, tend to remain in the crystallization mother liquor and are eliminated when the crystals are collected and washed.

motif n モチーフ syn *figure, pattern*

▶ Two examples containing the rare structural motif of a zig-zag chain of four phosphorus atoms were amongst the products structurally characterized.

motile adj 運動性の syn *active, mobile*

▶ The bacteria that are motile usually are propelled by one or more hairlike appendages called flagella.

motility n 運動性 syn *mobility, movability*

▶ Some bacteria have long hollow tubes called pili which do not contribute to motility. Instead they contribute to the adhesiveness of bacteria.

motion n 運動 syn *action, movement, shift*

▶ The effects arising from the motions of the electrons will be different for each kind of hydrogen; therefore, the resonance signal produced for each kind of hydrogen will come at different field strengths.

motionless adj 静止した syn *inactive, still*

▶ Normal alkanes have been shown to be essentially motionless at low temperatures; as the temperature gradually increased, rotation of the terminal methyl groups occurs and, at a few degrees below the melting point, rotation of the entire hydrocarbon chain sets in.

motivate vt 動機を与える syn *activate, encourage, provoke, stimulate*

▶ Research is often motivated by the need to optimize a particular property such as enhancing the ionic conductivities of solid electrolytes and improving the activity and selectivity of heterogeneous catalysts.

motivation n 動機 syn *impetus, purpose*

▶ Motivation for this research includes the search for new materials with interesting new properties.

mount vt 備える syn *display, exhibit, install, set up*

▶ Occasionally, the body mounts an immune response against its own cells or molecules, producing an autoimmune disease.

move vi **1** 移動する syn *advance, proceed, shift*

▶ A dislocation line can move in a crystal, in two significantly different ways, glide and climb.

vt **2** 動かす syn *affect, disturb*

▶ In permanent magnets, the domain walls are not easily moved.

movement n 運動 syn *migration, motion, shift, transfer*

▶ If a liquid is supercooled below its melting point, the rate of atomic and molecular movement decreases until a temperature known as the glass transition temperature is reached, at which internal equilibrium can no longer be maintained.

moving phase n 移動相

▶ In paper partition chromatography, the adsorbed water is the stationary phase and the organic solvent the moving phase.

much pron **1** 多量 much of には不可算名詞が続き，動詞は単数形を用いる．

▶ Much of our knowledge of organic chemistry has resulted from investigations of the composition of naturally occurring substances.

adj **2 too much** 多量すぎる syn *excessive*

▶ Too much potassium iodide should not be used, as otherwise lead iodide will separate out.

adv **3** 大いに syn *considerably, far*

▶ Thallium metal is much less reactive than rubidium, but much more so than silver or mercury.

4 同じを意味する語句を修飾して，ほぼ syn *nearly*

▶ Grignard reagents add to nitrile groups in much the same way as they add to carbonyl groups.

5 as much as …と同じ程度に

▶ High yields are achieved because the rate of the reaction catalyzed by enzymes might be increased by a factor of as much as 10^{12}, making side reactions unimportant.

6 as much…as possible できるだけ多く syn *completely, fully*

▶ The usual procedure is to obtain as much spectroscopic and chemical information as possible about the kinds of groups that are present.

mud n 泥 syn *dirt, sludge, soil*

▶ Oil-well drilling muds are made alkaline to prevent flocculation of their components.

mull n ペースト状のもの

▶ Nujol, mineral oil, is used to prepare mulls for infrared analysis.

multicenter bond n 多中心結合

▶ A globally delocalized cluster has a multicenter core bond in the center of the cluster polyhedron.

multicentered adj 多中心の

▶ In conventional terminology, LaB_6 would be described as a rigid, covalently bonded network of B_6 clusters having multicentered bonding within each cluster and two-center covalent B−B bonds between the clusters.

multicomponent adj 多成分の

▶ It is permissible to equate $a_{O^{2-}}$ with a_{Na_2O} in a simple binary system, but the procedure is much less satisfactory in multicomponent mixtures.

multicomponent system n 多成分系

▶ The principles of the phase rule apply to all multicomponent systems, including solvent blends, glass, alloys, and plastics.

multidentate ligand adj 多座配位子

▶ The most common donor atoms for the multidentate ligands are the electronegative oxygen and nitrogen atoms.

multidisciplinary adj 学際的な

▶ Nanotechnology is a highly multidisciplinary area that describes a field of applied science and technology focused on the design, synthesis, characterization, and application of materials and devices on the nanometer scale.

multifaceted adj 多面的な

▶ Continued competitiveness in this multifaceted industry depends upon readiness to improve existing processes and to introduce new ones.

multifunctionalized adj 多機能性の

▶ In recent work directed toward more sophisticated enzyme models, a number of specifically bifunctionalized and multifunctionalized cyclodextrins have been prepared.

multilayer n 多分子層

▶ The Brunauer-Emmett-Teller, or BET, method is based on an extension of Langmuir monolayer concept of adsorption to multilayers.

multilayered adj 多層の

▶ These versatile techniques are being used to prepare a broad range of technologically important materials that include diamond films, glass optical fibers, and the multilayered structures of semiconductor diode lasers.

multimode n 多モード

▶ Optical fibers with a graded refractive index are called multimode and may carry more than one wavelength.

multimolecular adj 多分子の

▶ The present review is mainly concerned with organic multimolecular inclusion compounds, particularly those of the true clathrate or cage type.

multimolecular layer adsorption n 多分子層吸着

▶ Derivation of adsorption isotherm equations for multimolecular layer adsorption is carried out on the assumption that the same forces that produce condensation are also responsible for multimolecular layer adsorption.

multiphoton dissociation n 多光子解離

▶ Infrared multiphoton dissociation provides an alternative method for producing radical species.

multiphoton excitation n 多光子励起

▶ Sulfur hexafluoride gave the first convincing evidence that multiphoton excitation really occurred so rapidly that collisional energy transfer could be avoided.

multiple n **1** 倍数

▶ Terpenes have a diverse range of structures but have the basic formula $C_{10}H_{15}$ or a larger number of carbon atoms that exist in multiples of five.

adj **2** 多数の syn *many*

▶ Conjugation of multiple carbon–carbon double bonds leads to significant changes in λ_{max} and in the ε value.

3 多様の syn *complex, complicated, intricate, involved*

▶ Many of the essential trace elements have multiple biochemical roles.

multiple bond n 多重結合 syn *multiple bonding*

▶ The most important use of Grignard reagents is for the formation of new carbon–carbon bonds by addition to multiple bonds, particularly carbonyl bonds.

multiplet n 多重線

▶ Perfectly symmetrical multiplets are to be expected only when the separation between multiplets is very large to the separation within multiplets, that is, when the chemical shift is much larger than the coupling constant.

multiplication n 掛け算

▶ Most molecules yield surface area that require multiplication by a factor of between 1. 2 and 1.5 to give areas agreeing with those calculated from the nitrogen isotherms.

multiplicative adj 掛け算の

▶ Entropies are additive, but probabilities multiplicative, and a relation satisfying this condition is $S = k \ln W$.

multiplicity n **1** 多重度

▶ The multiplicity $2S+1$ is indicated in the term symbol as a left-hand superscript.

2 多様性 syn *diversity, variety*

▶ A good example of the multiplicity of factors which affect rates of reaction is the relative rate of hydrolysis of Cl_2, Br_2, and I_2.

multiply adv **1** 多重に

▶ Many transition metal atoms form complexes with carbon monoxide and other small, multiply bonded molecules, by a combination of σ donation from the ligand and π donation to it.

vi **2** 増殖する syn *breed, develop, grow*

▶ Once attached, bacterial cells can multiply and then produce sufficient toxins to cause illness.

vt **3** 掛け合わせる

▷ On multiplying the left sides and the right sides of these equations, we obtain the relation

$$K_a K_b = [H_3O^+][OH^-]/[H_2O]^2$$

multiply charge anion n 多価陰イオン

▶ Electrospray ionization allows robust quantities of multiply charged anions to be transferred directly from bulk solutions into gas phase for experimental interrogation.

multipole n 多極子

▶ The Coulomb interaction can be expanded as a multipole series. Usually the dipole–dipole coupling predominates.

multistep adj 多段階の

▶ Because dye molecules are complex, they are synthesized through multistep processes employing numerous intermediate compounds.

multistep reaction n 多段階反応

▶ Generally, a multistep chemical reaction will have a slow rate-determining step and other relatively fast steps which may occur either before or after the slow step.

multitude n 多数であること syn *many, plenty*

▶ The multitude of stoichiometries, the complexities of the structures, and the intermediate nature of the bonding make the classification of intermetallic compounds difficult.

multivalent adj 多価の

▶ Although multivalent constructs can successfully elicit immune responses, monovalent vaccine candidates have also shown good results in producing antibodies against tumor-associated carbohydrate antigens.

muscle n 筋

▶ The reverse conversion of ATP to ADP provides energy for muscle activity.

muscle contraction n 筋収縮

▶ Muscle contraction is accomplished by the interaction of filaments composed of two different proteins, actin and myosin.

must aux …でなければならない syn *ought to, should*

▶ In carrying out the analysis of alkali carbonate it is important to make sure that there is no loss of CO_2 before the first-end point is reached; the solution must be cold, the acid must be added slowly, and each portion must be stirred in well before fresh acid is added.

mutagen n 突然変異原

▶ Chemicals called mutagens, many of which mimic the structure of nucleotides, can mistakenly be introduced into DNA and change DNA structure.

mutagenesis n 突然変異誘発

▶ Mutagenesis of viral capsids is well-established technique and allows alteration of surface properties of the particles.

mutant n 突然変異体

▶ Mutants can be created by exposing bacteria to X-rays, ultraviolet light, and a variety of chemicals that mutate DNA.

mutarotate vt 変旋光する

▶ Monosaccharides commonly mutarotate, exist as α- and β-hemiacetals, and give α- and β-methyl glycosides.

mutarotation n 変旋光

▶ The slow change of optical rotation in solution is called mutarotation, and it can be

interpreted as involving the interconversion of hemiacetals through the aldehyde intermediate.

mutate vt 突然変異を起こさせる

▶ Exposure to X-rays or ultraviolet light from the sun can also mutate DNA.

mutation n 突然変異

▶ The damage can lead to mutations, changes in the sequence of bases.

mutational adj 突然変異の

▶ Deletions and duplications can occur in the same mutational event when two homologous DNA strands overlap, break at the same time at two different points, and then rejoin with the wrong strand.

mutual adj 共通の syn *common, communal, joint*

▶ If two molecules are to undergo a chemical reaction, they not only must collide with sufficient mutual energy but must come together with such a mutual orientation that the required bonds can be broken and made.

mutual coagulation n 相互凝結

▶ Usually for mutual coagulation two colloids should be of opposite sign.

mutually adv 相互に

▶ The Co−O−Co angle in $LiCoO_2$ is close to 90°; hence, neighboring cobalt ions overlap with mutually orthogonal 2p orbitals on the bridging oxygens.

myofiber n 筋繊維

▶ Exercise involves muscle contraction which is brought about by the relative movement of filaments known as myofibers within muscle cells.

myofibrillar adj 筋原繊維の

▶ The sudden rise in cytoplasmic Ca^{2+} concentration stimulates the molecular interactions that cause myofibrillar contraction.

myosin n ミオシン

▶ Under resting conditions, the interaction of actin and myosin filaments is normally prevented by the intervention of group of proteins that form the troponin complex.

myriad n **1** 多数 syn *a enormous number, enormous numbers, many*

▶ Real catalysts comprise a myriad of potential active centers.

adj **2** 多様な syn *countless, limitless, many, numerous, unlimited*

▶ The most important instance of pattern recognition in chemistry is the periodic table, which systematizes the myriad complexity of chemical behavior in a two-dimensional pattern.

mysterious adj 不可思議な syn *ambiguous, obscure, paradoxial*

▶ The thermodynamic definition of entropy often seems both mysterious and arbitrary.

mysteriously adv 不可思議に

▶ Solid-state chemistry is concerned in large measure with extended solid-state structures, many of which may seem to be rather mysteriously described in terms of closest-packed arrays.

mystery n 謎 syn *paradox, puzzle, question*

▶ The nature of the colored solutions obtained on dissolving sulfur in oleum has remained somewhat of a mystery ever since their discovery in 1804.

N

naive adj 単純な syn *natural, simple*

▶ On a rather naive view, solids that are similar might be expected to show similarities in melting, provided these are measured in appropriate units.

naked adj 裸の syn *exposed*

▶ Compared with a naked proton, a shielded proton requires a higher applied field strength to provide the particular field strength at which absorption occurs.

name n **1** 名称 syn *designation, term*

▶ The names of the over 1500 enzymes identified are obtained by adding the suffix ase to the name of the process catalyzed or to the name of the molecule on which the enzyme acts.

vt **2** 命名する syn *call*

▶ Clathrate compounds were so named in 1947 by Powell.

3 指定する syn *designate, specify*

▷ The *R* and *S* scheme of naming absolute configuration is merely a nomenclature device and must not be thought to relate to the sign and magnitude of optical activity.

4 挙げる syn *list, mention*

▶ We have named all of substances likely to deviate because of quantization of molecular translation.

5 to name but a few 少し例を挙げれば

▶ Thermodynamics is a reliable guide in industrial chemistry, in plasma physics, in space technology, and in nuclear engineering, to name but a few applications.

named adj 言及された

▶ The solid oxides and sulfides all show, to a greater or lesser extent, homogeneous phases which vary in composition from that represented by the stoichiometric formula of the named compound.

namely adv すなわち that is, i.e.と同様に，より正確な情報を付け加えるときに用いる．セミコロンとコンマの用法は however のときと同じ． syn *specifically, that is to say*

▶ Each proton in the methyl group in ethanol feels three slightly different magnetic fields; there are, therefore, three lines with intensities in the ratio of number of spin states for each energy; namely, 1:2:1.

nanochannel n ナノチャンネル

▶ Referring to an earlier study by Hansen et al., in which *n*-hexane diffused faster in narrower nanochannels, the authors proposed a new model for the motion of adsorbed molecules which falls somewhere between flow and adsorption.

nanocrystal n ナノ結晶

▶ If a cluster is sufficiently large that it adopts the essential structural aspects of the crystalline solid, then it is described more specifically as a nanometer-scale crystal, or nanocrystal.

nanocrystalline adj ナノ結晶性の

▶ Ionic nanocrystals will grow if allowed to come into contact, so that a protective layer is required, as is done in the colloidal methods used to prepare nanocrystalline CdSe and GaAs particles in narrow size distribution.

nanofibrillar adj ナノ繊維の

▶ In applications that require thin films of nanofibers, the quality of the nanofibrillar films is dictated by the quality of dispersions they are cast from.

nanoparticle n ナノ粒子

▶ Nanoparticles can exert a multitude of effects in biological systems and individual cells.

nanoporous adj ナノ細孔のある

▶ Chemists have at their disposal nanoporous carbon, many nanoporous oxidic solids, nanoporous metals such as Pt and Au, nanoporous flexible polymers, nanoporous organic solids, and nanoporous metal-organic frameworks.

nanoscale n ナノスケール

▶ Materials processing techniques are well developed for the fabrication of nanoscale layers of a wide range of inorganic materials.

nanosized adj ナノサイズの
▶ Biorecognition in nanostructured materials is influenced by surface curvature and diffusion within nanosized spaces.

nanosystem n ナノスケールの系
▶ Nanosystems may be produced by either microfabrication, making big structures smaller and/or by embedding smaller features into macroscopic materials, or by using the techniques of component assembly and/or supramolecular chemistry to make small molecules bigger.

nanotubular adj ナノチューブの
▶ The versatility of tobacco mosaic virus as a biotemplate for fabrication of a range of nanotubular inorganic materials via metal deposition has been demonstrated.

nanowire n ナノワイヤー
▶ Discrete Bi nanowires may be created from a porous alumina template following its chemical removal.

narrow adj **1** 細い syn *slim, thin*
▶ In the technique of zone refining, the sample is normally in the form of a narrow cylinder.
vi **2** 狭くなる syn *constrict, decrease, diminish*
▶ For the higher homologs, the width of the liquid-crystalline phase at first increases with chain length and then narrows again.
vt **3** 狭める syn *lessen, limit, reduce*
▶ The mass spectrum can give a molecular formula or at least narrow the possibilities to a very few.

narrowing n 狭くなること
▶ Narrowing of the hyperfine lines sometimes results from warming or annealing the sample.

narrowness n 狭いこと
▶ The narrowness of the absorption lines poses an interesting problem.

nascent adj 発生期の
▶ Because zinc in the presence of dilute acid will accomplish reductions which hydrogen in its normal condition was incapable of effecting, it was customary to consider the action of zinc and acid as being due to nascent hydrogen.

nasty adj 危険な syn *acute, dangerous, serious, severe*
▶ The chemical advantages of these rather nasty liquid mineral acids as catalysts are compelling.

native adj **1** 本来の syn *intrinsic*
▶ In the use of probe methods in biology, great advantage is obtained if the chemistry of the probe is also very like that of the native atom, so isomorphous replacement has to be made with still greater care.
2 天然のままの syn *natural*
▶ Silver occurs both native and combined.

native protein n 未変性タンパク質
▶ Denaturation involves rupture of hydrogen bonds so that the highly ordered structure of the native protein is replaced by a looser and more random structure.

natural adj **1** 普通の syn *normal, ordinary, typical, usual*
▶ The boundaries of the fragments of a molecular system are defined in a natural way.
2 生得の syn *genuine, real*
▶ In chemistry, more than in physics, the essential and deep natural complexity of the subject matter is conspicuous.
3 天然の syn *native*
▶ Some optically active compounds are obtained from natural sources, since living organisms usually produce only one enantiomer of a pair.

natural abundance n 自然存在比
▶ Knowing the relative natural abundances of isotopes, one can calculate for any molecular formula the relative intensity to be expected for each isotopic peak: M＋1, M＋2, etc.

naturally adv **1** 天然に
▶ Many of the most important naturally occurring minerals and ores of the metallic elements are sulfides.
2 本来 syn *actually, inherently, intrinsically, normally*
▶ ZnO and SnO_2 are naturally slightly deficient in oxygen, especially at high temperatures and low oxygen partial pressures.
3 当然 syn *certainly, logically, needless to say, of*

course

▶ Naturally, each solute lessens the sorption of the other, but in practice it has been found that a more strongly sorbed solute largely displaces a more weakly sorbed substance from the surface.

natural product n 天然物

▶ Various natural products are synthesized in living systems from acetic acid and are known as acetogenins.

nature n **1** 性質 syn *character, features, properties, quality*

▶ The mechanism of an S_N reaction and the reactivity of a given alkyl compound RX toward a nucleophile Y depends upon the nature of R, X, and Y and upon the nature of the solvent.

2 天然 syn *environment, universe*

▶ Tin does not occur free in nature but mostly in the form of the dioxide.

near adj **1** 近い syn *attached, close*

▶ The crystal structure of Al is fcc, typical of many metals, each Al being surrounded by 12 nearest neighbors at 286 pm.

adv **2** ほとんど syn *almost, nearly, virtually*

▶ An energetic preference for linear or near-linear A—H···B configurations, at least in the crystalline state, is confirmed by the experimental data.

nearby adj 近くの syn *accessible, adjacent, close*

▶ ESR signals show splitting from coupling with the spins of certain nearby nuclei.

near infrared adj 近赤外の

▶ In electronically insulating ionic solids the absorption edge may occur in the ultraviolet, but in photoconducting and semiconducting materials it may occur in the visible or even in the near infrared spectral regions.

nearly adv ほとんど syn *almost, approximately, practically, virtually*

▶ Nearly all aldehydes and most methyl ketones afford solid, water-soluble addition compounds with $NaHSO_3$.

near ultraviolet n **1** 近紫外

▶ Saturated aldehydes and ketones absorb weakly in the near ultraviolet.

adj **2** 近紫外の

▶ Ultraviolet spectra of proteins are usually divided into the near and far UV regions.

neat adj 適切な syn *precise*

▶ The data were taken on a heating cycle of mixtures crystallized from a neat melt.

nebulize vt 霧状にする ☞ atomize

▶ Ultrasound can be used to nebulize solutions into fine mists, emulsify mixtures, and drive chemical reactions.

necessarily adv **1** 必然的に syn *inevitably, naturally, surely*

▶ As soon as a layer of product has been formed at the interface between two reacting solids, further reaction necessarily depends on mass transfer of one reactant (or both) through the product layer.

2 否定文で，必ずしも…でない

▶ One should appreciate that the assumption that bond energies can be transferred from one molecule to another is not necessarily valid.

necessary adj 必要な syn *essential, indispensable*

▶ When it is necessary to separate potassium from sodium, it is preferable to have the potassium in the form of the chloride.

necessary condition n 必要条件

▶ Chirality is the necessary and sufficient condition for the existence of enantiomers.

necessitate vt 必要とする syn *call for, demand, require*

▶ Although the mechanism for these reactions has not been rigorously established, it is plausible that they proceed via the formation of an iminium species, which would necessitate the involvement of water in the catalytic cycle.

necessity n **1** 必要性 syn *need, requirement*

▶ One of the most difficult aspects of long-chain peptide synthesis is the necessity of isolating and purifying each intermediate peptide along the synthetic route.

2 of necessity 必然的に syn *necessarily*

▶ The early studies on the structure of crystals were of necessity morphological, that is, based on the external appearance of the crystal.

need n **1 必要** syn *constraint, necessity, requirement*

▶ The driving force for all reactions of carbocations is the need to provide electrons to the electron-deficient carbon.

vt **2 必要がある** syn *call for, demand, require*

▶ All we need to know are the tabulated standard molar Gibbs energies for the reactants and products at the temperature of interest, and these can be obtained from standard tables.

aux **3 need not 必要がない** 否定文で，助動詞として用いられる．

▶ The color of a solid and a solution of the same solid need not be identical or even similar, because of the difference in environment of the molecules in the solid and in solution.

needle n **針状結晶**

▶ The acetyl compound, washed with warm water and ethanol and recrystallized from nitrobenzene, afforded yellow needles.

needless adj **needless to say 言うまでもなく** syn *of course*

▶ Needless to say, the field of chemical kinetics has been completely transformed over the past 100 years, indeed over the past ten years.

negate vt **を無効にする** syn *disprove, reverse*

▶ At high temperature the electron wanders onto a metal ion, M^+, momentarily neutralizing it and thereby negating its source of strong interaction with the remaining cluster.

negative n **1 負数**

▶ The Gibbs-energy change for a reaction is equal to the negative of the useful work which can theoretically be obtained from the reaction.

adj **2 数値として，負の** ☞ positive

▶ If heat is evolved, then ΔU is negative and the reaction is exothermic.

3 電気的に，負の

▶ The carbon-magnesium bond of the Grignard reagent is a highly polar bond, carbon being negative relative to electropositive magnesium.

4 光学的に，負の

▶ Another phase seems to be hexagonal and is uniaxial negative.

5 否定的な syn *negating, opposing*

▶ Attempts to detect charge motion have given negative results to date.

6 消極的な syn *uninterested, unresponsive*

▶ Observation of a single peak in the vapor-phase chromatographic analysis of a material is only negative evidence for purity.

negative catalyst n **負触媒**

▶ So-called negative catalysts do not act in the same way as catalysts but instead bring about their action either by destroying catalysts already present or by removing active intermediates such as atoms and free radicals.

negatively adv **負に**

▶ Negatively charged colloids are precipitated by the action of positive ions, and, conversely, the positively charged colloids are precipitated by negative ions.

neglect n **1 無視** syn *disregard, unconcern*

▶ The crystal used was approximately cylindrical in shape and was thin enough to warrant neglect of absorption corrections.

vt **2 無視する** syn *disregard, ignore*

▶ At room temperature, diffusion is sufficiently slow for these dopants that it can be neglected.

3 おろそかにする syn *omit*

▶ The thermal decomposition of dolomite has been studied by a number of workers, but the study of the kinetics in vacuum has been somewhat neglected.

negligible adj **無視してよい** syn *insignificant, minor, trivial, unimportant*

▶ Although hydrogenation is an exothermic reaction, it proceeds at a negligible rate in the absence of a catalyst, even at elevated temperatures.

neighbor n **1 隣接するもの**

▶ In a trigonal bipyramidal AX_5 molecule, the equatorial positions have only two neighbors at 90° whereas an axial position has three close neighbors at 90°.

adj **2 隣接の**

▶ Next nearest neighbor interactions are of the anion-anion and cation-cation type and are repulsive.

neighborhood n **1 近隣** syn *surroundings, vicinity*

▶ In glasses, the permanence of molecular neighborhoods, characteristic of crystalline bodies, exists together with a certain disorder characteristic of the liquid state.

2 in the neighborhood of およそ syn *about, approximately, nearly*

▶ In any event, the frequency factors are most commonly in the neighborhood of 10^{13} sec^{-1} in agreement with the transition-state theory if the entropy of activation is zero.

neighboring adj **隣接した** syn *adjacent, nearby*

▶ Neighboring stacks of molecules are held together by van der Waals forces.

neither pron **1 どちらも…でない** 二つのものを対象とするのが原則で，より多くのものを対象とするときは none を用いる．

▶ Since heat is not a state function, neither is the heat capacity.

neither of の次の名詞が複数形であっても，動詞は単数形を用いる．

▶ Neither of these extreme situations is at all likely.

adj **2 単数名詞を修飾してどちらも…でない**

▶ In neither case is the structure known: zinc hydride, ZnH_2, decomposes to its elements at below 100 ℃, cadmium hydride, CdH_2, is unstable at room temperature.

adv **3 neither A nor B AもBもどちらも…ない** これが主語の場合は動詞の数はBの数に一致させる．

▶ In most chemical processes, neither the energy nor entropy is held constant.

AとBは文法上，対等の関係にあることが必要である．名詞とは限らない．

▶ This reaction requires extremely high temperatures and exotic containers that neither react with the melt nor themselves melt at extremely high temperatures.

nematic adj **ネマチックの**

▶ Extending the length of a terminal *n*-alkyl chain increases the smectic tendencies relative to the nematic tendencies of a system.

nematogen n **ネマチック液晶性分子**

▶ The requirement that the molecule of a nematogen possesses a fairly rigid core structure is quite general.

nephelauxetic effect n **電子雲膨張効果**

▶ Nephelauxetic effects in the outer s and p orbitals of the probe ions Tl^+, Pb^{2+}, and Bi^{3+} are used for setting up scales of Lewis basicity for oxide systems.

nephelauxetic series n **電子雲膨張系列**

▶ The nephelauxetic series is quite different from the spectrochemical series, and the two assess different properties of the ligand field.

nephelometry n **ネフェロ分析，比濁分析**

▶ Opalescence is the basis of nephelometry, long used in quantitative chemical analysis.

nerve impulse n **神経インパルス**

▶ Nerve agents are cholinesterase inhibitors; they chemically inhibit the transmission of nerve impulses.

net n **1 網** syn *network*

▶ Lengthening the linker ligand often leads to structures with the same net, and this can lead to a series of structures with the same crystallographic space group but different pore sizes and volumes.

2 ネット

▶ Nets are abstract mathematical entities, consisting of a collection of points or nodes with some clearly defined connectivity or topology.

adj **3 最終の** syn *eventual, ultimate*

▶ Hydrolysis of the glycosidic bond proceeds with either net retention or inversion of anomeric configuration.

4 正味の syn *gross, overall, total*

▶ The usual net reaction for photosynthesis coverts H_2O and CO_2 to carbohydrates and O_2.

vt **5 うまく捕まえる** syn *capture, catch, trap*

▶ Collaborations with Drs. A. and B. netted several candidates for unimolecular rectifica-

tion.

net charge n 実効電荷 syn *effective charge*

▶ If the ion pairs have no net charge, they make no contribution to the conductivity, thus accounting for the anomalously low conductivity.

network n 網目（構造） syn *structure*

▶ Hybrid networks provide a unique opportunity to create interesting enantiometrically pure, porous networks.

neutral adj **1** 中性の

▶ Salts of strong acids with strong bases are not subject to appreciable hydrolysis. The solution reacts neutral and contains only as many H^+ and OH^- ions as corresponding to the ionization of water.

2 電気的に，中性の

▶ The entropies of hydration of neutral sets of ions can be calculated as the difference between the entropies of the ions in the gas phase and in aqueous solution.

neutrality n 中性

▶ The important feature to notice is that pH=7.0 corresponds to neutrality.

neutralization n **1** 酸塩基の中和

▶ The titration of an acid with a solution of a base is generally called neutralization.

2 電荷の中和

▶ Agglomeration is usually achieved by neutralization of the electric charges which maintain the stability of the colloidal suspension.

neutralize vt **1** 酸塩基について，中和する

▷ In neutralizing the alkaline solution of the vanadate with nitric acid, the solution must on no account to be made acid.

2 電荷に関して，中和する

▶ In minerals containing two or more cations and monoatomic anions, such as the perovskites and spinels, the charge on the anion is neutralized by the cations with which it is in contact.

3 無効にする syn *negate, nullify*

▶ The LUMO of CH_3^+ is delocalized over the methyl substituents. Adding electrons to the LUMO neutralizes hyperconjugation.

neutron bombardment n 中性子衝撃

▶ Elements of atomic number greater than 92 have been made either by neutron bombardment of uranium in a reactor or by nuclear reactions between cyclotron-accelerated ions and heavy atoms.

neutron diffraction n 中性子回折

▶ Neutron diffraction has been much used to locate light atoms, especially hydrogen in hydrides and hydrates.

neutron irradiation n 中性子照射

▶ Tritium is made on a large scale by neutron irradiation of enriched ^{6}Li in a nuclear reactor.

neutron scattering n 中性子散乱 ☞ magnetic scattering

▶ There is no simple dependence of neutron scattering power on atomic number.

never adv 決してない syn *not at all, on no account*

▶ The choice of critical micelle concentration is never unambiguous, since the change in slope occurs over a more or less narrow range of concentrations.

nevertheless adv それにもかかわらず，コンマ，セミコロンの扱い方は however の場合に同じ． syn *however, in spite of, yet*

▶ In pyrrole the nitrogen contributes two of the six π electrons, and although it does by virtue of its superior electron affinity acquire more than its share of the common stock, it is, nevertheless, left with a residual positive charge.

new adj 新しい syn *fresh, novel*

▶ The field is very extensive but introduces no new concepts into the general scheme of covalent heterocyclic molecular chemistry.

newly adv 新たに syn *just, lately*

▶ Nuclei will only develop into stable entities if the energy released through the formation of bonds in the solid state is greater than that required to maintain the newly created solid-liquid interface.

next pron **1** 次のもの

▶ In liquid hydrogen fluoride, the molecules tend to aggregate in chains and rings so that the positive hydrogen on one molecule attracts a negative fluorine on the next.

adj **2** 次の syn *second, subsequent*

▶ Once assured of the purity of a certain compound, the next problem is one of identification.

adv **3** …について syn *closest, nearest*

▶ Corundum is next to the diamond the hardest mineral.

4 次に syn *immediately, shortly*

▶ Assuming that data have been obtained giving concentrations at various times, the problem next arises of determining the order of a reaction with respect to all participants.

nicely adv うまく syn *properly, well*

▶ Aryldiazonium salts can, in most cases, be isolated as nicely crystalline fluoroborate salts.

nickel arsenide structure n ヒ化ニッケル型構造

▶ The NiAs structure is adopted by a variety of intermetallic compounds and by some transition-metal chalcogenides.

nitrate vt ニトロ化する

▶ $Sn(NO_3)_4$ readily oxidizes or nitrates organic compounds, probably by releasing reactive NO_3 radicals.

nitrating agent n ニトロ化剤

▶ The nitration of alkanes is usually carried out in the vapor phase at elevated temperatures using nitric acid or nitrogen dioxide as the nitrating agent.

nitration n ニトロ化

▶ Aromatic nitrations are usually effected with mixed acid, a mixture of nitric and sulfuric acids, at 0–120 ℃.

nitrify vt 亜硝酸塩に変える

▷ The amount mineralized to ammonium and then nitrified each year depends on the soil type, the weather, and probably the forms of organic N.

nitrogenation n 窒化

▶ The basic chemical of the cyanamide industry is calcium cyanamide obtained by nitrogenation of CaC_2.

nitrogen fixation n 窒素固定

▶ In nature, nitrogen fixation is performed by bacteria located on the root hairs of plants, as a result, plants are able to synthesize proteins.

nitrogen-fixing bacterium n 窒素固定菌

▶ Nitrogen is essential for the synthesis of proteins by the plant with the aid of nitrogen-fixing bacteria.

nitrogenous adj 窒素を含む

▶ By heating nitrogenous organic substances with concentrated sulfuric acid, potassium permanganate, mercury(II) oxide, etc., the organic matter is destroyed, and the nitrogen is completely changed to ammonium and held as ammonium hydrogensulfate.

no adj **1** 一つもない 名詞の単数形，複数形いずれの前に付けてもよい．用いる動詞が単数形か複数形かは，名詞の数によって決まる．ただし，no data は単数形として扱われる．

▶ At room temperature, silicon, like aluminum, has a very thin protective coating of oxide and so reacts with no other element except fluorine.

2 決して…でない be 動詞の補語としての名詞の前で用いる．

▶ Definitions and classifications are artificial conveniences, and the S_N1 and S_N2 categories are no exception.

nobility n 貴重性 syn *excellence, goodness*

▶ The inertness of tantalum arises particularly because of both its greater nobility and its high melting point which leads to a reduced tendency for alloy or intermetallic compound formation with the solid elements of interest.

noble adj 不活性な syn *unreactive*

▶ For many years gold was considered too noble to be chemically interesting.

noble gas n 貴ガス syn *inert gas, rare gas*

▶ Since the lighter members are by no means rare and since the heavier ones are not entirely inert, noble gases seems a more appropriate name and has come into general use during the past two decades.

noble metal n 貴金属 syn *precious metal*

▶ Passive chromium retains its luster in the air, is insoluble in dilute acids, and resembles the noble metals.

no-bond adj 非結合の
▶ The wave function ψ_0(AD) has been termed the no-bond function by Mulliken.

nodal adj 節の
▶ The number of nodal lines is equal to the quantum number *n*.

nodal plane n 節面
▶ A nodal plane is a planar surface that separates a wave function into regions of negative and positive sign.

node n 節
▶ A node is a point at which a wave function passes through zero.

nodule n 団塊 syn *lump*
▶ The size of nodules averages approximately 4 cm, varying from small pellets to masses several meters in diameter.

nomenclature n 命名法 syn *terminology*
▶ All nomenclature should be consistent, clear, and unambiguous and should conform to the rules established by IUPAC.

nominal adj 名目の syn *so-called*
▶ A positive temperature coefficient of solubility is not enough to provide a suitable supersaturation; the solid also must have a nominal solubility.

nominally adv 名目上は syn *supposedly*
▶ It was discovered that at high pressure some insulating members of charge-transfer complexes undergo a remarkable electronic phase transition, from a nominally neutral to a nominally ionic ground state.

nonalternant hydrocarbon adj 非交互炭化水素
▶ Nonalternant hydrocarbons are defined as those conjugated hydrocarbons which contain odd-membered rings,

nonaqueous solution n 非水溶液
▶ Partly because of the great practical importance of aqueous solutions and also because of the greater experimental difficulties in working with other solvents, nonaqueous solutions have received relatively little attention.

nonaqueous solvent n 非水溶媒 複合形容詞として使用.
▶ In these reactions, BrF_3 serves both as a fluorinating agent and as a nonaqueous solvent reaction medium.

nonassociated adj 非会合性の
▶ Hydrogen fluoride boils 100 degrees higher than the heavier, nonassociated hydrogen chloride.

nonbonded atom n 非結合原子
▶ Nonbonded atoms tend to take positions that result in the most favorable dipole-dipole interactions.

nonbonded interaction n 非結合相互作用
☞ noncovalent interaction
▶ The energy difference between the gauche and the anti forms of butane is the result of the repulsion of the electronic clouds of the methyl groups in gauche butane, called a nonbonded interaction.

nonbonding adj 非結合性の
▶ One or more edges of a cluster may be broken by the addition of an electron pair, so that a bonding electron pair is replaced by two nonbonding pairs.

nonbonding electron pair n 非結合性電子対 syn *lone pair, unshared electron pair*
▶ Other factors may be the difference in atomic radius, the type of hybridization, the polarization of nonbonding electron pairs, and the presence of exceptionally mobile electrons as in multiple bonds.

noncaloric adj カロリーのほとんどない
▶ Increasing research has developed several new noncaloric sweeteners.

non-carbon skeleton n 非炭素骨格 non と元素, 化合物質名の間にはハイフンを用いる.
▶ After the silicones, the polyphosphazenes form the most extensive series of covalently bonded polymers with a non-carbon skeleton.

noncentrosymmetric adj 対称中心のない
▶ A necessary condition for a crystal to exhibit spontaneous polarization and be ferroelectric is that its space group should be noncentrosymmetric.

nonclassical adj 非古典的
▶ Norbornenyl radicals have been shown not

to exhibit the nonclassical behavior that is seen in the corresponding cation.

nonclassical carbonium ion n 非古典的カルボニウムイオン

▶ With the experimental differentiation of trivalent carbenium from pentacoordinated carbonium ions, the classical vs. nonclassical carbonium ion controversy should now be finally settled.

noncombustible adj 不燃性の syn *incombustible*

▶ For example, sodium chloride, carbon tetrachloride, and carbon dioxide are noncombustible; sugar, cellulose, and ammonia are combustible but nonflammable.

nonconducting adj 非伝導性の

▶ Ionic compounds in the solid state are generally considered to be electrically nonconducting.

noncorrosive adj 非腐食性の

▶ The electrodissolution of metals in ionic liquids is essentially the same as the sulfuric and phosphoric acid process currently operated, but the electrolyte is noncorrosive and nontoxic.

noncovalent interaction adj 非共有相互作用 ☞ nonbonded interaction

▶ Proteins that possess a quaternary structure are composed of several separate polypeptide chains held together by noncovalent interactions.

noncrossing rule n 非交差則

▶ Two connecting lines may cross if they refer to orbitals that differ in parity but must not cross if they refer to orbitals of the same type. This is the noncrossing rule.

noncrystalline adj 非晶質の

▶ The reverse phase change, isotropic liquid-to-liquid crystalline phase, yields a noncrystalline opaque material.

noncumulative adj 累積しない

▶ Ethanol is rapidly oxidized in the body and is, therefore, noncumulative.

nondegenerate adj 縮退していない

▶ For a nondegenerate ground state, the alteration of the sense of bond alternation costs electronic as well as lattice energy; hence, solitons cannot be supported because they would be energetically unfavorable.

nondirectional adj 非方向性の

▶ Metallic bonding is, for the most part, nondirectional.

nondestructive adj 非破壊の

▶ Nondestructive tests are usually carried out by radiographic methods on large, finished metal products to determine the presence of internal defects likely to cause operational failure.

none pron いずれも…ない 意味によって，単数形とも複数形ともみなされる．

▶ None of the phenomena mentioned so far are restricted to titania.

nonelectrolyte n 非電解質

▶ Dissolved salts increase the internal pressure of the aqueous solutions to such an extent that the nonelectrolyte is squeezed out（salting out）.

nonempricial adj 非経験的な ☞ empirical

▶ The Mulliken definition of electronegativity is the only nonempirical one available and has received wide attention and acceptance.

nonequilibrium n 非平衡

▶ Nonequilibrium or metastable products are often produced by a process of fractional crystallization.

nonequivalent adj 等価でない

▶ The environments of these two diasterotopic protons are neither identical nor mirror images of each other; these protons are nonequivalent, and we expect an NMR signal from each one.

nonessential amino acid n 非必須アミノ酸 ☞ essential amino acid

▶ Glucose is primarily the source of the carbon skeleton for most of the nonessential amino acids.

nonetheless adv それにもかかわらず syn *nevertheless, regardless of*

▶ The chemistry of vanadium, although too rich, complex, and colorful to be predictable, seems, nonetheless, to be logically understandable in terms of electronic structure.

nonexistence n 非実在 syn *absence, lack*

▶ It is difficult to account for the existence of $Ni(CO)_4$ and the nonexistence of $Zn(CO)_4^{2+}$ unless double bonding is a factor.

nonexistent adj 存在しない syn *imaginary, unreal*

▶ Little reliable information is available on bromites and still less is established for iodites which are essentially nonexistent.

nonexponentially adv 非指数的に

▶ There is often a slower component of the emission which decays nonexponentially over several microseconds.

nonferrous metal n 非鉄金属

▶ The most important nonferrous metal sulfides are those of Cu, Ni, Zn, and Pb.

nonflammable adj 不燃性の syn *noncombustible*

▶ Sulfur hexafluoride is a colorless, odorless, tasteless, unreactive, nonflammable, nontoxic, insoluble gas prepared by burning sulfur in an atmosphere of fluorine.

nonflammability n 不燃性

▶ The unique properties of emulsion paints are ease of application, absence of disagreeable odor, and nonflammability.

nonhygroscopic adj 非吸湿性の

▶ Sulfamic acid is dry, nonvolatile, nonhygroscopic, colorless, white, crystalline solid of considerable stability.

nonideal adj 非理想的な

▶ The fugacity of a nonideal gas was defined as numerically equal to its activity, since the standard state of fugacity is by definition 1 bar.

nonideality n 非理想的性質

▶ The significant nonideality of alkane/perfluoroalkane mixtures has been well documented for many years.

nonideal solution n 非理想溶液 ☞ ideal solution

▶ The concepts of activity and activity coefficient are introduced to handle nonideal solutions.

nonidentical adj 同一でない

▶ Infrared, nuclear magnetic resonance, and mass spectra are essentially fingerprints of molecules and practically always are individualistic enough for nonidentical compounds to provide positive differentiation.

noninfectious adj 無感染の

▶ Plant viruses are noninfectious toward other organisms, and they do not present a biological hazard.

nonintegral adj 非整数の

▶ It is assumed that the orbital charge may have either an integral or nonintegral value and that the energy may be expressed as a quadratic function of the orbital charge.

noninteracting adj 相互作用をしていない

▶ The principal difficulty with the bond-moment analysis is that the electron distribution that results in the molecular dipole moment cannot always be treated in terms of separate, noninteracting components.

nonionic adj 非イオン性の

▶ An important reason for the current, widespread interest in radical-ion salts and charge-transfer complexes is that, in the solid state, many of these compounds exhibit nonionic electric conduction.

nonionized adj イオン化しない

▶ The formation of a nonionized substance causes an ionic reaction to go to completion.

nonlabile adj 置換不活性な

▶ Ligand dissociation in octacoordinate species is generally a relatively easy process although there is one notable exception, the nonlabile $Mo(CN)_8^{4-}$ ion.

nonlinear adj **1** 非直線の

▶ The microwave spectrum of HOCl in the gas phase confirms the expected nonlinear geometry with H−O 97 pm, O−Cl 169.3 pm, and angle H−O−Cl 103±3°.

2 非線形の

▶ In the search for materials with strong nonlinear responses, interest has focused principally on inorganic substances.

nonlinear optical adj 非線形光学の

▶ A prototypical nonlinear optical chromophore was 4-(*N*,*N*-dimethylamino)-4'-nitrostilbene in which the two benzene rings and the double bond are the conjugated π system and provide the polarizable electrons; the dimethylamino group acts as the donor, and the nitro group acts as the acceptor.

nonlinear optics n 非線形光学

▶ Dyes and dye-like molecules with highly polarizable π-electron systems are currently of interest in nonlinear optics, which is relevant to the new developments in laser telecommunications.

nonlocal adj 非局在の

▶ In contrast to the bulk susceptibility, the anisotropy of tropone shows little nonlocal character in spite of the fact that this molecule is essentially planar.

nonluminous adj 非発光性の

▶ Sodium salt vapors color the nonluminous gas flame a monochromatic yellow.

nonmetal n 非金属

▶ In general, nonmetals have very low to moderate conductivity and high electronegativity.

nonmetallic adj 非金属性の

▶ The most fundamental classification of the electronic properties of materials is in terms of metallic vs. nonmetallic properties.

nonnative adj 本来のものではない syn *foreign, strange, unfamiliar*

▶ Difluorotoluene allows scientists to use it as a probe of electrostatic effects in protein-DNA interactions without fear of interference by nonnative structure.

nonoccurrence n 発生しないこと

▶ The occurrence or nonoccurrence of rearrangement is a striking difference, and it provides one more piece of evidence that there are two mechanisms for nucleophilic substitution.

nonoxidizing adj 酸化力のない

▶ Furan polymers are dark-colored and are resistant to solvents, most nonoxidizing acids, alkalis, and specific corrosives such as dinitrogen pentaoxide.

nonpenetrating adj 浸透しない

▶ The pellet density is determined by the usual methods using nonpenetrating inert liquids and a pycnometer.

nonplanarity n 非平面性

▶ The nonplanarity of the molecules in the photochromic crystals is brought about by rotation of the aniline ring about the exocyclic C-N bond with respect to the plane of the salicylideneamine residue.

nonpolar adj 無極性の syn *apolar*

▶ A soap molecule has a polar end, $-COO^-Na^+$, and a nonpolar end, the long carbon chain of 12 to 18 carbons.

nonpolar compound n 無極性化合物

▶ Nonpolar compounds possess only weak intermolecular attractions and, therefore, have relatively low melting points and high vapor pressure compared to ionic materials.

nonpolarizable adj 非分極性の

▶ Lithium salts of large, nonpolarizable anions such as ClO_4^- are much more soluble than those of the other alkali metals, presumably because of the high energy of solvation of Li^+.

nonpolar solvent n 無極性溶媒

▶ As the concentration of alcohols in nonpolar solvents is increased, more and more of the molecules become associated and the intensity of the infrared absorption band due to associated hydroxyl groups increases at the expense of the free hydroxyl band.

nonporous adj 細孔のない

▶ Mesoporous materials are normally metastable with respect to nonporous forms of the same mineral.

nonproportionality n 比例していないこと

☞ proportionality

▶ The nonproportionality between the curves relating ESR intensity and paramagnetic susceptibility to copper concentration is explained as follows.

nonradiative adj 無放射の

▶ Ions that have nonradiative transitions to the ground state and which must be avoided in the preparation of phosphors include Fe^{2+},

Co^{2+}, and Ni^{2+}.

nonradiatively adv 無放射で

▶ The excited states all decay nonradiatively to the $^4F_{3/2}$ level from which laser action occurs to the $^4I_{11/2}$ level with a wavelength of 10,600 Å for neodymium glass and 10,640 Å for Nd^{3+}: YAG.

nonreducible adj 還元できない

▶ This chemistry appears to extend to several other transition metal oxides and even to some that are nontransitional and nominally nonreducible, provided that surface reduction actually occurs.

nonrenewable adj 再生不可能な

▶ Nonrenewable energy sources are materials of geologic origin, i.e., petroleum, natural gas, coal, shale oil, and uranium, which cannot be replaced once their supply is exhausted.

nonresolvable adj 分割できない

▶ Optical activity cannot be observed if any chiral molecules are present only as nonresolvable racemic modifications.

nonspecific adj 非特異性の syn *broad, general, unspecified*

▶ Gelatinous alumina aged to become insoluble and inert is a nonspecific emulsifying agent acting as a solid and not through reducing surface tension.

nonspherical adj 球状ではない

▶ Just as for nonspherical, nonpolar molecules, the lowest energy of polar molecules arises only in the most favorable angular orientation.

nonstoichiometric adj 不定比の

▶ Cu_2O is nonstoichiometric with an excess of oxygen, while ZnO normally contains a small excess of the metallic constituent.

nonstoichiometric compound n 不定比化合物

▶ Tungsten bronzes are well-defined nonstoichiometric compounds of general formula M_xWO_3 where M is some other metal, most commonly an alkali, and x is a variable < 1.

nonstoichiometry n 不定比

▶ For a long time, it was known that certain transition metal oxides could be prepared with an apparent wide range of nonstoichiometry.

nonstrained adj ひずみのない syn *unstrained*

▶ There is reason to believe that an attempt to construct a system of local rules for the diagonal components of the susceptibility tensor might succeed for nonstrained, nonaromatic molecules.

non-superimposable adj 重ねられない

▶ Optical isomers, enantiomorphs or enantiomers, as they are also known, are pairs of molecules which are non-superimposable mirror images of each other.

nonswellable adj 膨張しない

▶ On drying, gels of bentonite form coherent, translucent films which can be made nonswellable by ion exchange.

nontrivial adj 意味のある syn *no-nonsense*

▶ With changing composition, the intermolecular forces in the particle alter which in turn changes their spectroscopic properties and thus their refractive index data in a nontrivial manner.

nontypical adj 非典型的な

▶ Curiously, these two most nontypical MX_2 structure types are derived from two of the most flexible molecules among the group 2 dihalides.

nonuniformity n 不均一

▶ Below a thickness of 0.8 μm, spin-coated polymer films often possess many nonuniformities in structure and thickness.

nonvolatile adj 不揮発性の syn *involatile*

▶ The amino acids are nonvolatile crystalline solids which melt with decomposition at fairly high temperatures.

nor conj …もまた…ない

▶ Nor is the volume of the solution equal to the sum of the volumes of the components. 否定の表現を結ぶのに用いられる.

▶ No organic compound can be prepared in pure form by direct reaction of its component elements nor can the reverse reaction be carried out except in a few particular cases, so that the heats of formation must always be obtained by indirect methods.

normal adj **1** 正常な syn *conventional, ordinary, regular, usual*

▶ Samples containing normal naphthalene in normal durene, normal naphthalene in perdeuterodurene, and perdeuteronaphthalene in perduterodurene were examined.

2 直交する syn *perpendicular, vertical*

▶ If a piece of polarizing film is placed on top of another with the polarization of the first normal to that of the second, no light from below will come through both of them.

3 規定の ☞ normal solution

▶ Normal sulfuric acid is very carefully added with constant stirring until the precipitate redissolves.

4 炭素直鎖をもつ

▶ Normal alkanes undergo a remarkable series of solid-solid phase transitions prior to melting.

normal dispersion n 正常分散 ☞ anomalous dispersion

▶ Most media refract light of short wavelengths more strongly, showing what is called normal dispersion.

normality n 規定度

▶ The normality of a solution of an acid or base is the number of equivalents of acid or base per liter.

normalization n 規格化

▶ Because there is only one particle and it must be found somewhere, the integral over all spatial elements must be unity. This is known as the condition of normalization.

normalize vt 規格化する syn *standardize*

▶ The wave functions for molecular orbitals are normalized by multiplying a normalization constant.

normal liquid n 正常液体

▶ Our object in this work was to synthesize stable liquid-crystalline materials with long mesomorphic ranges and high transition temperatures to normal liquids.

normally adv 普通は syn *generally, ordinary, usually*

▶ α-Amino acids normally exist as zwitterionic salts and, therefore, show bands characteristic of the ionized carboxyl groups and an amine salt.

normal mode n 基準振動

▶ The complex vibrations of a polyatomic molecule can be dissected into sets of vibrations of the individual atoms called normal modes of vibration.

normal solution n 規定液

▶ The weight required to make a liter of normal solution depends upon the nature of the reaction involved.

normal spinel n 正スピネル ☞ inverse spinel

▶ In a normal spinel the A^{2+} cations occupy one-eighth of the tetrahedral interstices, and B^{3+} cations occupy half the octahedral spaces.

not adv **1** 後に続く語句を否定して，…でない

▶ Borides are not easy to prepare pure, and subsequent purification is often difficult.

2 not at all 少しも…ない

▶ Aluminum is one of the few abundant elements in the earth's crust which plants do not require at all for their metabolism.

3 not only A but B Aだけでなくまた B これが主語である場合，動詞の数は B に一致させる．A と B は名詞であるとは限らないが，文法上，対等の関係にあることを要する．次の not only A but also B に比べると，より B を強調する表現とされる．

▶ Chymotrypsin catalyzes hydrolysis not only of proteins but of ordinary amides and esters.

4 not only A but also B Aだけでなくまた B を意味する． これが主語である場合，動詞の数は B に一致させる．A と B は名詞であるとは限らないが，等位接続詞と同様に，文法上，対等の関係にあることを要する．

▶ Hydrolysis of esters is promoted not only by base but also by acid.

not only の代わりに not just を用いることがある．

▶ Gold could be not just an outstanding catalyst but the best catalyst for a reaction.

but also の代わりに but even を用いることがある.

▶ CCl_4 and CBr_4 molecules are sufficiently compact to rotate not only in the liquid but even in the solid state as well.

notable adj 顕著な syn *distinctive, extraordinary, remarkable*

▶ There were also early examples of hybrid materials with extended inorganic connectivity, the most notable being the layered zirconium phosphates.

notably adv 特に syn *especially, markedly, particularly*

▶ Silicon is notably more volatile than carbon and has a substantially lower energy of vaporization.

notation n 表示法 syn *signs, symbols*

▶ The ↑ notation signifies a gaseous product, and the ↓ indicates a product that precipitates out of solution.

note n **1** 注釈 syn *comment, explanation, remark*

▶ Certain examples of calamitic materials are worthy of further note.

vt **2** 気付く syn *notice, observe, perceive*

▶ It is important to note that an individual amino acid can be titrated by either acid or base.

3 言及する syn *comment, mention, report*

▶ It can be noted that N−N single bonds in hydrazines are weaker than the corresponding P−P bonds.

noteworthy adj 注目すべき syn *exceptional, notable, unique, unusual*

▶ A noteworthy feature of sulfonium salts is that, when substituted with three different groups, they can usually be separated into optical enantiomers.

nothing n **1** 無 syn *none*

▶ There is nothing very unexpected about most of the physical properties of aryl halides.

pron **2** 何も…ない

▶ Nothing has yet been said about the interpretation of hydration energy on a molecular model.

3 nothing but ただ…のみ syn *just, merely, only, simply*

▶ If the mineral on ignition loses nothing but water, the amount of the water can be determined by the loss in weight.

4 nothing of 少しも…ない syn *in no way, not at all*

▶ The designations D and L tell us nothing of the configuration of the compound unless we know the route by which the configurational relationship has been established.

5 nothing other than …にほかならない

▶ The potential-energy surface for such a reaction is nothing other than the common potential-energy curve for a diatomic molecule.

adv **6** 少しも…でない

▶ Although the X-ray structures of the complex aggregates displayed nothing unusual, a number of intermolecular C−H···Cl experimental contacts that were shorter than the sum of their calculated van der Waals radii were observed.

notice n **1** 注目 syn *attention, awareness*

▶ Often a generalization that encompasses many facts has escaped notice until a scientist with unusual insight has discovered it.

vt **2** 注目する syn *observe, perceive, recongnize*

▶ We noticed that color of the supernatant became more intense after a couple of centrifugation cycles, indicating an increased concentration of nanofibers in the supernatant.

3 指摘する syn *note, pay attention*

▶ It is important to notice that the transference numbers show some concentration dependence, particularly for electrolytes with highly charged ions.

noticeable adj 顕著な syn *discernible, distinct, recognizable*

▶ Although the ratio between nickel and titanium atoms in NiTi is close to 1:1, even a slight deviation from this value can cause a noticeable change in the temperature of the transition.

noticeably adv 顕著に syn *especially, notably, quite*

▶ The germanes are all less volatile than the corresponding silanes and, perhaps surprisingly, noticeably less reactive.

notion n 観念 syn *concept, conception, idea, thought*

▶ The cornerstone of the BCS theory is a notion that at first seems counterintuitive: electrons are attracted to each other to form what are called Cooper pairs.

notoriously adv **1** 悪い意味で，有名に

▶ The enthalpy of sublimation of carbon is a notoriously difficult quantity to obtain.

2 周知に syn *notably, prominently*

▶ The lanthanide elements, because of their similarity in size, are notoriously good at forming solid solutions with each other in their oxides.

notwithstanding prep …にもかかわらず syn *against, despite, regardless of*

▶ Notwithstanding their obvious complexity, living organisms can still be viewed as physicochemical systems that interact with their surroundings.

nourishment n 栄養物 syn *food, nutrition*

▶ Plainly one of the major and increasing problems facing the human race will be providing itself with adequate food and nourishment and, ultimately, limiting its own population growth.

novel adj 新しい syn *new, original, unusual*

▶ The relativivistic treatment results in a number of novel effects, both descriptive and theoretical, most of which can usually be neglected with little loss of accuracy and a great gain in convenience.

now n **1** 今 syn *right now*

▶ The nature of these other factors has remained elusive until now.

adv **2** 今では syn *nowadays, today*

▶ The recent work demonstrates that rate enhancement in heterogeneous catalysis is now being studied at molecular level.

nowadays adv このごろは syn *now, today*

▶ Nowadays, most chemists would regard ionic conductivity as one of the classical areas of physical chemistry, in which exciting new developments are unlikely to occur.

nowhere adv どこにも…ない syn *not anywhere*

▶ The ability of carbon to catenate is nowhere better illustrated than in the compounds it forms with hydrogen.

noxious adj 有毒な syn *harmful, poisonous, toxic*

▶ Examples of noxious gases are carbon monoxide, nitrogen oxides, hydrogen sulfide, sulfur dioxide, and vapors evolved by benzene, carbon tetrachloride, and a number of chlorinated hydrocarbons.

nozzle beam n ノズルビーム

▶ The nozzle beam is created by expanding a gas through a pinhole, from a high pressure oven, into a series of rapidly pumped chambers.

n-type semiconductor n n 型半導体 ☞ p-type semiconductor

▶ If doping increases the conduction-band electron population, an n-type semiconductor is produced.

nuance n 意味の微妙な差異 syn *detail, distinction, refinement, subtlety*

▶ Whereas molecular chemists are more interested in the nuance of angular geometry possibilities, solid-state chemists are more prone to ask why one particular structure is favored over another which is topologically different.

nuclear charge n 核電荷

▶ The energy of electron promotion depends on nuclear charge, atomic radius, and shielding; all three terms are intimately related.

nuclear division n 核分裂

▶ Chromosomes become visible under the light microscope only at certain phases of the nuclear division cycle.

nuclear fission n 原子核分裂 ☞ spontaneous fission

▶ In the process of nuclear fission a large nucleus splits into two highly energetic smaller nuclei and a number of neutrons.

nuclear force n 核力

▶ Nuclear forces are only exerted over very small distances of the order of 10^{-13} to 10^{-12} cm.

nuclear fusion n 核融合 ☞ fusion

▶ Nuclear fusion yields large amounts of energy in which the nucleus of light atoms, chiefly the hydrogen isotopes, unite to form helium.

nuclear magnetic resonance n 核磁気共鳴

▶ To cause nuclear magnetic resonance of the protons in acetaldehyde, we would have to increase the strength of the applied magnetic field to compensate for the shielding effect about each of the different types of protons in the molecule.

nuclear magnetic resonance spectrum, NMR spectrum n 核磁気共鳴スペクトル

▶ A set of protons with the same environment are said to be equivalent; the number of signals in the NMR spectrum tells us how many sets of equivalent protons a molecule contains.

nuclear quadrupole coupling constant n 核四極結合定数

▶ The nuclear quadrupole coupling constant eQq is the product of the quadrupole moment eQ of an atomic nucleus and the field gradient q, which gives the deviation of the potential V from spherical symmetry at the position of the nucleus.

nuclear quadrupole moment n 核四極モーメント

▶ Since $I < 1$, there is no nuclear quadrupole moment and, hence, no quadrupole broadening of ^{19}F resonance.

nuclear quadrupole resonance n 核四極共鳴

▶ The quadrupole moments of Cl, Br, and I have been exploited successfully in nuclear quadrupole resonance of halogen-containing compounds in the solid state.

nuclear reaction n 原子核反応

▶ The recognition of the enormous energy set free in nuclear reaction has suggested the source from which the stars draw their energy.

nuclear spin n 核スピン

▶ The isotope ^{17}O is important in having a nuclear spin, and this enables it to be used in NMR studies.

nuclear transition n 核転移

▶ Mössbauer spectroscopy involves nuclear transitions resulting from absorption of γ radiation; the conditions for absorption depend upon the electronic environment and site symmetry of the nucleus.

nucleate vi **1** 核となる

▶ The high-temperature structures nucleate most readily when they can.

vt **2** 核とする

▶ It is generally accepted that bone crystals are nucleated in the interstices of a crystalline assembly of collagen fibrils.

nucleation n 核生成

▶ The seemingly simple reversible change of monoclinic to triclinic *p*-dichlorobenzene begins at a nucleation site with a front which can be readily followed with the polarizing microscope.

nucleic acid n 核酸

▶ Nucleic acids are polymers made up of nitrogenous bases, monosaccharides, and phosphate in a regular repeating structure.

nucleobase n 核酸塩基

▶ The long wavelength absorption and emission properties of these novel nucleobases were employed to eliminate interference from UV-active amino acid residues and natural nucleobases.

nucleophile n 求核試薬 syn *anionoid reagent, nucleophilic reagent*

▶ By definition, whatever reacts with an electrophile must be a nucleophile.

nucleophilic adj 求核性の ☞ electrophilic

▶ Whereas ethyl bromide reacts easily with sodium methoxide in methanol to form methyl ethyl ether, vinyl bromide and bromobenzene completely fail to undergo nucleophilic displacement under similar conditions.

nucleophilic addition n 求核付加 ☞ electrophilic addition

▶ Of special importance in synthesis is the

nucleophilic addition of carbanions to α, β-unsaturated carbonyl compounds known as the Michael addition.

nucleophilic attack n 求核攻撃

▶ Nucleophilic attack can take place at any time after the hydrolysis and, thus, can involve any species from the initially formed ion pair to the free carbocation.

nucleophilicity n 求核性 ☞ electrophilicity

▶ The nucleophilicity of a reagent does not always parallel its basicity, measured by its ability to donate an electron pair to a proton.

nucleophilic substitution n 求核置換

▶ The components required for nucleophilic substitution are: substrate, nucleophile, and solvent.

nucleoprotein n 核タンパク質

▶ The so-called nucleoproteins are important constituents of the genes that supply and transmit genetic information in cell division.

nucleoside n ヌクレオシド

▶ A base covalently linked to either D-ribose or 2-deoxy-D-ribose is known as a nucleoside.

nucleosome n ヌクレオソーム

▶ Acetylated nucleosomes are associated with regions of active transcription, while deacetylated nucleosomes are associated with transcriptionally repressed regions.

nucleotide n ヌクレオチド

▶ Nucleotides are phosphoric acid esters of nucleosides in which there is a phosphate group at position C-5'.

nucleus n **1** 原子核 syn *atomic nucleus, kernal*

▶ The actual mass of the nucleus is always slightly less than the sum of masses of the constituents, the difference being the mass equivalent of the energy of formation of the nucleus.

2 結晶核 syn *crystal nucleus*

▶ When a saturated solution that has been freed of all nuclei is cooled just below it saturation point, no solid separates from solution.

3 細胞核 syn *cell nucleus*

▶ The nucleus is an organelle that normally contains a complete set of genetic instructions in the form of multiple, linear structure called chromosomes.

nuclide n 核種

▶ A particular nuclide is designated by the name of the element followed by the mass number

nuisance n 厄介なこと syn *difficulty, inconvenience*

▶ Interest in the gas hydrates originated mainly because of the nuisance of such compound formation in gas pipelines.

nullify vt 無効にする syn *abolish, cancel, neutralize, offset*

▶ Some of these new materials show such significant electrochemical stability that application to smaller-scale devices may nullify current economic barriers.

number n **1** 数 syn *digit, figure, numeral*

▶ A disadvantage of the use of heats of combustion is that, if one is interested in the difference in energy contents of two compounds, one must make use of the small difference between two large numbers.

主語である number が実際に数を意味するときには，名詞の単数形として扱われ，その前には the, this または that が付く．the large number となっても，単数形である．

▶ The large number of valence orbitals in polynuclear metal—metal-bonded complexes, coupled with their highly delocalized character, leads to a breakdown of simple orbital descriptions of ground- and excited-state reactivities.

数字を文頭に用いることは避けて，数と単位の両方ともを綴る．文中では数と単位の組み合わせを用いて差し支えない．

▶ Twenty-three milliliters of ethanol and 40 mL of acetic acid were mixed.

2 a number 多数 syn *several, some*

▶ This unified understanding will continue to be extremely fruitful because it bears directly on important catalytic processes, including a number with commercial importance.

3 a number of 多数の その後の名詞は複数

形，動詞も複数形を用いる． syn *several, some*

▶ There are a number of ways to achieve unsaturation in a molecule.

4 in number 数の上で syn *numerically*

▶ Platinum(IV) complexes rival those of Pt(II) in number and are both thermodynamically stable and kinetically inert.

vt **5** 番号を付ける syn *count, enumerate*

▶ The parent hydrocarbon is numbered starting from the end of the chain, and the substituent groups are assigned numbers corresponding to their positions on the chain.

number-average molecular weight n 数平均分子量 ☞ weight-average molecular weight

▶ Osmotic pressure gives a number-average molecular weight, because every ion, molecule, or particle gives the same osmotic effect.

numbering n 番号を付けること syn *counting*

▶ The atom numbering used in the analysis and in the subsequent discussion is shown in Figure 1.

numeral n 数字 syn *figure, number*

▶ The oxidation state of each P is shown as a superscript numeral.

numerator n 分母に対して，分子 ☞ denominator

▶ If an expression obtained by differentiation of a function is of such fractional form that the numerator is the differential coefficient of the denominator, the original function differentiated was the logarithm of the denominator.

numerical adj 数値の

▶ In order to obtain the maximum benefit from this high-quality experimental data, it is necessary to develop a volume-computation method whose numerical error is significantly less than the experimental error.

numerically adv 数の上で syn *in number*

▶ At low concentrations, the solution obeys Henry's law, and so the activity is then numerically equal to the molality.

numerical value n 数値

▶ When viscosity is measured under constant conditions, a definite numerical value can always be obtained even though this may vary greatly with the method employed.

numerous adj 多数の syn *many, substantial*

▶ Numerous reactions are available for the artificial production of tritium.

nutrient n **1** 栄養素 syn *nourishment*

▶ Bioreactors containing anaerobic bacteria can convert food waste and other trash into soil nutrients and methane gas.

2 補給物質

▶ A constant temperature difference is maintained between a nutrient and a growth chamber.

adj **3** 栄養分の

▶ Storage of inorganic compounds in early organisms may have led, by adaptation, to functional use, rather than simply to overcome external nutrient deficiencies.

4 補給の

▶ Growth from the vapor customarily involves a transport of vapor from a region containing solid nutrient material at a temperature t_1 to a second region of crystal growth at a temperature t_2 that is slightly below t_1.

nutrition n 栄養 syn *nourishment*

▶ Nutrition requires a steady supply of required chemicals, and storage is then a valuable buffer of steady-state use.

nutritional adj 栄養上の

▶ Food antioxidants are effective in very low concentration and not only retard rancidity but protect the nutritional value by minimizing the breakdown of vitamins and essential fatty acids.

O

obedience n 従うこと syn *agreement, conformity, observance*

▶ The empirical criterion for ideal behavior is obedience of the system to some relatively simple and general law.

obey vi 従う syn *agree to, follow, observe*

▶ When two liquids are sufficiently alike to obey Raoult's law, it is evident that they must be miscible in all proportions, because only where the internal forces are sufficiently unlike could there be separation into two liquid phases.

object n **1** 物 syn *entity, item, thing*

▶ Dissymmetric objects can be characterized by their chirality, or handedness.

2 目的 syn *aim, goal, objective, purpose*

▶ The object of chemistry is to discover the laws which govern the union or the decomposition of substances and to determine the limits within which such changes are possible.

objection n 異論 syn *argument, doubt, opposition, question*

▶ Several further objections can be raised to the above mechanism.

objective n **1** 目的 syn *aim, goal, object, target*

▶ Many applications of vibrational spectroscopy have as their objectives only the identification of a molecule or a functional group.

2 対物レンズ ☞ objective lens

▶ If the substance has a lower refractive index than the liquid, a slight lowering of the objective from good focus on the fragment causes brightness within the particle.

objective lens n 対物レンズ ☞ ocular lens

▶ The microscope usually has several objective lens of different magnification, which are readily interchangeable.

oblate adj 扁平な ☞ prolate

▶ An oblate ellipsoid will cause an elongation of the tetrahedral geometry, but this type of distortion has not been observed.

obligatory adj 必須の syn *indispensable, necessary, required, requisite*

▶ The brain has an obligatory requirement for glucose, so the maintenance of a blood glucose concentration of 5 mM is essential for brain function.

oblige vt 余儀なくさせる syn *demand, force, require*

▶ Because electrons with parallel spin are obliged by the Pauli principle to avoid each other, the electrostatic repulsion between them is less.

oblique adj 斜めの syn *angled, diagonal, inclined*

▶ All crystals of the triclinic system are referred to three unequal axes whose intersections are all oblique.

obscure adj **1** 不明瞭な syn *ambiguous, indistinct, unclear, vague*

▶ The origin of the rate enhancement effects observed in these complex systems remains obscure mainly because the precise structure of the active site is unknown.

vt **2** 不明確にする syn *becloud, cloud*

▶ The writing of the hydrated proton, i.e., of H_3O^+, and at the same time the use of symbols such as Na^+, Cl^-, OH^-, and so forth, tends to obscure the fact that all ions are associated with solvent molecules.

observable n **1** 原理的に観測可能な物理量

▶ Every observable must have a corresponding operator in the quantum theory.

adj **2** 観察できる syn *discernible, noticeable, perceptible, visible*

▶ In practice, the change of an equilibrium constant with pressure is fairly small, and quite high pressures have to be used in order to produce an observable effect.

observance n 観察 syn *examination, inspection, observation*

▶ This structural motif seemed ideal for the observance of thermotropic mesomorphism.

observation n 観察 syn *examination, inspection, observance, survey*

▶ The lifetime of the transition state is so short relative to the time scale of the majority of our investigation techniques that direct observations are normally not possible.

observational adj 観察上の

▶ All our knowledge, both factual and interpretive, is rooted ultimately in our observational and measuring skill.

observe vt **1** 観察する syn *examine, inspect, monitor, study*

▶ In considering any structure, it is important to observe the coordination, or the relationship between an ion and its nearest neighbors.

2 守る syn *conform to, follow, keep, obey*

▶ It must be remembered that other factors can also modify these physical properties; hence, caution must be observed in their interpretation.

observer n 観察者 syn *watcher*

▶ Even most casual observer is surely impressed by the remarkable synthetic capability of organisms to produce materials such as seashells, pearls, bone, coral, sea-urchin tests, etc.

obsolete adj すたれた syn *old, out of date*

▶ Vitriol is an obsolete term once used to refer to a number of sulfates because of their glassy appearance.

obstacle n 障害 syn *barrier, hindrance*

▶ With present computer-automated data interpretation, molecular complexity is not a great obstacle.

obstruct vt さえぎる syn *block, preclude, prevent*

▶ Since the product of an enzymic reaction resides in the active site at the end of the reaction, it is not surprising that a high concentration of it might obstruct the active site in some way and thus inhibit the reaction.

obtain vt 得る syn *acquire, get, seize*

▷ Commercially available acetonitrile, hydroiodic acid, hydrobromic acid, trimethylamine (25% in methanol), and bromine were used as obtained.

obtainable adj 購入できる syn *accessible, available, ready*

▶ All of the hydrocarbons used were of the best grade obtainable from Eastman Kodak Co.

obvious adj 明白な syn *apparent, clear, evident, perceptible, plain*

▶ One of the obvious problems in studying polymorphism is the determination of the melting point for a metastable polymorph.

obviously adv 明らかに syn *apparently, clearly, evidently, plainly*

▶ Along with a more fundamental understanding of the mechanism of acid-base reactions, the Lewis approach obviously offers a much greater generality than any previous approach.

occasion n **1** 場合 syn *circumstance, opportunity*

▶ Anhydrous sodium sulfate can be used on almost all occasions, but the drying action is slow and not thorough.

vt **2** 引き起こす syn *bring about, cause, give rise to, provoke*

▶ The statistical replacement of gold atoms by the smaller atoms of copper occasions a slight reduction in the cell side, which varies nearly linearly with composition, but otherwise no alteration in the structure occurs.

occasional adj 時折の syn *incidental, irregular*

▶ In addition to occasional compounds which feature low coordination numbers, there are many examples of 6, 8, and 12 coordination.

occasionally adv 時折 syn *sometimes*

▶ Only occasionally is ^{13}C near enough to another ^{13}C for ^{13}C–^{13}C spin-spin coupling to occur.

occlude vt 気体，液体を吸蔵する

▷ Care must be taken that liquid does not become occluded during solidification.

occupancy n **1** 占有 syn *occupation*

▶ Variations in local structure such as site occupancies and vacancies can be observed directly with high-resolution electron microscopy.

2 占有率

▶ In compound **3**, the asymmetric unit consists of one palladium complex molecule, one molecule of dichlorometane, and one half of a pentane molecule, the latter exhibiting 50% occupancy.

occupation n 占有 syn *occupancy*

▶ The X-ray diffraction experiment measures the average occupation of the tetrahedral sites.

occupied adj 被占の syn *involved*

▶ The loss of electrons by graphite leads to the production of holes in its almost fully occupied band, with a consequent increase in the number of charge carriers.

occupy vt 占める syn *fill up, take up, use up*

▶ If the d orbitals were occupied equally, the resulting electron density would have spherical symmetry.

occur vi 起こる syn *arise, come about, happen*

▶ When a chemical reaction takes place in a crystalline material, it invariably occurs heterogeneously.

occurrence n 存在 syn *appearance, existence*

▶ Despite their chemical similarity, Li, Na, and K are not closely associated in their occurrence, mainly because of differences in size.

oceanic adj 海洋の syn *marine*

▶ There are unlimited supplies of NaCl in natural brines and oceanic waters.

octagon n 八角形

▶ Ordinary crystalline sulfur consists of S_8 molecules which have the form of a puckered octagon.

octahedral adj 八面体の

▶ In an octahedral environment, the five d orbitals on a transition metal atom are no longer degenerate but split into two groups: the t_{2g} group of lower energy and the e_g group of higher energy.

octahedral complex n 八面体錯体

▶ Dealing with an octahedral complex, we shall assume the presence of six ligands octahedrally coordinated about the metal atom.

octahedrally adv 八面体形に

▶ In the structure of $CuSO_4 \cdot 5H_2O$, four molecules of water coordinate the copper atoms, each of which is also coordinated by two oxygen atoms of two SO_4 groups and is, therefore, octahedrally surrounded by neighbors.

octahedron adj 八面体 複数形は octahedra

▶ TiO_6 octahedra are linked by sharing edges and corners to form a three-dimensional framework.

octet n 八隅子

▶ There is a considerable tendency for boron to acquire an additional electron pair to fill the fourth orbital and so attain an octet of electrons.

ocular lens n 接眼レンズ ☞ objective lens

▶ The ocular and objective lenses serve the purpose of magnifying the object.

odd adj **1** 奇数の ☞ even

▶ The solubilities of cucurbit[*n*]uril (CB[*n*]) are intriguing; CB[*n*] with odd *n* are nicely soluble in neutral water whereas CB[*n*] with even *n* are poorly soluble.

2 ふたつ一組の片方の syn *unpaired*

▶ Since the magnetic susceptibility of radicals is almost entirely due to the spin of the odd electron, the magnetic susceptibilities of radicals are similar and approximately equal to the theoretical value for a free electron.

oddly adv 奇妙に syn *curiously, surprisingly*

▶ Oddly enough, there has been no exact definition of hardness.

odd number n 奇数 ☞ even number

▶ No stable elements heavier than nitrogen have an odd number of both protons and neutrons.

odd-numbered adj 奇数の

▶ Magnetic properties are always found with nuclei of odd-numbered masses, ^{1}H, ^{13}C, ^{17}O, ^{19}F, ^{31}P, etc., and nuclei of even mass but odd atomic number, ^{2}H, ^{10}B, ^{14}N, etc.

odor n 臭気 syn *aroma, fragrance, perfume, smell*

▶ Many compounds have a characteristic odor that is an effective means of identification.

odoriferous adj **1** 芳香を放つ syn *aromatic*

▶ It has been known since antiquity that the odoriferous constituents of a plant could be greatly concentrated in the form of an essential oil by gentle heating of the plant material.

2 悪臭のある syn *offensive*

▶ Occasionally these odoriferous characteristics can be turned to good use, as in the doping of odorless natural gas with ethanethiol as a warning against leaks.

odorless adj **無臭の**

▶ N_2O is a colorless, odorless gas, which has the power of supporting combustion, by giving up its atom of oxygen, leaving molecular nitrogen.

of prep **1 …からなる** 物質，材料などを示す．質に変化は伴わない． syn *constitute by*

▶ The balance is made of quartz.

2 …に関して syn *about, concerning*

▶ A considerable number of measurements have been made of the enthalpy and entropy changes that occur on adsorption and desorption.

3 …の中の

▶ Of the two products, methanol and benzoate, only the methanol was found contain ^{18}O.

4 原因を表して，…で

▶ The total amount of glycogen synthase in the cell does not change in a few minutes or hours, but the proportion of the inactive and active forms does, and this is under the control of insulin.

5 of＋名詞 形容詞の働きをする．of と名詞の間にさらに形容詞を挿入してもよい．

▶ Of particular interest is the very high conductivity of the hydrogen ion.

off prep **場所から外れて** syn *apart from, away from, beside*

▶ In a ^{13}C NMR spectrum we cannot see the absorption by protons because these signals are far off the scale.

off-center adj **中心を外れた** syn *one-sided*

▶ The structure of WO_3 is a distorted version of the ReO_3 structure in which tungsten atoms are slightly off-center in adjacent unit cells such that the W–W distances are alternately long and short.

off-color adj **色の悪い** syn *inappropriate, incompatible, unfit*

▶ If the initial solution is yellowish or has a dull or off-color, it is treated with decolorizing carbon.

offensive adj **不快な** syn *disagreeable, unpleasant*

▶ H_2Se is a colorless, offensive-smelling, poisonous gas which can be made by hydrolysis of Al_2Se_3.

offer vt **提供する** syn *provide, submit*

▶ An alkylbenzene with a side chain more complex than methyl may offer more than one position for attack, and so we must consider the likelihood of obtaining a mixture of isomers.

off-gas n **排気** ☞ exhaust

▶ Sodium chlorite is used as an oxidant for removal of nitrogen oxide pollutants from industrial off-gases.

offset vt **1 相殺する** syn *cancel, compensate, counterbalance, neutralize, nullify*

▶ At high temperatures, the entropy of solid solution formation is able to offset the significant positive enthalpy term.

adj **2 ずれる** syn *place out of line*

▶ The diamond structure is built up of two interpenetrating fcc lattices which are offset along the body diagonal of the unit cell by one-quarter of its length.

often adv **1 しばしば** syn *frequently, regularly*

▶ The perchlorate ion is used often as inert anion in studies of metal ion complexes in aqueous solution.

2 more often than not 50%以上の頻度で，たいてい

▶ The surfaces of the single crystals of polyethylene more often than not show irregular corrugations and pleats.

ohmic adj **電気抵抗の**

▶ The relatively low electric conductivity of ionic liquids can lead to ohmic heating when significant currents are passed through the electrolytes.

oil n **油**

▶ The residue is a light yellow, rather mobile oil consisting of pure vitamin K_1.

oil-in-water emulsion n **水中油滴型エマ**

ルション☞ water-in-oil emulsion

▶ Sodium stearate has very limited solubility at room temperature but within that range it forms normal oil-in-water emulsion.

oil shale n オイルシェール

▶ The organic component of oil shale is a bitumen-like solid which is a mixture of aliphatic and aromatic compounds of humic and algal origin.

oil-soluble adj 油溶性の

▶ A combination of an oil-soluble substance with a water-soluble substance can produce a very stable emulsion.

oily adj 油性の syn *fatty, greasy*

▶ The oily residue was crystallized from ether–hexane to give 80 mg of white needles, mp 88–89°.

olation n オール化

▶ Olation is the name given to the formation of polynuclear complexes by splitting out of water between hydroxo groups.

old adj 古いもの the を付けて，名詞的に用い，単数扱いをする． syn *early, former*

▶ Many of the older of the vapor-pressure data represent determinations of doubtful validity on substances of questionable purity.

old-fashioned adj 古風な syn *obsolete, outdated*

▶ Alkanes are sometimes referred to by the old-fashioned name of paraffins.

oleum n 発煙硫酸 syn *fuming sulfuric acid*

▶ Oleum consists of a complex mixture of different polysulfuric acids, and it is viscous and has a high vapor pressure of SO_3.

oligomer n オリゴマー

▶ Oligomers of chromophores have attracted interest as light-harvesting antenna energy-transfer systems as well as models for electron and hole transport.

oligomeric adj オリゴマーの

▶ Generally, gel filtration buffers do not dissociate oligomeric proteins into their subunits so that the proteins elute from the column according to their native size.

oligomerization n オリゴマー化

▶ Oligomerization does not preclude attempts to crystallize a protein as long as the oligomers are all identical.

oligosaccharide n オリゴ糖

▶ The word oligosaccharide describes carbohydrates with few（up to ca. 10）monosaccharide units.

omit vt 省略する syn *exclude, leave out*

▶ Methods not generally applicable to the growth of organic crystals either have been omitted entirely or are referred to summarily.

on prep **1** 場所，…の上に on と upon はしばしば交換できるが，on は位置や静止状態を，upon は方向や動きを強調する．

▶ When a reaction takes place on a surface on which there is a variation of activity, the overall rate is the sum of the rates on the various types of site.

2 動作，…によって

▶ On exposure to the air, the benzene of crystallization evaporates and the mandelic acid is left as a white powder.

3 基礎，原因，条件など，…に基づいて

▶ Each surface is probably quite rough on the atomic scale; thus, the atoms of the two surfaces are collectively at a variety of distances from one another.

4 方法，手段，器具，…によって 受動的表現を用いる．

▶ Infrared spectra were recorded on a Nicolet 7199 Fourier transform spectrometer.

5 …すると 動名詞の前におくと副詞の役割を果たす句となる．

▶ On heating the Fe^{2+}-containing minerals in air, oxidation of Fe^{2+} to Fe^{3+} occurs first.

once n **1 at once** 直ちに syn *immediately, instantly, promptly*

▶ It follows at once from this equation that the equilibrium constant K_c is equal to the ratio of the rate constants, k_1/k_{-1}.

adv **2** かつては syn *formerly, previously*

▶ A short time later, a wide variety of other medically important proteins that were once difficult to obtain became readily available as a result of recombinant DNA technology and expressing proteins in bacteria.

3 once again もう一度

▶ Strain-hardened metals can be rendered malleable and ductile once again by high-temperature annealing.

conj **4** ひとたび…すると syn *as soon as, at any time, if ever*

▶ Once there is a significant excited-state population, then stimulated emission will compete with absorption.

one n **1** 一つ

▶ There are a number of ways in which an aldose can be converted into another aldose of one less carbon atom.

2 one by one 一つずつ syn *individually, separately*

▶ In the method of molecular orbitals, the electrons are imagined to be fed one by one into the molecule.

pron **3** 限定語をともなって，**特定のもの** syn *a particular, a specific*

▶ A ferromagnetic material is one possessing a spontaneous magnetic moment.

one of ＋ 名詞の複数形の場合も，続く動詞は単数形を用いる．

▶ X-ray diffraction is one of the most important characterization tools used in materials science.

4 one another 二者以上の間で，**互いに** each other と区別する必要はない．

▶ An example of a solid solution that permits any degree of substitution is cupronickel, Cu_xNi_{1-x}: both Cu and Ni have the face-centered cubic structure, and the atomic radii are within 3% of one another.

adj **5** 一つの syn *single*

▶ It is not possible to obtain a comprehensive understanding of the various spectroscopic features observed using one particular method alone.

one-component system n 一成分系

▶ The system MgO is a one-component system, at least up to the melting point 2700 ℃, because the composition of MgO is always fixed.

one-dimensional adj 一次元の

▶ The structural properties of the resulting one-dimensional crystals are principally dictated by the structural chemistry of the bulk material.

one-dimensional solid n 一次元固体

▶ The original idea of Peierls was simply that a one-dimensional solid with a half-filled band would spontaneously undergo a transition to an insulator.

one-electron orbital n 一電子軌道関数

▶ The basic assumption is that the wave function can be well represented as a product of one-electron orbitals.

one-pot n ワンポット

▶ Reliable, one-pot syntheses of *p-tert*-butylcalixarenes are now available from *p-tert*-butylpenol, formaldehyde, and a base.

one-to-one adj 一対一の

▶ The carbon skeletons and structural formulae for alkanes are in one-to-one correspondence so that either can be used to deal with classes of alkanes.

ongoing adj 進行中の syn *continual, continued, continuous, running*

▶ The combination of several ongoing debates and a number of unanswered questions leaves much justification for continuing to study difluorotoluene and other nucleoside analogues as substrates for polymerases.

onium compound n オニウム化合物

▶ The ammonium salts constitute only one of several different classes of substance, which are known collectively as the onium compounds.

on-line adj オンライン式の

▶ On-line corrosion measurements are widely undertaken in production and process plants and in the laboratory.

only adj **1 the only** 唯一の syn *exclusive, lone, single, sole*

▶ The system $FeCl_3$–H_2O is binary as long as the only solid phases are ice and $FeCl_3$ or hydrates of $FeCl_3$ and as long as the vapor, if present, is pure H_2O or H_2O and $FeCl_3$.

adv **2** ただ…だけ 修飾する語，句に可能な限

り近接させて用いる. syn *exclusively, just*

▶ Chlorobenzene is converted into phenol by aqueous sodium hydroxide only at temperatures over 300 ℃.

onset n 開始 syn *beginning, initiation, start*

▶ Within 10 min of the onset of the reaction the whole solution was deeply colored.

onto prep …の上に upon と同様に方向や動きを示唆する.

▶ A pulsed laser beam is directed onto the surface of a rotating or translating graphite disk.

onward adv 先へ syn *forward*

▶ Simple oxo acids are formed mainly by the elements from group 13 onward.

opacity n 不透明 syn *opaqueness*

▶ Metals are sharply distinguished from other types of solid matter by a number of physical properties, of which the high electrical and thermal conductivity and the optical opacity are among the most obvious.

opalescence n 乳光

▶ The opalescence of aqueous solutions of salts of the heavy metals shows the presence of colloidal products due to hydrolysis.

opalescent adj 乳白色を発する syn *lustrous, pearly*

▶ On gently rubbing the inner walls of the tube with a glass rod, the sulfur crystallizes in the opalescent form.

opaque adj 不透明な syn *cloudy, obscure, turbid*

▶ Fine powders with particle sizes in the range 10 to 100 μm may often be transparent whereas they would be opaque in bulk form.

open adj **1** 開いた syn *unclosed*

▶ In most chemical systems we are concerned with processes occurring in open vessels, which means that they occur at constant pressure rather than at constant volume.

2 空の syn *available, unfilled, vacant*

▶ Five-eighths of the tetrahedral holes and all of the octahedral holes formed by the iodide ions are vacant, and these open sites provide possible pathways for the smaller copper cations to move through the crystal, carrying charge.

3 白抜きの ☞ filled circle, hatch, shaded

▶ The excellent fit from this treatment is shown with the open circles and curves in Figure 1.

4 隙間のある syn *unfilled, vacant*

▶ The zeolites have very open structures formed by joining silicate chains or cages such as the truncated octahedron.

5 未決定の syn *questionable, unsolved*

▶ The question of what constitutes "one kind of molecules" is left open.

6 公開の syn *accessible, available, free*

▶ The changes in ionic-covalent character of chemical bonding brought about by temperature are very much open to discussion among molten salt chemists.

7 open to 免れない syn *susceptible*

▶ This explanation seems to be open to question.

vt **8** 啓発する syn *furnish, introduce, launch, offer, present, provide*

▶ The introduction and development of annulated bis(imidazolium), bis(imidazolylidene)s and their respective polymeric materials have opened several new and exciting opportunities in macromolecular chemistry.

9 open up (a) 広げる syn *expand, extend, spread*

▶ Like charges can repel each other to cause the protein chain to open up, while opposite charges can act to pull the chain together.

(b) 開発する syn *develop, establish, launch, pioneer*

▶ The field was opened up by G. Wilkinson and his group in 1966 when they showed that [$Pt(PPh_3)_3$] reacts rapidly and quantitatively with CS_2 at room temperature to give orange needles.

open-chain adj 開鎖の

▶ Being α,β-unsaturated ketones, quinones are expected to have the possibility of forming 1,4-addition products in the same way as their open-chain analogs.

open-chain compound n 開鎖化合物

▶ The typical aldehyde reactions of D-(+)-glucose are presumably due to a small amount of open-chain compound, which is replenished as fast as it is consumed.

open circuit n **開路** 複合形容詞として使用

▶ Polarization of dry cells is the difference between the open-circuit voltage and the closed-circuit voltage.

opening n **1 始まり** syn *beginning, start*

▶ The latter issue is critical because generation of an alkylidene from the dienes could result in sequential opening of the ring and cross metathesis.

2 穴 syn *gap*

▶ The openings to hold the wires are cut wedge-shaped, so that any shape of wire can be inserted.

openly adv **公然と** syn *frankly, plainly*

▶ Even only a few years ago people would openly question whether gold could be the key component for such a wide number of catalysis discoveries.

openness n **開放状態** syn *freedom*

▶ In natural rubber, there is a certain openness of structure associated with $=CH-CH_2-$ linkages which in the cis isomer permits the molecule considerable freedom to undergo conformational changes.

open system n **開いた系** ☞ closed system

▶ The introduction of partial molar quantities is designed to handle open systems.

operate vi **働く** syn *act, function, serve, work*

▶ Other types of cells that are finding applications are miniature primary cells which operate at room temperature and which have a long life rather than a high power output.

operating adj **操作の** syn *operational*

▶ Metal organic chemical vapor deposition is a useful technique for growing layers, and new developments, allowing lower operating temperatures, have extended the scope of this approach.

operation n **1 操作** syn *manipulation*

▶ Rotation about this axis by $360/n$ degrees gives an identical orientation, and the operation is repeated n times before the original configuration is regained.

2 操業 syn *business, procedure, project*

▶ High-temperature chlorination of a number of paraffin hydrocarbons is carried out in large-scale industrial operations.

3 in operation 実施中で syn *in progress, ongoing*

▶ Many methods of treating wastes, either by converting them to useful by-products or by disposing of them, are in operation or under experimentation.

operational adj **操作上の** syn *actual, effective, real, working*

▶ An operational definition that is perhaps more useful is that a substance is pure when a given physical property is not changed by repetition of a purification process.

operationally adv **操作上**

▶ Operationally we define oxidation–reduction reactions in terms of changes in oxidation states or oxidation numbers.

operative adj **作用する** syn *active, efficient, working*

▶ Comparison of the para positions, where no steric effect can be operative, shows that this position is actually more activated in *tert*-butylbenzene than it is in toluene.

operator n **演算子**

▶ Quantum mechanics is concerned with observables, that is, with operators corresponding to physical properties.

opinion n **1 意見** syn *idea, perception, theory, viewpoint*

▶ Whereas there is no consensus of opinions as to the cause of overvoltage, its existence helps to explain quite a number of electrochemical phenomena.

2 in one's opinion …の意見では

▶ In the authors' opinion, molecular orbital theory is much less suited to qualitative treatment of organic molecules than is the resonance method.

opportune adj **適切な** syn *advantageous, bene-*

ficial, helpful

▶ It is opportune to mention the existence of a further type of polyanion.

opportunity n 機会 syn *chance, occasion, possibility*

▶ Microwave methods offer the opportunity to synthesize and modify the composition, structure, and morphology of materials, particularly composites via differential heating.

oppose vt **1** 反対する syn *resist, withstand*

▶ Lavoisier opposed the phlogiston theory.

2 妨害する syn *hinder, interfere, prevent, thwart*

▶ Thus, two rate-controlling factors tend to oppose each other so that the overall rate pattern is not predictable.

opposed adj **as opposed to** 対立するものとして syn *as contrasted with, rather than*

▶ Small particles as opposed to bulk substance have a much larger surface and thus show an improved solubility even in media in which the bulk substance hardly dissolves.

opposing adj 対立する syn *conflicting, contradictory, opposite*

▶ Often a delicate balance between opposing factors controls the stability or instability of a compound.

opposite n **1** 正反対なもの syn *converse, reverse*

▶ Even though the signs of rotation of the acids are thus the opposite of those of the nitriles from which they are obtained, no changes in configuration are presumed to have occurred in the hydrolyses.

adj **2** 正反対の syn *conflicting, contrary, different, opposing*

▶ The melt and growing crystal are usually rotated in opposite directions during pulling to maintain a constant temperature and uniform melt.

oppositely adv 反対に syn *contrarily, in opposition to*

▶ Intermediate-type bonds occur when a large ion, usually an anion, has its electron distribution distorted by an oppositely charged ion.

opposition n **in opposition** 対立して syn *competitive, conflicting*

▶ Sometimes ΔH and $T\Delta S$ are of similar importance in determining the sign of ΔG, sometimes they are in opposition, and sometimes they reinforce each other.

optical adj 光学の

▶ By optical examination, the thickness per layer and refractive index can be ascertained with great accuracy.

optical activity n 光学活性

▶ Optical activity is due to the different refractive indices of the medium for left- and right-handed circularly polarized light.

optical fiber n 光学繊維

▶ In its simplest form, an optical fiber consists of a core of an amorphous material surrounded by another material called the cladding.

optical glass n 光学ガラス

▶ Lanthanum oxide is an additive in high-quality optical glasses to which it imparts a high refractive index.

optical isomer n 光学異性体 syn *enantiomer, optical antipode*

▶ Optical isomers differ only in the direction in which they rotate the plane of polarization of plane-polarized light.

optical isomerism n 光学異性

▶ The only isomerism for tetrahedral complexes is optical isomerism.

optically adv 光学的に

▶ Two opposite faces of the crystal are highly polished, optically flat, and almost parallel, to form the ends of the resonant cavity where the stimulated process occurs.

optically active adj 光学活性な，旋光性の

▶ Some substances, of which quartz, sodium chlorate, and magnesium sulfate may be mentioned as typical examples, are optically active only in the solid state.

optically active substance n 光学活性体

▶ An optically active substance is one that rotates the plane of polarized light.

optically inactive adj 光学不活性な

▶ An equimolar mixture of two enantiomorphs is optically inactive; such a mixture is known as a racemic modification.

optical microscope n 光学顕微鏡 syn *light microscope*

▶ With optical microscopes, particles down to a few micrometers in diameter may be seen under high magnification.

optical pumping n 光ポンピング

▶ Population inversion can be achieved by optical pumping, by electron beam excitation, or by electron injection at a p–n junction.

optical rotation n 旋光 ☞ rotation

▶ For colored coordination compounds, the magnitude and sign of optical rotation in the visible region is often largely dependent upon the wavelength of the light source.

optical rotatory dispersion n 旋光分散

▶ The variation in sign and magnitude of rotation with wavelength is known as optical rotatory dispersion.

optical stability n 光学的安定性

▶ The optical stability of methylpropylphenylphosphine contrasts very strikingly with the behavior of the corresponding amines which have not yet been resolved, apparently because they undergo inversion too rapidly.

optic axis n 光軸

▶ The angle that the optic axes make with each other may vary from zero to 90°, but it is constant for any particular substance at a given temperature and pressure and for a given wavelength.

optimal adj 最適の syn *best, excellent, ideal, perfect*

▶ The original base choice was found to be optimal.

optimally adv 最適に

▶ The chemistry of living systems is for all practical purposes optimally adapted to function.

optimistic adj 楽観的な syn *bright, confident, positive*

▶ It is too optimistic to expect that a single, simple expression will account for the *p*–*V*–*T* relations of all systems.

optimization n 最適化

▶ The vast scale production of nitric acid requires the optimization of all the reaction conditions.

optimize vt 最適化する

▶ Further work in this area is needed to identify an energy efficient system in which all steps are optimized.

optimum adj 最適の

▶ Optimum zone length depends on the nature of the impurities being segregated.

option n 選択 syn *choice, selection*

▶ The rapid-scan option allows us to observe time-dependent processes with a time resolution of 30 ms at a resolution of 2 cm^{-1}.

optoelectronics n 光電子工学

▶ The field of liquid crystal science has found numerous applications ranging from optoelectronics to medicine.

optoelectronic adj 光電子工学の

▶ The core of any optoelectronic device must be material of high optical transmission, i.e., a highly perfect single crystal.

or conj **1 または** 二つまたはそれ以上の選択すべき同性質の語，句，節を対等につなぐ．列挙するときは，A, B, or C の形をとる．

▶ If a substance, neither a reactant nor a product, affects the rate, it is called an inhibitor, retarder, sensitizer, or a catalyst, depending on the nature of the effect.

or を重ねて用いることもある．

▶ A high melting point implies either a high enthalpy of melting or a low entropy of melting or both.

2 すなわち コンマで区切った中に類義語をおく．

▶ The forces operating in ionic crystals are very predominantly the electrostatic, or ionic, forces between the charged ions.

3 さもないと syn *otherwise*

▶ The temperature during working on soda-lime glass should not be too high, or the surface may become frosted as the result of volatilization of alkali.

oral adj 経口の syn *taken by the month*

▶ With oral administration of 0.1 or 0.2 grams

of a 0.25 percent solution of the sulfa drug daily over fourteen days, rabbits showed no pathological changes in either blood or urine.

orally adv 口頭で syn *verbally*

▶ Dalton's first paper on the relative masses of the different kinds of atoms was presented orally in 1803, and it appeared in print in 1805.

orbit n 軌道 syn *circuit, path, track*

▶ Bohr suggested that electrons revolve around the nucleus at certain fixed distances in a set of orbits.

orbital n **1 軌道関数**

▶ The 1s and 2s orbitals are spherical, while the p orbitals are dumbbell-shaped.

adj **2 軌道の**

▶ The orbital moment is not completely quenched, and the μ_M values are larger than given by the spin-only formula with $g = 2.00$.

orbital angular momentum n 軌道角運動量

▶ The magnetic moments of ions and atoms in crystals result from unpaired electron spins, from net orbital angular momentum, or from both.

orbital energy n 軌道エネルギー 複合形容詞として使用

▶ Orbital-energy diagrams that show correlation of reactant orbitals with product orbitals are called correlation diagrams.

orbital symmetry n 軌道対称性

▶ The principle of conservation of orbital symmetry assists us in making qualitative correlations of orbitals.

order n **1 秩序** syn *orderliness, pattern, regularity, system*

▶ A glassy material at absolute zero, for example, will not have the necessary molecular order to guarantee an entropy of zero at absolute zero.

2 状態 syn *condition, state*

▶ A water pump in good order gives suction nearly corresponding to the vapor pressure of water at the temperature of flow.

3 反応次数

▶ For the hydrogen-bromine reaction, the concept of order does not apply since its rate expression is not of the restricted form required for this concept.

4 順序 syn *sequence*

▶ The reducing action of these agents generally decreases in the order named.

5 程度 syn *category, class, level, rank, scale*

▶ A protein molecule will have a molecular weight of the order of ten to hundreds of thousands.

6 in order 望ましい syn *appropriate, fitting, right, suitable*

▶ A word of caution is in order regarding the use of such lists.

7 in order that (conj) …するために syn *with the purpose that*

▶ Emphasis may be laid on the fact that two solid phases are necessary in order that the dissociation pressure at a given temperature shall be definite.

8 in order to (prep) …するために syn *for the purpose of*

▶ Upon completion of the desired ligand-centered transformations, oxidative demetallation is often required in order to liberate the organic.

9 of high order 高度な syn *higher, superior, upper*

▶ The torsion type of balance has been developed to a high degree of sensitivity and stability but requires technical skill and ability of very high order in construction and in manipulation.

vt **10 配列する** syn *arrange, classify, organize*

▶ On annealing the alloy compositions AuCu and $AuCu_3$ at lower temperature, the gold and copper atoms order themselves.

order-disorder transition n 秩序-無秩序転移 syn *order-disorder transformation*

▶ A small anomaly observed in the heat capacity of naphthalene-tetracyanobenzene near 75 K with an entropy of transition of 0.71 cal K^{-1} mol^{-1} is believed to arise from an order-disorder transition involving the reorientation of the naphthalene molecules.

ordered adj 規則正しい syn *periodic, regular, systematic*

▶ The most common form of TiO_2, rutile, consists of a distorted hcp array of oxide ions with half of the octahedral interstices filled in an ordered way by Ti^{4+} ions.

ordering n **1** 整理 syn *classification, grouping*

▶ The orderings of softness and hardness are established purely empirically.

2 配列 syn *arrangement, disposition*

▶ Lithium ion ordering occurs involving occupancy of one half of the tetrahedral sites.

orderliness n 規則性 syn *order, pattern, regularity*

▶ So far as minerals and many binary and ternary oxides, hydroxides, and oxyhydroxides are concerned, the topotaxy and orderliness seem to stem from the maintenance of the organization of the oxygen and/or hydroxy ions within the structure as reaction proceeds.

orderly adj 規則的な syn *arranged, regular, systematic, uniform*

▶ Nearly all solid substances have an orderly arrangement of their constituent atoms.

order of magnitude n 桁 ☞ power

▶ It is the purpose of this series of papers to present a correlative and predictive scheme which is only slightly more complex but which yields about one order of magnitude more accurate.

order of reaction n 反応次数

▶ The order of reaction simply indicates how the rate depends on concentration.

ordinarily adv 普通に syn *customarily, generally, normally, usually*

▶ The term hydrochloric acid, as ordinarily used in analytical chemistry, refers to an approximately 12 M aqueous solution of density 1.2 containing about 39.1 percent of hydrogen chloride by weight.

ordinary adj 普通の syn *common, expected, general, normal, usual*

▶ No one conformation constitutes a discrete isolable substance under ordinary conditions.

ordinary ray n 常光線 ☞ extraordinary ray

▶ The ordinary ray is the fast ray in positive crystals and the slow ray in negative crystals.

ordinary temperature n 常温

▶ At ordinary temperatures, 18,045 parts of water dissolves 1 part of $SrCO_3$.

ordinate n 縦座標 ☞ abscissa

▶ The dielectric constant is plotted as the right-hand ordinate, and on the same figure is included a graph of the heat capacity against temperature; the heat capacity is plotted as the left-hand ordinate.

organ n 器官

▶ Cells appear to be able to recognize cells of like kind and thus to unite into coherent organs.

organelle n 細胞小器官，オルガネラ

▶ Very many cells and organelles contain precipitates of calcium and iron salts.

organic n **1** 有機物

▶ In organic-based devices the electronic or optical properties can be built into the molecular structure instead of being produced by the processing technique.

adj **2** 有機物の

▶ The key step in the production of ordered mesoporous materials involves formation of a stable emulsion in which the inorganic precursor surrounds a three-dimentional template of organic micelles.

organically adv 有機的に

▶ The range of synthetic methods grew to include organically modified surfaces, in which organic groups were covalently grafted to an inorganic surface of high surface area.

organic metal n 有機金属

▶ Although metallic in nature, organic metals have low conductivity compared to a metal such as copper.

organic reagent n 有機試薬

▶ Various organic reagents may be used to distinguish metals by selective precipitation, extraction, or elution from chromatographic column.

organic solvent n 有機溶媒

▶ Removing organic solvents in chemical synthesis is important in the drive toward benign chemical technologies.

organism n 生物体 syn *living thing*

▶ Organisms need magnesium in most phosphate enzymes: for example, those in carbohydrate metabolism and those for nerve and muscle action.

organization n 組織 syn *arrangement, configuration, pattern, structure*

▶ Organization exists in all living systems since they are composed of one or more cells that are the basic units of life.

organize vt 組織化する syn *arrange, structure, sort out*

▷ Vesicles are capable of organizing a large number of guest molecules in their compartments.

organized adj 組織された syn *ordered, orderly, regular, systematic*

▶ Materials exhibiting uniform particle size, polymorph selectivity, tailored morphology, oriented nucleation, organized assembly, and composite inorganic-organic structures are realistic areas for investigation.

organometallic compound n 有機金属化合物 syn *organometallic*

▶ The scope of application of organometallic compounds for the deposition of electronic materials, other than compound semiconductors, is large.

organosilane n オルガノシラン

▶ The organosilanes tend to be much more reactive than their carbon analogues, particularly toward hydrolysis, ammonolysis, and alcoholysis.

orient vt 方向付ける syn *conduct, direct, guide, lead, steer*

▶ The methyl substituent apparently orients the entering substituent preferentially to the ortho and para positions.

orientation n **1** 配向 syn *alignment, arrangement*

▶ By reason of thermal collisions, the orientation of magnetic molecules in the direction of an applied magnetic field is never completely attained in either a gaseous or a liquid system.

2 配向性

▶ A given group causes the same general kind of orientation, predominantly ortho, para or predominantly meta, whatever the electrophilic reagent involved.

3 方向 syn *location, position*

▶ In order to assign the R or S configuration to one of the enantiomers of a dissymmetric compound, we must view that enantiomer from a particular orientation.

orientational adj 配向の

▶ It is possible to have a discotic nematic phase in which there is orientational correlation of the short molecular axis in the absence of any positional ordering.

orientational order n 配向秩序

▶ Liquid-crystalline phases are characterized by the presence of long-range orientational order which in real systems extends over many millions of molecules.

orientation effect n 配向効果

▶ While orientation and induction effects are undoubtedly relevant in solids in which molecules are polar, neither can play any part in the numerous structures in which the atoms or molecules possess no permanent dipole moment.

orientation polarization n 配向分極 ☞ atomic polarization, distortion polarization, electronic polarization

▶ Orientation polarization involving rotation of dipolar molecules is a process requiring usually 10^{-12} to 10^{-10} sec.

oriented adj 配向した

▶ The only oriented molecular crystals met with in technology are those occurring in oriented films and fibers.

orifice n 穴 syn *hole, opening, pore*

▶ Under these conditions, the solid and vapor remain in equilibrium, while the vapor effuses through a carefully designed orifice of about 1-mm diameter

origin n **1** 起源 syn *base, basis, foundation,*

source

▶ The forces that hold ionic crystals together are entirely electrostatic in origin and may be calculated by summing all the electrostatic repulsions and attractions in the crystal.

2 原点 syn *beginning, start*

▶ The origin is at the top left-hand rear corner.

original adj 原型の syn *first, initial, primary, starting*

▶ On acidification, an aci isomer of the original compound is obtained which forms salts rapidly with bases but slowly reverts to the original nonacidic compound.

originally adv 最初に syn *early, initially*

▶ The volume of the carbon dioxide formed is equal to the volume of the carbon monoxide originally present.

originate vi 起こる syn *arise, emerge, stem*

▶ The repulsion energy which originates in the interpenetration of the electron clouds of ions presents a problem, since it is not well defined theoretically.

orthogonal adj 1 直交する

▶ In reduced Nb_2O_5, the crystallographic shear planes occur in two orthogonal sets, and the regions of perfect structure are reduced in size from infinite sheets to infinite columns or blocks.

2 矩形の

▶ If a crystal of rock salt is dropped, it will tend to fracture along cleavage planes to produce small cubes or orthogonal prisms.

orthogonally adv 直交して

▶ In κ-$(BEDT\text{-}TTF)_2(Cu(SCN)_2)$, the stacks of cations are absent and are replaced by orthogonally arranged dimers of BEDT-TTF cations, again resulting in strong S–S interactions.

ortho-para-directing adj オルト-パラ配向の

▶ Halogens are unusual in their effect on electrophilic aromatic substitution: they are deactivating yet ortho-para-directing.

orthorhombic adj 斜方晶系の

▶ Isostructural with YF_3 are the orthorhombic lanthanide trifluorides, samarium to lutetium inclusive.

orthorhombic system n 斜方晶系

▶ In the orthorhombic system, there are four space lattices, one lattice is a primitive cell, one lattice is base-centered, one is body-centered, and one is face-centered.

orthosilicate n オルトケイ酸塩 ☞ metasilicate

▶ Orthosilicates have relatively simple structures in which the SiO_4^{4-} grouping exists as a discrete entity in the crystal lattice.

oscillate vi 振動する syn *fluctuate, vibrate*

▶ The reaction is not oscillating between the reactants and final products, but rather, it is the concentrations of the intermediates that are involved in the various component elementary reactions, that are fluctuating.

oscillation n 1 振動 syn *vibration*

▶ Most physical properties of crystals are affected by the frequency of the thermal oscillations of atoms or molecules above their equilibrium lattice positions.

2 複数形で，変動 syn *fluctuation*

▶ In order to obtain the molecular scattering curve from a trace, a smooth background curve is drawn through the oscillations on the basis that there must be equal amounts of positive and negative area defined by the molecular scattering curve and the background line.

oscillator n 振動子

▶ Unlike mechanical and electrical oscillators, chemical oscillators never pass through their final equilibrium configuration in the course of an oscillatory cycle.

oscillator frequency n 振動周波数

▶ The chemical shift is easily recognizable as such by the fact that the spacing between the main groups is directly proportional to the oscillator frequency.

oscillator strength n 振動子強度

▶ The calculated oscillator strength of the first band of the polyenes is the same on the free-electron and LCAO approximations because of the similar forms for the molecular orbital wave functions.

oscillatory adj 振動する syn *oscillating*
▶ The oscillatory behavior observed is not only chemical in origin but also arises from the time scales of reaction induction, ignition, and throughput of the reactants and products.
oscillatory reaction n 振動反応
▶ Central to oscillatory reactions is a negative feedback mechanism.
osmosis n 浸透
▶ The phenomenon of osmosis is the tendency of solvent to flow into a more concentrated solution from which it is separated by a semipermeable membrane, a membrane that permits passage of the solvent molecules but not the solute.
osmotically adv 浸透圧で
▶ If cells are placed into water or dilute buffer, they swell due to the osmotically driven influx of water.
osmotic coefficient n 浸透係数
▶ The most convenient method to ascertain the effective number of particles is to determine the osmotic coefficient.
osmotic pressure n 浸透圧
▶ Osmotic pressure is a sensitive measure of the kinetic energy of the particles, since a molal solution has an osmotic pressure of about 25 atm and pressure may be measured in millimeters of water.
other pron **1** 通常複数形で，その他のもの
▶ An important characteristic of enzymes is their specificity; i. e., a given enzyme can catalyze one particular reaction and no others.
2 the を伴って，二つのうち先行する one に対して他方
▶ In sulfonation, the energy barriers on either side of the carbocation must be roughly the same height; some ions go one way, some go the other.
3 one from the others 両者を区別して
▶ The conformers interconvert extremely rapidly, and it would be impossible to separate one from the others.
adj **4** ほかの syn *different, distinguishable*
▶ While transition elements are responsible for the colors of most minerals, there are other mechanisms causing absorption of light in the visible region.
5 other than …とはほかの syn *apart from, except, except for*
▶ The alkali metals other than lithium form colored compounds of the formula MO_2, containing the anion O_2^-, and termed superoxides.
otherwise adj **1** 異なって syn *different*
▶ The loss of identity of particular protons by rapid rate process makes the NMR spectra of ethyl derivatives much simpler than they would otherwise be.
adv **2** 別のように syn *differently*
▶ All reactions, unless otherwise stated, were carried out under an atmosphere of dry, oxygen-free nitrogen, using standard Schlenk techniques.
3 さもなければ syn *if not, on the other hand*
▶ For ferro- and antiferromagnetic materials, the effect of temperature is to introduce disorder into the otherwise perfectly ordered parallel or antiparallel arrangement of spins.
4 or otherwise …かあるいはその逆で
▶ It is assumed and understood, implicitly or otherwise, that all other effects are absent, ignored, constant, or completely determined by P and T.
ought aux **ought to** はずである syn *have to, must, should*
▶ Phenol ought to be a stronger acid than an aliphatic alcohol since the conjugate base (phenoxide ion) is resonance stabilized.
out adv **1** 外へ syn *outside*
▶ The product will either accumulate for a time as metastable solid solution or diffuse out and form its own crystals.
2 十分に syn *completely, entirely, thoroughly*
▶ If particular intermolecular M—M or M—L interactions were important for mesophase formation, these should be diluted out at high percentages of the organic compound leading to unexpected features in the phase diagram.
outcome n 結果 syn *consequence, effect, re-*

sult

▶ Ill-defined, amorphous precipitates seemed much more likely outcomes than the single crystals suitable for the X-ray diffraction studies.

outer adj 外部の syn *external, exterior*

▶ It is evident that the outer regions of the electron distributions of the ions are considerably distorted from spherical shape.

outermost adj 最も外側の syn *distant, extreme, ultimate*

▶ The color in ruby is produced by light interacting with outermost 3d electrons of Cr^{3+}.

outer orbital n 外部軌道 ☞ inner orbital

▶ If a suitable metal ion is used as an acid probe for this effect, then through the effects of central-field and symmetry-restricted covalency there should be an expansion of the outer orbitals of the probe ion.

outer shell n 外殻 ☞ inner shell

▶ For electronic transitions involving outer shells, the associated energy usually lies in the visible and ultraviolet regions.

outer-shell electron n 外殻電子 ☞ valence electron

▶ In metals, a significant proportion of the outer-shell or valence electrons are effectively free to move throughout the structure and are completely delocalized.

outer-sphere mechanism n 外圏機構 ☞ inner-sphere mechanism

▶ Where both reactants are nonlabile, a close approach of the metal atoms is impossible, and the electron transfer must take place by a tunneling or outer-sphere mechanism.

outgas vt 気体を除去する

▷ When a gas is admitted to an outgassed solid, the unsaturated pseudovalence forces at its surface attract and retain some of the gas molecules in an adsorbed layer.

outgrowth n 副産物 syn *by-product,*

▶ The results to be described were an outgrowth of studies of reduction of molten rare earth metal trihalides with the respective metals in tantalum containers.

outlet n **1** 液体，気体などの出口 syn *opening*

▶ To monitor the physical properties of the nanoparticles produced, the outlet of the micromixer was coupled to a quartz cell through which UV-vis spectra were obtained.

2 販路 syn *market*

▶ Unlike Cl_2 and Br_2, iodine has no predominant commercial outlet.

outline n **1** あらまし syn *overview, summary*

▶ In this article, we present a brief outline of the important structural and chemical characteristics of the Y-Ba-Cu-O 90 K superconducting material.

vt **2** 概説する syn *sketch*

▷ More complicated types of solvation can be treated according to the same principles as those just outlined.

outlook n 見通し syn *expectation, prospect*

▶ The ligand lability makes the outlook for isolation of optical isomers in octacoordinate chelates rather poor.

out of prep …の範囲外に syn *beyond the range of*

▶ Pure, powdered CuBr looks white, because of its absorption onset at 2.9 eV, which is nearly out of the visible region of the spectrum.

out of plane n 面外 ☞ in plane

▶ The large number of intermolecular contacts involving atoms of the nitro groups may indicate that a portion of the large rotations of the nitro groups out of the plane of the picryl ring may result from intermolecular forces.

out-of-plane bending n 面外変角振動

▶ For aromatic rings, out-of-plane C-H bending gives strong absorption in the 675-870 cm^{-1} region, the exact frequency depending upon the number and location of substituents.

out-of-plane vibration n 面外振動

▶ Satisfactory potential functions exist to describe the out-of-plane vibrations of simple π-electron molecules such as ethylene and benzene.

output n **1** 出力 syn *energy, power*

▶ The line width of the laser output is

essentially limited by mechanical vibration of the optical components and turbulence in the dye cell.

2 生産高 syn *production, yield*

▶ The output of journal articles has increased almost exponentially in the last 70 years.

outright adv 完全に syn *completely, entirely*

▶ Neither referee had been at all concerned with the subject matter of his paper, and both rejected it outright.

outset n **at the outset** 最初に syn *firstly, initially*

▶ We realized at the outset that it might be possible to form salts in which the anions consist of trapped electrons.

outside n **1** 外側 syn *exterior*

▶ The ions Ca^{2+} and Na^{+} are concentrated in body fluids outside of cells.

prep **2** …の範囲を超えて syn *beyond*

▶ Further discussion of these fascinating series of reactions falls outside our present scope.

outstanding adj **1** 傑出した syn *famous, prominent, renowned*

▶ The outstanding chemical characteristic of vinyl halides is their general inertness in S_N1 and S_N2 reactions.

2 未解決の syn *remaining, unsolved*

▶ The outstanding problems are concerned with a more detailed understanding of microstructure in the amorphous phase and the dynamics of the ion transport.

outward adj **1** 表面的な syn *exterior, external, superficial*

▶ Graphitic oxide retains the outward crystalline form of the original graphite but is found to have undergone considerable swelling.

adv **2** 外側に syn *away*

▶ A protein molecule, coiled up to turn its lipophilic parts outward, is dissolved in the bilayer, forming a part of the cell wall.

outweigh vt 勝る syn *prevail, overcome, surpass*

▶ Angle strain in cyclooctatetraene ring is large enough to outweigh any resonance energy that might be gained thereby.

oven n 乾燥器

▶ The tube is cleaned in an acid bath, washed thoroughly, and dried in an oven.

over adj **1** 終わって syn *concluded, ended, finished*

▶ When the reaction was about half over, it was interrupted and the unconsumed alkene was isolated.

adv **2 over and over** 何度も syn *frequently, often, repeatedly*

▶ Methods like this can be repeated over and over with the addition of a new unit each time.

prep **3** …の上方に syn *above, on, upon*

▶ Acetonitrile may be dried over anhydrous calcium sulfate or by distilling from phosphorus pentaoxide.

4 …を超えて syn *beyond, exceedingly, more than*

▶ The line width of DPPH crystallized from benzene is over five times as great as that of DPPH crystallized from carbon disulfide.

5 …の間 syn *across, beyond*

▶ The physical and chemical properties of organometallic compounds vary over an extraordinary wide range.

6 over time 次第に

▶ Loss of hydrogen from BNH_x ligands over time may yield a less soluble rhodium cluster.

7 …関して syn *for*

▶ There is some controversy over the classification of complex anions as ionic liquids.

8 …の隅々まで syn *through, throughout*

▶ In the same way, cations are stabilized by dispersal of their positive charge over the solvent cluster.

overall adj **1** 総体的な syn *complete, entire, total, whole*

▶ The overall Si:O ratio in a silicate structure depends on the relative number of bridging and nonbridging oxygens.

adv **2** 全体的に見れば syn *on the whole*

▶ Overall the solution is neutral, but in the vicinity of any given ion there is a predominance of ions of opposite charge.

overall reaction n 全反応

▶ The rate of overall reaction is determined by the slow breaking of the C−Br bond to form the carbocation; once formed, the carbocation reacts rapidly to form the product.

overall reaction rate n 総括反応速度 syn *overall rate of reaction*

▶ Almost certainly, substituents affect the overall reaction rate by affecting the rate of migration; hence, migration must take place in the rate-determining step.

overall yield n 全収率

▶ If 90 percent yields could be achieved in each step in a thirty-step synthesis, the overall yield would still be only 4 percent.

overcome vt 打ち勝つ syn *overwhelm, prevail, supress*

▶ In order that salts and other substances dissolve in polar solvents such as water, there must be strong solute−solvent interactions to overcome the large crystal or bonding forces.

overcrowding n 混雑

▶ If a planar model of structure was constructed with 120° exocyclic angles at C(2), C(3), and C(4), the greatest overcrowding would be between Cl(1) and O(6).

overemphasize vt 強調しすぎる syn *enlarge, magnify, overstate*

▶ The immense industrial importance of alkaline earth carbonates and sulfates can hardly be overemphasized.

overestimate vt 過大評価する syn *miscalculate*

▶ The contribution of charge-transfer resonance to the ground state of benzene−halogen complexes has been overestimated in the past.

overgrowth n 繁茂

▶ Interfacial polymerization represents an effective method of suppressing the overgrowth of polyaniline.

overheating n 過熱

▶ Acetylation of D,L-alanine in acetic acid−acetic anhydride requires a 1-min reaction period at 100 ℃, and overheating converts the product into the azlactone.

overlap n **1** 重なり syn *overlapping*

▶ As the overlap between orbitals increases, the most bonding orbital of the band is stabilized to a greater extent, the most antibonding orbital of the band is destabilized to a greater extent, and a relatively wide band results.

vt **2** 重なる syn *intersect*

▶ When an atomic orbital overlaps several atomic orbitals on different nuclei, then a more extended molecular orbital is obtained.

overlap integral n 重なり積分

▶ In practice, for the π orbitals of conjugated hydrocarbons, the overlap integral of nearest neighbor atomic orbitals is about 0.25.

overlayer n 被覆層

▶ Thus, an important aspect of TiO_x overlayer formation is the demonstration of the TiO_x−metal surface interaction strength.

overlook vt 見落とす syn *disregard, ignore, neglect, omit*

▶ The biological activity of calixarenes has been well documented, though it has been largely overlooked.

overly adv 大いに syn *exceedingly, extraordinarily*

▶ The theoretical approach to electrolyte solubility based on the Born equation is not overly successful.

overnight adv 一晩中

▶ After standing overnight, two types of crystals were deposited.

overoxidation n 過酸化

▶ Usually, the oxidation of sulfides to sulfoxides has to conducted carefully to avoid overoxidation to the sulfones.

overshadow vt 重要性を奪う syn *diminish, dominate, minimize*

▶ It is generally difficult to demonstrate the effect of dipoles on melting and boiling points of halides, for it is usually overshadowed by the larger effect of the London forces.

oversimplification n 単純化しすぎ

▶ The Kihara model assumes the same minimum energy for a pair of molecules at their optimum separation regardless of their orientation. This is probably an oversimplification.

oversimplify vt 単純化しすぎる

▷ In order to calculate the thermodynamic properties of a particular system, we assume, as a first approximation, an oversimplified model.

overstate vt 誇張する syn *enlarge, exaggerate, magnify*

▶ The importance of this information to organic chemistry can scarcely be overstated, because the results of most crystal structure determinations are precise, detailed, and unambiguous.

overtone n 倍音

▶ The glass must be dry, because O—H bonds give rise to overtone bands that appear in the spectral region of interest.

overturn vt くつがえす syn *convert, upset*

▶ The traditional view that gold, being a noble metal, has a very limited chemistry has been overturned.

overview n 概観 syn *outline, review, summary*

▶ The purpose of this feature article is to give an overview of developments in the field of hybrid inorganic-organic framework structures over recent years.

overvoltage n 過電圧 syn *overpotential*

▶ The formal definition of overvoltage is the change of potential of the electron-conducting phase when reaction rate across its interface with the ion-conducting phase, with which it is in contact, is changed from zero to a certain velocity

overwhelm vt 圧倒する syn *overcome, supress*

▶ Sometimes the proportion of products greatly overwhelms the remnant of reactants in the equilibrium mixture, and for all practical purposes the reaction is complete.

overwhelming adj 圧倒的な syn *formidable*

▶ The overwhelming majority of carbon fibers are the products of polymerization of acrylonitrile.

overwhelmingly adv 圧倒的に syn *eminently, prominently, utterly*

▶ A quantitative description, which captures overwhelmingly the main contribution, includes a sum of Coulomb attractions and repulsions between all pairs of ions, along with short-range repulsions preventing overlap of the atomic cores.

owe vt 負っている syn *be indebted to*

▶ Most semiconductors owe their conductivity to the presence of impurities in solid solution, and the actual value of the conductivity is very sensitive to the amount of these impurities present.

owing adj **owing to** （prep）…のために syn *because of, on account of*

▶ Owing to the large number of interactions involved, the theoretical treatment of an ion in aqueous solution is very difficult.

own pron **1 of its own** 自身の

▶ Among K, Rb, and Cs, K is the only abundant element and forms silicate minerals of its own.

adj **2** 自身の syn *particular, respective, specific*

▶ At the present time, the view is commonly held that no single mechanism can account for all the known examples of racemization but that several different mechanisms, each with its own field of application, are required.

oxidant n 酸化剤 syn *oxidizing agent*

▶ Sulfites can act as oxidants in the presence of strong reducing agents; e.g., sodium amalgam yields dithionite, and formates yield thiosulfate.

oxidation n 酸化

▶ The term oxidation originally meant a reaction in which oxygen combines chemically with another substance, but its usage has long been broadened to include any reaction in which electrons are transferred.

oxidation number n 酸化数

▶ A convenient, formal concept used to characterize the state of oxidation of an element in a compound and also in nomenclature is that of oxidation number.

oxidation-reduction potential n 酸化還元電位 syn *redox potential*

▶ Crystal field theory may be used to help correlate oxidation-reduction potentials of certain complex ions.

oxidation-reduction reaction n 酸化還元反応 syn *redox reaction*

▶ The oxidation-reduction reaction between ninhydrin and an amino acid produces a pigment.

oxidation state n 酸化状態

▶ The oxides with the perovskite structure contain two cations, whose combined oxidation state is six; several cation combinations are, therefore, possible.

oxidative adj 酸化的な

▶ For both fatty acid and glucose oxidation, the initial oxidative removal of hydride ions from the carbon backbone gives rise to acetyl groups, which are carried on CoA.

oxidative addition n 酸化的付加

▶ A large fractional number of electrons transferred means a lowering of the energy barrier to oxidative addition.

oxidative cleavage n 酸化的開裂

▶ In mammalian organisms, including man, β-carotene readily undergoes oxidative cleavage at the central double bond to give two equivalents of an aldehyde known as retinal.

oxidative coupling n 酸化的カップリング

▶ Conducting polymers, such as poly(pyrrole) or poly(3-methylthiophene), are formed by oxidative coupling of the respective monomers.

oxidic adj 酸化物の

▶ The oxidic form of the catalyst is converted to the catalytically active sulfidic form during the hydrodesulfurization process.

oxidizable adj 酸化できる

▶ Commercial acetic acid contains traces of acetaldehyde and other oxidizable contaminants, which can be destroyed by refluxing the acid with 2-5% (by weight) of potassium permanganate for 2-6 hours.

oxidize vi **1** 酸化する

▶ When heated in air, arsenic sublimes and oxidizes to As_4O_6 with a garlic-like odor.

vt **2** 酸化させる

▶ Cyclohexanol can be oxidized directly to adipic acid by nitric acid in about 50% yield.

oxoanion n オキソアニオン

▶ The best-known oxoanion of iron is the ferrate(VI) prepared by oxidizing a suspension of hydrous Fe_2O_3 in concentrated alkali with chlorine.

oxonium salt n オキソニウム塩 複合形容詞として使用

▶ Catalysis of the enolization of acetone by acids involves, first, oxonium-salt formation and, second, removal of an α proton with water or other proton acceptors.

oxygenation n 酸素化

▶ The chromophore (Cu^{I}/Cu^{II}) in colorless deoxyhemocyanin turns bright blue upon oxygenation.

oxygen saturation curve n 酸素飽和曲線

▶ The oxygen saturation curve for hemoglobin differs in two major respects from the one for myoglobin. In the first place it is sigmoidal.

ozonization n オゾン化

▶ $AsOCl_3$ was made by ozonization of $AsCl_3$ in $CFCl_3/CH_2Cl_2$ at −78 ℃.

P

pace n 速度 syn *rate, speed, velocity*

▶ If the pace of exercise increases from a walk to a light jog, the rate of fatty acid oxidation by the muscles also increases.

pack vt 充填する syn *fill, load, package*

▶ Secondary forces are effective in determining in the solid state both the shape of the protein molecule and the way in which neighboring protein chains are packed together.

packet n 小さな束 syn *bundle, package*

▶ The energy of an oscillator is not continuous but comes in packets, called quanta.

packing n **1** 充填

▶ The packing of molecules in the solid state has a crucial effect on their properties.

2 充填物

▶ The column has a packing that provides a cooling space where part of the vapor of the boiling mixture condenses and drains through the apertures in the packing.

packing coefficient n 充填係数 syn *packing factor*

▶ Salts with discoidal cations and anions, e.g., $[PtCl_4]^{2-}$, generally have higher packing coefficients (70–76%) than those involving spherical cations and anions (64–71%).

packing density n 充填密度

▶ The plausibility of the packing density calculated from the experimental zero-point volume and the van der Waals volume will fix at least the upper limit of the van der Waals radius.

packing efficiency n 充填効率 ☞ packing coefficient

▶ The body-centered cubic packing efficiency is only 68%, and the face-centered cubic packing efficiency is 74%, so that an increase in pressure favors the close-packed fcc structure.

pains n 骨折り syn *effort, trouble*

▶ There is usually a slight oxidation of the iron, unless special pains are taken to exclude air during the heating and cooling.

painlessly adv たやすく syn *comfortably, easily*

▶ It is interesting to speculate why certain compounds have resisted discovery for long periods of time, only to be synthesized quite painlessly once the initial breakthrough has been made.

pair n **1** 対 syn *couple*

▶ The principal intermolecular interactions within the stacks are two hydrogen bonds between pairs of adjacent molecules.

2 in pairs 二つ一組になって

▶ In a normal molecule, where all the valence shells of the atoms are filled, the electrons occupy the various orbitals in pairs with opposite spin.

vi **3** 対になる syn *team*

▶ The surface silicon atoms pair into dimers connected by a σ and a π bond, thus having essentially double bond character.

vt **4** 対にする syn *join, unite*

▷ There are fewer soluble salts of anions of higher charge unless paired with univalent cations.

pairing n 対形成 syn *association, linkage*

▶ The linear Cr–O–Cr bridge evidently permits pairing of the d electrons of the two metal atoms via d_π–p_π bonds.

pairwise adj 対をなして

▶ In general, the simplifying assumption is made that the total interaction between two molecules may be represented by the sum of all the pairwise additive atom–atom interactions between the two molecules.

pale adj **1** 淡い syn *faint, light*

▶ Antimony compounds impart to the flame a pale, greenish white color.

vi **2** 薄くなる syn *diminish, fade, lessen*

▶ As the temperature is lowered, the color of nitrogen dioxide pales until the liquid becomes colorless and then freezes into colorless crystals at −10.2 ℃.

panel n パネル syn *plate, sheet, tablet*

▶ Figure 1 shows in the upper panel the infrared signatures of a spherical particle with N_2O in the core and CO_2 in the shell.

paper chromatography n 沪紙クロマトグラフィー

▶ In paper chromatography, the aqueous phase is held stationary in the microporous structure of the paper.

parabolic adj 二次の

▶ By varying the temperature, all laws of growth from parabolic to cubic can, in principle, be obtained for the system $Cu-Cu_2O$.

paracrystalline adj 準結晶の

▶ Paracrystalline solids are those, such as partially graphitized carbons, in which the process of ordering has not been completed.

paradigm n 範例 syn *guideline, ideal, standard*

▶ Because of its relative simplicity, the hydrogen molecule is generally taken as the paradigm of the covalent bond.

paradox n 逆説 syn *contradiction, inconsistency, problem*

▶ Perhaps the greatest surprises would be elicited by the paradox that a ceramic, which by definition was hitherto regarded as synonymous with an insulator, could be so fashioned, as with $YBa_2Cu_3O_{6+\delta}$, as to exhibit zero electric resistance.

paradoxical adj 相反する syn *contradictory*

▶ It may seem somewhat paradoxical, but metals are both strong absorbers and strong reflectors of electromagnetic radiation.

paradoxically adv 逆説的に

▶ Paradoxically, this good agreement was unfortunate, because it led to undue confidence in the hard-sphere kinetic theory of gases and delayed the development of the subject for many years.

paragraph n 節 syn *part, portion, section*

▶ The first paragraph of the results section requires some knowledge of the behavior of electronically excited iodine.

parallax n 視差

▶ Parallax errors were avoided by aligning the coross-hair and its image in the mirror adjacent to the centimeter rule.

parallel n **1** 類似 syn *resemblance, similarity*

▶ Very often factors that stabilize products also stabilize the transition state leading to those products, that is, that often there is a parallel between ΔH and E_{act}.

2 in parallel 平行して

▶ Two-directional strategies for polyether construction in which several synthetic operations are performed in parallel greatly minimize the total number of steps required to assemble large fused polyether fragments.

adj **3** 平行な

▶ The index of refraction of the extraordinary ray varies from its minimum value for the vibration parallel to the optic axis to its maximum value for the vibration normal to the optic axis.

vt **4** 類似する syn *comform, compare to, follow*

▶ The rich oxo acid chemistry of sulfur is not paralleled by the heavier elements of the group.

parallelism n **1** 平行

▶ Crystals frequently grow in groups or aggregates in which the various crystals show more or less tendency to parallelism.

2 対応 syn *correspondence, resemblance, similarity*

▶ Little parallelism was observed between the values of internal viscosity and those of the macroscopic viscosity.

parallelepiped n 平行六面体

▶ Unit cells are parallelograms in two dimensions and parallelepipeds in three dimensions.

parallel line n 二本の平行線,

▶ In structural formulas, unsaturation may be represented by parallel lines joining the carbon atoms.

parallel reaction n 並発反応

▶ We generate CH from the laser flash photolysis of $CHBr_3$, and so the parallel reaction, $CH + CHBr_3 \rightarrow$ products, will also be taking place.

parallel spin n 平行スピン

▶ Ferromagnetic substances, which are invariably solids, have structures composed of atoms possessing electrons with parallel spins.

paramagnetic adj **常磁性の**

▶ The breakdown of the Curie law is due to the effect of neighboring paramagnetic atoms, so that we would not expect the Curie law to be obeyed in the solid state, where the density of paramagnetic atoms will be high.

paramagnetic susceptibility n **常磁性磁化率**

▶ The simple Curie law states that paramagnetic susceptibility is inversely proportional to temperature.

parameter n **1 パラメーター** syn *contact factor*

▶ The empirical procedure for deriving the repulsion constant simply adjusts the parameter to make any reasonable equation give good results.

2 要因 syn *characteristic, feature*

▶ A large number of parameters affect the magnetic properties of materials.

paramount adj **最高の** syn *chief, main, predominant*

▶ Purity is of paramount importance in identifying compounds through comparison of their properties with those of known substances.

parasitic adj **寄生的な**

▶ For many chemical applications, the intensity will be of particular importance if the design specifications attempt to optimize radiation in the vaccum ultraviolet spectral region rather than regard it as a parasitic use.

para substitution n **パラ置換**

▶ If either the otho-para-directing substituent or the electrophile is bulky, then para substitution is favored for steric reasons.

parent adj **母体の** syn *original*

▶ The powerfully electron-withdrawing fluorine atoms help to stabilize the triflate anion, $CF_3SO_2O^-$, and make the parent acid CF_3SO_2OH one of the strongest Lowry–Brønsted acid known.

parenthesis n **丸括弧** 記号（ ）を指すときは複数形 parentheses を用いる． ☞ bracket, curly bracket, square bracket

▶ In the mica known as biotite, Mg^{2+} or Fe^{2+}, together with hydroxides, act to hold the sheets together, generating the formula $[(Fe,Mg)_3(OH)_2(AlSi_3O_{10})]^-$. The parentheses around Fe and Mg indicate that they form a solid solution.

parenthetically adv **付加的にいえば** syn *incidentally*

▶ Parenthetically we note the good structural sorting of AB_2O_4 compounds in general if crystal radii are used as indices and the data base is judiciously chosen.

parity n **偶奇性，パリティ**

▶ States with even and odd parity are designated by the subscripts g and u, respectively.

part n **1 構成要素の一つである部分** syn *component, constituent, portion*

▶ The bond system in phosphonitrilic derivatives is composed of two parts in both the ring and the ligands.

2 a part of …の要素 可算名詞の単数形または不可算名詞が続く．複数形の前では some of, many of を用いる． syn *component of, element of*

▶ When potassium cyanide is added to silver nitrate solution, a white precipitate of silver cyanide is formed, but if an excess of potassium cyanide is used, the precipitate dissolves and the silver has become a part of the anion.

3 part of …の一部分 syn *piece of, portion of*

▶ As part of our continuing studies on conducting organic charge-transfer complexes, we have recently examined the properties of the electron acceptor molecule 2-dicyanomethyleneindane-1,3-dione.

4 for the most part 大体は syn *generally, mainly, mostly, usually*

▶ For the most part, the negative charge in ClO^- must reside on the oxygen atom.

5 in part ある程度 syn *partially, partly, somewhat*

▶ The reactivity of a given alkyl derivative, RX, in either S_N1 or S_N2 reactions is determined in part by the nature of the leaving group, X.

6 in large part 大部分は syn *by and large, chiefly, largely, mainly*

▶ The properties of solutions have been

extensively studied, and it has been found that they can be correlated in large part by some simple laws.

7 書物の部分

▶ Page 706, reference 2, the part '1939, 9, 379' should be transferred to the end of reference 3.

8 割合 syn *percentage, portion, share*

▶ One part in 10^4 of tetracene in anthracene completely replaces the blue anthracene fluorescence with a green fluorescence due to tetracene.

9 役割 syn *function, role*

▶ We have already seen enough to realize that basicity plays an important part in our understanding of nucleophiles and leaving groups.

10 take part in に参加する syn *engage, partake, participate*

▷ Added substances frequently catalyze reactions by taking part in chain-propagation processes.

partake vi **1 partake in** …に加わる syn *participate, share, take part*

▶ Relatively small, highly symmetrical ions are unlikely to partake in strong secondary interactions such as hydrogen bonds and are thus often disordered.

2 partake of …の性質がある syn *possess*

▶ We distinguish between two limiting kinds of chemical bond, the purely ionic and purely covalent, though recognizing that any actual bond may partake of some of the qualities of each.

partial adj 部分的な syn *fragmentary, imperfect, incomplete*

▶ Attempts to correlate the entropy of fusion of an ionic crystal with the number of ions required to define its molecular structure in crystal and in melt meet with only partial success.

partial hydrolysis n 部分加水分解

▶ (+)-Maltose can be obtained, among other products, by partial hydrolysis of starch in aqueous acid.

partially adv 部分的に syn *in part, partly, to some extent*

▶ Elements with atoms or ions that have partially filled ($n-1$) orbitals are known as transition metals, since they come in between those using only their ns orbital and those that have np electrons.

partial molar volume n 部分モル体積

▶ Partial molar volumes (ϕ_m) are determined by accurate density measurements from equation

$$\phi_m = 1000(d_m - d)/mdd_m + M/d$$

where m is the molality of the electrolyte, d the density of the solvent, d_m the density of the solution, and M the molecular weight of the electrolyte.

partial pressure n 分圧

▶ An ideal solution is one in which all the components follow Raoult's law:

$$p_i = p_i^\circ n_i$$

in which p_i is the partial pressure of the component i from a solution in which n_i is the mole fraction of the same component, and p_i° is the vapor pressure of the pure substance i.

participant n 関係者 syn *contributor*

▶ Lattice imperfections are recognized as being important participants in the chemical reaction of solids.

participate vi **1** 関与する syn *contribute*

▷ When a set of hybrid orbitals is constructed by a linear combination of atomic orbitals, the energy of the resulting hybrids is a weighted average of the energies of the participating atomic orbitals.

2 participate in …に関与する syn *engage, partake, take part*

▶ The large variety of metal species M are known to participate in tungsten bronze formation.

participation n 関与 syn *involvement, share*

▶ Unless special procedures are employed, the kinetic results will not reveal the participation of the solvent.

particle n 粒子 syn *fragment, grain*

▶ Powdered lithium can be prepared by heating the metal in mineral oil at 250°under nitrogen with vigorous stirring; this gives fine, shiny particles.

particle accelerator n 粒子加速器

▶ The neutrons are obtained using particle accelerators to bombard a heavy metal target with high-energy particles such as protons.

particle size n 粒度

▶ Estimation of particle size may be formed from the distribution of intensity in powder diffraction patterns, the lines becoming diffuse with particles below 10^{-4} mm.

particular n **1** 通常複数形で，事項 syn *element, fact, item, information*

▶ The particulars below detail especially the steps in isolating these solids.

2 in particular 特に syn *especially, exactly, particularly, precisely*

▶ Certain crystalline oxides in particular, perhaps most notably those of transition metals, can occur in a remarkably wide range of composition.

adj **3** 特定の syn *certain, precise, specific*

▶ A system may be studied at some arbitrary, constant value of a particular condition, such as constant *P* or constant *T*.

4 特別の syn *notable, outstanding, special*

▶ The arylmethyl halides of particular interest are those having both halogen and aryl substituents bonded to the same saturated carbon.

particularly adv 特に syn *especially, notably, specially*

▶ Hydrogens attached to carbons adjacent to doubly bonded carbons are particularly reactive toward substitution.

particulate n **1** 微粒子 syn *granule*

▶ If the mineral acid concentration is reduced to below 0.001 M, submicron-sized particulates start to dominate the product morphology.

adj **2** 微粒子の syn *granular*

▶ No particulate materials were observed in the liquid flow, suggesting that the catalytic plates were mechanically stable.

parting n 分割 syn *division, separation*

▶ Parting with nitric acid does not dissolve platinum in the absence of silver, but if three parts by weight of silver are present for each part of gold and platinum, the platinum gradually dissolves.

partition n **1** 分配 syn *distribution*

▶ Small differences between partitioning of the components can be magnified by the large number of repetitive partitions possible in a long column.

vt **2** 分配する syn *distribute*

▶ When a given substance is partitioned between ether and water, the ratio of concentrations in the ether and water layers is a constant defined as the distribution ratio, *K*, specific to the solute, solvent pair, and temperature in question.

partition function n 分配関数

▶ The complete partition function for an isolated polyatomic molecule is

$$Q = Q^{T}Q^{R}Q^{E}Q^{IR}Q^{V}$$

provided that the energies for the separate motions are, to an adequate approximation, independent of one another.

partitioning n **1** 区分 syn *division, separation*

▶ A partitioning of the crystal structure on the basis of coordination numbers and/or structure type separates solids for which the associated monomers are linear from those for which the associated monomers are bent.

2 分配 syn *partition*

▶ A classic measure of polarity in organic molecules is partitioning between octanol and water.

partly adv ある程度は syn *in part,*

▶ The failure to isolate the cis complex may be due partly to the probable greater solubility of this isomer and possibly also to greater stability of the trans form.

parts per million, ppm n 百万分率

▶ Seawater contains approximately 35,000 parts per million of dissolved salts consisting mostly of sodium chloride.

pass vi **1** 進む syn *move, proceed, run*

▷ The large increase in dipole moment on passing from formaldehyde to acetoaldehyde is almost certainly due to the polar effect of the methyl group.

2 pass by 通過する syn *move, proceed*

▶ Free, or at least rapid, rotation is possible around all single bonds, except under special circumstances, as when the groups attached are so large that they cannot pass by one another.

3 pass through 通過する syn *enter, penetrate*

▷ Solvents were spectrograde and dried by passing through columns of activated basic alumina.

4 pass over 通過する syn *traverse*

▶ Upon steam distillation of the mixture of nitrophenols, the ortho isomer passes over in a substantially pure form.

vt **5** 渡す syn *deliver, give, transfer*

▶ Protons, although not free in solution, can be passed from one water molecule to another.

passage n 通過 syn *passing, transit*

▶ Dinitrogen was purified by passage through columns containing 4A molecular sieves and either manganese(II) oxide suspended on vermiculite or BASF catalyst.

passing n **in passing** ついでにいえば syn *incidentally*

▶ In passing, it can be pointed out that this expression is the basis for a quantity, presumably characteristic of the molecular volume, proposed by Sugden.

passivate vt 不動態化する

▶ Concentrated HNO_3 passivates Th, U, and Pu, but the addition of F^- ions avoids this and provides the best general method for dissolving these metals.

passivation n 不動態化

▶ The photoluminescence of porous silicon depends strongly upon the surface passivation.

passive adj 不動態の syn *inactive, inert*

▶ Aluminum readily replaces the hydrogen of hydrochloric acid, but it dissolves less readily in dilute sulfuric acid and becomes passive when treated with nitric acid.

passive film n 不動態皮膜

▶ The local breakdown of a passive film is promoted by the presence of aggressive ions, in particular halides, whose influence may depend on the metal being in tension.

passive state n 不動態

▶ In the passive state, the metal behaves as if it is more noble, i.e., has a position lower in the electromotive series, than in the active condition.

passivity n 不動態 syn *inactivity, inertness*

▶ The most obvious explanation of the facts concerning passivity is that it is due to a protecting film.

past n **1** 過去 syn *former times*

▶ In the past, the crystal classes were given names which were often long and cumbersome.

adj **2** 過去の syn *previous*

▶ The possibility of a scheme of this type has been mentioned on several past occasions.

adv **3** …を通り過ぎて syn *nearby*

▶ As the electrons responsible for conductivity travel past the positive ions, the ions are drawn toward the path of the electron by electrostatic attraction.

paste n ペースト

▶ Poly(tetrafluoroethylene) is mixed with light petroleum to form a paste which is then extruded, heated to remove the carrier, and sintered.

pasteurization n 殺菌

▶ Complete sterilization of milk requires ultrahigh pasteurization at from 94 ℃ for 3 sec to 150 ℃ for 1 sec.

pasty adj ペースト状の

▶ During this treatment, a change can be noted as the sulfonyl chloride undergoes transformation to a more pasty suspension of the amide.

path n **1** 行路 syn *route, track, trajectory*

▶ A path in a graph is an alternating sequence of distinct points and lines in the graph, beginning and ending with points, such that each line in the sequence is incident with the point immediately preceding it and with the point immediately following it.

2 筋道 syn *approach, direction, method, procedure*

▶ Attempts to account theoretically for the

growth of chain-folded lamellae from dilute solution have thus far followed on one or other of two paths.

pathogen n 病原体

▶ The aim of chemotherapy was not to effect a cure alone but to come to the aid of the host in its struggle with the pathogen.

pathogenic adj 病原性の

▶ The immune system protects the body from pathogenic organisms such as bacteria and viruses.

pathological adj 病理学の

▶ It now appears that elemental levels, even that of Sr, can vary with the choice of bone, with portions in a single bone, with pathological conditions, with the age of the individual at death, and possibly with the history of maternity and lactation in females.

pathway n 経路 syn *path, track*

▶ The importance of the nature and properties of the electrode surface in determining kinetic pathways and rates has become increasingly apparent.

pattern n **1** 様式 syn *model, prototype*

▶ Markovnikov pointed out that the orientation of addition follows a pattern which we can summarize as: in the addition of an acid to the carbon–carbon double bond of an alkene, the hydrogen of acid attaches itself to the carbon that already holds the greater number of hydrogens.

2 図形，模様 syn *figure, motif*

▶ There is a geometric pattern repeated over and over within a crystal.

vt **3** 手本とする syn *duplicate, follow, imitate*

▶ The study of metallomesogens is relatively young, and many of the types of compounds examined so far have been patterned after organic analogues.

patterning n 様式

▶ It became clear that the morphology and patterning of the inorganic structures only occurred for a specific choice of reagents and reaction conditions.

paucity n 少数 syn *absence, lack, shortage*

▶ This paucity of metal–metal bonding in halides was in striking contrast with the behavior of chalcogenides.

pay vi **1 pay off** 割に合う syn *be advantageous, profit*

▶ As M gets larger and softer, bending of the MX_2 monomer becomes less expensive and even pays off for the larger cations.

vt **2** 払う syn *deliver, extend, give*

▶ In this article, we will pay particular attention to spherical core–shell type micelles.

peak n **1** ピーク

▶ The M^+ peak is due to molecules containing only the commonest isotope of each element.

2 頂点 syn *apex, extreme, vertex*

▶ The molecular structure of P_4O_{10} is based on a tetrahedron of four phosphorus atoms, and it is convenient to envisage one of them forming the peak of the tetrahedron with the other three making the base.

vi **3** 最高になる syn *culminate, rise*

▶ As the localization of the particle is made more definite, the wave function peaks more strongly and its outer wings are annihilated more completely.

peculiar adj **1** 異常な syn *abnormal, curious, exceptional, unusual*

▶ Red blood cells are small compared with most other cells and are peculiar because of their biconcave disk shape.

2 peculiar to 特有の syn *characteristic of, typical of*

▶ Utilization of solid-state chemistry to achieve results not possible in solution demands a quantitative understanding of the effects on reaction rates of those features peculiar to the solid state, such as nucleation.

peculiarity n 特色 syn *characteristic, feature, hallmark, specialty*

▶ Many of the peculiarities of chromium(III) complexes are a result of the fact that the Cr^{3+} ion is not only small and highly charged but also has three, rather than two or four, 3d electrons.

peculiarly adv 独特に syn *especially, parti-*

cularly

▶ The ease with which the ionic model leads to quantitative predictions makes it peculiarly attractive as a basis for the discussion of a wide variety of inorganic problems.

Peierls transition n パイエルス転移

▶ A Peierls transition in an exactly half-filled band causes a lattice distortion in which the molecules dimerize.

pellet n **1** 小球 syn *ball, bullet, shot*

▶ When pellets of the starting materials were pressed in a strong magnetic field, the crystallites of $BaO \cdot 6Fe_2O_3$ became oriented with their *c* axes parallel, while the Fe_3O_4 remained unoriented.

vt **2** 小球状にする

▶ The nickel-silica catalyst was pelleted without the addition of any pelleting agent, weighed, and sealed into the sample tube.

pelletizing n 小球状にすること

▶ Thus, regrinding and pelletizing helps to ensure complete reaction.

pendant adj 側鎖の

▶ The cross-metathesis reactions of the substrates bearing a pendant 1,1 -disubstituted alkene were investigated, but only starting material was recovered in each case.

pendant group n 側鎖基 syn *pendant side group*

▶ By restricted rotation about chain bonds, pendant groups may hinder molecular rearrangement and retard the process of crystallization even in polymers with short chemical repeat units.

pending prep …を待ちながら syn *depending on*

▶ Pending development of models and theories which present all of nature as a unified whole, we often explain various aspects of the same system by using logically incompatible models.

penetrate vt 浸透する syn *diffuse, percolate, permeate*

▶ When monomers such as methyl methacrylate are added to a micelle, some of it becomes solubilized, which means that monomer penetrates into the micelles; the remainder stays in the aqueous phase.

penetrating adj **1** 貫通する

▶ For the study of liquids, as with that of solids, a more penetrating radiation is necessary so that the interior rather than surface structure can be studied.

2 突き刺すような syn *pungent, strong*

▶ The vapor of *p*-benzoquinone has a penetrating odor and attacks the eyes.

penetration n 浸透 syn *percolation, permeation*

▶ Adsorption is distinguished from the penetration of one component throughout the body of a second, called absorption.

pentagon n 五角形

▶ The five $-CH_2-$groups of cyclopentane can form a regular pentagon with only a little bending of the normal carbon bond angles.

pentagonal bipyramidal adj 五角両錐の

▶ The seven-coordinate $K_4[[V(CN)_7] \cdot 2H_2O$ has been shown to have a pentagonal bipyramidal structure.

penultimate adj 後から二番目の syn *next to the last*

▶ Any nonbonding electrons in molecules of the transition metals are d electrons from the penultimate shell.

peptide n ペプチド

▶ Peptides and proteins are composed of several amino acids joined by amide bonds; that is, the carboxy group of one amino acid is linked to the amino group of a second amino acid by an amide, or peptide, bond.

peptide chain n ペプチド鎖

▶ Proteins are made up of peptide chains, that is, of amino acid residues joined by amide linkages.

peptide linkage n ペプチド結合 syn *peptide bond*

▶ The peptide linkage is formed in a condensation reaction that requires the input of energy, while its hydrolysis to yield free amino acids is a spontaneous process that is normally very slow in neutral solutions.

peptide synthesis n ペプチド合成

▶ In peptide synthesis, the amino function of one of the compounds must be protected so that it cannot react with the carboxyl group of another compound of the same structure.

peptidoglycan n ペプチドグリカン

▶ The peptidoglycan, a mixed peptide-sugar complex, layer forms a net around a bacterial cell and helps maintain the structure and shape of the cell

peptization n ペプチゼーション

▶ In the peptization of globulin proteins, low lyotropic numbers or high heat of hydration gives a favorable orientation, good sorbabiity, and good peptizing action.

peptize vt 解こうする syn *deflocculate*

▶ Cellulose is peptized by the addition of organic solvents, such as ethanol-ether mixtures, leading to the familiar collodion sol.

peptizing agent n 解こう剤 syn *deflocculant, peptizer*

▶ In one procedure known as peptization, the disintegration into particles of colloidal dimensions is brought about by the action of a substance known as a peptizing agent.

per prep …につき

▶ It is important for the solid-state chemist to design intercalation hosts which can reversibly insert a large amount of lithium per formula unit.

perceive vt 認める syn *identify, observe, see*

▶ When viewed in the continuous radiation of daylight, alexandrite is perceived to be green because of the peculiar physiology of the human eye which is most sensitive to green light.

percent, % n パーセント 数字と記号の間は詰める.

▶ Sulfuric acid is more polar than water; hence, in going from 90 percent acid to 100 percent acid the polarity is actually increasing.

percentage n 百分率 名詞の複数形が続くときは，動詞も複数形とする. syn *proportion*

▶ Carbonyl groups in ketones, aldehydes, and esters are weakly basic, so that only a small percentage of the molecules are protonated even in strong aqueous acid.

perceptible adj 認識できる syn *detectable, discernible, noticeable*

▶ Some tetramer may be present but not in any readily perceptible proportion.

percolate vt 浸透する syn *penetrate, permeate*

▶ Nitrogen dioxide is passed through a tower packed with pieces of broken quartz through which water is percolating.

percolation n 浸透 syn *penetration, permeation*

▶ This low percolation threshold made sense after an interpenetrating network-like structure was revealed, composing of polyaniline nanofiber linkers with diameters of a few tens of nanometers.

perfect adj **1** 最適の syn *best, correct, proper,*

▶ The most perfect example of scattering of blue light is the blue sky, which otherwise would be black.

2 完全な syn *absolute, complete, ideal*

▶ The overall extent of thermal decomposition in sodium chlorate monocrystals depends very much on the sample perfection, the lower value being typical of the more perfect specimens.

vt **3** 完成する syn *accomplish, achieve, complete, finish, realize*

▷ The ammonia synthesis catalyst based on magnetite perfected by Haber and Mittasch is still centerpiece of fertilizer industry.

perfect diamagnetism n 完全反磁性

▶ Superconducting materials exhibit perfect diamagnetism and expel a magnetic field, provided the field is below the critical field strength.

perfection n 完全 syn *achievement, attainment, realization*

▶ A high degree of crystal perfection is needed: dislocations or defects in the crystal can be disastrous from a device performance aspect and can seriously reduce yields from a given wafer.

perfectly adv 完全に syn *accurately, exactly, precisely*

▶ The technique of thermal analysis, through cooling or heating curves, even if perfectly standardized in respect to thermal problems, has the inherent difficulty of the problem of the internal equilibrium of the sample itself.

perform vt 実行する syn *accomplish, complete, execute*

▷ One feature common to virtually all research performed in surfactant solution chemistry is the electrostatic complexity of the molecular interactions within association aggregates.

performance n **1** 性能 syn *behavior*

▶ There is no single physicochemical parameter that is adequate to predict clinical performance.

2 実行 syn *accomplishment, completion, execution*

▶ The successive performance of symmetry operations is the product of the operations.

perfume n 香料 syn *aroma, extract, fragrance, oder, smell*

▶ Coumarin itself occurs in clover and is widely used as a perfume and a flavoring agent.

perhaps adv たぶん syn *conceivably, possibly*

▶ Oxygen can form loose addition compounds with aromatic molecules, and the subsequent perturbation of the organic molecule leads, perhaps through an excited triplet state of the aromatic system, to oxidation.

perimeter n 周辺 syn *border line, boundary*

▶ The porphyrin ring consists of a macrocyclic pyrrole system with conjugated double bonds and various groups attached to the perimeter.

period n **1** 期間 syn *duration, span, time*

▶ A solution of $KMnO_4$ (1.11 g) in water was added to a mixture of 6-bromoveratraldehyde (1.22 g) and water (20 mL) at 75 ℃ over a period of 20 min.

2 周期

▶ The periodic table has the advantage over some of affording quick reference to the periods (horizontal) and the groups (vertical).

3 段階 syn *phase, stage, step*

▶ The specific conditions that exist during the period of growth will strongly affect the shape of the resulting crystal.

periodic adj 周期的な syn *cyclic, repeated, repetitive*

▶ The subject of periodic precipitation has a voluminous literature because such phenomena are a common occurrence in plants and other organisms and in mineralogical and geological formations.

periodically adv 周期的に syn *regularly, systematically*

▶ Diffraction patterns constitute evidence for the periodically repeating arrangement of atoms in crystals.

periodicity n 周期性

▶ Long-range periodicity, the characteristic property of the crystalline state, is produced by short-range interactions, resulting in optimal space filling.

peripheral adj **1** 周辺の syn *external*

▶ The electronic nature of these peripheral atoms has been the focus of much of the mechanistic debate.

2 末梢の syn *minor, secondary, unimportant*

▶ It is hoped that some impression of recent major aspects of liquid crystal research can be conveyed without giving too much attention to interesting but currently peripheral topics.

peripherally adv 周辺に syn *circumferentially, externally*

▶ The calixarenes bear some resemblance to the naturally occurring cyclodextrins which also possess repeating structural subunits with several hydroxyl groups arranged peripherally about a central cavity.

periphery n 周辺 syn *boundary, edge, perimeter, rim*

▶ Apparently, the Fe atoms at the center of the particle on titania are attracted to the periphery.

peritectic point n 包晶点

▶ When a compound is formed which undergoes decomposition with formation of another solid phase at a temperature below the congruent melting point of the compound, this temperature is spoken as an incongruent melting point or peritectic point.

permanence n 永続性 syn *durability, longevity, persistence*

▶ Mere permanence of an emulsion left completely undisturbed must be distinguished from stability under conditions of shaking, vibration, temperature change, centrifuging, or other stress.

permanent adj 永久的な syn *durable, lasting*

▶ Signals may induce transitory or permanent changes in cells.

permanent dipole n 永久双極子 ☞ induced dipole

▶ When the molecule is a permanent dipole, it leads to a tendency for the dipole to be aligned by the field or, rather, to be rotated slightly in the direction of alignment.

permanent dipole moment n 永久双極子モーメント ☞ dipole moment

▶ For many substances, the interaction of their permanent dipole moments constitutes only a small perturbation to the London force already present.

permanent hardness n 永久硬度 ☞ temporary hardness

▶ Permanent hardness of water is caused by alkaline-earth salts of the strong acids, usually calcium sulfate.

permeability n 透磁率

▶ Magnetically soft materials have low permeability and, hence, a hysteresis loop that is of narrow waist and of small area.

permeability barrier n 透過障壁

▶ For cell membranes to be effective permeability barriers, they must be flexible and allow relatively free motion of proteins that are embedded in or linked to them.

permeability coefficient n 透過係数

▶ In work with membrane, it is convenient to define a quantity known as the permeability coefficient, which is defined as the flux through unit area when there is a unit concentration difference.

permeable adj 浸透性の syn *penetrable*

▶ Dialysis is a differential diffusion where the membrane is permeable to molecules and ions but is too fine to allow colloid particles to pass.

permeate vt 行き渡る syn *spread through*

▶ A three-dimensional network of interconnected channels permeates the structure along which Na^+ may migrate.

permeation n 浸透 syn *penetration, percolation*

▶ Asymmetric polymer membranes combine the high selectivity of a dense membrane with the high permeation rate of a very thin membrane.

permeation chromatography n 浸透クロマトグラフィー

▶ Gas permeation chromatography involves the use of highly cross-linked gels which have a distribution of pore sizes and, therefore, can separate a polymer sample into fractions, on the basis of molecular volumes.

permissible adj 許容される syn *acceptable, allowable, allowed, tolerable*

▶ It is permissible to dry in a desiccator only those substances which will not lose water of crystallization.

permissible dose n 許容線量

▶ The permissible dose is the quantity of radiation which may be received by an individual over a given period with no detectable harmful effects.

permit vt …を許す syn *admit, allow, enable, tolerate*

▷ Hydroxylation of alkenes is the most important method for the synthesis of 1,2-diols, with the special feature of permitting stereochemical control by the choice of reagent.

permutation n **1** 互換

▶ The total molecular eigenfunction must be antisymmetric with respect to proton and neutron permutation.

2 順序の交換 syn *conversion, transformation, variation*

▶ The arene linker, N-substituents, and incorporated transition metals could be modified through simple permutations of the basic synthetic protocol.

perovskite structure n ペロブスカイト型構造

▶ Most known superconducting oxides are closely related to the perovskite structure (ABO_3) or one of its variants, the layered perovskite (A_2BO_4) or K_2NiF_4 structure.

peroxidase n ペルオキシダーゼ

▶ Hydrogen peroxide produced by reduction of O_2^- or O_2 can be handled by enzymes known as peroxidases by oxidizing a substrate.

peroxide effect n 過酸化物効果

▶ The reversal of the orientation of addition caused by the presence of peroxides is known as the peroxide effect.

peroxidic adj 過酸化物の

▶ Most alkenes react readily with ozone, even at low temperatures, to cleave the double bond and yield cyclic peroxidic derivatives known as ozonides.

perpendicular adj 垂直の syn *normal, vertical*

▶ Our own work has concentrated primarily on ortho-substituted tertiary benzamides, in which the amide group lies more or less perpendicular to the plane of the aromatic ring.

perpendicularly adv 垂直に syn *upward, vertically*

▶ The spectrum is uniquely explained by assuming that the bond is aligned perpendicularly to the applied field and that the molecule rotates rapidly about this direction.

perpetuate vt 永続させる syn *continue, extend, maintain, preserve*

▶ There is the related problem of how the information as to the amino acid sequences is perpetuated in each new generation of cells.

per se adv 本来 syn *in itself*

▶ It is not high concentrations of cholesterol per se that are necessarily damaging to blood vessels via atherogenesis but a high ratio of LDL : HDL.

persist vi 持続する syn *continue, last, remain*

▶ The violet color persists when solid iodine is dissolved in a noncomplexing solvent such as tetrachloromethane.

persistence n 持続性 syn *constancy*

▶ We use the N–I value for a mesogen as an index of the thermal persistence or thermal stability of that nematic phase.

persistent adj 持続的な syn *constant, continuous*

▶ The complex of methyl parathion with β-CD has useful and persistent activity against cotton insects.

perspective n **1** 考え方 syn *prospect, viewpoint*

▶ For perspective, pure water, for which hydrogen ions and hydroxide ions would be the charge carriers, would have a carrier concentration of only 10^{14} ions/cm^3.

2 in perspective 総合的視野で

▶ Considerable caution should be taken in handling compounds of Se and Te, but the hazards should be kept in perspective.

3 透視画

▶ The projection used in nearly all cases is not a true perspective but assumes that the observer is at infinite distance, so that all lines, which are parallel on the crystal, are also parallel in the drawing.

persuade vt させる syn *dispose, entice, induce, influence, prompt*

▶ In ferroelectric $BaTiO_3$, the individual TiO_6 octahedra are polarized all of the time; the effect of applying an electric field is to persuade the individual dipoles to align themselves with the field.

persuasive adj 説得力のある syn *convincing, impressive, influential*

▶ A more persuasive contribution was made by the German physician J. R. Mayer.

pertain vi 関係する syn *concern, relate*

▶ The original Hammett equation pertained to a limited set of equations.

pertinent adj 妥当な syn *appropriate, fitting, relevant, suitable*

▶ The designation organic is still very pertinent because the chemistry of carbon compounds is more important to everyday life than that of any other element.

perturb vt 変動を起こさせる syn *agitate, disturb*

▶ The ions surrounding an anion vacancy will all be drawn in slightly to new equilibrium positions with the result that the first excited anion levels (exciton levels) in the neighborhood of the vacancy are perturbed.

perturbation n 摂動 syn *agitation, disturbance*

▶ Relaxation techniques are widely employed for solution phase reactions, where a number of different types of perturbation can be employed.

perturbation theory n 摂動論

▶ Another type of transition, which is forbidden in first order time-dependent perturbation theory, is a transition involving several quanta simultaneously.

pesticide n 殺虫剤

▶ The pesticides emerged as weapons to control disease in man and animals caused by insect vectors, plant diseases transmitted by insects, and the ravaging of crops by insects, weeds, and fungi.

pestle n 乳棒 ☞ mortar

▶ Extensive grinding in a mortar and pestle produces particles with a radius of about 10^{-3} cm.

petroleum coke n 石油コークス

▶ Graphite is produced synthetically by heating petroleum coke to approximately 3000 ℃ in an electric resistance furnace.

petroleum ether n 石油エーテル

▶ The residue was purified by flash chromatography, eluting with petroleum ether, and recrystallized from petroleum ether to yield the *trans*-stilbene as needles

pH n 水素イオン指数 syn *hydrogen ion exponent*

▶ The color change of Methyl Red is at pH 4.4–6.0.

pharmaceutical n **1** 薬剤 syn *drug, medicine, pill*

▶ Pharmaceuticals are ingested intentionally.

adj **2** 製薬の

▶ Currently much of the great interest in the cyclodextrins arises from their pharmaceutical applications.

pharmacological adj 薬理学の

▶ Oxidative detoxification reactions either modify functional groups or add oxygen to the foreign compound, forming metabolites which lack pharmacological activity and are more rapidly eliminated in the urine because of increased water solubility.

phase n **1** 段階 syn *period, stage, step*

▶ Micas occur as a late crystallization phase in igneous rocks.

2 相 syn *form, state*

▶ The number of phases in a sample at equilibrium is the number of physically distinct and mechanically separable portions, each phase being itself homogeneous.

3 in phase 同位相で

▶ In plastic crystals, the only radiation scattered in phase comes from an electron distribution concentrated around the centers of gravity of the molecules.

4 out of phase 位相を異にして

▶ The heavy background in X-ray reflection from plastic crystals is due to the radiation scattered from displaced atoms that are out of phase.

phase boundary n 界面

▶ More than one phase would exist if the phase boundaries in the solid can be discerned by using a microscope.

phase diagram n 状態図

▶ Conventionally, temperature is the vertical scale and composition the horizontal one in binary phase diagrams.

phase equilibrium n 相平衡

▶ We have considered phase equilibrium between solid and liquid, in the case where the components were soluble in each other in any proportions, in both liquid and solid.

phase separation n 相分離

▶ Liquids are compatible if they are miscible and do not undergo phase separation on standing.

phase space n 位相空間

▶ In Bose-Einstein statistics, identical parts are supposed to be distinguishable only if they have different energies, i. e., are in different cells in the phase space.

phase-transfer catalysis n 相間移動触媒作用

▶ The power of phase-transfer catalysis lies in the fact that it minimizes the two chief deactivating forces acting on the anion: solvation and ion pairing.

phase transition n 相転移 syn *phase transformation*

▶ At lower temperatures, solid electrolytes may undergo a phase transition to give a polymorph with a low ionic conductivity and a more usual type of crystal structure.

phase transition temperature n 相転移温度

▶ The pertinent phase transition temperatures of the compounds were determined, and, in addition, the enthalpies and entropies of the mesomorphic transitions were measured by differential scanning calorimetry.

phenomenological adj 現象論的な

▶ A phenomenological approach which describes the observed properties is a sounder initial basis for discussion.

phenomenologically adv 現象論的に

▶ From an examination of the behavior of about 100 of the compounds, it was concluded that at least seven types of transformations in solid phase could be distinguished phenomenologically.

phenomenon n 現象 複数形は phenomena syn *event, fact, happening*

▶ Precipitation phenomena in crystalline systems are widespread and often technologically important.

pH gradient n pH 勾配

▶ If a mixture of proteins is placed on a gel that contains a stable pH gradient in which the pH increases steadily from the anode to the cathode, each protein will migrate to a position in the gel corresponding to its own isoelectric point.

phonon n フォノン

▶ If an electron and a hole finally coalesce with mutual annihilation, there is a release of energy which may result in the production of light or phonons.

phospholipid n リン脂質

▶ The esterification may be only partial, giving diacylglycerols that are intermediates in phospholipid synthesis.

phospholipid bilayer n リン脂質二重層

▶ Because the phospholipid bilayer of the inner mitochondrial membrane is impermeable to protons, the ejection of protons from the matrix by proteins establishes an electrochemical gradient.

phosphor n 蛍光体

▶ A phosphor that is used extensively in fluorescent lamps is an apatite, doubly doped with Mn^{2+} and Sb^{3+}.

phosphorescence n りん光

▶ Intersystem crossing to the lowest triplet state gives rise to phosphorescence.

phosphorescence yield n りん光収率

▶ For naphthalene the phosphorescence yield stayed constant throughout the S_1 and S_2 excitation regions.

phosphorescent adj りん光を発する

▶ Material that continues to emit light for a period after removal of the exciting energy is said to be phosphorescent.

phosphorylate vt リン酸エステル化する

▶ ATP simply phosphorylates, that is, transfers a phosphoryl group, $-PO_3H_2$, to some other molecule.

phosphorylation n リン酸エステル化 ☞ dephosphorylation

▶ The best example of molecular conversion is enzyme phosphorylation and dephosphorylation that occur in the control of glycogen synthesis and degradation.

photoactive adj 光活性

▶ In many cases, adsorption of a photoactive molecule on the clay introduces new features into the photophysics and consequently into the photochemistry also.

photoaddition n 光付加

▶ It is not required that intramolecular photoadditions occur at both ends of one double bond.

photocatalysis n 光触媒作用

▶ In photocatalysis, energy encouragement in the reactions is provided by absorbed light.

photocell n 光電セル syn *photoelectric cell*

▶ CdS absorbs visible light and is being used in the development of photocells, as a means of converting sunlight into other forms of energy.

photochemical chlorination n 光化学的塩素化

▶ The most satisfactory evidence for the proposed mechanism of chlorination is doubtless that obtained from a kinetic study of the photochemical chlorination of chloroform.

photochemical degradation n 光化学的分解 syn *photochemical decomposition, photodecomposition, photolysis*

▶ Dye-sensitized and pigment-sensitized photochemical degradations result from visible and near-ultraviolet irradiation and are generally oxidative.

photochemically adv 光化学的に

▶ The synthesis of new precursors is a challenge, for they must be volatile, highly pure, and photochemically reactive, while at the same time being thermally stable.

photochemical primary process n 光化学初期過程

▶ The process in which electromagnetic radiation is absorbed by a molecule, with the formation of species that undergo further reaction, is known as the photochemical primary process.

photochemical reaction n 光化学反応

▶ Clay systems either as aqueous colloids or as dry clays act as catalysts for photochemical reactions.

photochemical reduction n 光化学的還元

▶ Experimental and theoretical work on the photochemical reduction of solids has been dominated by the attention given to silver halides.

photochemical smog n 光化学スモッグ

▶ It is the simultaneous presence of nitrogen oxides and hydrocarbons that lead to the dramatic photochemical smogs now seen in many urban areas.

photochromic adj ホトクロミックな

▶ Photochromic systems are based on the ability of certain compounds to undergo reversible molecular rearrangements under the influence of light which results in a color change.

photochromic glass n ホトクロミックガラス

▶ One type of photochromic glasses is a silicate glass containing dispersed crystals of colloidal silver halide which is precipitated within the melt during cooling.

photochromism n ホトクロミズム syn *phototropy*

▶ In molecular materials, photochromism can be the result of a reversible change in charge distribution or in molecular structure upon irradiation.

photoconductive adj 光伝導性の

▶ Some semiconductors are photoconductive, i. e., their conductivity increases greatly on irradiation with light.

photoconduction n 光伝導 syn *internal photoelectric effect*

▶ Photoconduction along the *c* axis of Magnus' green salt has been observed with a threshold of about 4500 cm^{-1}.

photoconductor n 光伝導体

▶ Amorphous selenium is an excellent photoconductor and forms an essential component of the photocopying process.

photocurrent n 光電流

▶ The photocurrents observed during irradiation of solid methanol are explained in terms of free electrons resulting from ionization of the molecules.

photocycloaddition n 光環状付加

▶ The photocycloaddition reactions can be made to occur by direct irradiation, and under these circumstances an excited singlet state of one alkene reacts with the ground state of the other alkene.

photodimer n 光二量体

▶ 9-Cyanoanthracene, upon UV irradiation, yields the trans photodimer rather than the expected cis dimer.

photodiode n ホトダイオード

▶ The light pulse was monitored by a silicon photodiode, and the response was displayed simultaneously with the transient pulse on the dual beam oscilloscope.

photodissociation n 光解離

▶ The photodissociation of acetone at 193 nm has been a convenient source of methyl radicals for a number of kinetic studies.

photoelectric effect n 光電効果

▶ The collisional interpretation of the photoelectric effect has led to the idea that light is corpuscular.

photoelectric emission n 光電子放出 syn *photoemission*

▶ Even with the highly insulating paraffin hydrocarbons, it is possible to observe photoelectric emission.

photoelectron spectroscopic adj 光電子分光の

▶ The conclusion from the photoelectron spectroscopic studies is that the carbon monoxide is essentially dissociated when it is chemisorbed on these metals.

photoelectron spectroscopy n 光電子分光 ☞ ultraviolet−, X-ray−

▶ Photoelectron spectroscopy is the examination of the energy levels of molecules by determining the kinetic energies of electrons ejected by the absorption of high-frequency monochromatic radiation.

photoelectron spectrum n 光電子スペクトル

▶ Comparisons have been made between the photoelectron spectra of gaseous benzene, liquid benzene, and benzene chemisorbed on surfaces and have provided useful detailed information about the nature of the chemisorption process.

photogalvanic cell n 光ガルバニ電池

▶ There are formidable difficulties that make it difficult to make effective use of solar radiation with photogalvanic cells.

photographic emulsion n 写真乳剤

▶ It was stated that, during the exposure of chemically nonsensitized photographic emulsions, mainly an internal latent image is formed.

photography n 写真

▶ Photography makes use of the bromide or iodide, and the minute amount of photoreduction that occurs when light falls on the film is then amplified by developing it.

photoionization n 光イオン化

▶ Although photoionization may produce lower concentrations of ions, less ion fragmentation occurs, simplifying product identification.

photoisomerization n 光異性化

▶ Photoisomerization in the solid state need not be reversible: both internal redox reactions and cis–trans rearrangements are known which led to stable products.

photoluminescence n ホトルミネセンス

▶ Photoluminescence uses photons or light, often UV, for excitation

photolysis n 光分解 syn *photodecomposition,*

▶ Continuous generation of hydrogen by photolysis of water has been achieved using platinum catalyst in conjunction with ruthenium and rhodium.

photolytic adj 光分解の

▶ The motion of silver ions is involved in the mechanisms responsible for the formation of latent image and for the separation of photolytic silver.

photolyze vt 光分解する

▷ Methylene probably is formed on photolyzing diazomethane trapped in inert-gas matrices.

photometric adj 光度の

▶ Photometric analysis includes such techniques as spectrophotometry, spectrochemical analysis, Raman spectroscopy, colorimetry, and fluorescence measurements.

photomultiplier tube n 光電子倍増管 syn *photomultiplier*

▶ The unabsorbed photons hit a photomultiplier tube which produces a current whose magnitude depends on the number of incident photons.

photon n 光子

▶ Photons are discrete concentration of energy that seem to have no rest mass and move at the speed of light.

photopolymeric adj 光重合性の

▶ Development of the image in photopolymeric systems is based on many chemical and physical properties that can differentiate between polymerized and unpolymerized areas.

photoreceptor n 光レセプター

▶ Chlorophyll is a photoreceptor for wavelengths up to 700 nm.

photoresist n ホトレジスト

▶ There are many materials that in LB film form are virtually pinhole-free and display some of the properties associated with photoresists.

photosensitive adj 感光性の

▶ In a special type of photosensitive glass, the metal particles of the photographic image within the glass serve as nuclei for the growth of nonmetallic crystals.

photosensitization n 光増感

▶ Finally, mention should be made of the photoelectric properties of dyes, which are of importance in such areas as photosensitization phenomena, organic photoconductors, electrophotography, and photovoltaic cells.

photosensitizer n 光増感剤

▶ Since the formation of aggregates modifies the absorption spectrum and photophysical properties of a dye, it affects its ability to emit at a certain wavelength or to act as a photosensitizer.

photosynthesis n 光合成

▶ Most life on earth is dependent upon a series of redox reactions in photosynthesis by which solar energy is used to produce ATP and O_2 and to synthesize carbohydrates from CO_2.

photosynthetic adj 光合成の

▶ The absorption spectra of photosynthetic systems fall nicely within the visible region of the sun's spectrum that reaches the earth.

photosynthetically adv 光合成で

▶ Photosynthetically generated oxygen gas temporarily enters the atmosphere and is recycled about once every 2000 years at present rates.

photosynthetic bacterium n 光合成細菌

▶ In photosynthetic bacteria, the species oxidized is not H_2O but instead an organic molecule or H_2S.

photosynthetic organism n 光合成生物

▶ The reaction centers of photosynthetic organisms are undoubtedly the most sophisticated examples of photochemical energy conversion systems.

photosystem n 光化学系

▶ About 200 to 300 pigment molecules form a cluster called a photosystem; many such clusters exist in each membrane engaged in photosynthesis.

phototoxic adj 光毒性の

▶ Multiple exposures to external light from lasers and/or ambient light sources can induce phototoxic cell killing.

phototransformation n 光変換

▶ One of the most important problems is to find a method of decomposition or phototransformation of organic compounds that, although only sparingly soluble in water, are very dangerous when introduced into food chain and eventually into the human body.

phototropy n ホトトロピー syn *photochromism*

▶ Phototropy is the ability of a compound to undergo a light- or UV-induced reversible color change.

photovoltaic cell n 光起電力セル ☞ photocell

▶ Certain materials, such as selenium, produce electricity when they are irradiated, and devices that employ this effect are known as photovoltaic cells.

phrase vt 表す syn *describe, express, term*

▶ To phrase it differently, the four-coordinated oxo acids and ions in the last short period are substantially more stable to reduction than are their heavier homologs.

physical adj 物理的な

▶ Physical methods give molecular weights for cellulose ranging from 250,000 to 1,000,000 or more.

physical adsorption n 物理吸着 syn *physisorption*

▶ Physical adsorption consists in the binding of molecules to the surface of the adsorbent by essentially van der Waals' forces.

physically adv 物理的に

▶ Some materials that are physically rigid, such as glass, are regarded as highly viscous liquids because they lack crystalline structure.

physicochemical adj 物理化学の

▶ The perchlorate ion has for long been considered to be a noncoordinating ligand and has frequently been used to prepare inert ionic solutions of constant ionic strength for physicochemical measurements.

physiological adj 生理的な

▶ The extreme sensitivity to environment of the relative stabilities of both the +2 and +3 oxidation states of iron is the basis of many delicate physiological processes.

physisorb vt 物理吸着する

▶ Carbon monoxide is physisorbed on rhodium, as shown both by its ease of removal on warming and its vibrational frequency on the surface, which is close to that of gaseous carbon monoxide.

pick vt **1 pick out** 選び出す syn *choose, select*

▶ Any pieces of alkali metal which had not reacted with anthracene could be picked out, as they remained bright.

2 pick up 獲得する syn *catch, get*

▶ An example is provided by nickel oxide, NiO, which when heated in oxygen picks up oxygen to give a solid solution of formula, $Ni_{1-x}O$.

pictorial adj 描写の syn *clear, explicit, expressive, graphic, plain*

▶ In X-ray studies, the X-ray diffraction pattern from a given material is modeled by using the diffraction of visible light from a suitably scaled pictorial representation of the distribution of atoms and molecules in the original material.

pictorially adv 描写して

▶ The orbitals are commonly represented pictorially in two dimensions by a locus of points of equal electron density such that some given fraction, say 90%, of the total electronic charge is contained within figure.

picture n **1** 事態 syn *situation, state, status*

▶ In connection with orientation in the substitution of naphthalene, the picture is often complex, although the 1-position is most reactive.

2 描写 syn *model, representation*

▶ The paramagnetism of the superoxides suggests an ionic picture featuring the O_2^- anion; the moment of 2.04 μ_B observed for KO_2 is characteristic of one unpaired electron.

vt **3** 描写する syn *depict, display, illustrate, represent*

▶ It is sufficient to picture a molecular dipole as two equal and opposite charges $q^{\pm}$ separated by a distance r.

piece n **1** 断片 syn *fragment, lump, particle, portion*

▶ An ordinary piece of the metal copper does not consist of a single crystal of copper but of an aggregate of crystals.

2 一個 syn *article, component, item*

▶ The relatively long H···O distances and the small O−H···O angles are among several pieces of evidence which suggest that the hydroxyl hydrogen atoms are involved in a bifurcated hydrogen bond to the ortho chlorine atoms.

piecewise adv それぞれの部分に関して
▶ This conclusion has been arrived at slowly and piecewise.

piezoelectric adj 圧電の
▶ Materials like quartz are piezoelectric because they have the ability to develop a net dipole moment if they are mechanically deformed in particular directions relative to the arrangement of atoms in the crystal.

pigment n **1** 顔料 syn *paint*
▶ The distinction between a dye and a pigment is that a dye is actually absorbed by the materials to be colored, whereas a pigment is applied to the surface.
2 色素 syn *color, dye*
▶ Most pigments are insoluble in organic solvents and water; exceptions are the natural organic pigments, such as chlorophyll.
vt **3** 彩色する syn *paint*
part ▶ All types of carbon black are characterized by extremely fine particle size, which accounts for their reinforcing and pigmenting effectiveness.

pile n **1** 積み重ね syn *accumulation, stack*
▶ The top of the conical pile is flattened out and divided into quarters. Opposite quarters are taken for the next crushing.
2 電池 ☞ battery, cell
▶ Voltaic pile consisted of plates of different metals and offered a convenient source of electricity that soon attracted the attention of chemists.
3 原子炉 ☞ atomic pile
▶ Attempts are being made to increase the number of neutrons available from a pile and to reduce the dimensions of the beam and specimen.
vi **4 pile up** 蓄積する syn *accumulate, collect*
▶ Work hardening of metals is a process whereby many of the dislocations pile up and become immobilized.
vt **5** 積み重ねる syn *accumulate, cumulate, stack*
▶ In the crystal of CdI_2, the layers are piled on top of one another and held together only by the weak van der Waals forces.

pilot adj 予備的な syn *exploratory, provisional,*
▶ In the practical use of the sensitized photochemical equilibrium of cis and trans isomers, it is normally necessary to carry out pilot experiments to find what sensitizers are useful and to determine the equilibrium point which each gives.

pin vt 留める syn *affix, fix, hold, secure*
▶ When the period of the charge-density wave (CDW) is commensurate with the lattice periodicity, the CDW is pinned and stops sliding so the material becomes an insulator.

pinchcock n ピンチコック
▶ The pinchcock is carefully opened after the system has been evacuated until there is a steady stream of bubbles.

pinhole n 針で突いて作った小穴 syn *hole*
▶ The expansion of the solution takes place through the pinhole nozzle located in the expansion chamber.

pinpoint vt 特定する syn *identify, locate, specify*
▶ Finally, we must pinpoint the predominant conformation in which (+)-glucose exists.

pioneer vt 開拓する syn *begin, initiate, launch*
▶ Stock not only pioneered the chemistry of boron and silicon hydrides but he developed, from scratch, the chemical high-vacuum techniques which enabled him to study these extremely reactive, spontaneously flammable materials.

pioneering adj 先駆となる syn *advanced, new*
▶ The vast amount of research since this pioneering work has been discussed elsewhere.

pipet, pipette n ピペット ☞ transfer pipet
▶ The naphthalene and phenanthrene solutions were sampled with a pipet whose tip was covered with a small piece of filter paper to prevent particles of solid from entering.

pit n エッチピット ☞ etch pit
▶ It cannot be inferred that every pit represents a dislocation.

pitch n **1** ピッチ syn *tar*
▶ Pitch is brittle under shock and flows slowly but steadily under low shearing stress.

2 勾配 syn *inclination, slope, tilt*

▶ As a succession of layers is passed through, the director turns through 360° and this thickness represents the pitch length for the helix.

pitfall n 落とし穴 syn *difficulty, trap*

▶ One potential pitfall is that tiny solid particles may pass through the filter and continue to catalyze the heterogeneous reactions.

pitting n 孔食

▶ The fact that aluminum readily undergoes electrochemical pitting is used commercially in the production of lithographic plates, which require a very fine pit distribution for good definition in the final printing stage.

pivot vi 旋回する syn *revolve, spin*

▶ With less interference from the ortho substituents, the rings can more easily pivot about the central bond.

pivotal adj 重要な syn *critical, crucial, important, significant*

▶ It seems likely that metal ions will provide a pivotal role in future developments of liquid crystal technology just as they play key roles in nature's macromolecular assemblies.

pivot bond n ピボット結合 syn *rotational axis*

▶ The lack of rotation about the pivot bond is caused by steric hindrance between the bulky ortho substituents.

place n **1 in place** 決まった位置に syn *in situ, in (proper) position*

▶ One or more atoms may be removed from a cluster leaving the bonding electron pairs in place as nonbonding pairs.

2 in place of (prep) …の代わりに syn *instead of*

▶ In place of aluminum chloride, other Lewis acids can be used, in particular BF_3, HF, and phosphoric acid.

3 in the first place 第一に

▶ In the first place, it is important to know what is being measured.

4 out of place 不適当に syn *inappropriate, wrong*

▶ A brief deviation into its early history is here not entirely out of place.

5 take place 行われる syn *arise, happen, occur*

▶ If such a reaction appears not to take place, that can mean only that it takes place too slowly for measurement or that it is reversible, the equilibrium greatly favoring the reactants.

vt **6** 位置に置く syn *arrange, lay, locate, situate*

▶ A suitable amount of the solute, e. g., chrysene is mixed with boric acid, and the mixture is placed in a 10×2 cm boiling tube.

7 状態に置く syn *assign, give, put, set*

▶ These requirements place severe constraints on the selection of starting materials and the processing that can be used.

8 考える syn *identify, recall, recognize*

▶ There are insufficient structural data to place the symmetrically bicapped trigonal antiprism among the established octacoordinate geometries.

plain adj 明白な syn *clear, evident, obvious, unmistakable*

▶ It is plain that no significant inaccuracy in ΔH springs from this source.

plainly adv 明白に syn *apparently, clearly, evidently, obviously*

▶ The experimental evidence shows plainly that kinetic restrictions play an important role and that states of thermodynamic equilibrium are not easily achieved.

plan n **1** 計画 syn *program, project*

▶ Our plan in evaluating the reasonableness of these steps is to determine how much energy is required to break the bonds.

vt **2** 計画する syn *aim, envisage, intend, propose*

▶ Since we plan a general survey of the use of liquid crystals as solvents, we have been concerned with the choice of suitable materials for this work.

planar adj 平面の

▶ The UO_2 group in $UO_2(NO_3)_2{\cdot}6H_2O$ is surrounded by a near-planar hexagon of four oxygen atoms from two bidentate nitrate groups and two equivalent water oxygen atoms.

planarity n 平面性

▶ Planarity is common for molecules of second-period elements in which resonance stabilization is important.

planar structure n 平面構造

▶ The planar structures of Se_4^{2+} and P_6^{4-} correspond to the aromatic structure of the six- and 10-electron systems, respectively.

plane n 面 syn *(flat) surface*

▶ A crystal is symmetrical with respect to a plane when for each face or edge on one side of the plane there is a similar face or edge directly opposite on the other side, so that one side is the mirror image of the other.

plane figure n 平面図形

▶ Since the only variables in a one-component system are T and P, a plane figure with these coordinates is sufficient for the phase diagram of such a system.

plane of polarization n 偏光面

▶ Anisotropic substances cause a partial rotation of the plane of polarization of light as its passes through them.

plane of symmetry n 対称面 syn *mirror plane*

▶ A paraffin chain with an odd number of carbon atoms possesses a plane of symmetry perpendicular to its axis passing through the central atom.

plane-polarized light n 面偏光 ☞ polarized light

▶ With plane-polarized light, the direction of the electric field remains constant and only its magnitude varies.

planned adj 計画された syn *deliberate, intentional, systematic*

▶ These reactions cannot be described as planned syntheses as the product usually cannot be predicted with any certainty and indeed have often proved to be completely unexpected.

plant n **1** 植物

▶ Plants offer certain advantages over bacteria for the production of recombinant proteins.

2 工場 syn *works*

▶ A plant is any large-scale manufacturing unit including pipelines, reaction equipment, machinery, etc.

plant growth regulator n 植物成長調節物質

▶ A large number of chemicals that tend to increase the yield of certain plants, as well as plant-produced hormones, are included in the term plant growth regulator.

plaque n プラーク

▶ High-density lipoprotein in the plasma is antiatherogenic and removes cholesterol from macrophages, preventing the growth or even regression of the atherosclerotic plaque.

plasma n プラズマ

▶ A particle plasma is a neutral mixture of positively and negatively charged particles interacting with an electromagnetic field, which dominates their motion.

plasma membrane n 形質膜

▶ The plasma membrane is a fluid, highly dynamic, and complex double-layered barrier composed of lipids, proteins, and carbohydrates.

plasmid n プラスミド

▶ A strand or fragment of genetic material existing outside the chromosomes in certain types of bacteria is called a plasmid.

plastic n **1** プラスチック syn *resin*

▶ Addition of such modifying agents as fillers, colorants, etc. to high polymers yields an almost infinite number of products collectively called plastics.

adj **2** プラスチックの

▶ The scintillation efficiency of plastic scintillators deteriorates under irradiation, but it does not show any simple dependence on dose.

plastic crystal n 柔粘性結晶

▶ The plastic crystals are crystals composed of molecules of a globular nature and have rather remarkable properties.

plastic deformation n 塑性変形

▶ Plastic or irreversible deformation of crystalline substances occurs primarily by the movement of dislocations within grains.

plastic flow n 塑性流動

▶ Above the yield point, plastic flow begins to occur, and the crystal suffers an irreversible elongation.

plasticity n 塑性 ☞ elasticity

▶ Plasticity may be considered the reverse of elasticity.

plasticize vt 可塑化する

▶ Polyvinyl chloride can be plasticized blending it with substances of low volatility, which, when dissolved in the polymer, tend to break down its glass-like structure.

plasticizer n 可塑剤

▶ Camphor is a natural product and is also manufactured synthetically on a large scale for use as plasticizer for nitrocellulose to produce celluloid.

plate n **1** 板 syn *panel, sheet, slab*

▶ Plate and needle-shaped crystal forms were observed.

2 極板 ☞ electrode

▶ Electrodes are sometimes called plates or terminals.

vt **3** めっきする syn *coat, laminate, overlay*

▶ Chemical plating is more expensive than normal electrolytic plating but is competitive when intricate shapes are being plated and is essential for nonconducting substrates.

plateau n 平坦域 複数形は plateaus または plateaux syn *leveling off*

▶ In all cases, the electromotive force at high iodine concentrations reaches a plateau at 680 ±3 mV. This plateau corresponds to the appearance of elemental iodine.

plating n めっき syn *coating, lamination*

▶ Chromium plating has, to a considerable extent, replaced nickel plating as a protective coating for less resistant metals.

platinize vt 白金をかぶせる

▷ For the growth of the inorganic salts, platinized electrodes were attached to the cell in order to determine the ionic conductance, which can be used as a measure of the concentration of the solute in the growth zone.

platinum black n 白金黒

▶ The hydrogen electrode consists of a platinum electrode coated with platinum black in contact with a solution saturated with hydrogen.

plausibility n もっともらしさ

▶ With so many factors involved, it is not difficult to provide facile explanations of varying degrees of plausibility for most experimental observations, therefore, it is prudent to treat such explanations with caution.

plausible adj もっともらしい syn *conceivable, imaginable, probable*

▶ Since the rate of dissociation of $Fe(phen)_3^{2+}$ in methanol solution is also accelerated by chloride ion, it would appear that a more plausible interpretation of this behavior in nonaqueous systems is the formation of ion pairs.

play vt 果たす syn *perform*

▶ The multitude of roles that carbohydrates and their glyoconjugates play in biological processes has stimulated great interest in determining the nature of their interactions in both normal and diseased states.

pleasant adj 気持ちのよい syn *attractive, good, satisfying*

▶ Organic compounds having a pleasant odor are broadly designated as aromatic, regardless of chemical nature.

pleasingly adv 満足なことに syn *comfortably*

▶ Using 1-nitropentane, a mixture of diastereomers again resulted, although pleasingly the products were obtained in good yield and excellent enantioselectivity.

pleat n プリーツ syn *fold, overlap*

▶ Pleats running predominantly across the *b* diagonals of the single crystals of polyethylene represent drastic puckering and do not disappear on standing.

pleated sheet n プリーツシート

▶ A slight contraction of the peptide chains results in a pleated sheet, with a somewhat shorter distance between alternate amino acid residues.

plenty pron たくさん syn *abundance*

▶ There is plenty of evidence that Hg_2Cl_2 is linear and should have a zero dipole moment.

pleochroism n 多色性 ☞ dichromism

▶ In biaxial crystals, the amount of absorption may vary with the direction of vibration of the light, and the variation may reach a maximum or a minimum for light vibrating parallel to any one of the three ellipsoidal axes. This phenomenon is called pleochroism.

plot n **1** プロット syn *chart, graph*

▶ An accurate value of k cannot be obtained from a plot of ln v against ln c.

vt **2** プロットする syn *chart, draw, map*

▶ To demonstrate the versatility of the charges from this electronegativity equalization procedure, we have plotted proton NMR chemical shifts for CH_3X compounds against hydrogen atom charges.

plug n **1** 充填物

▶ The resulting mixture was diluted with ethyl acetate and filtered through a Celite plug.

vt **2** ふさぐ syn *block, clog, close, seal*

▶ If solid condensate threatens to plug the system, it can be dislodged by brief interruption of the cooling.

plunge n **1** 落下 syn *descent, drop, fall*

▶ Instead of a single reaction with a long plunge from energy level of carbohydrates and oxygen to that of carbon dioxide and water, there are long series of chemical reactions in which the energy level descends in gentle cascades.

vt **2** 突っ込む syn *immerse, sink, submerge*

▶ Pyrex glass tubes up to a diameter of 10 mm can be plunged directly into the flame of the blast lamp without cracking.

plurality n 多数 syn *a large number*

▶ This potential plurality provides one of the reasons why there is no agreement on the nature of the active sites or the reaction mechanism.

pluripotent adj 多能性の

▶ Embryonic stem cells are pluripotent, which means they can develop into any cell type of an organism.

plus prep …を加えて syn *added (to), increased (by)*

▶ The atomic number of iron is 26, and this plus five pairs of electrons gives the iron atom in $Fe(CO)_5$ an effective atomic number of 36.

pocket n 空洞 syn *cavity, hollow, pit*

▶ The empty pockets of space in the nematic lattice of the oxygen esters are reduced in size in the thioesters, allowing closer packing and leading to increased values for the both $\Delta H_{N\to I}$ and $\Delta S_{N\to I}$.

point n **1** 点 syn *dot*

▶ The points are experimental, while the curves are those calculated from the theories, as indicated.

2 ある時点 syn *moment, time*

▶ Up to this point, all of the molecules we have treated by the VB approach have required only electron-pair bonds in describing their electronic configuration.

3 問題点 syn *aspect, issue, matter*

▶ Another point worth noting is that ν_{CO} is lower for coordinated than for free CO, except in $[Ag(CO)]^+$, indicating that back-bonding populating the π^* orbitals actually does occur.

4 a case in point 適例 syn *appropriate example*

▶ The well-known oxygen-carrying property of hemoglobin and the hemocyanins is a case in point.

vi **5 point to** 示す syn *indicate*

▶ The addition of alcohols to water is accompanied by large negative excess entropies, a fact which points to a considerable structural enhancement.

6 point toward 向いている syn *direct*

▶ X-ray studies of solid ammonia indicate that three hydrogen atoms from different nitrogen atoms point toward the lone pair of electrons.

vt **7 point out** 指摘する syn *emphasize, mention, remind, stress*

▶ It should be pointed out that not only do exceptions occur but that the order of permeability of cations for any one case may change with concentration.

point charge n 点電荷

▶ Coulomb's law states that the force F between two point charges q_1 and q_2 a distance s apart is given by $F = q_1q_2/r^2$.

point defect n 点欠陥

▶ The term point defect refers to a site imperfection such as an interstitial atom or a vacancy.

point of reference n 基準 syn *reference point*

▶ The nonpolar covalent radii of atoms can be measured accurately and represent a point of reference with which to compare other radii.

point of view n 観点 syn *approach, outlook, perspective, viewpoint*

▶ From stereochemical point of view, nitrogen is second in importance only to carbon.

poised adj 用意が十分にできて syn *ready*

▶ It has been suggested that the metal in the enzyme is peculiarly poised for action and that this lowers the energy of the transition state.

poison n 1 毒物 syn *toxin*

▶ Any substance that is harmful to living tissues when applied in relatively small doses is called a poison.

2 触媒毒

▶ Thiophene may act as a poison for some catalysts.

vt 3 毒する syn *destroy, kill*

▷ The choice of a catalyst depends on the job to be done and the danger of it being poisoned by by-products or impurities in the reaction mixture.

poisonous adj 有毒な syn *deadly, lethal, noxious, toxic*

▶ Cyanogen, $(CN)_2$, is a colorless poisonous gas.

polar adj 極性の

▶ Some applications of LB films depend upon the presence of a polar assembly of molecules.

polar axis n 極性軸

▶ The presence of a polar axis in a molecular crystal has chemical implications that have only recently begun to be recognized.

polar bond n 極性結合

▶ The dipole moment of nitrogen trifluoride, with its polar bonds and pyramidal structure, would be expected to be quite high, but actually it is very small.

polar coordinate n 極座標

▶ If the equation of the circle is given in polar coordinates: $r = 2a \cos \theta$, the area is given by letting the radius vector r move over the angle from $-\pi/2$ to the angle $+\pi/2$.

polar effect n 極性効果

▶ The orientation of further substitution in an aromatic nucleus by a group initially present depends essentially on the differential electron density and availability at the various positions in the ring caused by the polar effects of the substituent group.

polarity n 1 極性

▶ Being compounds of low polarity, alkynes have physical properties that are essentially the same as those of alkanes and alkenes.

2 方向性

▶ The two strands of nucleotides in a double helix are considered antiparallel because the polarity of each strand is reversed relative to each other.

polarizability n 分極率

▶ Another feature common to ferroelectric materials is a high degree of polarizability which results in a high permittivity.

polarizable adj 分極性の ☞ nonpolarizable

▶ Cobalt occurs mainly with sulfur and arsenic rather than with oxygen; with its relatively high nuclear charge and small size it favors more polarizable atoms.

polarization n 1 分極 ☞ atomic－, electronic－, orientation－

▶ If a molecule is considered as an elastic body, deformable in an electric field, the total polarization can be written

$$P = P_o + P_e + P_a$$

where P_o is the orientation polarization, and P_e and P_a correspond with the displacements of electrons and atoms, respectively, in the molecule.

2 電気化学的分極

▶ Polarization is attributed to the simple resistance and other kinetic factors of the oxide reduction reaction.

3 偏光 ☞ depolarization

▶ The process of resolving the extremely complex vibrations of ordinary light into vibrations taking place in definite directions or in definite planes is called polarization.

polarize vt **1 分極させる**

▶ When electromagnetic radiation passes through a substance, the oscillating electric field of the radiation polarizes the material.

2 偏光させる

▶ A small sheet of polarizing film will polarize the light, letting through only that light which is vibrating in one particular direction.

polarized light n **偏光** ☞ plane-polarized light

▶ Enantiomers have identical physical properties, except for the direction of rotation of the plane of polarized light.

polarizer n **偏光子** ☞ analyzer

▶ The polarizer is held below the stage. It is generally left in place all the time, customarily with its polarization direction lying in the front-to-back plane of the microscope.

polarizing microscope n **偏光顕微鏡** ☞ optical –

▶ The polarizing microscope is a microscope especially made for viewing objects in crossed polarized light.

polar molecule n **極性分子**

▶ As already mentioned, the technique can be applied to nonpolar as well as to polar molecules so long as they have anisotropic polarizabilities.

polarogram n **ポーラログラム**

▶ Polarograms of complex TCNQ salts consist of two waves which are additive curves of neural TCNQ and $TCNQ^-$.

polarograph n **ポーラログラフ**

▶ A device for the automatic electroanalysis of a solution by means of the dropping mercury electrode is called polarograph.

polarographic adj **ポーラログラフによる**

▶ Polarographic studies of solution of Li^+ $TCNQ^-$ reveal two waves; the first is anodic and characteristic of the reversible, one-electron oxidation, and the second wave is cathodic and characteristic of the thermodynamically irreversible, one-electron reduction of $TCNQ^-$.

polarography n **ポーラログラフィー**

▶ An important analytical technique which is related to the behavior of the deposition of substances at electrodes is polarography.

polaron n **ポーラロン** ☞ bipolaron

▶ A polaron is an excited state in which the presence of an electron or hole causes a significant local distortion of the lattice.

polar plot n **極座標によるプロット**

▶ The polar plot of the velocity of growth is quite unsymmetrical, in keeping with the low symmetry of sucrose crystal.

polar reaction n **極性反応**

▶ In a polar reaction, electron-pair bonds are regarded as being broken in a heterolytic manner as contrasted to the free-radical processes.

polar solvent n **極性溶媒**

▶ Polar solvents, such as water and low-molecular-weight alcohols, are capable of effectively attenuating the generated field and, therefore, have relatively high dielectric constants.

polar structure n **極性構造**

▶ Of particular interest are those transformations in which a fundamental change of symmetry occurs, particularly a change from a centrosymmetric to a chiral or polar structure.

pole n **極**

▶ If the poles of a magnet are inclined toward each other, there is produced a nonhomogeneous field with an axis of symmetry.

polish n **1 光沢** syn *brightness, brilliance, luster*

▶ Sodium remained brightly reflecting for at least 2 min after cutting but then slowly lost its polish.

vt **2 磨く** syn *clean, rub, smooth*

▷ The most widely used solid polishing agent

is fine-ground red iron oxide, applied to the surface of plate glass, backs of mirrors, and optical glass.

polished adj 磨き上げた syn *bright, perfect, shiny, smooth*

▶ Some organic materials with good nonlinear properties are chemically stable over wide temperature ranges and have polished surfaces that do not degrade with time.

pollutant n 汚染物質 syn *contaminant, impurity*

▶ Numerous chemicals from many different sources are common pollutants in the environment.

pollute vt 汚す syn *contaminate, soil*

▷ The difficulty in applying classical methods to the purification of water polluted with traces of PCBs stems from their low concentration.

pollution n 汚染 syn *contamination*

▶ Two of the central concerns of environmental chemistry, pollution and sustainability, have long been matters of concern in our own culture.

polyatomic molecule n 多原子分子

▶ One very important application of infrared spectroscopy of polyatomic molecules in condensed phases is in chemical analysis.

polycation n ポリカチオン

▶ Various bismuth polycations have been identified, and in the process many confusing results have been rationalized.

polycondensation n 重縮合 syn *condensation polymerization*

▶ The hydrolysis and polycondensation reactions are initiated at numerous sites within the $Si(OCH_2CH_3)_4$ and H_2O solution as mixing occurs.

polycrystalline adj 多結晶質の

▶ Metals are usually polycrystalline: they are composed of many small crystals called grains.

polycyclic adj 多環式の syn *polynuclear*

▶ The ultimate polycyclic aliphatic system is diamond which is, of course, not a hydrocarbon at all but one of the allotropic forms of elemental carbon.

polycyclic aromatic hydrocarbon n 多環芳香族炭化水素

▶ Polycyclic aromatic hydrocarbons with both angular and linear types of ring fusion show absorption curves of a similar profile to that of benzene but with the absorption maxima shifted to longer wavelengths; the greater the number rings the more pronounced the shift.

polydentate ligand n 多座配位子

▶ A combination of letters in parentheses is used to designate polydentate ligands, e. g., (AA) is a symmetrical bidentate group such as ethylenediamine, (AB) is an unsymmetrical bidentate group such as glycinate, and (AABBAA) is a sexadentate group such as ethylenediaminetetraacetate ion.

polydisperse adj 多分散の ☞ monodisperse

▶ If the substance consists of macromolecules of different sizes, it is said to be polydisperse.

polyelectrolyte n 高分子電解質 syn *polymer electrolyte*

▶ The driving force behind the current interest in polymer electrolytes is because an ionically conducting polymer has significant advantages over conventional electrolytes for a range of practical applications, in terms both of fabrication and of subsequent use.

polyhedral adj 多面体の ☞ dodecahedral, icosahedral, octahedral, tetrahedral

▶ Many minerals and also synthetic periodic lattices have a network of holes that may be described as polyhedral.

polyhedron n 多面体 複数形は polyhedra

▶ The description of structures in terms of polyhedra does not necessarily imply that such entities exist in the structure as separate species.

polyhydric alcohol n 多価アルコール syn *polyol*

▶ Glycerin and the alkene oxide-derived diols are the most important polyhydric alcohols.

polymer n 高分子, ポリマー, 重合体 ☞ macromolecule

▶ The free-radical polymerization of butadiene in solution gives a polymer with a mixture of about 20% 1,2 -butadiene units, 40% cis-1,4 units, and 40% trans-1,4 units.

polymerase n ポリメラーゼ

▶ The thymidine analogues having varied size were studied as substrates for DNA polymerase.

polymeric adj 重合体の

▶ There are a number of naturally occurring polymeric substances that have a high degree of technical importance.

polymerizability n 重合可能性

▶ The opportunity of combining properties normally found only in organic molecular crystals, e. g., polymerizability and mesomorphism, with those usually associated with inorganic solids, e.g., conductivity and magnetism, is a fascinating possibility for molecular composites.

polymerizable n 重合可能な

▶ In telomerization reactions, a polymerizable unsaturated compound, taxogen, is reacted under polymerization conditions in the presence of radical-forming catalysts or promoters with a so-called telogen.

polymerization n 重合 ☞ depolymerization

▶ Polypropylene, made by polymerization of propylene with Ziegler catalysts, appears to be isotactic and highly crystalline with a melting point of 175°.

polymerize vi **1** 重合する

▶ Formaldehyde in water solution polymerizes to a solid long-chain polymer called paraformaldehyde or polyoxymethylene.

vt **2** 重合させる

▶ Tetracyanoethylene can be polymerized at 200° in the absence of metal by the use of a variety of organic catalysts.

polymorphic form n 多形相 syn *polymorph*

▶ Iron exists in three polymorphic forms: body-centered cubic α, stable below 910 ℃; face-centered cubic γ, stable between 910 and 1400 ℃; and body-centered cubic δ, stable between 1400 ℃ and the melting point 1534 ℃.

polymorphism n 多形

▶ When a compound exhibits polymorphism, it may be important to obtain a particular polymorph under controlled and reproducible conditions.

polynomial n 多項式

▶ The virial coefficients are represented by polynomials as functions of temperature.

polynucleotide n ポリヌクレオチド

▶ DNA is made up of two polynucleotide chains wound about each other to form a double helix 20 Å in diameter.

polyol n ポリオール syn *polyhydric alcohol*

▶ Pentaerythritol is an important polyol used in resins, plastics, and drying oils.

polypeptide n ポリペプチド

▶ Depending upon the number of amino acid residues per molecule, peptides are known as dipeptides, tripeptides, and so on, and finally polypeptides.

polyphosphate n ポリリン酸塩

▶ Many bacteria can store phosphate as polyphosphate, storage in a polymer which requires energy to make but which is of low osmotic activity.

polyprotic acid n 多塩基酸

▶ Qualitatively, the successive pK_a's of a polyprotic acid should increase as the protons are removed, simply on a charge basis.

polyribonucleotide n ポリリボヌクレオチド

▶ Ribonucleic acid is universally present in living cells and has a functional genetic specificity because of the sequence of bases along the polyribonucleotide chain.

polysaccharide n 多糖 syn *glycan*

▶ Polysaccharides are compounds made up of many, hundreds or even thousands, monosaccharide units per molecule.

polytype n ポリタイプ

▶ In the polytypes, the structures are the same in two dimensions, i.e., within the layers, and differ only in the third dimension, i.e., in the sequence of layers.

polyurethane foam n ポリウレタンフォーム

▶ The desirable features of polyurethane

foam, for example, good long-term performance as a thermal insulator and as an air filter, are linked to the network of pores in the foam.

polyvalency n 多価

▶ The propensity to self-assemble into monodisperse nanoparticles of discrete shape and size with high degree of symmetry and polyvalency makes viral capsids unique bionanoparticles.

polyvalent adj 多価の ☞ divalent, monovalent, trivalent

▶ Polyvalent cations form complexes with water and thereby are large hard spheres, whereas polyvalent anions do not form such complexes.

pool n **1** プール

▶ Polar molecules may move about relatively freely in vesicle-entrapped water pools.

2 蓄え syn *collection, reserve*

▶ The reaction releases CoA, thus regenerating the mitochondrial pool of this vital carrier.

poor adj **1** 質の劣った syn *inadequate, unsatisfactory*

▶ In the continuous range of compounds from WO_3 to $NaWO_3$, the compounds are initially poor semiconductors but become metallic in character at quite low sodium content.

2 わずかな syn *low, insufficient, sparse*

▶ Aromatic species tend to exhibit poor solubility in ionic liquids consisting of aliphatic cations and vice versa.

poorly adv 不十分に syn *badly, inadequately, insufficiently*

▶ Hindered amines performed poorly in this reaction.

poor solvent n 貧溶媒

▶ In a poor solvent the rate of transformation of a metastable to a more stable polymorph is slower.

popular adj 普通の syn *conventional, ordinary, standard*

▶ One of the most popular techniques for achieving atomic-scale synthetic control is chemical vapor decomposition.

popularly adv 普通に syn *generally, normally, ordinarily, routinely*

▶ Accurate electrochemical measurements are somewhat difficult because of the limited solubility of tetrathionaphthacene in the solvents popularly employed for anodic oxidations.

populate vt 存在する syn *occupy, reside*

▶ Other partly overlapping signals suggest that other conformers are populated but in ratios that cannot be determined.

population n **1** 集団

▶ The experimental enthalpy of formation for butane is about 0.27 kcal/mol more positive than the value that can be estimated for a hypothetical population consisting of 100% of the extended form.

2 占有数

▶ There are more transitions from the lower energy state than from the upper state on account of the population difference.

3 個体群

▶ Under favorable growth conditions in the laboratory, a small population of bacteria can divide rapidly to produce millions of identical cells.

population analysis n 占有数解析

▶ The electronic structure was evaluated in terms of the Mulliken population analysis and natural population analysis to calculate the gross atomic charges for the atoms.

population inversion n 反転分布

▶ The lifetime of the ^2E excited state of Cr^{3+} ion is fairly long, which means that a considerable population inversion has time to build up.

porcelain dish n 磁製皿

▶ The nickel solution in a porcelain dish is heated with bromine water and an excess of pure potassium hydroxide; the nickel is precipitated as brownish black $Ni(OH)_3$.

pore n 細孔 syn *aperture, hole, opening, orifice*

▶ The development of routes to larger pore materials has led to some significant extensions to the catalysts which can be immobilized in-pore.

porosity n 多孔度

▶ The density of a compressed powder differs from the true density of the material on account of the porosity of the pellet.

porous adj 多孔性の syn *permeable, spongy*

▶ When the solid is porous, the information that is required may comprise not only the surface area but in addition the size and distribution of the pores.

porphyrin n ポルフィリン

▶ Porphyrin complexes are biologically accessible compounds whose functions can be varied by changing the metal, its oxidation state, or the nature of the organic substituents on the porphyrin structure.

portable adj 携帯用の syn *mobile, movable*

▶ The reaction of water with CaH_2 or LiH forms a convenient portable source of hydrogen.

portion n 部分 syn *division, part, piece, section, segment*

▶ The isopoly anions may be considered to be portions of a closest-packed array of oxide ions with metal ions occupying the octahedral holes.

portionwise adv 分割して ☞ dropwise

▶ After sodium cyanide（23 g; 472 mmol）was dissolved in 2 L of methanol, *p*-bromanil（25 g; 59 mmol）was added portionwise over 1 h, the temperature was allowed to rise to 34 ℃, and the mixture turned red.

portray vt 描写する syn *depict, characterize, picture, represent*

▷ The electronic transitions portrayed in the absorption spectra of ruby are explained by a simple electrostatic model.

portrayal n 描写 syn *characterization, description*

▶ The NMR spectrum is, as a result of chemical shifts, a portrayal of the chemical environment of the various hydrogen atoms of the material.

pose vt 提起する syn *ask, postulate, present, submit*

▶ The problem of the determination of individual values of ionic entropies poses much the same difficulties as ionic enthalpies.

position n **1** 位置 syn *location, site*

▶ The chlorine atoms in $GaCl_2$ occupy positions at corners of an irregular dodecahedron with four at 3.18 Å and four at 3.27 Å from the Ga^+ ion.

2 in position 所定の場所に

▶ The water molecules in the hydration layers are more or less fixed in position and are oriented with respect to the ion by its intense electric field.

3 out of position 所定の場所をはずれて

▶ Defects may be atoms or ions that are missing from the structure, out of position, or replaced by impurity atoms or ions.

4 単数形で，立場 syn *circumstances, condition, situation, state, status*

▶ We are not yet in a position to discuss the thermochemistry of most of these elements in any detail.

vt **5** 置く syn *arrange, dispose, fix, situate*

▷ Certain pairs of elements positioned diagonally to one another in the periodic table have similar ionic size.

positional adj 位置の

▶ Only the positional parameters were allowed to vary for the hydrogen atoms, the isotropic thermal parameters being held constant at 4.0 $Å^2$.

positioning n 位置 syn *arrangement, distribution*

▶ The three pyrimidines, thymine, cytosine, and uracil, vary in the types and positioning of chemical groups attached to the ring.

position isomerism n 位置異性

▶ When there are two or more substituents on a benzene ring, position isomerism arises.

positive adj **1** 肯定的な syn *beneficial, favorable, useful* ☞ negative

▶ Ceramics are valuable clinical materials because they invoke either a neutral or positive biological response in the body.

2 正の

▶ While the *I* values must be positive, the *F*

values given by $I^{1/2}$ may be positive or negative.

positive electrode n 正極 syn *cathode*

▶ Many intercalation compounds have been studied as possible positive electrodes for rechargeable lithium batteries.

positive hole n 正孔 ☞ hole

▶ The positive holes that are left behind in the valence band are able to move, and gallium-doped silicon is a positive hole or p-type semiconductor.

positive ion n 正イオン syn *cation*

▶ In most cases, the intense peak of highest mass number corresponds to the positive ion (M^+) formed by removal of just one electron from the molecule (M) being bombarded.

positively adv **1** 肯定的に syn *certainly, definitely, surely, unquestionably*

▶ In the case of confirming the identity of a protein from an organism for which the genome has been sequenced, a short stretch of peptide sequence is often sufficient to positively identify the parent protein.

2 正に

▶ Electric current is conventionally regarded as the flow of a positively charged substance, and so the current flow is opposite to the flow of the electrons.

positive sign n プラス記号 syn *plus sign*

▶ By convention, ΔH is given a negative sign when heat is evolved and a positive sign when heat is absorbed.

possess vt 所持する syn *have*

▷ Other nucleophilic reactions that lead to the formation of P–C bonds occur with phosphorus compounds possessing P–H bonds.

possibility n 可能性 syn *opportunity, potential, potentiality*

▶ The problem of identifying glucose as a particular one of the sixteen possibilities was solved by Emil Fischer during the latter part of the 19th century.

possible n **1** 可能性 syn *feasibility*

▶ Scientific progress depends very much on combining the theoretical probable with experimental possible.

adj **2** 可能な syn *conceivable, feasible, plausible*

▶ There are several processes for obtaining ultrapure aluminum. Impurities as low as 0.2 ppm are possible.

3 if possible もしもできるなら syn *if at all possible, possibly*

▶ The conventional unit cells are chosen to exhibit the full lattice symmetry, if possible.

possibly adv ことによると syn *if possible, perhaps*

▶ Red tin(II) oxide is stable in air up to about 270°, possibly owing to the presence of a thin protective coating of tin(IV) oxide.

postpone vt 後回しにする syn *delay, suspend*

▶ Many solid substances with very symmetrical molecules can acquire sufficient entropy through the rotation or libration beginning at a transition point that melting is postponed to a higher temperature and the entropy of fusion and the normal liquid range are much diminished.

post-translational adj 翻訳後の

▶ The post-translational glycosylation pattern of natural glycoprotein is highly diverse , with the glycan not only ensuring the activity and stability of proteins but also adding additional functionalities.

postulate n **1** 前提 syn *hypothesis, supposition*

▶ Thermodynamics is a simple, general, logical science, based on two postulates, the first and second laws of thermodynamics.

vt **2** 仮定する syn *presume, speculate, suppose*

▶ From the NMR and IR studies, it has been postulated that the electron density of metal-bound ligand acting as hydrogen bond acceptor is the major factor in determining the strength of the hydrogen bonds formed.

postulation n 仮定 syn *assumption, hypothesis, speculation, supposition*

▶ The formation of chrysene and methylcyclopentanophenanthrene from selenium dehydrogenation of cholesterol led to postulation of the correct ring structure.

pot n 丸い容器 syn *kettle, pan*

▶ It is not possible to completely satisfy both sets of optimum reaction conditions simultaneously in a true one-pot process.

potency n 効力 syn *effect, influence, strength*

▶ The key to producing clinically relevant glycosidase inhibitors is not only high potency but also specificity over other enzymes which may be encountered.

potent adj 効力のある syn *powerful, strong*

▶ Ammonia is a potent neutral nucleophile which readily participates in substitution reactions.

potential n **1** 可能性 syn *capability, capacity, possibility, potency*

▶ These linear bimetallic species have potential for use as monomers in further copolymerization.

2 電位 ☞ electric potential

▶ As the frequency of the potential increases, the relaxation and electrophoretic effects will become less and less important, and there will be an increase in the molar conductivity.

3 電極電位 ☞ electrode potential

▶ The removal of one electron from a carboxylate anion can be effected by oxidizing agents of high potential, particularly persulfates.

adj **4** ポテンシャルの

▶ The potential valley due to such forces is always quite narrow.

5 潜在的な syn *implicit, possible*

▶ Ionization isomerism is possible in compounds which consists of a complex ion with a counter ion which is itself a potential ligand.

potential barrier n ポテンシャル障壁

▶ There seems to be a potential barrier to be overcome before a solute can take its final place in the surface and produce, after a lapse of time, a final adsorption and static surface tension.

potential difference n 電位差

▶ Application of a potential difference across a dielectric does lead to a polarization of charge within the material, although long-range motion of ions or electrons should not occur.

potential energy n ポテンシャルエネルギー ☞ kinetic energy

▶ The potential energy of the molecule is at a minimum for the staggered conformation, increases with rotation, and reaches a maximum at the eclipsed conformation.

potential-energy surface n ポテンシャルエネルギー曲面

▶ When the reaction is not a symmetrical one but involves three atoms that are not identical, the potential-energy surface is no longer symmetrical with respect to the two axes.

potentiality n 可能性 syn *capacity, opportunity, possibility, premise*

▶ Many chalcogenides have important technological potentialities for solid-state optical, electric, and thermoelectric devices.

potentially adv 潜在的に

▶ Silicon nitride is potentially a very useful high-temperature ceramic.

potential probe n 電圧端子 ☞ four-probe

▶ The assembly carries current electrodes and potential probes to eliminate electrode contact resistance.

potentiometric adj 電位差の

▶ Electrochemical systems may oscillate as changing electrode surfaces affect conductivity and/or potentiometric behavior.

potentiometrically adv 電位差によって

▶ To carry out an acid-base titration potentiometrically, one uses any convenient form of hydrogen electrode, such as a glass electrode.

potentiometric titration n 電位差滴定

▶ Potentiometric titration can be applied to acid-base and oxidation-reduction titrations.

pour vt 注ぐ syn *run*

▶ The reaction mixture was allowed to stir for 72 h, after which it was poured into NaOH(aq) (2 M, 75 mL), and the aqueous layer was extracted with dichloromethane(2×20 mL).

powder n 粉末 syn *dust*

▶ Ideal powders are characterized by a submicrometer size, an absence of particle aggregates, a narrow size distribution, and high chemical purity.

powdered adj 粉末にした syn *powdery, pul-*

verized

▶ Uranium hydride is finely powdered and extremely reactive.

powder pattern n 粉末回折像

▶ Comparison of powder patterns for nematic and isotropic phases shows that, with regard to the local molecular packing, these phases are little different.

powder X-ray diffraction n 粉末X線回折

▶ The powder X-ray diffraction method is very important and useful in qualitative phase analysis because every crystalline material has its own characteristic powder pattern.

powdery adj 粉末状の syn *powdered, pulverized*

▶ P_4O_{10} reacts vigorously with both wet and dry NH_3 to form a range of amorphous polymeric powdery materials.

power n **1** 能力 syn *ability, capability, capacity, potentiality*

▶ Most substances have the power to enter into many chemical reactions.

2 動力 syn *energy, fuel*

▶ Aerobic respiration is a much more efficient process for providing power for cells.

3 累乗 ☞ order of magnitude

▶ Rates of isomerization reactions are greater in solution than in the solid state by two to four powers of ten.

vt **4** 稼動させる

▶ The invagination of the plasma membrane is powered by the tightening of an actin–myosin belt underlying the membrane.

powerful adj 有力な syn *effective, important*

▶ In 1884, Fischer reported that the phenylhydrazine he had discovered could be used as a powerful tool in the study of carbohydrates.

powerfully adv 強力に syn *highly, intensely, profoundly, strongly*

▶ Interconversion of the enol and keto forms of ethyl acetoacetate is powerfully catalyzed by bases.

practicable adj 可能な（限りの） syn *achievable, attainable, feasible, possible*

▶ The minimum practicable power at most detectors is ca. 10^{-9} W.

practicability n 実用性 syn *usefulness*

▶ The feasibility, practicability, and wide scope of our general approach were supported by real, structurally characterized examples constructed with deliberate intent.

practical adj 現実的な syn *applicable, reasonable, usable, useful*

▶ From purely practical point of view, the prediction of an unknown bond length is best accomplished by a survey of analogous bonds in other molecules of similar structure.

practically adv ほとんど syn *almost, essentially, nearly, virtually*

▶ It has been shown that aluminum hydroxide will decolorize dilute solutions of alizarine and certain dyestuffs but has practically no adsorbing action on emerald green or chrysoidine.

practice n **1** 慣行 syn *custum, habit, procedure*

▶ A common practice in organic chemistry for the confirmation of the identity of a substance is to mix the substance with a pure sample of the substance with which it is suspected to be identical and determine the melting point.

2 実施 syn *pursuit, work*

▶ It is common observation in the practice of organic chemistry that when a solid substance is heated with a solvent, it appears to melt at a temperature much below the normal melting point.

3 in practice 実際問題として syn *actually*

▶ In practice, X-ray diffraction is the most widely used technique.

4 練習 syn *exercise, training*

▶ With a little practice, it is easy to tell when the end point is nearly reached by the fact the pink color is produced which fades slowly on stirring.

vt **5** 実践する syn *act, carry out, perform*

▶ It is important to practice drawing the possible isomeric structures that correspond to a single molecular formula.

practicing adj 現在活動している

▶ Practicing scientists must often be content

with a state of knowledge which does not reflect a unified version of reality.

practitioner n 従業者 syn *operator*

▶ Laddered polyether natural products of marine origin are amongst the largest and most complex targets to have confronted practitioners of natural product synthesis.

pragmatic adj 実際的な syn *empirical, practical, rational, realistic*

▶ The pragmatic criterion of the presence or absence of an inert pair effect can be taken as the tendency for the following reaction to proceed to the right:

$$MX_n \rightarrow MX_{n-2} + X_2$$

precarious adj 不確かな syn *delicate, unstable*

▶ A small metal particle placed on a TiO_2 surface finds itself in a precarious situation as the system is reduced in H_2.

precaution n 注意 syn *attention, care, caution*

▶ Precautions were taken as far as possible to prevent the entrance of moisture during the measurements.

precede vt 先行する syn *lead*

▶ Many examples of phosphorescence involve triplet → singlet transitions and are preceded by process in which an excited singlet state is converted into an excited triplet state.

precedence n 優先 syn *preference, priority, superiority*

▶ When two or more different hetero atoms are present, oxygen takes precedence over sulfur and sulfur over nitrogen for the number one position.

precedent n 先例 syn *example, model, paradigm, prototype*

▶ This is a somewhat unusual type of reaction for which, however, there is some precedent.

precedented adj 先例がある

▶ The failure of these reactions is surprising because the cross metathesis of allylic halides with other alkenes is precedented.

preceding adj 前述の syn *foregoing, previous, prior*

▶ Based on the preceding discussion, we predict that polyatomic solids like salts would have a high-temperature heat capacity limit of $3R \times p$, where p is the number of atoms in the chemical formula of the solid.

precipitant n 沈殿剤

▶ Sodium hexanitrocobaltate(III) is probably the most sensitive precipitant of potassium ions in aqueous solution.

precipitate n **1** 沈殿物

▶ The yellow precipitate was collected on a filter, washed with water, and dried to give 1.33 g of organic product.

vi **2** 沈殿する syn *sediment*

▶ When steel is slowly cooled below the austenite–ferrite phase-transition temperature, the excess carbon, which is no longer soluble in ferrite, precipitates as iron carbide.

vt **3** 沈殿させる syn *deposit*

▷ Nickel sulfide first precipitated is readily soluble in acid, but on standing it changes into a much more insoluble condition.

precipitation n 沈殿

▶ CaF_2 is a white, high-melting solid whose low solubility in water permits quantitative analytical precipitation.

precipitation hardening n 析出硬化 ☞ hardening

▶ Precipitation hardening occurs when the room-temperature solubility of an impurity is exceeded, and a fine-grained precipitate forms throughout the metal.

precise adj **1** 正確な syn *correct, definite, exact,*

▶ The precise degree of acidity, as well as the concentration of acid sites, is governed by the particular organic grouping that bears the sulphonic, carboxylic, or phenolic OH.

2 to be precise 正確に言うと syn *to be accurate*

▶ To be precise, we should also specify the temperature of the product, but we shall not do this unless it differs from 298 K.

precisely adv 正確に syn *absolutely, correctly, exactly, rigorously*

▶ The relative strengths of the intermolecular M···O interactions are inversely related to the strengths of the M–M bonds, though the latter are not known precisely.

precision n **1** 精密さ syn *accuracy, correctness, rigor*

▶ Density values were generally obtained with a precision of 0.0005 g/mL.

adj **2** 精密な syn *accurate, correct, exact*

▶ A fractional or percentage precision measure gives a definite idea of the precision of the measurement, as it involves both the value of the quantity and its average deviation.

preclude vt 除外する syn *avoid, exclude, rule out*

▷ Because the negatively charged group is resonance stabilized, carboxylate anions are unreactive toward nucleophilic attack, precluding the possibility of a reverse reaction.

preconceived adj 予想した syn *predetermined, prejudged*

▶ Fischer deliberately avoided any theoretical treatment because in his own words, "one must be urgently cautioned, while making observations, doing analyses, or making other determinations, not to be influenced by theory or preconceived ideas".

precursor n 前駆物質 syn *forerunner, predecessor*

▶ Proteins that are destined for secretion from cells are usually synthesized in a precursor form.

predecessor n 前のもの syn *forerunner, precursor*

▶ Each successive chemical element differs from its predecessor not only in having one more electron in its valence shell but also in having a higher charge on its nucleus.

predetermine vt 前もって決める syn *arrange, fix, destine*

▷ For this purpose, a method must permit the joining together of optically active amino acids to form chains of predetermined length and with a predetermined sequence of residues.

predict vt 予想する syn *anticipate, forecast*

▶ It is generally known that the diamagnetic susceptibilities of many organic compounds may be predicted theoretically by assuming that they are more or less additive.

predictability n 予測可能性 syn *certainty, conceivability*

▶ Many theoreticians have fallen back on simple predictability of observables as a criterion of success of a theory.

predictable adj 予想できる syn *foreseeable, probable*

▶ It seemed crucially important in early work in so uncertain a field to use bridging ligand that were as predictable as possible in their metal binding geometry.

predictably adv 予想通りに

▶ The only stable lead tetrahalide that can be made is, predictably, the fluoride, PbF_4.

prediction n 予想 syn *forecast, hint, suggestion*

▶ Experimental measurements of cation ordering are important to establish the validity of the predictions made by such lattice energy calculations for particular cations in particular structures.

predictive adj 予想する syn *inspired*

▶ This predictive capability can be achieved because the crystal properties of a molecular solid reflect primarily a combination of the molecular contributions.

predictor n 指標

▶ The valence shell electron pair repulsion model is a qualitative predictor of the angular geometry of an AH_n or AX_n system with a main group central atom.

predispose vt 傾向を与える syn *affect, influence*

▶ They do not possess a cyclic conformational constraint but instead rely on the conformational preferences of the open-chain system to predispose the precursor toward ring closure in high yield.

predisposition n 傾向 syn *preference, tendency*

▶ Despite this predisposition of a particular

substrate toward a particular mechanism, we can still control the course of reaction to a considerable degree by our choice of experimental conditions.

predissociation n 前期解離

▶ In predissociation, the absorption spectrum is discrete but broadened according to the uncertainty principle.

predominance n 優勢 syn *dominance, superiority*

▶ Stirring an aqueous solution of sodium chlorate leads to a predominance of crystals of one handedness, sometimes right, sometimes left, but not depending on the direction of stirring.

predominant adj 顕著な syn *dominant, main, primary, superior*

▶ There are exceptions in some special cases when more or less pure particles of a certain shape are predominant.

predominantly adv 顕著に syn *mainly*

▶ The difference in shape manifests itself very prominently.

predominate vi 優勢である syn *dominate, outweigh, prevail*

▶ Except for one case, the studies on Rhodamine 6G aggregation have been performed within concentration ranges where the monomer predominates.

preexponential factor n 前指数因子

▶ For reactions between ions of opposite signs, the preexponential factors are much higher than for reactions between neutral molecules.

prefer vt …のほうを取る syn *choose, select*

▶ The monomers that are bent strongly prefer the C_{3v} triply bridged dimer geometry and condense to form extended solids that have high coordination numbers.

preferable adj より望ましい syn *better, favorable, superior*

▶ Dimethyl phthalate is preferable to the diethyl and dibutyl esters both because the reaction proceeds best with the lowest homolog.

preferably adv なるべくなら syn *instead, rather*

▶ In order to be liquid at room temperature, the cation should preferably be unsymmetrical, e.g., R^1 and R^2 should be different alkyl groups in the dialkylimidazolium cation.

preference n **1** 選択 syn *choice, selection*

▶ A glass with all tetrahedral coordination will necessarily give a lower loss because Fe^{2+} ion, with its strong octahedral site preference, just cannot fit into the more ordered parts of the structure but only elsewhere.

2 in preference to …に優先して syn *instead of, preferably, rather than*

▶ Chlorine is also prepared by electrolysis, and aqueous solutions may be employed, since chlorine is commonly liberated in preference to oxygen because of the oxygen overvoltage.

preferential adj 優先的な syn *advantageous, better, favorable*

▶ An example of a preferential chemical combination is that of hemoglobin with carbon monoxide, with which it unites 200 times as readily as with oxygen when exposed to a mixture of the two.

preferentially adv 優先的に

▶ Neither Fe^{3+} nor Fe^{2+} form stable ammonia complexes in aqueous solution, as they combine preferentially with the hydroxyl ions present.

preheat vt 予熱する

▶ The pure alkane and a diluent gas are preheated and then sprayed into the reactor from a jet orifice.

preliminary n **1** 前置き syn *beginning, introduction, preparation*

▶ As a preliminary to the discussion, it is necessary to establish the concept of partial molar quantity.

adj **2** 序文の syn *introductory*

▶ Authors are requested to read the note on copyright published in the preliminary section of each issure.

3 予備的な syn *advance, initial, prior*

▶ Preliminary Communications report results

whose immediate availability to the scientific community is likely to be beneficial to the discipline.

premature adj 早まった syn *ill-timed, untimely*

▶ To avoid premature assumptions, the calibration data have been worked up by both equations, to provide a van der Waals calibration and a Joffe calibration.

premelting adj 融点直下の

▶ Analogous premelting transitions are found in a number of more complex systems which contain hydrocarbon chains as major constituents.

premise n 前提 syn *assumption, hypothesis, supposition*

▶ The procedure of dividing internuclear distance devised by Pauling is based on the premise that the size of an ion is determined by the distribution of the outermost electrons.

premium n **1 be at a premium** 貴重である syn *rare, scanty, scarce, sparse*

▶ Since ^{13}C is a low abundance mucleus with a small magnetic moment and long polarization times, sensitivity is always at a premium in ^{13}C NMR spectroscopy.

adj **2** 高品質の syn *costly, expensive*

▶ The development of premium fuels for airplanes and automobiles resulted from the use of catalysts in cracking.

premix vt あらかじめ混ぜる

▶ Reactants and precursors are premixed and flowed into the photolysis cell at the required pressure.

preoccupation n 関心事 syn *passion*

▶ The major preoccupation today of the study of biology by inorganic chemists is with catalysis.

preorganization n 前もっての組織化

▶ The secondary, tertiary, and quaternary structures of macromolecules are key features of the preorganization required for controlled nucleation.

preparable adj 調合できる

▶ Although not always preparable from N_2, many other nitrides are known.

preparation n **1** 製法 syn *prescription, recipe*

▶ Amine oxides decompose when strongly heated, and this reaction provides a useful preparation of alkenes.

2 製品 syn *material, product, substance*

▶ An Eastman Kodak preparation was washed with sodium hydrogencarbonate solution until effervescence ceased, then washed, dried, and distilled from a Claisen flask with fractionating side arm.

3 標本

▶ The helical structure gives rise to two unique identifying textures when microscope slide preparations of the cholesteric phase are examined using a polarizing microscope.

4 in preparation 準備中

▶ A detailed report concerning this class of compounds is in preparation.

preparative adj 調製の

▶ By using soft chemistry or other preparative techniques, it is often possible to prepare solid solutions that are much more extensive than those existing under equilibrium conditions.

preparatively adv 調製上

▶ A preparatively more useful form of this reaction is the crossed Cannizzaro reaction which ensues when a mixture of an aromatic aldehyde and formaldehyde is allowed to react under the influence of strong base.

prepare vt 調製する syn *make, modify, produce, transform*

▷ Vesicles prepared from relatively simple surfactants are the most sophisticated and versatile.

prepared adj **prepared for** 用意ができて syn *ready, willing and able*

▶ One must be prepared for the sea of electrons of a metal to play a role in the bonding of surface groups that has no direct counterpart in simple chemical systems.

preponderance n 優位 syn *bulk, majority*

▶ In spite of these differences in reactivity, chlorination rarely yields a great preponderance of any single isomer.

preponderate vi 優位を占める syn *predominate, prevail*

▶ In view of their chemical similarities, Nb and Ta invariably occur together, and their chief mineral, (Fe, Mn)M_2O_6 (M=Nb, Ta) is known as columbite or tantalite, depending on which metal preponderates.

preprogrammed adj あらかじめ設定された

▶ Smart materials have the capability to sense changes in their environment and respond to the changes in a preprogrammed way.

prepurified adj あらかじめ精製された

▶ For many purposes, prepurified nitrogen containing less than 0.1% oxygen is satisfactory as an inert gas.

prereduced adj あらかじめ還元した

▶ A 30-Å-thick Pt or Rh film on prereduced titania is covered with half a monolayer of TiO_x after 2 min at 400 ℃.

prerequisite n 必要条件 syn *condition, necessity, qualification, requirement*

▶ An essential prerequisite for the development of a particular sol-gel process is the availability of a stable, concentrated aqueous sol.

prescribe vt 定める syn *define, specify, stipulate*

▶ Williams and Hoard prescribed the capped octahedron for the $NbOF_6^{3-}$ ion from analysis of two-dimensional X-ray measurements.

prescribed adj 所定の syn *conventional, established, standard*

▶ Quantum mechanical calculations furnish not only the energy of a molecule for a prescribed nuclear conformation but also a wave function, whose square at any point in space gives an electron density.

prescription n **1** 方法 syn *direction, instruction*

▶ A useful description of the electronic structure of organic molecules may be obtained by the following prescription.

2 処方 syn *formula, recipe*

▶ The prescription of quantities is based upon considerations of stoichiometry as modified by the results of experimentation.

presence n 存在 syn *association, existence*

▶ The presence of acidic and basic groups in proteins means that there will generally be positive or negative charges on the protein molecule.

present n **1** 現在 syn *nowadays, today*

▶ Up to the present, structural studies on discontinuous transformations have seldom been correlated with thermodynamic considerations.

2 at present 今は syn *just now, today*

▶ At present, the majority of synthetic reactions are performed using techniques and apparatus that have been in place for decades.

adj **3** 存在して syn *remaining*

▶ In all cases, reactions proceed to an equilibrium, where both products and reactants are present but show no further tendency to change their concentrations.

4 現在の syn *contemporary, current, present-day*

▶ Much of our present understanding of the nature and biological role of proteins has come about through chemical and biological studies.

vt **5** 提供する syn *exhibit, give, offer, produce*

▶ A more extensive tabulation of sublimation data has been recently presented by Jones.

6 引き起こす syn *introduce, offer, produce, submit*

▷ Finely divided or microporous materials presenting a large area of active surface are strong adsorbents and are used for removing colors, odors, and water vapor.

7 present oneself 考えなどが浮かぶ syn *put forth*

▶ Usually several possibilities present themselves, and the choice can be narrowed further by preparation of derivatives.

present-day adj 現代の syn *current, modern*

▶ Much of the effort of present-day scientific research is directed toward an elucidation of the forces responsible for the chemical bond.

presently adv 目下 syn *currently*

▶ The complexes presently discussed are formed by the weak interaction of electron donors with electron acceptors.

preservation n 保存 syn *conservation*

▶ A sample for preservation is transferred with a capillary dropper to a small specimen vial wrapped in metal foil or black paper to exclude light.

preservative n 保存料

▶ Nitrites are used as preservatives and curing agents in meats.

preserve vt 維持する syn *keep, maintain, sustain*

▷ The linking unit usually contains multiple bonds about which freedom of rotation is restricted, so preserving the rigidity and elongation of the molecules.

press n **1** プレス

▶ Lithium in the form of small blocks is pressed into a wire with a sodium press equipped with a 3-mm die.

vt **2** 加圧する syn *compress, squeeze*

▶ Iodine can be pressed into being a metal.

pressing adj 緊急の syn *critical, crucial, dangerous, serious*

▶ Acid rain is one of the more obvious and pressing results of degradation of air quality.

pressure n **1** 圧力 syn *compression, force*

▶ Most of the experiments were conducted in a static gas-handling system in which the sample could be evacuated for several hours to a pressure of ca. 10^{-6} mm.

vt **2** 圧力をかけて…させる syn *contrain, force, press*

▶ By stimulating the immune system, the vaccine has pressured the immune system into stockpiling antibodies and immune memory cells.

pressure-sensitive adj 感圧の

▶ Some pressure-sensitive adhesives are chemically reactive, including many of those based on organofunctional acrylics and polyisobutylenes.

presumably adv おそらく syn *probably, seemingly, undoubtedly*

▶ Hydrogen selenide is a gas at room temperature; indeed, its melting and boiling points are lower than those of hydrogen sulfide, presumably because there is even less hydrogen bonding.

presume vt 推測する syn *assume, presuppose, surmise*

▶ The related molybdenum compound is presumed to have analogous packing on the basis of the similarity of the cell parameters determined by XRD powder patterns.

presuppose vt 前提とする syn *assume, presume, suppose, surmise*

▶ In the transition state theory, we assumed that the concentration of the activated complex could be calculated using statistical mechanical expressions which presuppose equilibrium.

pretransitional adj 転移に先立つ

▶ When selective reflections are in the visible range, the rapid pretransitional unwinding of the pitch is seen as a dramatic color change from blue to red on cooling.

pretreatment n 前処理

▶ The same solid can have totally different sorptive properties when given different pretreatments.

pretty adv かなり syn *fairly, moderately, quite, rather, somewhat*

▶ In fact, by this time chemists had pretty well convinced themselves that no oxo acids of fluorine were ever likely to be isolated.

prevail vi 支配的である syn *dominate, predominate*

▶ There is no assurance that the same geometry for a given species will prevail in all the physical states.

prevailing adj 支配的な syn *dominant, predominant*

▶ This coefficient is a function of temperature alone and serves as a measure of the prevailing intermolecular attractions.

prevalence n 普及 syn *acceptance, popularity, predominance*

▶ Another feature of E1 reactions is the prevalence of the initially formed carbonium ion to rearrange, especially if, in doing so, a more stable ion results.

prevalent adj 広く行き渡っている syn *common, dominant, widespread*
▶ There is a relationship between the structural trends in the MX_2 and M_2X_4 series of molecular structures and the prevalent structure types in the group 2 dihalide solids.

prevent vt 防止する syn *avoid, obstruct, obviate, preclude*
▷ Changing one or two amino acids in a protein can alter the overall shape of a protein, dramatically disrupting its function or in some cases preventing the protein from functioning at all.

prevention n 予防 syn *avoidance, inhibition*
▶ Vaccination is one of the most effective methods in the prevention and control of bacterial infections.

preventive adj 予防の syn *precautionary*
▶ A sunburn preventive agent must be nontoxic, nonstaining, nonvolatile, reasonably soluble in and stable to water, and able to filter out the burning but not the tanning component of sunlight.

previous adj 前の syn *earlier, preceeding, prior*
▶ The reactivity of individual specimens of a solid is highly dependent on their previous chemical and thermal history.

previously adv 以前に syn *before, earlier, formerly*
▶ A substance is said to be pure when its physical and chemical properties coincide with those previously established and recorded in the literature.

primarily adv 主として syn *chiefly, essentially, mainly, principally*
▶ On account of the strongly electronegative character of oxygen, many oxides AO_2 are primarily ionic and have symmetrical structures of the fluorite, rutile, and β-cristobalite types composed of A^{4+} and O^{2-} ions.

primary adj **1** 第一級の
▶ An alkyl group is described as primary if the carbon at the point of attachment is bonded to only one other carbon.
2 最初の syn *earliest, first, initial, original*
▶ These primary compounds may then react further, making the total reaction quite complicated.

primary amine n 第一級アミン
▶ In the laboratory, pure primary amines are best prepared by the reaction between potassium phthalimide and an alkyl halide to give an *N*-alkylphthalimide, which is then cleaved to give the corresponding primary amine.

primary cell n 一次電池
▶ The production of an electric current through chemical reaction is achieved in primary cells and storage cells.

primary metabolite n 一次代謝産物
▶ The organic constituents of living systems can be divided into two major groups: primary metabolites and natural products.

primary standard n 一次標準
▶ In volumetric analysis, potassium dichromate is preferred since it lacks the hygroscopic character of the sodium salt and may, therefore, be used as a primary standard.

primary structure n 一次構造
▶ The primary structure of a functional protein molecule consists of the linear sequence of amino acids in each of its polypeptide chains.

primary valence n 主原子価 syn *principal valence*
▶ The molecules of linear polymers consist of sequences of many small structural units, called monomers, which are joined together end-to-end by primary valence bonds.

prime adj 主要な syn *primary, principal*
▶ Silver salts are often soluble in aromatic and olefinic solvents. Here presumably π-bonding effects are of prime significance.

primitive adj 単純な syn *basic, simple*
▶ By functionally primitive we mean aprotic species which possess no permanent molecular dipole.

primitive lattice n 単純格子
▶ A cell that contains only one lattice point is a primitive cell, and a lattice made up of such cells is called a primitive lattice.

principal adj 主な syn *chief, dominant, main, primary, prime*

▶ Hydrogen bonding in water is the principal reason for the insolubility of nonpolar substances in this solvent.

principally adv 主として syn *chiefly, mainly, mostly, primarily*

▶ Variations in the composition of minerals principally result from the isomorphous replacement of one ion by another.

principle n **1** 法則 syn *law, rule*

▶ As a general principle, the sum of pK_a and pK_b for a conjugate acid-base pair must be equal to 14 in aqueous solution.

2 要素 syn *constituent, essence*

▶ The chief coloring principle of litmus, the azolitmin, is dark brown powder only slightly soluble in water and insoluble in alcohol and ether.

3 in principle 原則として syn *basically, fundamentally, theoretically*

▶ Seventeen asymmetric centers are present on the backbone of 26 carbon atoms of monensin, which means that, in principle, 131,072 different stereoisomers exist for the antibiotic.

principle of microscopic reversibility n 微視的可逆性の原理

▶ According to the principle of microscopic reversibility, a reaction and its reverse follow exactly the same path but in opposite directions.

prior adj **1** 先の syn *earlier, previous*

▶ If there were a prior equilibrium involving removal of a proton or tritium ion, a difference in the equilibrium constant would be expected.

2 prior to (prep) …より先に syn *before, preceeding, previous to*

▶ Decolorizing carbon is added to the hot solution prior to filtration, and the solution is kept hot for a brief period, shaken to wet the carbon, and filtered.

priority n 優先順位 syn *precedence, preference, superiority*

▶ Priority in scientific work can only be established by a proper publication in the form of a book or an article in a scientific journal.

prism n **1** プリズム ☞ antiprism

▶ Prisms are forms each of whose faces cuts the two lateral axes and is parallel to the vertical axis.

2 多角柱状結晶

▶ Crystals grown from chloroform at room temperature were orange six-sided prisms elongated on *b*.

prismatic adj 三稜形の ☞ antiprismatic

▶ The trans form begins to separate in prismatic needles, and after 20-25 min crystallization appears to have stopped.

privileged adj 特別扱いの syn *favored, special*

▶ Each sodium ion is symmetrically surrounded by six chlorine neighbors and stands in no privileged position relative to any one of them.

probability n 確率 syn *expectation, likelihood, possibility*

▶ Other things being equal, a particle has an equal probability of jumping in any direction.

probable adj ありそうな syn *apt, likely, plausible, possible*

▶ All that we can hope to obtain from experimental data is the most probable value of the quantity or quantities in question.

probably adv たぶん syn *likely, possibly, presumably,*

▶ The surfaces of anthracene and other organic crystals and plastics deteriorate on exposure to the atmosphere, probably owing to oxidation.

probe n **1** プローブ

▶ Because the probes made from large organic molecules cause structural perturbations which alter the system under examination, it cannot be easily shown whether an observed feature is due to the probe or is an intrinsic property of the system.

2 端子 ☞ four-probe, potential probe

▶ The measurement of a voltage drop by noncurrent-carrying probes is greatly superior to the crude two-probe method.

vi **2** 追求する syn *examine, explore, investigate*

▷ Thermal decomposition of solid H_3NBH_3 occurs in two steps, probed by differential scanning calorimetry and volumetric analyses.

problem n **1** 問題 syn *puzzle, question*

▶ The complexity of the problem even in the simplest cases makes the interpretation of the experimental results difficult.

2 難問 syn *complication, difficulty, trouble*

▶ One problem that remains to be solved is the sensitivity of polyacetylene to oxygen.

problematic adj 問題の多い syn *difficult, doubtful, questionable, uncertain*

▶ Isomerization after ring closure proved to be problematic.

procedure n 方法 syn *method, scheme, strategy, way*

▶ Excellent procedures are available for preparation of primary, secondary, and tertiary amines by the reduction of a variety of types of nitrogen compounds.

proceed vi 進行する syn *advance, continue, progress*

▶ The reaction usually proceeds readily in the cold, particularly when the methyl esters of the lower molecular-weight carboxylic acids are involved.

process n **1** 過程 syn *activity, development, function*

▶ Since all natural processes are spontaneous, they must occur with an increase of entropy; therefore, the sum total of the entropy in the universe is continually increasing.

2 処置 syn *course, mechanism, operation, technique*

▶ Products of the transformation-reduction process were characterized by analysis, by X-ray diffraction, and by electron diffraction and high-resolution electron microscopy.

vt **3** 加工する syn *alter, convert, modify, transform, treat*

▷ Steam distillation is useful in processing natural oils and resins, which can be separated into steam-volatile and nonsteam-volatile fractions.

processable adj 加工可能な

▶ A stable, processable, conducting polymer would have a major technological impact.

processibility n 加工の可能性

▶ The purification method affects their further processibility in solvents and ultimately their device performance.

processing n 加工

▶ In sol-gel processing, starting materials that include metal salts undergo thermal or chemical treatments, and the resulting dispersible oxides form stable colloidal dispersions when added to dilute mineral acid.

processivity n 漸進性

▶ The key features of DNA polymerase III are its catalytic potency, its fidelity, and its processivity.

processor n 処理装置

▶ Modern processors contain 40 million transistors or more in a device which can fit comfortably in a desktop computer, giving access to impressive computing power for even the average user.

prochiral adj プロキラル

▶ When glycine is placed in an asymmetric molecular environment such as a solution of stretched gelatin gel, each hydrogen atom gives a distinct ^{1}H NMR spectrum. The α-carbon is, therefore, said to be prochiral.

prochiral center n プロキラル中心

▶ A carbon to which a pair of enantiotopic ligands are attached is called a prochiral center.

produce vt **1** 作り出す syn *deliver, furnish, supply*

▶ The first step, which involves electrophilic attack on the double bond and heterolytic breaking of both a carbon-carbon and a bromine–bromine bond, produces a a bromide ion and a cation with the positive charge centered on a carbon atom.

2 引き起こす syn *cause, give rise to, yield*

▶ The alkyl groups of the open-chain compounds have considerably more freedom of motion and produce greater steric hindrance in transition states for addition.

3 提示する syn *disclose, display, reveal, show*
▶ Electrochemical studies have produced evidence for the reversible addition of up to five electrons to C_{60} in solution.

product n **1 成果** syn *outcome, output, result*
▶ The lattice theory of crystal structure was a product of theoretical physics.
2 生成物 syn *artifact, output*
▶ The normal products of combustion of the heavier alkali metals in air are KO_2, RbO_2, and CsO_2.
3 積
▶ The rate of a reaction must be proportional to the product of two concentrations [A] and [B] if the reaction simply involves collisions between A and B molecules.

production n **生成** syn *creation, formation, outcome, preparation*
▶ Needless to say, the production of shorter pulses requires more complex equipment and greater costs.

productive adj **有効な** syn *profitable, rewarding, valuable*
▶ The idea of adding molecular diluents to ionic liquids may seem counter productive.

proffer vt **提供する** syn *offer, present*
▶ It is realistic that ionic liquids should proffer an alternative for the deposition of metals that can only currently be applied using vapor deposition.

profile n **1 調査** syn *examination, survey, study*
▶ More detailed information about the molecular packing is contained in the intensity profile of the diffraction pattern.
2 輪郭 syn *outline*
▶ Allylic ethers have a similar reactivity profile to that of unfunctionalized alkenes.

profitable adj **1 有益な** syn *advantageous, helpful, valuable, useful*
▶ Clearly much profitable research needs to be done by inorganic and physical chemists on the absorption of hydrogen by the lighter elements and their alloys.
2 有利な syn *beneficial, effective, fruitful, productive, worthwhile*
▶ By polymerization of acetylene in the presence of nickel cyanide, cycloöctatetraene could be manufactured easily on a large scale; however, no profitable commercial uses of the substance have as yet been developed.

profitably adv **有益に** syn *favorably, statisfactory, well*
▶ The fragment formalism, now widely used to view the structures of inorganic and organometallic molecules, may profitably be used to examine the electronic structure of extended arrays.

profound adj **重大な** syn *deep, extreme, great, intense, overwhelming*
▶ The nature of the alkyl group of the substrate exerts a profound effect on which mechanism is to be followed.

profoundly adv **甚大に** syn *extremely, internsely, greatly, very*
▶ Odor and flavor are closely related, and both are profoundly affected by sub-microgram amounts of volatile compounds.

program n **1 プログラム** syn *plan, scheme*
▶ Programs are now available for the representation and storage of structural and other isomers.
vt **2 プログラムで設定する** syn *organize, plan*
▷ Differential thermal analysis is a technique in which the temperature of a sample is compared with that of an inert reference material during a programmed change of temperature.

programmable adj **プログラム化できる**
▶ DNA is extremely adaptable to nanoscience due to its stability, its programmable length, and the ease of chemical functionalization.

progress n **1 in progress 進行中** syn *ongoing*
▶ While this study was in progress, several reports of similar approaches were published.
vt **2 進む** syn *advance, continue, proceed*
▶ Monitoring the reaction by high-performance liquid chromatography over four days showed that the reaction did not progress significantly after three days.

progression n 進行 syn *advance, development, extension, progress*

▶ The progression from loops to helices can usually be associated with climb of screw dislocations brought about by the absorption of vacancies.

progressive adj 漸進する syn *continuous, gradual*

▶ As the temperature is raised, there is a progressive transition from the tetragonal superstructure toward the cubic arrangement of the disordered solid solution.

progressively adv 次第に syn *gradually*

▶ The diphenylpolyenes, $C_6H_5-(CH=CH)_n-C_6H_5$, absorb light at progressively longer wavelengths as *n* increases.

prohibitive adj 手が出ない syn *discouraging*

▶ The reaction is slow and an excess of amine must be used. If the amine is valuable, this is prohibitive.

prohibitively adv 法外に

▶ The very large triple bond energy in dinitrogen tends to make the activation energy prohibitively large.

project vi **1** 突出する syn *protrude*

▶ The polar ends of soap molecules project outward into the polar solvent, water.

vt **2** 投影する

▶ The atoms of three adjacent dimers are projected onto the best plane defined by the carbon and two oxygen atoms of a carboxyl group and the attached ring carbon in one of the molecules of the central dimer; this plane is shown in the figure by a dotted shading.

projection n **1** 投影 ☞ Fischer projection

▶ In Figure 1 is shown a projection down *c* of the hexagonal close-packed arrangement of anions.

2 突起 syn *extension*

▶ Absorptive cells have numerous hairlike projections called microvilli on their outer surface.

projection formula n 投影式 ☞ cross formula

▶ The so-called Fischer projection formulas are widely used for distinguishing between optical isomers.

prolate adj 扁長の ☞ oblate

▶ In the AX_5 square pyramidal molecules of the transition metals, interaction with a prolate ellipsoidal d subshell causes the apical bond to be longer than the equatorial bonds.

proliferation n 急増 syn *expansion, growth, increase, spread*

▶ The tremendous proliferation of synthetic materials in recent years was made possible by the increasingly sophisticated use of catalysts.

prolific adj 豊富な syn *abundant*

▶ For cobalt, rhodium, and iridium, +3 oxidation state is the most prolific one, providing a wide variety of kinetically inert complexes.

prolong vt 延長する syn *extend*

▶ If the electrophoresis is prolonged and the conditions are carefully controlled, a separate boundary develops for each protein.

prolonged adj 長期の syn *extended, extensive, lengthy, long*

▶ One example is sulfonation which is a reversible reaction leading to 1-naphthalene sulfonic acid at 120 ℃ but to the 2-isomer on prolonged reaction or at temperatures above 160 ℃.

prolongation n 延長 syn *continuation, maintenance*

▶ In a Soxhlet extractor, the same solvent is used over and over again, and even substances of very slight solubility can be extracted by prolongation of the operation for the necessary period of time.

prominence n 卓越 syn *distinction*

▶ Lithium soaps have come into prominence because to the properties of sodium soaps they add more resistance to water.

prominent adj 顕著な syn *famous, notable, noteworthy, well-known*

▶ The glassy condition is often very prominent even in binary and higher systems, both in organic and in inorganic substances.

prominently adv 顕著に syn *importantly, notably, significantly*

▶ In an electrospray ionization mass spectrum of the potassium salt of $IrBr_6^{2-}$, the parent divalent anion appears prominently along with $IrBr_5^-$, $IrBr_4^-$, and $K^+IrBr_6^{2-}$.

promise n 有望 syn *expectation, likelihood, potential, probability*

▶ Mass spectrometry is showing great promise in handling structural problems.

promising adj 有望な syn *encouraging, favorable, optimistic, positive*

▶ The adaptation of ideas and concepts derived from biomineralization research to the synthesis of inorganic materials with controlled properties appears to be a promising area of investigation.

promote vt **1** 昇位させる syn *raise*

▶ Thermal energy is sufficient to promote a large fraction of these donors' excess electrons into the conduction band.

2 促進する syn *assist, encourage, stimulate, support*

▷ While Os(II) proved to be a versatile dearomatization agent, capable of promoting a vast array of arenes and aromatic heterocycles toward electrophilic additions, cycloadditions, and related reactions, there are several drawbacks.

promoter n **1** 助触媒

▶ Aluminum oxide and potassium oxide are added as promoters to the iron catalyst used in facilitating a combination of hydrogen and nitrogen to form ammonia.

2 プロモーター

▶ The recombination of a proto-oncogene to an active promoter region may increase components of signal transduction pathways and disrupt their normal functioning.

promotion n **1** 昇位 syn *elevation*

▶ In the delocalized bonding picture, promotion of electrons produces conduction-band electrons and valence-band holes, in equal numbers.

2 触媒促進 syn *improvement*

▶ Rate enhancement, or promotion, of heterogeneous catalysts has long been viewed as an empirical subject.

prompt adj **1** 迅速な syn *fast, rapid, swift*

▶ If the solution remains supersaturated, addition of a seed crystal of the cis form causes prompt separation of the cis form in a paste of small crystals.

vt **2** 促す syn *encourage, induce, influence, provoke*

▶ These observation of clean dinitrogen hydrogenation with a family of side-on bound zirconocene- and hafnocene-N_2 complexes prompted more detailed studies into the mechanism of N-H bond formation.

promptly adv 迅速に syn *at once, immediately, instantly, quickly*

▶ The washing should be done promptly without allowing the precipitate to remain dry for any length of time.

prone adj しがちな syn *apt, liable, likely*

▶ Ethers are potentially hazardous chemicals since, in the presence of atmospheric oxygen, a radical-chain process can occur, resulting in the formation of peroxides which are unstable, explosion-prone compounds.

pronounced adj 目立った syn *distinct, noticeable, prominent*

▶ Changes in crystal structure and phase behavior, which often are caused by relatively minor differences in molecular structure, have pronounced effects on the heat capacity of organic solids, as shown by the results for the isomeric heptanes.

proof n **1** 証明 syn *evidence, validation, verification*

▶ The tremendous effort recently made in the isolation and proof of structures of antibiotics has led to the recognition of several new classes of naturally occurring amino sugars.

2 通常複数形で，校正刷り

▶ Authors will receive proofs by e-mail if an address has already been provided or by fax.

proofreading n 校正

▶ One important mechanism by which DNA polymerases increase fidelity of replication is in proofreading.

propagate vt 伝播する syn *develop, grow,*

spread

▷ To grow large crystals from the melt, the problem is to find a reliable method of initiating a single crystal nucleus and of propagating it through a large volume of material.

propagation n 伝播 syn *development, reproduction*

▶ Even with equilibrium constants obtained from careful spectroscopic studies, the propagation of errors can lead to standard deviation in ΔH° of 1–2 kJ mol^{-1}.

propagation reaction n 成長反応

▶ The propagation reactions occur in competition with chain-terminating steps, which result in destruction of atoms or radicals.

propellant n **1** 推進薬

▶ When powdered, aluminum combines explosively with liquid oxygen and has been used as a rocket propellant.

2 スプレー用の高圧ガス

▶ The strong possibility that chlorofluorocarbons contribute to depletion of the ozone layer of the upper atmosphere has resulted in prohibition of their use as propellants.

propensity n 性質 syn *habit, tendency*

▶ Ag_2HgI_4 crystal possesses a unique propensity for bulk diffusion of ions.

proper adj 適切な syn *adequate, fitting, suitable*

▶ One of the most useful methods of preparing an aromatic carboxylic acid involves oxidation of the proper alkylbenzene.

properly adv 適切に syn *appropriately, correctly, suitably*

▶ Because of the great variety of compounds that are properly considered coordination compounds, a classification is difficult, even for study purposes.

property n 性質 syn *attribute, characteristic, feature, hallmark, quality*

▶ The properties of one functional group are often greatly modified by the presence of another, particularly when the groups are joined by a common bond as in the carboxyl group.

proportion n **1** 比率の意味を含む部分．集団を構成する個々を強調するときには，動詞の複数形と組み合わせる． syn *division, part, portion*

▶ A large proportion of the metallic elements form alkoxides of the type $M(OR)_x$.

2 比率 syn *ratio, relationship*

▶ Gases can exist in only one phase at normal pressures since gases mix in all proportions to give a uniform mixture.

3 in proportion 比例して syn *proportional*

▶ The typical heat capacity curve has both a zero value and a zero slope at 0 K and initially increases to a first approximation in proportion to T^3.

proportional adj 比例する syn *proportionate, related*

▶ The hysteresis loss is proportional to the area inside the hysteresis loop.

proportional counter n 比例計数管

▶ The powder diffractometer has a proportional Geiger counter as the detector which is connected to a chart recorder.

proportionality n 比例

▶ The magnetic susceptibility is the constant of proportionality between the magnetization and the strength of the applied field.

proportionate adj 比例した syn *proportional*

▶ The two chief rules deduced are: first, that for every voltage applied a proportional hydrostatic pressure is produced and, second, that for every amount of current passed a proportionate volume of liquid is transported.

proportioned adj バランスの取れた

▶ Naphthalene can be crystallized efficiently from a suitably proportioned methanol–water mixture; the hydrocarbon is dissolved in excess methanol, and water is added little by little at the boiling point until the solution is saturated.

proposal n 提案 syn *presentation, suggestion*

▶ In 1848 Louis Pasteur made a set of observations which led him a few years later to make proposal that is the foundation of stereochemistry.

propose vt 提案する syn *present, submit, suggest*

▷ This sequence of reactions in the addition of water to alkenes is just reverse of that proposed for the dehydration of alcohols.

prospect n 予想 syn *anticipation, expectation, outlook*

▶ We explored some of the prospects for the synthesis of 1D polymers incorporating M−M quadrupole bonds.

prospective adj 予期される syn *anticipated, expected*

▶ The recent improvements in the processibility of conducting polymers has multiplied their prospective uses.

prosthetic adj 補欠分子の

▶ Adverse reactions between foreign or prosthetic surfaces and blood components are the pre-eminent factors restricting the use of certain biomaterials.

prosthetic group n 補欠分子族

▶ The prosthetic group of rhodopsin is 11-*cis*-retinal: an unsaturated aldehyde derived from vitamin A, which in turn is derived from β-carotene.

protect vt **1** 保護する syn *keep, preserve*

▶ The potassium iodate solution is very stable and can be preserved for years if protected from evaporation.

2 防御する syn *guard, shield*

▷ The amide hydrolysis proceeds with good yield so that acylation is useful for protecting an amino group.

protected adj 保護された syn *shielded*

▶ The next step in peptide synthesis is to activate the carboxyl function of the protected amino acid.

protecting group n 保護基

▶ After formation of the desired number of peptide bonds in the synthesis of a polypeptide, the protecting groups must be removed to give the final product.

protection n 保護 syn *screen, shield, security*

▶ The rules of stereochemistry provide an added value from the use of chiral surfactants by providing protection against one of the continuing problems of monolayer chemistry, the production of artifacts due to the intrusion of impurities into the tiny quantities of material in the monolayer film.

protective adj 保護的な syn *strong, tough*

▶ Iodine penetrates the protective layer of oxide on the surface of magnesium.

protective action n 保護作用

▶ Striking demonstration of protective action may be made by showing how a soluble soap can carry charcoal and finely divided iron(III) oxide through filter paper, when in the absence of the soap none will pass through.

protective colloid n 保護コロイド

▶ To make colloidal systems stable, a sufficient quantity of stabilizing agent must be introduced or some protective colloid added.

protective film n 保護膜 ☞ passive film

▶ A metallic glass without the protective film is more reactive than the corresponding crystalline material.

protein n タンパク質

▶ The distinction between proteins and peptides is arbitrary. Compounds which have molecular weights greater than 10,000 are generally referred to as proteins.

protein folding n タンパク質の折りたたみ

▶ Density variations in different regions of the protein interior and packing defects have been identified and related to conformational fluctuations, hydrogen exchange, and the protein folding problem.

proteolytic enzyme n タンパク質分解酵素

▶ Proteolytic enzymes disrupt peptide bonds of the primary protein structure, creating fragments of various lengths or degrading the protein completely to amino acids.

protic adj プロトン性の syn *protogenic*

▶ With both protic N–H and hydridic B–H bonds, three H atoms per main group element, and a low molecular weight, H_3NBH_3 has the potential to meet the stringent gravimetric and volumetric hydrogen storage capacity targets needed for transportation applications.

protic solvent n プロトン性溶媒 ☞ aprotic solvent

▶ As protic solvents, alcohols solvate anions especially strongly through hydrogen bonding.

protocol n 手順 syn *custum, practice, rule*

▶ In both experimental protocols, more than 1 equiv of H_2 was released when the reaction was allowed to proceed for more than 6 h.

protogenic adj プロトン性の syn *protic*

▶ Sulfuric acid is so strongly protogenic that most compounds of oxygen and nitrogen accept protons from it to some extent.

protolysis n プロトリシス 複数形は protolyses

▶ There is a general tendency for protolysis to be greater the higher the cationic charge.

protolytic reaction n 陽子移行反応

▶ With superacids, alkanes readily undergo protolytic reactions involving tertiary, secondary, and primary C−H as well as C−C bonds.

proton affinity n 陽子親和力

▶ It is often more convenient to rationalize the relative strengths of acids in terms of the proton affinities of their conjugate bases.

protonate vt プロトン付加する ☞ deprotonate

▶ Hydrolysis is fastest at the intermediate pH where the weaker base is mostly free and the stronger base is mostly protonated.

protonation n プロトン付加 ☞ deprotonation

▶ Protonation of neopentyl alcohol yields an oxonium ion.

proton exchange n プロトン交換

▶ Proton exchange between two molecules of alcohol is so fast that the proton cannot see nearby protons in their various combinations of spin alignments but in a single average alignment.

proton gradient n プロトン勾配

▶ Energized storage utilizes metabolism, and the generation of proton gradients must be included with other ion storage modes.

protonic adj プロトンの

▶ Various materials have been examined for protonic conductivity.

protonic acid n プロトン酸

▶ The protonic acid, NH_4^+, formed by ammonium salts in solution in ammonia, liberates hydrogen with such metals as calcium.

proton motive force n プロトン駆動力 ☞ electrochemical gradient

▶ The excess concentration of hydrogen ions on one side of a membrane creates an electric potential called a proton motive force.

proton NMR n プロトン NMR ☞ proton resonance

▶ Practically, obtaining a usable spectrum is more difficult for ^{13}C NMR than the proton NMR and requires more sophisticated instrumentation.

protonolysis n プロトノリシス

▶ After protonolysis, aldehydes yield primary alcohols, and ketones yield secondary alcohols.

proton resonance n プロトン共鳴 ☞ proton NMR

▶ The separation of proton resonances brought on by electron-shielding effects is called the chemical shift.

proton transfer n プロトン移動

▶ The comparatively high mobility of the HSO_4^- ion undoubtedly arises from successive proton transfers to the ion from an adjacent H_2SO_4 molecule.

prototropic adj プロトトロピーによる

▶ The majority of prototropic reactions can be catalyzed by acids as well as by bases.

prototropy n プロトトロピー ☞ tautomerism

▶ When two or more tautomeric structures differ in the position of a hydrogen atom, as well as in the distribution of the valence bonds, the tautomerim is frequently described as prototropy.

prototype n 典型 syn *example, illustration, model, sample, standard*

▶ The well-studied prototype of tetrahedral systems would be the organic compounds containing four groups bound to carbon.

prototypical adj 典型的な syn *authentic, basic, genuine, primary*

▶ Prototypical electric insulators and semiconductors are found in the nonmetal portion of the periodic table.

protrude vt 突き出す syn *extend, project*

▷ The complex NbF_7^{2-} has been assigned a trigonal prism structure with a fluoride ion protruding from one of the tetragonal faces.

prove vi **1** 判明する syn *be shown, turn out*

▶ Predictions based solely on crystal-field stabilization energy considerations sometimes prove incorrect.

vt **2** 確かめる syn *certify, confirm, establish, support, validate, verify*

▶ With respect to the infrared spectra of aryl halides, correlations between structure and absorption bands of aromatic carbon–halogen bonds have not proved to be useful.

proven adj 立証された syn *rational, reasonable, sound, valid*

▶ Layered $LiMnO_2$ is potentially attractive since it combines the proven performance of the $LiCoO_2$ structure with the low cost and low toxicity of manganese.

provide vt 提供する syn *equip, furnish, supply*

▶ A very pure inert gas is used to provide a nonreactive atmosphere for the Czochralski process.

provided conj もし…ならば syn *as long as, only if, providing that*

▶ Very pure samples can be produced by repeating the zone-refining process several times, provided the molten sample does not react and pick up impurities from the container or from the atmosphere.

providing that conj もし…ならば syn *as long as, only if, provided, provided that*

▶ Providing that the components are adequately volatile, gas-liquid chromatography (GLC) is perhaps the most powerful technique for the rapid and convenient analysis of the composition of mixtures of organic compounds.

provision n **1** 準備 syn *arrangement, measures*

▶ In addition to minimizing background impurity levels, provision has to be made for the controlled introduction of specific impurities, to give the required electric properties in each layer deposited.

2 make provision for …に備える syn *make ready, prepare*

▶ Any general treatment of complex formation must make provision for covalent bonding between metal and ligand.

provisional adj 暫定的な syn *temporary*

▶ This general consistency provides full confirmation of the provisional statement that one or more definite polymers higher than the dimer exist in the acid vapor.

provoke vt 引き起こす syn *activate, motivate, prompt, stimulate*

▶ The nature of the active site has provoked considerable debate.

proximity n 近接 syn *neighborhood, vicinity*

▶ In solution where molecules are always in such close proximity that mutual electrostatic interactions have to be taken into account, neutral bond fission is not usually favored.

proximity effect n 近接効果

▶ Proximity effects are important in the enhancement of reaction rates.

prudent adj 分別のある syn *reasonable*

▶ The prudent course is to conserve resources, minimize wasteful practices, and improve efficiency.

pseudo-first-order adj 擬一次の

▶ The relative concentration of ethyl radicals is detected by the mass spectrometer at various injector positions, giving the pseudo-first-order rate coefficient for the reaction.

pseudohalogen n 擬ハロゲン

▶ The pseudohalogens react with various metals to give salts containing X^- anions.

pseudomorph n 仮像

▶ A pseudomorph is a transformed crystal: the external shape of the original crystal may be discernible, even though the internal structure is that of the new form.

psychological adj 心理的な

▶ Whatever the mechanism for psychological response to color, the color of a substance that does not emit light itself is due to the light it transmits or reflects.

p-type adj p 型の

▶ Holes must be forced to move under the barrier from p-type side to the n-type side during current flow.

p-type semiconductor n p 型半導体 ☞ n-type semiconductor

▶ A decrease in the conductivity of a p-type semiconductor indicates an electron transfer from chemisorbed gas to semiconductor.

publicity n 世間の注目 syn *attention, prominence*

▶ There is little doubt that publicity associated with these experimental findings stimulated the wide interest and excitement in this class of materials.

publicize vt 宣伝する syn *advertise, promote*

▶ The effects of acidic rainfall are most evident and highly publicized in Europe and the northwestern United States.

publish vt 発表する syn *disclose, report*

▶ Some mechanistic considerations on related problems were published previously.

pucker vt 襞を取る syn *contract, wrinkle*

▶ Since most polyethylene crystals are puckered in some manner after drying, it must be supposed that a hollow pyramid represents their most common growth habit.

puckered adj 襞のある

▶ The structure of C_4F is like that of the monofluride, but the layers are planar and not puckered, and the distance between them is smaller (5.5 Å).

puckering n 襞になること

▶ Puckering of the pentagonal ring may be observed for the solid-state pentagonal bipyramidal molecules or ions.

pull vt **1** 引っ張る syn *drag, draw*

▶ As the seed is pulled from the melt, the melt adheres to the crystal and is pulled into the cooler region of the furnace.

2 pull together 統合する syn *establish, form, organize, set up*

▶ On this basis, it can be expected that a small amount of free liquid will pull itself together to form a more or less spherical drop.

pulsation n 脈拍 syn *beat*

▶ The beating mercury heart phenomenon involves pulsation of a mercury globule near a corroding electrode not necessarily touching the mercury.

pulse n **1** パルス syn *vibration*

▶ A pulse of light is used to produce a transient species, an atom, radical, or excited state, whose concentration is then monitored as a function of time.

vt **2** パルス状にする syn *vibrate*

▷ A pulsed electric discharge creates excited Kr^+ and F^- ions from a mixture of $Kr/F_2/He$.

pulse radiolysis n パルス放射線分解

▶ It might be argued that hexane is an unlikely solvent for the study of ionic reactions, but charged transient species can readily be generated in alkanes by the technique of pulse radiolysis.

pulverize vt 粉状にする syn *crush, grind, mill, powder*

▶ At low temperatures and at about 200 ℃ metallic zinc is so brittle that it can be pulverized, but at 110–150 ℃ it is ductile and can be drawn out into wire and rolled into foil.

pump n **1** ポンプ

▶ Any pump radiation must be lower than the InSb band gap, ca. 1900 cm^{-1}, to avoid direct absorption by the crystal.

vt **2** 送り込む syn *deliver, force, push, send*

▶ Much energy is required to pump Ca^{2+} ions into the sarcoplasmic reticulum against a concentration gradient.

3 pump out 汲み出す syn *drain*

▶ The transport of glucose and amino acids into the cell is coupled with Na^+ transport. The Na^+ entering the cell in this way must be pumped out again.

4 励起するために放射線を当てる

▷ The excitation process involves pumping the active centers into an excited state that has a reasonably long lifetime; a situation can then be reached in which a population inversion occurs.

pumping n **1** ポンプを使うこと

▶ Pumping away the volatiles is also essential, because OF_2 is formed in amounts comparable to the HOF, and only by pumping can it be removed.

2 ポンピング

▶ By pumping the band gap of a semiconductor, electrons are promoted to the conduction band leaving a hole in the valence band.

pungent adj 刺激する syn *bitter, caustic*

▶ Tetrahydronaphthalene is a colorless liquid with a pungent odor.

purchase vt 購入する syn *acquire, get, obtain*

▶ All solvents were distilled before use; dry solvents were purchased from Fluka.

pure adj 純粋な syn *clean, clear, uncontaminated*

▶ Ordinary phosphorus is poisonous and is colorless when pure.

purely adv **1** 純粋に syn *completely, entirely, perfectly, totally*

▶ Tetraiodomecrurate(II) ion, HgI_4^{2-}, is the most stable purely inorganic complex known.

2 単に syn *basically, merely, only, simply*

▶ Distinctions between ionic, coordinate, and covalent structures in the halides and their complexes frequently depend on purely arbitrary demarcations.

purge n **1** 浄化 syn *elimination, expulsion, removal*

▶ An oven-dried round-bottom flask was cooled to room temperature under a nitrogen purge and charged with *N*,*N*-dialkylbenzenesulfonamide.

vt **2** 浄化する syn *clean, purify, wash*

▶ Both solutions were purged with pure gaseous nitrogen during the electrolysis.

purification n 精製 syn *refinement, refining*

▶ The usual purpose of distillation is purification or separation of the components of a mixture.

purify vt 精製する syn *clean, cleanse*

▶ Bromine trifluoride and chlorine trifluoride were purified by several distillations in a Monel apparatus.

purine n プリン

▶ The purines are the derivatives of the substance purine and constitute another important class of nitrogen heterocycles.

purity n 純度 syn *perfection*

▶ The enantiomeric purity of this compound was determined by chiral high performance liquid chromatography.

purple adj 紫色の

▶ If a colored KCl crystal is cooled in liquid nitrogen, the purple color observed at room temperature changes to a deep pink color at 77 K.

purplish adj 紫色がかった

▶ I_2Cl^+ compounds are dark brown or purplish black.

purport vt 主張する syn *claim, profess, state*

▶ Shortly after the successful preparation of perbromates, Cox and Moore published the results of calculations that purported to disprove all the earlier theoretical arguments and to show that there was no theoretical reason why perbromates should not exist.

purpose n 目的 syn *aim, goal, motivation, objective*

▶ The purpose of qualitative analysis is not simply to find out what elements are contained in a given substance, but the aim should also be to get a good idea of the relative amounts that are present.

purposeful adj 意図的な syn *deliberate, intentional, planned*

▶ Very recently more purposeful synthetic routes have been devised.

purposefully adv 意図的に syn *deliberately, intentionally*

▶ Sulfur is unique in the extent to which new allotropes can now be purposefully synthesized using kinetically controlled reactions.

purposely adv 故意に syn *deliberately, intentionally*

▶ The study of polymorphism often depends on purposely avoiding thermodynamic equilibrium.

pursue vt 追求する syn *continue, conduct, follow*

▶ The matter appears not to have been pursued further, and today it seems virtually certain that Cady's interpretation was correct.

pursuit n 研究 syn *search*

▶ This is an area of current pursuit.

push vt **1** 押しやる syn *drive, press*

▷ In crystals such as KI, where I–I distance is greater than that found in LiI, it is assumed that the anion and cation are in contact with each other, with the I^- ions pushed apart.

2 push aside 押しのける syn *set aside*

▶ Since a larger solute molecule has to push aside more solvent molecules during its diffusion, it will move more slowly than a smaller molecule.

3 push out 押し出す syn *displace, expel, remove*

▶ Acting as a nucleophile, the alkene attaches itself to one of the bromines and pushes the other bromine out as bromide ion.

put vt **1** 課する syn *send, subject*

▶ The fact that in a crystal the pattern of arrangement of the atoms is repeated in all directions puts certain restrictions on the kind of symmetry crystals can have.

2 投入する syn *place, position*

▶ If one deliberately puts peroxides into the reaction system, HBr adds to alkenes in exactly the reverse direction.

3 見積もる syn *offer, present, propose, submit*

▶ The experimental result that CN^- is much softer than OH^- puts limits on the unknown bond strength of silyl cyanide.

4 put forth 提起する syn *advance, offer, propose*

▷ The arguments were necessarily qualitative or, at best, semiquantitative, and each theorist tended to discount the proposals put forth by others.

puzzle n 難問 syn *paradox, problem, question*

▶ One remaining puzzle is the role that HOF plays in the formation of OF_2 from the reaction of fluorine with water or base.

puzzling adj 困らせる syn *ambiguous, bewildering, confusing, contradictory*

▶ The reaction is generally thought to involve carbonium-ion intermediates but several puzzling features remain.

pycnometric adj 比重測定による

▶ With increasing excess of oxygen in wüstite, both the pycnometric density and the lattice parameter were found to decrease progressively.

pyramid n 角錐

▶ The crystals of polyethylene grow, not as flat plates, but as hollow pyramids which collapse subsequently under the influence of surface-tension forces when being dried down.

pyramidal adj 錐形の

▶ Phosphine has a pyramidal structure, as expected, with P–H 142 pm and the H–P–H angle 93.6°.

pyrimidine n ピリミジン

▶ The major pyrimidines found in DNA are thymine and cytosine; in RNA they are uracil and cytosine.

pyroelectric n **1** 複数形で，焦電体

▶ Materials that possess a temperature-dependent spontaneous polarization and respond to a rate of change of temperature are known as pyroelectrics.

adj **2** 焦電性の

▶ Pyroelectric crystals are related to ferroelectric ones in that they are noncentrosymmetric and exhibit a net spontaneous polarization.

pyroelectric effect n 焦電効果

▶ The pyroelectric effect in a crystal becomes apparent only when the crytal is heated, thereby changing the spontaneous polarization.

pyrolysis n 熱分解 syn *thermal decomposition*

▶ A chemical transformation produced by applying heat to a single compound is called a pyrolysis reaction.

pyrolytic adj 熱分解の

▶ Though the term pyrolysis implies decomposition into smaller fragments, pyrolytic change may also involve isomerization and formation of higher molecular weight compounds.

pyrolytic graphite n 熱分解黒鉛

▶ In these specimens of pyrolytic graphite, the carbon hexagon networks lie parallel to one another but with turbostratic disorder.

pyrolyze vt 熱分解する syn *thermolyze*

▷ Stoichiometric and homogeneous $BaTiO_3$ can be prepared by recrystallizing and pyrolyzing barium titanium citrate, $BaTi(C_6H_6O_7)_3 \cdot 6H_2O$.

pyrophoric adj 自然発火しうる

▶ Nickel tarnishes when heated in air and is actually pyrophoric if very finely divided.

Q

Q switch n Qスイッチ
▶ At the other end of the ruby crystal rod is a device known as a Q switch which may either allow the laser beam to pass out of the system or may reflect it back through the rod for another cycle.

quadratic function n 二次関数
▶ The X-ray data for a, b, c, and β in the neutral phase can be fitted by a quadratic function of T and P.

quadruple bond n 四重結合
▶ Experimentally determined electron densities are consistent with the quadruple bond picture.

quadruply adv 四重に
▶ The complex anion $[Sb_4Cl_{12}O]^{2-}$ contains two square-pyramidal $Sb^{III}Cl_5$ units sharing a common edge and joined via unique quadruply bridging Cl atom to two pseudo-trigonal bipyramidal $Sb^{III}Cl_3O$ units.

quadrupole adj 四極子の
▶ By use of microwaves, transitions among the various energy levels associated with the quadrupole orientation can be induced.

quadrupole mass filter n 四重極マスフィルター
▶ The products are usually detected as a function of scattering angle by means of a mass spectrometer which incorporates a quadrupole mass filter.

quadrupole moment n 四重極子モーメント
▶ The quadrupole moment of an atomic nucleus interacts with the electric field gradient near the nucleus, such that if this field gradient is unsymmetrical, absorption of microwave radiation can take place.

quadrupole splitting n 四極子分裂
▶ A rather different application of quadrupole splittings is the detection of minute distortions from ideal symmetry and the determination of the energy separation between the various orbital states.

qualification n 条件 syn *condition, prerequisite, restriction*
▶ The assumption of complete dissociation sometimes needs qualification for strong electrolytes.

qualitative adj 定性的な
▶ Stretched-film spectrum and fluorescence polarization measurements provide essentially the same qualitative information.

qualitatively adv 定性的に
▶ The factors that govern whether or not solid solutions, especially the more complex ones, form are understood only qualitatively.

quality n **1** 性質 syn *characteristic, property*
▶ Temperature is the quality that determines the direction in which thermal energy flows.
2 品質 syn *grade*
▶ Obtaining crystals of suitable quality for an X-ray structure study proved to be exceedingly difficult and time-consuming.

quantify vt 定量化する syn *count, estimate, evaluate*
part ▶ It would be useful here if data were provided with their uncertainty quantified.

quantitative adj 定量的な
▶ Some idea of the quantitative significance of the atomic charges calculated by our electronegativity equalization procedure can be obtained by comparing the charges with those calculated by other methods.

quantitative analysis n 定量分析
▶ Quantitative analysis treats of the methods for determining in what proportion the constituents are present in any compound or mixture of compounds.

quantitatively adv 定量的に
▶ The favorite and most practical method for quantitatively describing electron distribution in a molecule is the assignment of partial charges to the atoms.

quantity n **1** 分量 syn *amount, volume, weight*
▶ Decaborane（14）is obtainable in research

quantities by the pyrolysis of B_2H_6 at 100–200 ℃ in the presence of catalytic amounts of Lewis bases.

2 in quantity 多量に syn *in abundance*

▶ Free methyl could easily be obtained in quantity by heating azomethane to about 400 ℃.

quantization n 量子化

▶ Quantization of the electronic structure occurs as the length scale of a material is reduced toward the low nanometer range.

quantize vt 量子化する

▶ Since the radii of the Bohr orbits are quantized, it follows naturally that only certain energies exist for that atom: the energy of the atom is also quantized.

quantum n 量子 複数形は quanta

▶ When the dimensions of the tailored features are sufficiently small, quantum size effects can be observed.

quantum effect n 量子効果

▶ The quantum effects are obviously so large for helium and hydrogen that we can immediately exclude them from consideration.

quantum efficiency n 量子効率 syn *quantum yield*

▶ The quantum efficiencies of carrier generation in the crystals studied were found to be temperature-independent over the temperature region studied.

quantum mechanical adj 量子力学的な

▶ In chemistry, resonance is a mathematical concept based on quantum mechanical considerations.

quantum mechanically adv 量子力学的に

▶ What is indeed available through these quantum mechanically determined density distributions is a classical description of the chemical bond.

quantum-mechanical tunneling n 量子力学的トンネル効果

▶ As far as kinetic isotope effects are concerned, quantum-mechanical tunneling need be considered only when H or muonium atoms are involved.

quantum state n 量子状態

▶ The ability to separate particles in different quantum states has made the Stern–Gerlach experiment particularly important in providing a conceptual basis for quantum mechanics.

quantum yield n 量子収率 syn *quantum efficiency*

▶ If molecular iodine vapor is exposed to light of wavelength less than 499 nm, the quantum yield for dissociation is unity.

quarter n 四分の一 syn *fourth*

▶ The first and last quarters of the distillate were discarded.

quasilinear adj 準直線形の

▶ The so-called quasilinear molecules are those found experimentally to be linear, but having a relatively low potential energy barrier to bending.

quasiperiodic adj 準周期的

▶ Quasiperiodic crystals are a new class of aperiodic atomic structures which have noncrystallographic point symmetries, but which nevertheless give diffraction patterns with sharp diffraction peaks indicative of long-range order.

quaternary ammonium ion n 第四級アンモニウムイオン

▶ Quaternary ammonium ions can carry permanganate ions from an aqueous layer into a nonaqueous layer where the substrate awaits.

quaternary structure n 四次構造

▶ If two or more polypeptide chains spontaneously associate, they form a quaternary structure.

quaternization n 四級化

▶ The quaternization is carried out usually using an alkyl halide.

quench vt **1** 消光する syn *extinguish*

▶ The effect of the tetracene impurity is to quench the anthracene emission.

2 抑制する syn *repress, supress*

▶ After 20 h the reaction was quenched with saturated NH_4Cl, diluted with ether, and poured into a separatory funnel.

3 冷却する syn *cool*

▶ Samples that are rapidly quenched from high temperatures, or vacuum annealed, are characterized by large concentrations of oxygen defects.

quenchable adj 失活可能な

▶ At absolute zero, all quenchable energy has been quenched.

quenching n 消光 syn *extinction, extingishing*

▶ Simple atoms and molecules are very susceptible to the quenching by collisions and rarely exhibit fluorescence except in the gas phase at pressures below 10 mmHg.

quest n 探求 syn *exploration, pursuit, search*

▶ The quest for potent and selective inhibitors is extremely active at present.

question n **1** 問題点 syn *issue,matter, point*

▶ Many questions remain as a central challenge.

2 疑問 syn *doubt, uncertainty*

▶ There are many questions to be answered, but what is clear from the plastocyanin results is that the distance and the driving force are not the sole determining factors for reaction.

3 beyond question 確かに syn *certainly, surely, undoubtedly*

▶ The tetrahedral structure of methane has been verified by electron diffraction, which shows beyond question the arrangement of atoms in such simple molecules.

4 in question 問題の syn *under consideration, under discussion*

▶ Directions in crystals and lattices are labeled by first drawing a line that passes through the origin and parallel to the direction in question.

5 out of the question 問題にならない syn *impossible*

▶ Under drastic conditions, any diene will condense with any olefin; these reactions occur in the gas phase and so an ionic mechanism is out of the question.

vt **6** 疑う syn *doubt, suspect*

▶ Evidence for the true solid-state reactions is convincingly presented by Hedvall, and their occurrence in many systems is now never questioned.

questionable adj 問題のある syn *doubtful, uncertain*

▶ The geometry of minimum energy should lie close to the experimental geometry otherwise the value of that particular method for calculating energy surfaces is questionable.

quickly adv 急速に syn *fast, rapidly*

▶ Reactions proceeded more quickly in dichloromethane relative to chloroform but with slightly lower enantioselectivity.

quietly adv そっと syn *mildly*

▶ A drop of diglycol laurate placed quietly on water spreads horizontally to give myelin forms.

quite adv **1** 非常に syn *completely, entirely, fully, totally, very*

▶ Quite generally throughout inorganic and organic chemistry, there is almost everywhere qualitative interpretation of phenomena in terms of molecular and electronic theories. 形容詞＋名詞を修飾する場合には，冠詞に先行する．

▶ Quite a different result was obtained on heating H_3NBH_3 in ionic liquid solvents.

2 quite a number of 相当な数の

▶ Quite a number of statistical quantities can be defined whose behavior more or less resembles that of thermodynamic entropy.

quote vt 引用する syn *cite, mention, reference, refer to*

▶ It has proved convenient in thermodynamic work to define certain standard state and to quote data for reactions involving these standard states.

quotient n 割り算の商

▶ The quotient M/H, known as the volume susceptibility, is given the symbol χ and is dimensionless.

R

racemate n ラセミ体 syn *racemic modification*

▶ A racemate exhibiting a nematic phase can be resolved into its enantiomers, each of which will exhibit a cholesteric phase.

racemic mixture n ラセミ混合物 syn *conglomerate*

▶ Conversion of the molecules of one enantiomer into a racemic mixture of both is called racemization.

racemization n ラセミ化

▶ Any preparation of an alkyl halide from an alcohol must involve breaking of the carbon-oxygen bond and is, hence, accompanied by the likelihood of stereochemical inversion and the possibility of racemization.

racemize vt ラセミ化する

▶ Optically acitive biphenyl derivatives are racemized if the two aromatic rings at any time pass through a coplanar configuration by rotation about the central bond.

radial distribution function n 動径分布関数

▶ The radial distribution function is related to the probability of finding a particular internuclear distance in the molecule, so that the calculated function consists of a number of peaks, each of which represents an internuclear distance present in the sample molecule.

radially adv 放射状に

▶ The molecules in the stream of the sample gas are randomly oriented so that the diffraction pattern that is produced must be radially symmetric.

radiant energy n 放射エネルギー

▶ The various forms of radiant energy are characterized by their wavelength, and together they comprise the electromagnetic spectrum.

radiate vi **1** 放射状に伸びる syn *spread*

▶ Six hydrogen atoms, one associated with each carbon, radiate out from the benzene ring at distances of 1.09 Å.

vt **2** 放射する syn *emanate, emit*

▷ The total power radiated from the sun in all directions is estimated at 3.8×10^{26} watts.

radiation n 放射線 syn *light*

▶ In order to generate the characteristic monochromatic radiation, the voltage used to accelerate the electrons needs to be sufficiently high so that ionization of the copper 1s electrons may occur.

radiation damage n 放射線障害

▶ We must add to the list of complicating factors the strong radioactivity of plutonium, which causes self-heating and radiation damage and results in further change of the properties with time.

radiation dosimetry n 放射線線量測定

▶ Much of the experimental evidence regarding coloring of polymers stems from the interest in these centers as a basis for radiation dosimetry.

radiation effect n 放射線効果

▶ Polymers tend to occupy a unique position in the study of radiation effects because their mechanical properties are highly sensitive to radiation damage.

radiation-induced reaction n 放射線化学反応

▶ Numerous investigators have studied the radiation-induced reactions in methanol, in both the liquid and the solid phase.

radiationless transition n 無放射遷移 syn *nonradiative transition*

▶ Like singlet-singlet internal conversion, triplet-triplet radiationless transitions are apparently so rapid that phosphorescence only occurs from the lowest triplet state.

radiative adj 発光する ☞ nonradiative

▶ The emission of multicomponent systems is determined by a balance between the radiative properties of the components and the transfer of energy between the components.

radiative decay n 放射性崩壊

▶ Only a minute fraction of the molecules survive long enough to undergo radiative decay.

radiatively adv 発光して ☞ nonradiatively

▶ The enhancement in light-emitting diode intensity at low temperature reflects the competition between whether the return of electrons from the conduction band to the valence band occurs radiatively or non-radiatively.

radiator n 放熱体

▶ MgO is unusual in being both an excellent thermal conductor and a good electric insulator, thus finding widespread use as the insulating radiator in domestic heating ranges and similar appliances.

radical n **1** 遊離基，ラジカル ☞ free radical

▶ The acid-base character of radicals has been recognized for some time, and it is customary to speak of electrophilic radical, such as $Cl^{\cdot}$, and nucleophilic radicals, such as $(CH_3)_3C^{\cdot}$.

adj **2** 極端な syn *drastic, extreme*

▶ If impurities are present, they often will impede crystallization by being adsorbed on the growing crystal faces, resulting in some instances in radical habit changes of the crystal.

radical anion n ラジカル陰イオン syn *anion radical*

▶ The blue color formed when alkali-metal polysulfides are dissolved in highly polar solvents such as dimethylformamide or hexamethylphosphoramide is now known to be due to the S_3^- ion, a radical anion with an S−S−S angle of 105° and $\lambda_{max} = 600$ nm, $\varepsilon_{max} = 10^4$ M^{-1} cm^{-1}.

radically adv 根本的に syn *completely, entirely, totally, wholly*

▶ A technological breakthrough which could lead to underground gasification or hydrogenation to produce liquid hydrocarbons would certainly radically effect coal usage.

radical mechanism ラジカル機構

▶ It is very difficult, if at all possible, to prove conclusively whether these polymerizations proceed by a radical or by an ionic mechanism, since the usually employed diagnostic tools are not applicable to solid-state processes.

radical recombination n ラジカル再結合

▶ Depending on the fuel/air ratio and conditions, hydrocarbon radical recombination may compete with abstraction and subsequent degradation, leading to the formation of higher hydrocarbon species.

radioactive adj 放射性の

▶ Tritium differs from the other two isotopes of hydrogen in being radioactive.

radioactive disintegration n 放射性壊変

▶ Radioactive disintegrations follow first-order kinetics and, therefore, have half-lives that are independent of the amount of radioactive substance present.

radioactive nuclide n 放射性核種 syn *radionuclide*

▶ Elements with radioactive nuclides amongst their naturally occurring isotopes have a built-in time variation of relative concentration of their isotopes.

radioactive tracer n 放射性トレーサー

▶ Paper chromatography has proved very useful in following the mechanisms of biological processes using radioactive tracers.

radioactivity n 放射能

▶ Radioactivity can be caused artificially in many stable elements by irradiation with neutrons in a nuclear reactor or by charged particles from an accelerator.

radiocarbon dating n 放射性炭素年代測定

▶ The decay of ^{14}C isotope in dead organic matter is the basis of the radiocarbon dating procedure.

radiochemical adj 放射化学の

▶ Bismuthine is extremely unstable and was first detected in minute traces by F. Paneth using a radiochemical technique involving $^{212}Bi_2Mg_3$.

radiofrequency n ラジオ周波数

▶ At a given radiofrequency, all protons absorb at the same effective field strength, but they absorb at different applied field strengths.

radioimmunoassay n 放射線免疫検定

▶ Radioimmunoassays are highly sensitive tests that rely on detection of protein by a radioactively labeled antibody.

radioisotope n 放射性同位体

▶ Disposal of waste containing radioisotopes and of spent nuclear reactor fuel presents a serious problem for which there is as yet no completely satisfactory solution.

radiolysis n 放射線分解

▶ Because the concentration of solvated electrons produced by radiolysis is low and the concentration of other solute species can be controlled, radiation chemistry provides the most reliable identification of the spectrum of the isolated solvated electron.

radiolytic adj 放射線分解の

▶ Polonium dissolves readily in acids to yield pink solutions of Po^{II} which are then rapidly oxidized further to yellow Po^{IV} by the products of radiolytic decomposition of the solvent.

radio wave n ラジオ波

▶ Virtually all parts of the spectrum of electromagnetic radiation, from X-rays to radio waves, have found some practical application for the study of organic molecules.

radius n 半径 複数形は radii

▶ The electron density of an atom or ion extends indefinitely in space, so that it is not possible to assign a precise radius to either.

radius ratio n 半径比

▶ If the cation-to-anion radius ratio shrinks to below 0.414, the anions are too crowded relative to the smaller cation in an octahedral geometry; their repulsions can be reduced by a smaller coordination number.

radius-ratio rule n 半径比則

▶ Use of radius-ratio rules to view the structures of even the alkali halides is very unsatisfactory.

raft n スラブ syn *slab*

▶ The transformation of large, globular Ag particles into small, thin rafts takes place only in the vicinity of Pt particles.

raftlike adj スラブ様の

▶ If the above geometry is reversed by depositing a metal onto titania, the interaction is sometimes sufficient to bring about or stabilize a raftlike configuration of the metal.

rainwater n 雨水

▶ Rainwater is acidic due to atmospheric CO_2, SO_2, and nitrogen oxides; its pH is typically 5.6.

raise vt **1** 高める syn *lift*

▶ In the direction of the applied field the barrier is lowered by an amount $eaF/2$, and against the field it is raised by the same amount.

2 提出する syn *mention, present, suggest*

▶ Against the second method of standardization, objection has been raised that it has been found difficult to prepare absolutely pure iron.

3 raise to the *n*-th power *n* 乗する

▷ It is permissible to speak of the order of a composite reaction, provided that the rate is proportional simply to concentrations raised to certain powers.

Raman active adj ラマン活性 ☞ vibrational mode

▶ For a vibration to be Raman active, there must be a change of polarizability of the molecule as the transition occurs.

Raman effect n ラマン効果

▶ A different type of scattering of light, usually called the Raman effect, arises when a tube of substance irradiated from above with powerful light of frequency is examined by a spectroscope directed toward the side of the tube.

Raman spectrum n ラマンスペクトル

▶ Raman spectra are produced by exposing a gas, liquid, or solid to bright light and examining the spectrum of the light scattered laterally.

ramification n **1** 複数形で効果 syn *consequence, effect, implication, result*

▶ Only a tiny mismatch in the energy balance equation can have quite large ramifications on whole body mass, in the long term.

2 小区分 syn *branch, subdivision*

▶ The ultramicroscope fails to resolve the small, solvated primary particles in a jelly and, therefore, does not show the ramification of their loose aggregates.

ramify vi 小区分される syn *divide, separate, split*

▷ The formation of ramifying aggregates giving an apparently high viscosity may proceed to such an extent that the system begins to resemble an elastic solid; a sol becomes a jelly.

ramp n **1** 傾斜 syn *descent, gradient, slope*

▶ Specimen temperature was regulated with a programmed control system connected to the furnace, capable of multiple heating and cooling ramps at various rates.

vi **2** 勾配をなす syn *ascend, decend, incline*

▷ With temperature ramping, the onset of TiO_x attack is placed at 200 ℃.

rancidity n 酸敗

▶ Rancidity is due to the presence of volatile, bad-smelling acids and aldehydes.

random n **1** **at random** 行き当たりばったりに syn *erratically, irregularly*

▶ If we start with a specimen of pure gold and add to it progressively more and more copper, the atoms of this latter element replace those of gold at random at the sites of the cubic face-centered cell.

adj **2** 無秩序の syn *arbitrary, indefinite*

▶ The phenomenon of diffusion that is observed in pure liquids and gases is a direct consequence of the ordinary random motion of the particles.

random copolymer n ランダム共重合体

▶ If more than one type of monomer is present, linear copolymers are formed which may either be random copolymers or block copolymers.

randomization n 無作為化

▶ As the pressure is raised, the time period during which dissociation can occur becomes smaller and smaller, and eventually energy randomization cannot occur.

randomize vt 無作為化する

▶ Collisions with the other molecules of the sample ensure that the accelerated motion of one molecule is rapidly randomized in direction and shared between all the other molecules.

randomly adv 無秩序に syn *at random*

▶ Eventually, the oxidation of FeO leads to γ-Fe_2O_3 in which all the cations are Fe^{III} which are randomly distributed between octahedral and tetrahedral sites.

randomness n 無秩序

▶ The randomness of structure usually causes the density of a liquid to be somewhat less than that of the corresponding crystal.

random walk n ランダム歩行

▶ The molecules are assumed to perform random walks through the solution.

range n **1** 変動の範囲 syn *extent, scope, span*

▶ The purity of as-received organic materials is usually in the range 95-99 percent.

n **2** 一連 range of に続く名詞は複数形、動詞は全体を強調するときは単数形、個々を強調するときは複数形を用いる. syn *series, variety*

▶ A range of Pt and Pd complexes exists with structures analogous to Magnus' green salt.

vi **3** 及ぶ syn *extend, fluctuate, spread, vary*

▷ It is possible to prepare stable polymerized vesicles ranging in diameters between 300 and 3000 Å.

rank n **1** 等級 syn *class, level, status*

▶ For anions and polyatomic cations, it is possible to obtain rank orderings of hardness but not absolute values.

vi **2** …に位する syn *stand*

▶ The investigation of the nature of acids and bases ranks as one of the earliest preoccupation of chemistry.

ranking n 順位 syn *class, grade, rank*

▶ Although classification of a reagent as hard or soft is fairly straightforward as a function of charge and position in the periodic table, the relative ranking on the hardness or softness scale is often not obvious from these considerations.

rapid adj 速い syn *fast*

▶ The migration of the reaction front in the long direction of the crystal is more rapid than that perpendicular to the long axis.

rapid-flow method n ラピッドフロー法
▶ The rapid-flow methods are limited by the speed with which it is possible to mix solutions.

rapidity n 急速 syn *speed*
▶ The rapidity of growth often differs for different crystal planes, and the slowest growing faces dominate the appearance of the crystal.

rapidly adv 速やかに syn *quickly, speedily*
▶ As would be expected, the inductive effect falls off rapidly with increasing distance of the substituent from the carboxyl group.

rare adj まれな syn *exceptional, scarce, uncommon, unusual*
▶ Ruthenium is extremely rare, being only slightly more abundant than natural technetium.

rare element n 希元素
▶ Because pegmatites represent the end product of the crystallization of a magma, rare elements become concentrated and the conditions are right for further differentiation and growth of large crystals.

rare gas n 希ガス syn *inert gas, noble gas*
▶ The principal characteristics of the heavier rare gases are spherical shape and an inverse sixth power attractive potential.

rarely adv めったに…しない syn *hardly, infrequently, scarcely, seldom*
▶ In practice, reaction rarely reaches completion in a mixture of powdered solids, because a certain proportion of the solid particles are not in contact with another reactant.

rarity n 希少さ syn *scarcity*
▶ Direct recovery of Se and Te is not usually economically feasible because of their rarity.

rate n **1** 速度 syn *pace, speed, velocity*
▶ The number of oxidizing agents available to the organic chemist is growing at a tremendous rate.

2 割合 syn *proportion, scale*
▶ The initial heat of swelling is at the rate of hundreds of joules per gram, soon falling off toward an immeasurably small value.

3 at any rate 少なくとも syn *anyway*
▶ Most of the molecules would be so oriented that their ϕ's are equal to, or at any rate near, this favored value.

rate constant n 速度定数
▶ The rate constant increases with increasing pressure if there is a volume decrease on forming the transition state.

rate-determining step n 律速段階 syn *rate-limiting step*
▶ The rate-determining step in the precipitation of barium sulfate is the formation of a cluster of four barium ions and four sulfate ions.

rate law n 反応速度式 syn *rate equation*
▶ A rate law is the empirically determined relation between the rate of a reaction and the concentrations of the species that occur in the overall chemical reaction.

rate of conversion n 変換速度
▶ The reason why the enol can be isolated, even though it is apparently the less stable form, is of course that the rate of its conversion into the keto form is exceptionally small.

rate of effusion n 流出速度
▶ The rate of effusion of a gas through a small hole is inversely proportional to the square root of its molecular weight.

rate of flow n 移動率
▶ When a liquid is forced by electroosmosis through the pores of a membrane, the rate of flow is determined by two opposing factors: the force of electroosmosis on the one hand and the frictional force between the moving liquid layer and the wall on the other.

rate of shear n ずり速度
▶ The apparent viscosity at very low rates of shear is far greater that at very high rates of shear, the difference being attributed to effects of structural viscosity.

rather adv **1** かなり syn *quite, somewhat, very*
▶ The absorption spectrum of CsI consists of a series of rather sharp peaks.

2 それどころか syn *alternatively, instead, preferably*
▶ Energy can be transformed from one form to

another, but it cannot be created or destroyed; rather, energy is conserved.

3 rather than むしろ syn *instead of*

▶ In the acid-catalyzed bromination of ketones the enol, rather than the enolate, undergoes the bromination.

rating n 格付け syn *grade, rank*

▶ Mohs scale of the hardness of mineral or mineral-like materials consists of 10 values, ranging from talc, with a rating of 1, to diamond, with a rating of 10, the rating being based on the ability of each mineral to scratch the one directly below it in the series.

ratio n 割合 syn *proportion*

▶ Ethyl acetoacetate ordinarily exists at room temperature as an equilibrium mixture of keto and enol tautomers in the ratio of 92.5 to 7.5.

rational adj **1** 合理的な syn *logical, reasonable*

▶ We suggest a systematic classification of hybrid frameworks that places new and existing materials in a simple rational context.

2 有理の

▶ If a reference face makes intercepts *a*, *b*, and *c* on three axes, then the intercepts made by any other face must be in the proportion of rational multiples of these intercepts.

rationale n 理論的解釈 syn *explanation, reason*

▶ It is important to realize that the above rationale was established after the fact.

rationality n 合理性 syn *reasonableness*

▶ Rationality means no more than that explanations have to be commonly reasonable in the circumstances.

rationalization n 正当化 syn *explanation, reasoning*

▶ Any kind of a theoretical prediction or rationalization of the rate of reactions must inevitably take into account the details of how the reactants are converted to the products, in other words, the reaction mechanism.

rationalize vt 合理的に説明する syn *account for, explain, justify*

▶ The activating and ortho, para-orienting influence of alkyl substituents can be rationalized on the basis of inductive effects.

rationally adv 合理的に syn *practically, realistically, reasonabley, sensibly*

▶ Thermodynamics, the study of the transformations of energy, enables us to discuss all these matters rationally.

rattle vi ガタガタする syn *vibrate*

▶ A situation in which a cation may rattle inside its anion polyhedron is assumed to be unstable.

raw adj 生の syn *crude, unrefined*

▶ The raw energy value computed in a molecular mechanics calculation may be designated as the steric energy.

raw material n 原料

▶ Bauxite is the raw material for aluminum.

ray n 光線 syn *beam*

▶ In a beam of unpolarized light, each ray is vibrating normal to the propagation direction, and any orientation of the vibration plane around the beam axis is possible.

reach n **1** 届く範囲 syn *range, scope*

▶ Chemical conversions and electron-transfer processes passing through cationic intermediates could not be exploited with C_{60} and C_{70} but are now in reach with the larger carbon spheres.

vt **2** 達する syn *accomplish, achieve, attain*

▷ When reaching a certain concentration, the osmotic coefficient falls off sharply, proving the disappearance of ions to form colloidal particles or micelles.

react vi **1 react with** 化学的に…と反応する

▶ Lithium metal does not react at an appreciable rate with hydrogen gas at room temperature; the complete reaction requires temperatures in excess of 700 ℃.

2 刺激などに反応する syn *respond*

▶ The human body reacts immediately whenever it is subjected to the introduction of any foreign substance, including larger polypeptides.

reactant n 反応物

▶ If the equilibrium constant *K* were greater

than 1, then on mixing together equal volumes of each of the reactants, the reaction would proceed to the right.

reaction n 反応

▶ In the few studies of reactions of single crystals, the observable change has been found to begin at one or more nucleation sites and to spread through the crystal.

reaction center n 反応中心

▶ The chlorophyll pigment acts as a natural molecular antenna in photosynthesis by absorbing light and transferring energy to the reaction center.

reaction coordinate n 反応座標

▶ The reaction coordinate can be any desired measure of the progress of the reaction.

reaction dynamics n 反応動力学

▶ Reaction dynamics deals with the intermolecular motions that occur and with the relationships between the quantum states of the reactant and product molecules.

reaction intermediate n 反応中間体 syn *intermediate, intermediate product*

▶ Carbenes, carbonium ions, carbanions, and free radicals are the four most important classes of organic reaction intermediates containing carbon in an unstable valence state.

reaction kinetics n 反応速度論 syn *chemical kinetics*

▶ Radicals are highly reactive intermediates with fast reaction kinetics yet can be generated under mild conditions without the need for strongly acidic or basic conditions.

reaction mechanism n 反応機構

▶ The study of the kinetics of single-phase reactions led to a considerable understanding of the details of reaction mechanisms.

reaction path n 反応経路

▶ The reaction path is represented by the dashed line shown in the contour diagram, and for energetic reasons the majority of the reaction systems will follow this path.

reaction product n 反応生成物

▶ Corrosion is strongly influenced by the rate of supply of reagents to the interface and the rate of removal of reaction products.

reaction rate n 反応速度 syn *rate of reaction*

▶ To understand and model detonation processes, we must be able to assess what will be the effects of large increases in pressure on these reaction rates.

reaction time n 反応時間

▶ Transmission electron microscopy analysis of an aliquot of the solution collected after a reaction time of 6 h at 25 ℃ revealed the presence of 2-nm rhodium nanoparticles.

reactivate vt 再活性化する

▶ Glycogen synthase is reactivated by protein phosphatase which catalyzes the hydrolysis of the phosphate groups from the enzyme.

reactive adj 反応性の

▶ Iron is highly reactive chemically, oxidizes readily in moist air, and reacts with steam, when hot, to yield hydrogen and iron oxides.

reactive oxygen species n 活性酸素種

▶ Uncapped or inadequately capped quantum dots are not very stable and can produce reactive oxygen species.

reactive scattering n 反応性散乱

▶ The angular pattern of the reactive scattering is sensitive to the reaction mechanism.

reactive site n 反応点

▶ Weak-acid cation-exchange resins have weak fixed reactive sites such as the carboxylic group.

reactivity n 反応性

▶ Reactivity and orientation in electrophilic aromatic substitution are determined by the rates of formation of the intermediate carbocations concerned.

reactor n 反応器

▶ In a stirred-flow reactor, in which the concentrations are maintained constant within the reactor, it is not necessary to consider a thin slab; instead, we consider the reactor as a whole.

read vt **1** 誤植を…と読み替える syn *understand*

▶ Page 108, last line, for 'function' read 'functions.'

2 read into 推断する syn *assume, conclude*

▶ This diagram is only schematic, and no meaning should be read into the finer details of line shapes.

readily adv 容易に syn *easily, smoothly*

▶ Azoxybenzene is readily prepared by reduction of nitrobenzene in an alkaline medium with a variety of mild reducing agents.

readiness n 容易さ syn *ease*

▶ The importance of ethylene oxide lies in its readiness to form other important compounds.

reading n 読み syn *indication, measure*

▶ Alkali errors are well known and lead to unreliable pH readings in test solutions under alkali conditions, pH > 9, in the presence of significant amounts of sodium ions.

read-out n 読み出し syn *display*

▶ Quenching the photoluminescence could serve as a visible wavelength read-out for sensing of chemical vapors or gaseous molecules in general.

ready adj すぐ間に合う syn *immediate, prompt, quick*

▶ Despite its ready availability in the atmosphere, nitrogen is relatively unabundant in the crustal rocks and soils of the earth.

reagent n 試薬

▶ Since the heat of a reaction depends on whether a reagent is solid, liquid, or gas, it is necessary to specify the state of the reagents.

real adj **1** 真の syn *genuine, true, valid*

▶ The lack of a firm descriptive base in a great many systems means that synthesis is often an exploration in the real sense of the word.

2 現実の syn *actual*

▶ All real materials possess defects or imperfections in the regular repeat pattern of the crystal.

realistic adj 現実的な syn *practical, pragmatic*

▶ The basic problem for all interfaces is to get a realistic model of the chemical composition and molecular structure without which one is stuck with formal thermodynamics.

realistically adv 現実に syn *practically*

▶ Realistically we can probably only expect an increase of one or two orders of magnitude in the current if we use multilayer enzyme coverage on the electrode.

reality n **1** 実体 syn *actuality, fact, truth*

▶ Just as the observed geometry of a molecule corresponds to a local energy minimum, then also the calculated geometry should correspond to a calculated energy minimum, otherwise the energy surface is seriously distorted from reality.

2 in reality 実は syn *actually, really*

▶ This catalyst called solid phosphoric acid is, in reality, liquid phosphoric acid, present as a liquid layer on a high-area silica or silica–alumina gel.

realizable adj 実現できる syn *feasible, possible*

▶ The simplest physically realizable model of an ionic bond is that of two oppositely charged ions in contact, each necessarily polarized to a slight extent by the electric field of the other.

realization n **1** 認識 syn *conception, recognition, understanding*

▶ Realization of the importance of surface polarity and hydrophobicity on the adsorption properties of a material has led to the development of a number of methods to estimate and quantify these properties.

2 実現 syn *achievement, establishment*

▶ One crucial development in the application of atomic force microscopy to relevant biological systems was the realization that biological samples could be analyzed at high resolution under solution.

realize vt **1** 理解する syn *appreciate, understand*

▶ It is important to realize that the solvent itself is an impurity in growth from solution and can be incorporated into the crystal.

2 実現する syn *accomplish, achieve, produce*

▶ Both endo- and exovesicular catalytic reactions have been realized in surfactants and polymerized surfactant vesicles.

really adv 実際に syn *actually, definitely, surely, truly*

▶ Any process that really occurs must be irreversible in the thermodynamic sense.

realm n 領域 syn *area, territory*

▶ The realm of chemistry on surfaces takes on new dimension when two substances are adsorbed on the same surface.

real solution n 実在溶液

▶ Real solutions are formed from liquids in which the A···A, A···B, and B···B interactions are different.

real-time adj 即時応答の

▶ Shock tubes, especially when combined with flash photolysis and real-time detection, provide a method of studying elementary reactions up to ca. 3000 K.

rearrange vi **1** 転位する

▶ Addition of an oxygen atom from hydrogen peroxide to a primary or secondary amine would be expected to lead to amine oxide-type intermediates, which then could rearrange to hydroxylamines.

vt **2** 再配列する syn *adjust, arrange, shift*

▶ To calculate the spacing between the diffraction planes in the copper metal, rearrange the Bragg equation for the unknown spacing d:

$$d = n\lambda/2\sin\theta$$

where $\theta = 5.64°$, $n = 1$, and $\lambda = 0.711$ Å; therefore, $d = 3.62$ Å.

rearrangement n **1** 再配列 syn *reorganization*

▶ The glassy phase is capable of internal atomic and molecular rearrangements prior to the onset of crystallization.

2 転位

▶ The process involves the rearrangement of an amine cyanate and is analogous to Wöhler's classical synthesis of urea from ammonium cyanate.

reason n **1** 理由 syn *ground, rationale*

▶ The preparation of aryl halides from diazonium salts is more important than direct halogenation for several reasons.

2 by reason of …のために syn *on account of, because of, owing to*

▶ By reason of the nonrecognition of the importance of the solid dissociation product for the definition of the dissociation pressure of a salt hydrate, many of the older determination lose much of their value.

vt **3** 推論する syn *conclude, deduce, think*

▶ From a knowledge of the similarity in the first ionization potentials of molecular oxygen and xenon, Bartlett reasoned, correctly, that the xenon compound, $XePtF_6$, should be stable.

reasonable adj 合理的な syn *logical, rational, sound*

▶ We now have a reasonable understanding of the main factors influencing reaction rates; however, important problems remain unsolved and attention is directed to many of them.

reasonableness n 合理性 syn *rationality*

▶ The stability of a structure can often be roughly estimated from its reasonableness.

reasonably adv 適度に syn *practically, pretty*

▶ Tertiary and aromatic nitroso compounds are reasonably stable substances, which, although usually blue or green in the gas phase or in dilute solution, may be isolated as colorless or yellow solids or liquids.

reasoning n 推論 syn *analysis, rationalization*

▶ We have pointed out that these results can be explained by steric reasoning.

reassociation n 再会合

▶ The inhibitory or stimulatory activity of the α subunit is blocked by the spontaneous hydrolysis of the bound guanosine triphosphate and its subsequent reassociation with the β and γ subunits.

recall vt 思い出す syn *remember*

▶ It is important to recall that the early $LiMn_2O_4$ spinels, although attractive in many ways, did not demonstrate sufficient capacity retention on cycling.

recapture vt 奪い返す syn *recover, regain*

▶ The empty cavities in activated molecular sieve crystals have a strong tendency to recapture the water molecules that have been driven off.

receive vt 1 受ける syn *experience, undergo*

▶ One of the first examples of a tautomeric compound to receive careful and detailed study was ethyl acetoacetate.

2 受け入れる syn *accept, acquire*

▷ α-Glucose and β-glucose, containing ca. 5% and 2% of the other anomer, respectively, were used as received.

receiver n 受器

▶ Equilibrations occur in all parts of the column, and the vapor that eventually passes into the receiver is highly enriched in the more volatile component.

recent adj 最近の syn *current*

▶ Much of the recent interest in magnetism at low temperatures arises from a more basic interest in the general phenomenon of phase transitions.

定冠詞を付して名詞に転用された例.

▶ Hypofluorous acid is the most recent of the halogen oxo acids to be prepared.

recently adv 最近に syn *lately, now*

▶ Recently, magnetic materials based on the garnet structure have begun to replace traditional ferromagnetic materials.

receptive adj 受容性のある syn *amenable, flexible*

▶ Most bacteria do not take up DNA easily unless they are treated to make them more receptive.

receptivity n 1 感受性 syn *sensitivity*

▶ ^{19}F NMR can be observed with high receptivity at a frequency fairly close to that for ^{1}H.

2 受容性

▶ In biological systems, such interactions are regarded as allosteric when binding at one active site induces conformational changes which alter the receptivity of a remote site.

receptor n レセプター

▶ These light receptors stimulate special proteins in the plasma membrane, and these, in turn, transduce the signal.

rechargeable adj 再充電可能な syn *renewable*

▶ The lithium battery can store more than twice the energy compared with conventional rechargeable batteries of the same size and mass.

recipe n 方法 syn *approach, method, plan, procedure*

▶ DNA provides the genetic recipe that permits cell division to produce identical cells.

reciprocal n 1 逆数

▶ The reciprocal of the bulk modulus is the compressibility.

adj 2 逆数の

▶ Abscissa units are reciprocal Å.

3 相互の syn *complementary, corrective, mutual*

▶ Spin–spin coupling is a reciprocal affair, and the effect of the secondary protons on the tertiary proton must be identical with the effect of the tertiary proton on the secondary protons.

reckon vt 考える syn *assume, imagine, suppose, think*

▶ On account of the mechanism of crystal growth, one must reckon that crystal growth faces are always heterogeneous.

reclaim vt 再生利用する syn *recover, restore*

▷ Among the materials widely reclaimed in industry are aluminum, steel, paper, rubber, glass, etc.

reclamation n 再生利用 syn *recovery, restoration*

▶ Reclamation of eutrophied lakes can best be effected by addition of soluble A^{III} salts to precipitate the phosphates.

recognition n 認識 syn *acceptance, awareness*

▶ Stereochemical recognition can be achieved by regulating the spatial disposition of functional groups across the matrix surface.

recognizable adj 認識できる syn *discernible, distinct, noticeable*

▶ In general, the absorption bands observed in the solution spectra are still recognizable in the solid-state spectra, but they undergo significant shifts in frequencies.

recognizably adv 認識できるほどに

▶ The encapsulated crystal has a structure recognizably related to that of the bulk material.

recognize vt **1** 識別する syn *distinguish, identify*

▷ In solution chemistry, one of the principal goals of research is that of clearly recognizing the factors that are involved and fitting the data into a quantitative theory.

2 認知する syn *accept, realize*

▶ The unique behavior of colloids is now recognized to be exhibited by particles in the size range of about 100 to 10,000 Å.

recoil vi 跳ね返る syn *jump back*

▷ When a molecule undergoes rapid bond dissociation after absorbing a pulse of polarized light, the directions along which the recoiling fragments separate will be correlated with the direction of polarization.

recoilless adj 無反跳の

▶ Under certain conditions of recoilless emission, all of the energy change in the nuclei is transmitted to the emitted γ-rays, and this gives rise to a highly monochromatic beam of radiation.

recombinant DNA adj 組み換え DNA

▶ The transfected bacteria can be grown to produce large amounts of the recombinant DNA and/or recombinant protein encoded by the DNA.

recombination n **1** 原子の再結合

▶ Many metals react with atomic hydrogen, in addition to acting as catalysts for its surface recombination to hydrogen molecules.

2 荷電体の再結合

▶ It is possible to prepare a diode laser from a semiconductor p-n junction by making the structure in the shape of a laser cavity and electrically stimulating the recombination of such large numbers of carriers that lasing action results.

3 組み換え

▶ Recombination represents the formation of new gene combinations resulting from the breakage and reunion of chromosomes.

recombine vi 再結合する

▶ The minority carriers can recombine with the majority carriers that are in abundant supply.

recommend vt 推奨する syn *advice, guide, propose, suggest*

▶ It is not advisable to attempt to compensate the effect of iron(III) sulfate by adding phosphoric acid, although this practice has been recommended.

reconcile vt 両立させる syn *bring together, unite*

▶ In describing a liquid, the problem is to reconcile the random motion with the presence of a high degree of local order.

reconsider vt 再考する syn *reexamine, review*

▶ The thermodynamics of aqueous sodium chloride solutions derived from electromotive force measurements has been reconsidered.

reconstruction n 改造 syn *reconversion*

▶ The semiconductor surface has undergone a systematic preparation, usually including anneal to generate specific reconstructions.

reconvert vt 転換する

▶ Energy is expended to reconvert the pyruvate back into oxaloacetate for further metabolism.

record n **1** 記録 syn *recording*

▶ If a reaction is rapid, it is necessary to employ analytical techniques that allow records to be obtained instantaneously.

2 on record 記録されて

▶ Other similar examples are on record, e.g., the large increase in the dipole moment of H_3C-CCl_2-Cl compared with that of H-CCl_2-Cl has been ascribed to hyperconjugation.

vt **3** 記録する syn *register*

▶ ^{1}H, ^{13}C, and ^{31}P NMR spectra were recorded on a Varian Mercury 400, a Bruker Avance 300, or a JEOL GX-270 spectrometer.

recorder n 記録計

▶ Conductivity is measured using a pair of platinum black electrodes, and the conductivity meter is connected directly to a chart recorder.

recording adj 記録する

▶ Recording infrared spectrometers with excellent resolution and reproducibility are commercially available and are widely used in organic research.

recourse n 依存 syn *access*

▶ The ΔH° value is obtained from calorimetric results at a single temperature and without recourse to the van't Hoff equation.

recover vt 回収する syn *reclaim, restore*

▶ The anilide could be recovered unchanged from boiling for 15 min with 10% aqueous acetone.

recoverability n 回収の可能性

▶ A significant amount of work has been carried out on such systems with a view to improving the stability of enzymes under a wider range of conditions and their recoverability.

recovery n 回収 syn *restoration*

▶ Recovery of uranium from the manufacture of phosphoric acid and other phosphate chemicals is expected to become an important source of this metal.

recruit vt 補充する syn *mobilize*

▶ For translation to be initiated, the ribosome must be recruited to mRNA and correctly positioned over the start codon.

recrystallization n 再結晶

▶ Recrystallization was effected from a minimum amount of acetonitrile.

recrystallize vt 再結晶する

▶ The solid was recrystallized from ether at -78 ℃, forming thin yellow-orange needles.

rectangular adj **1** 長方形の syn *four-sided*

▶ The d^8 ions, Ni^{2+}, Pd^{2+}, and Pt^{2+}, commonly have square planar or rectangular planar coordination in their compounds.

2 直交する syn *right-angled*

▶ All crystals which are referred to the isometric system have three equal rectangular axes.

rectification n 整流

▶ The form of the current-voltage curve and even the direction of rectification depend on the nature of the contact.

rectify vt **1** 調整する syn *correct, improve, revise*

▶ Model systems have a place in our attempts to understand how transmembrane ion transport occurs, how it sometimes impaired, and how impairment of function may be rectified.

2 整流する

▷ The nature of the electrode contact and grain boundary resistance should be investigated closely with respect to any rectifying and non-ohmic characteristics.

3 精留する

▷ Column efficiency is improved by insulating the column and by introducing a small condenser at the very top of the column to effect partial condensation of the already highly rectified vapor.

recur vi 繰り返される syn *repeat*

▶ Certain metal ions form well-defined and robust clusters that recur in many hybrid materials.

recurrence n 再現 syn *repetition*

▶ While recognizing the recurrence of certain bond lengths, we must at the same time accept the fact that a given pair of atoms may apparently enter into more than one type of bonding arrangement.

recurrent adj 周期的に起こる syn *periodic, regular, repeated*

▶ Polytype can be rationalized in terms of recurrent stacking sequences.

recurring adj 繰り返し発生する syn *regular, repeated*

▶ A polymer is defined as a long-chain molecule with recurring structural unit.

recyclable adj 再生利用できる

▶ The catalyst is only recyclable with addition of pyridine.

recyclability n 再生利用できること

▶ Recyclability over several runs is quite impressive; the enantioselectivity and conversion drop off to some extent, although this cannot be due to leaching as the reaction supernatants are inactive.

recycle n **1** 再生利用

▶ The catalyst can be easily recovered and reused over eight reaction recycles with little loss in activity.

vt **2** 再生利用する syn *cycle, recur, return, rotate*

▶ In the petroleum refining industry, some of the product stream may be recycled and blended with the fresh input materials to obtain a product of maximum value.

red blood cell n 赤血球

▶ Vitamin B_{12} is a coenzyme in a number of biochemical processes; the most important of which is the formation of red blood cells.

reddish adj 赤みがかった

▶ The reddish color of copper is produced by Cu_2O, which forms easily in air, and a further dark hue comes from black CuO.

red heat n 赤熱

▶ At a dull red heat, Fe_2O_3 is reduced to metal, but in such cases black, pyrophoric iron is formed which cannot be exposed to the air. By heating to a bright red heat, however, the iron becomes gray and is no longer pyrophoric.

redifferentiate vi 再分化する

▶ Callus cells have the capability to redifferentiate into shoots and roots, and a whole flowering plant can be produced at the site of the injury.

redispersible adj 再分散できる

▶ Most of the sols dehydrate to hard, glassy, redispersible gels simply by evaporation in air at room temperature.

redissolve vt 再び溶かす

▶ Sufficient addition of gelatin to the colloidal gold enabled the latter to be evaporated and then redissolved in water.

redistill vt 再蒸留する

▶ When the total water-insoluble layer is separated, dried, and redistilled through the dried column, the chaser again drives the cyclohexene out of the column.

redistribution n 再分布

▶ Infrared absorption occurs on a long time scale in comparison to intramolecular energy redistribution, and so fragmentation can occur in bonds other than those targeted for dissociation.

redness n 赤熱状態 syn *burning*

▶ In attempting to reduce Fe_2O_3 to iron by means of hydrogen, it is very important to heat the oxide to bright redness.

redox adj 酸化還元の syn *oxidation-reduction*

▶ The great majority of redox enzymes or coenzymes such as NADH and NADPH do not undergo rapid, direct electrochemical reaction at electrode surfaces.

redox cell n 酸化還元電池 syn *oxidation-reduction cell*

▶ There is an important class of oxidation-reduction cells, known a redox cell, in which both the oxidized and reduced species are in solution.

redox potential n レドックス電位 syn *oxidation-reduction potential*

▶ Depending upon the ligands present, the redox potential of a given cytochrome can be tailored to meet the specific need in the electron-transfer scheme.

redox system n レドックス系 syn *oxidation-reduction system*

▶ To carry out a titration of an oxidizing agent against a reducing agent, one measures the potential of the redox system, which varies sharply at the end point.

redraw vt 描き直す

▶ A face-centered tetragonal cell can be redrawn as a body-centered tetragonal cell; the symmetry is still tetragonal but the volume is halved.

redress vt 取り戻す syn *adjust, improve, satisfy*

▶ There are many other important solvents, and it is convenient to discuss them generally at this point to redress the balance.

red shift n レッドシフト ☞ blue shift

▶ Ample evidence exists to show that applied pressures of a few thousand atmospheres cause large red shifts and intensity enhancements in the charge-transfer bands of molecular complexes.

red-shift vt レッドシフトさせる

▷ For head-to-tail dimers, the transition to the higher-energy excited state is forbidden, and the spectrum shows a single band red-shifted with respect to the monomer.

reduce vt **1** 減らす syn *decrease, diminish, lessen*

▷ Crystals were obtained by reducing the solvent volume by a third and storing the flask under a nitrogen atmosphere in the fridge.

2 還元する

▶ Manganese dioxide is reduced by concentrated hydrochloric acid, giving chlorine, and by concentrated sulfuric acid, liberating oxygen.

3 簡単な形にする syn *convert, modify*

▶ The complicated concentration dependence of this equation cannot be reduced to a simpler expression.

reduced adj **1** 還元された

▶ Partially reduced ring compounds are often referred to as dihydro or tetrahydro derivatives of the parent unsaturated compound.

2 換算した

▶ The most convenient empirical quantity is the reduced vapor pressure at a point well removed from the critical point.

reduced mass n 換算質量

▶ Because of the inverse dependence of vibration frequencies on the square roots of reduced masses of molecules, isotopically substituted molecules have different vibrational spectra.

reduced pressure n **1** 減圧

▶ Filtration, followed by removal of solvents under reduced pressure, afforded an oily orange solid.

2 換算圧力

▶ As an illustration, two gases having the same reduced temperature and reduced pressure are in corresponding states and should occupy the same reduced volume.

reduced temperature n 換算温度

▶ For a simple fluid, e.g., A, Kr, Xe, CH_4, the reduced vapor pressure is almost precisely 0.1 at a reduced temperature of 0.7.

reducible adj 還元可能な

▶ In polarography, the observed current is usually governed by the rate of diffusion of the reducible substance to the cathode.

reducing agent n 還元剤 syn *reductant*

▶ Hydrogen sulfide is in general a good reducing agent, being itself oxidized to sulfur, sulfur dioxide, or even sulfuric acid.

reducing equivant n 還元等量

▶ Biologically, nitrogen fixation is accomplished by the nitogenase family of enzymes at ambient temperature and pressure, but this process requires 16 reducing equivalents of ATP.

reducing flame n 還元炎

▶ The borax bead which has been colored pale blue by copper becomes a transparent ruby-red in the reducing flame if a trace of tin is added.

reducing power n 還元力

▶ In neutral or alkaline solution, the reducing power of some of the usual reducing agents toward nitrobenzene is less than in acid solution.

reduction n **1** 減少 syn *decrease,*

▶ The reduction in size to nanostructured materials has resulted in dramatic changes in properties.

2 還元

▶ Tetrahydrofuran has high solvent power and is a useful solvent for reductions with lithium aluminum hydride that proceed slowly in ether.

reduction potential n 還元電位

▶ For most of these transformations, differences in reduction potentials for various intermediates make cyclic voltammetry a particularly convenient method for monitoring the progress of the reaction.

reduction wave n 還元波

▶ No electrochemical reduction of tetrathionaphthacene was observed in dichloromethane, but reduction waves were observed when benzonitrile was employed as solvent.

reductive adj 還元的な

▶ Alternative wet routes to hydrolytically stable halides are metathetical precipitation and reductive precipitation reactions.

reductively adv 還元で

▶ It is fortunate that unstrained cyclic ethers are not reductively cleaved except under very forcing conditions.

redundancy n 不必要な重複 syn *excess*

▶ Cross-referencing is employed to avoid unnecessary redundancy.

redundant adj 重複した syn *superfluous, unnecessary*

▶ It is necessary to supplement the algorithm with a procedure to check for and eliminate redundant structures.

re-emission n 再発光

▶ If the potential energy curves for the ground and excited states do not overlap, then the electron may revert to the ground state only by re-emission of radiation.

re-emit vt 再発光する

▶ If a substance has absorbed radiant energy and does not re-emit this as fluorescence, then it follows that the photochemical energy has been dissipated by some other means.

reenter vt 再び入る

▶ Protons are pumped out of the matrix during the operation of the electron-transport chain, and they reenter the matrix driving ATP synthesis.

reentrant adj **1** 凹入の

▶ Any preferred evaporation or dissolution near the dislocation terminus would produce a small reentrant dimple.

2 液晶について，リエントラント

▶ The reentrant behavior is often associated with the tendency of polar molecules to associate.

reesterfication n 再エステル化

▶ Reesterfication requires glycerol 3-phosphate that is synthesized via glycerol kinase, but most peripheral tissues do not contain this enzyme.

reesterify vt 再エステル化する

▶ The activated fatty acids are either oxidized after transport into the mitochondria or reesterified with glycerol to give triacylglycerols that are stored in lipid droplets in the cytoplasm.

re-examine vt 再吟味する syn *reconsider, review*

▶ Since our experimental conditions coincide quite closely with those of Ritter and Simons, their data have been re-examined.

refer vi **refer to** (a) ことを指す 受身も可 syn *denote, mean, signify*

▶ Indeed, the very labels ionic and covalent refer only to idealized situations.

(b) 呼ぶ 受身も可 syn *name, quote, specify*

▶ Many carbohydrates have a sweet taste, so they are often referred to as sugars.

(c) 参照する 受身も可 syn *make reference to*

▶ Since only differences are significant, one electrode may arbitrarily be assigned the value zero and all others referred to it.

reference n **1** 参照 syn *connection, relation, specification*

▶ The behavior of copper toward acids can be understood by reference to the electromotive series.

2 with reference to (prep) …に関して syn *concerning, regarding*

▶ The emf determination of equilibrium constants may be illustrated with reference to the measurements of the dissociation constant of an electrolyte such as acetic acid.

3 参考文献

▶ Details of these magnetic measurements and their interpretation are given in the references cited.

adj **4** 基準となる

▶ Despite the fact that absolute energies and enthalpies cannot be determined, it is desirable for purposes of concise tabulation to establish an arbitrary reference scale.

vt **5** 参照する syn *make reference to, refer to*

▶ Proton NMR spectra were recorded on a Brucker AC 200 spectrometer and referenced to the residual protons of the solvents; shifts are given with respect to Me_4Si.

reference electrode n 参照電極

▶ An accurate and stable reference electrode is essential, and this is not easy to fabricate as a disposable structure.

reference point n **1 目印** syn *point of reference*

▶ The presence of a few heavy atoms in a macromolecule allows these reference points to be located, and these permit a description of the remainder of the molecule to be attempted.

2 参照点

▶ The standard state is an important concept because it provides a reference point upon which to establish all numerical values of thermodynamic quantities.

refine vt **1 精製する** syn *decontaminate, purify*

▶ Copper is refined electrolytically, the silver and gold present separating as an anode sludge.

2 改良する syn *improve, perfect*

▶ Coordinates and anisotropic thermal parameters of all non-hydrogen atoms were refined.

refined adj **1 精巧な** syn *exact, precise, sophisticated*

▶ To provide the refined data required for such an investigation, a new method for the precise determination of vapor densities has been devised.

2 精錬した syn *pure, purified*

▶ Whereas traditional ceramics are prepared from naturally occurring minerals, advanced ceramics are either derived from highly refined minerals or made by chemical synthesis.

refinement n **改良** syn *improvement*

▶ As the structure refinement proceeds, the quality of the electron density map improves.

refining n **精製** syn *purification*

▶ Aromatic hydrocarbons and alkenes have much higher solubilities in liquid SO_2 than paraffins, a distinction which is of great practical value in the refining of petroleum.

reflect vt **反映する** syn *demonstrate, illustrate, show*

▷ Synthetic methods for the preparation of materials are extremely varied reflecting the importance not only of composition and structure but also morphology, defects, and particle size distribution.

reflectance n **反射率** syn *reflectivity* ☞ diffuse−, specular−

▶ Reflectance spectra of pellets of the sodium tungsten bronzes were recorded by Brown in which more than 95% of the incident light was absorbed.

reflected light n **反射光**

▶ If natural light is incident on a metal surface, the reflected light is partly but never completely polarized.

reflection n **1 反映** syn *evidence, result, sign*

▶ The values of melting points and boiling points depend upon the strength of intermolecular forces and are thus often a reflection of the solid structures adopted by the halides.

2 反射 ☞ selective−, total−

▶ When the electric vector of the incident light is parallel to the chain direction, we find typical metallic reflection which can be described by the Drude theory.

3 鏡映

▶ The π orbitals are all antisymmetric with respect to reflection in the *xy* plane.

reflective adj **反射性の**

▶ Doped indium and tin oxides have interesting property that they are transparent to visible light, although highly reflective in the IR.

reflux n **1 還流**

▶ When 2.5 g of picryl ether **1** was heated in 50 mL of freshly distilled benzene for 6 hr under reflux, the pale yellow solution became deep orange and some crystallization occurred.

vt **2 還流させる**

▶ The reaction mixture was left to reflux for 3 h.

reflux condenser n **還流冷却器**

▶ The acetone is boiled and the hot condensate from the reflux condenser flows back through the porous thimble over the solid barium hydroxide contained therein and comes to equilibrium with diacetone alcohol.

refolding n **再生**

▶ Several proteins and enzymes, when completely unfolded by urea and with disulfide bridges reduced, are capable of refolding to give the active, native state on removal of the urea.

reform vi 再生する syn *rebuild, reorganize*

▶ As ZnS reacts with gaseous iodine, the gases, which are at equilibrium at 900 ℃, diffuse down the tube to cooler end, where the equilibrium conditions are different, and ZnS reforms, generally as single crystals.

reformation n 再構成 syn *conversion, transformation*

▶ Brown iodine solutions frequently turn violet on heating and brown again on cooling due to the ready dissociation and reformation of the complex.

reforming n 改質

▶ Decomposition of hydrocarbon gases or low-octane petroleum fractions by heat and pressure, either without a catalyst or with a specific catalyst, is called reforming.

reformulation n 再公式化

▶ In several cases, reformulation in terms of only moderately delocalized orbitals is helpful.

refract vt 屈折させる

▶ When a ray of homogeneous light passes from a vacuum into an isotropic homogeneous medium, unless the ray is incident normally on the surface of separation, it is refracted toward the normal in the medium.

refraction n 屈折 syn *refringence*

▶ The ratio of the sine of the angle of incidence to the sine of the angle of refraction is the index of refraction of the second medium.

refractive adj 屈折する

▶ Nearly all organic crystals are doubly refractive, although a few are isotropic and do not give a lighted image.

refractive index n 屈折率 syn *index of refraction*

▶ In diamond, the refractive index is appreciably larger for violet light than it is for red.

refractory n **1** 耐火物

▶ Transitions such as α- ⇄ β-quartz or quartz ⇄ cristobalite have a deleterious effect on silica refractories because volume changes associated with each transition reduce the mechanical strength of the refractory.

adj **2** 耐熱性の syn *heat-resistant*

▶ Boron is an extremely hard refractory solid of high melting point, low density, and very low electric conductivity.

refrigerant n 冷媒 ☞ coolant

▶ For studies above 90 K, liquid nitrogen is used as refrigerant.

refueling n 燃料補給

▶ Compressed hydrogen lacks the necessary volumetric capacity to drive 500 km without refueling.

regain vt 回復する syn *recover, restore*

▶ A substance is said to be elastic if it spontaneously regains its former shape after being deformed.

regard n **1** 考慮されるべき事項　通常は単数形 syn *aspect, matter, point*

▶ In such a regard, selective inhibitors can be invaluable for modulating effects in cells or in vivo.

2 with regard to …に関しては syn *about, concerning, regarding*

▶ With regard to the center of the molecule, there are two possible ways of finding the dimer, with Ru(II) on one side or on the other.

vt **3** みなす syn *consider, imagine*

▶ The solid hydrates probably are more properly regarded as clathrate compounds.

4 as regards (prep) …の点では syn *about, concerning, regarding*

▶ In the description of the form of a crystal, especially as regards the position of its faces, it is convenient to assume certain axes of reference, passing through the center of the ideal form.

regarding prep 関心の的に関して syn *about, concerning*

▶ Apart from the selection rules, the analysis of Raman spectra is very similar to that of infrared spectra, and the same considerations apply regarding information as to symmetry and dimensions.

regardless adj **regardless of** (prep) …にかかわらず syn *despite, in spite of*

▶ Dibenzo-18-crown-6 forms 1:1 compounds with salts of many metals regardless of the oxidation state of the metal or of the number of anions in the salt.

regenerate vt 再生する syn *recover, renew, restore*

▶ Molecular sieves are regenerated by heat removal of the water, followed by treatment with an inert gas.

regeneration n 再生 syn *restoration, resurgence*

▶ The cholesterol dibromide that crystallizes from the reaction solution is collected, washed free of the companion substances or their dehydrogenation products, and debrominated with zinc dust, with regeneration of cholesterol in pure form.

region n 区域 syn *area, part, zone*

▶ Opposite spin permits two electrons to occupy the same region.

regioselective adj 位置選択的な

▶ Oxymercuration–demercuration is highly regioselective and gives alcohols corresponding to Markovnikov addition of water to the carbon–carbon double bond.

register vt **1** 示す syn *display, manifest, show*

▶ If the cell registered a negative electromotive force, we would know that the tendency of the reaction is to proceed in the opposite direction.

2 記載する syn *note, record, report*

▶ Sometimes significant advances in materials science are registered, not as a result of inspired new preparations but rather as a consequence of elegant variation in the bulk processing of the material.

regrettable adj 残念な syn *unfortunate*

▶ It is regrettable that general usage names the trivalent planar ions of the CH_3^+ type as carbonium ions, for if the name is considered analogous to other onium ions, then it should relate to the highest valency state carbocation, that is the pentacoordinated cation of the CH_5^+ type.

regrettably adv 残念にも syn *sadly, unfortunately*

▶ We note the regrettably poor agreement between data of different investigations, a situation which occurs constantly in the literature.

regrinding n 再粉砕

▶ Regrinding the sample exposes unreacted portions of the reactants.

regular adj **1** 規則正しい syn *periodic, systematic, uniform*

▶ By disorder we mean any deviation from the regular arrangement of all-trans chains found in the low-temperature solid.

2 通例の syn *customary, normal, routine, usual*

▶ Regular consumption of fish and shellfish contaminated with mercury and methylmercury poses serious health threats to humans including birth defects and brain damage.

regularity n 規則性 syn *constancy, order, ordrliness*

▶ There appears to be little regularity in the magnitude of the effect of *o*-methyl groups upon nematic clearing points as the linkage group and molecular length-to-breadth ratio vary.

regularly adv 定期的に syn *consistently*

▶ Organic reactions in the solid state have been reported regularly since before the turn of the century.

regulate vt 調節する syn *adjust, control, modify*

▷ Many signaling pathways are controlled by regulating the degradation of key protein components in them.

regulation n 調節 syn *adjustment, control, modification*

▶ The regulation of the partial pressure of the solvent above the surface to insure controlled evaporation is done by means of one or more thicknesses of filter paper or other porous material placed over the crystallizer.

regulator n 制御因子

▶ The quantity of the sterol lipid cholesterol in animal membranes is a key regulator of membrane fluidity.

regulatory adj 調節する syn *regulative*

▶ The endogenous chemicals that are present everywhere in plants or animals and that exert regulatory actions are called hormones.

regulatory enzyme n 調節酵素

▶ Heterotropic effects are frequently observed with regulatory enzymes.

regulatory sequence n 調節配列

▶ Many genes that are tightly regulated by cells contain regulatory sequences called enhancers.

rehydrate vi 再水和する

▶ The two most striking general properties of the zeolites are their capacity for base exchange and the ability of the hydrated materials to lose water or rehydrate without any change in optical or crystallographic properties.

reign vi 支配する syn *dominate*

▶ The importance of the environmental atmosphere is well illustrated by the studies on Teflon, where confusion reigned for several years owing to oxygen effects.

reinforce vt 強化する syn *strengthen, support*

▶ Plastics can be reinforced, usually with glass or metallic fibers for added strength.

reinforcement n 強化 syn *augumentation, strengthening*

▶ A mixture of stabilizing agents may reveal their incompatibility, or conversely, it may result in a strong reinforcement of the stabilizing action of either material alone.

reinvestigation n 再調査

▶ The whole problems of line widths and exchange interactions in free radicals is worthy of careful reinvestigation.

reiterate vt 反復していう syn *repeat, restate*

▶ We can only reiterate the caution given previously against attaching too much significance to the crystal-field stabilization energy term in isolation.

reject n **1** きずもの syn *damaged or defective merchandise*

▶ If a particular gelation is unsatisfactory, the reject gel can be redispersed and recycled without significant loss.

vt **2** 排除する syn *dismiss, refuse, turn down*

▶ Bacteria store P, K, Mg, Fe and Cu amongst other elements, and they reject mainly Na, Ca and Si.

rejection n **1** 排除 syn *exclusion*

▶ Once cells had compartments, storage became much more complicated since pumping out of cytoplasm may no longer mean only rejection but also retention in special retained vesicle solutions.

2 拒絶 syn *refusal, turndown*

▶ The body of the receptor looks upon the skin or kidney cells of the donor as foreign matter and immediately sets up an antibody-type rejection mechanism.

relate vt 関係させる syn *associate, connect, correlate*

▶ Copper and gold are chemically closely related, have a similar electronic configuration, and have the same crystal structure.

related adj 関連した syn *allied, corelated*

▶ Separation of a single compound from a mixture of related substances is very time-consuming and frequently does not yield material of the required purity.

relation n **1** 関係 syn *association, connection, link, relationship*

▶ A relation between C_P and C_V is given by the thermodynamic expression

$$C_P = C_V + \alpha VT/\beta$$

where α is the temperature coefficient of expansion, β is the coefficient of compressibility, and V is the molar volume.

2 in relation to …と比較して syn *referring to*

▶ The results have been considered in their relation to other electromotive force measurements.

relationship n 関係 syn *relation*

▶ This is just one example of a general relationship: in a set of similar reactions, the less reactive the reagent, the more selective it is in its attack.

relative adj **1** 相対的な

▶ We may define relative stability as the difference in the standard Gibbs energy between the product and reactant molecules.

2 relative to 文法上，対等な関係にあるもの…と比較して syn *in reference to*

▶ The calculated energy at the col, relative to the energy of $H+H_2$, was 40.21 kJ mol^{-1}.

relative configuration n 相対配置 ☞ absolute configuration

▶ Most applications of stereochemistry are based upon the relative configurations of different compounds, not upon their absolute configurations.

relative error n 相対誤差

▶ For exact work in absolute weighing it is necessary to determine not only the relative errors of the weights among themselves but also the absolute error of the set by comparing one of the largest pieces with a standard weight.

relatively adv 比較的 syn *comparatively, more or less*

▶ Relatively large anions will be in contact with one another when packed around a small cation such as Li^+ in LiI.

relax vi **1** 緩和する syn *ease, moderate, relieve*

▶ In metals, the atoms surrounding the vacancy appears to relax inward by a few percent, i.e., the vacancy becomes smaller.

2 平衡状態に戻る

▶ The system is initially at equilibrium under a given set of conditions. These conditions are then suddenly changed; the system is no longer at equilibrium, and it relaxes to a new state of equilibrium.

relaxation n 緩和 syn *easing, moderation, relief*

▶ Relaxations can occur to several vibrational levels in the ground state; hence, the fluorescence can be detected at a different wavelength from excitation.

relaxation process n 緩和過程

▶ In contrast to the paramagnetic susceptibility measurement, which is equally sensitive to any type of domain size, the detectability of the ESR spectrum is dependent on relaxation processes.

relaxation time n 緩和時間

▶ If a labile system is perturbed from equilibrium by a small amount, it will approach equilibrium in a first-order process characterized by a relaxation time τ which is the time needed for the system to traverse a fraction $1/e$ of its path to equilibrium.

relaxed adj 緩和された syn *at ease, easy, free*

▶ A good elastomer should not undergo plastic flow in either the stretched or relaxed state.

relay adj 中継の

▶ Using NMR, we demonstrate a degree of conformational interaction between the Ar-CO and N-CO axes, suggesting their potential application in stereochemical relay systems.

release n **1** 放出 syn *liberation*

▶ The prolonged exposure of dry crystals of silver halides to light is accompanied by the release of halogen and the separation of silver.

vt **2** 放出する syn *liberate*

▶ These electro-optical devices are optimized to release as much of the energy as possible as photons.

relevance n 関連性 syn *bearing, connection*

▶ The properties of simple amides are particularly important in their relevance to the chemistry of peptides and proteins.

relevant adj 問題に関連した syn *associated, related*

▶ To convert the pairwise interactions into the heat of sublimation, we must sum over all relevant distances in the crystal.

reliability n 信頼度 syn *dependability, stability*

▶ The sensitivity and reliability of the ninhydrin test is such that 0.1 μM of amino acid gives a color intensity which is reproducible to a few percent.

reliability factor n *R* 因子 syn *R factor*

▶ A useful measure of the agreement between the observed and calculated structure factors is given by the reliability or *R* factor.

reliable adj 確実な syn *certain, sound, sure*
▶ The only commercial calcium compound with reliable composition is $CaCO_3$.
reliably adv 確実に syn *dependably*
▶ Charge-transfer absorption bands provide a valuable diagnostic tool for recognizing inner-sphere electron-transfer reactions and then for predicting reliably their reaction rates.
reliance n 依存 syn *dependence*
▶ The green chemistry credentials of such systems are complex; on one hand palladium can be catalytic at remarkably low levels, but sustainability is questionable given the scarcity of the precious metal and the reliance on supply from a small number of geographical regions.
relief n **1** 緩和 syn *easing, relaxation*
▶ The relief of strain by cracking appears to provide a catalytic effect on the rearrangement.
2 起伏 syn *contrast*
▶ Some phases etch away more quickly than others, and this gives relief to the initially flat surface.
relieve vt 軽減する syn *ease, moderate, reduce*
▶ Before making measurements, new pressed samples were cycled to above and below the transition temperature two or three times in order to relieve strains and improve their homogeneity.
reluctance n 抵抗 syn *unwillingness*
▶ The reluctance of benzene to undergo chemical transformations has long been understood in terms of its resonance stabilization.
reluctant adj …しにくい syn *opposed*
▶ Another marked difference between oxygen and sulfur is that sulfur, like most of the elements in the second and higher rows of the periodic table, is reluctant to form double bonds.
rely vi **rely on** …に頼る syn *depend on*
▶ Many processes in bioremediation rely on applications of microbial biotechnology.
remain n **1** 複数形で残骸 syn *residue*
▶ Petroleum is thought to be derived from the remains of tiny sea animals laid down in past geologic ages.
vi **2** 残る syn *be left*
▶ Obviously much more remains to be done, and experiments are in progress to identify the intermediates in the gas phase or in inert gas matrices.
3 とどまる syn *continue, persist, stay*
▶ If no appreciable ranges of solid solubility exist among the reactants and products of a solid-solid reaction, their thermodynamic activities remain constant throughout the course of the process.
4 …の手に帰する syn *stay*
▶ It remained for W. Gibbs to draw the correct conclusion that the work done in an electrochemical cell is equal to the decrease in what is now known as the Gibbs energy.
remainder n 残り syn *balance, remains, rest*
▶ Much of the remainder of this book will deal with galvanic cells.
remaining adj 残りの syn *left-over, unused*
▶ Except when the substituting ligands are very good π acids, substitution usually stops after two or three carbonyls have been replaced, because the remaining COs become saturated in their ability to withdraw electron density from the metal.
remark n 所見 syn *comment*
▶ A similar remark may be made with respect to the effect of pressure.
remarkable adj 著しい syn *extraordinary, outstanding, significant*
▶ The difference in chemical shift between the two diastereotopic benzylic protons of the sulfoxide is remarkable.
remarkably adv 非常に syn *exceedingly, really, very*
▶ Since microemulsions are self-organizing systems, they are remarkably easy to prepare.
remedy n **1** 救済手段 syn *prescription, treatment*
▶ Chemical scrubbers that prevent the emission of the pollutants offer one of the possible remedies.

vt **2 改善する** syn *improve, rectify, reform*

▶ Some workers remedied the situation by retaining the kinetic theory of gases but introducing into the preexponential factor a steric factor.

3 治療する syn *cure, restore*

▶ DNA damage, in which bases acquire bulky substituents or they are displaced from their normal hydrogen-bonded positions, is remedied by nucleotide excision repair.

remember vt **記憶する** syn *recall*

▶ Some typical solvents are arranged in an approximate order of dielectric constant at 25 ℃, but it should be remembered that dielectric constants vary with temperature.

remind vt **注意する** syn *cause to remember, put in mind of*

▶ It is well to be reminded that the actual stabilizing agent in the interface is not necessarily identical either chemically or physically with the emulsifying agent or agents that were added to the bulk liquids.

reminder n **注意** syn *note*

▶ The example is a salutary reminder of the influence of pH, solubility, and complex formation on the standard reduction potentials of many elements.

reminiscent adj **reminiscent of …を暗示する** syn *indicative, suggestive*

▶ Obviously, in contrast to mixtures of chiral liquids, large diastereomeric interactions reminiscent of those in the crystalline state can be expressed in monolayers, and they may be highly sensitive to temperature.

remote adj **遠く離れた** syn *distant, faraway*

▶ The acid-strengthening effect of an alkyl group may show up in positions remote from the alkyl substituent.

remote control n **遠隔操作**

▶ Irradiated nuclear fuel is one of the most complicated high-temperature systems found in modern industry, and it has the further disadvantage of being intensely radioactive so that it must be handled exclusively by remote control.

remote-sensing adj **遠隔探査の**

▶ Optical fibers are applicable to remote-sensing devices which permit analysis of samples at widely separated locations.

removable adj **除去できる** syn *separable*

▶ Since ethers have no reactive π system or removable proton, they lack reactivity and are useful solvents for organic reactions.

removal n **除去** syn *elimination, removing*

▶ In a molecule of acetaldehyde, CH_3CHO, the proton on the carbonyl carbon, being closer to the electronegative oxygen, will experience a greater electron removal by the oxygen than the three methyl protons which are at a greater distance .

remove vt **除去する** syn *separate, take out*

▶ The solvent was removed on a steam bath under reduced pressure, and the residue was recrystallized from ethyl acetate to give 0.45 g of TCNQ.

removed adj **隔たった** syn *distant, remote*

▶ The contribution of each C–C and C–H bond is supposed to be influenced not only by its nearest neighbor bonds but also by neighbors once, twice, and further removed.

renaissance n **復活** syn *restoration, resurgence, revival,*

▶ Zinc is now experiencing a dramatic renaissance in its chemistry due to the important roles that it plays in biological systems.

render vt **…にする** syn *cause, make*

▶ A crystal of KCl may be rendered nonstoichiometric with excess metal by heating in potassium vapor.

rendering n **表現** syn *presentation, version*

▶ When both Sb and Mn are present in alkaline earth halophosphates, energy absorbed by the defect associated with the Sb is shared between the Sb and Mn emitting centers with the result that white light of adequate color rendering is obtained.

renewable adj **再生可能な**

▶ Renewable energy sources are those that can be replenished on a predictable time basis known collectively as biomass.

renewed adj 復活した syn *fresh, restored*
▶ In recent years, there has been a renewed effort to identify high performance thermoelectric materials for applications in power generation and device cooling.

renowned adj 有名な syn *famous, well-known*
▶ The penicillin antibiotics are probably the most renowned sulfur-containing natural products.

reorganization n 再編成 syn *modification*
▶ Reorganization energy for going from an octacoordinate square antiprism or dodecahedron to nonacoordnate trigonal prism or monocapped square antiprism should be very small.

reorient vt 再配向させる
▷ Rotation in plastic crystals can be described adequately in terms of orientational disorder with the molecules vibrating about discrete, distinguishable positions in the crystalline lattice and reorienting with very high frequency.

reorientation n 再配向
▶ In water, the structure is less ideal, and the rate-determining step is now reorientation of the water molecules in the tertiary structure.

repair n **1** 修復 syn *restoration*
▶ Most bacterial cells have DNA repair mechanisms aimed at excising the UV-damaged DNA and replacing the excised segment with an undamaged one.
vt **2** 修復する syn *restore*
▶ Telomerase repairs telomere length at the ends of chromosomes by adding DNA nucleotides to cap the telomere after each round of cell division.

repair enzyme n 修復酵素
▶ These repair enzymes, in marked contrast to A family replicative enzymes, may require Watson–Crick hydrogen bonding to incorporate a nucleotide.

repeat n **1** 繰り返し syn *repetition*
▶ Defects in crystals are imperfections in the regular repeat pattern of the crystal and may be classified in terms of their dimensionality.
vt **2** 繰り返す syn *duplicate, reproduce*
▶ Another increment of hydrogen was then added, and the measurements was repeated.

repeated adj 繰り返された syn *repetitive*
▶ Substitution of hydrogen by halogen raises the viscosity by an amount which increases with the atomic weight of the halogen; repeated substitution of hydrogen by chlorine in the same molecule produces different effects.

repeating unit n 繰り返し単位 syn *repeat unit* ☞ monomer unit
▶ All polymeric substances are characterized by the presence in the molecule of a repeating structural unit.

repel vt 反発する syn *repulse*
▶ We know that molecules repel one another on close approach, that this repulsion arises very suddenly, and that it is a manifestation of the Pauli exclusion principle.

repertoire n 能力範囲 syn *repertory*
▶ More recent contributions expand the repertoire to include other potential advantages to explain the common evolutionary selection of multimeric proteins.

repertory n 情報などの集積 syn *collection, repertoire, reservoir*
▶ A versatile procedure for the direct conversion of ozonides to alcohols would be a potentially valuable addition to the synthetic repertory.

repetition n 繰り返し syn *duplication, repeat*
▶ Before repetition of a run on the identical sample, it was, of course, necessary to evacuate the sample again at 350° and to cool it in vacuo.

repetitious adj 繰り返す syn *repetitive*
▶ Up to this point, we have examined the orderly nature of crystals, their repetitious structure and their symmetry.

repetitive adj 繰り返す syn *continuous, periodic, persistent*
▶ A crystal is a solid composed of atoms arranged in an orderly repetitive array.

replace vt 取り替える syn *substitute*
▶ Aluminum can replace Si in the SiO_4 tetrahedra, requiring the addition of another

cation or the replacement of one by another of higher charge to maintain charge balance.

replaceable adj 取り替えられる

▶ The equivalent weight of an acid is determined by the number of replaceable hydrogen atoms in the acid molecule.

replacement n 置換 syn *substitution*

▶ Carboxylic acids react with phosphorus trichloride, phosphorus pentachloride, or thionyl chloride with replacement of OH by Cl to form acid chlorides.

replenish vt 補充する syn *fulfill, supply*

▶ Our store of carbohydrates is constantly replenished by the recombining, in plants, of carbon dioxide and water.

replenishment n 補充

▶ A static foam is one that has no mechanism of replenishment as it subsides.

replica n レプリカ syn *copy, duplicate, reproduction*

▶ Replicas taken from these preparations reveal to some extent the roof-like shapes of uncollapsed pyramids.

replicate vt 複製する syn *copy, duplicate, reproduce*

▶ Shifting the cube in all three mutually perpendicular directions, by the length of a side of the cube, will generate six identical unit cells, which can be further replicated to create the entire structure.

replication n 複製 syn *copy, duplicate*

▶ Replication requires a single copy to be made of the whole genome once in the life of the cell.

replicative adj 複製の

▶ The smallest analogue was much less efficiently replicated than the best, and larger analogues were poorly replicated, apparently owing to steric clashes in the replicative polymerases.

report n **1** 報告 syn *article, communication, record*

▶ A preliminary report of the structure of NbS_2Br_2 and NbS_2Cl_2 indicates nonacoordination for the niobium atoms.

vt **2** 報告する syn *mention, note, state*

▶ Although the measurements were made in different ranges of temperature, neither investigator reports evidence for the large variation of ΔH with temperature required to explain so great a difference.

3 report back 折り返し報告する

▶ Mostly, these approaches involve the use of spectroscopic probe molecules that are sensitive to their surroundings and can report back via some measurable change in their behavior.

reportedly adv 報告によれば

▶ Reportedly isostructural with La_2O_3 are La_2O_2S, Ce_2O_2S, and Pu_2O_2S with four oxygen and three sulfur atoms bonded to the metal atoms.

repository n 貯蔵所 syn *storehouse*

▶ The double helix of DNA is the repository of the hereditary information of the organism.

reprecipitate vt 再沈殿させる

▶ The precipitate is filtered off, dissolved, and reprecipitated a second and a third time.

reprecipitation n 再沈殿

▶ 4-Nitrobenzoic acid, after purification by reprecipitation from aqueous alkaline solution with dilute HCl, was crystallized from a hot solution in ethyl acetate by slow cooling to give plates with {001} as the prominent faces.

represent vt **1** 表す syn *illustrate, stand for*

▶ The small spheres represent Na^+ ions, and large spheres denote halide ions, Cl^-.

2 象徴する syn *depict, reflect, show*

▶ The development of polymerized vesicles has represented a major breakthrough in membrane mimetic chemistry.

representation n **1** 表現 syn *model, picture, reproduction*

▶ The quantitative representation of the volumetric behavior of fluids over both gas and liquid regions has proven to be an unusually difficult problem.

2 群の表現

▶ For a representation, the trace of each matrix is called a character.

representative n **1** 代表 syn *example, sam-*

ple

▶ The protonations of aromatic hydrocarbons are representative of equilibria of the carbonium ion type.

adj **2 代表的な** syn *characteristic, illustrative, typical*

▶ A representative, but by no means complete, selection of standard heats of formation is tabulated in Appendix.

repress vt **抑制する** syn *suppress*

▷ Addition of water to nitric acid at first diminishes its electric conductivity by repressing the autoprotolysis reactions.

repression n **抑制** syn *suppression*

▶ Metabolic pathways are controlled via enzyme repression and derepression that occur in response to variations of metabolite levels.

reprocessing n **再処理**

▶ Serious radiation hazards are involved in reprocessing of fission products that require use of appropriate shielding and remote-control handling procedures.

reproduce vi **1 繁殖する** syn *multiply, regenerate*

▶ Viruses can reproduce only within living cells.

vt **2 再現する** syn *match, simulate*

▶ Despite uncertainty as to the true value of net resonance energy in conjugated systems, several empirical schemes have been devised which reproduce standard heat of atomization of aromatic systems reasonably well.

reproducibility n **再現性**

▶ The reproducibility of the boiling point elevations was found to be ±0.0002°.

reproducible adj **再現性がある**

▶ The *d* spacings are reproducible from sample to sample unless impurities are present to form a solid solution or the material is in some stressed, disordered, or metastable condition.

reproducibly adv **再現性をもって**

▶ The synthesis reproducibly yielded only two surface-patterned shapes, that is, solid spheroids and hollow shells.

reproduction n **1 再現** syn *copying, duplication*

▶ Color reproduction is achieved by the formation of varying amounts of cyan, magenta, and yellow dyes in three separate superimposed layers sensitive to red, green and blue light, respectively.

2 複製 syn *copy, duplicate, replica*

▶ In biochemistry, replication refers to reproduction of the DNA molecule, which is composed of two interlocking chains of nucleotides.

repulsion n **反発** syn *resistance*

▶ The repulsion energy is inversely proportional to the *n*th power of the internuclear separation.

repulsive adj **1 反発する** syn *repellent*

▶ In interactions involving contacts between pairs of atoms on the peripheries of different molecules, the force is repulsive.

2 胸が悪くなるような syn *disagreeable, unpleasant*

▶ Hydrogen azide has a repulsive, intensely irritating odor and is a deadly poison.

repulsive force n **斥力**

▶ The bonding of close-packed ions to give a stable configuration implies an equilibrium in which attractive forces are balanced by repulsive forces.

require vt **必要とする** syn *desire, need, want*

▶ If there is to be extensive incorporation of molecules of one substance into a crystal of another, there must be similarity in molecular size and shape, but no particular relationship of functional groups is required.

required adj **必要な** syn *indispensable, necessary*

▶ The water required for steam distillation of 1 g of benzene is only 0.10 g.

requirement n **1 必要条件** syn *condition, prerequisite, qualification*

▶ High purity and high sensitivity are essential requirements of laboratory reagents.

2 必要 syn *demand, need*

▶ The rapid preparation of oxide materials can be beneficial with respect to processing and potential reduction in energy requirements.

requisite adj 必須の syn *necessary, obligatory*

▶ Low-temperature crystallization is necessary when the solubility of the compound in the requisite solvent is too high at ordinarily obtained temperatures for recovery to be economic.

research n 研究 syn *exploration, investigation*

▶ Recent research on micelles formed by fluorocarbon/hydrocarbon surfactant mixtures offers compelling evidence of a high degree of segregation of the two surfactants.

researcher n 研究者

▶ Subsequent investigations by other researchers have found a variety of outcomes when attempting to reproduce this work.

resemblance n 類似 syn *conformity, correspondence, similarity*

▶ The reactions of activated aryl halides bear a close resemblance to S_N2-displacement reactions of aliphatic halides.

resemble vt …に似ている syn *approximate, be similar to*

▷ The stability of $H_5O_2^+BrO_4^-$ is very limited, resembling that of anhydrous perchloric acid.

reservation n 留保 syn *exception*

▶ Agreement between six quite different sets of assumptions was remarkably good, with a few reservations about one assumption when water is solvent.

reserve n **1** 蓄え syn *stock, store, supply*

▶ Fats are the main constituents of the storage fat cells in animals and plants and are one of food reserves of the organism.

2 留保 syn *limitation, reservation*

▶ Conclusions based on the hard sphere approximation for atoms in condensed phases need to be treated with reserve.

adj **3** 予備の syn *alternate, auxiliary*

▶ Myoglobin is a major reserve supply, not a primary supply, and that accounts for its high abundance in diving species.

vt **4** 残しておく syn *keep, preserve, retain*

▶ The prefix iso- is reserved for substances with two methyl groups at the end of an otherwise straight chain.

reservoir n 貯蔵所 syn *stock, store, supply*

▶ There are two large reservoirs of organic material, coal and petroleum, and aromatic compounds are obtained from both.

reside vi …に存する syn *be, lie, rest*

▶ There are numerous examples of substrate-controlled asymmetric conjugate additions of nitroalkanes to α,β-unsaturated carbonyls, in which the chirality resides in either the donor or the acceptor.

residence time n 滞留時間

▶ On raising the temperature further, exchange becomes sufficiently rapid to reduce the residence time in any environment far below the time scale of the measurement.

residual adj 残余の syn *remaining*

▶ For rings from C_7 to C_{12} there appears to be a residual strain of 1 to 1.5 kcal per CH_2.

residual magnetization n 残留磁化 syn *remnant magnetization*

▶ Materials that are magnetically hard are those with a high coercivity and a high residual magnetization.

residual resistance n 残留抵抗

▶ In a normal metal, a smooth decline in resistance with temperature ceases at very low temperatures when scattering becomes limited by fixed defects; a limiting or residual resistance exists.

residue n **1** 残留物 syn *remainder, remains*

▶ The residue was dissolved in 0.2 M ethanolic NaOEt and heated at reflux for 10 h.

2 アミノ酸残基

▶ The specific function performed by each of these zinc enzymes is dictated by both the nature of the residues which bind zinc to the protein and the amino acid spacer lengths between the active site residues.

resin n 樹脂

▶ A resin with active sulfonic groups can be converted to the sodium form and will then exchange its sodium ions with the calcium ions present in hard water.

resinous adj 樹脂質の

▶ The fine carbon particles present a large, active surface for adsorption of dissolved substances, particularly the polymeric, resinous, and reactive by-products that appear in traces in most organic reaction mixtures.

resist vt …に耐える syn *block, hinder, prevent, restrain*

▶ Platinum resists the action of all common acids except aqua regia and solutions containing chlorine.

resistance n **1** 抵抗 syn *obstruction, opposition*

▶ The resistance to compression decreases with increased internuclear distance and, significantly, decreases by nearly an order of magnitude over the range of ion separations considered.

2 電気抵抗 ☞ electric resistance

▶ The contact between a metal and semiconductor has a resistance which varies, in many cases considerably, with the direction of the current.

3 耐性 syn *tolerance*

▶ The resistance of lead to corrosion is mainly a result of the inert surface layer of oxide, chloride, or carbonante that forms if it is exposed to air.

resistanceless adj 抵抗のない

▶ As the magnet approaches the pellet, it will induce a resistanceless supercurrent in the surface of the superconductor.

resistant adj **1** 抵抗力のある syn *stubborn*

▶ Normally, C–H bonds are highly resistant to attack by basic reagents.

2 複合語として，…に耐える

▶ Extensive work by Heller resulted in the development of a series of fatigue-resistant, thermally stable photochromic compounds.

resistant bacterium n 耐性菌

▶ The biosynthetic production method has become important because of the necessity of generating new antibiotics that will attack natural-penicillin-resistant bacteria, of which there are many.

resistive adj 抵抗性の

▶ Most often energy is applied as heat, and commonly utilized procedures include resistive furnace heating, hydrothermal synthesis, and gas phase deposition.

resistively adv 抵抗によって

▶ Control experiments comparing the rate of reaction for microwave-induced plasma and resistively heated samples showed that in some cases microwave-induced plasma products had experienced temperatures well in excess of 1100 ℃.

resistivity n 抵抗率 ☞ electric resistivity

▶ Resistivity is a function of thermal vibrations of the crystal lattice as well as of the degree of localization of valence electrons.

resolution n **1** 解像度 syn *accuracy, precision, sharpness*

▶ In many circumstances, mass spectrometry is the method of choice, particularly since resolution is routinely achieved for masses as high as 600.

2 分割 syn *separation*

▶ The solid-state resolution of crystalline binaphthyl, which undergoes a transition from a centrosymmetric to a chiral structure when heated, has been studied extensively.

resolvable adj 分解できる

▶ The absorption spectra of solids consist of a short series of very broad peaks with no resolvable fine structure.

resolve vi **1** 分解する syn *decompose, separate*

▶ At high resolution the spectrum of acetaldehyde resolves into multipets, the low field proton splitting into a quartet and the high field resonance showing up as doublet.

vt **2** 解像する syn *separate, take apart*

▶ The human eye can resolve objects of 100 microns in any dimension.

resolving power n 分解能

▶ The electron microscope is characterized by extremely high resolving power due to the ultrashort wavelength of electronic radiation.

resonance n **1** 共鳴

▶ Strictly speaking, resonance is less important for the acid because the contributing structures are of different stability, whereas the equivalent structures for the ion must necessarily be of equal stability.

2 分光学における共鳴

▶ In the terminology of spectroscopy, resonance is the condition in which the energy state of the incident radiation is identical with that of the absorbing atoms, molecules, or other chemical entities.

3 共鳴線 ☞ resonance line

▶ The ratio of the intensity of the four resonances will be 1:3:3:1 since there are three times as many ways of having methyls with net spins of $+1/2$ and $-1/2$ as there are of having methyls with net spins of $+3/2$ and $-3/2$.

resonance effect n 共鳴効果

▶ The resonance effect involves delocalization of electrons, typically, those called π electrons.

resonance energy n 共鳴エネルギー

▶ Stabilization and resonance energies are not necessarily equal, since the former is the net of all effects, both stabilizing and destabilizing.

resonance fluorescence n 共鳴蛍光

▶ Resonance fluorescence is a well-used technique for monitoring the concentration of atomic species.

resonance frequency n 共鳴周波数

▶ An oscillating electric field applied to a quartz crystal makes the quartz crystal oscillate at a natural resonance frequency that is dependent on the size of the crystal.

resonance hybrid n 共鳴混成体

▶ The contribution of each structure to the resonance hybrid depends upon the relative stability of that structure: the more stable structures make the larger contribution.

resonance line n 共鳴線 ☞ resonance

▶ The resonance lines centered at 5.07 ppm are indicative of a hydrogen on a double bond, and integration of this multiplet reveals that it is due to just one proton.

resonance stabilization n 共鳴安定化

▶ The electronic excitation energy required for butadiene or any conjugated polyene is relatively small primarily because of resonance stabilization of the excited state.

resonance structure n 共鳴構造

▶ Resonance confers additional stability on a covalent bond so that a resonance structure is energetically more favored than any of the single configurations which contribute to it.

resonant adj 共鳴の

▶ If the pump frequency is close to the band gap, then resonant enhancement of the stimulated Raman process occurs.

resort n **1** 頼ること syn *backup, resource*

▶ Esterification of aromatic carboxylic acids with phenols cannot be accomplished by a direct esterification procedure, and resort must be made to the greater reactivity exhibited by the acid chlorides.

vi **2 resort to** …に頼る syn *employ, use*

▶ Most radical are stable in a sufficiently rigid matrix, and it is seldom necessary to resort to temperatures below 77 K.

resource n 方策 syn *asset, means, possession*

▶ The potential for some computation in chemistry can be met only with the greater capacity and capability of the largest scientific computers coupled with specialized resources, such as software libraries and graphics systems.

respect n **1** 点 syn *aspect, matter, point, property*

▶ Rubidium differs markedly from potassium in only one respect: the formation of metallic oxides.

2 in all respects すべての点で syn *extirely*

▶ Since aromatic amino and hydroxy compounds have special stabilization as enamines and enols, their behavior is not expected to parallel in all respects that of the less stable vinylamines and vinyl alcohols.

3 with respect to (prep) …に関して syn *about, concerning, regarding*

▶ In their positive oxidation states, the halogens are nearly always thermodynamically unstable with respect to oxygen and elemental halogen, the only marked exception being iodine in the +5 state.

respectability n 社会的地位

▶ The idea of using molecules as electronic devices has gained attention and respectability in the past quarter century.

respectable adj かなりの syn *fair, moderate, satisfactory*

▶ For Fe, Ru, and Os derivatives, pyrolysis under carefully defined conditions produces respectable though unaccountable yields.

respective adj それぞれの syn *particular, relevant, specific*

▶ Although in general both types of defect will occur, the respective energies necessary for their formation will usually be sufficiently different in any given crystal to make one type of disorder predominate over the other.

respectively adv それぞれ 複数の語の間の対応が記載の順序によることを示す．通常，文末にコンマに続いて置く．対応が明らかなときには用いない．

▶ In the crystal of **1**, the intramolecular O(1)···O(9) and O(4)···O(10) distances are 2.566 and 2.576 Å, respectively; the corresponding H···O contacts are 1.67 and 1.81 Å, and the O—H···O angles are 146 and 145°. In **2**, the intramolecular O(1)···O(9) and O(4)···O(10) distances are 2.545 and 2.546 Å, while the corresponding H···O contacts are 1.88 and 1.79 Å, with O—H···O angles of 143 and 150°.

respiration n 呼吸 syn *breath*

▶ Respiration is a complex system of reactions in which molecules provided by glycolysis are converted into carbon dioxide and water.

respiratory chain n 呼吸鎖

▶ The complex sequence, bringing about the oxidation of NADH indirectly by O_2, is known as the respiratory chain.

respire vi 呼吸する syn *breathe*

▶ Fruits and vegetables continue to respire after harvest, a fact must be taken into account in transportation and storage.

respond vi 応答する syn *answer, react, reply*

▷ Cells are continuously receiving information from their surroundings and, in turn, responding in a way that is determined by their genes and epigenetic factors.

response n **1** 応答 syn *answer, reaction, reply*

▶ At high glucose concentrations or in situations where the supply of oxygen is restricted, the response can be limited by the oxygen supply rather than by concentration of glucose.

2 in response to …に応じて

▶ These molecules change their cation binding ability in response to the pH of their environment.

responsibility n 責任 syn *obligation*

▶ In our generation, we have had to take more responsibility for our products than had ever been the case before.

responsible adj 原因となる syn *accountable*

▶ Polymorphic forms of explosives may have widely differing rates of chemical reaction; the less stable forms are responsible for the high shock sensitivity of the material.

responsive adj 敏感な syn *reactive, receptive, sensitive*

▶ The first examples reported of a class of responsive macrocyclic receptor molecules are the pH-responsive crown ethers which contain one or more carboxylic acid functional groups attached to the periphery of crown ether macrocycles.

rest n **1** 残り syn *balance, remainder, residue*

▶ Only about 5 percent of hydrogen manufactured today is sold as hydrogen, the rest being used in the chemical industry for the manufacture of such chemicals as ammonia and methanol.

2 at rest 静止して syn *motionless*

▶ The assumption that the atoms were at rest can never been true under experimental conditions,

vi **3** 静止している syn *lie, remain, stay*

▶ If a radical can be completely isolated in an inert matrix, it presumably rests in a stable state and can exist indefinitely.

4 rest upon …に基づく syn *depend on (upon)*

▶ The alignment of the M–M axis with respect to the polymer axis rest upon the well-defined coordination chemistry of the M_2 subunit and the choice of organic bridge.

vt **5** 置く syn *be placed, be situated, lie, reside*

▶ For final chilling the flask is rested in a beaker of appropriate size containing ice and water.

restate vt 再び述べる syn *explain, reword*

▶ Many of these modifications have been reviewed recently by D. and R. and will not be restated here.

restatement n 言い換え syn *rephrasing*

▶ The Schrödinger postulate can be set down in terms of a restatement of the Hamiltonian in which the latter is converted to an operator.

resting state n 静止状態

▶ Rapidly metabolizing cells require increased amounts of oxygen relative to the usage during their resting state.

restoration n 復元 syn *recovery, reestablishment*

▶ The most common instance of restoration of a material to its original condition is that of cellulose for rayon production.

restorative adj 補強の syn *fortifying*

▶ The major applications are in restorative dentistry which is concerned with the repair of teeth and oral structures damaged by disease or trauma.

restore vt 回復する syn *bring back, renew*

▶ When oxygen is admitted, the triphenylmethyl rapidly reacts to form the peroxide, and the yellow color disappears. More dimer dissociates to restore equilibrium and the yellow color reappears.

restoring force n 復元力

▶ When a particle is subject to a restoring force linearly proportional to its displacement from some point, it undergoes simple harmonic motion about the point.

restrain vt 抑制する syn *hinder, limit, restrict*

▶ During normal growth, these self-destructive enzymes are restrained.

restraint n 束縛 syn *control, limitation, restriction*

▶ In the absence of primary cross links between molecules, the weak van der Waals forces and the restraint due to tangling or an occasional branch cannot prevent the linear molecules from slipping past each other.

restrict vt 制限する syn *confine, limit, regulate*

▶ The orientation of one or more layers of water molecules around an ion restricts their motion compared with their freedom in the bulk of the solvent.

restricted adj 特定の syn *specific*

▶ Despite considerable advances in the past decade, understanding even in this restricted area is still fragmentary.

restricted rotation n 束縛回転 syn *hindered rotation*

▶ As a consequence of the special geometry of the carbon–carbon double bond and restricted rotation about this bond, the possibility arises of geometric isomerism in appropriately substituted alkenes.

restriction n 制限 syn *condition, qualification*

▶ It is known that an ion has a negative contribution of entropy which is due to the restriction of degrees of freedom of solvent molecules in the vicinity of the ion and that this effect is greater the greater the charge of the ion.

restriction enzyme n 制限酵素

▶ Restriction enzymes are extremely important tools for the molecular biologists as they are used in many techniques that require DNA manipulation and have applications in diagnostics and recombinant protein production.

restructure vi 再構築する

▷ Restructuring knowledge in response to the rapid growth of knowledge is a central theme.

result n **1** 結果 syn *consequence, development, outcome*

▶ The net result is that, other things being equal, a high concentration of nucleophile favors the S_N2 reaction and that a low concentration favors the S_N1 reaction.

2 as a result 結果として syn *consequently, therefore, thus*

▶ As a result, the replacement of a proton by a deuteron removes from an NMR spectrum both the signal from that proton and the splitting by it of signals of other proton.

vi **3** 結果として生じる syn *emerge, occur*

▶ When 2.5 mL of 0.6% tetrachloroauric(III) acid and 3–3.5 mL of 0.09 M potassium carbonate in 120 mL of water are treated with 0.5 mL of a solution of phosphorus in ether, a red sol of gold results whose particles are too small to be seen in the ultramicroscope.

4 result from 起因する syn *arise, develop, emerge, follow*

▶ It is commonplace in chemistry to find that similar values of free energy result from widely differing but compensatory values of enthalpies and entropies.

5 result in 結果的に終わる syn *conclude, end, terminate*

▶ Chemisorption of hydrocarbons may result in formation of chemisorbed hydrogen atoms and hydrocarbon fragments.

resultant n **1** 結果

▶ The overall charge distribution in a molecule is a resultant of a number of factors, of which the bond polarity is only one.

2 ベクトル和

▶ The net polarization of a piece of ferroelectric material is the vector resultant of the polarizations of the individual domains.

adj **3** 結果として生じる syn *consequent, subsequent*

▶ It is significant that those atoms and molecules which have an odd number of electrons have an unbalanced electron spin and, therefore, must have resultant magnetic moments.

resume vt 再開する syn *carry on, continue*

▶ Our modeling of tricritical phenomena started with van Knoynenburg's work, but recently we have resumed making similar calculations and in much greater detail.

resurgence n 復活 syn *regeneration, restoration, revival*

▶ The past twelve years have seen a resurgence of interest in heterocycle carbonylation, much of it stimulated by one patent.

retain vt 保持する syn *hold, keep, maintain, preserve*

▷ The crystals underwent ready anion exchange while retaining their crystallinity.

retard vt 遅らせる syn *slow, thwart*

▶ The transformation of white tin to grey tin is greatly retarded by traces of Bi, Pb, Sb, Cd, Au, and Ag and is accelerated by traces of Al, Zn, Co, Mn, and Te.

retardation n 遅延 syn *delay, prevention*

▶ When an ion moves through the solution as a result of an applied potential field, the atmosphere lags behind and causes a retardation of the motion.

retardation effect n 遅延効果

▶ The addition of salt has only very slight retardation effect on the rate of racemization of $Ni(phen)_3^{2+}$, but the rate is slightly increased in the presence of alkali.

retention n 保持 syn *conservation, preservation*

▶ Preparation of an alkyl sulfonate does not involve the breaking of the carbon–oxygen bond and, hence, proceeds with complete retention.

retention of configuration n 立体配置の保持

▶ Electrophilic substitution at carbon normally proceeds by front-side attack with retention of configuration.

retention time n 保持時間

▶ A provisional identification of the components of the mixture may be made from a comparison of the retention times with those obtained for the pure components.

retentive adj 保持力のある syn *tenacious*

▶ Capillary chromatography uses an open capillary tube with a thin retentive layer on its inner wall.

rethinking 再考 syn *reconsideration, reexami-*

nation

▶ These molecular mechanics calculations led to a rethinking of what controlled metal ion selectivity in macrocycles and to some ideas that are applicable to metal ion selectivity in all ligands.

retina n 網膜

▶ In the rod cells of the retina of a mammal, there is a conjugated protein called rhodopsin.

retinal n レチナール

▶ Retinal can be biochemically converted to its 11-cis isomer which in turn reacts with a protein called opsin to give the Schiff base rhodopsin, the principal photosensitive pigment of the retina.

retract vi 引っ込む syn *pull back, withdraw*

▶ When the liquid and the material of the capillary repel each other, the liquid in the tube retracts from the walls.

retrieval n 検索

▶ Many of these publications are available for computerized information retrieval.

retrieve vt **1** 取り戻す syn *recover, regain*

▶ When an acid gives up its proton, the residue must be a base, since it cam retrieve a proton.

2 情報を引き出す

▶ The crystallographic data used in the survey were retrieved from the Cambridge Structural Database.

retrospect n 回想 syn *hindsight*

▶ In retrospect, the coal tar dyestuff industry had created two essential prerequisites for the production of synthetic medicines.

return n **1** 復帰 syn *restoration*

▶ High-density lipoprotein is involved in the return of cholesterol to the liver where it is excreted.

vi **2** 戻る syn *come back, revert*

▷ Returning to the question of crystallinity, it seems reasonable that molecules which can fit together in close proximity will show a higher degree of order than molecules with irregular branches which interfere with the close approach required for effective van der Waals interaction.

vt **3** 戻す syn *bring back, restore*

▶ The enzyme methemoglobin reductase returns the oxidized heme to the +2 state ordinarily.

reunification n 再統一

▶ Organometallic chemistry is another approach to the reunification of chemistry limited only by the periodic table.

reusable adj 再利用できる

▶ Other interesting examples of reactions catalyzed by modified mesoporous silicas include the chlorination of organic acids using reusable silica immobilized phosphazenium chloride catalysts.

reusability n 再利用性

▶ Mesoporous silica supported $AlCl_3$ showed excellent reusability and improved selectivity to the important 2,6-dialkylnaphthalene compared to other catalysts.

reuse n **1** 再利用

▶ Efficient recovery and reuse of palladium is extremely important both to preserve the metal and to avoid contamination of drugs.

vt **2** 再利用する

▶ The encapsulated *cis*-$[Mn(bpy)_2]^{2+}$ catalyst can be reused following drying at 47 ℃, and it is reported that repeated catalyst regeneration is possible.

reveal vt 明らかにする syn *disclose, display, expose, show*

▷ X-ray diffraction patterns obtained from bulk polymers are diffuse and exhibit a considerable proportion of background scattering, revealing the presence both of small crystallites and of regions in which the molecules are severely disordered.

revealing adj 意義深い syn *illuminating, informative*

▶ The reactions of ICl frequently parallel those of the parent halogens but with subtle and revealing differences.

reversal n 逆転 syn *reverse, turn*

▶ The reversals that occur in donor strength for a series of donors with change in the reference Lewis acid have interested many chemists.

reverse n **1** **the reverse** **逆** syn *contrary, converse, opposite*

▶ Contributing structures in which negative charge resides on an electronegative element and positive charge resides on an electropositive element are more important than those in which the reverse is true.

adj **2** **逆の** syn *contrary, converse, inverse, opposite*

▶ The relaxation time is a function of the forward and reverse rate constants and the concentration of ions.

vt **3** **反対にする** syn *invert, overturn*

▶ An electron has a much larger magnetic moment than the nucleus of a proton, and more energy is required to reverse the spin.

4 **逆方向に動かす** syn *move backward*

▶ It is sometimes stated that iodination fails because the reaction is reversed as the result of the reducing properties of the hydrogen iodide that is formed.

5 **逆転させる** syn *upset*

▶ Flocculation can often be reversed by agitation, as the cohesive forces are relatively weak.

reverse osmosis n **逆浸透** ☞ osmosis

▶ Reverse osmosis involves application of pressure to the surface of a saline solution, thus forcing pure water to pass from the solution through a membrane that is too dense to permit passage of sodium and chloride ions.

reverse reaction n **逆反応** ☞ forward reaction

▶ All chemical reactions are potentially reversible, although in many cases the reverse reaction is so slow that it may be neglected.

reversibility n **可逆性**

▶ Sulfonation is unusual among electrophilic aromatic substitution reactions in its reversibility.

reversible adj **可逆的な** ☞ irreversible

▶ For satisfactorily reversible platinum electrodes one finds experimental results in agreement with this expression.

reversible change n **可逆変化**

▶ In a reversible change, it is supposed that the system can move continuously through a succession of such equilibrium states.

reversible electrode n **可逆電極**

▶ Because ion polarization is very rapid unless ion-reversible electrodes are available, simple DC conductivity measurements are less common for these materials.

reversible process n **可逆過程** ☞ irreversible process

▶ On studying reversible processes, which are observed on both heating and cooling, it is common to observe hysteresis.

reversible reaction n **可逆反応** ☞ irreversible reaction

▶ The aldol condensation is a reversible reaction, and the dehydration may be needed to drive the process to completion.

reversibly adv **可逆的に**

▶ Cu_2HgI_4 is thermochromic; it reversibly changes color with temperature.

reversion n **逆戻り** syn *return, reverting*

▶ The reason for the slow and relatively inefficient reversion to X_2 is the need for a three-body collision in order to dissipate the energy of combination.

revert vi **戻る** syn *return*

▶ All metallic glasses will revert to a polycrystalline phase or phases on heating to elevated temperatures.

review n **1** **総説** syn *review article*

▶ It is beyond the scope of this review to discuss in detail the synthesis of ionic liquids, and the reader is referred to excellent review for many details.

vt **2** **論評する** syn *examine, inspect, survey*

▶ Millero has comprehensively reviewed the whole subject of apparent molal volumes; thus we make no attempt to do so here.

revise vt **改定する** syn *correct, improve, modify*

▷ Revised values for the atomic, structural, and group parachors and refractivities so far determined are tabulated.

revision n 訂正 syn *correction, modification*
▶ The extremely short length of the oxygen–sulfur bond in sulfones and similar compounds has led to a revision of the earlier view that it is a semipolar bond in favor of a double bond formulation.

revival n 復活 syn *restoration, resurgence*
▶ For many years little was done to develop fuel cells for commercial purpose, but since the 1960s there has been a considerable revival of interest in this problem.

revolution n **1** 革命 syn *drastic change, transformation*
▶ In its various forms phase-transfer catalysis has started a revolution in the technique by which organic reactions are carried out, in the laboratory and in industry.
2 回転 syn *rotation*
▶ Light traveling along the axis of revolution of the ellipsoid will have a single velocity.

revolutionary adj 画期的な syn *creative, new, novel, original*
▶ Porous silicon is a potentially revolutionary variant of crystalline silicon because of its tunable electro-, photo-, and chemoluminescent properties.

revolutionize vt 大変革を起こす syn *reform, revise*
▶ The scanning tunneling microscope is an instrument that is revolutionizing the ability to image and manipulate matter.

revolve vi **revolve around** 中心題目として，考慮する syn *consider, contemplate, think about*
▶ Many of the properties and applications of crystalline inorganic materials revolve around a surprisingly small number of structure types.

reward n **1** 報酬 syn *compensation, return*
▶ The work illustrates the integrity of the statement that the more closely we enquire into the nature of things the greater is our reward.
vt **2** 賞を与える
▶ A complete synthesis of hemin was achieved by H. Fischer in 1929, and his contributions were rewarded with a Nobel Prize.

rewarding adj する甲斐のある syn *advantageous, profitable, worthwhile*
▶ Preparation of such plots using readily available compilations of data can be a revealing and rewarding exercise.

reword vt 言い換える syn *paraphrase, rephrase*
▶ Bent's rule might be reworded: The p character tends to concentrate in orbitals with weak covalency, and s character tends to concentrate in orbitals with strong covalency.

rewrite vt 書き改める syn *change, convert, transform*
▶ The Langmuir isotherm can be rewritten

$$C_e/(x/m) = 1/aK + C_e/a$$

so that a plot of $C_e/(x/m)$ vs. C_e is a straight line.

***R*f value** n R_f値
▶ R_f values vary with the solvent system and type of filter paper used.

rheological adj レオロジーの
▶ The rheological properties of a liquid can often be controlled by adding small amounts of high-molar-mass polymers.

rhodopsin n ロドプシン
▶ Rhodopsins interact with special proteins in the cytoplasmic disc membranes of rod cells or in the plasma membranes of cone cells.

rhombic adj 斜方晶系の
▶ This crystalline form of sulfur is monotropic with respect to rhombic and monoclinic sulfur.

rhombohedron n 菱面体
▶ In the trigonal system, a rhombohedron is usually chosen as the unit cell.

ribonucleic acid, RNA n リボ核酸
▶ RNA differs from DNA in that every thymine is replaced by uracil and that the backbone sugar is ribose instead of deoxyribose.

ribosome n リボソーム
▶ All cells have ribosomes made partly of protein and partly of ribonucleic acid molecules. Ribosomes function in the synthesis of proteins.

rich adj **1** 豊富な syn *abundant*
▶ The charge deficit in the bond region between electron-rich atoms can be attributed to the exclusion principle, which obviously, for

these atoms, works against excessive accumulation of electron density in this region and hence against chemical bonding.

2 rich in …の含有量の多い syn *abundant in*

▶ At this point, the mother liquor should be rich enough in the more soluble cis form for its isolation.

3 豊かな syn *fertile, fruitful, productive*

▶ The rich chemistry of substituted bis(cylopentadienyl) zirconium and hafnium complexes bearing side-on coordination dinitrogen ligands is highlightened in this perspective.

richness n 潤沢 syn *abundance, wealth*

▶ Chemists have exploited the richness of silicates to create and extend another very varied region of silicon chemistry.

ridge n うね syn *crest, strip*

▶ When orthorhombic $C_{36}H_{74}$ crystals were grown from petroleum ether and heated to temperatures between 45 and 60 ℃, roof-shaped ridges appeared.

ridiculously adv 途方もなく syn *absurdly*

▶ The acidity and basicity constants of amino acids are ridiculously low for $-COOH$ and $-NH_2$ groups.

rife adj …に富む syn *abundant, rich*

▶ Many of these intermetallic compounds exist over a range of composition, and non-stoichiometry is rife.

right n **1** 右

▶ Within each main-group family to the right of the transition series, the pairs Al and Ga, Si and Ge, P and As, S and Se, and Cl and Br display the greatest similarities.

2 in one's own right それ自身の資質によって syn *through one's own position*

▶ A knowledge of equilibrium electrochemistry forms a basis for understanding nonequilibrium processes, and it has many important applications in its own right.

adj **3 適切な** syn *accurate, correct, exact, proper*

▶ If the molecules are complex, in only a small fraction of the collisions will the molecules come together in the right way for reaction to occur.

right-handed adj 右回りの ☞ left-handed

▶ Circular dichroism spectrophotometry measures as the difference in absorption of left- and right-handed circularly polarized light.

right angle n **at right angles to** …に直角に syn *perpendicular to*

▶ The sample is illuminated, at right angles to the spectrometer, by an intense source of monochromatic radiation, usually the 4,358 Å visible line of a mercury-vapor lamp.

rightly adv 当然のことながら syn *appropriately, properly*

▶ We rightly expect lanthanum to be an electropositive metal, less so than barium but more so than yttrium.

rigid adj **1 剛性の** syn *hard, stiff, strong*

▶ Under certain conditions, especially in rigid glasses, the population of the lowest triplet state may become large enough to permit observation of a new characteristic absorption spectrum.

2 厳密な syn *exact, precise, rigorous*

▶ The classification of oxides as acidic and basic is not rigid.

rigidify vt 堅くする syn *stiffen*

▶ A thin film of blood is smeared on a glass slide, which is then placed in methanol to fix the cells; this process rigidifies the cells and preserves their shape.

rigidity n 剛性

▶ Rigidity is necessary to prevent the collisional deactivation of the triplet state that occurs so rapidly in fluid solution.

rigidly adv 厳格に syn *firmly, precisely*

▶ If contact with acidic and basic substances is rigidly excluded, then interconversion is slow enough that it is possible to separate the lower-boiling enol from the keto form by fractional distillation under reduced pressure.

rigor n 厳密さ syn *precision*

▶ Although the conclusions may be correct, the analyses possess little rigor.

rigorous adj 厳密な syn *exact, precise, strict*

▶ An interesting case of a specific interaction

is represented by the reversible reaction of sodium phenoxide with carbon dioxide under rigorous exclusion of moisture.

rigorously adv 厳密に syn *precisely*

▶ Lanthanide hydration numbers have not been rigorously established for the solution state.

rim n **1** 曲線状のへり syn *edge, perimeter, periphery*

▶ Each with two ends of horizontal fourfold axes is marked by black squares on the rim of the projection.

vt **2** へりを付ける

▶ Monolayer polyethylene crystals or the basal layers of terraced crystals may be rimmed by thin borders which might typically be 100 Å thick.

ring n 環 syn *circle, loop*

▶ The borazine molecule has a planar ring with 120° bond angles and six equivalent B–N bonds of length 1.44 Å.

ring closure n 閉環 syn *cyclization, ring formation*

▶ A concerted ring closure of butadiene to cyclobutene is thermally allowed in a conrotatory manner.

ring current n 環電流

▶ The diamagnetic ring current in tetranitrogen tetrasulfide is due to mixing with the delocalized system of the orbitals involved in the S–S bond.

ring formation n 環形成 syn *cyclization, ring closure*

▶ Ring formation confers rigidity on the molecular structure such that rotation about any of the C–C ring bonds is prevented.

ring opening n 開環 複合形容詞として使用.

▶ The most studied ring-opening carbonylations of epoxides are hydroformylations, the net additions of CO and H_2.

ring puckering n 環パッカリング

▶ Gross physical properties of condensed phases are affected by the low-frequency torsional oscillations and internal rotations of chain molecules as well as the ring puckering and conformation changes of alicyclic ring structures.

ring system n 環系

▶ The chlorophyll ring system is a porphyrin in which a double bond in one of the pyrrole rings has been reduced.

rinse vt ゆすぐ syn *clean, cleanse, wash*

▶ Burets should be rinsed with at least three 10-mL portions of the liquid before filling them.

ripple n 波形 syn *wave*

▶ The X-ray absorption spectra in the region of an absorption edge usually shows a ripple from which information on local structure and bond distances may be obtained.

rise n **1** 上昇 syn *elevation*

▶ An exothermic reaction tends to nullify the rise in temperature by shifting back toward the reactants.

2 give rise to もとである syn *cause, generate, produce*

▷ It is important to point out that a crucial feature is the structural anisotropy of the molecules, giving rise to an anisotropy of electronic polarizability which generates the weak, anisotropic dispersion forces which stabilize the liquid-crytal phases.

vi **3** 上昇する syn *be elevated*

▶ In general, boiling points and melting points rise with an increase of intermolecular forces.

4 増大する syn *increase, grow, swell*

▶ In water containing a little hydrochloric acid, AgCl is less soluble than in pure water, but as the quantity of hydrochloric acid is increased, the solubility of AgCl rises rapidly.

5 空に昇る syn *climb*

▶ Warm, moist air rises into cooler regions higher in the atmosphere.

6 rise to 対処する syn *react, respond*

▶ Future development will rely heavily on the ability of solid-state chemists to rise to the challenge of designing new intercalation compounds for lithium which optimize the many different properties that are important for a positive electrode in a rechargeable lithium battery.

rise time n 立ち上がり時間

▶ In a binary or ternary system, the emission may have a finite rise time or be slightly lengthened in duration owing to the finite time of intermolecular energy transfer.

rising adj **1** 上昇する syn *increasing*

▶ With rising temperature the conductivity of PbF_2 increases smoothly and rapidly until at ca. 500 ℃ a limiting value of ca. 5 ohm^{-1} cm^{-1} is reached.

2 増大する syn *growing*

▶ The rising interest in such compounds comes from several developments in organic chemistry, physics, engineering, biochemistry, and biology.

risk n 危険度 syn *danger, hazard*

▶ High-density lipoprotein concentrations in the blood are inversely related to the risk of heart disease.

rival n **1** 匹敵するもの syn *competitor*

▶ The bond additivity scheme here presented works as well as any rivals.

vt **2** …に匹敵する syn *match*

▷ A cadmium-containing hybrid with large channels and over 50% void space gave 93% enantiometic excess for the addition of diethylzinc to 1-naphthaldehyde, rivaling the results obtained from homogeneous analogs.

road n …への道 syn *approach, method, procedure, way*

▶ Given that sulfuryl chloride is a chlorinating agent this was scarcely a surprise, but it set us the road to success.

roast n **1** 焙焼

▶ The duration of the roast is determined by the kinetics of the gas–solid reactions.

2 焙焼した物

▶ In the absence of soda ash, SeO_2 can be volatilized directly from the roast.

vt **3** 焙焼する syn *burn, heat*

▷ Beryllium is extracted from beryl by roasting the mineral with Na_2SiF_6 at 700–750 ℃.

roasting n 焙焼

▶ Combustion of elemental sulfur or roasting of pyrites yields sulfur dioxide.

robust adj 強靱な syn *strong, tough*

▶ One concern has been whether the chiral ligands are sufficiently robust to survive the reaction conditions that are required for hybrid framework formation without racemization.

robustly adv 強靱に syn *strongly*

▶ The metal–metal interactions are very weak relative those in the robustly bonded examples to be described.

robustness n 強靱さ syn *strength*

▶ The key to successful reverse osmosis lies in the reliability and robustness of the membrane.

rock salt structure n 岩塩型構造

▶ Most halides and hydrides of the alkali metals and silver have the rock salt structure, as do also a large number of the chalcogenides of divalent metals such as alkaline earths and divalent transition metals.

rod n 棒 syn *bar, stick*

▶ A molten zone is passed slowly along a solid rod of the material to be purified by moving either the sample or the heating element.

rod cell n 桿細胞

▶ The human eye can detect the absorption of as few as five photons of light by five rod cells.

rodlike adj 棒状の

▶ The point is that by choosing highly colored metal complexes, rodlike dyes can be obtained for dissolution in rodlike hosts, leading in principle to higher solubility and larger dichroic ratios.

role n 役割 syn *function, task*

▶ Iron is the most abundant transition element and serves more biological roles than any other metal.

roll vi **1** 進む syn *go, pass*

▶ If the Gibbs energy of A＋B is higher than that of C＋D, then the reaction will have a tendency to roll from left to right.

2 転がる syn *rotate*

▶ A drop of water or other volatile liquid rolls about on a hot metal plate or on the surface of a boiling liquid.

vt **3 roll up** 丸める syn *coil, curl, wrap*

▶ Micelles may be considered as sections of liquid monolayer that have been rolled up into small aggregate units and dispersed through the aqueous subphase.

Roman numeral n ローマ数字

▶ On the matter of convention, the various polymorphic forms are best designated by Roman numerals: I, II, III, IV, etc.

roof n 頂部

▶ The floor and roof of the cavity are formed by hexagons of hydrogen-bonded oxygen atoms which are nearly planar.

room temperature n 室温 複合形容詞として使用

▶ Room-temperature ionic liquids are not new. Ethylammonium nitrate, which is liquid at room temperature was first described in 1914.

rooted adj …に起源をもつ syn *inherent*

▶ Not all of the limitations of crystal-field theory are rooted in the neglect of metal–ligand electron exchange.

rooting n 発根

▶ Hormons can stimulate the rooting of plant cuttings and encourage more rapid growth of meat animals.

root-mean-square velocity n 根平均二乗速度

▶ The term root-mean-square velocity implies that each of the molecular velocities is squared, then the average of the squared velocities is taken, and finally, the square root of this average is determined.

rotamer n 回転異性体 syn *rotational isomer*

▶ Molecular mechanic calculations suggest that in benzanilides with a bulky ortho substituent on the benzanilide ring the Z rotamer becomes preferred to the E rotamer.

rotate vt 回転させる syn *revolve, turn*

▶ Regardless of the direction in which they rotate polarized light, all monosaccharides are designated as D or L on the basis of the configuration about the lowest chiral center, the carbonyl group being at the top.

rotating electrode n 回転電極

▶ Nearly saturated solution of tetrathionaphthacene in dichloromethane was investigated by voltammetry at the rotating platinum electrode.

rotation n **1** 回転 syn *revolution*

▶ The rotation of the molecules of the yellow form around the Cl–Cl vector should disrupt the molecules within the stack and in adjacent stacks in the *c* direction but have little effect on neighboring stacks in the [110] direction.

2 旋光 ☞ optical rotation

▶ These reactions merely provide further examples of the fact that a change in the sign of rotation does not necessarily imply a change in configuration.

rotational adj 回転の

▶ The two methyl groups of ethane may take up various rotational dispositions with respect to each other.

rotational axis n 旋回軸 syn *pivot bond*

▶ If rotation of a molecule by 360°/n results in an indistinguishable configuration, the molecule is said to have an n-fold rotational axis.

rotational energy n 回転エネルギー

▶ One observes, in fact, that molecules like H_2, N_2, and CO_2, which are linear, gives rise to no absorptions that can be attributed to changes in the rotational energy of the molecules.

rotational energy level n 回転エネルギー準位

▶ Rotational energy levels are normally very closely spaced so that rather low-energy radiation suffices to change molecular rotational energies.

rotational spectrum n 回転スペクトル

▶ Structural parameters of the HOF molecule have been determined from its rotational spectrum.

rotational symmetry n 回転対称 ☞ translational symmetry

▶ Crystals may display rotational symmetries of 2, 3, 4 and 6.

rotation axis n 回転軸

▶ One-, two-, three-, four-, and sixfold rotation axes are permissible, corresponding to rotations by 360°, 180°, 120°, 90°, and 60°.

rotator n 回転子

▶ The rotator phases of the tetrasubstituted methanes were observed to be waxy.

rotatory inversion n 回反

▶ Possession of an *n*-fold axis of rotatory inversion means that the molecule is brought into coincidence with itself by rotation through $360°/n$ and inversion through its center.

rotatory power n 旋光能

▶ The phenomenon of the rotation of the vibration direction of plane polarized light is known as rotatory power or optical activity.

rough adj **1** ざらざらした syn *coarse, irregular, uneven*

▶ When tetrachloroauric(III) acid was used in rapidly mixed reactions, thicker nanofibers of polyaniline with rougher surfaces were produced, with randomly distributed gold nanoparticles as a by-product of the polymerization reaction.

2 概略の syn *approximate, cursory*

▶ As a rough guide, any reaction which occurs in an inert solvent at a reasonable rate at a temperature 60–100° below the melting points of reactants can probably be made to occur in the solid.

roughen vt 粗くする

▶ The difference in the size of the atoms roughens the slip plane on the atomic scale, making it harder for dislocations to move.

roughly adv おおよそ syn *approximately, nearly*

▶ This is a case in which accurate data can only roughly be interpreted because of a lack of sufficient knowledge.

roughness n でこぼこ

▶ Very few studies have addressed the physical properties of the films such as hardness and roughness which are vital.

round n **1** 循環 syn *cycle, series*

▶ For fatty acid to be subdivided into acetyl-CoA units, several rounds of the sequence of reactions are required.

prep **2** 回りに syn *about, around*

▶ The molecule of $OSCl_2$ is pyramidal and owes its shape to the use of sp^3 hybrid orbitals round a S atom with one lone pair.

round-bottom flask n 丸底フラスコ

▶ The Friedel–Crafts reaction is carried out in a 500-mL round-bottom flask equipped with a short condenser.

route n 経路 syn *course, direction, road, way*

▶ The low-temperature routes yield material of lower crystallinity as is evident from the broad peaks in the powder X-ray diffraction patterns, but the particle size remains in the micron range.

routine adj 日常的な syn *customary, familiar, ordinary, usual*

▶ The application of the scientific method does not consist solely of the routine use of logical rules and procedures.

routinely adv 日常的に syn *ordinarily*

▶ Reactions were routinely carried out under pure dry nitrogen or argon, using dry oxygen-free solvents unless otherwise noted.

rovibronic transition n 回振電遷移

▶ Dye lasers are used as the excitation source as their output frequency can be precisely tuned to match a particular rovibronic transition.

row n 横の列 syn *line, series*

▶ The absolute electronegativity of the transition metals increases as we go from left to right in any row across the periodic table.

rub vt こする syn *polish, scratch*

▷ The oxime usually separates as an oily layer; on very thorough cooling and rubbing the walls of the tube with a stirring rod the material can be caused to solidify.

rubberlike adj ゴム状の

▶ Copolymers of ethylene and propylene made with Ziegler catalysts have highly desirable rubberlike properties and are potentially the cheapest useful elastomers.

rubber stopper n ゴム栓

▶ Dihydropyran (50 g, 0.6 mole) and Raney nickel catalyst (8 g) are mixed in a bottle which has a rubber stopper with a tube leading to a hydrogen gas source.

ruby laser n ルビーレーザー
▶ A ruby laser contains a ruby crystal rod, several centimeters long and 1 to 2 cm in diameter.

rudimentary adj 基本的な syn *basic, elementary, fundamental*
▶ The boranes formed two rudimentary classes distinguished by their formulas, viz. B_nH_{n+4} and B_nH_{n+6}, which we now know as nido and arachino boranes.

rule n **1** 規則 syn *guide, law, principle*
▶ The justification of this rule comes partly from theory and partly from experimental facts.
2 the rule 通例 syn *fact, practice, routine, standard*
▶ Inversion of configuration is the general rule for reactions occurring at chiral centers, being much commoner than retention of configuration.
3 as a rule 概して syn *generally, normally, usually*
▶ Titanium unites with hydrogen forming what is usually referred to as TiH_2 but is, as a rule, nearer $TiH_{1.75}$.
4 rule of thumb 経験則 syn *practical approach*
▶ The rule of thumb that ionic substances dissolve in polar solvents is quite generally used, and it certainly seems to be valid in most instances.
vt **5 rule out** 除外する syn *eliminate, exclude*
▶ We may rule out the presence of a hydroxyl group in a material that does not exhibit an absorption in the 3600 cm^{-1} region.

run n **1** 作業時間 syn *course, period*
▶ Care was taken to hold the temperature of the sample constant within a degree throughout a run.
vi **2** 継続する syn *cover, go*
▶ The third series, $[Cl(Cl_2PN)_nPCl_3]^+PCl_6^-$, runs from $n=0$; i. e., $PCl_4^+PCl_6^-$, through P_3NCl_{12}, $P_4N_2Cl_{14}$, and $P_5N_3Cl_{16}$ to $P_6N_4Cl_{18}$.
3 延びる syn *extend, stretch*
▶ The two strands of DNA run in opposite directions.
4 にじむ syn *flow*
▶ Thixotropic paints do not drip or run.
5 …のようになっている syn *keep, maintain, sustain*
▶ In every case the argument runs in the same way as the rudimentary calculations outlined above.
6 run out 尽きる syn *end, cease*
▶ According to geologists, fossil fuel reserves are being rapidly depleted and may well run out after the end of the century.
vt **7** 実行する syn *operate, perform*
▶ All proton and carbon NMR spectra were run on a Brucker WM-250 spectrometer.
8 処理する syn *handle, manupulate*
▶ Solids are often run as finely ground suspensions (mulls) in various kinds of oils or are ground up with potassium bromide and compressed by a hydraulic press into wafers.
9 流す syn *pour*
▷ The purest silver halides for experimental work with large crystals are produced by running solutions of silver nitrate and hydrogen halides into a common vessel through concentric jets.

runaway n **1** 暴走
▶ Alumina is known to exhibit a phenomenon called thermal runaway where microwave heating increases with increasing temperature.
adj **2** 止めどもない syn *uncontrolled*
▶ Detergents containing phosphates contribute to the pollution of lakes and streams by fostering the runaway growth of algae, often followed by its decay and the consequent depletion of oxygen.

rung n はしごの横木
▶ The rungs of this twisted ladder consist of the complementary base pairs, and the sides of the ladder consist of sugar and phosphate molecules, creating backbone of DNA.

running adj 延びている syn *continuous, ongoing*
▶ The presence of chains of metal atoms running throughout the structure of single crystals of a compound can only be established by a full X-ray structure determination.

runoff n 地中に吸収されないで流れる廃水

▶ Following heavy rains, these chemicals may create runoff that can contaminate adjacent surface water supplies such as ponds, lakes, streams, and rivers.

rupture n **1** 破断 syn *division, fracture,*

▶ Denaturation is ascribed to the rupture of certain linkages in the native protein and probably involves partial combination between the ionizable groups of the protein.

vt **2** 破断する syn *break, divide, fracture, separate, split*

▷ Most inorganic acids generate their protons by rupturing an O–H bond: common examples are oxo acids of nonmetals, such as sulfuric and hypochloric acids.

rust n **1** さび syn *decay, deterioration*

▶ There may be some unobserved rust on the iron wire.

vi **2** さびる syn *decay, deteriorate*

▶ If the sample of iron is in contact with moist air, it rusts quickly.

rust-colored adj 赤さび色の

▶ Tetracynoquinodimethane is a rust-colored, crystalline solid which can be sublimed at temperatures above 250 ℃ at atmosphere pressure or at 20 ℃ under vacuum.

rusting n 腐食

▶ The rusting of iron is a familiar example of corrosion which is catalyzed by moisture.

rutile structure n ルチル型構造

▶ A coordination of 6:3 occurs in several AX_2 structures, of which the commonest is the tetragonal rutile structure, named after one of the mineral forms of TiO_2.

Rydberg series n リュードベリ系列

▶ By and large, there is remarkably good agreement between the ionization potential values obtained from direct photoionization experiments and those obtained from ultraviolet Rydberg series measurements.

S

sacrifice n 犠牲 syn *loss*

▶ While thermoplastic fluoropolymers with increased solubility are almost as chemically inert as poly(tetrafluoroethylene), some sacrifice in thermal and chemical resistance results when reactive groups replace inert fluorinated groups.

sacrificial adj 犠牲的な syn *sacrified*

▶ Our work on porous materials began with the selective leaching of a sacrificial phase from dense bulk composites.

sacrificial anode n 犠牲アノード

▶ The advantage in using sacrificial anodes, zinc, magnesium, and aluminum are commonly used, is that little maintenance is required, apart from the need to replace anodes.

saddle point n 鞍部 syn *col*

▶ The vibrational frequencies may be determined from the shape of the potential energy surface at the saddle point.

safe adj 確かな syn *secure, sound, sure*

▶ It is safe to venture that the interest in organic crystals will continue to increase and that many new uses will be found for them in the years ahead.

safety n 安全 syn *safeness*

▶ Gaseous diazomethane may be handled with safety by diluting it with nitrogen.

sake n **for the sake of** …のために syn *for*

▶ For the sake of simplicity, the mechanism is illustrated here as a one-step cyclic concerted process.

salience n 顕著なこと syn *prominence*

▶ Because of its uniqueness and salience in reaching clusters related those in nitrogenase, the pathway of core conversion is a matter of interest.

salient adj 顕著な syn *distinctive, outstanding, significant*

▶ Another salient feature of colloidal matter is the enormous development of surface relative to the amount of matter present.

saline adj 塩分を含んだ syn *salty*

▶ The usefulness of sea water and other saline sources to produce fresh water is also of increasing importance.

salt n 塩

▶ The red color of the hydrated chromium(II) acetate is in sharp distinction to the blue of the simple salts.

salt bridge n 塩橋

▶ The salt bridge could be a tube containing saturated potassium chloride solution.

salt effect n 塩効果

▶ Some idea of the salt effect is shown by the fact that a solution of 0.01 M KNO_3 will dissolve 12 percent more AgCl and 170 percent more $BaSO_4$ than the same volume of pure water.

salt in vi 塩溶する ☞ salt out

▶ In most cases, the initial addition of a salt helps to salt out even where further additions help to salt in.

salting in n 塩溶 ☞ salting out

▶ The volume of the system is increased by perchloric acid to give a negative electrostriction; salting in occurs.

salting out n 塩析 ☞ salting in

▶ All the electrolytes, except perchloric acid, cause electrostriction of water and a salting out of the organic species.

salt out vt 塩析する ☞ salt in

▶ Less soluble soaps are readily salted out from aqueous solution by electrolytes.

salvage vt 再利用する syn *reclaim, recover, retrieve*

▶ The de novo synthesis of pyrimidines and purines is energetically expensive so most of the purines and pyrimidines obtained from the degradation of nucleic acids are salvaged for reuse.

same pron **1** 同様のこと the をつけて用いる

▶ The same is also true of the process of oxidation of the element.

adj **2** 同じ syn *identical*

▶ Carbon monoxide is poisonous because of its ability to combine with the hemoglobin in the blood in the same way that oxygen does.

3 other things being the same ほかの条件が同じとして ☞ other things being equal

▶ Other things being the same, the amount of inversion decreases as the stability of the carbonium-ion intermediate increases.

4 the same as …と同じ syn *equal, identical*

▶ An important limitation for MX (1:1) compounds is that the coordination number of the cation is the same as that of the anion.

sample n **1** 標本 syn *specimen*

▶ Room-temperature resistivities are reported to be of the order of 1000 μΩ cm but vary widely from sample to sample.

vt **2** 標本として抽出する syn *test*

▶ The polyaniline product is sampled periodically with a pipette from the reaction bath for examination under an electron microscope.

sampling n 抽出見本 syn *sample*

▶ The above examples constitute only a sampling of the variety of possible pressure-tuning studies of interest to chemists.

sandwich n **1** サンドイッチ状のもの

▶ The CdI_2 structure may be regarded as a sandwich structure in which Cd^{2+} ions are between layers of I^- ions, and adjacent sandwiches are held together by weak van der Waals bonds between the layers of I^- ions.

vt **2** …の間に挟む

▶ In the 2:1 complex of phenol and *p*-benzoquinone, each quinone molecule is sandwiched between two phenol molecules which are parallel to the quinone.

sandwich structure n サンドイッチ構造

▶ Nickelocene has the sandwich structure of ferrocene and is similarly susceptible to ring-addition reactions.

saponification n けん化

▶ Saponification was thus a chemical action in which an inorganic base replaced an organic base combined with a fatty acid.

saponify vt けん化する

▶ The stable liquid materials in the composition range 30–40% SiO_2, 10–20% Na_2O, 60–40% H_2O find extensive use in industrial and domestic liquid detergents because they maintain high pH by means of their buffering ability and can saponify animal and vegetable oils and fats.

satellite n サテライト

▶ From the ^{13}C satellite peaks detected in ^1H NMR spectrum, we found the coupling constant between these hydrogen atoms was 8.6 Hz.

satisfactorily adv 満足に syn *adequately, nicely, well enough*

▶ Fluorine, being the most electronegative element, cannot be satisfactorily prepared by displacement from its stable compounds with another element.

satisfactory adj 満足な syn *acceptable, adequate, sufficient*

▶ On the whole, the Rice–Ramsperger–Kassel–Marcus theory has proved very satisfactory.

satisfy vt **1** 満たす syn *fulfill, meet*

▶ A suitable material for such applications must satisfy stringent molecular and macroscopic requirements.

2 満足させる syn *content*

▶ Most scientists are satisfied when things are beyond reasonable doubt.

satisfying adj 満足を与える syn *pleasing, satisfactory*

▶ Although it can be rationalized in terms of the character of the bond, a more intuitively satisfying explanation comes from the results of NMR studies on the HOF molecule, which show the fluorine to have a charge of about −0.5 *e* and the hydrogen to have a charge of about +0.5 *e*.

saturate vt 飽和させる

▷ Cholestanol, being saturated, does not react with bromine and remains in the mother liquor.

saturated adj **1** 原子価の飽和した ☞ unsaturated

▶ Alkanes and alkane-like saturated groups will give upfield peaks in both ^{13}C NMR and ^1H NMR: absorption by sp^3-hybridized carbons and the protons attached to them.

2 溶液の**飽和した** syn *well-supplied*

▶ Whereas water dissolves in ether to the extent of 1.5%, water saturated with sodium chloride (36.7 g/100 g) has no appreciable solubility

saturated solution n **飽和溶液**

▶ The concept of a saturated solution is very important to the crystal grower, since it is from this point that almost all crystal-growing procedure begins.

saturated vapor pressure n **飽和蒸気圧**

▶ The pressure of the system corresponding to the tie line pressure is the saturated vapor pressure of the liquefied gas.

saturation n **1** 原子価の**飽和** ☞ unsaturation

▶ The state in which all available valence bonds of an atom are attached to other atoms is called saturation.

2 蒸気圧の**飽和**

▶ A single isotherm taken to equilibrium pressures approaching the saturation value for both adsorption and desorption should be sufficient for the determination of the capillary volume and the size distribution.

save vt **保存する** syn *conserve, keep, preserve*

▶ The product is collected and washed with a very small quantity of water, and a few crystals are saved for seed.

saw vt **のこぎりで切る**

▶ Unlike alkali halides, silver halides do not cleave readily, and specimens have to be sawed from the crystals with the appropriate orientations.

say vt **1** …**という** syn *affirm, remark, state*

▶ There is a useful rule in organic chemistry which says that the milder a reagent, the more selective it is.

2 論じる syn *disclose, mention, report*

▶ Little can be said about the nature and behavior of surface molecules.

scaffold n **1 骨格** syn *framework, support*

▶ Tabacco mosaic virus has been used as a scaffold for the selective attachment of fluorescent dyes and other small molecules.

vt **2** 足場で**支える**

▷ The conception of a scaffolding structure accounted for a large sedimentation volume and increased apparent viscosity.

scalable adj **拡大・縮小可能な**

▶ Reliable and scalable synthetic methods for conducting polymer nanostructures must be developed in order to provide the necessary materials base for both research and applications.

scale n **1 規模** syn *range, scope*

▶ On a molecular scale, the processes are complex, and on a macroscopic scale they provide an almost puzzling variety of phenomena.

2 目盛 syn *graduation*

▶ To convert temperatures from one scale to another, we need only remember that the Fahrenheit degree is 5/9 of the centigrade degree and that 32 degrees is the freezing point of water.

3 to scale 一定の縮尺で

▶ Actually, atomic radii are so large relative to the lengths of chemical bonds that when a model of molecule like methyl chloride is constructed with atomic radii and bond lengths, both to scale, the bonds connecting the atoms are not clearly evident.

4 薄片 syn *flake*

▶ Free iodine forms scales resembling graphite in appearance; its density is 4.94 at 17 ℃.

vt **5 scale up 拡大する** syn *enlarge*

▶ Its preparation was easily scaled up to 100 g level.

scan n **1 走査** syn *sweeping search*

▶ The cyclic voltammogram for this solution was recorded at a scan rate of 50 mV s^{-1} and exhibited a reduction wave with $E_p = -1.13$ V.

vt **2 走査する** syn *survey, sweep*

▶ By recording the voltage applied to this piezocrystal as the tip is scanned across the surface by additional *x* and *y* piezotranslators, one obtained line scans showing the movements of the tip needed to maintain a constant current.

scanning electron microscopy n 走査型電子顕微鏡法

▶ Scanning electron microscopy can be very helpful to visualize the shape of nanoparticles and their degree of agglomeration.

scanning tunneling microscope, STM n 走査型トンネル顕微鏡

▶ The imaging of atoms and molecules adsorbed on substrates is one of the most interesting STM applications.

scant adj わずかな syn *limited, meager, sparse*

▶ Mössbauer spectroscopic studies in iodine-containing species suggest rather scant s-orbital participation.

scanty adj わずかな syn *insufficient, scanty, short*

▶ Comparatively little is known of the chemical properties of metal borides; results are scanty and unsystematic.

scarce adj 数が少ない syn *deficient, insufficient, lacking, scanty*

▶ The aromatic ring may tend to seek out the scarce unrearranged ions because of their higher reactivity.

scarcely adv ほとんど…ない 否定の意味を持つ. syn *barely, hardly*

▶ Anhydrous strontium chloride dissolves scarcely at all in absolute alcohol.

scarcity n 払底 syn *lack*

▶ Although liquid helium could serve as a coolant, its scarcity and processing costs made it expensive.

scarlet adj 緋色の

▶ Halides are usually colorless unless the metal ion itself has a characteristic color. The principal exceptions are certain anhydrous iodides: AgI, yellow; PbI_2, bright yellow; BiI_3, dark brown; HgI_2, scarlet.

scatter n **1** 撒き散らされた状態

▶ The experimental scatter of these data, particularly those for high temperatures and low pressures, is not inconspicuous.

vt **2** 散乱させる syn *disperse*

▶ The electrons in the beam are scattered by the potential field of the sample molecules.

scattered adj 散乱された syn *diffuse*

▶ The wavelength of the scattered light is increased by a single, definite amount which depends only on the angle through which the light is scattered.

scattering n 散乱 syn *dissipation*

▶ Scattering is the result of fluctuations in density or in concentration due to Brownian motion of the molecules.

scattering angle n 散乱角

▶ The two beams of reactants cross at a certain angle, and the products are detected as a function of their scattering angle.

scavenge vt 除去する syn *scour*

▶ The photozymes can scavenge the organic contaminants present in polluted water at very low concentrations.

scavenger n スカベンジャー

▶ In chemistry, the term scavenger is applied to any substance added to a system or mixture to consume or inactivate traces of impurities.

scenario n 筋書き syn *outline, plan, scheme*

▶ One can envisage various scenarios for catalysis in and/or by ionic liquids.

scene n **come on the scene** 登場する syn *appear*

▶ The advance of science inevitably accelerates when incisive diagnostic and measuring techniques come on the scene.

schematic adj 図式的な syn *graphical, pictorial*

▶ Schematic phase relations are given in Figure 1 for a one-component system in which the axes are the independent variables, pressure and temperature.

schematically adv 図式的に

▶ The various processes of absorption, fluorescence, and phosphorescence as well as the competing radiationless processes are depicted schematically in Figure 1.

scheme n 体系 syn *program, system*

▶ The enthalpies of formation of alkyl radicals $C_nH_{2n+1}^{\cdot}$ can be reproduced satisfactorily by means of an empirical additive bond energy scheme previously applied to alkanes and other hydrocarbons.

Scherrer's formula n シェーレルの式

▶ Scherrer's formula assumes that the material is composed of particles of the same size and shape, which is probably never the case, and the results are only approximate.

Schlieren texture n シュリーレン組織

▶ In thinner layers, the threaded nematic texture changes to the Schlieren texture with point-like disclinations.

Schottky barrier n ショットキー障壁

▶ Hot carriers are emitted from the Schottky barrier of the semiconductor and move into the metal coating

Schottky diode n ショットキーダイオード

☞ diode

▶ Schottky diodes are formed by contact between a semiconductor layer and a metal coating.

Schrödinger equation n シュレーディンガー方程式

▶ The Schrödinger equation is a second-order partial differential equation and, depending on the form of the potential, the magnitude of energy, the shape of the container, and so on, has many possible solution.

scientific adj 科学的な

▶ One aspect of scientific progress is the accumulation of new scientific instruments.

scientifically adv 科学的に

▶ Life forms on earth are scientifically amazing, since they are made up of easily oxidizable organic molecules but depend on O_2 for their existence.

scintillation n **1** 火花を発すること syn *spark*

▶ All nitrates deflagrate on being heated on charcoal, i.e., the charcoal burns at the expense of the oxygen of nitric acid, with vivid scintillation.

2 シンチレーション

▶ When ionizing radiation impinges on a fluorescent material, it causes the emission of a short flash of luminescence known as a scintillation.

scintillator n シンチレーター

▶ Anthracene is a more suitable crystal scintillator, and, for a given energy of incident radiation, it gives scintillation pulses about five times the amplitude of those from naphthalene.

scissile adj 切れやすい syn *separable*

▶ These specificity pockets on the surface of the enzyme accommodate the side chain of the amino acid residue located on the carbonyl side of the scissile bond of the substrate.

scission n 切断 syn *division, separation, splitting*

▶ The changes observed in an irradiated polymer may result from a combination of crosslinking and chain scission, especially if radiation encompasses a wide range of wavelengths.

scope n **1** 可能性 syn *capacity, opportunity*

▶ There is great current interest in this field since it offers scope for the reproducible synthesis of structures having cavities, tunnels, and pores of precisely defined dimensions on the atomic scale.

2 視野 syn *area, field, range*

▶ The details of this are beyond the scope of this account.

scrambling n スクランブリング

▶ Hydrogen-deuterium scrambling was observed in superacid solutions of deuteriated methane.

scrap n 廃物 syn *waste*

▶ Glass is made by melting a mixture of sodium carbonate, limestone, and sand, usually with some scrap glass of the same grade to serve as a flux.

scrape vt 削る syn *damage, scratch*

▶ Potassium was scraped as free as possible from oxide, filtered through a glass capillary in vacuo, and then distilled with the precautions previously described.

scratch n **1** かき跡 syn *abrasion, damage*

▶ Care is required in the initial operation of making a fine, straight scratch extending about a quarter of way around the tube.

vt **2** 引っ掻く syn *damage, rub*

▶ Laboratory chemists often scratch the walls of a glass vessel with a glass rod to encourage a solute to crystallize.

screen n 1 遮へい物 syn *cover, protection, shield*

▶ In water, the solvent molecules arrange themselves around the ion, providing an efficient screen and reducing the Coulomb forces.

vt 2 遮へいする syn *conceal, guard, protect, shield*

▶ The earth is screened from far-UV radiation by oxygen in the atmosphere.

3 検査する syn *evaluate, examine*

▶ Two of the chiral imidazolidinone catalysts developed by MacMillan and Austin were screened but gave none of the expected product under these particular reaction conditions.

screening n 1 スクリーニング

▶ Many catalysts have been designed on the basis of large catalyst screening programs.

2 遮へい ☞ shielding

▶ Within a condensed phase environment, the solvent or counter ions act to stabilize the excess charges and reduce the extent of Coulombic repulsion via dielectric screening.

screening constant n 遮へい定数

▶ Calculations of screening constants were made by Slater who found that outermost electrons are much less efficient at screening nuclear charge than are electrons in inner shells.

screw n 1 らせん syn *helix, spiral*

▶ Where a dislocation that is wholly or partly screw emerges at a crystal surface, this surface can grow indefinitely at low supersaturations.

vt 2 ねじで締める syn *rotate, turn, twist*

▷ The gold electric contacts were held firmly against the electrode faces by screwing the brass end pieces into the main chamber and thus pressing the Teflon insulators on to the back of the contacts.

screw axis n らせん軸

▶ As deposition continues, the ledge rotates around screw axis.

screw dislocation n らせん転位 ☞ edge dislocation

▶ If a screw dislocation emerges at a close-packed surface, growth proceeds more favorably than if the crystal is perfect.

scrub vt 洗い落とす syn *rub, wash*

▶ The condensate that continually drops back into the flask is scrubbed of the volatile liquid and enriched in the less volatile one.

scrubbing n 洗浄 syn *washing*

▶ Much of coal is too high in sulfur content to meet desirable pollution standards unless sulfur is removed by scrubbing.

sea n 海

▶ Metals are a special type of crystalline array because they are built from an assembly of cations immersed in a sea of electrons.

seal vt 密閉する syn *close, shut*

▷ Indium-rich solders are valuable in sealing metal–nonmetal joints in high vacuum apparatus.

sealed adj 密閉された syn *impermeable, tight*

▶ To avoid interaction with water, a lipid bilayer tends to form self-sealed hollow pancake or sphere called vesicle.

sealed tube n 封管

▶ Through glass there is no diffusion, so that gases may be kept unchanged in sealed tube for years.

seam n 継ぎ目 syn *joint, junction*

▶ All seams of the calorimeter are force-fitted and sealed vacuum tight by silver alloy welding.

search n 探索 syn *examination, exploration*

▶ The importance of achieving such an understanding is particularly acute in the search for acceptable smectic C phases for use in ferroelectric devices.

searching n 1 探索 syn *pursuit, search*

▶ The exact choice of conditions for a preferential extraction of one metal from others requires a great deal of searching by trial and error.

adj 2 綿密な syn *detailed, systematic*

▶ This admittedly limited but fairly searching

test indicates the true vapor density of carbon tetrachloride as a value intermediate between the predictions of the two theoretical equations of state.

seat n **1** 源 syn *base, center, site*

▶ The interface is a seat of Gibbs energy which we shall define as the work necessary to enlarge the interface reversibly by 1 cm^2 at constant temperature and pressure and at constant amount of matter in the system.

vt **2** 位置する syn *place*

▶ The hole in the crown ether can be smaller than the cation; in this case, the cation is simply seated in the cavity on one face or the other of the crown.

seawater n 海水 syn *salt water*

▶ The ratio of bromine to chlorine in seawater is only about 1:300.

second n **1** 秒

▶ Because phosphorescence is a process with low probability, the triplet state may persist from fractions of a second to many seconds.

adj **2** 二番目の syn *next, subsequent*

▶ Of the first row transition metals, zinc is second only to iron in its importance in enzyme systems.

adv **3** 次に syn *next, secondly*

▶ Second, X-ray crystallography may be used to determine the structure.

secondary adj **1** 二次的な syn *minor, subsidiary*

▶ It should be mentioned that secondary forces, which are not covalent chemical bonds, can also operate to bind protein chains together.

2 第二級の

▶ The oxidation of secondary alcohols with sodium dichromate in dilute sulfuric acid gives acceptable yields of ketones since these do not normally undergo extensive further oxidation under the reaction conditions.

secondary amine n 第二級アミン

▶ Secondary and tertiary amines, particularly those with different alkyl groups, are advantageously prepared by lithium aluminum hydride reduction of substituted amides.

secondary battery n 二次電池

▶ The conventional method of storing energy is by means of primary and secondary batteries.

secondary ion mass spectrometry n 二次イオン質量分析法

▶ The surface layer can be sampled with secondary ion mass spectrometry. Because this method strips away the surface during the measurement, continued bombardment then samples underlying layers for depth profiling.

secondary metabolite n 二次代謝物

▶ Secondary metabolites are compounds produced in larger quantities and function as growth regulators but with no recognized specific activity related to the life cycle of the host plant.

secondary structure n 二次構造

▶ In the secondary structure, peptide chains are arranged in space to form coils, sheets, or compact spheroids, with hydrogen bonds holding together different chains or different parts of the same chain.

second moment n 二次モーメント

▶ The most useful parameter derived from the solid-state NMR spectra is the second moment, the mean square width of the line.

second-order reaction n 二次反応

▶ The hydrogen－iodine reaction is a second-order reaction, and the order with respect to each reactant separately is one.

second-order transition temperature n 二次転移温度

▶ As the temperature of the supercooled melt is decreased, the motion of the molecules become increasingly sluggish and a second-order transition temperature is reached at which the times required for molecular rearrangement become in effect infinitely long.

secrete vt 分泌する syn *generate, release, yield*

▶ About 20 g of bile salts are secreted by the liver into the duodenum each day.

secretion n 分泌 syn *release*

▶ Certain proteins such as insulin are released from the cell in a process called secretion.

secretory protein n 分泌タンパク質
▶ The sequence of events involved in the synthesis and transport of secretory proteins from glands can be followed using autoradiography.

section n **1** 部分 syn *part, portion*
▶ The Langmuir trough is filled with pure water and divided into two sections by a fixed barrier placed about one-third of the length of the tank from one end.

2 断面 syn *cross section*
▶ All sections normal to the axis of revolution are circular.

3 書物，文章の節 syn *part*
▶ The experimental section is largely composed of an account of how the light absorbed by the reaction mixture was determined.

secular determinat n 永年行列式
▶ By using the symmetry orbitals in each irreducible representation, the secular determinant is set up and solved for the energies and coefficients.

secure adj **1** 確実な syn *certain, sure*
▶ The data obtained for simple molecular complexes rest on a secure theoretical foundation.

vt **2** 手に入れる syn *acquire, obtain*
▶ Pyrite and related sulfur minerals are roasted to secure SO_2 gas which is then usually used directly for the manufacture of H_2SO_4.

securely adv 確実に syn *definitely*
▶ Iodoform can be recognized by its odor, by its yellow color, and, more securely, by the melting point.

security n 機密保護 syn *protection*
▶ F_3ClO was discovered in 1965 but not published until 1972 because of US security classification.

sediment n **1** 堆積物 syn *deposit, precipitate*
▶ Sediment in lakes and oceans is rich in organic materials from the breakdown of decaying materials such as leaves and dead organisms.

vi **2** 沈降する syn *precipitate*
▶ The nuclei tend to be the first to sediment to the bottom of the sample tube at forces as low as 1000*g* for ca. 15 min in a tube 7 cm long.

sedimentation n 沈降
▶ The sedimentation of protein molecules in a centrifuge depends directly on the mass of each molecule.

sedimentation volume n 沈降体積
▶ The small addition of a surface active polar substance decreases the sedimentation volume greatly.

see vt **1** 認める syn *observe, perceive, recognize*
▶ The NMR spectrometer can often see the proton either in each environment or in an average of all of them, depending upon the temperature.

2 気付く syn *appreciate, realize, understand*
▶ We can see the usefulness of the molecular orbital approach if we examine the water molecule.

3 遭遇する syn *experience, undergo*
▷ If the electron density is less than one, the net positive charge means that the screening seen by any one electron is reduced and that the Coulombic attraction to the nucleus is increased.

Seebeck coefficient n ゼーベック係数
☞ thermoelectric power
▶ The Seebeck cofficient is perhaps more useful than the Hall coefficient in the greater ease of application.

seed n **1** 種 syn *grain*
▶ (+)-Glucose molecules can be combined to form the large molecules of starch, which is then stored in the seeds to serve as food for a new growing plant.

2 種晶 ☞ seed crystal
▶ A seed that promotes formation of a crystallization nucleus need not necessarily be composed of the same molecules as the compound that is to be crystallized.

seed crystal n 種晶 ☞ seed
▶ Seed crystals may be prepared from the melt by cooling the molten substance in a thin tube and initiating crystallization at one end; the last part to crystallize is often a single crystal.

seeded adj 種晶添加した

▶ For the growth of some materials, it is sufficient to place the saturated, seeded solution covered with a cloth in a basement for a few weeks.

seeding n 種晶添加

▶ Seeding may also occur if small amounts of the crystalline material are present as contaminants: unintentional seeding.

seek vt 求める syn *look, search*

▶ Scientists seek to control the arrangement of matter on a smaller and smaller scale.

seem vi 思われる syn *appear, look, sound*

▶ The various chemical measures are not as directly informative as they might seem.

seemingly adv 見たところ syn *apparently, evidently*

▶ The attack of a solvent or chemical reagent on a seemingly uniform solid surface is frequently localized in etch pits.

segment n セグメント syn *part, portion, section*

▶ Linear O–Hg–O segments occur in HgO and may be rationalized on the basis of sp hybridization of mercury.

segmental adj セグメントの

▶ Poly(methyl methacrylate) and polystryrene are linear polymers, and segmental motion can occur despite the overall solid appearance.

segregate vt 分離する syn *isolate, partition, separate*

▶ As a rule, the outside of an ingot solidifies first, and some of the impurities are likely to be concentrated or segregated in the interior.

segregation n 偏析 syn *isolation, separation*

▶ Segregation takes place when particles are of different sizes and densities.

seldom adv めったに…ない syn *infrequently, rarely, scarcely*

▶ Normally, boiling points are recorded at 101.325 kPa (1 atm), whereas they seldom are obtained experimentally under this exact pressure.

select adj **1** 選んだ syn *chosen, preferred, selected*

▶ The plasma membranes are permeable to water and a select group of small molecules but are impermeable to large molecules and many ions.

vt **2** 選ぶ syn *choose, prefer*

▶ For this investigation, compounds of the general formula **1** were prepared, where X and Y were selected to be ethoxy and *n*-propyl, respectively.

selection n **1** 選択 syn *choosing, selecting*

▶ The selection and preparation of representative samples for analysis are matters of great importance.

2 選ばれたもの syn *group, series*

▶ Several hundred oxides and halides form the perovskite structure. A selection is given in Table 1.

selection rule n 選択則

▶ The IR and Raman spectra of a particular solid are usually quite different since the two techniques are governed by different selection rules.

selective adj 選択的な syn *discriminative, particular*

▶ Metal ion reactions with simple anions and complexing agents are moderately selective.

selective adsorption n 選択吸着

▶ Separation in adsorption chromatography depends on the selective adsorption of the components of a mixture on the surface of the solid.

selectively adv 選択的に syn *discriminatingly*

▶ The more soluble, lighter colored crystals were selectively dissolved by addition of a small amount of dichloromethane.

selective reaction n 選択反応

▶ Differential heating is potentially possible allowing selective reaction at a particular surface or modification, e.g., chemically using microwave-induced plasma.

selective reflection n 選択反射 ☞ total reflection

▶ The selective reflection and thermochromic properties of the cholesteric phase have stimulated imaginative ideas for applications.

selectivity n 選択性

▶ Toward carbocations an aromatic ring is a reagent of low reactivity and hence high selectivity.

self-antigen n 自己抗原

▶ Tumor-associated carbohydrate antigens are self-antigens and can be tolerated by the immune system.

self-assembly n 自己集合

▶ An alternative approach is to take advantage of the self-assembly properties of liquid crystals and to combine the active molecules with substituents that will give rise to liquid crystal properties.

self-associate vi 自己会合する

▶ Amides have a very strong tendency to self-associate by hydrogen bonding, and the appearance of the spectrum is very much dependent on the physical state of the sample.

self-condensation n 自己縮合

▶ Decarbonylation and dehydration of the α-hydroxy acid forms formylacetic acid which then undergoes self-condensation.

self-consistency n 首尾一貫性

▶ The self-consistency of X-ray measurements are expressed by an *R* value, a form of standard deviation for the calculated and observed intensities.

self-consistent adj 自己矛盾のない

▶ Taking the radius of each of two adjacent ions as the distance from the nuclear center to the electron density minimum between the ions yields radii that are self-consistent.

self-diffusion n 自己拡散

▶ Coefficients for self-diffusion can be measured isotopically labeling some of the molecules and determining how the labeled gas diffuses into the unlabled.

self-sustaining adj 自立した syn *self-suffiient*

▶ The reaction of Na_2SiF_6 with metallic Na is highly exothermic and is self-sustaining without the need for external fuel.

semicircular adj 半円形の

▶ Because of the difficulty of forming continuous ingots in tubes with internal diameters less than 2 mm, the samples are placed in open boats of semicircular cross section.

semiconducting adj 半導性の

▶ The conductivity of semiconducting transition-metal compounds is often increased when the transition element is present in more than one oxidation state.

semiconductivity n 半導性

▶ The semiconductivity of crystalline organic π–π molecular complexes with nonionic ground states has been interpreted as intrinsic or extrinsic involving impurity acceptor ions.

semiconductor n 半導体

▶ A semiconductor has a smaller band gap, and electrons are more easily thermally promoted into the conduction band.

semiconductor-to-metal transition n 半導体–金属転移

▶ The conductivity of polyacetylene increases extremely rapidly as the dopant is added, and a semiconductor-to-metal transition occurs at about 1 to 5 mol% added dopant.

semiempirical adj 半経験的な

▶ It has proved possible to construct empirical and semiempirical sets of ionic radii, such that the sums of such radii reproduce with a reasonable accuracy the observed interionic distances found in ionic crystals.

semimetallic adj 半金属の

▶ $CsSnBr_3$ forms black lustrous crystals with a semimetallic conductivity of ca. 10^3 ohm^{-1} cm^{-1} at room temperature.

semipermeable membrane n 半透膜

▶ The phenomenon of reverse osmosis is the flow of water through a semipermeable membrane from a more concentrated into a less concentrated solution.

semiquantitative adj 半定量的な

▶ The parameters assigned agree with qualitative chemical intuition regarding the acid–base interactions and also with semiquantitative estimates regarding the amount of covalency.

semitransparent adj 半透明な syn *translucent*

▶ After removing the anion-rich supernatant liquor, the flocculated hydrous cerium(IV) oxide readily disperses in water to give stable semitransparent sols.

send vt 送る syn *transmit*

▶ MeV X-rays supply the driving force to send electrons through the dielectric sample and the electrometer circuit without benefit of an applied electric field.

sensation n 知覚 syn *impression, sense*

▶ Light of 5000 Å gives the sensation of blue-green, 5600 Å yellow-green, etc.

sense n **1** 意味 syn *meaning*

▶ The very strong acid, sulfuric acid, causes nitric acid to ionize in the sense, $HO^- \cdots NO_2^+$, rather than in the usual way, $H^+ \cdots NO_3^-$.

2 in no sense 決して…でない

▶ Each positive ion is surrounded by a number of negative ions and each negative by a number of positives, at equal distances, so that it is in no sense correct to say that one ion is bound to one or two neighbors more than to others.

3 make sense 意味をなす syn *be logical*

▶ When a new protein with a role in signal transduction is discovered, it is given a name, which makes sense to members of the laboratory concerned.

4 感覚 syn *impression, intuition, sensation*

▶ One must keep a sense of balance about the extent to which simple rules of structure are applied.

5 互いに反対方向のうちの一方の向き

▶ All ionic crystal surfaces will show a mode of distortion described as puckering in which there occur small displacements of anions and cations in opposite senses perpendicular to the plane of the surface.

vt **6** 感知する syn *detect, feel, perceive*

▶ The scanning tunneling microscope does not really give the physical positions of the atoms but senses their electrons and bonds.

sensible adj 賢明な syn *logical, realistic, reasonable*

▶ A diamond-like network seemed the simplest and most sensible target.

sensitive adj 敏感な syn *delicate, reactive, subtle*

▶ Since the discovery of air-sensitive chloroaluminate ionic liquids, subsequent research concentrated on their use in electrochemical applications.

sensitively adv 敏感に

▶ The decomposition of chlorous acid depends sensitively on its concentration, pH, and the presence of catalytically active ions such as Cl^-.

sensitiveness n 敏感さ syn *acuteness, sensitivity*

▶ In order to carry out an analysis with certainty, it is necessary to understand not only the reactions of the different elements but also the sensitiveness of each reaction.

sensitivity n 感受性 syn *delicacy, sensitiveness*

▶ One further characteristic of charge-transfer bands should be their wavelength sensitivity to the polarity of the solvent.

sensitize vt **1** 感光性を与える

▶ Pinacyanol sensitizes silver bromide emulsions powerfully to red light.

2 敏感にする

▶ Antenna chromophores can selectively sensitize certain types of reactions in a complex mixture of substrates by properly matching the energy levels of antenna chromophore to the substrate molecule.

sensitized luminescence n 増感ルミネセンス

▶ As noted for the tetracene–anthracene system, a good test of purity is the absence of impurity-sensitized luminescence.

sensitized photolysis n 増感光分解

▶ Mercury-sensitized photolysis of N_2O is a more convenient route to ground state O atoms.

sensitizer n 増感剤

▶ Color sensitizers are dyes added to silver halide emulsions to broaden their response to various wavelengths.

sensor n センサー syn *probe, tester*

▶ The enzyme immobilized in a membrane converts the substrate to be detected into NH_3 which is then detected by the ammonia sensor to give a quantitative measure of the substrate concentration.

separable adj 分離できる syn *removable*

▶ Butylene is easily separable by distillation, but the other substances are in the same boiling point range as the product.

separate adj **1** 別々の syn *different, independent, unrelated*

▶ Separate solutions, A and B, of the starting metals or compounds were prepared.

2 離れた syn *isolated*

▶ In the violet isomer of $CrCl_3 \cdot 6H_2O$, the six water molecules are coordinated to the chromium, and the three chloride ions are separate ions in the solid and in solution.

vi **3** 分離する syn *segregate*

▶ The tetraacetate $Ge(OAc)_4$ separates as white needles, when $GeCl_4$ is treated with TlOAc in acetic anhydride and the resulting solution is concentrated at low pressure and cooled.

4 separate from 分離する syn *segregate from*

▶ The particular isomer that separates from solution is largely determined by the relative solubilities of the isomeric salts.

5 separate into …に分かれる

▶ Some stable colloidal sols separate into two liquid layers, each containing some of colloid when to the sol is added either another colloid or an electrolyte or a nonelectrolyte.

vt **6** 分離する syn *divide, split*

▶ Dissolved sodium chloride separates 1-butanol and water into two layers.

7 切り離す syn *partition*

▶ The glass membrane of the glass electrode separates two different solutions, as does the KCl salt bridge.

8 分割する syn *classify, distinguish*

▶ Frequently we separate the internal energy into its components of rotational, vibrational, and electronic energy; but we recognize that this kind of separation is an approximation that does not always hold.

separately adv 別々に syn *independently, individually*

▶ A spectacular example is afforded by benzene and tetracyanoethylene, each of which separately is colorless, but which give a bright-orange complex when mixed.

separation n **1** 分離 syn *division*

▶ The separation of hafnium from zirconium was effected by repeated recrystallizations of the complex fluorides.

2 距離 syn *distance*

▶ In K_3CrO_8, the oxygen–oxygen bond length in the peroxide group is 1.405 Å, and the chromium–oxygen separations are 1.846 and 1.944 Å.

separation factor n 分離係数

▶ Starting with normal water and a separation factor of 5, the deuterium content rises to 10% after the original volume has been reduced by a factor of 2400.

separatory funnel n 分液ロート

▶ To remove traces of sulfurous acid from the product, the distillate is transferred to a separatory funnel and about 10 mL of 10% sodium hydroxide solution is added.

sequence n **1** 一連 syn *chain, series*

▶ The cleavage of a carbon–carbon double bond by oxidation with ozone followed by hydrolysis to yield carbonyl compounds is a reaction sequence of considerable importance.

2 配列 syn *arrangement, organization*

▶ A short polypeptide of 100 amino acid residues may have any one of 20^{100} different sequences.

3 順序 syn *order*

▶ The polymorphism at atmospheric pressure can be summarized by the following sequence of reactions on heating: α-quartz → β-quartz → β-tridymite → β-cristobalite → liquid.

4 in sequence 次々と

▶ Cytochromes are electron-transferring proteins that act in sequence to transfer electrons to O_2.

vt **5** 配列を決める

▶ Many bacterial promoters have been sequenced.

sequence isomer n シークエンス異性体

▶ Glycylalanine and alanylglycine are examples of sequence isomers; they are composed of the same amino acids, but they are combined in different sequences.

sequential adj 続いて起こる syn *continuous, successive*

▶ Signal transduction is followed by sequential changes in the activities of proteins inside the cell.

sequentially adv 連続して syn *in succession, successively*

▶ It is possible sequentially to transfer the layers to produce an LB film of several hundreds of layers.

sequester vt 封鎖する syn *insulate, isolate, separate*

▷ Ferritin lowers the concentration of free iron by sequestering it.

sequestering agent n 金属イオン封鎖剤

▶ Two groups of organic sequestering agents of economic importance are the aminopolycarboxylic acids, such as EDTA, and the hydroxycarboxylic acids, such as gluconic, citric, and tartaric acids.

sequestration n 金属イオン封鎖作用

▶ The term sequestration may be used for any instance in which an ion is prevented from exhibiting its usual properties because of close combination with an added material.

serial adj 連続的な syn *sequential*

▶ Rectangular cells may be used for serial titration experiments as about 60% of a cell can be empty for the first spectrum and gradually filled.

series n **1** 系列 syn *chain, order, sequence*

▶ In the series glycine, alanine, valine, and leucine, R_f increases with increasing molecular weight; the larger the alkyl group the more the acid tends to move along with the organic solvent.

2 a series of 一連の 続く名詞は複数形，集団を強調するときは動詞の単数形，個々を強調するときは動詞の複数形を用いる．

▶ A series of closely related structures are found for complex fluorides $A_mB_nF_p$ in which $p = 3m$.

3 in series 直列に

▶ In the first case an elastic element and a frictional resistance have been connected in series.

4 級数

▶ The exponential series is special cases of a general form of the expansion of a function in a power series which is called Taylor's theorem.

serious adj 重大な syn *crucial, important, significant*

▶ A serious problem was that the terms hard and soft were not well defined, either theoretically or experimentally.

serotonin n セロトニン

▶ Serotonin affects blood pressure, promotes intestinal peristalsis, and even appears to be involved in mechanisms of psychic phenomena in the brain.

serum n 漿液

▶ In the case of milk, the serum is a true solution of sugars, proteins, and mineral compounds in water.

serve vi 役に立つ syn *function, perform, suffice*

▷ Metal complexes can complement purely organic molecules by serving as structural scaffolds for the design of compounds with bioactivity.

service n 器具のひとそろい

▶ The method chosen for small-scale laboratory preparations of oxygen depends on the amount and purity required and the availability of services.

set n **1 set of** …のひと組 続く名詞は複数形，全体を強調するときは動詞の単数形，個々を強調するときは動詞の複数形を用いる． syn *group, collection*

▶ Many reactions studied kinetically can be explained by a particular set of simple processes which are so reasonable and so in accord with all chemical experience that we accept them as essentially true.

adj **2** 所定の syn *established, predetermined*

▶ The particular spatial arrangement for any dissymmetric molecule when it is viewed according to a set convention is called its absolute configuration.

vi **3 凝固する** syn *coagulate, harden, solidify, stiffen*

▶ The conductivity of an electrolyte in a jelly does not change appreciably when the jelly sets.

4 set in (a) 始まる syn *begin, initiate*

▶ At a temperature of about 67 ℃, disorder sets in, and the Cu^+ and Hg^{2+} cations are randomly distributed about all of the tetrahedral holes in the structure of Cu_2HgI_4.

(b) …となる syn *arrive, come*

▶ A frictional force sets in to balance the diffusion force when some constant velocity is reached.

5 set off 出発する syn *depart, move, start*

▶ When an electric field is applied, the ions set off to the oppositely charged electrodes.

6 set up 生じる syn *begin, start*

▶ When the solid corrosion product occupies a larger volume than the metal from which it is formed, then large stresses can be set up if the corrosion occurs in a confined region.

vt **7 決める** syn *indicate, specify*

▶ When the glassy condition occurs, it simply sets a limit to the phase rule investigation of the system and will for that reason not be considered further.

8 問題などを宛がう syn *assign*

▶ Kopp set himself the problem of determining boiling points with all possible accuracy.

9 活字に組む syn *arrange*

▶ The names of point groups are set in italic type.

10 …の状態にする syn *place, put*

▶ If a solution of potassium iodide is treated with chlorine water, iodine is set free, and the solution turns yellow to brown.

11 set against …と比べる syn *compare, contrast*

▶ Set against this, it is easy to show that even a small rise in daily energy expenditure can have significant consequence for body weight reduction in the long term.

12 set aside 脇に置く syn *set apart, store*

▶ The melted preparation should be carefully set aside for several minutes without physical shock before examination.

13 set off 爆発させる syn *detonate, explode*

▶ 2,4,6-Trnitrotoluene is not set off easily by simple impact and even burns without exploding.

14 設置する syn *assemble, construct*

▷ A number of physical measurements are conveniently made by setting up appropriate electrochemical cells.

setting n 環境 syn *environment*

▶ Nanocrystals of NaCl or NaBr are important in natural settings including marine atmospheric regions, where they crystallize from salt-water droplets that are swept up from ocean waves and dried.

setting point n 凝結点

▶ The setting point of jellies is greatly dependent upon the presence of other substances.

settle vi **1 沈む** syn *fall, precipitate, sink*

▶ Fe_2O_3 and other solid impurities may be removed by allowing them to settle.

2 定着する syn *remain, reside, stay*

▶ As the process of settling into ledges and kinks on a surface continues, there comes a time when the entire lower terrace has been completed.

3 settle down 落ち着く syn *come, move, stay*

▶ If a system is maintained at constant temperature and pressure, it settles down at an equilibrium state in which its Gibbs energy is a minimum.

vt **4 解決する** syn *conclude, decide, resolve*

▶ The question whether an inversion occurs in the allyl group when it migrates to the para position is still far from settled.

settled adj 片付いた syn *concluded, ended*

▶ The question is as yet far from settled; perhaps a detailed study of the very slow proton transfers afforded by proton cryptates can provide an answer.

settling n 沈降 syn *precipitation*

▶ After settling, the precipitate is washed by

decantation, dried, and melted in vacuum.

several pron **1** いくつか syn *some*

▶ Several of the commercially important magnetic oxides have the spinel structure.

adj **2** 不定数を強調して，いくつかの syn *a few, some*

▶ Corrosion may be controlled in several ways, including modification of the anodic and cathodic processes by adding inhibitors or removing reactants.

severe adj 厳しい syn *harsh, rigorous, strict*

▶ One severe restriction in the applicability of the electron microscope to the examination of colloid and, more especially, biological systems is the fact that at present the specimens are almost always exposed to high vacuum.

severely adv 厳しく syn *rigorously, strictly*

▶ Solvolytic reactions are often of synthetic value, but they also severely limit the range of reactions which may be conducted in a particular solvent.

severity n 激しさ syn *rigor, strictness*

▶ The severity of conditions required for comparable reaction to occur within the same period of time can be observed.

sewage n 汚水 syn *effluent*

▶ Techniques for using bioremediation to degrade human wastes at a sewage treatment plant are quite different than degrading the variety of chemicals that exist in the environment.

sextet n セクステット ☞ octet

▶ One reason for believing this is simply the anticipated difficulty of forming a highly unstable intermediate in which an electronegative element like nitrogen has only a sextet of electrons.

shade n 濃淡 syn *color, hue, intensity, tint*

▶ If the solutions are colored exactly the same shade, then the amounts of manganese which they contain are the same.

shaded adj 周囲より色を濃くした syn *dark, hatched, filled, shadowed*

▶ Shaded circles represent water molecules.

shading n 陰影 syn *expression*

▶ The central dimer is shown by line shading, the dimer above that plane by heavy lines, the dimer below by dashed lines.

shadow n シャドウ syn *darkness*

▶ One method of obtaining the heights of surface irregularities is by means of the so-called shadow technique.

shake vt ふりまぜる syn *agitate, stir*

▶ When alkenes are shaken under a slight pressure of hydrogen gas in the presence of a small amount of the catalyst, alkenes are converted smoothly and quantitatively into alkanes of the same carbon skeleton.

shaking n ふりまぜ syn *agitation*

▶ Immediately after shaking the thixotropic system solidifies again, a tube containing it may be inverted without flow occurring.

shallow adj 浅い

▶ As oil droplets move through water, they frequently sweep others along with them so that creaming is sometimes more effective in deep vessels than in shallow pans.

shape n **1** 形 syn *form*

▶ A great variety of adsorption-isotherm shapes are found.

vt **2** 形作る syn *form, make*

▶ Many methods have been described in the literature trying to shape polyaniline into nanostructures.

shaped adj しばしば複合語で，…の形をした

▶ Typically, the molecule of a smectic compound is rod-shaped with polar groups along most of its length providing the lateral intermolecular attractive forces.

shape-memory alloy n 形状記憶合金

▶ NiTi is probably the most well-known of the shape-memory alloys.

shape-memory effect n 形状記憶効果

▶ The processes underlying the shape-memory effect illustrate links between the structure, microstructure, and composition of NiTi and general thermodynamic principles of phase transformations.

share n **1** 分け前 syn *allocation*

▶ The boron halides and the organoboranes are Lewis acids and may accept an electron pair from a base to form tetracovalent boron compounds in which the boron atom has a share of eight electrons.

vt **2 共有する** syn *allot, partition, split*

▶ In NaCl, each octahedron edge is shared between two octahedra, resulting in an infinite framework of edge-sharing octahedra.

sharing n 共有 syn *allotting, dividing*

▶ Many different polyanions can be formed by the sharing of corners or edges of a given set of tetrahedra or octahedra.

sharp adj **1 はっきりした** syn *marked*

▶ Sharp distinction between poisons and nonpoisons is not always possible, because many variables must be taken into consideration in each case.

2 鋭い syn *acute*

▶ The surface free energy per unit area is a function of the crystal face and is a minimum for a polyhedron with flat faces and sharp edges.

sharpen vi **1 尖る** syn *intensify*

▶ A single broad resonance gradually sharpens to a narrow singlet at higher temperatures.

vt **2 尖らせる** syn *thin*

▷ The surface probe is a metal needle sharpened to 0.1 μm radius or better by electrochemical etching.

sharply adv **1 はっきりと** syn *distinctly*

▶ Biological systems generally discriminate sharply between stereoiomers.

2 荒々しく syn *severely*

▶ In extreme cases, synthetic plastic materials may be as flexible as a sheet of rubber, but when hit sharply with the fingernail it breaks off in pieces.

3 素早く syn *abruptly, quickly, suddenly*

▶ This will cause the liquid to move along the capillary; immediately it reaches the mark, the filter paper is sharply removed.

sharpness n 鋭さ syn *fineness, resolution*

▶ The C≡N stretching band in cyano compounds not only appears in the range 2040–2170 cm^{-1} but is also identifiable by its sharpness and high intensity.

shatter vi **1 粉々になる** syn *disintegrate, fracture, pulverize*

▶ The crystals shattered on cooling to 100 K; thus, despite the large thermal vibration, the present analysis had to be carried out at room temperature.

vt **2 粉砕する** syn *disintegrate, fracture*

▷ The enthalpy change in shattering a molecule entirely into its component atoms is called the enthalpy of atomization.

shear n **1 ずり**

▶ During shear, the atoms to either side of the slip plane move, relatively, in opposite directions.

vi **2 せん断される**

▶ Under the action of a shearing stress, the top row shears over the bottom row by one unit.

shearing stress n ずり応力，せん断応力

▶ Many crystals that contain tetrahedral groups, e.g., ZnO, ZnS, are piezoelectric since application of a shearing stress distorts the tetrahedra.

shear plane n ずり面

▶ Octahedra within the crystallographic shear planes share some edges whereas in unreduced regions of WO_3, the corresponding linkages are by corner-shearing only.

shear structure n ずれ構造 ☞ block structure

▶ For studying defects in crystallographic shear structures, the electron microscope is indispensable.

shed vt 放つ syn *emanate, emit, radiate*

▶ The atom or molecule can shed its energy only in discrete steps.

shed light on 解明する syn *clarify*

▶ Chemical theory has been notably unsuccessful in its attempt to shed light on the perbromate problem.

sheet n **1 薄板などの一枚** syn *plate, slab*

▶ Glass is normally thought of as transparent, but if the edge of a sheet of ordinary window glass is examined, it appears dark green.

2 層 syn *lamina, layer*

▶ The mineral talc has the formula $Mg_3(OH)_2Si_4O_{10}$ and, as expected for an Si:O ratio of 1:2.5, the structure contains infinite silicate sheets.

adj **3** 薄板状の

▶ Novel sheet aluminum phosphate anions bearing ionizable OH groups have recently been characterized.

shelf-life n 貯蔵寿命

▶ Polymerized vesicles have shelf-lives of several months and remain stable in solutions of up to 25 percent alcohol.

shell n **1** 殻 syn *framework, skeleton*

▶ The liquid crystal droplets are dispersed in a continuous polymer matrix without precoating with a polymer shell.

2 電子殻 ☞ inner shell, outer shell

▶ Electrons in the second shell are, on the average, farther away from the nucleus, therefore, of higher energy than the electrons in the first shell.

shield n **1** 遮へい syn *protection, screen*

▶ The valence electrons act as shields against an applied magnetic field, and the magnitude of the shielding effect depends upon the electron density at a particular site in the molecule.

vt **2** 遮へいする syn *guard, protect, screen*

▶ When the field felt by the proton is diminished, the proton is said to be shielded.

shielding n 遮へい syn *screening*

▶ Because of the low electronegativity of silicon, the shielding of the protons in the silane is greater than in most other organic molecules.

shielding effect n 遮へい効果

▶ The net result of the shielding effect is that various set of protons in a molecule will experience nuclear magnetic resonance at different applied magnetic field strengths.

shift n **1** シフト syn *movement, transfer* ☞ blue shift, red shift

▶ A 1,2-shift of hydrogen can convert initially formed secondary cation into a more stable tertiary cation.

vi **2** 移動する syn *move, transfer*

▶ The wavelength of the emission peak shifts to smaller values as the temperature is raised.

vt **3** 移動させる syn *displace, move*

▶ In chloranil complex with hexamethylbenzene, the rather large substituents cause the aromatic nuclei of adjacent molecules to differ in orientation by about 16°, and their centers are also shifted through about 0.9 Å relative to each other.

4 重点を移す syn *change, switch*

▶ In dilute solution, the center of interest is largely shifted to the little region of modified solvent immediately surrounding each ion.

shine n **1** 輝き syn *brightness*

▶ Highly polished specimens of beryllium retain their shine indefinitely.

vt **2** 照らす syn *glow, radiate*

▶ If light having an energy above the band gap of the semiconductor shines on the junction, the electron-hole pairs created in the junction region are separated by the electric field therein.

shining adj 輝く syn *bright, brilliant*

▶ Zinc is a shining white metal with bluish-gray luster.

shiny adj 輝く syn *bright*

▶ At room temperature, iodine is a dark, shiny, volatile solid with a layer structure containing pairs of iodine atoms.

shock n 衝撃 syn *impact*

▶ Dry silver acetylides may be quite shock-sensitive and can decompose explosively.

shock tube n 衝撃波管

▶ In a shock tube, well-mixed reactants and precursors are subjected to a very rapid increase in pressure which causes rapid heating of the mixture to several thousand kelvin and dissociation of the precursor species.

shoot vi **1** 突き出る syn *develop, grow*

▶ A crystal which grows very rapidly may shoot out in various chief directions like frost of one orientation on a window pane, and then the skeleton form may be filled in during a period of slow growth.

vt **2** 発射する syn *throw, transmit*

▶ Electrons are shot into a good insulating plastic until the electric field of the trapped charge exceeds the dielectric strength of the solid.

short n **1 in short** 要するに syn *briefly*

▶ In short, electrostatic arguments are generally not applicable to ligands which show a pronounced tendency to form π bond.

adj **2** 距離が短い syn *little, small*

▶ The normal coordination environment of the Cu^{2+} ion is distorted octahedral with four short and two long bonds.

3 時間が短い syn *brief, limited, short-lived*

▶ Fullerene-derived materials have had an enormous impact on the materials chemistry community in the short time they have been available for study.

4 簡単な syn *brief, concise*

▶ Even a short survey of the literature makes it quite clear that there is a surprising variety of fluorescent DNA base replacements.

5 不足している syn *deficient, insufficient*

▶ If electrons are in short supply, then a pair of electrons can bond three atoms in a triangular array, rather than just two atoms.

6 fall short of 達しない syn *lack, need, require*

▶ The rigor of the analysis in most of the X-ray structural determinations of the polymeric lattices falls short of establishing unequivocally the geometry of the coordination polyhedron.

7 short of 前置詞的に，…を除いて syn *except for*

▶ Short of being boiled with nitric acid or putting in a flame Neoprene filter adaptors should last indefinitely.

short circuit n 短絡

▶ These silver iodide derivatives may be used as solid electrolytes without any danger of short circuits through the electrolyte.

shortcoming n 欠点 syn *drawback, weakness*

▶ A major shortcoming of the LB technique is the lack of long-term chemical, thermal, and mechanical stability of the films.

shortcut n 近道

▶ The theory of metals is based on the more difficult and involved wave mechanics; there seems to be no shortcut in understanding it.

shorten vi 距離が短くなる syn *diminish*

▶ The contraction of a stretched polymer is frequently an entropic process; the enthalpy change is usually small, and the driving force is the tendency of the long molecules to shorten, with an entropy increase.

shortening n 短縮 syn *contraction*

▶ In linear conjugated systems, the increase in the bond order of the acceptor bond causes a marked shortening of the bond.

short-lived adj 短寿命の syn *fleeting, temporary, transitory*

▶ Photochemical, as well as thermal, decompositions have been used for the production of short-lived free radicals.

shortly adv まもなく syn *quickly, soon*

▶ Shortly after insulin became available, growth hormone was cloned in bacteria and became available for human use.

short-range adj 短距離の ☞ long-range

▶ In pure crystalline solids, the repetition of short-range coupling between adjacent, similar molecules in a regular crystal lattice causes a single energy level of the isolated molecule to split into a band of levels in the crystal.

short-range order n 短距離秩序 ☞ long-range order

▶ Short-range order is a more fundamental concept than long-range order since the principal atomic interactions in a crystal are those between close neighbors.

short-term adj 短期の syn *brief, transient* ☞ long-term

▶ Using less nitrogen fertilizer makes little difference to nitrate leaching in the short-term but will reduce leaching in the long-term.

should aux **1** …すべきである syn *have to, must, ought to*

▶ Lithium metal should not be warmed in a nitrogen stream for it tends to form the nitride.

aux **2** …のはずである

▶ Electron-releasing substituents should intensify the negative charge, destabilize the anion, and thus decrease acidity.

shoulder n スペクトルの肩 syn *edge, side*

▶ A prominent shoulder of lower energy allows for longer wavelength excitation at 3000 cm^{-1}.

show vi **1 show up** はっきり現れる syn *appear, be noticeable*

▶ Formation of intramolecular hydrogen bonds shows up clearly in NMR spectra.

vt **2** 明らかにする syn *confirm, prove*

▶ Mass spectroscopic analysis showed that it contained almost no deuterium.

示す syn *demonstrate,display, indicate, manifest*

▶ To show how the intensity of the diffracted beam is affected by the arrangement of the atoms, we can take cesium chloride structure as an example.

shrink vi 縮む syn *contract*

▷ The ever shrinking sizes of electronic devices means that more and more information has to be put onto each silicon chip.

shrinkage n 収縮 syn *diminution, shrinking*

▶ The shrinkage is evidently the result of the compressive effect of the surface tension of the evaporating liquid.

shuffling n 組み換え syn *rearrangement*

▶ The extensive shuffling of atoms has to accompany the conversion of the starting materials, Ni or Co with Al, which have the fcc crystal structure, to the CsCl structure of the products, NiAl or CoAl.

shutter n シャッター syn *blind, cover, shade*

▶ Ga, As, and Al are heated in separate furnaces that are equipped with shutters, and the heated beams strike a suitable substrate.

shuttle n **1** シャトル

▶ Homogeneous solution mediators are presumably able to diffuse into and out of the enzyme's active site and thus to act as a shuttle carrying electrons from the active site to the electrode.

vi **2** 往復する syn *commute*

▶ This series of enzyme and transporter reactions has effectively shuttled reducing equivalents in NADH from the cytoplasm to the matrix.

shuttle system n シャトル系

▶ As cytoplasmic NADH cannot cross the inner mitochondrial membrane, a redox shuttle system is required.

side n **1** 側面 syn *face, surface*

▶ Rubbing the sides of the beaker with a glass rod hastens the formation of the precipitate.

2 辺 syn *edge*

▶ A monoatomic gas is contained in a rectangular vessel, the sides of which have lengths *a*, *b*, and *c*.

3 side by side 並んで syn *together*

▶ Almost all metallic films and mirrors consist of colloidal particles lying side by side but not necessarily touching.

side arm n 器具の枝

▶ When the neck of the flask has been warmed sufficiently, the vapors begin flowing out through the side arm into the condenser.

side chain n 側鎖 複合形容詞として使用.

▶ Cholesterol undergoes a series of reactions in the liver involving double-bond reduction, ring oxygenation, and partial side-chain destruction to give several closely related acids, known as the bile acids because they are secreted in the bile.

sided adj 複合語として，面をもった

▶ Five-sided prisms and more-than-six-sided prisms cannot be packed together to fill space.

side effect n 副作用

▶ Because of their toxicity and often serious side effects, use of sulfa drugs in treating disease is limited.

side-on adj サイドオンの ☞ end-on

▶ The side-on structure has been established in two dinickel complexes

side-on coordination n サイドオン配位

▶ Our studies in this area were initially motivated by the desire to understand side-on vs. end-on dinitrogen coordination in bimetallic zirconocene- and hafnocene-N_2 compounds.

side product n 副成物 syn *by-product*

▶ The use of pyrrolidine resulted only in

significant formation of unidentified side products.

side reaction n 副反応

▶ This side reaction was effectively suppressed when mildly basic conditions were employed.

siderophile element n 親石元素

▶ Siderophile elements are typically those which tend to exist in the native state, whether associated with Fe or not.

sieve vt ふるい分ける syn *filter, separate*

▷ The sieving or screening action, which makes it possible to separate smaller molecules from larger ones, is the most unusual characteristic of molecular sieves.

sight n **1** 照準 syn *view, eyeshot*

▶ It did seem wise initially to lower my sights a little and aim for the simplest possible tetrahedral and square planar components.

2 at first sight 一見して

▶ The situation is somewhat more complicated than it appears at first sight to be.

sigmatropic adj シグマトロピーの

▶ The mechanism presumably involves a 1,2-sigmatropic shift to contract the seven-membered ring with the resulting aldehyde being trapped by the ethylene glycol.

sigmatropic reaction n シグマトロピー反応

▶ A concerted reaction in which a group migrates with its σ bond within a π framework is called a sigmatropic reaction.

sigmoidal adj S字状の

▶ The kinetics of typical chemical bath deposition processes follows a sigmoidal profile similar to those observed for autocatalytic reactions.

sign n **1** 記号 syn *symbol*

▶ The sign of rotation is sometimes written as d for (+) and l for (−), or dl for (±).

2 兆候 syn *evidence, indication*

▶ The crystal at this stage still showed no signs of having melted or softened.

signal n **1** 信号 syn *indication, sign*

▶ The presence of an aldehyde group is apparent from the appearance of a signal at 9.7 ppm relative to tetramethylsilane.

2 きっかけ syn *impetus, stimulus*

▶ Electromagnetic radiation is an important nonchemical signal that affects plants and animals through many of the same signal pathways that the chemical signals utilize.

vt **3** 示す syn *indicate, notify*

▶ The associated contraction of the *a* axis of the unit cell signals a shorter Co···Co distance, reduced from 2.83 Å in the case of $LiCoO_2$ to 2.81 Å for $Li_{0.9}CoO_2$.

signal-to-noise ratio n 信号雑音比

▶ The high powers and signal-to-noise ratios of tunable lasers greatly simplify the recording of spectra under adverse conditions where very weak signals are obtained.

signature n 特徴 syn *characteristics, features*

▶ Not all spectral signatures arise from the interaction of all the molecules in a particle as it is the case for the exciton coupling.

significance n 重要性 syn *importance, value*

▶ The cyclobutadience ring system has fundamental significance as the first cyclic polyene for which Kekulé structures can be written.

significant adj **1** 重要な syn *important, valuable*

▶ From an empirical point of view, this distance of closest approach is obviously very significant in determining the packing in molecular crystals, and it is conveniently tabulated as the van der Waals radius.

2 著しい syn *considerable, substantial*

▶ The time that takes to mix reactants or to bring them to a specified temperature may be significant in comparison to the half-life of the reaction.

significant figure n 有効数字

▶ For most chemical work, four significant figures are sufficient.

significantly adv 著しく syn *distinctly, notably, remarkably*

▶ While the center of gravity motion of the polymer chains is very small, at the atomic level local relaxational processes may still provide liquid-like degrees of freedom which are in some ways not significantly different from those in an ordinary molecular liquid.

signify vt 意味する syn *indicate, represent, symbolize*

▶ A positive electromotive force signifies a deficiency of electrons on the right-hand electrode.

silent electric discharge n 無声放電

▶ The white crystalline cyclic tetramer [O=PF−O]$_4$ was obtained by subjecting equimolar mixtures of PF_3 and O_2 to a silent electric discharge at −70°.

silica gel n シリカゲル

▶ Silica hydrosol, on standing, polymerizes into a white jelleylike precipitate, which is silica gel.

silica glass n 石英ガラス

▶ The vapor-phase deposition process for silica glass is inherently slow, which in turn makes the fiber a high-cost product.

silica hydrosol n シリカヒドロゾル

▶ Mixing a sodium silicate solution with a mineral acid such as sulfuric or hydrochloric acid produces a concentrated dispersion of finely divided particles of hydrated SiO_2, known as silica hydrosol or silicic acid.

siliceous adj シリカを含む

▶ The so-called HSM-5 uniform heterogeneous catalyst is a protonated form of a highly siliceous molecular sieve with a composition close to SiO_2.

silver vt 銀めっきする

▶ The edges may be silvered to redirect the light back to the cells.

silvery adj 銀白色の syn *shining, shiny*

▶ Chromium in the massive state is lustrous, silvery, and fairly soft.

similar adj **be similar to** 類似した syn *resemble*

▶ A measurement of how similar one molecule is to another can be a useful parameter in studies of relationships between molecular structure and activity.

similarity n 類似性 syn *resemblance*

▶ This similarity in behavior suggests a similarity in mechanism.

similarly adv **1** 類似して syn *alike, equally, identically*

▶ Cadmium carbonate behave similarly to calcium carbonate and, on being heated, gives rise to the equilibrium $CdCO_3 \rightleftharpoons CdO + CO_2$.

adv **2** 同様に syn *in like manner, likewise*

▶ Similarly, in oxides the full band corresponds to the p electrons of the oxide ion, O^{2-}.

simple adj 簡単な syn *clear, lucid, plain*

▶ A increasing number of Hartree–Fock wave functions are being made available for simple molecules.

simple-minded adj 単純な syn *naive, simple*

▶ The available data can be described by a rather simple-minded procedure.

simplicity n 簡単 syn *clarity* ☞ complexity

▶ In view of the simplicity of the argument, it is not surprising that there are many exceptions to this radius-ratio rule.

simplify vt 簡単化する syn *clarify, explain*

▶ For large systems, the approach has to be simplified by reducing it to the dominant contributions.

simplified adj 簡単化された syn *popular*

▶ A simplified picture of the mechanism of Grignard addition involves coordination of the carbonyl oxygen with magnesium, followed by transfer of the organic portion of the Grignard reagent to the positive carbon of the aldehyde or ketone.

simply adv **1** 単に syn *barely, merely, only*

▶ The reaction is carried out simply by mixing together the two reactants, usually in an inert solvent like carbon tetrachloride.

2 否定文で，全然 syn *absolutely, totally, wholly*

▶ Many of the methods that are commonly used for the preparation of alkyl halides simply do not work when applied to the preparation of aryl halides.

simulate vt 模倣する syn *approximate, imitate, mimic, reproduce*

▷ High-pressure xenon arcs with appropriate

filters are a practical means of simulating the spectral distribution of daylight with a single light source.

simulated adj 模擬の syn *artificial, assumed*

▶ In the absence of single crystals for structure determination, the structures were solved by the ingenious use of Monte Carlo simulation with simulated annealing.

simulation n シミュレーション syn *imitation*

▶ All of the (C–)H atoms in the sample were included in the simulation.

simultaneous adj 同時の syn *coincident, concurrent*

▶ The preparations were made by simultaneous precipitation from solutions containing stoichiometric amounts of reactants.

simultaneous equation n 連立方程式

▶ By adding or subtracting the simultaneous equations, simple expressions can be derived containing the concentration of a single radical.

simultaneously adv 同時に syn *at the same time, together*

▶ In contrast to single metal centers, metal surfaces and clusters present possibilities for binding small molecules to more than one atom simultaneously.

since adv **1** 以来

▶ The validity of Fischer's chain and ring formulae has since been proved by the rigorous investigations of many other researchers.

2 since then その時以来

▶ The idea of using optical analogues to aid in the interpretation of X-ray diffraction patterns originated with Sir L. Bragg around 1938, and the method has developed considerably since then.

conj **3 だから** 理由が複数あるときにはその都度，since を使用する．動詞が過去形の場合は以来の意味と混乱する恐れがある． syn *as, because, for*

▶ Since indeterminate experimental errors may effect the accuracy of the data and since some error is present in the literature values used in the calculations, we estimate an error in the $\Delta G^{\cdot}$ values of ±2 kcal/mole metal.

single adj **1** 一個の syn *isolated, one*

▶ The phase boundary is represented by a single vertical line separating the components of the phases.

2 個別の syn *individual*

▶ An activating group activates all positions of the benzene ring; even the positions meta to it are more reactive than any single position in benzene itself.

3 single out 選抜する syn *choose, select*

▶ Titania was singled out in early studies because titanium cations had been shown to bond to various other transition-metal cations in a group of oxides known as hexagonal barium titanates.

single crystal n 単結晶 複合形容詞として使用.

▶ Confirmation of this assignment was provided by single-crystal X-ray diffraction and a series of NMR experiments.

single strand n 一本鎖 複合形容詞として使用.

▶ In the secondary structure of RNA, helixes are again involved, but this time nearly always single-strand helixes.

singlet n **1** 一重項 ☞ singlet state

▶ If the ground state is the singlet, as in most molecules, the molecule will be diamagnetic.

2 一重線

▶ A hydroxyl proton ordinarily gives rise to a singlet in the NMR spectrum.

singlet oxygen n 一重項酸素

▶ Singlet oxygen, $^1O_2$, can readily be generated by irradiating normal triplet oxygen, $^3O_2$, in the presence of a sensitizer.

singlet state n 一重項状態 ☞ singlet

▶ The energies of the excited π-electron singlet states can be obtained from its ultraviolet absorption spectrum.

single-valued adj 一価の

▶ The energy of a system is a single-valued function of its macroscopic state.

singly adv 一倍に

▶ Even singly charged cations obviously form

a highly ordered primary hydration sphere and thereby virtually saturate the possible orientation polarization of the surrounding medium.

singular adj 並外れた syn *extraordinary, uncommon, unusual*

▶ The abundance and complexity of varieties of organic compounds make the problem of organic nomenclature one of singular importance.

singularly adv 並外れて syn *especially, particularly*

▶ As a class, alkanes are singularly unreactive.

sink n **1** 溜め

▶ The electrodes consist of conductors that introduce the source and sink of electrons into the solution.

vi **2** 沈む syn *descend, fall*

▶ In the float-sink method, a few crystals of the material are suspended in liquids of a range of densities until a liquid is found in which the crystals neither sink nor float.

sinter n **1** 焼結物 ☞ sintered-glass filter

▶ The cloudy solution was filtered through sinter, the filtrate was evaporated in vacuum, and the resulting solid was redissolved with heating in acetonitrile (15 mL).

vt **2** 焼結させる

▷ The vast majority of work to date has focused on the synthesis and study of sintered, polycrystalline samples obtained by sintering and annealing pellets composed of the starting materials Y_2O_3, CuO, and $BaCO_3$.

sintered-glass filter n ガラス沪過器 ☞ sinter

▶ Other samples of the solvent-free sodium complexes were washed on a sintered-glass filter under nitrogen with successive aliquot portions of a solvent.

sintering n 焼結 ☞ fritting

▶ Sintering involves heating just below the melting point; this process promotes bonding between the grains composing the pellet, thereby increasing the density and strength of the pellet.

sinusoidal adj 正弦曲線の

▶ The modulated structure derives from a modulation, often sinusoidal in nature, of the average atomic positions and arises from electron-phonon coupling.

siphon n **1** サイホン

▶ The direction of electrophoresis is conveniently observed by placing the liquid in a siphon the ends of which are closed by a membrane, dipping into a cathode and anode vessel, respectively.

vt **2** サイホンで吸う syn *drain, pump out*

▶ After some 48 h or so the stirring was stopped, and the ethereal layer was siphoned off.

sit vi 位置する syn *rest, stay*

▶ The Mg^{2+} ion sits at the middle of the porphyrin ring that forms the skeleton of chlorophyll molecules.

site n **1** 位置 syn *location, place, position*

▶ The softest sites of a molecule are those where the electron density can be changed most easily, and attack by reagents will occur at these sites.

2 on site その場で

▶ Because of the reactivity, instability, and hazardous nature of O_3, it is always generated on site.

situate vt 位置する syn *locate, place, position*

▶ If two homogeneous bulk phases are in contact, the atoms or molecules at the interface are situated in an environment different from that in the interiors of the phases.

situated adj 位置して syn *fixed, settled*

▶ Molecules situated near the point of emergence of the dislocation are likely to have a higher chemical potential and a greater tendency to evaporate or dissolve than those distant from it.

situation n 事態 syn *case, circumstance, condition, state*

▶ The concept of light that one uses in a particular situation depends somewhat on the situation.

six-membered ring n 六員環

▶ Cyclohexane is a typical cycloalkane and has

six methylene groups joined together so as to form a six-membered ring.

sizable adj かなり大きな syn *considerable, substantial*

▶ At elevated temperatures, the atoms in a crystal undergo sizable displacements from the averaged positions in the course of molecular translational and especially rotational vibrations.

size n サイズ syn *dimensions, extent, magnitude*

▶ A typical size crystal was 6.5×6.5×3 mm.

skeletal adj 骨格の

▶ Long-range order or a rigid skeletal structure is not essential features of a solid electrolyte.

skeleton n 骨格 syn *framework*

▶ The structures show enhanced thermal stability on account of the inertness of the inorganic skeleton.

skeptical adj 懐疑的な syn *doubtful*

▶ Ostwald remained skeptical of the real existence of atoms until 1909, when he was finally convinced by the experiments of Thomson on cathode rays and of Perrin on the Brownian movement.

skepticism n 疑念 syn *distrust, doubt*

▶ Little's theory was based on electron-phonon interactions, and it was greeted with some criticism and skepticism.

sketch vt 写生する syn *depict, draw, outline*

▷ The faces of many crystals with cubic unit cell show the eight-sided crystal form, octahedron, sketched in Figure 1.

sketchy adj 皮相な syn *crude, ill-defined, rough, superficial*

▶ The aqueous solution chemistry of arsenic is very sketchy.

skew configuration n スキュー配座

▶ In the gas phase, H_2O_2 molecule adopts a skew configuration with a dihedral angle of 111.5°.

skill n 技能 syn *ability, experience, technique*

▶ Those who practice in any area of chemistry have a large body of knowledge and particular skills in an activity.

skillful adj 熟練した syn *able, capable, professional*

▶ Skillful use of the microscope can give useful supporting information about structural changes in transformations in solids, though the conclusions are necessarily more superficial.

skim vt 液体からすくい取る syn *remove, separate*

▶ The Pb bullion is melted and held just above its freezing point when Cu rises to the surface as an insoluble solid which is skimmed off.

slab n 厚板 syn *plate, sheet*

▶ When a drop of liquid is placed on a slab of solid with a gas above, there are three surfaces of contact.

slag n スラグ syn *scoria*

▶ Slag is often the medium by means of which impurities may be separated from metal.

Slater determinant n スレイター行列式

▶ A convenient way of formulating the product wave function that embodies the Pauli requirements is as a Slater determinant.

slender adj 細長い syn *narrow*

▶ The capillary should extend to the very bottom of the flask, and it should be slender and flexible so that it will whip back and forth in the boiling liquid.

slice n 薄片 syn *piece, slab*

▶ A thin slice of tissue can be fixed to a microscope slide and incubated with a specific labeled probe molecule that hybridizes only to the transcript or protein of interest.

slide n **1** すべること

▶ Along a uniform slide wire, there is a linear potential drop.

2 スライド

▶ A microarray is a glass slide onto which DNA representing up to 30,000 genes is bound in distinct spots, with each spot consisting of a different gene.

vi **3** すべる syn *glide, slip*

▶ Layers are held together primarily by weak van der Waals forces, permitting them to slide relative to one another.

slight adj わずかな syn *little, minor, negligible, small*

▶ If a very little of fluorescein is present, the silver chloride precipitate assumes a reddish tint as soon as there is a very slight excess of Ag^+ present.

slightly adv わずかに syn *marginally, somewhat*

▶ Ammonium molybdate in neutral or slightly acid solutions of nickel salts causes a greenish white precipitation on heating to 70 ℃.

slip n **1** すべり

▶ The process of movement of dislocations is called slip, and the pile-up of half-planes at opposite ends gives ledges or slip steps.

vi **2** すべる syn *glide, slide*

▶ As a consequence of the motion of the dislocation, the top half of the crystal slips to the right by one plane of atoms, yet the entire set of bonds between the planes that slip need not be broken simultaneously.

slip plane n すべり面

▶ The behavior of a metal under stress depends very much on the direction of the applied stress relative to the direction and orientation of the slip directions and slip planes.

slit n スリット syn *aperture, opening*

▶ With good resolution, only the ions of a single mass number pass through the slit and impinge on the collector.

slope n 勾配 syn *inclination, tilt*

▶ The slope of the vapor pressure curve can be obtained with enough precision to determine the enthalpy of sublimation with fair accuracy.

slow adj **1** 遅い syn *sluggish*

▶ Silver-white tin changes to brittle gray tin at temperature of 18 ℃, but the transition is normally very slow.

vt **2** 遅くする syn *hamper, oppose*

▶ The overall rates of chain reactions are usually very much slowed by substances which can combine with atoms or radicals and convert them into species incapable of participating in the chain-propagation steps.

3 slow down 速度を落とす syn *reduce speed, retard*

▶ If we could sufficiently slow down rotations about single bonds by lowering the temperature, we would expect an NMR spectrum that reflects the instantaneous environments of protons in each conformation.

slowly adv ゆっくり syn *carefully, gradually*

▶ The sample was then cooled slowly and stored for several days at room temperature to help ensure equilibration.

slowness n 遅さ

▶ Because of the slowness with which equilibrium is established, the variation of composition may not taken place immediately.

slow neutron n 低速中性子

▶ Moderators are used in nuclear reactors, because slow neutrons are most likely to produce fission.

sludge n スラッジ syn *precipitate, residue*

▶ Selenium is also recovered from the sludge accumulating in sulfuric acid plants.

sluggish adj 緩慢な syn *inactive, slow*

▶ Because transitions among the three modifications of SiO_2 are very sluggish, each of these polymorphic forms can be studied at temperatures outside its range of stability.

sluggishly adv 緩慢に

▶ Because the ions are much more massive than the electrons and move more sluggishly, this track of displaced positive charge persists long enough to attract a second electron and thus forms the Cooper pair.

sluggishness n 緩慢さ syn *inactivity, passivity*

▶ Many carbon compounds are unaffected by air and unhydrolysed by water, but their apparent stability is a consequence of kinetic sluggishness.

slurry n **1** スラリー

▶ The insoluble hydrous cerium(IV) oxide can be deaggregated by treating an aqueous slurry with nitric acid at a carefully controlled H^+:Ce(IV) mole ratio.

vt **2** 泥状にする

▷ The synthesis and processing of the working hydrodesulfurization catalyst involves slurrying the high-surface-area alumina support with solutions of molybdenum and cobalt, followed by firing to yield a mixture of the oxides.

slush n どろどろのもの

▶ The ordinary ice bath should consist of a slush of crushed ice covered with water.

small adj **1** 小さい syn *little, tiny*

▶ Magnus' green salt crystallizes as small tetragonal needles containing $[Pt(NH_3)_4]^{2+}$ cations and $[PtCl_4]^{2-}$ anions stacked alternatively above one another in the direction of the needle axis.

2 わずかな syn *limited, minor, negligible, slight*

▶ Small additions of alcohol to a solution of iodine in chloroform suffice to shift the color stepwise from violet to brown.

small-angle scattering n 小角散乱

▶ The small-angle scattering around the primary X-ray beam is just of the right order of magnitude to determine particle dimensions.

small ring n 小員環 複合形容詞として使用

▶ When small-ring cyclic glycols are oxidized by lead tetraacetate, the cis isomer is transformed at a greater rate than the trans isomer.

small-scale adj 小規模の ☞ large-scale

▶ Solutions in a substance like camphor, which gives a very large freezing point depression, are particularly suitable for small-scale operations.

smart adj 高性能な syn *ingeneous, intelligent*

▶ Electrorheological fluids are smart materials, whose viscosity can be tuned by an applied electric field.

smear vt にじませて不鮮明にする syn *spread*

▶ A relatively large dipole-dipole interaction between the two unpaired electrons leads to a fine structure splitting which, in a polycrystalline sample, smears out the resonance to the point of undetectability.

smectic phase n スメクチック相

▶ If we introduce partial positional order in addition to the orientational order, a family of smectic phases can be generated, which are characterized by having some layering of the molecules.

smectics n スメクチック液晶物質

▶ A feature of the liquid-crystal polymers is that in common with low-molar-mass smectics, but in contrast with low-molar-mass nematics, materials may remain aligned for months to years in the liquid-crystal state following the removal of an aligning electric or magnetic field.

smell n **1** 匂い syn *aroma, fragrance, odor*

▶ Ozone can be detected by its smell in concentrations as low as 0.01 ppm.

vi **2** 匂う

▷ Tricarbon dioxide is an evil-smelling gas with boiling point 7 ℃ and melting point −111 ℃.

smelt vt 製錬する

▷ Furnaces are used for steel production, for smelting iron and other ores, and for manufacture of furnace carbon black, etc.

smelting n 製錬

▶ Sulfur is recovered in the smelting process for some of the metal sulfides.

smog n スモッグ syn *fog, fume, mist*

▶ Dusts, smokes, and smog are produced as the result of many industrial operations and combustions.

smoke n **1** 煙 syn *cloud, smog*

▶ The tars occurring in cigarette smoke can lead to lung cancer.

vi **2** 煙を出す

▶ The oil smokes somewhat at this temperature, but use of a fume hood takes care of this.

smooth adj **1** 平坦な syn *even, flat, plane*

▶ Atomically smooth faces of a crystal are formed from close-packed planes of atoms of low Miller indices.

2 円滑な syn *easy, even, free*

▶ The reactivity of aryl halides such as the halobenzenes and halotoluenes is exceedingly low toward nucleophilic reagents that normally effect smooth displacements with alkyl halides and activated aryl halides.

vt **3 ならす** syn *even, flatten, level*

▷ The smoothed experimental data from Figure 1 are included for comparison.

smoothly adv **円滑に** syn *easily, freely, readily*

▶ In the presence of a small amount of phosphorus, aliphatic carboxylic acids react smoothly with chlorine or bromine to yield a compound in which α-hydrogen has been replaced by halogen.

S_N reaction n S_N反応 ☞ nucleophilic substitution

▶ S_N reactions carried out using the solvent as the nucleophilic agent are called solvolysis reactions.

S_N1 reaction n S_N1 反応

▶ S_N1 reactions of neutral substrates go faster in water than in ethanol; they go faster in 20% ethanol than in 80% ethanol.

snugly adv **ぴったりと**

▶ Models indicated that the tetramethylammonium cation might fit snugly into the adamantine-like cavities formed by the diamond net.

so adv **1 そのように**

▶ The most widely used example of halogen exchange is provided by the preparation of alkyl iodides from chlorides or bromides using sodium iodide in a solvent, such as acetone, in which sodium iodide is soluble but sodium chloride or bromide is relatively less so.

2 非常に

▶ Since a transition state is so unstable and since it has such an extremely short lifetime, there is no way to determine its precise structure.

3 接続詞的に, **それゆえ** syn *accordingly, consequently, therefore*

▶ The chemistry of a molecule is a property of the electrons, so the electron density is perhaps the most obvious property to use in quantitative comparisons.

4 and so 接続詞的に, **それゆえに**

▶ Pyridine was added to remove the HCl and so prevent the reverse reaction.

5 so much (adj) **その程度までの**

▶ It has been demonstrated that in the formation of quinhydrones by mixing the solid quinone and hydroquinone reactants, the solids have sufficient mobility to react completely to form a new product but not so much mobility that there is rearrangement of that product to its more stable isomer by a hydrogen-transfer reaction.

6 or so 期間の表現に続いて, **ほど** syn *approximately*

▶ The crystals should be melted completely by holding the melt for 30 sec or so about 10 to 20 ℃ above the melting point.

7 so that (a) **…となるように**

▶ It is difficult to control the alkylation of a primary amine so that it will yield the secondary amine exclusively.

(b) **そのために** 結果を述べる副詞節の前にはコンマを要する.

▶ For a given alkyl group, the boiling point increases with increasing atomic weight of the halogen, so that a fluoride is the lowest boiling, an iodide the highest boiling.

8 so…that 程度, 結果を表して, **非常に…なので**

▶ The structure of α-AgI is so suited for easy motion of Ag^+ that the ionic conductivity actually decreases slightly on melting at 555 ℃.

soak vt **浸す** syn *immerse, wet*

▷ The release of halogen from dry crystals of silver halides may be demonstrated chemically with indicators such as filter paper soaked in a solution of fluorescein and dried.

so-called adj **いわゆる** 引用符を併用しない. syn *misnamed, supposed*

▶ For shorter *n*-alkanes, C_{19}—C_{23}, we found only the so-called rotator transition.

soda lime n **ソーダ石灰**

▶ Soda lime is a mixture of calcium oxide with sodium hydroxide or potassium hydroxide intended for the absorption of carbon dioxide and water vapor.

sodium pump n **ナトリウムポンプ** ☞

pump

▶ Maintenance of these large concentration gradients requires a sodium pump.

soft adj **1 軟らかい** syn *compressible, flexible*

▶ The sulfur end of thiocyanate ion is assumed to be much softer than the nitrogen end and, hence, to prefer soft Lewis acids.

2 磁性が軟質の

▶ Magnetically soft materials have a high permeability and a low coercive field and tend to have low hysteresis losses.

3 穏和な syn *gentle, mild*

▶ Soft chemistry involves modifying an existing compound under relatively mild conditions to produce a closely related material.

soft acid n **軟らかい酸** ☞ hard acid

▶ A soft acid is one in which the acceptor atom is large, carries a low positive charge or none at all, and has electrons in orbitals which are easily distorted.

soft base n **軟らかい塩基** ☞ hard base

▶ A soft base is one in which the valence electrons are easily polarized.

soften vi **1 軟らかくなる** syn *melt*

▶ Laboratory tubing made of soda-lime glass softens at a comparatively low temperature and can be worked most satisfactorily using an air-gas blast lamp.

vt **2 硬水を軟化する**

▶ Most boiler feed water is softened before being used.

3 金属材料を軟らかくする syn *moderate, temper*

▶ A work-hardened metal can be softened again by annealing at high temperatures, generally at temperatures above one-half of the melting point on the absolute temperature scale.

softener n **軟化剤**

▶ The *cyclo*-polyphosphates are used as water softeners. Sodium *cylco*-hexaphosphate, $Na_6P_6O_{18}$, is especially effective for this purpose.

soft glass n **軟質ガラス**

▶ At room temperature, soft glasses are somewhat permeable to protons almost to the exclusion of all other ions.

softening n **1 軟化**

▶ Annealing is important and requires care and skill, for the annealing point is not much below the temperature of softening.

2 硬水の軟化

▶ Both natural and artificial zeolites are used extensively for water softening.

softening point n **軟化点**

▶ Pyrex or other borosilicate glass has a higher softening point than soda-lime glass and a lower coefficient of expansion.

softness n **軟らかさ**

▶ Pearson's rule, now known as the HSAB principle, is used as an operational definition of hardness or softness when the assignment of hard and soft to a reference pair of acids (or bases) seems relatively unambiguous.

soil n **1 うわ土** syn *humus*

▶ Humic acid is a brown, polymeric constituent of soils, lignite, and peat.

vt **2 表面を汚す** syn *contaminate, smear*

▷ Glass soiled with oleic acid and rinsed with dilute sodium oleate solution shows a zero receding angle of contact.

sol n **ゾル**

▶ Sols are dispersions of colloidal particles in a liquid.

solar cell n **太陽電池**

▶ The essential component of a solar cell is a thin sheet or wafer of crystalline or amorphous silicon.

solder n **はんだ**

▶ The solder acts as an adhesive and does not form an intermetallic solution with the metals being joined.

sole adj **単独の** syn *only, particular*

▶ Lithium alkyls can be prepared not only in the presence of ether but also with low-boiling petroleum ether as sole solvent.

solely adv **ただ** syn *merely, only, simply*

▶ Crystals held together solely by London dispersion forces melt at comparatively low temperatures, and the resulting liquids vaporize easily.

sol-gel process n ゾルゲル法

▶ The capability for preparing high-density spherical particles with controlled size range using colloid precursors established the sol-gel process as a practical and versatile technique.

sol-gel processing n ゾルゲル加工

▶ Sol-gel processing techniques using colloidal hydrous oxide intermediates can be used to prepare inorganic oxides of very high density.

solid n **1** 固体

▶ Solids like Ge_xSi_{1-x} often form disordered substitutional solid solutions; that is, germanium atoms randomly replace silicon atoms in elemental silicon.

adj **2** 固体状態の

▶ The stoichiometry of solid compounds can be determined by counting the atoms in their unit cells.

3 一様な syn *consistent, homogeneous, uniform*

▶ The solid curve shows qualitatively how the potential energy varies as a function of interatomic distance.

solid acid n 固体酸

▶ Mesoporous silica is a popular support material for solid acids designed for liquid phase applications.

solid acid catalyst n 固体酸触媒

▶ Solid acid catalysts that could serve to minimize the content of aromatics, carcinogenic benzene in particular, from gasoline and other fuels are required.

solid electrolyte n 固体電解質

▶ In solid electrolytes, one component of the structure, cationic or anionic, is not confined to specific lattice sites but is essentially free to move throughout the structure.

solidification n 固化

▶ During solidification the part that solidifies last is usually different from that which first separates on cooling.

solidify vi 固化する syn *crystallize, freeze,*

▶ Some substances solidify to the amorphous state if they are cooled very rapidly, because time is necessary to permit the atoms to change from the molecular groupings into regular arrangement, or lattices, in which the interatomic forces are in equilibrium.

solid line n 実線 ☞ dashed line, dotted line

▶ The solid line shows the transverse displacement y at a position x at time t. The dashed line shows the same curve at a later time $t+\Delta t$.

solidly adv すっかり syn *firmly, rigidly, tightly*

▶ Space can be filled solidly with cubic cells that are exactly alike.

solid phase n 固相

▶ The dodecahedral structure of $H_3O^+(H_2O)_{20}$ may carry over into the solid phase.

solid-phase peptide synthesis n 固相ペプチド合成法

▶ A major breakthrough came with the development of solid-phase peptide synthesis by R. B. Merrifield.

solid solution n 固溶体

▶ The ability of two substances to form solid solutions is determined primarily by geometrical rather than chemical considerations.

solid state n 固態

▶ The characteristic of the solid state is that the substance can maintain itself in a definite shape that is little affected by changes in temperature and pressure.

solid-state laser n 固体レーザー

▶ The solid-state laser is basically a luminescent solid in which certain special requirements have been met.

solid-state polymerization n 固相重合

▶ The most intriguing problem in solid-state polymerization concerns the extent to which the geometry of the crystal lattice controls the characteristics of the reaction.

solid-state reaction n 固相反応

▶ Solid-state reactions usually occur between apparently regular crystal lattices in which motion of lattice units is very restricted and depends in a complex manner on the presence of lattice defects.

solidus n 固相線

▶ Below the solidus temperature, homogenization of the metal, with the elimination of coring, occurs rapidly.

solid-state synthesis n 固相合成

▶ Many solid-state syntheses are based simply on heating a mixture of solids together with the intent of producing a pure, homogeneous sample of desired stoichiometry, grain size, and physical-chemical properties.

soliton n ソリトン

▶ A soliton is a topological excited state in which the sense of bond alternation is reversed in the vicinity of a charge carrier.

solubility n 溶解度

▶ If the system in equilibrium contains a solution and another phase that is one of the components of the solution in the form of a pure substance, the concentration of that substance in the solution is called the solubility of the substance.

solubility curve n 溶解度曲線

▶ The solubility curve is, for any given substance, continuous, so long as the solid phase remains unchanged.

solubility product n 溶解度積

▶ The emf method is a valuable one for measuring solubility products for salts of very low solubility, for which direct solubility measurements cannot be made with high accuracy.

solubilization n 可溶化

▶ Solubilization supplies a means of bringing into solution any substance in any desired solvent provided there is a suitable detergent or colloidal electrolyte available.

solubilize vt 可溶性にする

▶ The most useful property of micelles is their ability to solubilize, i.e., to dissolve hydrophobic material in their interiors.

soluble adj 可溶性の

▶ Osazones are much less soluble in water than the parent sugars, since the molecular weight is increased by 178 units and since the number of hydroxyl groups reduced from five to four.

solute n 溶質 ☞ solvent

▶ Although there is no fundamental difference between components in a solution, we call the component that constitutes the larger proportion of the solution the solvent. The component in lesser proportion is called the solute.

solution n **1** 溶液

▶ Aqueous solutions containing mixtures of copper(II) and silver nitrates with total concentration about 0.4 M were added to boiling solution of approximately 0.1 M K_2HgI_4.

2 溶解 syn *dissolution, mixing*

▶ There is no intrinsic reason why the sign of the integral molar enthalpy of solution at infinite dilution should be related to the temperature dependence of the solubility of a solute.

3 解答 syn *answer, explanation*

▶ The solution to the problem was found by using the interesting solvent formamide in which (cetyltrimethylammonium)$_4Ge_4S_{10}$ is soluble and self-assembles into a germanium cluster lyotropic liquid crystal.

solvate n **1** 溶媒和物 ☞ hydrate

▶ The reference molecule A has the phenyl rings shaded as has the reference benzene solvate molecule.

vt **2** 溶媒和する

▷ For highly solvated macromolecules in water, the solvating layers of water help to prevent the individual particles from agglomerating.

solvated electron n 溶媒和電子

▶ The preparation of salts of the alkali metal anions has its origins in the study of metal-ammonia solutions and solvated electrons.

solvation n 溶媒和 ☞ hydration

▶ Solvation should not be ignored in the liquid phase: reactants are normally surrounded by a shell of solvent molecules and will diffuse as a solvent-solvate assemblage.

solvation effect n 溶媒和効果

▶ Solvation effects can play an extremely important role in the rates and paths of chemical reactions, and it is here that our interest should probably be the greatest.

solvation energy n 溶媒和エネルギー

▶ Solvation energy in water is very closely

linked with crystal radii in keeping with a relatively simple ion-dipole model.

solvation sheath n **溶媒和のさや**

▶ Conductance measurements suggest that cations carry a large solvation sheath in dipolar aprotic solvents, while anions are relatively naked.

solvation shell n **溶媒殻**

▶ The first solvation shell of an ion is highly ordered, because of the strong ion-dipole forces, and the solvent molecules are unable to reorient.

solvatochromic adj **溶媒和発色による**

▶ The solvent strength of supercritical fluids has most often been quantified by studying the solvatochromic shifts in the UV-vis absorption bands of organic dyes.

solve vt **解明する** syn *answer, clarify, elucidate, explain*

▶ The parallels and differences among hemoglobin, hemerythrin, and hemocyanin illustrate the ways in which evolution has often solved what is basically the same problem in different ways in different groups of animals.

solvent n **溶媒** ☞ solute

▶ Dimethyl sulfoxide is a versatile solvent which provides a particularly useful medium for reactions between polar and nonpolar reagents.

solvent cage n **溶媒かご**

▶ The encounter pair of reactants is trapped inside the solvent cage for a period in which they can bound on and off each other before one partner finds a route out of the cage.

solvent effect n **溶媒効果**

▶ Solvent effects are particularly marked for bimolecular reactions involving ions.

solvent extraction n **溶媒抽出** syn *liquid-liquid extraction*

▶ Because many uncharged complexes are insoluble in water but soluble in organic solvents, they are used for separations based on solvent extraction procedures.

solvolysis n **加溶媒分解** syn *solvolytic reaction*

▶ It has been suggested that there is a continuous spectrum of mechanisms for solvolysis ranging from classical S_N1 reaction at the one end to the single-step S_N2 reaction at the other.

solvolytic adj **加溶媒分解の** ☞ hydrolytic

▶ Quantitative solvolytic data for octacoordinate species are lacking.

solvolyze vi **加溶媒分解する** ☞ hydrolyze

▶ When *t*-butyl chloride solvolyzes in 80% aqueous ethanol at 25 ℃, it gives 83% of *t*-butyl alcohol by substitution and 17% of isobutylene by elimination.

some pron **1** …のうちいくらか 単数形としても複数形とも見做される syn *part*

▶ When monochromatic light is passed through a material, some of it is scattered.

adj **2 いくらかの** 複数名詞の前に some を置いたときには，いくらかは特定されない．もし，それが主語であれば，動詞は複数形を用いる．syn *several*

▶ An interesting phenomenon found in some crystals is the tendency to be pyro- and piezoelectric.

不可算名詞の前に some を置いて特定されない量，種類を表す．この場合には，動詞は単数形を用いる．

▶ Some evidence for p_π-d_π bonding in silicon compounds is provided by the shapes of the molecules $N(SiH_3)_3$ (planar) and H_3SiNCO and H_3SiNCS (linear).

3 単数形の可算名詞を伴って，**ある** syn *unknown, unspecified*

▶ A large number of alloys and intermetallic compounds show some kind of magnetic ordering.

4 数詞とともに用いて，**約** syn *about, approximate*

▶ Significantly, the Si-Si bond is weaker than the C-C bond by some 30 kcal/mole, whereas the Si-O bond is stronger than the C-O bond by some 22 kcal/mole.

somehow adv **どうにかして** syn *in some way*

▶ The observation of millimeter-sized inorganic forms bearing micrometer-dimension surface patterns suggested to us that vesicles were somehow involved in the templating process.

something pron あるもの

▶ Physically, the restrictor is a piece of stainless steel or fused silica capillary with a very small bore inside diameter, something on the order of 30–50 μm.

sometimes adv 時には syn *occasionally*

▶ Oxalic acid and the acid oxalats are used sometimes as acids and sometimes as reducing agents.

somewhat adv 多少 syn *more or less, moderately, rather*

▶ Esters usually show the carbonyl band at somewhat higher frequencies than ketones of the same general structure.

somewhere adv どこかに

▶ As a result of the law of conservation of energy, this energy must go somewhere since it cannot simply cease to exist.

sonication n 音波処理

▶ Dispersal of surfactants in water by ultrasonic irradiation using either a bath or a probe type sonicator is called sonication.

sonolysis n ソノリシス

▶ Sonolysis in pure water produces hydrogen atoms, hydroxyl radicals, molecular hydrogen, oxygen, and hydrogen peroxide.

soon adv **1** 速やかに syn *immediately, promptly, quickly*

▶ When a mixture of cyclohexanol and phosphoric acid is heated in a flask equipped with a fractionating column, the formation of water is soon evident.

2 as soon as (conj) …するとすぐに syn *immediately, when*

▶ As soon as one glycine molecule has been converted into its acid chloride, it might well react another molecule of glycine.

3 sooner or later 遅かれ早かれ syn *eventually, ultimately*

▶ The concepts mentioned here must sooner or later find application in many of the low-temperature phenomena of ionic crystals.

soot n 煤

▶ Soots formed by polymerization of gaseous hydrocarbons have low bulk density and low electrical and thermal conductances; they are useful for thermal insulation at high temperatures.

sophisticated adj 精巧な syn *advanced, elaborate, refined*

▶ Most devices are made from single-crystal layers with specific material requirements based on sophisticated preparative techniques and good knowledge of chemical and physical processes involved.

sophistication n 高機能化 syn *complexity, refinement, subtlety*

▶ When we compare functions of metal ions with organic groups, the use of the former as catalysts is essential and the latter is a sophistication to gain selectivity.

sorb vt 収着する ☞ absorb, adsorb

▶ Silver halides can sorb dye molecules, the first layer of which have the hydrophilic ends adhering to the silver halide surface with the hydrophobic ends exposed to the aqueous solution.

sorbate n 収着質 ☞ adsorbate

▶ Stronger interactions involve direct electron transfer between the sorbate and the sorbent.

sorbent n 収着媒 ☞ adsorbent

▶ Significant progress is being made on complex metal hydrides such as Ti-doped $NaAlH_4$ and new porous sorbent materials for hydrogen storage.

sorption n 収着 ☞ absorption, adsorption

▶ The distinction between adsorption and absorption is not always clear-cut, and the noncommittal word sorption is sometimes used.

sorptive adj 収着の

▶ Clay is important for its sorptive properties in the bleaching and refining of many materials.

sort n **1** 種類 syn *class, kind, type*

▶ Intramolecular hydrogen bonds of the sort in *o*-nitrophenol are encountered throughout organic chemistry.

2 sort of (adv) …のようなもの syn *rather, somewhat*

▶ Polymers have a range of molecular weights, and any data for the size or weight of molecules of a polymer must represent some sort of average value.

vt **3 分類する** syn *categorize, classify*

▶ The positive ions produced by electron bombardment are sorted as to their mass-to-charge ratio by the analyzing magnet.

4 sort out えり分ける syn *choose, select, separate*

▶ With thermodynamics, we can sort out the reactions that need to be driven and calculate the extra driving force of reactions that occur spontaneously.

sorting n 区分

▶ The gross structural sorting is then explicable in principle on these grounds, although, at present, we do not understand the exact location of the dividing lines.

sought-after adj **珍重される** syn *desirable, desired*

▶ In addition to being a more potent dearomatization agent, the Re(I) fragment also possessed many of the sought-after qualities absent in the Os(II) system.

sound adj 確実な syn *dependable, firm*

▶ The hypothesis of hyperconjugation has been given a sound physical basis by the application of the molecular orbital method to the problem.

sound wave n 音波

▶ A sound wave is a longitudinal wave, one in which the particles of the medium move back and forth in the direction of propagation.

sour vi 酸っぱくなる syn *curdle, ferment*

▶ Milk sours when lactose is converted into lactic acid by bacterial action.

source n **1 源** syn *cause, origin, root*

▶ Since both ring and double bond are good sources of electrons, there may be competition between the two sites for certain electrophilic reagents.

2 複数形で，**情報源** syn *documentation*

▶ The references to Table I include the sources of data for the respective substances for the other tables of this paper.

space n **1 空間** syn *place, room*

▶ X-ray analysis shows that the amylose chains are coiled in the form of a helix, inside which is just enough space to accommodate an iodine molecule: the blue color is due to entrapped iodine molecules.

2 紙面

▶ Far more space would be needed to describe all their extraordinary properties, but a final mention should be made of the most remarkable superconductivity.

vt **3 間隔を置いて並べる** syn *align, arrange, array*

▷ Modern cells employ arrays of anodes and cathodes spaced 3-mm apart and carrying current at 2700 A m^{-2} into brine at 60–80°.

space charge n 空間電荷

▶ Mott's theory provides theoretical justification for numerous empirical laws of growth when oxide layer is so thin that effects of space charges may be neglected.

space-charge-limited current n 空間電荷制限電流

▶ Although space-charge-limited currents were indicated in the case of the electron steady-state photocurrents, the electron transient photocurrents were found to depnd linearly on the applied voltage.

space filling n 空間充填

▶ Many of the other metallic elements can be described in terms of a different type of extended structure that is not as efficient at space filling, called body-centered cubic.

space-filling model n 空間充填模型

▶ Space-filling model is widely used to determine the possible closeness of approach of groups to each other and the degree of crowding of atoms in various arrangements.

space group n 空間群

▶ In the space group $P\bar{1}$, there is one unique molecule with a crystallographically imposed center of inversion present in the unit cell.

space lattice n 空間格子
▶ The high-temperature phase of CsCl has the NaCl space lattice.

spacer n スペーサー
▶ Often, one material is used to form spacer layers and has no purpose other than to allow the molecular dipoles within the active layers to be orientated in a common direction.

spacing n 面間隔 ☞ lattice spacing
▶ The normal practice in using powder patterns for identification purposes is to pay most attention to the *d* spacings.

spallation n 原子核の破砕
▶ The efficiency of the spallation process, yielding about 30 neutrons per proton, gives a high neutron flux suitable for diffraction experiments.

span n **1** 全範囲 syn *extent*
▶ Use of transition-metal complex catalysts affords a much wider span of control over rates, selectivity, and extent of hydrogen release from ammonia-borane.
vt **2** 及ぶ syn *extend, reach*
▷ The first Bohr radius typically expands from 0.53 Å in hydrogen to tens of angstroms, spanning many atoms, for common semiconductor dopants.

sparingly adv わずかに
▶ Calcium oxide reacts rapidly and exothermally with water to give the sparingly soluble hydroxide, $Ca(OH)_2$.

spark n 火花 syn *flash*
▶ Metallic cerium is strongly pyrophoric; it gives sparks when struck or rubbed against steel.

sparking n 点火
▶ Either rapid squeezing of the piezoelectric crystal or rapid release of the pressure applied to the crystal can cause sparking in the nearby combustible gas and ignite it.

sparse adj 乏しい syn *few, meager, rare*
▶ The chemistry of oxidation states above +4 is sparse for cobalt, rhodium, and iridium.

sparsely adv 希薄に syn *sparely*
▶ Sodium ions are accommodated rather sparsely in separate Al_2O_3 layers where they are free to move and so make γ-alumina a good electric conductor.

spatial adj 空間的な
▶ Stereochemistry deals with those properties of molecules that result from the spatial relationships of their constituent groups.

spatially adv 空間的に
▶ The solid acid catalyst ZSM-5 is a so-called uniform heterogeneous catalyst in the sense that the active sites are distributed in a spatially uniform fashion throughout the bulk of the solid.

spattering n はね
▶ The reaction is extremely exothermic and can result in explosive spattering of the mixture if water is added to sulfuric acid.

speak vi **speak of** …のことをいう syn *comment, mention, refer to*
▶ When radiation is emitted in a chemical reaction, we speak of chemiluminescence.

special adj 特別の syn *extraordinary, rare, unconventional, unique*
▶ By Langmuir's atomic hydrogen torch refractory metals, such as tungsten and tantalum, can be melted and worked, and satisfactory joints can be made in special alloy steels which are difficult to weld in an ordinary flame on account of oxidation.

specialized adj 専門化した syn *particular, specific*
▶ When each membrane became specialized in pumping, either chemically selective or directive, storage was made selective within vesicles within cells.

specially adv 特に syn *especially, particularly*
▶ The stabilization of the anions of these specially activated esters is greater than for simple esters because the negative charge can be distributed over more than two centers.

specialty n 特別なこと syn *area, field, subject*
▶ Solid-state chemistry has frequently been regarded as a specialty of little interest to the

practicing organic chemist.

species n 種 syn *kind, sort, type*

▶ Many biological minerals exhibit species with specific structures and shapes and uniform particle sizes.

specific adj **1** 限定的な syn *distinct, limited, restricted*

▶ Lead tetraacetate and periodic acid have a specific action on 1,2-glycols and related compounds, oxidizing them to two molecules of carbonyl compounds.

2 具体的な syn *proper, typical*

▶ As a specific example of a calculation of ΔS for a phase transition, we consider the data for the fusion of ice at 0 ℃.

3 単位量あたりの意味で，比

▶ The word specific before the name of any extensive physical quantity refers to the quantity per unit mass.

specifically adv **1** 特に syn *notably, particularly*

▶ Among vanadates the V_{10} anion, among tungstates the W_{12} anion, and among molybdates the Mo_7 anion are specifically stabilized.

2 具体的にいうと syn *in depth, in detail*

▶ Specifically, it was proposed that these molecules with the composition C_{20+2m} can take the stable form of hollow closed nets composed of 12 pentagons and *m* hexagons.

specification n **1** 記述 syn *description*

▶ The specification that the hot solution saturated with solute at the boiling point be let stand undisturbed means that crystallization is allowed to proceed without subsequently moving the flask, disturbing the bench on which it is resting, or inserting a thermometer or stirring rod.

2 仕様 syn *qualification, requirement*

▶ Using the concentrated aquasols, procedures have been developed that enable gel products to be prepared to a predetermined specification in terms of composition, shape, and size.

specific conductivity n 比導電率

▶ The specific conductivities of these true solid electrolytes are several orders of magnitude higher than those that have their origin in point defects.

specific heat n 比熱

▶ The specific heat will rise sharply near the critical temperature and then, all order being destroyed, fall again to its normal value.

specificity n 特異性

▶ Each enzyme is characterized by specificity for a narrow range of chemically similar substrates and also other molecules that modulate their activities; they are called effectors and they can be activators, inhibitors, or both.

specific rotation n 比旋光度

▶ Specific rotation is the number of degrees of rotation observed if a 10-cm tube is used and the compound being examined is present to the extent of 1g mL^{-1}.

specify vt 指定する syn *identify, indicate, mention*

▶ Since K_c° changes only slightly with pressure, it is usually not necessary to specify constant pressure conditions.

specimen n 試料 syn *sample*

▶ Carefully pressed samples were a uniform yellow-orange color and were semitranslucent, e. g., ordinary typewriting could be read through specimens 1-mm thick.

speck n 点 syn *spot*

▶ Vapor is normally evolved from specks of foreign material in the glass of a clean vessel.

spectacular adj 目覚しい syn *dramatic, splendid*

▶ Specutacular aluminophosphate mesh patterns might arise from distortions of close-packed arrays of polydispersed vesicles or cells, subjected to boundary and/or shear effects.

spectacularly adv 目覚しく syn *marvelously, well*

▶ Ammonia-borane releases hydrogen gas when heated, but the reaction rates and products formed depend quite spectacularly on the reaction conditions.

spectral adj 分光学的な

▶ All of the new compounds exhibited satisfactory spectral properties and elemental analyses.

spectrally adv スペクトルに

▶ Pyroelectric crystals can be made spectrally sensitive by coating the surface of the crystal with appropriate absorbing material.

spectrochemical series n 分光化学系列

▶ The order of ligand-field strength for common ligands is called the spectrochemical series.

spectrophotometric adj 分光測光による

▶ A calorimetric procedure for obtaining *K* and ΔH has recently been reported which is inherently more reliable than the spectrophotometric procedure.

spectrophotometrically adv 分光測光によって

▶ The rates of rearrangement of nitrito complexes of cobalt(III) to the corresponding nitro compounds were determined spectrophotometrically, and in all cases the data were found to give a first-order plot.

spectrophotometry n 分光測光

▶ The speed with which the system relaxes can be measured, usually by spectrophotometry, and we can then calculate the rate constants.

spectroscopic adj 分光の

▶ Because of all the special features, laser investigations have many advantages over conventional spectroscopic studies.

spectroscopically adv 分光学で

▶ The absorption of light by molecular chlorine has been shown spectroscopically to result in the production of the atomic chlorine.

spectroscopic method n 分光法

▶ The distances between atoms and the angles formed by the atoms in many simple molecules have been determined by spectroscopic methods.

spectroscopic term n スペクトル項

▶ The proper crystal-field theory explanation of the spectra requires an understanding of the way an electrostatic field splits the spectroscopic terms of a metal ion.

spectroscopy n 分光学 不可算名詞である.

▶ Both infrared and Raman spectroscopy have selection rules based on the symmetry of the molecule.

spectrum n **1** スペクトル 複数形は spectra

▶ The ^{1}H NMR spectrum of this solution at low temperature showed a single conformer corresponding to the major conformer in the spectrum at room temperature.

2 範囲 syn *range, scale*

▶ Crystalline materials exhibit the complete spectrum of bond types, ranging from ionic to covalent, van der Waals, and metallic.

specular adj 鏡面的な

▶ If there were no transfer of translational energy with the surface, then the scattering would be specular with the angle of reflection equal to the angle of incidence.

specular reflectance n 鏡面反射率 ☞ diffuse reflectance

▶ In modulated specular reflectance spectroscopy, an electrode is irradiated and the reflected light is detected.

speculate vi **1** 推測する syn *consider, ponder, think, wonder*

▶ It is interesting to speculate on how different human history might have been if reduction of iron ores had never been performed.

vt **2** 推測する syn *evaluate, judge, postulate*

▶ Given this polymorphism, it is difficult to speculate which crystal is the most stable form.

speculation n 確実な根拠のない推測 syn *guess, hypothesis, postulation, supposition*

▶ The precise mechanism is complicated and has been the subject of much speculation and controversy.

speed n **1** 速度 syn *velocity*

▶ The speed with which molecules and ions can pass through membranes is a matter of great importance, particularly in biology.

vt **2** 速度を速める syn *accelerate, advance, promote*

▶ Where the solvent speeds up an S_N1 reaction enormously, it slows down the S_N2 reaction.

speedily adv 速やかに syn *promptly, quickly, rapidly*

▶ A catalyst might be found which can operate at low temperatures so that the favorable equilibrium can be established speedily.

spend vt 費やす syn *allot, assign, devote*

▶ Several sections must be spent considering some aspects of ionic solutions.

spent adj 使用済みの syn *used*

▶ A high proportion of activity could be restored by treating the spent catalyst with an ethanolic solution of propylamine.

sphere n **1** 球

▶ Titanium dioxide spheres produced by alkaline hydrolysis of an alkoxide which have a colloidal size can be considered as ideal powders.

2 領域 syn *area, discipline, field, range*

▶ In most systems of interest in the general sphere of solid-state chemistry, the vapor pressure remains low for large variations in temperature.

spherical adj 球状の syn *globular, round, spheric*

▶ The micelle for horse ferritin is approximately spherical, with an outer diameter of 10,000–11,000 pm and an inner diameter of ca. 6000 pm.

spherically adv 球状に

▶ Being spherically symmetrical, the *ns* electron pair then loses it stereochemical effects.

spheroidal adj 回転楕円体形の

▶ For O and N singly bonded to a carbon atom the shapes are virtually spherical, but the remainder, F, S, Cl, Se, Br, and I, are spheroidal, always having the shorter radius along the atom-to-carbon vector.

spherulite n 球晶

▶ Globular spherulites are less convenient subjects for morphological study than spherulites grown in thin films.

spherulitic adj 球晶の

▶ High polymers crystallize from the melt characteristically with a spherulitic habit.

spill n **1** こぼれ syn *flood, leak, leakage*

▶ Most large chemical spills such as oil spills occur in marine environments often far away from populated areas.

vi **2** あふれる syn *overflow*

▶ The hydrogen molecule dissociates on the metal into atoms, and the atoms spill into the surrounding support to provide highly active reducing species.

vt **3** こぼす syn *throw out*

▶ Petrol vapor and propane gas are heavier than air and only disperse slowly if they are spilt.

spillage n こぼすこと

▶ Spillage of mercury may be a toxic hazard due to droplet proliferation.

spillover n 流出

▶ Platinum particles catalyze the reduction of the surface by the hydrogen spillover mechanism.

spin n **1** スピン

▶ There are four net spins possible for the methyl protons: +3/2, +1/2, −1/2 and −3/2.

vi **2** 回転する syn *rotate, revolve*

▷ Nuclear magnetic resonance is possible because the nuclei of certain atoms like hydrogen behave as though they were charged spinning bodies.

vt **3** 回転させる syn *rotate, turn*

▶ The sample is spun in a centrifuge.

spin crossover n スピン交差

▶ Iron（III）*N,N*-dialkyldithiocarbamates provide probably the best-documented examples of the high-spin–low-spin crossover.

spin decoupling n スピンデカップリング

▶ The spin decoupling is achieved by irradiating the substrate with a strong radiofrequency signal corresponding to the resonance frequency of one of the nuclei; the spectrum resulting from the remaining nuclei is scanned to observe any simplification which results.

spinel n スピネル ☞ normal spinel, inverse spinel

▶ The spinels are minerals with the empirical

formula AB_2O_4. There are 2:3 spinels containing A^{2+} and B^{3+} ions and 4:2 spinels containing A^{4+} and B^{2+} ions.

spin-lattice relaxation time n スピン-格子緩和時間

▶ The spin-lattice relaxation time measurements have confirmed the existence of large in-plane motions for the naphthalene and pyrene molecules in all the complexes studied.

spin multiplicity n スピン多重度

▶ The wave function $\psi_1(A^+-D^-)$ corresponds to a dative structure in which one electron has been transferred from the donor to the acceptor molecule, while maintaining the overall spin multiplicity of the state ψ_0.

spinning n **1** 自転 syn *revolution, rotation*

▶ The spinning of this nuclear charge produces a magnetic moment.

2 紡糸

▶ Serpentine asbestos is the mineral chrysotile, magnesium silicate. The fibers are strong and flexible, so that spinning is possible with longer fibers.

spin-off n 技術開発の副産物 syn *by-product*

▶ We have flexible, fused-silica capillaries with a polymer overcoat for chromatography; these columns are a spin-off of fiber-optics technology.

spin-only formula n スピンだけの式

▶ If there are *n* unpaired electrons, the magnetic moment is given by the spin-only formula.

spinodal decomposition n スピノーダル分解

▶ Vycor glass is formed through the spinodal decomposition of a borosilicate glass to borosilicate-rich and borosilicate-poor regions.

spin-spin coupling n スピン-スピン結合

▶ Spin-spin coupling patterns are characteristic of a particular grouping of protons and may be used to uniquely define a structure.

spin-spin coupling constant n スピン-スピン結合定数

▶ Another complexity arises when two sets of chemically equivalent nuclei have spin-spin coupling constants that are unequal.

spin-spin interaction n スピン-スピン相互作用

▶ Ntrobenzene has different chemical shifts for its ortho, meta, and para hydrogens and six different spin-spin interaction constants.

spin-spin relaxation time n スピン-スピン緩和時間

▶ The spin-spin relaxation time T_2 may often be obtained very conveniently from line widths for quadrupolar nuclei.

spiral n **1** らせん syn *helix*

▶ Since the dislocation source remains fixed while the outer part of the step advances at a steady rate, the step winds itself up into a spiral.

adj **2** らせんの syn *helical*

▶ Since these etch pits are formed under nonequilibrium conditions of undersaturation, their form depends upon the rates of retreat of the spiral step and the evaporation or dissolution kinetics in different crystallographic directions.

spite n **in spite of** (prep) …にもかかわらず syn *despite, regardless of*

▶ In spite of the apparent success of the dielectric constant in correlating electrolyte behavior, we should not get the impression that solution properties can be understood in terms of this parameter alone.

spitting n 泡を吹き出すこと

▶ At the melting point, silver absorbs more than 20 times its own volume of oxygen, and on cooling, the oxygen is liberated and causes the characteristic spitting.

splashing n 飛散

▶ A certain amount of stirring that does not cause splashing is beneficial, as it circulates fresh solution to the growing crystal faces and helps to prevent high local supersaturations at the surface of the solution.

splaying n 外に開くこと

▶ Decachloropyrene is an overcrowded molecule; the strain is relieved by a splaying of angles and by large out-of-plane displacements to give a saddle-shaped molecule.

splendid adj すばらしい syn *exceptional, extraordinary, magnificent, spectacular*

▶ Iodine has a splendid variety of polyanions, and in this it differs from bromine and chlorine.

splice vt 接合する syn *bind, unite*

▶ Certain genes used to produce antibodies are alternatively spliced to produce some antibody proteins that attach to the surface of cells as well as other antibodies with different structures causing them to be secreted into the bloodstream.

split vi **1** 分裂する syn *cleave, divide, separate*

▶ As cyclohexane is cooled down, the single sharp peak observed at room temperature is seen to broaden and then, at about −70 ℃, to split into two peaks, which at −100 ℃ are clearly separated.

vt **2** 分裂させる syn *break, rupture*

▶ The absorption of a photon of blue light by a chlorine molecule splits the molecule into two chlorine atoms.

3 分割する syn *divide, separate*

▷ The occurrence of an appropriate lattice distortion results in the partly filled band of the molecular metal splitting into filled and empty bands.

splitting n **1** 結合の分裂 syn *break, rupture*

▶ Free radicals are formed only by the splitting of a covalent bond.

2 状態の分裂 syn *division*

▶ For thallium (I), splitting of the $^3P_1 \leftarrow {}^1S_0$ band occurs in all the glasses. Such splitting may be due to Jahn-Teller splitting in the excite state.

3 分離 syn *parting, separation*

▶ Polymerization may occur by condensation, involving the splitting out of water molecules by two reacting monomers.

spoilage n 食物の腐敗 syn *damage*

▶ Removal of terpenes is necessary to inhibit spoilage, particularly of oils derived from citrus sources.

spongy adj 海綿状の syn *porous*

▶ The spongy lead is the reducing agent and the lead dioxide the oxidizing agent in the chemical reaction that takes place in the storage cell.

spontaneity n 自発性

▶ Spontaneity is not a complete criterion for irreversibility.

spontaneous adj 自発的な syn *automatic, natural*

▶ A negative ΔG° for a reaction means that the process is spontaneous; a compound having a negative $\Delta_f G^\circ$ is, therefore, thermodynamically stable with respect to its elements.

spontaneous fission n 自発核分裂 ☞ nuclear fission

▶ The dominant feature of the actinides is their nuclear instability, as manifest in their radioactivity and tendency to spontaneous fission.

spontaneously adv 自発的に syn *freely, readily*

▶ That many elements react spontaneously with oxygen is remarkable when it is realized that the process of converting molecular oxygen into oxide ions is strongly endothermic.

spontaneous polarization n 自発分極

▶ Ferroelectric substances are an important class of polar materials in which the direction of spontaneous polarization can be reversed by applied electric field.

spool n フィルムのリール

▶ Eastman succeeded in attaching the light-sensitive emulsions to strips of paper and later to celluloid, which could be rolled on spools.

spore n 胞子

▶ Complete disinfectants destroy spores as well as vegetative forms of microorganisms.

spot n **1** 斑点 syn *fleck, speck*

▶ The identities of the amino acids that produce the various spots are established by comparison with the behavior of known mixtures.

2 on the spot 現場で syn *at the place of*

▶ Ozone is usually manufactured by electronic irradiation of air on the spot, as it is too expensive to ship.

spot test n 斑点試験

▶ The separated or enriched fractions may be identified by spot tests, microscopy, spectrophotometry, emission spectrography, mass spectroscopy, etc.

spray n **1** 噴霧 syn *mist*

▶ Fog and smoke are common examples of natural aerosols; fine sprays are man-made.

vt **2** 噴霧する syn *atomize, disperse, sprinkle*

▶ One way to prevent particle growth through agglomeration and coagulation is to spray the supercritical solutions directly into an aqueous surfactant solution.

spread n **1** 広まり syn *expansion, extension*

▶ The energy spread of the band reflects the atomic overlap.

2 分布 syn *difference, extent, range*

▶ The spread of reactivities of halogens toward methane is so great that only chlorination and bromination proceed at such rates as to be generally useful.

adj **3** 広がった

▶ The oxidation of a film of oleic acid spread on a dilute solution of potassium permanganate is stopped by reducing the surface area and crowding the oleic acid molecules closer together so that their double bonds are inaccessible to the permanganate.

vi **4** 広がる，展開する syn *diffuse*

▶ If a drop of thin lubricating oil is put on the surface of cold water in a glass, it does not spread out and stays too thick to show interference colors.

vt **5** 分布させる syn *distribute*

▶ In ClO_4^- the negative charge is spread over four equivalent oxygen atoms.

spreading n 広がり syn *expansion, extension*

▶ No spreading is observed with a pure liquid paraffin, but if a fatty acid or similar polar compound is first added to the paraffin, it spreads out immediately to form a very thin film.

spreading coefficient n 拡張係数

▶ Wettability of solid powders, spreading coefficients of liquids, and protective action of colloidal substances are intimately associated with interfacial behavior.

spring n **1** スプリング

▶ The McBain-Baker balance comprises a helical spring of fine quartz carrying light bucket containing the absorbent.

vi **2** 跳ね返る syn *bound, hop, jump*

▶ If one removes the load at time *t*, then the system springs back slowly with decreasing velocity.

sprinkle vt 散らす syn *scatter, spatter, spray*

▷ Samples for diffractometry are thin layers of the fine powder sprinkled onto a glass slide.

spurious adj 偽の syn *false, fictitious, unreal*

▶ There have been many spurious claims of structural proof by demonstration of optical rotation.

square n **1** 正方形

▶ The distance from any oxygen to the center of the square is 2.04 Å.

2 二乗

▶ The square of the velocity measures only the magnitude of this quantity.

3 平方

▶ One milligram of pure Martius Yellow will dye to full strength three 3×3-cm squares of silk weighing 56 mg each.

vt **4** きちんと整頓する syn *adjust, arrange, modify*

▶ The end of a glass tube can be squared by sprinkling a little 200- or 400-mesh grinding powder onto a flat piece of glass, moistening it with water, and grinding with a circular rubbing motion until the end is planar and smooth.

square-antiprismatic adj 正方逆プリズムの

▶ Barium hydroxide octahydrate has a slightly distorted square-antiprismatic array of eight water molecules about the barium atom at a distance of 2.69–2.77 Å.

square bracket n 角括弧 ☞ bracket, curly bracket, parentheses

▶ The square brackets imply, as is customary, the active mass of the reagent, and for solutes the active mass is taken as the molar concentration.

square complex n 正方形錯体

▶ It should be pointed out that more careful examination of these square complexes shows that they have a marked tendency to coordinate a fifth and/or a sixth group at a slightly longer distance from the central metal ion than are the ligands in the original square.

square planar adj 平面正方形の

▶ A prolate ellipsoid causes the more commonly observed elongation of the octahedron, which in the limit gives a square planar AX_4 geometry with the loss of two ligands.

square pyramidal adj 正方錐形の

▶ ［VO(acac)$_2$］ has the square pyramidal C_{4v} structure with the ＝O occupying the unique apical site.

square root n 平方根

▶ The conductivity multiplied by the square root of the concentration, when plotted against the concentration for any colloidal electrolyte, yields a straight line.

squash vt 押しつぶす syn *crush, squeeze*

▶ In chromium-doped apatites, the CrO_4^{3-} tetrahedron is distorted in such a way as to be squashed or compressed along one of the fourfold rotatory inversion axes of the tetrahedron.

squeeze vt **1** 圧搾する syn *compress, press, squash*

▶ The effects of temperature and pressure on the size of the halide vacancy suggest that as the crystal is cooled or squeezed and the ions move closer together, the F-center absorption peak should shift to higher energies.

2 squeeze out しぼり出す syn *expel, extract*

▷ Internal hydrostatic pressures created in gels are responsible for squeezing out nonelectrolytes from the gels.

stability n 安定度

▶ In dealing with equilibria, we shall compare the stability of the reactants with the stability of the products.

stability constant n 安定度定数

▶ The selectivity of one cation relative to another can be conveniently expressed as the ratio of the stability constants.

stabilization n 安定化

▶ Working with molecular models, Watson and Crick assembled a structure in which all the building blocks fitted without crowding and, of prime importance, which permitted the greatest stabilization by hydrogen bonds.

stabilization energy n 安定化エネルギー

▶ The stabilization energies of pyrrole, furan, and thiophene obtained from experimental and calculated heats of combustion are only about half of the stabilization energy of benzene.

stabilize vt 安定させる

▷ Anion vacancies occur in cubic, lime-stabilized zirconia of formula, $Zr_{1-x}Ca_xO_{2-x}$.

stabilizer n 安定剤 syn *stabilizing agent*

▶ Hydrolysis is the usual method of preparing colloidal iron(III) hydroxide. However, the preparation fails unless an appreciable proportion of iron(III) salt is retained as stabilizer.

stable adj 安定な syn *lasting, longstanding, steady, unchanged*

▶ The easy way to determine which of two forms is stable at a given temperature is to observe the relative solubility of the two in a solvent.

stable form n 安定形

▶ Although calcite is the most stable form of calcium carbonate at the ordinary temperature, the metastable modification, aragonate, nevertheless exists under the ordinary conditions in an apparently very stable state.

stable isotpe n 安定同位体

▶ Using a variety of nuclear syntheses, it is now possible to account for the presence of the 273 known stable isotopes of the elements up to ^{209}Bi and to understand, at least in broad outline, their relative concentrations in the universe.

stack n **1** 積み重ね syn *accumulation, heap, pile*

▶ Infinite stacks of alternate donor and acceptor molecules appear to be a common feature of all solid 1:1 charge-transfer complexes in which there is no participation through hydrogen bonding.

vt **2** 積み重ねる syn *accumulate, pile up*

▶ In these complexes, planar or nearly planar monomer units are stacked above one another to form metal-atom chains.

stacking n 積み重ね

▶ In general, stacking of aromatic systems in DNA correlates well with size and hydrophobicity.

stacking disorder n 積み重ねの乱れ

▶ Stacking disorder occurs when the normal stacking sequence is interrupted at irregular intervals by the presence of wrong layers.

stage n **1** 段階 syn *level, step*

▶ Urea and formaldehyde are united in a two-stage process in the presence of pyridine, ammonia, or certain alcohols with heat and control of pH to form intermediates that are mixed with fillers to produce molding powders.

2 台

▶ The micromethod of Kofler consists in placing a minute fragment of material on a cover glass centered over a small hole in a metal stage of controllable temperature and establishing the point of melting by observation under the microscope.

stagger vt ずらして配列する syn *alternate, vary*

▶ In Magnus' green salt, alternate ions are staggered by 28° allowing a close approach of the platinum atoms along the chain.

staggered conformation n ねじれ形配座

▶ Any ethane-like portion of a molecule tends to take up a staggered conformation.

stain n **1** 汚れ syn *discoloration, spot*

▶ The protease subtilisin, derived from *Bacillus subtilis*, is a valuable component of many laundry detergents, where it functions to degrade and remove protein stains from clothing.

vt **2** 染色する syn *dye*

▶ The treatment stains nuclei blue, cell cytoplasm pink, and some subcellular organelles either pink or blue.

stand vi **1** そのままでいる syn *remain, stay*

▶ Then a solution of 1.5 mL of triethylamine in 15 mL of acetonitrile was added, and the dark green solution was allowed to stand at room temperature overnight.

2 ある関係にある syn *exist, persist, remain*

▶ In much of its chemistry, phosphorus stands in relation to nitrogen as sulfur does to oxygen.

3 stand for …を意味する syn *designate, mean, represent*

▶ The name laser stands for light amplification by stimulated emission of radiation.

vt **4** 受ける syn *experience, withstand*

▶ The classical mechanisms for glycoside hydrolysis have stood the test of time and a vast amount of biochemical investigation and remain largely unchanged.

standard n **1** 基準 syn *criterion, measure, model*

▶ Water is the primary viscosity standard with an accepted viscosity at 20 ℃ of 0.01002 poise.

adj **2** 普通の syn *ordinary, typical, usual*

▶ α-Halogeno acids are converted into α-hydroxy acids by hydrolysis; the standard conditions usually are to boil with aqueous sodium carbonate solution.

standard cell n 標準電池

▶ The potentiometer wire can be calibrated by use of a standard cell, such as the Weston cell.

standard deviation n 標準偏差

▶ Best straight lines and standard deviations were in all cases computed with a computer using a standard root-mean-square program.

standard electrode potential n 標準電極電位

▶ By combining the standard electrode potentials for two electrodes, we can deduce the emf of a cell involving the two electrodes.

standard electromotive force n 標準起電力

▶ The standard electromotive force is that potential associated with a reaction conducted in a cell in which the reactants and products are in their standard states, which for solutions of *n-n* electrolytes is represented by an ideal solution of unit molality.

standard enthalpy n 標準エンタルピー

▶ If in any reaction the products and reactants are in their standard states, we may denote the enthalpy change as the standard enthalpy of the reaction.

standard heat of formation n 標準生成熱 syn *standard enthalpy of formation*

▶ The standard heats of formation of a great number of chemical compounds have been determined, often with considerable accuracy, not only by calorimetric means but by many diverse methods.

standardization n 標定 syn *normalization*

▶ A titration error is compensated when the same error is made in the standardization and in the analysis.

standardize vt **1** 標定する syn *normalize*

▶ It is possible to standardize a solution of permanganate by a number of different methods.

2 標準化する syn *systematize*

▶ Nuclear magnetic resonance data have been standardized by reporting the chemical shifts observed in terms of the ppm variation in field strength between the resonance under observation and the resonance of a reference compound, usually tetramethylsilane.

standard solution n 標準溶液

▶ The concentrations of the standard solutions used in volumetric analysis are usually referred to the so-called normal concentration.

standard state n 標準状態

▶ Specifications of the standard state of a chemical species vary from system to system, i. e., it may be different for a pure substance as a gas, liquid, or solid and for a component in solution.

standing n 持続 syn *duration*

▶ Sparingly soluble amides crystallize out from the reaction mixture upon standing, as in the case of succinamide.

starch n デンプン

▶ In any one starch, there are two structurally different polysaccharides.

start n **1** 開始 syn *beginning, onset, outset*

▶ The reaction is exothermic at the start when the concentrations of benzene and nitric acid are maximal and becomes sluggish toward the end as the concentrations of reactants decrease.

vi **2** 起こる syn *get going*

▶ The reaction with lithium starts more readily and proceeds at a greater rate than that with magnesium.

3 …から手を付ける syn *commence, begin*

▷ When one carries out an S_N1 reaction starting with a single pure optical isomer of a tertiary compound, the product is usually a mixture of both optical isomers.

vt **4** 運転を開始する syn *set in motion*

▶ The stirrer is started, even though very little liquid is present, for the crushing of pieces of light magnesium foil in contact with the liquid is effective in initiating reaction.

starting material n 出発物質

▶ The starting materials for zeolite synthesis are solutions of salts of silicate and aluminate anions in aqueous alkali.

starting point n 出発点

▶ The calculations provide a useful starting point for further modeling work on the these polymers.

startling adj 驚くべき syn *amazing, astonishing, surprising*

▶ There are no startling differences in the molecular geometry of the two conformers.

starvation n 飢餓 syn *famine, hnger*

▶ The beginning of starvation is normally considered to be ca. 5 h after the last meal.

starved adj **starved of** …に欠乏して syn *in need of, lacking*

▶ The reaction mixture rapidly becomes starved of oxygen, and so the reaction slows down.

state n **1** 状態 syn *condition, situation*

▶ Any change in physical state will necessarily change the value of ΔH, since all changes of state involve changes in energy.

vt 2 述べる syn *affirm, assert, express*

▷ The position of any face of a crystal is conveniently expressed by stating the lengths on the axes at which the face cuts these axes.

state function n 状態関数

▶ Another important characteristic of a state function is that when the state of a system is changed, the change in any state function depends only on the initial and final states of the system and not on the path followed in making the change.

state of the art n 科学，技術の最先端

▶ The state of the art in chemical documentation is the sophisticated and extensive information bank assembled by Chemical Abstracts Service, a division of ACS.

statement n 陳述 syn *account, expression, report*

▶ If the solid substance can exist in different allotropic modifications, the particular form of substance which is in equilibrium with the solution must be known, in order that the statement of the solubility may be definite.

static adj 静的な syn *fixed, stationary* ☞ dynamic

▶ Measurments of static vapor pressure by means of manometers are used most commonly in the pressure range near or above 1 mm.

static surface tension n 静的表面張力 ☞ dynamic surface tension

▶ For solutions, the dynamic surface tension is almost always very different from the static surface tension.

station vt 位置に付く syn *locate, place, position*

▶ The four outer electrons of a carbon atom would station themselves at the vertices of a regular tetrahedron inscribed in the sphere, for they would then be as far as possible from each other, thereby minimizing repulsive interactions between the negative charge.

stationary adj 静止した syn *fixed, static*

▶ Because the detection position is stationary in the stopped-flow system, a greater variety of detection systems can be utilized as detector mobility is no longer an experimental criterion.

stationary phase n 固定相 ☞ mobile phase

▶ Chemically modified mesoporous silica may be used as a stationary phase in high performance liquid chromatography.

stationary state n 定常状態

▶ The atom can exist in various stationary states characterized by the value of the quantum number *n*.

statistical adj 統計的な

▶ Very recent work in our group has focused on the directional nature of hydrogen bonds, and a statistical method was developed to analyze a variety of hydrogen bonds, which reinforced the linear tendency of many types of hydrogen bonds.

statistically adv 統計的に

▶ A polyethylene chain may adopt the energetically favored, but statistically unlikely, zigzag shape.

status n 事態 syn *condition, situation, state*

▶ Because of the very special status of water as a solvent, the terms hydrophilic and hydrophobic are used in reference to water solubility and water insolubility.

stay vi とどまる syn *remain, stand*

▶ Electrons tend to stay as far apart as possible because they have the same charge and also, if they are unpaired, because they have the same spin.

steadily adv 着実に syn *consistently, constantly, regularly*

▶ In the past decade, the properties of nanocrystals have attracted steadily growing interest from chemists, physicists, and other materials scientists and engineers.

steady adj 一様な syn *continuous, regular, uniform*

▶ Down the alkaline-earth series, there is a steady increase in the heat of reaction which is paralleled by a steady increase in the decomposition temperature of the carbonate.

steady state n 定常状態

▶ After ca. 3 days of starvation, whole-body metabolism reaches a new steady state.

steady-state approximation n 定常状態近似

▶ The kinetics may be deduced by applying the steady-state approximation to the reactive intermediates.

steam vt 蒸す syn *boil*

▶ After soaking in aqueous chrome alum or acetate, the fabric is steamed to precipitate colloidal hydrolysis products which subsequently fix the dye.

steam bath n 蒸気浴

▶ The water of crystallization may be eliminated by drying on the steam bath or in an oven at 120 ℃.

steam distil vi 水蒸気蒸留する

▶ Nitrobenzene steam distils at 99 ℃ and requires 4.0 g of water per gram.

steam distillation n 水蒸気蒸留

▶ Steam distillation offers the advantage of selectivity, since some water-insoluble substances are volatile with steam and others are not, and some volatilize so very slowly that sharp separation is possible.

steam reforming n 水蒸気変成

▶ Steam reforming of natural gas is an important method of producing hydrogen by the reaction $CH_4 + H_2O \rightarrow 3H_2 + CO$.

steel-blue adj はがね色の

▶ Tantalum is a steel-blue-colored metal when unpolished.

steel-gray adj 鉄灰色の

▶ Arsenic forms brittle steel-grey crystals of metallic appearance.

steep adj 急勾配の syn *abrupt, nearly vertical, sharp*

▶ Cycling does not affect the steep portion of the temperature-increasing curve to any great extent, but the point of intersection with the α-modification line does vary as the latter moves about.

steeped adj 染まった syn *soaked*

▶ As for all concepts there are limitations to its use, but it has the advantage of being conceptually simple and readily applied to solid-state problems by those not steeped in the culture of the subject.

steepest decent n 最急降下

▶ The dividing surface is constructed perpendicular to the minimum-energy path, which is the path of steepest decent from the lowest point of the col into the reactant and product valleys.

steer vt 向ける syn *conduct, direct, guide*

▶ Evidently the rigid environment of the clamped organic reactant steers both chlorine atoms to attack the double bond from the same side.

stem n **1** 脚

▶ If the filter fits tightly against the side walls of the funnel, the stem should fill with liquid.

vi **2** 生じる syn *arise, come, result*

▶ The difference in structure between NH_4F (wurtzite structure) and NH_4Cl (CsCl structure) stems from the greater strength of the N–H⋯F bond than the N–H⋯Cl bond.

stemless adj 脚のない

▶ The filters should be stemless to avoid crystallization in the stem and supported merely by resting in the mouth of the receiving flask.

step n **1** 段階 syn *stage*

▶ The temperature of the sample was increased in steps with spectra being measured at each step.

2 表面欠陥のステップ syn *ledge*

▶ Since organic crystals often have lattice spacings several hundred angstroms high, unimolecular spiral steps are much easier to observe on these compounds.

stepped adj 段のある

▶ The stepped faces invariably grow very rapidly by means of addition of material to the steps.

stepwise adj 段階的な

▶ With the successive formation of a series of hydrates, such a diagram has a characteristic stepwise appearance, in which each horizontal marks the pressure required for the formation of a higher hydrate or for its decomposition.

stereoblock n ステレオブロック

▶ A stereoblock polymer is made of comparatively long sections of identical stereospecific structure, these sections being separated from one another by segments of different structure.

stereochemical adj 立体化学的な

▶ In addition to these geometric relationships, the stereochemical arrangement of the surfactant head groups is of fundamental importance.

stereochemically adv 立体化学的に

▶ We had to decide whether certain ligands that were equivalent in chemical composition and location on a chain or ring were or were not stereochemically equivalent.

stereochemical nonrigidity n 立体化学的柔軟性

▶ These studies point to, but do not define, the streochemical nonrigidity of octacoordinate structures.

stereochemistry n **1** 立体化学

▶ Tartaric acid has played a key role in the development of the stereochemistry and particularly the stereochemistry of the carbohydrates.

2 立体化学的性質　この意味では可算名詞である.

▶ Most Fe^{II} complexes are octahedral but several other stereochemistries are known.

stereographic projection n ステレオ投影

▶ The stereographic projection may be derived from the spherical projection by projecting each pole on the upper hemisphere onto the equatorial plane of the sphere by means of a straight line from the pole of the face to the south pole of the sphere.

stereoisomer n 立体異性体

▶ In dealing with those aspects of stereochemistry that depend on isolation of stereoisomers, we can ignore the existence of easy-to-interconvert isomers, which means most conformational isomers.

stereoisomeric adj 立体異性の

▶ The three stereoisomeric tartaric acids consist of a pair of enantiomorphs, known as (+)- and (−)-tartaric acid, and an optically inactive form, known as mesotartaric acid.

stereoisomerism n 立体異性

▶ Stereoisomerism may be defined broadly as isomerism of compounds with the same structural formulas but having different spatial arrangements of the various groups.

stereoregular adj 立体規則性の

▶ There are many other possible types of stereoregular polymers, one of which is called syndiotactic and has the side-chain groups oriented alternately on one side and then the other.

stereoregularity n 立体規則性 syn *tacticity*

▶ Isomers which are completely lacking in stereoregularity have the R groups arranged on both sides at random and called atactic polymers.

stereoscopic view adj 立体図

▶ Figure 1 is a stereoscopic view of the packing in compound **1** looking down the *b* axis.

stereoselective adj 立体選択的な

▶ A large part of the research in the field of pheromones involves development of new, highly stereoselective ways to introduce the carbon–carbon double bond or other structural elements into a molecule.

stereoselectivity n 立体選択性

▶ The stereospecificity of their action demands an equal stereoselectivity in their synthesis: enantioselectivity to match enantiospecificity and diastereoselectivity to match diastereospecificity.

stereospecific adj 立体特異性の

▶ Using stereospecific catalysts, one can prepare either isotactic polymers in which all of the pendant R groups lie on one side of the plane of the zigzag, or syndiotactic polymers in which they alternate regularly from one side to the other.

stereospecifically adv 立体特異的に

▶ Studies of picryl ethers derived from compound **1** by introduction of a single substituent into the para position of one of the phenyl rings had shown the rearrangement to proceed stereospecifically with migration of that aryl ring trans to the picryloxy group.

stereospecific reaction n 立体特異反応

▶ A stereospecific reaction is one in which stereochemcially different molecules or streochemically different parts of molecules react differently.

stereospecificity n 立体特異性

▶ In biological systems, such stereospecificity is the rule not the exception.

steric adj 立体的な

▶ Formal steric enthalpy is a precisely defined measure of the steric component of the enthalpy of formation.

sterically adv 立体的に

▶ In sterically hindered molecules, overlapping of the van der Waals surface occurs between atoms which are not bonded to each other.

steric effect n 立体効果

▶ It is apparent that the smaller total velocity of nitration of *tert*-butylbenzene is due to entirely to the inhibition of ortho substitution by a large steric effect.

steric factor n 立体因子

▶ The probability factor is sometimes called a steric factor, since it can be thought of as the fraction of collisions with sufficient energy that also have the proper orientation for reaction.

steric hindrance n 立体障害 ☞ interference

▶ The tendency for the chains to orient can often be considerably reduced by random introduction of methyl groups which, by steric hindrance, inhibit ordering of the chains.

steric repulsion n 立体反発

▶ An anomalous enthalpy of formation, attributable perhaps to steric repulsion, or to resonance stabilization, will often dramatically affect reactivity.

steric strain n 立体ひずみ

▶ It is somewhat surprising that the relief of the intramolecular steric strain in the yellow form takes the form of bond angle deformation rather than a greater out-of-plane bending or greater rotation of the substituent.

sterilize vt 消毒する syn *clean, disinfect, purify*

▶ The characteristic chlorine odor of water that has been sterilized with hypochlorite is due to chloramines produced from attack on bacteria.

steroid n ステロイド

▶ Although steroids occur in all plant and animal organisms, the most important ones are those of animal origin. The most abundant animal steroid is the crystalline alcohol cholesterol.

stick n **1** 棒 syn *pole, rod*

▶ The sticks of lithium metal were washed with successive portions of benzene and ether under nitrogen.

vi **2** 付着する syn *adhere, attach*

▶ The cations in the solution tend to withdraw electrons from the electrode, and the resulting neutral M atoms stick to its surface.

sticking coefficient n 付着係数

▶ Bulk Pt(100) exists as a square (1×1) lattice, with a high sticking coefficient for dissociative adsorption of molecular oxygen.

sticky adj べとべとする syn *viscous*

▶ If dried at room temperature, slightly sticky crystals of a hydrate are obtained.

stiff adj 曲がらなくて硬い syn *rigid, tough*

▶ Double-helical DNA, although a long molecule, is relatively stiff, and it imparts high viscosity to a solution.

stiffen vi 堅くなる syn *crystallize, harden, solidify*

▶ The melt formed on mixing the reagents stiffens within minutes to yield a sticky solid that hardens further on standing.

stiffness n 硬さ syn *inflexibility, rigidity*

▶ There are two fundamental elastic moduli for isotropic materials: the bulk modulus, which measures volumetric stiffness, and the shear modulus, which measures shape stiffness.

still n **1** 蒸留器

▶ The term reflux refers to the liquid that has condensed from the rising vapor and allowed to flow back down the fractionating column toward the still.

adv **2 まだ** syn *even now, until this time*

▶ If the substance is still not quite pure, the same process of recrystallization is repeated until the presence of no impurity can be detected.

3 接続詞的に，**それでも** syn *even then, yet*

▶ Even when two hydrogen molecules and an oxygen are in the intimate contact of a collision, still it is such a rare thing for their atoms to rearrange themselves to form two water molecules that for all practical purposes it never happens.

4 比較級を強めて，**なおいっそう** syn *even*

▶ In general, dyes aggregate more strongly in water than in organic solvents and more generally still in solutions of high ionic strength.

stimulate vt **刺激する** syn *activate, excite*

▶ Even a single foreign protein molecule contains numerous antigenic sites, each of which can stimulate a specific antibody response.

stimulated emission n **誘導放出**

▶ Stimulated emission of radiation of frequency $\nu = E_g / h$ may be produced by recombination of these electron-hole pairs.

stimulation n **刺激** syn *agitation, impetus*

▶ The physiological stimulation derived from beverages such as tea, coffee, and many soft drinks is due to the presence of caffeine.

stimulatory adj **刺激の** syn *stimulating*

▶ G proteins are important intermediates in signal transduction pathways because they determine whether the signal will be stimulatory or inhibitory.

stimulus n **刺激** 複数形は stimuli syn *impetus, signal*

▶ Stimuli external to a cell, such as hormone signals or light, are detected by specific proteins, receptors, and photosystem.

stipulate vt **規定する** syn *prescribe, specify*

▷ Pauling's first rule, stipulating a connection between the cation-to-anion radius ratio and cation coordination number, is in line with that analysis.

stipulation n **条件** syn *condition, requirement*

▶ The hydrogen-bonding requirements can be satisfied between the chains by the stipulation that a thymine side group lines up opposite an adenine group and a guanine group opposite a cytosine group.

stir vt **撹拌する** syn *agitate, shake*

▶ The mixture was stirred at room temperature under nitrogen overnight.

stirrer n **かきまぜ機**

▶ Stirrers operated magnetically are recommended, since they do not protrude from solution and, therefore, do not act as nucleation points on the surface.

stirring n **撹拌** syn *agitation, shaking*

▶ Temperature, rate of growth, extent of stirring, and the presence of impurities all play an important role in the determination of the crystal shape.

stirring bar n **撹拌子**

▶ The hot plate contains a rotating electromagnet that stirs the solution by means of the encapsulated stirring bar.

stitch vt **縫い綴る** syn *connect*

▷ The structure of C_{60} resembles a soccer ball, stitched from 20 hexagons and 12 pentagons.

stock n **在庫品** syn *reserve, store*

▶ The hydrocarbons under investigation were taken from stock.

stoichiometric adj **化学量論的な**

▶ Because reactions can occur in steps, it is found very frequently that the order of a reaction is less than corresponds directly to the stoichiometric equation.

stoichiometric coefficient n **化学量論係数**

▶ For the reaction

$$N_2 + 3H_2 \rightarrow 2NH_3$$

the stoichiometric coefficient for the product NH_3 is 2, that for N_2 is -1 and that for H_2 is -3.

stoichiometric compound n **定比化合物**

☞ nonstoichiometric compound

▶ A zinc chromium spinel, $ZnCr_2O_4$, which is used as a magnetic recording medium in cassette recorders, results from the decomposition of the stoichiometric compound ammonium zinc chromate, $(NH_4)_2Zn(CrO_4)_2$.

stoichiometric number n 化学量数

▶ The stoichiometric number is introduced into the equations for the kinetics of complex reactions to allow for the possibility that the rate-determining step may occur more or less than once in the completion of the overall reaction as written.

stoichiometry n **1** 化学量論

▶ All stoichiometry essentially is based on the evaluation of the number of moles of substance.

2 組成 この意味では可算名詞である.

▶ As oxygen is progressively eliminated, a whole series of M_nO_{3n-1} stoichiometries are feasible between the MO_3 structure containing corner-shared MO_6 octahedra and the rutile structure consisting of edge-shared MO_6 octahedra.

Stokes' line n ストークス線

▶ Besides the Rayleigh line of frequency ν_0, there are fainter lines of frequency ν on each side, called Stokes lines when ν is smaller than ν_0, and anti-Stokes lines when ν is larger than ν_0.

stop vi **1** 停止する syn *cease, terminate*

▶ The chlorination of methane does not have to stop with the formation of methyl chloride, and it is possible to obtain the higher chlorination products.

vt **2** 止める syn *halt, terminate*

▶ The product is a weaker base than the starting material and, therefore, less susceptible to acid-catalyzed enolization, so that the reaction can be stopped after a single bromination.

stopcock n 活栓

▶ Joule's apparatus consisted of two containers, connected through a stopcock, the whole apparatus being placed in a water bath, the temperature of which could be measured very accurately.

stopped-flow spectrophotometry n ストップトフロー分光光度法

▶ Stopped-flow spectrophotometry has been employed to study the kinetics of binding of Cu^{II} to α- and β-CD.

stopper n **1** 栓 syn *plug*

▶ The salt solutions, contained in 100-mL flasks with ground-glass stoppers, were shaken mechanically with excess solute in a constant temperature bath at 25.0 ± 0.05℃.

vt **2** 栓をする syn *plug*

▷ The reaction is completed by either refluxing the mixture for one half hour or stoppering the flask with the calcium chloride tube and letting the mixture stand overnight.

storage n **1** 貯蔵

▶ Complete removal of lithium from $LiNiO_2$ corresponds to a charge storage capacity of 275 mA h g^{-1}.

2 記録装置

▶ The demand for increased memory storage capacity has led to the need for submicron resolution.

storage battery n 蓄電池

▶ Almost half the Pb produced is used for storage batteries, and the remainder is used in a variety of alloys and chemicals.

store n **1** 蓄え syn *accumulation, stock, supply*

▶ A important goal of chemistry is to be able to predict what will happen on the basis of our store of chemical information.

vt **2** 貯蔵する syn *accumulate, amass, stock*

▶ Fish in certain circumstances do take up mercury from their environment and do store it in their flesh.

stow vt 詰め込む syn *load, pack, store*

▶ The Cr^{2+} ion is slightly smaller in low-spin complexes, where the d electrons are stowed between the ligands, than in high-spin ones, in which the d electron clouds are more extended.

straight adj まっすぐな syn *linear*

▶ By convention, a single straight line connecting the atomic symbols is used to represent a single bond, two such lines to represent a double bond, and three a triple bond.

straight chain n 直鎖 複合形容詞として使用

▶ Of the straight-chain primary alcohols, the

C_8 and C_{10} members are used in the production of high-boiling esters used as plasticizers.

straighten vi straighten out まっすぐになる syn *disentangle, uncurl*

▶ When tension is applied and the material elongates, the chains in the amorphous regions straighten out and become more nearly parallel.

straightforward adj 簡単な syn *lucid, plain, simple*

▶ It is a comparatively straightforward matter to determine standard enthalpies and Gibbs energies of formation and absolute entropies, for pairs of ions in solution, for example, for a solution of sodium chloride.

strain n **1** ひずみ ☞ stress

▶ Strain in the system of bonds of reactants and the release of the strain as the transition state converts to products can provide rate enhancement of chemical reactions.

2 株

▶ Advances in biochemistry and cell biology made it possible to purify large amounts of antibiotics from many different strains of bacteria.

strained adj ひずみのある syn *forced, unnatural*

▶ Epoxides owe their importance to their high reactivity, which is due to the ease of opening of the highly strained three-membered ring.

strain energy n ひずみエネルギー

▶ The extra strain energy in the neighborhood of a dislocation can lead to faster dissolution, making a pit at the end of each dislocation.

strain-free adj ひずみのない syn *strainless*

▶ The very large ring compounds like cyclopentadecane are essentially strain-free.

strainless adj ひずみのない syn *strain-free*

▶ There has been considerable interest for many years in the synthesis of conjugated cyclic polyalkenes with a large enough number of carbons in the ring to permit attainment of a strainless planar structure.

strand n 鎖 syn *chain, string*

▶ A lower proportion of mercury gives a gold solid with metallic luster and formula $Hg_x^{2+}(AsF_6)_2$ where x may be nonintegral and varies between 3 and 8; it contains long strands of Hg_x^{2+} ions perpendicular to each other, with the anions in between.

strange adj 不思議な syn *curious, peculiar, uncommon*

▶ Since adsorption is primarily a surface phenomenon, it is not strange to find that the extent of the adsorption is determined largely by the amount of exposed surface.

strategic adj 計画上重要な syn *critical, crucial, vital*

▶ Strategic research, by definition, focuses upon areas of scientific activity which are perceived to be particularly promising in so far as their likelihood to yield worthwhile practical applications is concerned.

strategy n 方策 syn *procedure, scheme, tactic*

▶ Nitrogen oxide control strategies focus either on preventing NO formation by gradually controlling the addition of oxidant or by removing NO from the exhaust gases.

stratosphere n 成層圏

▶ The ClO radical in particular is implicated in environmentally sensitive reactions which lead to depletion of ozone and oxygen atoms in the stratosphere.

straw-colored adj 淡黄色の

▶ Pure nitrobenzene is a straw-colored liquid boiling at 206–207 ℃.

stray light n 迷光

▶ The luminescence is very weak and may easily be swamped by stray light.

stray radiation n 迷光放射

▶ Lead is a very effective material for shielding X-ray equipment and absorbing stray radiation.

streak n **1** 線 syn *line, stripe*

▶ The ionizing particles produce a streak of nucleation centers which result in localized condensation.

2 条痕 syn *mark*

▶ Pyrite is a brass-yellow or brown tarnished

mineral with greenish or brownish-black streak.

stream n 1 流れ syn *current, flow*

▶ Solid anhydride in a flask was maintained under a stream of ammonia gas for 14 days at which time the increase in weight showed that more than the stoichiometric amount of ammonia had been absorbed.

vi 2 流れる syn *flow*

▶ When eight volts is applied between mercury electrodes in an aqueous solution of tetraethylammonium hydroxide, the surface tension is so reduced that the mercury streams into the water as a brown cloud of colloidal particles.

strength n 1 長所 syn *ability, asset, strong point*

▶ It is a great strength of thermodynamics that we can make use of the concept of internal energy without having to deal with it on a detailed molecular basis.

2 強度 syn *robustness, toughness*

▶ All ammonium salts are relatively unstable compounds, the degree of stability depending, in general, upon the strength of the acid which is combined with the ammonium.

3 濃度 syn *concentration*

▶ An oxalic acid solution of this strength can be prepared by dissolving exactly 6.303 g of pure, crystallized oxalic acid in water and diluting to a volume of 1 L in a calibrated flask with water at the laboratory temperature.

strengthen vt 強める syn *fortify, reinforce*

▶ The liquid crystal-to-liquid transition occurred only for M = Mo and was raised compared to the nonfluorinated complex, suggesting that the $M_2 \cdots O$ interaction was strengthened by the electron-withdrawing ligand.

stress n 1 応力 ☞ strain

▶ Although glass is isotropic, sometimes glasses that have not been annealed properly show stress birefringence.

vt 2 強調する syn *emphasize, highlight, underline*

▶ It must be stressed that laser-induced fluorescence only provides a relative measure of concentration and not an absolute concentration.

stress relaxation n 応力緩和

▶ The experimental methods used to study and investigate the viscoelastic properties of solid materials are essentially of three types which can be termed creep tests, stress relaxation tests, and dynamic tests.

stress-strain curve n 応力-ひずみ曲線

▶ The stress-strain curves of vulcanized rubber display hysteresis, because strain persists when the deformation stress is removed, thus producing a hysteresis loop instead of a reversible pathway of the curves.

stretch n 1 伸展 syn *span, spread*

▶ One of the unique features of DNA polymerase III is its ability to continuously synthesize very long stretches of DNA.

vt 2 引き伸ばす syn *elongate, lengthen*

▶ When wool is stretched, the hydrogen bonds within the helical chain are broken and are replaced by hydrogen bonds between adjacent chains.

stretching n 1 伸張 syn *expansion, extension*

▶ Natural rubber is amorphous when unstretched but has oriented crystalline structure on stretching.

2 伸縮振動 ☞ stretching vibration

▶ Bands due to carbon-carbon stretching may appear at about 1500 and 1600 cm^{-1} for aromatic bonds, at 1650 cm^{-1} for double bonds, and at 2100 cm^{-1} for triple bonds.

stretching vibration n 伸縮振動 ☞ stretching

▶ In the infrared spectrum of 1-butene, the absorption band near 1650 cm^{-1} is characteristic of the stretching vibration of the double bond.

stretching vibrational adj 伸縮振動の

▶ The hydroxyl group has a stretching vibrational frequency about 3500 cm^{-1}, somewhat greater than that for C−H, about 3000 cm^{-1}.

strict adj 1 厳密な syn *exact, precise, rigorous*

▶ A cell making use of a salt bridge is not one

that can be analyzed by strict thermodynamic arguments.

2 in a strict sense 厳密な意味で

▶ In a strict sense, the entropy of ordinary crystalline chloromethane is not zero at 0 K.

strictly adv **1** 厳密に syn *exactly, precisely*

▶ Bond energies are strictly applicable to molecules in which the bonds are of the normal lengths.

2 strictly speaking 厳密にいえば syn *in the narrow sense, in the strict sense*

▶ Strictly speaking, in constant current mode, the tip follows the contours of constant local density of states.

3 完全に syn *completely, perfectly*

▶ Most crystals are very poor conductors of the electric current, but strictly opaque crystals are commonly good conductors.

strike vt **1** 衝突する syn *collide with, hit*

▷ In an X-ray tube, electrons from a hot filament are accelerated by a voltage and then brought to rest by striking a solid target.

2 叩く syn *hit, knock*

▶ Heavy metal chlorites tend to explode or detonate when heated or struck.

3 行き着く syn *attain, conclude, reach*

▶ Water appears to strike the best compromise with regard to the structural features that make up ionizing power, that is, dielectric constant and solvating ability.

striking adj 顕著な syn *amazing, exceptional, remarkable*

▶ The diamond is probably the most striking example of dispersion in optically isotropic crystals.

strikingly adv 顕著に syn *especially, notably, particularly*

▶ There are no properties of the ammonia-water system which behave strikingly differently at the composition NH_4OH from what they do at neighboring concentrations.

string n 一続き syn *chain, series*

▶ A string of nucleosomes is wrapped in a cylindrical manner to form a solenoid-shaped fiber.

stringent adj 厳重な syn *severe, strict*

▶ As the device complexity increases in terms of different compositions, so the demands on the material properties become more stringent.

strip n **1** 細長い一片 syn *band, belt*

▶ Field effect transistors consist of a conducting channel formed by a strip of n- or p-type semiconductor adjacent to a gate.

2 クロマトグラムの細長い一片

▶ The amino acids are all colorless, and the strip bears no indication of their distribution until it is sprayed with a solution of ninhydrin.

vt **3** 奪う syn *remove, take away*

▶ At the exceedingly high temperatures prevailing (10^8 K) the nuclei are stripped of electrons and attain velocities comparable with those of particles from the cyclotron.

stroke n ストローク syn *dash, mark*

▶ The phase boundaries in individual electrodes were denoted by a vertical stroke.

strong adj **1** 濃い syn *concentrated*

▶ For most purpose a normal solution is too strong, and 0.1 or 0.5 N solutions are commonly used.

2 強い syn *powerful*

▶ *m*-Nitrobenzoic acid is only 5.1 times stronger as an acid than benzoic acid and only 1.2 times weaker than *p*-nitrobenzoic acid.

3 顕著な syn *conclusive, sound*

▶ The greater the number of physical properties measured, the stronger the evidence.

strong acid n 強酸 ☞ weak acid

▶ Strong acids such as $HClO_4$, HNO_3 and H_2SO_4 are completely ionized in water, and their relative strengths cannot be differentiated.

strong base n 強塩基 ☞ weak base

▶ Strong bases, NaOH, KOH, or $Ca(OH)_2$, added to an ammonium salt in the presence of a little water cause the evolution of ammonia on heating.

strong electrolyte n 強電解質 ☞ weak electrolyte

▶ There are many substances which behave as a strong electrolyte when dissolved in one solvent but as a weak electrolyte when dissolved in another solvent.

strongly adv 強く syn *effectively, powerfully*

▶ A highly electronegative atom will strongly induce electrons away from its neighbors.

structural adj **1** 構造上の syn *constitutional*

▶ The structural ideas on nucleic acids are of immense interest to biochemists and biologists as well as to physical chemists.

2 構造用の

▶ The principal advantage of Mg as a structural metal is its low density.

3 組織の syn *essential, fundamental*

▶ Proteins are found as structural material in such widely diverse forms as skin, hair, muscle, and horn.

structural formula n 構造式

▶ A structural formula indicates the location of the atoms, groups, or ions relative to one another in a molecule as well as the number and location of chemical bonds.

structural isomer n 構造異性体

▶ Structural isomers have different structural frameworks in which the bonding arrangements for the component atoms are different.

structural isomerism n 構造異性

▶ Efforts have been made by various chemists to subdivide the field of structural isomerism into smaller fields.

structurally adv 構造的に

▶ X-ray diffraction studies have established that the 1-2-3 compound is structurally similar to the perovskite family.

structural transformation n 構造変換

▶ During the structural transformation from austenite to martensite upon cooling, these particular planes of atoms in the austenite structure slide relative to one another.

structural viscosity n 構造粘性

▶ The structural viscosity depends upon how long it has had to build up its structure undisturbed.

structure n **1** 構造 syn *framework, shape*

▶ With the finest particles of bentonite 0.01% produces a structure, and 0.85% a soft transparent jelly.

vt **2** 組み立てる syn *arrange, organize*

▶ This article is structured to give a brief introduction to liquid crystals for the first-time reader.

structure analysis n 構造解析

▶ They made a very careful structure analysis of the crystals in the lower ferroelectric region.

structured adj 構造化された syn *integrated, organized*

▶ Structured organic surfaces have the potential to regulate the crystallization of inorganic materials.

structure determination n 構造決定

▶ Oligomerization often precludes structure determination with NMR spectroscopy.

structure factor n 構造因子 syn *crystal structure factor*

▶ The intensity of reflection *I* from a particular plane is proportional to the square of the quantity called the structure factor *F*, which is a function of the positions of atoms present and their scattering powers.

structureless adj 構造のない

▶ At concentrations above 10^{-4} mol/L, a broad structureless band appears in the fluorescence of pyrene solutions, and the intensity of this band is proportional to the square of the concentration of the solution.

structure-sensitive adj 構造敏感性の

▶ The process of melting is less akin to atomization than is boiling, and melting points are usually much more structure-sensitive.

struggle n 努力 syn *attempt, endeavor, effort*

▶ In textbooks, the imaginative steps, the testing of suggestions, the struggle to clarify ideas, the hesitant steps forward, and the breakthrough are largely lost, and instead there is a clear systematic development of the subject.

stubborn adj 頑固な syn *persistent, tenacious*

▶ To clean up some of these stubborn and particularly toxic pollutants, we may need to use bacteria and plants that have been genetically altered.

stubbornly adv 頑固に

▶ Nitrogen stubbornly resists ordinary chemical attack, even under stringent conditions.

study n **1** 研究 syn *exploration, investigation, research*

▶ The study may be concerned with a large number of individual components that comprise a macroscopic system.

vt **2** 調べる ☞ examine, investigate

▷ The most thoroughly studied phototropic solids are the *N*-benzylideneanilines, especially the derivatives of salicylaldehyde.

subcellular adj 細胞レベル以下の

▶ Centrifugation continues until the subcellular particles achieve density equilibrium with their surrounding solution.

subdivide vi **1** 細分される syn *separate, split*

▶ A pure liquid such as xylene or styrene spontaneously subdivided itself into innumerable small globules when placed upon an aqueous solution of a soap or detergent.

vt **2** 再分割する syn *categorize, classify*

▶ Nucleic acids are subdivided into two types: ribonucleic acid, containing the sugar D-ribose, and deoxyribonucleic acid, containing the sugar D-deoxyribose.

subdivision n 細分 syn *fragment, piece, portion, segment*

▶ The rate and ease of combustion may depend as much on their state of subdivision as on their chemical nature.

subgrain boundary n サブ粒界

▶ The interfaces between grains are called subgrain boundaries and can be treated in terms of dislocation theory.

subgroup n **1** 亜族

▶ In the A subgroup metals, the coordination is so irregular that the concept of a precise radius ceases to have any very definite meaning.

2 部分群

▶ An important subgroup of transcription factors regulates expression of genes that determine the identity of body regions.

subject n **1** 題目 syn *issue, matter*

▶ Autoxidation is one of the most important known to chemistry and has been the subject of a very large number of studies.

adj **2 subject to** …の支配を受ける syn *dependent on*

▶ In general, melting points and particularly the ranges of temperature over which substances are liquids are very subject to precise details of solid and liquid structures.

vt **3 subject to** 受けさせる syn *expose, submit*

▶ We have subjected the earlier data to a more recently developed technique of calculation with the result that better agreement with the results of other measurements is obtained.

sublattice n 副格子

▶ Intercalation beyond $x = 3$ would require the C_{60} sublattice to adapt a symmetry different from fcc in order to provide more than three interstitial sites per molecule.

sublevel n 副準位

▶ In a transition to a higher electronic level, a molecule can go from any of a number of sublevels, corresponding to various vibrational and rotational states, to any of a number of sublevels; as a result, untraviolet absorption bands are broad.

sublimable adj 昇華する

▶ TeF_4 can be obtained as colorless, hygroscopic, sublimable crystals by controlled fluorination of Te with F_2/N_2 at 0 ℃.

sublimate n 昇華物

▶ Microscopic rodlike crystals of sublimate formed after long reaction times or at higher temperatures.

sublimation n 昇華

▶ Crystals which had been prepared by sublimation had less well-formed faces.

sublime adj **1** 最高の syn *great, high*

▶ The sublime goal of this work is of course the ability to predict the products and rates of reactions for any oxidation.

vi **2** 昇華する

▶ Each constituent sublimes and resublimes until it reaches a point in the temperature gradient at which it vapor pressure is negligible.

vt **3** 昇華させる

▶ Any organic material that can be sublimed without excessive decomposition is a possible candidate for growth from the vapor.

sub-micrometer n マイクロメーター以下 すでに接頭辞を含む語の前にはハイフンを要する.

▶ For sub-micrometer-sized particles, it is essential to use electron microscopy.

submultiple n 約数

▶ Complete elemental analyses permit computation of empirical formulas which are equal to, or are submultiple of, the molecular formulas.

subordinate adj 副次的な syn *secondary, subsidiary*

▶ Chemically, coal is a macromolecular network composed of groups of polynuclear aromatic rings, to which are attached subordinate rings connected by oxygen, sulfur, and aliphatic bridges.

suboxide n 亜酸化物

▶ Cesium forms an even more extensive series of suboxides: Cs_7O, bronze-colored; Cs_4O, red-violet; $Cs_{11}O_3$, violet crystals.

subsection n section をさらに分割した小区分

▶ It will be convenient to divide the discussion into five subsections dealing in turn with the trihalides, the pentahalides, other halides, halide complexes of M^{III} and M^{V}, and oxohalides.

subscript n 下付き文字

▶ In formulas showing composition or molecular structure, the numerical subscript of the symbol of an element gives the number of atoms of the element in the molecule.

subsequent adj **1** その後の syn *following, later*

▶ Subsequent investigations have made it evident that, in thermal decompositions of organic molecules, free radical formation often comprises only a small proportion of the total reaction and that simple molecular products, such as methane and water, can be formed directly.

2 subsequent to …の後で syn *after, following, succceeding*

▶ Azo–hydrazone tautomerism of arylazonaphthols has been the topic of many investigations subsequent to Zincke's original observation in 1884.

subsequently adv その後に syn *afterward, later*

▶ It is possible that an *N*-nitroso compound is formed first, which subsequently isomerizes to the *p*-nitroso derivative.

subshell n 副殻

▶ The irregularity associated with the filled d subshell still occurs.

subsidiary adj 補助的な syn *auxiliary, secondary, supplemental*

▶ There are certain reactions, involving nonpolar molecules, in which it seems that the solvent plays a relatively subsidiary role.

substance n 物質 syn *material, matter*

▶ Sugars are neutral, combustible substances, and these properties distinguish them from other water-soluble compounds.

substantial adj 相当な syn *considerable, significant, sizable*

▶ If the temperature of a gas is changed by a substantial amount, it will be necessary to take into account the variation of the heat capacity with the temperature.

substantially adv かなり syn *amply, fully, sufficiently, well*

▶ Salts are likely to be soluble if the cation and anion differ substantially in size.

substantiate vt 実証する syn *affirm, confirm, prove, verify*

▶ The existence of free triarylmethyl radicals has been substantiated in a number of ways.

substituent n 置換基

▶ The properties characteristic of these and other various types of organic compounds are largely determined by the substituent, or functional group, common to each.

substituent constant n 置換基定数

▶ The substituent constant (σ) is a number (+ or −) indicating the relative electron-withdrawing or electron-releasing effect of a particular substituent.

substitute n **1** 代役 syn *alternative, replacement, substitution*

▶ Calcium silicate can be effectively reduced to very fine particles by passing it through the transition temperature; such a phenomenon can be turned to advantage as a substitute for grinding a substance to achieve small particle size.

vt **2 substitute for** 置換する syn *displace, exchange, replace*

▶ An important feature of the silicate minerals is that aluminum may substitute for silicon in the tetrahedra.

substitution n **1** 置換 syn *displacement*

▶ The tendency to aggregate increases with increasing halogen substitution and also with the presence of more polarizable halogens.

2 代用 syn *alternative, replacement, substitute*

▶ In its beginning stages, virtually all chemical innovation begins as a process of substitution.

substitutional alloy n 置換型合金

▶ Substitutional alloys are obtained by starting with metal X and gradually replacing its atoms with those of metal Y.

substitutional solid solution n 置換型固溶体

▶ In substitutional solid solutions, the atom or ion that is being introduced directly replaces an atom or ion of the same charge in the parent structure.

substitutionally adv 置換型に

▶ Since the paramagnetic ion is present only in small amounts, it is assumed that its site symmetry is identical to that of the host ion that it substitutionally replaces.

substitution reaction n 置換反応 syn *displacement reaction*

▶ Elimination reactions are generally found to compete with substitution reactions.

substoichiometric adj 不足当量の

▶ Whereas substoichiometric amounts of diethylamine resulted in improved enantioselectivities but diminished yields, excess diethylamine resulted in enhanced yields but decreased enantioselectivities.

substrate n **1** 基板 syn *basis*

▶ Reactive metal alkyls and nonmetal hydrides are co-reacted by thermally decomposing them onto a heated substrate.

2 基質

▶ Each of these reactions requires the presence of acid to convert the alcohol into the actual substrate, the protonated alcohol.

subtle adj 微妙な syn *delicate, fine, refined*

▶ The distinction between hcp and ccp is subtle, and several metals adopt both structures under different conditions.

subtlety n **1** 複雑さ syn *intricacy, sophistication*

▶ The interest of nickel(II) stereochemistry is its subtlety; but it is the organometallic chemistry of nickel that is more surprising.

2 微妙 syn *delicacy*

▶ Second-order transitions will usually be missed completely on heating under the microscope because of their subtlety.

subtly adv 微妙に syn *delicately*

▶ It is not surprising that sulfur can replace oxygen in a large number of organic compounds, subtly altering their acid–base, redox, and complexing properties by virtue of its lower electronegativity and increased size.

subtract vt 引く syn *deduct* 動詞として minus は用いられない.

▶ If the digitized spectrum before photolysis is subtracted from the spectrum after photolysis, only the features that change are seen.

subtraction n **1** 引き算 syn *deduction*

▶ Subtraction of the last two equations, with term-by-term cancellation of the exponential series, gives $N[1 - \exp(-x)] = n_0$.

2 削減 syn *removal*

▶ The subtraction of the monolayer from the solid surface is rendered visible by the movement of the thread over an equal area.

subunit n サブユニット

▶ One heme with its protein chain is known as a subunit.

success n 成功 syn *achievement, attainment*

▶ An enormous number of host-activator combinations have been studied for luminescence, with a fair degree of success.

successful adj **成功した** syn *fruitful*

▶ They have reported the successful encapsulation of Mn(III) salen complexes into zeolite Y using a ship-in-a-bottle approach.

successfully adv **首尾よく** syn *adequately, nicely,*

▶ In the case of defect conductors, the apparent activation energy has been successfully split into terms representing the creation and migration of the point defects.

succession n **1 連続** syn *chain, sequence, series*

▶ PrO_{2-x} forms a disordered nonstoichiometric phase at 1000 ℃, but at lower temperatures this is replaced by a succession of intermediate phases with only very narrow composition ranges of general formula Pr_nO_{2n-2} with $n = 4$, 7, 9, 10, 11, 12 and ∞.

2 in succession 連続して syn *succesively*

▶ Various solvents can be tried in succession for crystallizing the substance, for a solvent which proves unsatisfactory can be evaporated quickly and replaced by another one.

successive adj **連続する** syn *continuous, uninterrupted*

▶ The family of alkanes forms a homologous series, the constant difference between successive members being CH_2.

successive reaction n **逐次反応** syn *consecutive reaction*

▶ This polymerized product may be considered the result of successive reactions.

successively adv **引き続いて** syn *in succesion*

▶ The product was purified for analysis by washing successively in cold water, 95% ethanol, and petroleum ether followed by recrystallization from acetone-petroleum ether to give nearly colorless, well-formed prisms.

such pron **1 そのようなもの**

▶ Only true chemical substances or definite combinations of such, can be components of a system.

2 as such そのようなもの

▶ Quartz was the first optically active substance to be recognized as such.

adj **3 そんな**

▶ Although isomerization reactions are quite common in organic chemistry, such reactions do not appear to take place commonly in the solid state.

4 such and such これこれの

▶ Put simply, knowing the enthalpy change of reaction is the essential first stage in deciding whether such and such a proposed process is likely to go or not.

5 such as 同種類のものを例示して，**…のような** syn *like*

制限的な用法では，such as の前にコンマを用いない．

▶ In structures such as ice, each oxygen is attached tetrahedrally to four hydrogens, because this is required for the maximum number of hydrogen bonds to be formed.

such as の後に，文法上，対等な関係にあるものを並べるときには，それらを and でつなぐか，or でつなぐかを検討する．

▶ Natural fibers such as cotton and human hair can feel very rough after washing.

補足説明をする非制限的な用法では，コンマで区切る．

▶ Such defects have considerable effect on certain physical properties, such as the color and conductivity of crystals.

6 such as to do …するような

▶ The normal development of a crystal is such as to obscure its symmetry.

7 such…as… …のような…

▶ The ruby has absorbed light of such wavelengths as to leave the remaining light colored red.

8 such that …のようなもの

▶ The frequency of reorientation is such that nuclear magnetic resonance and X-ray measurements cannot be expected to distinguish between it and free rotation.

suck vt **1 吸収する** syn *draw*

▶ When the reaction is endothermic, the principal driving force is the increasing entropy of the system, because the entropy of the surroundings decreases as enthalpy is sucked into the system.

2 suck in 吸い込む syn *inhale*

▶ The flask on cooling sucks in dry air through the calcium chloride.

suction n 吸引

▶ The suction is likely to draw small particles of the precipitate into the pores of the paper, and then filtration is usually as slow as or slower than it would have been without the suction.

suction filtration n 吸引沪過

▶ Small quantities of solids can be collected and washed by centrifugation rather than by suction filtration.

suction flask n 吸引フラスコ ☞ filtering flask

▶ Round-bottom Pyrex ware and thick-walled suction flasks are not liable to collapse.

suction pump n 吸引ポンプ

▶ A filtering flask is connected to a suction pump with special suction tubing made with thick walls so that it will not collapse under vacuum.

sudden adj 急な syn *abrupt, quick, rapid*

▶ The perturbation from equilibrium may be the result of a sudden change in temperature, pressure, or strong electric field.

suddenly adv 急に syn *abruptly, quickly, rapidly*

▶ Heating of the flask is begun only after the system has been evacuated; otherwise the liquid might boil too suddenly on reduction of the pressure.

suffer vt こうむる syn *experience, undergo*

▷ A comparison of the crystal structure of trioxane and its polymer shows that both are hexagonal, with the dimension along the *c* axis suffering only 0.5% elongation during polymerization, whereas dimensions at a right angle to it are contracted by 4%.

suffice vi 十分である syn *be adequate, be enough, be sufficient*

▶ It should be understood that the mere magnitudes of dipole moments do not suffice to afford an understanding of their influence upon intermolecular forces.

sufficient adj 十分な syn *adequate, enough*

▶ Absorption of light by acetone results in the formation of an excited state which has sufficient energy to under go cleavage of a C–C bond and form a methyl free radical and an acetyl free radical.

sufficiently adv 十分に syn *adequately, enough, satisfactorily*

▶ When a nanocrystal is heated, e. g., by absorption of a sufficiently large energy, it melts and therby becomes a finite molten salt.

sugar charcoal n 砂糖炭

▶ Arrested growth in the recrystallization of nongraphitizing carbons leads to products with low densities, for example, 1.79 g/cm^3 for sugar charcoal after heating to 3,000 ℃.

suggest vt **1** 提案する syn *propose, recommend*

▶ The apparent contradictions suggest a closer examination of the conflicting statements.

2 示唆する syn *imply, indicate*

▶ The early results suggest that additional exploration of this molecular strategy is warranted.

suggestion n 提案 syn *advice, proposal, recommendation*

▶ Gomberg's suggestion that the triphenylmethyl radical could exist as a fairly stable species was not well received.

suggestive adj 示唆に富む syn *indicative, reminicent*

▶ The electronic spectrum of the tetrakis(acetylacetonato)europium(III) anion in alcohol is grossly altered on addition of dimethylformamide, suggestive of complex formation.

suit vt 適する syn *be acceptable, be convenient, be suitable*

▶ If a nitrate solution is boiled with zinc dust and an alkali, a considerable evolution of ammonia takes place. This reaction is particularly suited for the detection of nitric acid in the presence of chloric acid.

suitable adj 適当な syn *applicable, appropriate, fitting, proper*

▶ A certain molecular complex can be prepared by crystallization of a mixture of the components from a suitable solvent.

suitably adv 適当に syn *appropriately, properly*

▶ Substituents in macrocyclic ligands are not, in general, suitably placed to have a profound impact on the sterics of the coordination pocket.

suite n ひと組 syn *collection, series, set*

▶ The liver contains both the enzyme galactokinase and a suite of enzymes that are necessary to convert galactose 6-phosphate into glucose 6-phosphate.

sulfonate vt スルホン化する

▶ To sulfonate we use a large excess of concentrated or fuming sulfuric acid; high concentration of sulfonating agent and low concentration of water shift the equilibrium toward sulfonic acid.

sulfonation n スルホン化 ☞ desulfonation

▶ The particular shape of potential energy curve that makes sulfonation reversible also permits an isotope effect to be observed.

sum n **1** 和 syn *total*

▶ The Br···O distance of 3.30 Å is 0.15 Å less than the sum of the van der Waals radii.

vt **2** 合計する syn *add*

▶ In these simple perovskites, the metal oxidation states must sum to +6 in order to yield an electrically neutral compound.

3 sum up 要約する syn *digest, review, summarize*

▶ To sum up this section, the deformation energies play a decisive role in dimer formation and are responsible in large part for the strong preference for the D_{2h} structure in Be_2X_4 and Mg_2X_4.

summarize vt 要約する syn *condense, sum up*

▶ The more important physical properties of the elements are summarized in Table 1.

summary n **1** 概要 syn *brief, outline, review*

▶ A discussion of the stereochemistry of coordination compounds is not complete without a summary of the different types of isomerism.

2 in summary 要約すると syn *briefly, concisely, in brief*

▶ In summary, a selection of novel modular, hybrid ligands has been synthesized containing the sulfone moiety.

summation n 合計 syn *addition*

▶ This summation counts each pair interaction twice, hence, the introduction of the numerical coefficient 1/2.

sum total n 総計 syn *sum, total*

▶ The law of conservation of energy, simply stated, is that the sum total of energy in the universe is constant.

sunlight n 日光

▶ Benzopinacol crystallizes in dramatic fashion on exposure of a solution of benzophenone in isopropyl alcohol to bright sunlight.

superacid n 超酸

▶ Superacids are protonated solvent molecules of the already strong acids.

supercoiled adj 高次コイルの

▶ Most plasmid DNAs in bacteria exist as supercoiled circular structures.

supercoiling n スーパーコイル化

▶ The torsional stress introduced by this supercoiling is relieved by DNA gyrase.

superconducting adj 超伝導性の

▶ C_{60} was found to react with some of the alkali metals to form solids of composition M_3C_{60} that are superconducting with critical temperature on the order of 30 K.

superconduction n 超伝導 ☞ superconductivity

▶ Currently it is not clear whether or not BCS theory can adequately explain organic superconduction.

superconductivity n 超伝導 ☞ superconduction

▶ A large-scale shift to superconducting technology will hinge on whether wires can be prepared from the ceramics that retain their superconductivity at 77 K, while supporting large current densities.

superconductor n 超伝導体

▶ Alloys of niobium, particularly Nb–Ti, had been responsible for the greatest advances in superconductor technology

supercool vt 過冷却する syn *undercool*

▶ Isolated metal droplets 50–100 microns in diameter could be supercooled by more than 100 degrees, while in bulk the same metals are difficult to supercool more than a few degrees.

supercooled liquid n 過冷液体

▶ On cooling, the liquid does not recrystallize but becomes supercooled; as the temperature drops, so the viscosity of the supercooled liquid increases until eventually it becomes a glass.

supercooling n 過冷却

▶ Ordinarily the supercoolings observed are small because nucleation catalysts are present; these may be foreign particles or the walls of the containing vessel on or at which nucleation of the new solid phase is facilitated.

supercritical adj 超臨界の

▶ Supercritical CO_2 has proven effective as a solvent for the synthesis of fluoropolymers which are generally insoluble in conventional solvents.

supercritical extraction n 超臨界抽出

▶ Much of the research currently being conducted using supercritical fluid extraction centers on the reduction of organic and aqueous waste streams from industrial processes.

supercritical fluid n 超臨界流体

▶ The low viscosity of supercritical fluids allows for considerable narrowing of proton NMR signals compared to the signals observed in conventional solvents.

superexchange n 超交換

▶ In order to get efficient energy transfer by superexchange, it appears to be important that the metal–oxygen–metal bond should be approximately linear.

superficial adj **1** 表面の syn *external, exterior, surface*

▶ There is a likelihood that the formation of a superficial WO_3 layer may interfere with the intrinsic bronze spectra by reflectance method.

2 見かけの syn *apparent, cursory*

▶ The resemblance is more than just superficial.

3 表面的な syn *cursory, nominal, quick*

▶ Superficial inspection of the structures of oxime picryl ethers 1**a**–**c** and their isomerization products suggest that they must have very different molecular shapes.

superficially adv 見かけ上 syn *apparently, seemingly*

▶ Superficially, the crystal structure of $YBa_2Cu_3O_7$ may be regarded as a perovskite with 2/9 of the oxygens missing.

superfluid adj 超流動性の

▶ The superfluid phase of helium interferes with the cooling process by creeping around the apparatus.

superfluous adj 余計な syn *excessive, needless, unnecessary*

▶ The prefix di- to denote disubstitution is sometimes omitted as superfluous, but most current opinion regards this form of redundancy desirable to help prevent errors.

superheat vt 過熱する

▷ For substances of extremely low vapor pressure it may be necessary to resort to distillation with superheated steam or distillation in vacuum, or both.

superheating n 過熱

▶ In solid-state phase changes, superheating and supercooling are common, because the new phase generally has to nucleate and grow within the phase that is initially present.

superimposable adj 重ねられる ☞ nonsuperimposable

▶ Molecules that are not superimposable on their mirror image are chiral.

superimpose vt 重ねる syn *overlay*

▶ In talc and kaolinite, the layers are electrically neutral, and they are loosely superimposed on one another to form the crystalline material.

superimposition n 重ね合わせ

▶ An illustration of the probable superimposition of inductive and hyperconjugative electron release is afforded by the values of the dipole moments of monoalkylbenzenes in the vapor state.

superior adj 優れた syn *better*

▶ While exponential terms are believed to be somewhat superior expressions of the repulsive potential, inverse power terms are much easier to use and have proven almost as satisfactory.

superiority n 優位 syn *distinction, importance*

▶ Water owes its superiority as a solvent for ionic substances not only to its polarity and its high dielectric constant but to another factor as well: it contains the –OH group and thus can form hydrogen bonds.

superlattice n 超格子

▶ Another system which illustrates the transition from the random arrangement of the solid solution to the ordered structure of the superlattice is that of iron and aluminum.

supermolecule n 超分子

▶ The transition from a plastic which exhibits flow, through molecular slip, to a rigid three-dimensional supermolecule can be visualized in terms of an increase in the cross-link concentration.

supernatant n **1** 上澄み ☞ supernatant liquid

▶ The soluble proteins and other solutes remain in the supernatant from this step.

2 adj 上澄みの

▶ The mixture is allowed to settle for several minutes, the supernatant solution is decanted, and the dense precipitate is washed with one or two portions of fresh ether.

supernatant liquid n 上澄み液 ☞ supernatant

▶ On standing, the systems leave an excess of supernatant liquid above a compact sediment.

supernumerary adj 過剰な syn *excess, extra*

▶ The condensation of MoO_6 and MoO_4 polyhedra to produce these large anions requires large quantities of strong acid, as the supernumerary oxygen atoms are removed in the form of water molecules.

superoxide n 超酸化物

▶ The superoxides MO_2 contain the paramagnetic ion O_2^- which is stable only in the presence of large cations such as K, Rb, Cs.

superpose vt 重ねる

▶ The monochromatic peaks are superposed on a background of white radiation which is produced by the general interaction of high-velocity electrons with matter.

supersaturate vt 過飽和にする

▷ A solution, containing more solute than it ordinarily would at that temperature when in equilibrium with excess solid, is termed supersaturated.

supersaturated solution n 過飽和溶液

▶ From very dilute solutions, a white crystalline precipitate of $Mg(NH_4)PO_4$ separates only after standing some time, owing to the tendency to form supersaturated solutions.

supersaturation n 過飽和

▶ A metastable polymorph can be prepared either by achieving a high degree of supersaturation in the vapor or solution state or by suprcooling in the melt state.

superscript n 上付き文字 ☞ subscript

▶ A radical is indicated by a dot as right superscript to the symbol of the element or group.

supersede vt 取って代わる syn *replace, substitute*

▶ The term secondary valence has now been superseded by the term coordination number.

superstructure n 超構造

▶ When face-centered cubic NiO is examine by neutron diffraction, extra peaks are observed which indicate the presence of a superstructure.

supplement vt 補足する syn *add, complement*

▶ For unsaturated ligands in general, the electrostatic viewpoint must be replaced or supplemented with the molecular orbital or valence bond approach.

supplier n 供給者 syn *dealer, donor*

▶ Starting materials sourced from commercial suppliers were used as received.

supply n 1 供給 syn *delivery, furnishing, provision*

▶ In air arsine burns to arsenic oxides and water, but if the oxygen supply is limited or the flame is cooled, only the hydrogen burns and the arsenic is liberated.

vt 2 供給する syn *deliver, furnish, provide*

▷ Metal dichlorides were used as supplied from Fluka without further purification.

support n 1 支持 syn *backup, encouragement, reinforcement*

▶ The stereochemistry of halogen addition gives powerful support for a two-step mechanism.

2 担体 syn *carrier*

▶ Xerogels can be used as catalyst supports and precursors for a wide range of glasses, ceramics, films, and fibers, depending upon the method of preparation.

vt 3 支持する syn *advance, promote*

▷ Of all the evidence supporting the mechanism, the strongest single piece is the occurrence of rearrangements.

4 支える syn *carry*

▷ The typical catalyst used in the hydrodesulfurization process is MoS_2 supported on alumina.

supporting electrolyte n 支持電解質

▶ The supporting electrolyte for electrochemical experiments in dichloromethane was tetrabutylammonium hexafluorophosphate.

suppose vt 仮定する syn *assume, presume, surmise*

▶ Grain boundaries are supposed to consist of suitable arrays of edge dislocations.

supposed adj 想像された syn *assumed, hypothetical*

▶ An intimate mixture of ice and salt containing 23.6 percent of sodium chloride melts at a definite and constant temperature and exhibits, therefore, a behavior supposed to be characteristic of a pure chemical compound.

supposedly adv 推定上 syn *presumably, theoretically*

▶ Since osmotic pressure is supposedly proportional to the number of particles present in solution, the value for any solution is dependent on the temperature, solvent used, pH, or any other factor which influences aggregation or dissociation.

supposition n 仮定 syn *assumption, postulate, presumption*

▶ The solubility of the several hydrocarbons in aqueous solutions increased with increasing silver ion concentration to a greater degree than was consistent with the supposition that only a 1-1 complex, $AgAr^+$, was formed.

suppress vt 抑制する syn *block, preclude, prevent*

▶ When single crystals are subjected to a finite external pressure, the insulating phase is suppressed, leading to metallic behavior and eventually a superconducting ground state.

suppression n 抑止 syn *prevention, repression*

▶ The suppression of H_2 and CO chemisorption was initially attributed to an electronic perturbation of the metal atoms, caused by their interaction with titanium cations at the oxide surface.

supramolecular adj 超分子の

▶ In many cases, the control of particle size is achieved by confining the syntheses to discrete localized volumes through the use of supramolecular assemblies of organic molecules.

sure adj **make sure** 確かめる syn *assure, confirm, secure*

▶ It is necessary to make sure that sufficient acid is present and to dilute the solution.

surely adv 確かに syn *certainly, definitely, undoubtedly*

▶ To the extent that the entropy of mixing is a function of random configurations and is assumed independent of heat interactions, it is surely independent of pressure.

surface n 1 表面 syn *exterior, interface*

▶ The isotherm for two substances adsorbed on the same surface is of importance in connection with inhibition and with the kinetics of surface reactions involving two reactants.

adj **2 表面の** syn *external, exterior, interfacial*

▶ The most stable surfaces tend to be flat and close-packed with each surface atom surrounded by a large number of nearest neighbors.

vt **3 滑らかにする**

▶ A stopcock which will hold without lubricant or solvent is surfaced carefully in a final grinding with rouge as the abrasive.

surface-active agent n **界面活性剤** syn *detergent, surfactant*

▶ Water and oil do not mix, but on addition of an appropriate surfactant, otherwise known as a detergent or surface-active agent, a microemulsion may form as one of several possible systems, depending on the relative concentrations of the three components.

surface analysis n **表面分析法**

▶ Surface analysis is quite different from bulk analysis; frequently, factors important for surface analysis are not important to bulk analysis.

surface area n **表面積**

▶ Many attempts have been made to determine the effective surface area of porous bodies.

surface charge n **表面電荷**

▶ The surface charge of colloid particles must be balanced by an equal excess of opposite charge in the liquid.

surface charge density n **表面電荷密度**

▶ Surface charge density changes the surface tension of the mercury and, hence, the shape of the globule.

surface coverage n **吸着率**

▶ At a higher surface coverage, the scattered intensity increases substantially, demonstrating a reduced sticking coefficient.

surface defect n **表面欠陥**

▶ When surface defects have been removed, ice is essentially an insulator with an immeasurably small conductivity.

surface energy n **表面エネルギー** ☞ interfacial energy

▶ From the theoretical viewpoint, the surface energy is identified with the difference of potential energy brought about by splitting a crystal along a specific crystallographic plane and separating the two parts to infinite distance.

surface-enhanced Raman effect n **表面増強ラマン効果**

▶ The surprising discovery of the million-fold intensification involved in the surface-enhanced Raman effect provides a technique with applicability yet to be determined.

surface excess n **表面過剰**

▶ If a component tends to accumulate at the interface, its surface excess is positive.

surface film n **表面膜**

▶ Minute impurities in the water can appreciably affect the properties of a surface film.

surface mobility n **表面移動度**

▶ If α represents the ratio of the absolute temperature of an ionic solid to its melting point on the absolute scale, surface mobility of lattice units is believed to become appreciable at about $\alpha = 0.3$, whereas lattice diffusion requires $\alpha = 0.5$ or higher.

surface phenomenon n **表面現象**

▶ The goal of modern physical-chemical studies of surface phenomena is the understanding of these phenomena by means of a molecular model.

surface potential n **表面電位**

▶ In studies on germanium, various gases were found to affect the space-charge layer near the surface, the surface potential, and the surface conductance.

surface pressure n **表面圧**

▶ It is possible to extrapolate the solid region of the curve to zero surface pressure to determine the area occupied by a single molecule in the compact film.

surface reaction n **表面反応**

▶ Varying the temperature of the surface in

the molecular beam experiments permits the activation energy of the surface reaction to be determined.

surface state n 表面準位

▶ Surface states occur with energies within the forbidden band.

surface stress n 表面応力

▶ In the case of a solid, the surface tension is the work spent in forming a unit area of surface while the surface stress depends upon the work spent in stretching the surface.

surface structure n 表面構造

▶ Doping TaS_2 with a small amount of Ti can controllably perturb the electronic distribution, allowing imaging of the surface structure.

surface tension n 表面張力 ☞ dynamic surface tension, static surface tension

▶ Mercury with the highest surface tension of any liquid does not flow but disintegrates into droplets.

surface viscosity n 表面粘性

▶ Surface viscosity is another relevant rheological property which may be measured by several devices.

surfactant n 界面活性剤 syn *detergent, surface-active agent*

▶ If a very dilute solution of surfactant molecules is dropped on the surface of pure water and the solvent evaporates or dissolves, the surface-active molecules are restricted to the interface with their polar head groups bound to the aqueous subphase and their fatty tails assuming various orientations relative to the surface plane, depending upon the available area per molecule.

surmise vt 推測する syn *assume, imagine, speculate, suppose*

▶ Since the forces opposing compression of the ionic crystal are those of repulsion between the ions, we can surmise that these repulsive forces have a much stronger dependence on the internuclear distance than do the attractive forces which resist thermal expansion.

surmount vt 乗り越える syn *exceed, surpass*

▶ Practical methods of producing titanium metal from oxide ores have to surmount two problems.

surpass vt 上回る syn *exceed, excel*

▶ The structural complexity of borate minerals is surpassed only by that of silicate minerals.

surprise n 驚き syn *shock, wonder*

▶ It is no surprise to find that most aryl halides are usually much less reactive than alkyl or allyl halides toward nucleophilic reagents in either S_N1- or S_N2-type reactions.

surprising adj 驚くべき syn *amazing, extraordinary, remarkable*

▶ The pairs of major elements Na^+–K^+ and Ca^{2+}–Mg^{2+} are so similar chemically that it is surprising that they differ so greatly in their biological functions.

surprisingly adv 驚くほどに syn *notably, particularly*

▶ The alkali metals were listed in a single column because it was observed that the members of this family had quite surprisingly similar properties.

surround vt 囲む syn *enclose, encompass*

▷ Many reductions, such as from MnO_4^- to MnO_2, involve a decrease in the number of oxygen atoms surrounding a metal ion, and since these are removed by H^+ to form water, this type of change is favored by high acidity.

surrounding n **1** 複数形で外界 syn *environment, neighborhood*

▶ Reaction in a mixture of solids is exothermic; as such a mixture is heated, a point is reached at which the heat of reaction can no longer be dissipated to the surroundings under approximately isothermal conditions, and the temperature within the mass of mixture rises appreciably.

adj **2** 周囲の syn *adjacent, nearby, neighboring*

▶ Unstable colloidal particles primarily consist of materials which are insoluble in the surrounding medium.

survey n **1** 調査 syn *appraisal, evaluation, examination, study*

▶ A survey of the literature shows several very interesting investigations in the field of catalytic condensations of aldehydes in which acidic, basic, and amphoteric catalysts are used.

vt **2 概観する** syn *appraise, evaluate, examine,*

▶ The number of such reactions is so large that no attempt can be made here to survey the entire field.

survival n 生存 syn *existence, living, subsistence*

▶ Maintenance of the integrity of chromosomal DNA is essential for the survival of all forms of life.

survive vt より生き残る syn *persist, subsist*

▶ The strong W–N multiple bond of the precursor is likely to survive the deposition process and facilitate incorporation of nitrogen into the film.

susceptibility n **1 感受性** syn *sensitivity, tendency*

▶ A common property of coordinated alkenes is their susceptibility to attack by nucleophiles such as OH^-, OMe^-, $MeCO_2^-$, and Cl^-.

2 磁化率 ☞ magnetic susceptibility

▶ The susceptibilities of the different kinds of magnetic materials are distinguished by their different temperature dependences as well as by their absolute magnitudes.

susceptible adj **1** …を受けやすい syn *affected by, subject to*

▶ Heme is always susceptible to oxidation when in the presence of dioxygen.

2 感受性の syn *influenceable*

▶ Bacterial adhesion is mediated by specific lectins on bacteria that bind to complementary carbohydrates on susceptible host cells.

suspect n **1** 考えられるもの

▶ The usual suspects of elemental sulfur, thionyl chloride, sulfuryl chloride, and SCl_2 itself were tried.

vt **2 感づく** syn *feel, imagine, sense, surmise*

▶ When the miscibility gap is close to the solubility curve, its presence may be suspected from the typical S shape of this curve.

suspected adj 疑われる syn *doubtful, questionable*

▶ It is possible to synthesize high-purity mineral phases and then intentionally dope them with a suspected impurity center to see whether a particular emission band can be induced or enhanced.

suspend vt 懸垂する syn *retain, withhold*

▶ If oil is suspended as independent drops in a heavier liquid such as water, the drops will tend to rise under the influence of gravity.

suspended adj 懸濁した

▶ An emulsion consists of droplets of one liquid suspended in another continuous liquid.

suspension n **1 吊り糸（針金）**

▶ Low-conductivity suspensions minimize thermal conduction between the various parts of the cryostat.

2 懸濁液

▶ The reactions of metal ions with colloidal suspensions are of great interest in a variety of chemical fields including catalysis, electrochemistry, and soil chemistry.

suspicion n 疑い syn *doubt, skepticism*

▶ The possibility of a solvent effect always leaves a suspicion that a dipole moment that is observed to be finite should really have been zero.

sustain vt **1 維持する** syn *continue, maintain, preserve*

▷ The two processes central to sustaining life on earth, respiration and photosynthesis, involve electron-transfer reactions of metalloproteins and metalloenzymes.

2 承認する syn *admit, approve*

▶ The accuracy of any data presently available is an order of magnitude removed from that which would be required to sustain such a treatment.

sustainability adv 持続性

▶ Sustainablity is increasingly an important issue in the wider context dealing with population, health, the environment, energy, and renewable resources.

sustainable adj 持続できる syn *feasible, possible, workable*

▶ This situation is highly sustainable as very little carbohydrate is lost, and there are usually large reserves of triglycerides in the body.

sustained adj 持続する syn *continued, prolonged, unchanged*

▶ This rapid and sustained increase in computing power has facilitated advances across the whole spectrum of scientific and engineering fields.

swamp vt 圧倒する syn *overcome, overwhelm*

▶ The relatively small changes in mobility with temperature are completely swamped by the much larger changes in the number of mobile electrons.

swarm n 群 syn *flock, mass*

▶ The initial process in condensation can be imagined as a sticking together of a swarm of water molecules into a microscopic droplet.

sweep n **1** 掃引 syn *scan*

▶ The most familiar electrochemical techniques are cyclic voltammetry or, if only one sweep is involved, linear sweep voltammetry.

vt **2** 運び去る syn *carry, remove, take away*

▶ When the methyl radical is swept along in a current of hydrogen, more or less methane is formed.

sweet adj 甘い syn *sugary*

▶ A polypeptide from a tropical fruit called the serendipity berry is said to be 3000 times sweeter than sucrose.

swell vi **1** 膨潤する syn *distend, expand*

▶ In early European cathedrals, organ pipes made of tin were found to swell and then to disintegrate in cold weather.

vt **2** 膨潤させる syn *enlarge, expand*

▷ The addition of a small amount of water serves to turn a plastic thixotropic mass of completely swollen clay into a readily flowing slurry.

swelling n **1** 膨潤 syn *enlargement*

▶ The swelling of gelatin in water is much affected by acids, bases, and electrolytes.

adj **2** 膨潤する

▶ The swelling power of liquids for rubber lies in the following decreasing series: carbon tetrachloride, chloroform, etc.

swift adj 迅速な syn *fast, quick, speedy*

▶ The chemist's contribution toward a deeper understanding of the natural world consists partly in developing better methods of analysis and swifter, more efficient means of synthesis.

swirl vt 渦を巻かせる syn *circulate*

▷ Since both benzene and nitrobenzene are insoluble in the acid mixture and form an upper layer, thorough agitation must be maintained by swirling throughout the entire reaction period in order to obtain a successful result in the time indicated.

switch n **1** スイッチ

▶ The Q switch may simply be a rotating mirror timed to allow out the laser beam when it has reached its optimum intensity.

vi **2** 切り替わる syn *change, shift*

▶ Thin films of Cu(TCNQ) and Ag(TCNQ), when subjected to an electric field, switch at a certain threshold potential from a high resistance state to a low resistance state.

vt **3** 切り替える

▷ Other important switching and recognition mechanisms take place in the transport of ions through ion channels in the cell membrane.

4 switch off スイッチを切る

▶ Ferroelectric materials are distinguished from ordinary dielectrics by their extremely large permittivities and the possibility of retaining some residual electric polarization after an applied voltage has been switched off.

switching n 転換 syn *exchange, interchange*

▶ Nonlinear optical materials can be used in devices for fast switching, with modulating waves carrying information and amplifying signals.

symbol n 記号 syn *abbreviation, code*

▶ The symbol HA implies only that the species can act as a proton donor: it can be a neutral species, an anion, or a cation.

symbolic adj 記号の syn *typical*

▶ In the common symbolic representation for the atomic orbitals, the numerical value of n is followed by a letter representing l in which s, p, d, and f refer to $l = 0, 1, 2, 3$, respectively.

symbolism n 記号体系 syn *notation, sign*

▶ The nomenclature and symbolism employed throughout are those developed and employed by Ingold and his research school.

symbolize vt **1** 象徴化する syn *express, represent*

▷ Only those crystals free from pockets of occuluded water and those of the morphological type symbolized in Figure 1 were selected for study.

2 記号で表す syn *denote, stand for*

▶ The approximate temperature below which glass-like behavior is apparent is called the glass transition temperature and is symbolized by T_g.

symbiotic adj 共生の

▶ Some insects have symbiotic nitrogen-fixing bacteria in their intestines.

symbiotic relationship n 共生関係

▶ Many plants, particularly legumes, have a symbiotic relationship with nitrogen-fixing bacteria, which live in special nodules on the roots.

symmetric adj 対称な ☞ antisymmetric

▶ The interaction of enantiomers with a symmetric chemical or physical environment is identical in all respects.

symmetrical adj 対称的な syn *symmetric, well-balanced*

▶ The highly symmetrical neopentane boils at a so much lower temperature than does the isomeric and still nonpolar but much less symmetrical *n*-pentane.

symmetrically adv 対称的に

▶ Only part of the cubic face of sodium chloride is illustrated, since the rest of it is symmetrically related.

symmetric top n 対称こま syn *symmetric rotor*

▶ If two of the three moments of inertia are equal, the molecules is known as a symmetric top.

symmetry n 対称性

▶ Accordingly, substances with molecules of high symmetry crystallize more readily than those of low symmetry; that is, they have higher melting point.

symmetry element n 対称要素

▶ The cube and octahedron have the same symmetry elements and belong to the O_h group.

symptom n 症状 syn *indication, syndrome*

▶ Sickle cells block blood vessels, and patients suffer from poor oxygen delivery to tissues, causing joint pain and other symptoms.

syn addition n シン付加 ☞ anti addition

▶ The reaction of 1,2-dimethylcyclopentene illustrates the stereochemistry of the synthesis: hydroboration–oxidation involves overall syn addition.

synchronously adv 同時に

▶ After diffraction, the electrons strike a photographic plate which is exposed synchronously with the introduction of the sample gas for a second or less.

synchrotron n シンクロトロン

▶ With synchrotron sources, time-resolved X-ray structure determinations are feasible in situ.

synchrotron radiation n シンクロトロン放射光

▶ The most intense, currently available source of tunable radiation in the extreme ultraviolet and X-ray region is synchrotron radiation.

syn conformation n シン配座

▶ They are flipped into the syn conformation, where they are likely to have some shape similarity to cytosine.

syndiotactic polymer adj シンジオタクチックポリマー

▶ Syndiotactic polymers have asymmetric centers alternating in pattern between cis and trans positions.

syneresis n シネレシス

▶ Syneresis is promoted by all influences which promote coagulation, such as increasing concentration of particles, of electrolyte, etc.

synergic adj 相乗的な syn *collective, cooperative*

▶ It is very difficult to interpret shifts in s–p spectra so as to distinguish between the effect of stereochemical change and the effect of changing basicity which is synergic with it.

synergistically adv 相乗的に

▶ The two olefin-metal bond components reinforce one another synergistically.

synergy n 相乗効果 syn *cooperation*

▶ The synergy between mechanistic studies and new reaction types will continue to drive progress in this field.

syn form n シン形 ☞ anti form

▶ Because of the geometric isomerism of the C=N bond, there are three possible forms of the dioximes of the symmetrical 1,2-diketones: the syn, anti, and amphi forms illustrated below.

synonym n 類義語

▶ Dioxins is a synonym for polychlorinated dibenzo-*p*-dioxins and polychlorinated dibenzofurans.

synonymous adj 同じことを意味する syn *equal, equivalent*

▶ The hydrogen-ion concentrations are synonymous with those of the tetrahedral Al in the zeolite framework.

synovial fluid n 関節液

▶ Hyaluronic acid is a polysaccharide of 2-amino-2-deoxyglucose and glucuronic acid; it is an important component of animal connective tissue and of the synovial fluid which is the natural lubricant of joints.

synthesis n 合成 複数形は syntheses

▶ The term synthesis refers to creation of a substance that either duplicates a natural product or is a unique material not found in nature, by means of one or more chemical reactions.

synthesis gas n 合成ガス

▶ With transition-metal catalysts, synthesis gas yields alcohols, aldehydes, acrylic acid, etc.

synthesize vt 合成する

▶ Barbituric acids are readily synthesized by the reaction of urea with substituted malonic esters.

synthetic adj **1** 合成の

▶ There were no precedents for the successful ring-closing metathesis of enol ethers at that time; we began to explore the synthetic sequence.

2 人工的な syn *artificial, man-made*

▶ Synthetic methanol is suitable for most purposes without purification but usually contains traces of acetone and formaldehyde.

synthetically adv 合成によって

▶ Methyl salicylate occurs in many plants, but it is also readily prepared synthetically by esterification of salicylic acid, which in turn is made from phenol.

synthetic detergent n 合成洗剤

▶ Many of the synthetic detergents have become very valuable for specific applications, but soap is still the cheapest and most efficient of all the known surface-active agents.

synthetic dye n 合成染料

▶ Recognition of the selective and specific affinity of synthetic dyes for certain cells and cell constituents had made synthetic dyes the first working theme of the new science of chemotherapy.

synthetic polymer n 合成高分子

▶ Synthetic polymers include a wide variety of materials having properties ranging from hard and brittle to soft and elastic.

synthetic resin n 合成樹脂

▶ A distinction should be made between a synthetic resin and a plastic, the former is the polymer itself, whereas the latter is the polymer with such additives as filters, colorant, plasticizers, etc.

synthetic rubber n 合成ゴム

▶ The materials called synthetic rubber are not really synthetic rubber, since they are not identical with the natural product.

syringe n 注射器

▶ The plungers of the syringes can be forced down rapidly, and a rapid stream of two solutions passes into the mixing system.

syrup n シロップ

▶ It is easier to isolate an osazone than to isolate the sugar, and sugars that are syrups often give crystalline osazones.

syrupy adj シロップのような syn *oily, slimy*

▶ Extensive hydrogen bonding persists on

fusion, and phosphoric acid is a viscous syrupy liquid that readily supercools.

system n **1** 系

▶ The name system as used in thermodynamics denotes, generally, any finite body or collection of bodies or any region possessing energy, every part of which has a definite temperature and pressure and is the object of investigation.

2 体系 syn *scheme*

▶ A system, as logical as it is simple, is to number the polymorphic forms in the order of their discovery, which should in general follow their order of stability

3 組織 syn *organization, setup*

▶ The material capacity targets are system requirements and not just material requirements.

systematic adj 系統的な syn *organized, systematized*

▶ Thus, the systematic analysis of large numbers of related structures is a powerful research technique, capable of yielding results that could not be obtained by any other method.

systematic absence n 結晶回折の消滅則

▶ In some simple cases, the systematic absence of certain reflections corresponding to particular values of the Laue indices suffices to determine the structure completely.

systematically adv **1** 規則正しく syn *accurately, correctly, exactly, precisely*

▶ The unit cell of an extended solid is a collection of its atoms, and sometimes fractions of its atoms, that can be systematically translated through space to create the entire solid.

2 系統的に

▶ The effects of branching of terminal alkyl chains in mesogens have been investigated systematically.

systematic name n 系統名 ☞ trivial name

▶ Ideally, every organic substance would have a completely descriptive and systematic name to permit only one structural formula to be written for it.

systematics n 分類法

▶ Cesium forms at least nine crystalline oxides whose structures can be rationalized in terms of general bonding systematics.

systematization n 体系化 syn *organization*

▶ It is the systematization that makes the textbook an effective way of communicating knowledge, but it is achieved at the expense of losing a sense of chemistry as an activity.

T

table n 表 syn *chart, list, tabulation*

▶ The following table gives the reduction potentials of some of the common substances which easily undergo oxidation or reduction.

tablet n 平板 syn *plate, slab*

▶ K_2SiF_6 crystals may form in thin basal tablets which have $c = 1.60$, $d = 3.08$.

tabular adj **1** 表の

▶ A comprehensive summary of available thermodynamic data for crystals, regardless of source, will be included in tabular form.

2 平板状の

▶ Relatively slow crystallization as, for example, from 0.01% solution in xylene results in diamond-shaped tabular crystals.

tabulate vt 表に作る syn *classify, categorize, list, systematize*

▷ A second problem arises from the use of tabulated refractive index data.

tabulation n 表 syn *chart, list, table*

▶ A tabulation of bond energies for various types of bonds to phosphorus is given in Table 1.

tacit adj 暗黙の syn *implicit, implied, unexpressed, unspoken*

▶ This observation brings us to one of the tacit assumptions of this inquiry, the invariance of the van der Waals radius of an atom under the most drastic environmental changes.

tacitly adv 暗黙裡に syn *implicitly*

▶ All previous treatments of this model have tacitly or explicitly accepted the assumption that equivalent electrostatic radius in both phases is the crystal radius.

tackle vt 取り組む syn *confront, pursue*

▶ In this review, we tackle the question of why specific oligomeric structures may have evolved for a very well-characterized enzyme family.

tactic n 方策 syn *procedure, scheme, strategy*

▶ Replacing groups of atoms by different but chemically similar groupings in the form of bioisosteres is a widespread tactic.

tactoid n タクトイド

▶ The tactoids are swarms of elongated particles such as vanadium pentaoxide, tobacco mosaic virus, myosin, fibrin, and benzopurpurin.

Tafel's equation n ターフェルの式

▶ The logarithmic relationship between the overvoltage and the current density, as expressed by the Tafel equation, reflects the logarithmic relationship between the rate of reaction and the activation energy.

tag n **1** 標識 syn *label, marker, sticker*

▶ The tag protein allows for the isolation and purification of the recombinant protein as a fusion protein.

vt **2** 標識を付ける syn *identify, label, mark*

▶ Compounds containing tracers are often said to be tagged or labled.

tailor vt 条件に適合させる syn *adapt, adjust, fit, modify, suit*

▶ The normal IR emission frequency is set by the energy of the band gap and may be tailored by altering the chemical composition of the semiconductor.

tailored adj 適合させた syn *custom-made, fitted*

▶ Chemical vapor deposition permits the growth of atomically abrupt interfaces and thus of countless tailored solid structures.

tailoring n 適合

▶ The kinetics of the phase transformation depends strongly on temperature, a fact that allows careful tailoring of the size distributions and types of second-phase precipitates within steels.

tailor-made adj 特別あつらえの syn *custom-made, fitted*

▶ A more general approach to the construction of infinite network structures that might provide tailor-made materials of the future was proposed.

take vt **1** 運ぶ syn *bring, lift, transport*

▶ Absorption takes the molecule to several levels of S_1, the relative probabilities being determined by the Franck–Condon principle and the relative configurations of the nuclei in the two states.

2 要する syn *demand, necessitate, need, require*

▶ It takes 16 to 18 kcal more energy to break the carbon–halogen bond in a vinyl halide than in the corresponding ethyl halide.

3 みなす syn *accept, assume*

▶ If the radius of the hole-forming spheres is taken as 1.000, the cubic, octahedral, and tetrahedral holes will accommodate spheres with radii of 0.732, 0.414, and 0.225, respectively.

4 測定する syn *acquire, obtain*

▷ The spectrophotometric data taken during these experiments reveal that even these simple systems decompose in a complex manner.

5 行使する syn *exercise, use*

▶ With phenols, as with amines, special precaution must often be taken to prevent polysubstitution and oxidation.

6 取り扱う syn *accept, consider, view*

▶ As in all such cases which are really concerned with the differing polarities of covalent bonds, such formalism should not be taken literally.

7 受け取る，理解する syn *understand*

▶ Internal conversion is usually taken to mean a nonradiative transition between two states of like multiplicity.

8 take up （a）**吸収する** syn *accept, adopt*

▶ A monolayer can reversibly take up an organic vapor to form a mixed film.

（b）**取り上げる** syn *accept, pick up*

▶ According to the Lewis definition, an acid is a substance that can take up an electron pair to form a covalent bond.

（c）**…を取る** syn *assume, follow*

▶ Liquids have no definite shape but a definite volume; a portion of liquid takes up the shape of that part of a vessel which contains it.

tandem n **1 in tandem with　連合して**

▶ We investigated whether the more nucleophilic character of the rearranged radical could be exploited in electrophile trapping in tandem with the deoxygenation rearrangement.

adj **2 縦に並んだ**

▶ Especially noteworthy are the processes developed involving tandem intermolecular radical addition followed by homoallylic rearrangement and homoallylic rearrangement followed by intermolecular radical trapping.

tandemly adv **縦列に**

▶ Examination of replicating eukaryotic DNA by electron microscopy shows the presence of many tandemly arranged bubbles clustered in groups of 20–80 in various regions of the DNA.

tandem mass spectrometry n **タンデム型質量分析法**

▶ In tandem mass spectrometry, one mass spectrometer feeds ions of a selected mass into a collisional zone that induces fragmentation into a new set of fragment ions for analysis in a second mass spectrometer.

tangent n **1 接線** syn *tangent line*

▶ The initial velocity is measured as the slope of the tangent of the progress curve at $t = 0$.

adj **2 接する**

▶ The respective axes passing through the tangent spheres of the three p orbitals lie at right angles to one another.

tangential adj **接線の，正接の**

▶ We define the coefficient of viscosity as the numerical value of the tangential force on unit area of either of two parallel planes at unit distance apart when the space between these planes is filled with the fluid in question and one of the planes moves with unit velocity in its own plane, relative to the other.

tangentially adv **接線として**

▶ The curves of liquid and vapor composition must touch tangentially at the point of maximum or minimum pressure.

tangle n **1 もつれ** syn *entanglement*

▶ Stringlike molecules like the cellulose derivatives curl up more or less into a tangle, depending upon their relation to the solvent.

vi **2 もつれる** syn *intertwine, intertwist*
▶ Acid-catalyzed polymerization of $Si(OCH_2CH_3)_4$ gives linear molecules that are occasionally cross-linked and tangle to form additional branching and gelation.

tangled adj **もつれた** syn *complicated, intricate, involved*
▶ A number of developers of high activity reduce silver halide to tangled bundles of filaments of silver.

tape n **テープ** syn *band, ribbon, strip*
▶ Heating tapes were wound around the rest of the system and were heated to temperatures of 140–150 ℃ to avoid condensation of iodine.

tapping n **軽く叩くこと** syn *knocking*
▶ Filling of the cell was accomplished with the precautions already described, aided by gentle tapping.

tar n **タール**
▶ It is often possible in one vacuum distillation to remove foreign coloring matter and tar without appreciable loss of product.

tare vt **…の風袋を量る**
▷ The absorption cells used were a pair of tared, stoppered quartz cells of 1-cm path length.

target n **1 目標** syn *goal, object, objective*
▶ Large fused polycyclic ether natural products of marine origin are some of the most complex and formidable synthetic targets found in nature.

vt **2 目標に定める** syn *aim, focus*
▷ Inhibitors targeting sialidases have received a great deal of interest over the past decade or so.

tarnish vi **変色する** syn *degrade, spoil, stain*
▶ Silver resists oxidation but tarnishes in air through reaction with atmospheric sulfur compounds.

tarnishing reaction n **造膜反応**
▶ A tarnishing reaction will not occur if a compact layer of oxide or salt between the two reacting species is an insulator.

task n **仕事** syn *job, work*
▶ The systematic enumeration of all geometrical possibilities for each stoichiometry is an extremely complex task even for simple systems.

taste n **1 味** syn *flavor*
▶ Carbon dioxide is a colorless, odorless gas with a weakly acid taste, owing to the formation of some carbonic acid when it is dissolved in water.

2 好み syn *preference*
▶ The choice of origin of the repeating unit is to a certain extent a matter of personal taste, even though the size and shape or orientation of the cell are fixed.

vi **3 味がする** syn *savor*
▶ Thallium is very toxic and was formerly used to kill ants and rats, but since its compounds neither taste nor smell, it is very dangerous and so is little used nowadays.

tasteless adj **無味の** syn *flavorless, watery*
▶ Water is a colorless, odorless, tasteless liquid.

tautomer n **互変異性体**
▶ Compounds whose structures differ markedly in arrangement of atoms, but which exists in easy and rapid equilibrium, are called tautomers.

tautomeric adj **互変異性体の**
▶ In a fluid state, a tautomeric compound is generally considered to be a mixture of two or more different substances.

tautomerism n **互変異性**
▶ Tautomerism is a type of isomerism in which migration of a hydrogen atom results in two or more structures, called tautomers.

tautomerization n **互変異性化**
▶ The construction of thermodynamic cycles has provided insight into tautomerization equilibria and acidities for Os(III) species too unstable to isolate.

taxogen n **タキソゲン**
▶ Organic compounds containing an olefinic double bond, such as ethylene, propylene, hexene, octene, or styrene, are normally employed as taxogens.

tear n **1 引き裂き** syn *rupture, split*
▶ Strength, elongation, tear resistance, and

flexibility of polyvinyl alcohol improve with increasing molecular weight.

vt **2 tear apart** 引き裂く syn *split*

▷ Finding substrates that will hold biological molecules rigidly without tearing them apart has been a problem.

technical adj **1** 工業的な syn *industrial, technological*

▶ A much-investigated reaction of great technical importance is the oxidation of sulfur dioxide to sulfur trioxide.

2 専門の syn *specilized*

▶ The paper was published in a rather obscure technical journal and lay there quietly for many months.

technically adv 工業的に

▶ It is possible to correlate the properties of many of the technically important thermoplastic and elastic polymers with their chemical structures.

technique n 手法 syn *approach, method, procedure*

▶ The discovery of new materials requires the development of a diversity of synthetic techniques.

technological adj 科学技術的な syn *technical*

▶ Many of the technological applications of semiconductors are associated with doped or extrinsic materials.

technologically adv 科学技術的に

▶ One technologically important eutectic mixture is used for solder.

tedious adj 退屈な syn *laborious, monotonous, uninteresting*

▶ This method of titration is tedious, but the results obtained are accurate.

tell vt **1** 示す syn *disclose, indicate*

▷ The infrared spectrum helps to reveal the structure of a new compound by telling us what groups are present in the molecule.

2 知る syn *ascertain, distinguish, understand*

▶ We cannot tell from the rotations alone whether the acid and ester have the same or a different arrangement of groups about the asymmetric center.

telogen n テロゲン ☞ telomerization

▶ Many different types of compounds can be employed as telogens, e.g., halogenated hydrocarbons, such as chloroform or carbon tetrachloride, halogen derivatives of cyanogens, aldehydes, alcohols, and the like.

telomerization n テロメル化

▶ During telomerization reactions, the telogen is split into radicals that attach to the ends of the polymerizing taxogen.

temper vt 焼き戻す syn *anneal, harden, toughen*

▷ Tempering the steel allows the carbon to diffuse short distances and allows some of the internal stresses to relax, yielding a greater resistance to fracture.

temperature coefficient n 温度係数

▶ From the vapor pressure of the crystal and the corresponding temperature coefficient, the enthalpy of sublimation may be determined.

temperature factor n 温度因子

▶ In addition to the positional variables, atoms are vibrating, either isotropically or anisotropically, and the so-called temperature factors are additional variables which must be determined in any good quality structure refinement.

temperature gradient n 温度勾配

▶ Zone refining occurs when a long sample is slowly drawn through a temperature gradient that is sufficiently narrow that only a small section of the sample is melted at any point.

temperature-jump method n 温度ジャンプ法

▶ The temperature-jump method is one of a group that depend upon a rapid perturbation followed by a shift of the composition of the solution toward the new position of equilibrium.

tempering n 焼き戻し syn *annealing, hardening*

▶ In tempering, the sample is gently heated after it has been quenched by sudden cooling.

template n **1** 鋳型 syn *die, model*

▶ A condensation or crosslinking reaction occurs around the template to produce the ordered structure, and the template may subsequently be removed by washing or thermal treatment.

vt **2 鋳型となる**

▷ The appearance of chiral templated materials should lead to some exciting new chemistry, both in enantiomeric separation and chiral catalysis.

template synthesis n **鋳型合成**

▶ In the synthetic zeolites, the aperture and channel size may sometimes be controlled by a sort of template synthesis.

temporal adj **一時的な** syn *temporary*

▶ In collisional cooling cells, particles can be observed over minutes to hours which allows us to study the temporal evolution of the aggregates.

temporarily adv **一時的に** syn *briefly*

▶ The reactants are adsorbed on the enormous surface of a finely divided solid metal or bonded temporarily to a soluble metal ion.

temporary adj **一時的な** syn *fleeting, short-lived, transient*

▶ If the molecule has no permanent dipole moment, the applied field leads only to a temporary distortion of the molecule.

temporary hardness n **一次硬度** ☞ permanent hardness

▶ The hardness of water produced by the presence of alkaline-earth hydrogencarbonates disappears on boiling and is designated as temporary hardness.

tempting adj **心をそそる** syn *appealing, attractive*

▶ It is tempting to see these changes as products of modern science, particularly modern chemistry.

ten n **1** 複数形で，**何十倍もの**

▶ Tens of thousands of organosillicon compounds have been synthesized.

adj **2 10 の**

▶ Sufficiently high pressure may stop solid particles to pass through larger pores even when the pore is ten times the diameter of the particles.

tenable adj **主張できる** syn *plausible, possible, rational, workable*

▶ The approximations made in the Debye–Hückel calculation are tenable only at very low concentrations.

tenability n **可能性**

▶ The ability to vary the composition of Lewis or Brønsted acid adds an additional dimension to the tenability of the eutectic-based ionic liquids.

tenacious adj **堅固な** syn *determined, persistent, stubborn*

▶ Some colloids may form tenacious, plastic, even visible films at the interface.

tenaciously adv **頑強に**

▶ After the crystalline product is collected, it should be washed thoroughly with ether to remove the nitrobenzene, which otherwise adheres tenaciously to the crystals.

tend vi **傾向がある** syn *favor, incline, lean*

▶ The long linear molecules tend to coil up and exhibit considerable flexibility with little force between molecules, allowing plastic flow in unvulcanized rubbers.

tendency n **傾向** 動詞が続くときには，前置詞 to を用いる． syn *trend*

▶ Reactivity in electrophilic aromatic substitution depends upon the tendency of a substituent group to release or withdraw electrons.

tensile adj **張力の** syn *elastic, extensible*

▶ High-tensile carbon fibers or whiskers are made from rayon, polyacrylonitrile, or petroleum pitch.

tensile strength n **引張り強度**

▶ An oriented crystalline polymer usually has a much higher tensile strength than the unoriented polymer.

tension n **張力** syn *stress, tightness*

▶ At elastic limit, a semicrystalline state is reached, which is different from the one produced by cold drawing a crystalline polymer in that it is stable only while under tension.

tentative adj **暫定的な** syn *experimental, pro-*

visional

▶ The tentative inference that the higher-melting isomer has the more symmetrical trans configuration eventually was found to be correct.

tentatively adv 暫定的に

▶ These observations on C_8K or C_8Br can be tentatively interpreted on the basis of a model in which the layers of graphite behave amphoterically and take up electrons from the potassium or give up electrons to the bromine.

tenth n 1 10分の1

▶ The process is characteristically slow, and 3–6 weeks may be required to obtain a few tenths of a gram of product.

adj 2 10分の1の

▶ H_3PO_4 corresponds to a moderately strong acid and is about 35 percent ionized in tenth-normal solution.

term n 1 用語 syn *designation, expression, name*

▶ Other terms with the same general meaning as destructive distillation are pyrolysis and thermal decomposition.

2 項

▶ The total vibrational energy of the *N* molecules can now be obtained by adding up the terms for the number of molecules in each level times the energy of that level.

3 in terms of …の見地から syn *concerning, regarding*

▶ For given amount of material, it is then generally possible to write an equation that describes the state in terms of intensive variables.

adj 4 期間の

▶ The optimum dye should have a long-term photo and thermal stability.

vt 5 呼ぶ syn *call, designate, entitle*

▶ The mechanisms by which bacteria fix nitrogen at ambient temperatures and pressures has been termed one of the most conspicuously challenging, unsolved problems in chemistry.

terminal n 1 端子 syn *connector*

▶ Both terminals of the cell are of the same material.

adj 2 末端の syn *ending, terminating*

▶ Terminal substituents are important in their effect on the dielectric anisotropies of liquid crystal materials.

terminate vi 1 終わる syn *cease, stop*

▶ The tip often terminates in a small cluster of atoms or the desired single atom.

vt 2 終える syn *end, stop*

▷ Hydrosilylation of alkenes and alkynes at the hydride-terminated porous silicon surface results in the formation of a monolayer of alkyl chains linked to the surface by a very stable Si–C bond.

termination reaction n 停止反応 syn *termination*

▶ The overall radical concentration is determined by a balance between the initiation reactions which form them and the termination reactions which remove them.

terminology n 用語 syn *nomenclature, term*

▶ Before proceeding to describe the volume computation method, it is necessary to define the terminology used in this work.

terminus n 末端 syn *end*

▶ A leaving group departs from the migration terminus, taking the bonding electrons with it.

ternary adj 三成分の

▶ If four substances are stoichiometrically related in the double decomposition AB + CD $\rightleftarrows$ AD + CB, the system consisting of these four substances is ternary because all possible compositions may be expressed in terms of any three of them.

terrace n 表面欠陥のテラス

▶ A typical type of surface defect is a ledge between two otherwise flat terraces.

tertiary adj 第三級の

▶ Tertiary phosphine oxides, $R_1R_2R_3P{=}O$, have nonplanar configurations, since they can normally be separated into enantiometric forms.

tertiary alcohol n 第三級アルコール

▶ Tertiary alcohols undergo dehydration the most rapidly of the alcohols, because they form the most stable carbocations and then, once formed, these cations yield the most stable alkenes.

tertiary structure n 三次構造

▶ The protein chain coils are folded into tertiary structures by additional hydrogen bonding and by covalent bonding of sulfur atoms of cysteine units.

test n **1** 試験 syn *assay, examination, trial*

▶ Whenever possible, we select for a characterization test a reaction that is rapidly and conveniently carried out and that gives rise to an easily observed change.

vt **2** 試す syn *assay, assess, examine*

▶ The neutral-ionic transition offers us a unique opportunity to test and compare some of the most general factors involved in charge transfer.

testable adj 試験できる

▶ In its application, the crystal-field theory has provided a range of qualitatively novel and testable expectations, despite all the limitations of that theory.

testify vi 証拠となる syn *attest, give evidence*

▶ Sulphonic acids RSO_3H, especially those in which the moiety R contains fluorinated carbon atoms as close as possible to the sulfur, possess high acidic strength, as triflic acid amply testifies.

testing n 試験 syn *examination, trial*

▶ Polycrystalline materials, whether pure or impure materials, are things of practical interest and have mostly been used in experimental testing.

test tube n 試験管 ☞ in vitro

▶ Nonpolar nucleoside analogues have been shown to have a useful application for in vitro transcription to produce RNAs in the test tube.

tether n つなぎの鎖 syn *chain*

▶ Typical tethers for attaching fluorophores to DNA involves several to as many as 1-2 dozen bonds and often make use of simple linear alkyl chains.

tetragonal adj 正方晶系の

▶ At room temperature, Cu_2HgI_4 crystallizes in a tetragonal unit cell with unit cell lengths $a = b = 6.09$ Å and $c = 12.24$ Å. All of the angles in a tetragonal cell are 90°.

tetragonally adv 正方晶系に

▶ Tetragonally distorted octahedral complexes are intermediate between the octahedral and square-planar extremes.

tetrahedral adj 四面体の，四面体形の

▶ A tetrahedral field also splits the d orbitals into two groups, but in the opposite manner to an octahedral field.

tetrahedrally adv 四面体形に

▶ Each H_2O molecule in $LiOH \cdot H_2O$ is tetrahedrally coordinated by 2Li from the same chain and 2OH from other chains.

tetrahedron n 四面体 複数形は tetrahedra

▶ The metal atoms in M_2B are arranged in layers of fused tetrahedra, leaving square antiprismatic holes between the layers in which the boron atoms are accommodated.

tetramer n 四量体

▶ Crystalline $SeCl_4$ and $TeCl_4$ are isotypic, and the structural unit is a cubane-like tetramer.

tetramereric adj 四量体の

▶ $SeBr_4$ has a cubane-like tetrameric unit.

tetravalent adj 四価の

▶ The building block of structural organic chemistry is the tetravalent carbon atom.

texture n 組織 syn *character, consistency, feature, structure*

▶ The mesophase shows very different textures depending upon whether it is accessed by heating or cooling.

than prep **1** …よりも syn *in comparison with*

▶ An important feature of solid-state phase changes that distinguish them from many other phase changes is that superheating and supercooling are extremely common, being the rule than the exception.

conj **2** …よりも，比較級の形容詞または副詞の後に使用する．比較は文法上，対等な関係にある語について行う．この条件を満たすに必要な語は省略しない．

▶ The calixarenes are characterized by much higher melting points and lower solubilities in common organic solvents than their acyclic counterparts.

3 rather than　**むしろ**　syn *instead of*

▶ The features discussed tend to be salient rather than detailed.

that pron **1** 反復のときに，**それ**

▶ The most important class of molecular rearrangements is that involving 1,2-shifts to electron-deficient atoms.

2 制限的な節を導いて，**するところの**　前置詞の目的語とはしない．

▶ A monochromatic beam of X-ray strikes a finely powdered sample that has crystals randomly arranged in every possible orientation.

3 in that　**…だから，…という点で**　syn *because, in the fact that*

▶ Conductivity is an unusual property in that it spans roughly 30 orders of magnitude: one of the largest ranges known for a physical property.

4 that is　**すなわち**　syn *namely, that is to say*

▶ That is, as the concentration approaches zero, the activity approaches the concentration.

節と節の間で用いるときは，前にセミコロン，後にコンマを用いる．

▶ In arithmetic and ordinary algebra, multiplication commutes; that is, $2 \times 3 = 3 \times 2$.

5 that is to say　**すなわち**　syn *namely, that is*

▶ That is to say, the entering group does not always take up the same position on the ring as that vacated by the halogen substituent.

adj **6** **その，あの**

▶ The basic idea is that the mole fraction solubility of a substance in an ideal solution only depends on the properties of that substance.

conj **7** **…ということ**　主語節を導いて

▶ That organic crystals are more difficult to grow than inorganic crystals is due mainly to their great supercooling tendencies and to their low thermal conductivity.

目的語節を導いて

▶ The conditions for transfer are simply that the absorption of the acceptor overlap the fluorescence spectrum of the donor and that the medium between be transparent to the wavelengths of radiation involved.

the article **その，問題の**　可算名詞の複数形，不可算名詞を含め，すでに言及したものに付ける．

▶ In any scheme involving the packing of spheres, there will be unoccupied spaces between the spheres.

はじめに用いた語句がそのままの形で繰り返されるとは限らない．

▶ When a saturated solution of magnesium chloride is mixed with magnesium oxide, the mixture soon solidifies, forming a mass hard as stone, known as magnesium cement, consisting of magnesium chloride hydroxide.

名詞が省略されているとわかる叙述用法の形容詞の最上級に付ける．

▶ Of the 22 distinct structural arrangements of tetrachlorodioxin, the most toxic is three orders of magnitude more toxic than the second most toxic.

前置詞句や関係詞節によって特定されたものや人には the を付ける．

▶ The fact that the boiling point always increases, that the melting point often increases, and that the solubility in water always decreases as the number of CH_2 groups increases in a homologous series is too familiar to require comment.

可算名詞（普通名詞，集合名詞）の場合，すべてに通じる一般的なことを述べるときには，the を付けた単数形を主語に用いる．

▶ The H_2 molecule reacts with a large number of other elements.

あるいは，the を付けないで複数形を用いる．

▶ Alkenes are readily converted by chlorine or bromine into saturated compounds that contain two atoms of halogen attached to adjacent carbons; iodine generally fails to react.

人名を冠した法則，方法，反応などには，定冠詞を付ける．あるいは，'sを付した所有格を用いて，冠詞は付けない．

▶ A wide range of liquids give approximately the same entropy of vaporization: this is Trouton's rule.

構文 the comparative adjective…, the comparative adjective は…**であればあるほど，それだけ余計に…する**の意味で使用する．この構文は…すればするほどの節を必ず前にする．簡潔であること，二つの部分が緊密であることが肝要である．

▶ The lower the energy required for ionization of the molecule and the more stable the molecular ion, the more intense will be the peak in the mass spectrum.

theme n **主題** syn *concept, idea, subject, topic*

▶ The concept of the localized electron pair has been a central theme in the development of bonding theories since Lewis first postulated that a chemical bond was a consequence of a shared electron pair.

themselves pron **自ら**

▶ If a substance like sugar lies as a solid on the bottom of a beaker filled with water, the molecules of sugar tend to distribute themselves throughout the solution; in other words, the sugar dissolves.

then adv **1 その時には**

▶ If muscle glycogen is depleted completely, then it is impossible to sustain contraction and it is likely that liver glycogen will also reduced.

2 次に

▶ The solution of 2,4-diamino-1-naphthol dihydrochloride is treated with acetic anhydride and then sodium acetate.

3 そのうえ syn *additionally, also*

▶ There is then a repulsive force, and the energy rises as the intermolecular distance is reduced.

theorem n **定理** syn *principle, rule*

▶ The reason for the deviations from the Nernst heat theorem in its original form was that substances are often not in a state of true equilibrium at low temperatures.

theoretical adj **1 理論上の** syn *ideal*

▶ No theoretical explanation of this regularity is assigned at this time.

2 仮定上の syn *hypothetical*

▶ The concept of an ideal gas is theoretical as no actual gas meets the ideal requirement.

theoretically adv **理論的に** syn *fundamentally, ideally*

▶ The permanganate ion is strongly oxidizing and is theoretically able to oxidize water to oxygen, but in practice the reaction is very slow.

theorize vi **学説を立てる** syn *guess, hypothesize, speculate*

▶ In 1678, the Dutch physicist Christiaan Huygens theorized that the existence of cleavage planes in crystals could be accounted for by the regular packing of spheroidal particles in layers.

theory n **理論** syn *guess, hypothesis, speculation*

▶ Theories can be accepted or rejected if an appropriate experiment can be designed to test their validity.

therapeutic adj **治療の** syn *medicinal*

▶ Many other therapeutic proteins with valuable applications for treating medical illness in humans have been expressed in and isolated from bacteria.

there adv **1 そこで**

▶ There the standard state was that of unit fugacity.

2 there is …がある 文法上の主語の単数，複数に動詞を一致させる．

▶ With the change in shape of the retinal moiety, there is a change in shape of the entire rhodopsin molecule; the protein portion must adjust its conformation to accommodate this altered guest.

thereafter adv **その後に**

▶ The synthesis of the corresponding hexafluorophosphate followed shortly thereafter.

thereby adv **それによって**

▶ By suitable crosslinking in the resin, this swelling can be made more difficult, and selectivity of the resin for different cations is thereby altered.

therefore adv **したがって** syn *accordingly, consequently, hence, thus*

導入の語として文頭で使用した場合には，次にコンマを要する．

▶ Therefore, atoms are not likely to combine in such a bimolecular process but must do so in a termolecular collision where the third molecule removes energy in the form of translation or vibration.

二つの節を結ぶときは，前をセミコロンで，後をコンマで区切る．

▶ Michler's Hydrol Blue is a normal cyanine dye. The rings are so placed that in the classical structure one is benzenoid, the other quinoid; therefore, cyanine mesomerism reduces the symmetry of the former but increases that of latter.

その他の場合，前後ともコンマで区切る．

▶ In spite of these advances in chemistry, the details of the atom and, therefore, of many aspects of the molecule, remained a complete mystery.

therein adv そこに

▶ Aromatic diazonium salts can be converted into a variety of compounds and therein lies their utility.

thereof adv それの

▶ A third point connected with the identity of the metal ions concerns their kinetic stability or lack thereof.

thereon adv そのうえに

▶ The total effective charge on any particle or a surface is the net algebraic sum of all the charges thereon.

therewith adv それとともに

▶ The ability to bond an interstitial element Z within all Zr_6X_{12} clusters and therewith to vary the charge of the core over a greater range affords a much greater versatility in composition and structure and a far better basis on which to develop systematics regarding cluster synthesis, bonding, and structure.

thermal adj 熱の

▶ The number of cation vacancies present in alkali halides depends on the purity and thermal history of the crystals.

thermal agitation n 熱運動 syn *thermal motion*

▶ Constant thermal agitation ensures that the molecules of a sample are distributed among the energy levels available to them.

thermal conductivity n 熱伝導率

▶ Detection is usually achieved by measuring changes in thermal conductivity of the effluent gas.

thermal decomposition n 熱分解

▶ With a higher alkylammonium hydroxide, thermal decomposition leads to the formation of an alkene.

thermal diffusion n 熱拡散

▶ Artificial enrichment of ^{17}O and ^{18}O can be achieved by several physical or chemical processes such as the fractional distillation of water, the electrolysis of water, and the thermal diffusion of oxygen gas.

thermal energy n 熱エネルギー

▶ Because of the minimal thermal energy needed to ionize the dopants, extremely low temperatures are required to observe these transitions spectroscopically.

thermal equilibrium n 熱平衡

▶ Thermal equilibrium in the spin system is maintained through energy exchange between the normal thermal motions of the molecule and those of the surrounding molecules.

thermal expansion coefficient n 熱膨張率

▶ The measurements of thermal expansion coefficients on organic crystals have been made primarily by X-ray diffractional analysis and, rarely, by goniometric studies.

thermal insulator n 断熱材

▶ Air is particularly effective as a thermal insulator when trapped within a solid network.

thermalize vt その熱による状態にする

▷ The ejected electrons become thermalized and solvated in just a few picoseconds.

thermally adv 熱的に

▶ Rutile is the most thermally stable of the three naturally occurring crystal forms of titanium(IV) oxide.

thermal motion n 熱運動 syn *thermal agitation*

▶ Low-melting-point solids, if not cooled, often

exhibit some form of disorder due to their high thermal motion.

thermal stability n 熱安定性

▶ Triethylphosphine was found to be particularly suitable owing to its high thermal stability; it remains largely undecomposed in a system where the residence time of the species over the heated substrate is short.

thermal vibration n 熱振動

▶ As a piece of metal is heated, the increased thermal vibrations allow atoms to rearrange and to go lower energy states by diffusion, thereby reducing the number of dislocations present.

thermionic emission n 熱電子放出

▶ Electrons from the metal may readily penetrate a thin oxide layer either by the quantum-mechanical tunneling effect or by thermionic emission and so react with oxygen to form adsorbed oxygen ions.

thermistor n サーミスター

▶ With a $BaTiO_3$ thermistor, when a current is passed through it, the resisitivity increases dramatically as it heats and consequently the current switches off.

thermochemical cycle n 熱化学サイクル ☞ Born–Haber cycle

▶ The heat of formation of a crystal may be related to lattice energy by constructing a thermochemical cycle known as a Born–Haber cycle.

thermochemical radius n 熱化学半径

▶ By examining salts of alkali metals, where the thermochemical radius of cation is simply its ionic radius, the thermochemical radii of complex anions may be determined.

thermochromic adj サーモクロミックな

▶ A thermochromic ink is produced with the liquid crystal by encapsulating it in a polymer.

thermocouple n 熱電対

▶ Thermocouples are essential for determinations of extreme temperatures that are beyond the range of liquid-in-glass thermometers.

thermodynamic adj 熱力学的な

▶ A thermodynamic analysis may rule out a given reaction for the synthesis of some substance by indicating that such a transformation cannot proceed spontaneously under any set of available conditions.

thermodynamically adv 熱力学的に

▶ For bromates and idodates, disproportionation to halide and perhalate is not thermodynamically feasible.

thermodynamic control n 熱力学支配 ☞ kinetic control

▶ When the ratio of the products is determined by the ratio of their equilibrium constants, we say the overall reaction is subject to thermodynamic control.

thermodynamic equilibrium n 熱力学的平衡

▶ It is necessary in preparing the metastable forms to prevent thermodynamic equilibrium by working fast and with small quantities.

thermodynamic stability n 熱力学的安定性

▶ There is an important distinction to be made between thermodynamic stability expressed in terms of the equilibrium constant of a reaction and kinetic stability of a substance.

thermoelectric adj 熱電気の

▶ Zintl phases are prime candidates for applying this concept to obtain high zT thermoelectric materials.

thermoelectric power n 熱電能 ☞ Seebeck coefficient

▶ The thermoelectric power is defined either as the mean value for 1° over a specified range of temperature or as the value for 1° difference at a specified temperature.

thermogravimetry n 熱重量分析法

▶ In thermogravimetry, the sample, usually a few milligrams in weight, is heated at a constant rate, typically in the range 1 to 20 ℃ min^{-1}.

thermoluminescence n 熱ルミネセンス

▶ The energy that is stored in the F center can be released through thermoluminescence, the process of using heat to generate luminescence from a compound.

thermolysis n 熱分解 syn *thermal cracking*
▶ Thermolysis studies of B_5H_{11} in the temperature range 40–150 ℃ established first order kinetics with a rather low activation energy of 72.6 ± 2.4 kJ mol^{-1}.

thermolyze vt 熱分解する syn *pyrolyze*
▶ At 80 ℃ the time needed to thermolyze half of the neat H_3NBH_3 was 290 min.

thermonuclear reaction n 熱核反応
▶ The sun's energy, on which life on the earth depends, is produced by a series of thermonuclear reactions deep in the solar interior.

thermoplastic adj 熱可塑性の
▶ The thermoplastic type can be resoftened to its original condition by heat; the thermosetting type cannot.

thermosetting adj 熱硬化性の
▶ One of the oldest-known thermosetting synthetic polymers is made by condensation of phenol with formaldehyde using basic catalysts.

thermostat n **1** サーモスタット
▶ Sample temperature was determined by appropriately positioned thermocouples and by a mercury-in-glass thermometer in the thermostat bath.
vt **2** サーモスタットで調温する
▷ The reaction is carried out in a thermostatted and well-stirred reaction vessel.

thermotropic adj サーモトロピック
▶ Planar molecules such as metal-containing phthalocyanines, porphyrins, and β-diketonates with long attendant side chains show thermotropic mesogenic behavior, as do related planar organic molecules.

thermotropic liquid crystal n サーモトロピック液晶 syn *lyotropic liquid crystal*
▶ Thermotropic liquid crystals are formed from compounds whose molecules are mainly either rod-shaped or disc-shaped, either by heating the crystalline solid or by cooling the isotropic liquid.

thermotropic liquid-crystalline adj サーモトロピック液晶性の
▶ Two apparently different types of cubic mesophase are known to occur in thermotropic liquid-crystalline systems.

these pron これら
▶ The exciting radiation promotes sensitizer ions into an excited state. These then transfer energy to neighboring activator ions.

thick adj **1** どろどろした syn *dense, viscous*
▶ The thick white precipitate was collected and washed with ethanol and ether giving bicyclo[2,2,2]octane-1,4-dicarboxylic acid as a white powder.
2 数量を示す名詞の後に用いて，厚さの
▶ A typical experiment involves the deposition of a thin layer of *o*-chloranil on a film ca. 5 × 10^{-4} cm thick of phthalocyanine covering an electrode system.

thicken vi 厚くなる syn *develop, expand, swell*
▶ The parabolic law is followed as the oxide layer thickens.

thickness n 厚さ syn *breadth, span, width*
▶ Typical crystals were hexagonal plates of thickness 0.5 to 1 mm and 0.3- to 0.3-cm^2 area.

thimble n 円筒沪紙
▶ For exhaustive extraction of dried leaves or seeds, the solid is packed into a filter paper thimble placed in a Soxhlet extractor.

thin adj **1** 薄い syn *slim*
▶ For examination in transmission, samples should usually be thinner than ca. 2000 Å.
2 希薄な syn *fluid, watery*
▶ Lubricating oils are too thin when hot and too nearly solid when chilled.
vi **3** 薄くなる syn *diminish, reduce*
▶ Within a few minutes the film thins, and the reflected light exhibits interference colors that ultimately turn black.

thin film n 薄膜
▶ In order to delineate an appropriate reaction system for quantitative study, we first studied reactions of thin films of hydrocarbons with bromine vapor.

thing n **1** 事柄 syn *aspect, factor, point*
▶ The unusual thing about alcohols is that they boil higher than many other compounds of considerable polarity.

2 for one thing 一つには

▶ For one thing, viscosity varies quite enormously.

3 有形の物 syn *item, matter*

▶ There is no such thing as a typical animal cell, since cells vary in overall size, shape, and contents of various subcellular organells.

think vt **1 考える** syn *expect, imgine, suppose*

▷ The rhenium chloride ion, thought to be $[ReCl_4]^-$, was shown to be the trimer $[Re_3Cl_{12}]^{3-}$.

2 think of …のことを考える syn *consider, think about*

▶ It is helpful to think of colligative as denoting properties bound together in a common explanation.

thinking n **思考** syn *idea, reasoning, viewpoint*

▶ This thinking led to the synthesis of a new class of tin sulfide liquid crystals with a novel structure based on a mesh mesophase.

thin layer n **薄層**

▶ The components are not uniformly distributed throughout each phase because thin layer at the interface may be rich or deficient in some of the species.

thin-layer chromatography n **薄層クロマトグラフィー**

▶ Frequently mixtures are obtained, and these can be separated by paper chromatography, paper electrophoresis, thin-layer chromatography, ion exchange, or gel chromatography.

thinly adv **1 まばらに** syn *sparsely*

▶ Scandium is very widely but thinly distributed.

2 薄く

▶ Many cells and parts of cells react strongly with colored dyes such that they can be easily distinguished in thinly cut sections of tissue by using light microscopy.

thiol n **チオール**

▶ Thiols have animal and vegetable origins; notably, butanethiol is a component of skunk secretion; propanethiol is evolved from freshly chopped onions.

third n **1 3番目**

▶ One group of phenothiazine derivatives possesses antihistamine activity, another counteracts certain forms of muscular rigidity, and a third has what are generically termed tranquilizing effects.

2 3分の1

▶ Another third of the sodium silicate produced serves as a source of silica for derived products such as silica gels and sols, alumina-silica cracking catalysts, zeolites, and precipitated silicas.

adj **3 3番目の**

▶ To every third atom of the peptide chain is attached a side chain.

this pron **1 このこと** 次の例のように，先行する全文を指すことができる．何を指すかが不明確なときには名詞で記載する．

▶ The Li^+ ion is small and this is the key to much of the chemistry of lithium.

adj **2 この**

▶ In this way, 1,4-naphthoquinone, 9,10-anthraquinone, and 9,10-phenanthraquinone are prepared from naphthalene, anthracene, and phenanthrene, respectively.

thixotropic adj **チキソトロピーを示す**

▶ The thixotropic system may appear to be a solid jelly and yet upon shaking appear as transparent liquid.

thixotropy n **チキソトロピー**

▶ Thixotropy is observed in some clays, paints, and printing inks which flow freely on application of slight pressure, as by brushing or rolling.

thorough adj **徹底的な** syn *careful, exhaustive, extensive*

▶ Thorough stirring of mixture is required to maintain the lowest possible temperature.

thoroughly adv **綿密に** syn *carefully, exhaustively, extensively*

▶ Diazonium tetrafluoborate is collected, thoroughly dried, and then carefully heated to the decomposition point, the products being an aryl fluoride, nitrogen, and boron trifluoride.

those pron **1 それらのもの**

▶ The precipitating action of ions of opposite

sign to those on the surface of the particle is greater the higher the valence of the ion.

adj **2** それらの

▶ Organic chemistry was originally defined as the chemistry of those substances which are formed by living matter.

though adv **1** けれども syn *but, however, nevertheless*

▶ Many, though by no means all, of these complexes are those obtained by the interaction of aromatic hydrocarbons with electron acceptors.

conj **2** …にもかかわらず syn *although, even though, while*

▶ It is convenient to recognize five limiting types of interaction in hydrates, though the boundaries between them are vague and undefined.

thought n 考え syn *anticipation, expectation, prospect*

▶ The thought that we can discuss 5 percent of a total interaction energy quantitatively but cannot make more than a rough estimate of the remaining 95 percent is a sobering one.

thousand n **1** 1000 複数形は多数を意味する.

▶ Charge-transfer transitions, being electric dipole allowed, result in absorption bands with peak linear absorption constants of some tens of thousands of cm^{-1} and also several thousand cm^{-1} wide.

adj **2** 1000 の

▶ A polymeric copper phthalocyanine with a resistivity one thousand times less than that of copper phthalocyanine is obtained by reacting pyromellitic anhydride with urea and a copper salt.

thread vt 縫うように通り抜ける syn *pass*

▶ Polymer molecules may thread their way many times through the same crystallite by folding sharply back and forth on themselves.

three-body collision n 三重衝突

▶ The recombination of hydrogen atoms can only take place in the gaseous phase upon the occurrence of a three-body collision between two hydrogen atoms and a third particle capable of absorbing the liberated energy.

three-center bond n 三中心結合

▶ A less common type of tetrahedral cluster is that having four three-center face bonds. An example is B_4Cl_4.

three-dimensional structure n 三次元構造

▶ All crystalline forms of SiO_2 have three-dimensional framework structures with sharing of all four oxygens of the SiO_4 tetrahedra.

III–V compound n III–V 化合物

▶ A particular advantage of III–V compounds as semiconductors is that the energy difference between the conduction and valence bands can be varied by using mixed compounds.

threefold adj 三つの要素の

▶ The adoption of semiempirical methods is a threefold compromise between rigor, experiment, and intuition.

three-membered ring n 三員環

▶ Epoxides are ethers, but the three-membered ring gives them unusual properties.

threshold n しきい syn *beginning, onset*

▶ For each electrolyte, these is a small threshold concentration which has to be exceeded in order to produce noticeable coagulation.

threshold energy n しきいエネルギー

▶ The threshold energy, an experimental quantity, is the minimum initial relative kinetic energy actually necessary for reaction.

threshold value n しきい値

▶ Most reactive collisions have energies only slightly higher than the threshold value.

thrice adv 三回 syn *three times*

▶ Picric acid was thrice recrystallized from ethanol and dried at 60 ℃.

throat n 細い通路

▶ Under certain conditions, small clusters of molecules may be generated in the throat of a molecular beam.

through prep **1** …を貫いて syn *into, to*

▶ If more than one ion can pass through the membrane, the situation is more complex.

2 …により syn *by means of*

▶ The condensation of polyhedra through square faces is not such a common aggregation process.

3 …じゅうに syn *in the middle of, throughout*

▶ Crystallographic shear planes may occur at random through the structure or may be spaced at regular intervals.

4 通じて syn *including, inclusive of*

▶ The oxidation states 0 and III through to VII are known for plutonium, so it is not surprising that numerous compounds exist.

throughout adv **1** 終始 syn *thoroughly*

▶ Single crystals of compounds **6** and **7** were analyzed on a Nonius Kappa CCD diffractometer using molybdenum radiation throughout.

prep **2** 全体にわたって syn *all over, everywhere*

▶ Throughout biological morphology, the most common arrangement of cellular monodispersions is the hexagonal pattern.

throughput n 一定期間内に処理される量

▶ The work led to the discovery of new phases, underlining the power of high throughput methods in materials discovery.

throw vt **1** 投じる syn *deliver, give, pass*

▷ Unstable colloids are illustrated by the pink suspension of selenium formed by throwing a true molecular solution of selenium in carbon disulfide into a large amount of ether, in which selenium is insoluble.

2 throw out 放り出す syn *eject, expel*

▶ The rotating collision complex will throw out product molecules in all directions; both forward and backward scattering will be observed.

thrust n **1** 要旨 syn *meaning, point*

▶ The thrust of the preceding argument is that the structure of a liquid is mostly, but not entirely, determined by the repulsion between molecules.

vt **2** 押しやる 過去形，過去分詞も同じ syn *force, push, shove*

▶ If clean glass is thrust into water covered by a monolayer of oleic acid, the glass remains clean, but when it is gradually raised out of the water it withdraws an equal area of monolayer.

thumbnail adj 簡潔な syn *brief, concise, short*

▶ We may compare this thumbnail sketch of the molecular scene to one of the solid state.

thus adv **1** したがって 導入の語として文頭で使用した場合には，次にコンマを用いる. syn *accordingly, consequently, therefore*

▶ Thus, each portion of the molecule affects the reactivity of the other portion and determines the orientation of attack.

2 このように syn *so*

▶ Aluminum chloride is moderately soluble in nitrobenzene, which is thus a good solvent for Friedel–Crafts reactions, particularly at temperatures ranging from −15 to 25 ℃.

tie n **1** 複数形で，つながり syn *bond, link*

▶ Of particular interest are the conformations of the polymer molecules, the extent and nature of molecular ties between crystallites, and the influence on these factors of chemical structures and crystallization conditions.

vt **2** 関係する syn *associate*

▶ The means by which this exo position is occupied is intimately tied to the number of additional halides, and this in turn mainly determines structure type.

tie line n タイライン

▶ The compositions of a pair of conjugate solutions are joined by a line, known as a tie line.

tight adj …の漏らない syn *sealed*

▶ Cellophane that has not been swollen is practically gas-tight.

tightly adv 堅く syn *firmly, rigidly, strongly*

▶ Fluorocarbons have unusually low boiling points because tightly held electrons in the fluorine atoms have a small polarizability.

tightness n 緊密さ

▶ A widely accepted figure for assessing tightness of packing is the packing coefficient, *k*, the ratio of the molecular volume to the volume available in the particular condensed phase we are interested in.

tilt n **1** 傾き syn *angle, inclination, slope*

▶ In addition to the orientational ordering, the smectic phases also possess varying degrees of spatial ordering and molecular tilt such that several smectic phases may exist.

vt **2** 傾ける syn *angle, incline*

▷ Smectic G and J phases have tilted quasi-hexagonal packing of the molecules, the tilts being toward a side and an apex of the hexagon, respectively.

tilting n 傾き syn *inclination, tilt*

▶ The relative tilting of the two molecules is undoubtedly due to the existence of two intermolecular hydrogen bonds.

time n **1** 期間 syn *duration, period*

▶ For a considerable time the tungsten bronzes were thought to be unique, but in recent years analogous compounds of molybdenum, vanadium, niobium, and titanium have been prepared and found to have similar properties.

2 時期 syn *chance, occasion*

▶ This is a suitable time to summarize the knowledge presently available from these structural studies.

3 時間 syn *period, stretch*

▶ The time that takes to make a measurement of concentration may be significant compared to the half-life.

4 回，度 syn *instance*

▶ This equation has been tested a considerable number of times.

5 複数形で，倍 ☞ times

▶ Because of the difference in activation energy alone, chlorine atoms are 375,000 times as reactive as brome atoms toward methane.

6 all the time いつも syn *always*

▶ Pathogens are changing all the time giving rise to both drug- and vaccine-resistant strains and new strains of disease-causing bacteria and viruses.

7 at a time 一度に syn *simultaneously*

▶ There are even numbers of carbons in fatty acids because the acids are built up, two carbons at a time, from acetic acid units.

8 at any time いつでも syn *as soon as, ever, once*

▶ If the solvent is of fixed composition, that is, dilute in the reactants, then the concentration of NO_2^+ is directly proportional to the concentration of molecular nitric acid at any time.

9 at one time かつて syn *formerly, once*

▶ In fact, the carcinogenic properties of certain aromatic hydrocarbons may be directly dependent on their chemical reactivity, which is far greater than was at one time believed.

10 at the same time 同時に syn *at once, simultaneously*

▶ The atom or group of atoms that defines the structure of a particular family of organic compounds and, at the same time, determines their properties is called the functional group.

11 at the time その間ずっと syn *in the event*

▶ Examples of 1-D coordination polymers are relatively common in the early literature, even though they were not seen at the time as part of a vast and remarkable family of materials.

12 for a time ちょっとの間 syn *briefly, fleetingly*

▶ A sol may persist for a time at temperatures well below the setting point while the particles are gradually linking.

13 for the first time 初めて

▶ It has become practical for the first time to relate the empirical information regarding the motions of atoms during a simple chemical reaction to the forces which are responsible for the observed motion.

14 for the time being さしあたり syn *for the moment, temporarily*

▶ For the time being, we shall concern ourselves with the mechanism of addition reactions.

15 in recent times 最近 syn *recently*

▶ Molecular magnetic materials have attracted significant attention in recent times.

16 in time やがては syn *eventually*

▶ If rigid particles suspended in a liquid in which they are insoluble are not prevented from cohesion, they will in time form aggregates.

time-consuming adj 時間のかかる syn *ineffective, time-wasting*

▶ Polonium can only be studied using stringent safety precautions and time-consuming techniques.

time course n 時間経過

▶ The different concentration maximum and overall time course of blood glucose that occur in a normal person in response to a standard dose of a particular food can be quantified and given a number; it is called glycemic index.

timeliness n 時を得ていること syn *opportune time*

▶ Papers describe completed research of timeliness and importance.

time of flight n 飛行時間 複合形容詞として使用.

▶ A recent advance which offers exciting prospects is the use of pulsed neutron sources coupled with time-of-flight analysis.

time-of-flight mass spectrometry n 飛行時間型質量分析計法

▶ The translational energy of the scattered molecule can be measured by time-of-flight mass spectrometry.

time-resolved adj 時間分解の

▶ The development of sub-picosecond laser has allowed us to probe the time-resolved behavior of processes which had previously to be considered as instantaneous.

times prep 掛ける syn *multiplied by*

▶ The solubility of the solute times the volume of the solution to be evaporated should be sufficient for the weight of the crystal or crystals to be grown.

time scale n **1** 完成に要する時間

▶ Time scales of reactions studied by discharge flow are limited to the millisecond range.

2 時間の尺度

▶ All of the Ar-CO, Ar-N, and N-CO bonds are expected to rotate slowly on the NMR time scale at 25 ℃, though fast on the laboratory time scale.

timing n 間合い syn *time*

▶ Depending upon the timing of bond-breaking and bond-making, the nucleophilic attack can be either S_N1-like or S_N2-like.

tint n 色合い syn *shade*

▶ The brick industry is another major user of MnO_2 since it can provide a range of red to brown or gray tints.

tiny adj とても小さい syn *microscopic, minute, small*

▶ Crystals form initially in tiny regions of the parent phase and then propagate into it by growth.

tip n **1** 先端 syn *apex, extremity, vertex*

▶ Tips are usually made by an electrochemical etching process, starting with fine wire.

vi **2** ひっくり返る syn *turn over*

▶ The suction tubing is so heavy that the 50-mL and 125-mL filter flasks will tip over easily unless provided with a support.

vt **3** ひっくり返す syn *upset*

▶ The Madelung constant for CN = 8 is only marginally larger than for CN = 6. Thus, small energies coming from other sources can tip the balance.

tissue n 組織

▶ (+)-Glucose is carried by the bloodstream to the tissues, where it is oxidized, ultimately to carbon dioxide and water, with the release of the energy originally supplied as sunlight.

title adj 表題と同名の

▶ Diphenyl sulfone and chlorodiphenylphosphine were coupled as stated by the general protocol with *n*-buthyllithium to generate the title compound as a colorless solid.

titrate vt 滴定する

▶ A solution of sodium chloride is titrated with a solution of silver nitrate; silver chloride, which is only slightly soluble, is precipitated.

titration n 滴定 ☞ acid-base titration, precipitation titration

▶ At the end point of an acid-base titration, there is a sharp change in the concentration of hydrogen ions, which is reflected in a sharp change in the electromotive force.

to prep **1** …のほうへ

▶ To a hot solution of 104 mg of TCNQ in 15 mL of tetrahydrofuran was added 100 mg of anthracene.

2 …に至るまで

▶ Oxidation of alcohols to the aldehyde or ketone stage is usually accomplished by the use of Cr(VI).

3 程度，限度を表して，…まで

▶ Solutions for magnetic susceptibility determinations were prepared by weighing appropriate quantities of solute and solvent to the nearest 0.1 mg in 5-ml stoppered vessels.

4 …のために

▶ To obtain accurate results a number of precautions must be taken.

5 …に対して

▶ Silica is chemically resistant to all acids except HF.

6 対

▶ Each kind of ion has a particular ratio of mass to charge, or m/e value.

7 適用の範囲を示し，…にとって

▶ To the author's knowledge, no one has studied the possible importance of other electrostatic interactions in the benzene–halogen complexes.

8 不定詞を導いて形容詞的に，…する

▶ What we can confidently predict is that this is a field which will remain highly active and exciting for years to come.

today adv **現今では** syn *at present, now*

▶ Today we produce energy inefficiently to produce goods that decay by corrosion.

together adv **共に** syn *jointly*

▶ All silicates contain tetrahedral SiO_4 units, which may be linked together by sharing corners but never by sharing edges or faces.

tolerable adj **耐えられる** syn *acceptable, allowable*

▶ The intramolecular overcrowding in the yellow form becomes progressively less tolerable as thermal vibrations increase with a rising temperature.

tolerance n **耐性** syn *resistance*

▶ The first point to note is the great variety of mechanisms of metal tolerance which have been demonstrated or suggested in plants.

tolerant adj **耐性のある**

▶ Analyses of a range of metal-tolerant plants provide further evidence of the diversity of mechanisms involved.

tolerate vt **許容する** syn allow, permit

▶ The purported benzene complex was not expected to tolerate exposure to the electrophilic reagents.

too adv **1** 一つのものを強調して，…もまた，通常は文尾に置く． ☞ also

▶ Generally the molecular oxides of the nonmetals are acidic, many being acid anhydrides, and so too are the higher oxides of transition metals.

2 すぎる syn *exceedingly, overly*

▶ Too much sulfur completely destroys the elastic properties and gives hard rubber of the kind used in cases for storage batteries.

3 to do と相関的に用いて，…にすぎる

▶ The available data are too few and too discordant to permit any generalizations about the dependence of the polymerization process on the acid strength and/or length and character of the hydrocarbon portion of the alkanoic acids.

tool n **手段** syn *means, way*

▶ NMR has become a fundamental research tool for structure determinations in organic chemistry.

top n **1** **頂上** syn *peak, summit*

▶ It is never safe to take samples from the top of a large pile of material, but portions should be selected from all parts.

2 こま

▶ Another interesting family of globular molecules, which are to be considered more nearly symmetrical tops than spherical rotators, has the basic skeleton of bicyclo[2,2,2]octane.

adj **3** **一番上の** syn *highest*

▶ The top layers of activated carbon will initially remove the pollutant from the water and will do so until that layer of carbon reaches its capacity.

top-down adj 綿密に組織された

▶ These techniques are known as top-down and bottom-up design, respectively.

topic n 話題 syn *issue, matter, subject*

▶ The transfer of excitation energy in solids and the mechanism of decay of phosphorescence appear to be fruitful topics for further study.

topochemical adj トポ化学的な

▶ The operation of topochemical and topotactic forces in the reactions of organic solids has been established for certain photochemical reactions and polymerization systems.

topochemical reaction n トポ化学反応

▶ Topochemical reactions are those in which the process is dependent on the geometric arrangement of reactive groups within the lattice.

topochemistry n トポ化学

▶ The most elegant application of the principle of topochemistry in the solid-state chemistry has been the polymerization of a number of colorless monomeric diacetylenes under heat, UV, X-ray, or γ-ray irradiation to form single crystalline, highly colored polymeric polydiacetylenes.

topographic adj 地形上の syn *topographical*

▶ Many research areas are gaining and will continue to gain relevant information by looking at the topographic structure of surfaces.

topography n 地形 syn *terrain*

▶ Scanning electron microscopy complements optical microscopy for studying the texture, topography, and surface features of powders or solid pieces.

topological adj 位相的な

▶ Techniques are available for transferring the monolayer onto a solid support and for building up organized multilayers in controlled topological arrangements.

topology n **1** トポロジー

▶ Graph theory may be regarded as a branch of topology concerned with the use of graphs to depict neighborhood relationships.

2 形態

▶ The topologies of chiral structures might be in some cases be quite different from those of their racemic analogues.

topotactic adj トポタクチックな

▶ The term topotactic refers to solid-state reactions in which the product crystal orientation is related to that of the reagent.

topotaxially adv トポタキシアルに

▶ It has long been accepted that the reaction of oxides or mixed oxides proceeds topotaxially by addition or removal of oxygen layers, with appropriate migrations of cations.

topotaxy n トポタキシー

▶ Topotaxy is concerned with the three-dimensional orientational relations between the crystallography of the product and reactant phases.

torch n バーナー

▶ The high heat of recombination of hydrogen atoms finds application in the atomic hydrogen torch.

torsion n ねじれ

▶ A simple method of determining rigidity is by torsion of a cylindrical rod or wire of isotropic material.

torsional adj ねじれの

▶ In order to model aromatic polymers theoretically, it is necessary to understand the energetics for torsional motion of the polymer about its bonds.

torsional angle n ねじれ角

▶ The assumption of equal torsional angles is an approximation, and indeed crystal data indicate that in some situations the two torsional angles may not be equal.

torsional strain n ねじれひずみ

▶ Any deviations from the staggered arrangement are accompanied by torsional strain.

total n **1** 合計 a total of に続く名詞は複数形，全体を強調するときは動詞は単数形，個々を強調するときは複数形を用いる． syn *sum, whole*

▶ A long-standing problem with respect to the description of many octahedral clusters is that they contain a total of 86 rather than the 84 cluster electrons that is appropriate for edge bonding.

adj **2** 全体の syn *entire, overall, whole*

▶ The total filtrate is then evaporated to the original optimal volume and let stand undisturbed at room temperature until the solution has acquired the temperature of the surroundings and crystals have ceased to increase in number or size.

3 完全な syn *complete, perfect, thorough*

▶ The enthalpy of combustion is the change of enthalpy for the total oxidation of a material.

totally adv すっかり syn *completely, fully, perfectly*

▶ At best, the problem can only minimized by these methods but never totally eliminated.

total reflection n 全反射

▶ In its simplest form, an optical fiber consists of a central core region in which the light transmission occurs, encased in a cladding with smaller refractive index so that light passed down the core undergoes total internal reflection at the core-cladding interface, and thus cannot escape.

total synthesis n 全合成

▶ In the past five years, several impressive total syntheses of fused polyether natural products of marine origin have appeared in the literature.

touch vt **1** 接触する syn *bring into contact with*

▶ The coordination number in the hexagonal close packed structure is 12: a central sphere is touched by six spheres in its close-packed plane, by three in the close-packed plane above it, and by three more in the close-packed plane below it.

2 変化させる syn *affect, disturb*

▷ By proper selection of conditions, it is quite easy to hydrogenate the side chain without touching the aromatic ring.

tough adj 強靭な syn *durable, hard*

▶ Aside from the color due to the pigments present, protection of a surface by a tough organic film is the chief purpose of paint.

toughness n 頑丈なこと syn *durability, reliability, stability*

▶ Cermets combine the strength and toughness of metal with the heat and oxidation resistance of the ceramic material.

toward prep **1** …をさして syn *for, to*

▶ If either the acid or the alcohol is cheap and readily available, it can be used in large excess to shift the equilibrium toward the products and thus to increase yield of ester.

2 …に対して syn *for, to*

▶ Bromine is generally much less reactive toward hydrocarbons than chlorine.

tower n 塔

▶ A current of air was freed from all traces of carbon dioxide by passing through concentrated KOH solution and then through a tower filled with soda lime.

toxic adj 有毒な syn *harmful, poisonous*

▶ Only Fe and Ni react directly with CO under mild conditions to give $Fe(CO)_5$ and $Ni(CO)_4$, both of which are very toxic because they decompose thermally to release very toxic CO.

toxicity n 毒性

▶ Carbon monoxide has a relatively high toxicity due to its ability to form a complex with hemoglobin that is some 300 times more stable than the oxygen-hemoglobin complex.

toxin n 毒素 syn *poison*

▶ It often happened that so much toxin was released into the bloodstream by disintegration of bacteria in the course of chemotherapy.

trace n **1** 微量 syn *bit, speck, spot*

▶ Most technologically important semiconductors are of the extrinsic type, in which the charge-carrier production is determined by trace amounts of impurities or by lattice imperfections.

2 跡 syn *mark*

▶ The intensity of the resulting curve is measured with a microphotometer, and, to ensure uniformity, it is advantageous to rotate the photographic plate rapidly about the center of the diffraction pattern, whilst the trace is being recorded.

3 記録 syn *record*

▶ A typical diffractometer trace is shown in Figure 1 for SiO_2.

4 マトリックスのトレース

▶ The sum of the diagonal terms of a matrix is called the trace of the matrix.

vt **5 たどりつく** syn *track*

▶ Bases show certain characteristic reactions which can be traced to the hydroxide, OH, that they contain.

trace analysis n **痕跡分析**

▶ Contaminants often are present at low concentrations and may be determined by one or more of the techniques known collectively as trace analysis.

trace element n **微量元素**

▶ The rapidity with which changes in trace element supply result, first, in detectable biochemical changes and, later, in tissue damage progressing to disease or death differs greatly between elements.

tracer n **トレーサー**

▶ The main uses of deuterium are in tracer studies to follow reaction paths and in kinetic studies to determine isotope effects.

tracing n **トレーシング**

▶ The measurements were made by first filling the cells with solvent and obtaining a blank tracing over the desired spectral region.

track n **1 走路** syn *course, path, route*

▶ Etching occurs preferentially at defect sites and thus enlarges the damage tracks.

vt **2 探知する** syn *monitor, record*

▶ As the scanning probe of an atomic force microscope tracks across a surface, the forces it experiences are translated into vertical deflections of the lever.

tractable adj **扱いやすい** syn *manageable, workable*

▶ The treatment is mathematically tractable, and the development is physically reasonable.

traditional adj **伝統的な** syn *customary, standard, usual, well-known*

▶ Chemical oxidative polymerization of aniline is the traditional method for preparing polyaniline in bulk.

traditionally adv **慣例上**

▶ Perovskite oxides traditionally have been prepared by high-temperature solid-state reaction of the binary oxides or suitable oxide precursors such as carbonates, nitrates, oxalates, citrates, etc.

train n **長い列** syn *column*

▶ In a quantitative combustion, a weighed sample of the organic compound is passed through a combustion train: a tube packed with copper oxide heated to 600–800 ℃, followed by a tube containing a drying agent and a tube containing a strong base.

trajectory n **軌跡** syn *course, track*

▶ A statement of both $x(t)$ and $p(t)$ is called a trajectory of the particle.

tranquilize vt **精神を安定にする** syn *calm, pacify, relax*

▷ The tranquilizing action is thought to be the result of a reduction in the concentration of brain serotonin.

trans adj **1 トランスの**

▶ The group which is in a trans relationship to the hydroxyl group is the one that migrates.

pref **2 トランス** 一般には，ローマン体を用いる．ただし，化学物質名の一部をなすときには，イタリック体とし，ハイフンを用いる． ☞ cis

▶ In *trans*-2-butene, the prefix trans indicates the two methyl groups are on opposite sides of the double bond.

trans addition n **トランス付加** ☞ cis addition

▶ There is ample evidence to show that bromine and many other reagents give trans addition.

transamination n **アミノ変換反応**

▶ Glutamate provides the amino group for the synthesis of many other amino acids through transamination reactions.

transannular adj **渡環の**

▶ S_8^{2+} and Se_8^{2+} consist of an eight-membered ring with an exo-endo conformation and a rather long transannular bond.

transcend vt 越える syn *exceed, surpass*

▶ Phosphorus has an extensive and varied chemistry which transcends the traditional boundaries of inorganic chemistry.

transcribe vt 転写する syn *copy, replicate, reproduce*

▶ A sequence of deoxynucleotides in DNA is first transcribed into a sequence of ribonucleotides in RNA.

transcript n 転写物 syn *duplicate*

▶ The introns are excised before the processed mRNA transcript is exported to the cytoplasm.

transcription n 転写 syn *transcript*

▶ DNA serves as a template on which molecules of RNA are formed in the process called transcription.

transduce vt **1** 変換する

▶ In vision, a stimulus from the outside world is transduced in the order: light → isomerization of retinal → conformational change in rhodopsin.

2 伝達する

▶ These special proteins transduce the signal to deeper levels within the cell.

transducer n トランスデューサー

▶ Chlorophyll can readily transfer radiant energy to its chemical environment and, thus, acts as a transducer in photosynthesis.

transduction n 変換

▶ Transduction of the biorecognition event is generally based on changes in the optical thickness caused by the adsorption or desorption of biological molecules from the surface of the pores of the porous silicon optical sensor.

trans effect n トランス効果

▶ The utility of an empirical rule such as that of the trans effect becomes apparent if we consider a few reactions of platinum(II) complexes.

transesterification n エステル交換反応

▶ Fats are sometimes converted by transesterification into the methyl esters of carboxylic acids; the glycerides are allowed to react with methanol in the presence of a basic or acidic catalyst.

trans fatty acid n トランス脂肪酸

▶ Trans fatty acids are formed inadvertently during the manufacture of margarines in a process called hydrogenation.

transfection n トランスフェクション

▶ Plasmids that contain recombinant DNA can be introduced into bacteria by transfection.

transfer n **1** 移動 syn *move, transport*

▶ Thus, the transfer of a hydrogen from one aldehyde molecule to another does not involve water in the transfer mechanism.

vt **2** 移す syn *deliver, give, pass*

▶ Alkali metal atoms may penetrate between the graphite sheets, to which they transfer electrons.

transferability n 移しうること

▶ Such schemes aim to express the atomization enthalpy of the substance as the sum of empirical bond-energy terms having a high degree of transferability.

transferable adj 移すことのできる syn *movable, transportable*

▶ This alone does not prove that susceptibility can be found from a sum of transferable group susceptibilities.

transference n 移動 syn *transit, transport*

▶ The uptake of both anions and cations by such resins is accompanied by the transference of considerable amounts of water, causing the exchange resins to swell.

transfer pipet n ホールピペット

▶ A transfer pipet has only one mark upon it and serves for measuring off a definite amount of liquid.

trans form n トランス形 ☞ cis form

▶ When NO condenses to a liquid, partial dimerizatiom occurs, the cis form being more stable than the trans form.

transform n **1** 変換 syn *change, conversion*

▶ There is a clear analogy between the mathematical transform and the production of images in optical and electron microscopes.

vt **2** 変換する syn *change, convert, modify*

▷ Denaturation is an important alteration in

the properties of proteins which tends to make them insoluble in water, transforming a hydrophile to a hydrophobe sol.

3 形質転換する

▶ Only the cells that were transformed with plasmid DNA containing the antibiotic-resistant genes will grow to produce colonies on an agar medium containing antibiotics.

transformation n **1 転移** syn *transition*

▶ The transformation of the yellow form to the white form seems superficially to involve only the change of a hydrogen position.

2 変形 syn *alteration, modification*

▶ By transformation, $(P/V)(1+P/P_s)$ is seen to be a linear function of P/P_s.

3 形質変換

▶ This process is called bacterial transformation because one can change the appearance or properties of a bacterial cell by introducing foreign genes.

transient n **1 過渡現象**

▶ In general, the importance of the transients arises from their possible contribution to our knowledge of the reaction details.

adj **2 過渡的な** syn *fleeting, short-lived, transitory*

▶ Cholesterol reacts with acetic acid containing sulfuric acid to give a transient purple color changing to blue and then green.

transiently adv **過渡的に** syn *briefly, fleetingly, temporarily*

▶ The disadvantage of X-ray powder technique is its dependence on the crystallinity of the material; many intermediate products in solid reactions are formed transiently and in a poorly crystallized condition, and they are consequently difficult to detect by means of X-rays.

transistor n **トランジスター**

▶ A transistor, or n–p–n junction, is built up of two n-type regions of Si separated by a thin layer of weakly p-type.

transit n **通過** syn *transfer, transport*

▶ A mixed film greatly affects transit of substances related to or soluble in either constituent.

transition n **1 遷移**

▶ Transitions between states of different multiplicity are forbidden.

2 転移 syn *transformation*

▶ Many compounds that are molecular metals at room temperature undergo a metal-to-semiconductor transition at low temperatures.

transitional adj **過渡的な** syn *intermediate, transient*

▶ The two eclipsed forms of butane are necessarily unstable species and have only a fleeting existence as transitional forms between the conformers found in the energy troughs.

transition dipole moment n **遷移双極子モーメント**

▶ The Franck–Condon principle may be put on a quantitative basis by analyzing the transition dipole moment.

transition element n **遷移元素**

▶ Many metals in the later transition element groups show substantial trends to increasing covalency leading either to lower coordination numbers or to layer-lattice structures.

transition-metal complex n **遷移金属錯体**

▶ The transition-metal complexes are of increasing importance to organic chemists today as catalysts of unprecedented power and selectivity.

transition moment n **遷移モーメント**

▶ For an all-trans polyene, the transition moment to the lowest electronic state is not quite directed along the length of the molecule but has a small transverse component.

transition probability n **遷移確率**

▶ The selection rules dictate what will be spectroscopically observable and usually arise as symmetry constraints on integrals representing transition probabilities.

transition state n **遷移状態**

▶ The transition state is located at the maximum of the minimum energy route between reactants and products, the saddle point.

transition state theory n **遷移状態理論**

▶ Transition state theory starts by assuming that the transition state is in equilibrium with the reactants.

transition temperature n 転移温度

▶ One of the most important problems in studying polymorphism in a given system is to determine the transition temperature.

transitorily adv 一時的に syn *transiently*

▶ All the higher oxidation states of manganese are strongly oxidizing, but Mn(V) exists only transitorily, while Mn(VI) and Mn(III) normally disproportionate.

transitory adj 一時的な syn *momentary, short-lived, transient*

▶ In spite of their transitory existence, free radicals are capable of initiating many kinds of chemical reactions by means of a chain mechanism.

translate vt 翻訳する syn *explain, interpret*

▶ At molecular level, the information has to be translated into proteins within the cell.

translation n 並進

▶ A distance of 3.11 Å between ester carbonyl atoms of molecules is related by a unit cell translation in *y* and *z*.

translational adj 並進の

▶ In considering translational periodicity generally, it is convenient to begin with a discussion of a planar pattern.

translational diffusion n 並進拡散

▶ Nuclear magnetic resonance provides a powerful and unique way of studying molecular rotational and translational diffusion in the solid state.

translational symmetry n 並進対称 ☞ rotational symmetry

▶ Every crystal structure has a lattice that is a representation of the translational symmetry of the structure.

translucent adj 半透明な syn *semitransparent*

▶ An emulsion is usually turbid, opaque, translucent, or opalescent, but if the refractive index of both liquids is made the same, it becomes transparent.

transmembrane adj 膜貫通型の

▶ Regulation of transmembrane ion transport is a vital aspect of bioinorganic chemistry.

transmission n **1** 透過 syn *forwarding, transport*

▶ The direction of vibration, and not the direction of transmission, determines the index of refraction.

2 伝導 syn *conveyance, transportation*

▶ The ability of a superconductor to support a resistanceless direct current can be exploited in the lossless transmission of electric power.

3 伝達 syn *communication*

▶ Ribonucleic acid plays major roles in the transmission of the genetic information encoded by DNA.

transmit vt 透過する syn *pass through*

▶ For compounds such as benzene and naphthalene whose strong absorptions lie below 3650 Å, lamps with quartz envelops which transmit the 2537 Å resonance line are used.

transmittance n 透過率

▶ The transmittance *T* is the ratio of the intensities of transmitted to incident light.

transmitted light n 透過光

▶ The characteristic ruby-red color of Al_2O_3—Cr^{3+} in transmitted light is actually a complex mixture of blue and orange-red light.

transmutation n 核変素 syn *nuclear transmutation*

▶ Radioactive decay of uranium can be regarded as a type of transmutation.

transparent adj 透明な syn *clear*

▶ When monochromatic radiation is incident on systems like dust-free, transparent gases and liquids, or optically perfect, transparent solids, most of it is transmitted without change, but some is scattered.

transport n **1** 輸送 syn *transportation*

▶ From the nature and the extent of observed deviations from the relation, much new knowledge has been gained of the details associated with the transport of materials in ionic crystals

vt **2** 輸送する syn *bear, bring, carry, take, transfer*

▷ Oxygen transported to cells is utilized for the reversible oxidation of cytochromes as a source of energy in mitochondria.

transporter n 輸送体 syn *carrier*

▶ The membrane exists as a two-dimensional fluid of freely diffusing lipids, dotted or embedded with proteins that may function as channels or transporters of solutes across the membrane.

transport number n 輸率 syn *transference number*

▶ Before attempting to correlate experimental data on electric conductivity and diffusion, it is necessary to determine the extent to which these component conductivities are participating, that is, to evaluate the transport numbers.

transverse adj 横方向の

▶ To obtain compounds which pass directly from the crystalline to the nematic phase, the molecules should not contain too many groups with a transverse dipole component and preferably not more than one long alkyl chain.

trap n **1** 気体，蒸気を捕捉するトラップ

▶ The trap is a suction flask half-filled with water and with a delivery tube inserted to within 1 cm of the surface of the water.

2 電子を捕捉するトラップ

▶ As phosphorescence proceeds, electrons are excited from the traps into the conduction band.

3 転位を捕捉するトラップ

▶ Added carbon atoms in iron can act as dislocation traps and halt the gliding motion.

n **4** 遊離基を捕捉するトラップ

▶ Iodine is sometimes employed as a free-radical trap or scavenger in the study of reaction mechanisms.

vt **5 捕捉する** syn *confine, hold, keep*

▷ Trapping an electron at a halide vacancy in an alkali halide crystal leads to the intense color of the so-called F center.

trapped adj 捕捉された

▶ The energy of excitation spreads rapidly through the crystal until it becomes trapped by an impurity molecule with lower energy levels or at some imperfection in the crystal lattice.

trapped electron n 捕捉電子

▶ The trapped electron is not wholly confined within the anion vacancy but interacts with the nearest and even next-nearest neighbors.

trapping n 捕捉

▶ Trapping will reduce the drift mobility but not the microscopic mobility or the Hall mobility.

travel n **1** 伝播 syn *motion, passage*

▶ No matter what the direction of travel from the center, the vibration direction of the ordinary ray will always be normal to that of the extraordinary ray.

vi **2 進む** syn *move, proceed*

▶ The ejected electron has a substantial kinetic energy and travels several nanometers before it is slowed down and eventually solvated.

traverse vt 横断する syn *cross*

▶ Because of the particle's x component of velocity, the molecule will traverse the container of length x in the x direction, collide with the wall yz, and then rebound.

treat vt **1 取り扱う** syn *analyze, deal with, examine, investigate, survey*

▶ In order to treat large numbers of compounds, it is necessary to have fast automatic procedures.

2 処理する syn *deal with, handle*

▶ A boiling solution of 1.0 g of TCNQ in 100 mL of tetrahydrofuran was treated with a boiling solution of 0.57 g of triethylammonium iodide in 7 mL of acetonitrile.

treatise n 論説 syn *monograph, paper*

▶ Well's treatise on structural inorganic chemistry provides a more exhaustive assay of ionic and metallic lattices.

treatment n **1** 扱い syn *manupulation*

▶ The treatment is satisfactory at all but very high pressures.

2 処理 syn *action, handling*

▶ Treatment of B–O species with strong reducing agent such as sodium hydride results in formation of B–H bonds.

treble vt 3倍にする
▶ Two millimoles per gram of methyl groups trebled the activity of a sulfonic acid-micelle-templated silica system.

tree n 樹形図
▶ Each alkane graph is a tree with *n* points of degree 4 (carbon points) and $2n + 2$ points of degree 1 (hydrogen points).

tremendous adj 途方もない syn *enormous, huge, immense, vast*
▶ Defects can exert tremendous influence over the physical and chemical properties of a material.

tremendously adv 途方もなく syn *greatly, highly, immensely*
▶ The relative proportions of the several substances present in the solution of a quinhydrone vary tremendously with the concentration of the solution, with its acidity or alkalinity, and with the nature of the particular quinone-hydroquinone pair that is involved.

trend n 傾向 syn *tendency*
▶ While in some instances the experimentally measured ΔH_{NI} is smaller for methyl-substituted derivatives than for the parent compound, this trend cannot be universally true.

triad axis n 三回軸 syn *threefold axis*
▶ Whereas a cube is distinguished by four triad axes, it also contains most of the symmetry elements observed in the other systems.

trial n 試験 syn *check, examination, test*
▶ Two complete series of trials were made, one preceding a carbon tetrachloride calibration and the other immediately following it.

trial and error n 試行錯誤
▶ By knowing how a reaction takes place, we can make changes in the experimental conditions, not by trial and error but logically, that will improve the yield of the product we want.

triangular adj 三角形の
▶ Crystals of sodium chloride, although usually found as cubes, often grow with their corners modified by triangular faces.

triatomic adj 三原子を有する ☞ monoatomic, diatomic
▶ Nitrous oxide is a moderately unreactive gas composed of linear unsymmetrical molecules, as expected for a 16-electron triatomic species.

triaxial adj 三軸の
▶ In the triaxial ellipsoid, there are three mutually perpendicular axes which are, in general, unequal in length.

tricapped trigonal prism n 三面冠三角柱
▶ The stereochemistry of most nine-coordinate complexes approximates to the tricapped trigonal prism, formed by placing additional ligands above the three rectangular faces of a trigonal prism.

tricky adj 微妙な syn *complicated, difficult, delicate, sensitive*
▶ The only tricky point about the application of the Boltzmann distribution is the determination of the number of states corresponding to a particular energy level.

triclinic adj 三斜晶系の
▶ The X-ray diffraction pattern is that of the triclinic phase.

triclinic system n 三斜晶系
▶ In the triclinic system, the simple space lattice has a primitive unit cell, with three axes of unequal lengths and unequal angles.

tridentate ligand n 三座配位子 syn *terdentate ligand*
▶ Acyclic tridentate ligands often bind in a T-shaped manner that has no significant biological relevance.

trigger vt 誘発する syn *activate, initiate*
▷ Humic acid is a natural stream pollutant and is thought to be capable of triggering the red tide phenomenon due to microorganisms in seawater.

triglyceride n トリグリセリド
▶ Inhibition of pancreatic lipase prevents the hydrolysis of triglycerides and reduces the absorption of fatty acids.

trigonal bipyramid n 三方両錐
▶ The two basic structures for five-coordination are the trigonal bipyramid and the square pyramid.

trigonal crystal adj 三方晶

▶ The low-temperature trigonal crystals of SbI_3 are usually in hexagonal basal plates with $d = 4.85$.

trigonal prism n 三角柱

▶ Metal diborides have the ultimate structure obtainable by the condensation of trigonal prisms and also the first type in which a discernible extended boron lattice occurs.

trigonal prismatic adj 三角柱の

▶ The two sets of nickel ions are superposed in projection down *c* and give the trigonal prismatic coordination for arsenic.

trigonal pyramid n 三方錐

▶ In contrast to BF_3, the NH_3 molecule, while still having a threefold axis, is not planar but has a configuration usually referred to as a trigonal pyramid.

trigonal pyramidal adj 三方錐の

▶ The X-ray analysis shows the solid to consist of trigonal pyramidal XeO_3 units, with the xenon atom at the apex.

trigonal system n 三方晶系

▶ The trigonal system is characterized by a single threefold axis.

trimer n 三量体

▶ By varying the conditions, yields of the cyclic trimer or tetramer and other oligomers can be optimized.

trimeric adj 三量体の

▶ Our experimental data are entirely consistent with the hypothesis that the vapors of acetic and trimethylacetic acids contain monomeric, dimeric, and trimeric species, the latter in small but appreciable concentrations.

trimerize vi 三量化する

▶ Colorless crystals of cyanamide result from the reaction of NH_3 on ClCN and trimerize to melamine at 150°.

trimolecular reaction n 三分子反応

▶ The first trimolecular gas reaction was discovered and investigated in 1914 by Trautz: it was the reaction

$$2NO + Cl_2 \rightarrow 2NOCl$$

and he investigated it from 8 to 283 ℃.

trimorphic adj 三形の

▶ NaO_2 is trimorphic, having the marcasite structure at low temperatures, the pyrite structure between −77 and −50 ℃, and a peudo-NaCl structure above this, due to disordering of the O_2^- ions by rotation.

triple bond n 三重結合 ☞ acetylenic linkage

▶ The high dissociation energy of CO and very short C-O bond distance suggest that the Lewis structure is better represented by a triple bond, including a C←O dative bond.

triple-decker n 三段重ね

▶ If the 1-2-3 compound had an idealized perovskite structure, it would possess nine oxygen atoms in its formula and would have the triple-decker structure.

triple point n 三重点

▶ The kelvin is strictly defined as the fraction 1/273.16 of the temperature interval between the absolute zero and the triple point of water.

triplet n **1** 三重線

▶ Because the peak is split into a triplet, the aldehyde group must be joined to a carbon which bears two hydrogens.

2 三重項

▶ When solutions containing both naphthalene and benzophenone were irradiated, the naphthalene triplet resonance was easily observed.

triplet state n 三重項状態

▶ A triplet state has two unpaired electrons and is normally more stable than the corresponding singlet state because, by Hund's rule, less interelectronic repulsion is expected with unpaired than paired electrons.

triply adv 三重に

▶ The Li_4C_4 cluster consists of a tetrahedron of 4Li with a triply bridging C above the center of each face to complete a distorted cube.

tripod ligand n 三脚配位子 syn *tripodal ligand*

▶ Tris(2-aminoethyl)amine is one of the so-called tripod ligands which are quite unable to give planar coordination but instead favor trigonal bipyramidal structures.

tritiated adj トリチウム化した

▶ Tritiated ammonium salts can be readily prepared by dissolving the salt in tritiated water and then removing the water by evaporation.

triturate vt すり砕く syn *grind, mill, pulverize*

▶ Sodium chlorate must not be triturated with any combustible substance.

trivalent adj 三価の syn *tervalent*

▶ Electrophilic reactions of hydrocarbons were long recognized to involve electron-deficient, trivalent hydrocarbon ions.

trivial adj ささいな syn *insignificant, meaningless, unimportant*

▶ Accurate synthetic analogues of zinc enzymes are not trivial to obtain.

trivial name n 慣用名 ☞ common name

▶ Semidescriptive and trivial names for organic compounds are being coined continually and their use is widespread.

troposphere n 対流圏

▶ NO in the troposphere is converted to NO_2 by reaction of hydrocarbon peroxy radicals.

trouble n 苦労 syn *concern, difficulty*

▶ One of the greatest sources of trouble for the analytical chemist is the presence of impurities in reagents and the action of solutions upon glassware.

troublesome adj 厄介な syn *annoying, difficult,*

▶ Because aldehydes are susceptible to further oxidation, the conversion of primary alcohols to aldehydes can be troublesome.

trough n 槽 syn *basin*

▶ Spreading is performed in the Langmuir trough, which controls the surface area and pressure.

true adj 真実の syn *actual, factual, genuine*

▶ The use of the ionic model does not necessarily imply that it is true, merely, that it is convenient and useful.

truly adv 本当に syn *actually, really*

▶ If the reaction is truly heterogeneous, there should be no further conversion of substrate when the catalyst is removed.

truncated adj 先端を切った形の

▶ The typical basketlike framework of ultramarines is a truncated octahedron, the polyhedron obtained by cutting off the apices of an octahedron.

truth n 事実 syn *fact, reality*

▶ The truth probably lies somewhere between these two extremes.

tube n 管 syn *pipe*

▶ The ordinary mercury thermometer consists of a quantity of mercury in a glass tube.

tubing n 管材料

▶ Reduced scandium chlorides as best synthesized in sealed tantalum tubing starting with $ScCl_3$ or a reduced mixture in the cooler end of tube and strips of Sc metal in the hot and central zones.

tubule n 細管

▶ The dentin that forms the tough inner core of the tooth is a vital tissue, made of a composite of collagen and hydroxyapatite, containing tubules that radiate from the pulp cavity at the center of the tooth.

tumor n 腫瘍 syn *cancer, saroma*

▶ Mutations in the proteins that negatively control signal transduction pathways may also lead to cellular transformation and the development of tumors and cancers because signal pathways are active that should not be.

tunability n 調整できること

▶ The tunability of the band gap in the GaP_xAs_{1-x} and $Al_xGa_{1-x}As$ families of solids is used extensively in the design of light-emitting diodes and diode lasers.

tunable adj 調整できる

▶ Dye lasers produce tunable radiation in the visible and near IR regions.

tunable laser n 波長可変レーザー

▶ Tunable lasers using electron injection have been the most extensively developed and are called diode lasers.

tune vt 調整する syn *adjust, regulate*

▶ The degree of porosity and the pore size may be tuned across a wide range.

tungsten bronze n タングステンブロンズ

▶ Tungsten and molybdenum bronzes owe

their name to their metallic luster and are used in the production of bronze paints.

tuning n **1** 同調

▶ The pressure tuning is the element that makes pressure effects a powerful tool for characterizing electronic states and excitations, testing theories, and uncovering electronic transitions to new ground states with different physical and chemical properties.

2 調整 syn *adjustment, alignment*

▶ Intracellular mineralization is characterized by enormous fine-tuning of the crystal chemistry.

tunnel n **1** トンネル syn *hole, passage*

▶ The Mn_2O_4 framework of the spinel structure in $LiMn_2O_4$ is highly stable and defines a series of tunnels formed by the face-sharing of tetrahedral lithium and empty octahedral sites.

vi **2** 進む syn *penetrate*

▶ An electron may tunnel from an atom in the probe tip to an atom on the surface.

tunneling n トンネル現象

▶ The motion of the electron through a classical barrier is called tunneling.

tunneling current n トンネル電流

▶ The electron probability distributions fall of exponentially with distance; therefore, the tunneling current provides a very sensitive probe of interatomic separation.

tunneling spectroscopy n トンネル分光

▶ Current imaging tunneling spectroscopy can provide electronic information both for filled and empty states, which are below and above the Fermi energy level.

turbid adj 濁った syn *cloudy, opaque*

▶ Acid soaps are insoluble in water and appear as very soft turbid flocs in hydrolyzed soap solutions.

turbidity n **1** 濁り

▶ Since the molecules of soap or detergent are very small compared with colloidal dimensions, the particles causing turbidity are groups of these molecules.

2 濁り度

▶ If the intensity of the incident radiation is I_0, and l is the length of the path through the scattering medium, the intensity of the transmitted radiation is given by

$$I = I_0 e^{-\tau l}$$

where τ is known as the turbidity.

turbulence n 乱れ syn *disorder, disturbance*

▶ In purely viscous flow, the movement of the liquid is devoid of turbulence and is lamellar.

turbulent adj **1** 乱流の

▶ Empirically, it has been found that for Reynolds numbers less than about 2,100 the flow is laminar, for values greater than 4,000 it is turbulent.

2 乱れた syn *chaotic, unsettled*

▶ The history of the thermo- and photochromism of bianthrone has been a long and quite turbulent one.

turbulent flow n 乱流

▶ At higher flow rates or for large-diameter tubing the flow type changes to what is known as turbulent flow.

turn n **1** 変わり目 syn *alteration, change*

▶ Some of these intriguing compounds have been known since the turn of the century but only recently have their structures been elucidated by single-crystal X-ray analysis.

2 回転 syn *cycle, revolution, rotation*

▶ One turn of the helix occurs each 3.6 amino acid residues.

3 in turn それがまた syn *in succession,*

▶ The tip is connected to a positioner that is in turn connected to the feedback controller.

vi **4** …に変じる syn *alter, convert*

▶ The aqueous solution of hydroiodic acid is even more difficult to keep than a solution of hydrobromic acid; it soon turns brown, owing to the separation of iodine.

5 向かう syn *move, shift*

▷ Turning now to the other possibility for oxidation-reduction reactions involving atom or group transfer, we are in a more familiar field as far as kinetic theory is concerned.

6 turn out であることが判明する syn *appear*

▶ Many proteins contain metals such as iron, zinc, and copper, and these metal atoms turn out to be intimately involved in the biological functions of the molecules to which they are bound.

vt **7** 変える syn *adapt, change, modify*

▶ The yellow dyestuff, fluorescein, is turned red by bromine liberated by the action of an oxidizing agent upon a bromide in the presence of acid.

8 向ける syn *take up*

▶ Next we turn our attention to the extended MX_2 crystal structures; there are, of course, the familiar forms in which these compounds commonly occur under ambient terrestrial conditions.

9 回転させる syn *rotate*

▶ In operation, the analyzer is turned to a position of minimum light transmission, indicating that the two prisms are optically crossed.

10 turn on 作動させる syn *activate, set in motion*

▶ At any given time in a cell, only certain genes are turned on or expressed to produce proteins while many other genes are silenced or repressed.

turnings n 削り屑

▶ When a solution of an alkyl halide in dry ether is allowed to stand over turnings of metallic magnesium, a vigorous reaction takes place.

turnover n 代謝回転

▶ There are too many differences between the amino acids for any useful generalization to be made about the processes except for the turnover of the amino group.

turnover number n 代謝回転数 regeneration と reuse は 1 つの概念とみなされている.

▶ Turnover numbers based on Mn are not high, and regeneration and reuse of catalysts is an issue that has yet to be confirmed.

turnover rate n 代謝回転速度

▶ Cationic gold(I) species gave excellent turnover numbers and turnover rates for the addition of alcohols to alkynes.

twice adv **1** 二回 syn *two times*

▶ Tetracyanoethylene from Eastman Kodak was twice recrystallized from chlorobenzene and twice sublimed.

2 二倍に

▶ Adsorption of oxygen proceeds to at least twice the extent that adsorption of carbon monoxide does.

twin n **1** 双晶

▶ One individual of the twin is always related to the other by a symmetry operation, and the twins share a plane of atoms that fits right into the structure of each.

adj **2** 双晶の

▶ Two or more individual crystals may grow in contact so that neither is complete, and the result is a twin crystal.

3 対をなす syn *matching, matched*

▶ Twin drawbacks of many metal-containing liquid crystal systems are the high temperatures of transition and the tendency to decomposition.

vt **4** 双晶にする

▶ If the crystal divides into strips which extinguish alternately as the crystal rotates, then the crystal is likely to be twinned.

twinned adj 双晶になった

▶ A twinned crystal is not a single crystal but contains crystal domains in two, or more, symmetry-related orientations.

twinning n 双晶形成

▶ Single crystals exhibit extensive twinning upon cooling through the phase transition but become single again when warmed to room temperature.

twist n **1** ねじれ syn *distortion*

▶ If a chiral molecule forms a nematic or tilted smectic liquid-crystalline phase, then the structure becomes chiral and contains a helical twist.

vi **2** ねじれる syn *curve*

▶ One phenyl ring twists in the opposite sense to the other, giving a helical conformation.

vt **3** ねじる syn *bend, curve*

▶ In some cases, the plane of the free radical is twisted with respect to the original plane of the carbon skeleton.

twist conformer n ねじれ配座異性体

▶ The twist conformers have greater energy than the chair form since they retain from the boat form some eclipsing of neighboring hydrogen atoms.

twist form n ねじれ形

▶ Manipulation of a molecular model of the twist conformers of cyclohexane shows us that one twist form may be converted into another via a third distinct conformer called a boat.

two pron **1** 二つ

▶ Sorption is a surface phenomenon that may be either absorption or adsorption, or a combination of the two.

adj **2** 二つの

▶ Progesterone exists in two crystalline forms of equal physiological activity.

two-dimensional adj 二次元の

▶ β-Alumina is a two-dimensional conductor.

two-dimensional chromatography n 二次元クロマトグラフィー

▶ Two-dimensional chromatography can give additional specificity, resolution, and sensitivity when coupled with techniques such as isoelectric focusing.

twofold adj 二重の（2-fold とはしない）

▶ The reason for this is twofold.

twofold axis n 二回軸 syn *diad axis*

▶ Molecules with only a twofold axis have transitions polarized either along that axis or perpendicular to it.

two-photon absorption n 二光子吸収

▶ Two-photon absorption is a process whereby molecules simultaneously absorb two photons and is inherently weak at normal light intensities.

two-photon process n 二光子過程

▶ Raman scattering is a two-photon process, with one photon going in and one coming out, and as a result the angular momentum of the photons can remain unchanged or change by two units.

two-thirds n 三分の二

▶ Pure Al_2O_3 has a rhombohedral crystal structure comprising a hcp array of oxide ions with Al ordered on two-thirds of the octahedral interstices.

Tyndall effect n チンダル効果

▶ The scattered light, or Tyndall effect, is observed when a light beam is passed through a colloidal solution.

type n 型 syn *class, category, kind, sort*

▶ Nitroso compounds may be of the C-NO or N-NO type.

typical adj 典型的な syn *characteristic, regular, representative*

▶ Hydrolysis of amides is typical of the reactions of carboxylic acid derivatives.

typically adv 一般的に syn *generally, ordinarily*

▶ Enthalpies of sublimation are typically evaluated from the temperature dependence of the vapor pressure over the crystalline phase.

typify vt 代表する syn *exemplify, represent*

▶ The formation of an alcohol by a substitution reaction is typified by the reaction of a primary or secondary alkyl halide with dilute aqueous base.

U

ubiquitous adj 遍在する syn *prevalent*

▶ Manganese is a particularly ubiquitous element in small amounts in naturally occurring minerals, and correlations between manganese concentration and the intensity of emission bands in certain silicates have been demonstrated.

ultimate n **1** 究極のもの syn *extreme, limit*

▶ Nanocrystal structures represent the ultimate in miniaturization of solid-state materials and devices.

adj **2** 基本的な syn *basic, elemental, fundamental*

▶ The electrode chosen as the ultimate standard is the standard hydrogen electrode.

3 究極の syn *conclusive, decisive, final*

▶ The ultimate goal of artificial photosynthesis is to combine oxidation and reduction half cells that would efficiently split water upon solar irradiation.

ultimately adv 結局 syn *eventually, finally*

▶ Molecules can hold charge and indeed may conduct electricity, but they have other potentially useful properties that could be used for storing information and ultimately providing switching capabilities.

ultracentrifuge n **1** 超遠心機

▶ The first, and by far the most usual measurement to be made with the ultracentrifuge, is rate of sedimentation.

vt **2** 超遠心機にかけて分離する

▶ If a solution of a substance of low molar mass is ultracentrifuged, equilibrium is established within a fairly short period of time, and there will be a density gradient in the solution.

ultrahigh vacuum n 超高真空

▶ Many of the most modern techniques of surface science require that the measurement be carried out in ultrahigh vacuum (i.e. below 10^{-9} torr).

ultrapure adj 純度が極めて高い

▶ Pyrolysis of SiH_4 is a commercial route to ultrapure Si.

ultrasonic adj 超音波の

▶ Ultrasonic dispersion of nonmetallic liquids such as benzene, toluene, water, or oils is sometimes used.

ultrasonic wave n 超音波 syn *ultrasound*

▶ Ultrasonic wave originates from a piezoelectric material to which a high-voltage alternating current has been applied at a frequency range from approximately 20 kHz to 10 MHz.

ultrastructure n 超微細構造

▶ Chemists have a key role to play in the fabrication of ultrastructures because properties characteristic of this level of organization are ultimately governed by the nature of molecular interactions occurring at solid–liquid and solid–solid interface.

ultrathin film n 超薄膜

▶ Hollow titanium tubes were generated by coating tobacco mosaic virus with an ultrathin film of titanium followed by removal of the organic template by oxygen plasma treatment.

ultraviolet (UV) light n 紫外光 syn *ultraviolet radiation*

▶ Scintillators such as pyrene partially protect methyl methacrylate polymer from degradation by UV light or ionizing radiations.

ultraviolet (UV) photoelectron spectroscopy, n 紫外光電子分光 ☞ *X-ray photoelectron spectroscopy*

▶ UV photoelectron spectroscopy, which probes the valence shell electrons in the adsorbate, has been used to study the modification of the molecular orbitals on adsorption.

ultraviolet spectroscopy n 紫外分光法

▶ Several species may be detected by ultraviolet and visible spectroscopy, including those containing unpaired electrons.

ultraviolet spectrum n 紫外スペクトル

▶ The ultraviolet spectrum can tell a good deal about the structure of carbonyl compounds: particularly, about conjugation of the carbonyl group with a carbon–carbon double bond.

unable adj することができない syn *not able,*

powerless, unfit, unqualified
▶ Antibodies are usually unable to pass from the intercellular fluid into living cells.
unabundant adj 豊富でない
▶ Boron is comparatively unabundant in the universe; it occurs to the extent of about 9 ppm in crustal rocks.
unacceptable adj 受け入れられない syn *undesirable, unsatisfactory*
▶ The main chemical problem in reducing SnO_2 comes from the presence of Fe in the ores which leads to a hard product with unacceptable properties.
unaffected adj 影響されない syn *uninfluenced*
▶ If the viscosity is structural due to a mechanical framework of colloidal dimensions, it may be much more nearly unaffected by large changes of temperature.
unaltered adj 変わらない syn *fixed, invariable, unmodified*
▶ The value of determinant is unaltered if rows are changed into columns or columns into rows.
unambiguous adj 明瞭な syn *definite, obvious*
▶ This result does not allow us to write an unambiguous structure for our unknown.
unambiguously adv 明瞭に syn *definitely, perfectly, simply*
▶ The prefixes *R* and *S* enable us to specify unambiguously the absolute configuration of a compound, because their use does not depend on a relationship to any other compound.
unanswered adj 答えのない syn *unsolved*
▶ There are still unanswered questions on nucleation and on the role of the rigid and diffuse double layer; therefore, many more experiments are required.
unanticipated adj 予期しない syn *accidental, surprising, unexpected*
▶ The unanticipated bending of the group 2 dihalides is the focus of continued interest in these molecules.
unassociated adj 非会合の
▶ Hydrogen bonding increases the boiling point of a liquid beyond that expected from comparison with comparable non-hydrogen-bonded liquids and causes a greater entropy of vaporization than that of unassociated liquids.
unavailability n 利用できないこと
▶ The unavailability of the d orbitals in Be and Mg helps to explain the lower four- and six-coordination in the Be and Mg systems.
unavailable adj 得られない syn *inaccessible, unatainable*
▶ Liquids can be supercooled, freezing substantially below their normal melting point, if a nucleation site for the growth of crystals is unavailable.
unavoidable adj 避けがたい syn *inevitable*
▶ The formation of a small amount of precipitate during the layering procedure is unavoidable and will not greatly affect results.
unaware adj 気付かないで syn *ignorant*
▶ Most chemists are almost completely unaware of the nature of polymorphism and the potential usefulness of the knowledge of this phenomenon in research.
unbound adj 他の物質と結合していない syn *untied*
▶ Transformation of the unbound portion of the aromatic ligand invariably alter the π acidity of the ligand and can dramatically shift the reduction potential.
uncertain adj 不確かな syn *doubtful, questionable, unreliable*
▶ Solid materials were obtained earlier by Ruff and by Peacock from the reaction of SbF_6 with sulfur, but the exact nature of these solids is uncertain.
uncertainty n 不確実性 syn *ambiguity*
▶ Reported equilibrium constants for a particular complex frequently differ by substantially more than the assigned uncertainties.
uncertainty principle n 不確定性原理
▶ The average time that the excitation is associated with one molecule is, from the uncertainty principle, of the order of $\triangle t \approx h/\triangle E$.
unchanged adj 不変の syn *stable, sustained*
▶ A mixture of ethanol and water boils

unchanged when the water content is 4 per cent and the temperature 78 ℃.

uncharged adj 電荷を帯びていない

▶ Formamide and dimethylformamide are particularly useful solvents for synthetic work in which both ionic and uncharged organic compounds are involved.

unclear adj 不明瞭な syn *ambiguous, obscure, uncertain*

▶ Many details of the preparative reaction mechanism remain unclear.

uncoil vt 巻いたものをほどく syn *unfold*

▶ Denaturation uncoils the protein, destroys the characteristic shape, and with it the characteristic biological activity.

uncommon adj めったにない syn *exceptional, rare, scarce*

▶ Simple reactions in the gas phase are uncommon, most mechanisms being complex.

uncomparable adj 匹敵するもののない

▶ The calorimetric method is one of a very few techniques that offer uncomparable criteria of overall purity, without regard to the number and identity of impurities.

unconsumed adj 消費されていない syn *remaining, unused*

▶ Sometimes, larger amounts of the reagent are used, and the relative amounts of the two reactants, ordinary and labeled, left unconsumed are measured.

uncontaminated adj 汚されていない syn *clean, pure, sterile*

▶ In the Sn^{II} series, $SnSO_4$ is a stable colorless compound which is probably the most convenient laboratory source of Sn^{II} uncontaminated with Sn^{IV}.

unconventional adj 型にはまらない syn *unfamiliar, unusual*

▶ The highly reduced halides of the early transition elements, with their strongly metal—metal bonded clusters, chains, and sheets in compounds such as Sc_7Cl_{12}, Sc_5Cl_8, and ZrCl, provide numerous examples of such unconventional chemistry.

uncoordinated adj 配位されていない

▶ The Pd–O distance is 1.997(1) Å, while the uncoordinated oxygen atoms reside 3.065(2) Å from the metal center.

uncorrected adj 補正されていない

▶ Melting points were determined using a Büchi 535 melting point apparatus and are uncorrected.

uncoupling n 脱共役

▶ Despite the medical dangers of chemically mediated uncoupling of oxidative phosphorylation, there are examples of deliberate and controlled uncoupling that occur naturally.

uncover vt 明らかにする syn *disclose, discover*

▶ Like other theories, mechanisms are subject to change as new information is uncovered or as new concepts are developed in related areas of science.

undeformable adj 変形しない

▶ The small, highly charged hard Al^{3+} ion can exist, provided that it is combined with similarly undeformable hard anions such as F^-, OH^-, and O^{2-}.

undemanding adj 多くを要求しない syn *tolerable*

▶ The open-chain ligand triethylenetetramine is flexible and undemanding.

undeniable adj 明白な syn *certain, obvious*

▶ Proton transfer is an elementary chemical reaction of undeniable importance.

under prep **1** …の下に syn *directly below*

▶ The *ab* crystal face is preferentially nucleated at the organic surface because the electrostatic accumulation of Ca ions under carboxylate headgroups results in a positively charged layer that is complementary to the hexagonally packed Ca layer of the vaterite *ab* face.

2 …のもとに

▶ Under the same conditions, chlorobenzene completely fails to react.

3 中の syn *subject to*

▶ The solid under investigation is compressed into flat, cylindrical pellets which after weighing are arranged in a small pile under pressure between a pair of metal electrodes.

underdeveloped adj 十分に開発されていない syn *imperfect, incomplete*

▶ In comparison, synthetic solid-state microwave chemistry is relatively underdeveloped.

underestimate vt 過小評価する syn *discount*

▶ The importance of the vapor-phase systems can hardly be underestimated in regard to increasing our understanding of the complexes in that they represent a reasonable approximation to the theoretical model of a single donor molecule interacting with a single acceptor molecule.

underestimation n 過小評価

▶ The first failure of the simple Lindemann mechanism arose because of an underestimation of the rate of activation.

undergo vt こうむる syn *be subjected to, experience, suffer*

▶ Borazine underwent a dehydrocoupling reaction to form polyborazylene ($[B_3N_3H_4]_n$) after ca. 48 h at 70 ℃, and polyborazylene lost H_2 to form ceramic BN above 900 ℃.

underlie vt 基礎となる syn *support*

▷ Underlying both chemical kinetics and thermodynamics are the more detailed theories of statistical mechanics and the kinetic-molecular theory.

underlying adj **1** 基礎をなす syn *basic, fundamental*

▶ The development of the faces of a crystal is determined by the underlying symmetry characteristic of that particular crystal.

2 下にある syn *lying below*

▶ We say that s electrons penetrate these underlying filled shells.

underneath adj 真下の syn *below, beneath*

▶ The degraded portions are removed by an oxygen plasma to expose a second layer underneath.

underpin vt 支える syn *sustain*

▶ This article describes the solid-state chemistry of intercalation compounds that underpins a revolutionary new rechargeable lithium battery.

underscore vt 強調する syn *emphasize, stress*

▶ The importance of the octahedron and tetrahedron in extended structures is underscored by the multitude of solids that can be constructed by sharing corners and edges of these shapes.

understand vt 理解する syn *comprehend, know, recognize*

▷ Understanding these systems becomes critically dependent on advanced surface probes, and there is perhaps no subject in contemporary catalysis in which these techniques have had a greater influence.

understandable adj 理解できる syn *clear, intelligible, unambiguous*

▶ Throughout our study of polynuclear hydrocarbons, we shall find that the matter of orientation is generally understandable on the basis of this principle: of the large number of structures contributing to the intermediate carbocation, the important ones are those that require the smallest sacrifice of resonance stabilization.

understandably adv 当然のことながら syn *clearly, discernibly, distinctly*

▶ Understandably a voluminous literature has accumulated, including volumes of catalogued properties and several recent books devoted entirely to the effects of radiation on plastics.

understanding n 知識 不加算名詞であるにもかかわらず，an understanding の表現も用いられる． syn *awareness, comprehension*

▶ One of the primary skills that a chemist brings to a materials research group is an understanding of synthesis.

understate vt 控えめにいう

▶ The importance of the reaction should not be understated.

undertake vt 着手する syn *begin, start*

▶ A more complete investigations of these *n*-alkanes is currently being undertaken.

undertaking n 通常単数形で，仕事 syn *effort, project, task, work*

▶ To open up a new field of research is a formidable undertaking in any area, but those involved in chemical education are subject to strong and unusual influences.

undesirable adj 望ましくない syn *unacceptable, unwanted*

▶ Disulfide bonds stabilize proteins, but they may also be undesirable, so cells contain reducing agents that prevent or reverse this reaction.

undesired adj 望ましくない syn *undesirable, unwelcome*

▶ An inhibitor retards or stops an undesired chemical reaction, such as corrosion, oxidation, or polymerization.

undetectable adj 検知不可能な syn *indiscernible*

▶ Although chlorine and hydrogen react with one another at room temperature, the rate is so slow as to be virtually undetectable.

undisputed adj 明白な syn *accepted, unmistakable*

▶ Despite the undisputed importance of organic chemistry in physical, biological, and medical sciences, few attempts have been made to summarize the thermodynamics of organic substances.

undissociated adj 解離していない

▶ Salts are partially but not completely dissociated into ions in aqueous solution and often are largely undissociated.

undissolved adj 溶解していない

▶ The presence of undissolved silica may not be noticed from a visual inspection of a glass sample, whereas it would be obvious immediately under the microscope.

undisturbed adj 平静な syn *motionless, stationary*

▶ Acetylsalicylic acid can be crystallized by dissolving it in ether, adding an equal volume of petroleum ether, and letting the solution stand undisturbed in an ice bath.

undoubtedly adv 疑いなく syn *certainly, unquestionably*

▶ The autoxidation of olefins usually produces a very complex mixture of substances, but the initial reaction product is undoubtedly a peroxide capable of liberating iodine from potassium iodide.

undue adj 過度の syn *excessive*

▶ We deliberately avoid higher temperature and undue exposure to light.

unduly adv 過度に syn *excessively*

▶ Some decomposition may occur if acetylsalicylic acid is recrystallized from a solvent of high boiling point or if the boiling period during recrystallization is unduly prolonged.

unequal adj 同等でない syn *unbalanced*

▶ All crystals of the monoclinic system have three unequal axes, one being at right angles to the other two, which are not at right angles to each other.

unequaled adj 無類の syn *excellent, notable, splendid*

▶ The sequence information coded into the base-pairing of the biopolymer provides unequaled structure-building capabilities based on complementary hydrogen bonding.

unequivocal adj 確実な syn *certain, definite, specific*

▶ Unequivocal classification of a crystal is generally not possible by means of morphological consideration alone.

unequivocally adv 確実に syn *clearly, definitely, perfectly*

▶ Only for $MnFe_2O_4$ and $NiCo_2S_4$ are there good, modern studies which unequivocally confirm the normal spinel structure.

uneven adj 一様でない syn *irregular*

▶ Increased loadings in grafted materials give reduced adsorption, which is attributed to pore blockage caused by the uneven distribution of silanes around the pore openings.

unexplained adj 原因不明の syn *mysterious*

▶ Although the combined steric and resonance factors can thus apparently account for the gross features of the dissociations of substituted ethanes into free radicals, many of finer details are as yet unexplained.

unexpected adj 予期しない syn *surprising, unanticipated, unforeseen*

▶ The reaction of certain substrates demonstrated an unexpected sensitivity to water.

unexpectedly adv 思いがけなく syn *extraordinarily, unusually*

▶ Like many other facts of nature, the insolubility of hydrocarbon liquids in water is easy to observe and describe but unexpectedly difficult to explain.

unexplored adj 未調査の syn *strange, unfamiliar, unknown*

▶ A relatively unexplored field of research in superconducting materials has been inorganic oxide system.

unfavorable adj 都合の悪い syn *adverse, demanding*

▶ While we can be sure from the laws of thermodynamics that an unfavorable energy change will not occur spontaneously, all we can say about a favorable one is that it might happen.

unfit adj 不適当な syn *inappropriate, undesirable*

▶ The referee said that the paper is nothing but nonsense, unfit even for reading before the Society.

unfortunately adv 不運にも syn *badly, regrettably*

▶ Unfortunately, many fundamental chemical techniques are not precise enough to be useful in establishing the purity of compound beyond about 0.1 mole % contamination.

uniaxial adj 一軸性の

▶ Smectic B phase is uniaxial with true hexagonal symmetry associated with dynamic rotational disorder about the sixfold axis.

unicellular adj 単細胞の

▶ Stem cells have many of the features of a primitive unicellular amoeba, so in some senses differentiation is like evolution.

unicellular organism n 単細胞生物

▶ The stability of Si–O–Si units in water gives rise to hydrated inorganic polymers that are molded into elaborate shapes by many unicellular organisms.

unidentified adj 未確認の syn *mysterious, unfamiliar, unknown*

▶ Most organic materials become colored when irradiated, but responsible centers are largely unidentified.

uniform adj 一様な syn *homogeneous*

▶ The hydrogen–chlorine system seems to behave in just the same way as the pure chlorine system, in that uniform temperature and pressure are established.

uniformity n 均等性 syn *homogeneity*

▶ Glasses intended for vision-correcting and such applications as lenses for cameras, microscopes, and other instruments must be of extremely high quality and uniformity to meet requirements for refractive index and light dispersion.

uniformly adv 一様に syn *consistently, steadily*

▶ DeCoursey undertook a very systematic effort to prepare perbomates by a variety of methods, with results that once more were uniformly negative.

unify vt 統一する syn *consolidate, unite*

▶ Electron-transfer theories from inorganic chemistry and charge-transfer views on organic systems have been unified to provide a basis for predicting oxidation–reduction reaction rates.

unimportant adj 重要でない syn *insignificant, meaningless*

▶ The geometry of the group 2 dihalide monomers is quite unimportant in determining the structure types of the corresponding solids.

unintentional adj 意図しない syn *accidental*

▶ The carrier concentration of ca. 10^{17} cm^{-3} in *trans*-polyacetylene results from unintentional doping, probably from catalyst residues that are chemically bound to the chain ends.

uninteresting adj つまらない syn *dull, monotonous, tedious*

▶ The history of chemistry contains many empirical rules, some of which have been incorporated and explained within theories, while others have been abandoned as accidental and uninteresting.

union n 結合 syn *combination, junction*

▶ When the two rings are joined through two

equatorial-type bonds, *trans*-decalin results, while an axial-equatorial union gives *cis*-decalin.

un-ionized adj イオン化していない

▶ The Fermi level position will dictate the percentage of acceptors that are ionized or un-ionized.

unique adj 独特な syn *unmatched, unparalled, unsurpassed*

▶ Nucleic acids are unique among the biopolymers in that they contain phosphate.

uniquely adv **1** 比類なく syn *especially, notably,*

▶ The Pariser-Parr-Pople self-consistent field molecular orbital method has emerged from this period as a uniquely useful aid to the chemist concerned with the molecular design of dyes.

2 唯一の結果だけを生じる syn *particularly*

▶ The space group was uniquely defined by the systematic absences as C_{2h}^{5}-$P2_1/n$.

uniqueness n 異常 syn *distinction, individuality*

▶ Water retains its uniqueness due to an open structure well past a temperature 35–45 ℃ where properties like compressibility and heat capacity show discontinuities.

unit n **1** 単位

▶ The quantum is the basic unit of electromagnetic energy.

2 構成単位 syn *component, constituent, part, section, segment*

▶ In contrast to chromium(II) chloride, which has an ionic structure similar to calcium chloride, the corresponding molybdenum and tungsten chlorides are based on $[M_6Cl_8]^{4+}$ units.

adj **3** 単位の

▶ B_2H_6 is spontaneously flammable; it has a higher heat of combustion per unit weight of fuel than any other substance except H_2, BeH_2, and $Be(BH_4)_2$.

unitary adj 単一の syn *individualistic, proper, unique*

▶ Although the other unitary systems, pure liquids and pure plastics, also scintillate, their efficiency is too low for them to be of practical use.

unit cell n 単位胞

▶ Two oxygens at $z = 1/2$ in the unit cell of rutile are collinear with titanium and form the axes of the octahedron.

unite vt 結合する syn *combine, unify*

▶ Sulfur unites directly with all elements except the noble gases, nitrogen, tellurium, iodine, iridium, platinum, and gold.

unity n 数量の単位としての **1**

▶ Air has a refractive index (1.000294) so nearly unity that it is commonly used as a convenient standard.

univariant adj 一変の

▶ The system water-steam is univariant because only one degree of freedom, either *P* or *T*, is needed to describe completely the system at equilibrium.

universal adj **1** 普遍的な syn *common, general, prevalent, widespread*

▶ This correlation between equilibrium constants and heats of reactions is neither rigorous nor universal.

2 万能の syn *comprehensive, extensive*

▶ Mass spectrometers act as sensitive universal detectors and are, hence, widely used in gas-phase kinetic experiments.

universality adj 普遍性 syn *generality, prevalence*

▶ Until salting out and in effects in solvents other than water have been examined, the universality of the theory cannot be judged.

universally adv 例外なく syn *invariably*

▶ Thus, in contrast to Ni(II), which forms tetrahedral, square planar, square pyramidal, trigonal bipyramidal, and octahedral complexes, Pd(II) and Pt(II) form complexes that are almost universally low-spin and square planar.

unknown n **1** 未知のもの

▶ Identification of an unknown is usually possible within 30 min of obtaining its measured X-ray powder pattern.

adj **2** 未知の syn *obscure, strange, unfamiliar, unidentified*

▶ Although compounds of Al[I] are stable as gaseous species such as AlCl(g), they are unknown as solids.

unlabeled adj 標識のついていない ☞ labeled

▶ The reaction of unlabeled molecules follows exactly the same stereochemical course as the reaction of the labeled ones.

unless conj …でない限り

▶ Infrared spectra were obtained with a Perkin-Elmer model 521 spectrophotometer unless otherwise indicated.

unlike adj **1** 同じでない syn *different, dissimilar, distinct*

▶ Like charges should not reside on atoms close together in a contributing structure, but unlike charges should not be greatly separated.

prep **2** …と違って これを用いた導入の句は，主文の主語を引き合いに出すものでなければならない． syn *different from, distinct from, in contrast with*

▶ Unlike Sr, Zn occurs in relatively high levels in meat.

unlikely adv ありそうもない syn *doubtful, improbable, unthinkable*

▶ It is unlikely that a compound that is sufficiently unstable to react at an appreciable rate in the solid state will survive the procedure necessary for its isolation.

unlimited adj 無限の syn *endless, immense, infinite, limitless*

▶ These examples illustrate the huge scope of this burgeoning area, with its unlimited permutations of metals and ligands.

unmarked adj 標識のついていない syn *unidentified*

▶ Around the perimeter of the ions, the electron density drops to the comparatively small value of 0.2 electrons/$Å^3$ and, in the unmarked regions, it is even less than this.

unmatched adj 並ぶもののない syn *unique, unparalled*

▶ Despite the great abundance of structural types and the widely varying stoichiometries which are unmatched elsewhere in chemistry, it is possible to classify the structures of silicate minerals on the basis of a few simple principles.

unmensioned adj 言及されていない

▶ Many important details of the behavior of conducting and semiconducting materials are left unmentioned in this brief section.

unmistakable adj 明白な syn *apparent, clear, evident, obvious, plain*

▶ A great many properties of both elements and their compounds exhibit an unmistakable periodicity based upon electron configuration.

unmodified adj 無修飾の syn *unaltered*

▶ In the absence of rate enhancement, the unmodified active sites are not less active than modified sites.

unnatural adj 人為的な syn *artificial*

▶ The L-enantiomer is unnatural and is not metabolized by animals.

unnecessarily adv 不必要に syn *unduly*

▶ The thermodynamic methods that are used to treat nonideal effects will, at first, appear to be unnecessarily devious.

unobservably adv 観察されないほどに

▶ The formation of water from gaseous hydrogen and oxygen is a spontaneous reaction at room temperature, although its rate may be unobservably small in the absence of a catalyst.

unoccupied adj 空の syn *empty, vacant*

▶ Energy absorptions in the ultraviolet and visible regions result in the transition of electrons from the low-energy bonding or nonbonding to the unoccupied high-energy antibonding orbitals.

unofficial adj 公式でない syn *unauthorized*

▶ A prevalent but unofficial procedure names alcohols as substitution products of carbinol, CH_3OH, a synonym of methanol.

unoriented adj 配向していない

▶ Above the melting temperature, unoriented crystalline polymers are amorphous and undergo plastic flow, which permits them to be molded.

unpaired electron n 不対電子 syn *odd electron*

▶ Magnetic susceptibility measurements have been used widely to determine the number of unpaired electrons in complexes.

unpleasant adj 不快な syn *disagreeable, unacceptable, unwelcome*

▶ Sometimes liquid hydrofluoric acid, a particularly unpleasant chemical to handle, is used in place of phosphoric acid because of their superior selectivity as a catalyst.

unprecedented adj 前例のない syn *uncommon, unexpected, unusual*

▶ Through modifications in the catalysts, hydrogenation can be carried out with unprecedented selectivity.

unpredictability n 予言できないこと syn *variability*

▶ Bridging ligands bearing terminal carboxylato donors, in contrast to those with N donor ligands, appeared much less attractive because of their unpredictability.

unpredictable adj 予知できない syn *indefinite, uncertain, unforeseeable*

▶ Different functional-group classes show clear preferences for specific hydrogen-bond patterns in their crystal structures, despite the presence of other unpredictable and nonspecific lattice forces.

unpredictably adv 予知できなく

▶ Ozonides, like most substances with peroxide bonds, may explode violently and unpredictably.

unprotected adj 保護されていない syn *exposed, naked, open, uncovered*

▶ The surface is exposed to the vapor of a group 13 element, and the impurity atoms diffuse into the unprotected area to form a layer of p-type Si.

unquestionably adv 明白に syn *clearly, certainly, evidently, undoubtedly*

▶ Borazine has often been referred to as inorganic benzene, while these is unquestionably some resemblance between the two compounds as far as bonding is concerned, there is little in the chemistry of the compound to suggest aromatic character as understood by organic chemists.

unravel vt **1** 解明する syn *explain, interpret, solve*

▶ Well-planned mechanistic experiments may help unravel the complexity of this catalyzed process.

2 ほぐす syn *disentangle, separate*

▶ Crystals of the fibrous minerals show the extraordinary property of being easily unraveled into fibers.

unreacted adj 反応していない

▶ Removal of solvent is expected to collapse the network of pores, gradually resulting in additional crosslinking as unreacted –OH and –OR groups come into contact.

unreactive adj 反応しない syn *inactive, inert*

▶ Despite both its stability and its kinetic sluggishness, dinitrogen is not totally unreactive.

unreal adj 架空の syn *artificial, unrealistic*

▶ Most of the small-scale work related to theories has been done with single crystals, which makes it somewhat unreal from the point of view of the engineer.

unrealistic adj 非現実的な syn *impractical, unrealizable, unreasonable*

▶ It is unrealistic and misleading to assign a radius to the chloride ion which is constant for all solid chlorides.

unrealizable adj 実現されない syn *impossible, inconceivable, unattainable, unrealistic*

▶ Slight distortions of the spherical charge distributions of the ions must be allowed, or the ionic bond becomes a physically unrealizable situation.

unreasonable adj 不当な syn *impractical, inappropriate, unrealistic, untenable*

▶ We can discard as unreasonable any dissociation reaction, if the ΔH's for breaking the bonds are greater than 30 to 35 kcal.

unrelated adj 無関係な syn *distinct, independent, separate*

▶ This is particularly dramatic example of the kind of evidence that gave rise to the idea that these apparently unrelated reactions proceed through the same intermediate.

unreliable adj 信頼できない syn *uncertain*

▶ Small amounts of manganese can be determined accurately and quickly by the colorimetric method, but if more than 1.5 percent of manganese is present the results are unreliable.

unremarkable adj 目立たない syn *ordinary, undistinguished, usual*

▶ $SnCl_4$, $SnBr_4$, and SnI_4 can be made by direct action of the elements and are unremarkable volatile liquids or solids comprising tetrahedral molecules.

unreported adj 報告されていない syn *undescribed, undisclosed, unpublished*

▶ It is important to recognize that the determination of the structural formula of a previously unreported compound is really a different type of problem from identifying it with some already known substance.

unresolved adj 未解決の syn *unanswered, uncertain, unsure*

▶ The more difficult question about the status of polymers higher than the dimer is almost completely unresolved.

unresponsive adj 鈍感な syn *dull, inert*

▶ Molecular nitrogen is so unresponsive to ordinary chemical reactions that it has been characterized as almost as inert as a noble gas.

unrivaled adj 無比の syn *leading, unique, unparalleled*

▶ CO has an unrivaled capacity for stabilizing metal clusters and for inserting M-C bonds.

unsatisfactory adj 不満足な syn *deficient, inadequate, insufficient*

▶ The state of our knowledge of compounds containing O-O links is still somewhat unsatisfactory, despite the fact that the compounds have been known and used analytically for a long time.

unsaturated adj 不飽和な

▶ Because of this property of readily combining with other substances such as the halogens, ethylene and related hydrocarbons are said to be unsaturated.

unsaturated bond n 不飽和結合 syn *unsaturated linkage*

▶ The position of the unsaturated bond is always indicated by citing the lower of the two numbers of the carbons between which the unsaturated bond is situated.

unsaturated compound n 不飽和化合物

▶ Alkynes behave in some ways as typical unsaturated compounds with two double linkages.

unsaturation n 不飽和 ☞ saturation

▶ Whereas the two tests above are specific to sterols, the test with tetranitromethane is a generally applicable method of detecting unsaturation.

unsensitized adj 増感されていない

▶ Unsensitized silver halide emulsions are most responsive in the blue region of the spectrum and, thus, do not correctly represent the light spectrum striking them.

unshared electron pair n 非共有電子対 syn *lone pair, nonbonding electron pair*

▶ The band at 2770 Å for methyl ethyl ketone is due to excitation of an electron from one of unshared electron pairs on oxygen.

unsolvated adj 溶媒和していない

▶ Unsolvated LiO_3 has not been prepared, but the ammine $LiO_3{\cdot}4NH_3$ is known.

unspecified adj 特に指示していない syn *indefinite, nonspecific*

▶ Just as ［*uvw*］ refers to a direction which is unspecified, so（*hkl*）refers to a plane which is unspecified.

unstable adj 不安定な syn *unpredictable, unreliable, variable*

▶ Chloric acid is an unstable acid that, like its salts, is a strong oxidizing agent.

unstrained adj ひずみのない syn *relaxed, steady*

▶ The simple polyethers are particularly stable, because they contain only strong C-O, C-C and C-H bonds, which are unstrained.

unstretched adj 引き伸ばされていない

▶ In the unstretched material, the polymer chains are twisted more or less at random.

unsubstantiated adj 実証されていない syn *questionable, uncertain*

▶ An early claim for KrF_4 remains unsubstantiated.

unsubstituted adj 未置換の

▶ Chloroacetic acid is 100 times as strong as acetic acid, dichloroacetic acid is still stronger, and trichloroacetic acid is more than 10,000 times as strong as the unsubstituted acid.

unsuccessful adj 不成功の syn *ineffective, unfortunate, useless, vain*

▶ All subsequent efforts to confirm his work were entirely unsuccessful.

unsuitability n 不適切 syn *impropriety*

▶ The general unsuitability of simple VB theory to deal with electronic spectra is relevant here as elsewhere.

unsurpassed adj 卓絶した syn *leading, supreme, unparalleled*

▶ Their unsurpassed advantage for the study of intermolecular forces is that it is possible to actually manipulate the orientation and degree of organization of molecules in a monolayer assembly and to control directly the approach of molecules to each other by variation of the surface area.

unsuspected adj 思いもよらない syn *unaware*

▶ The rewards of elegant and skillful synthetic programs are supplemented by an unusual number of chance discoveries and totally unsuspected reactions.

unsymmetrical adj 非対称的な ☞ asymmetric

▶ An interesting problem arises in the addition of unsymmetrical reagents, such as HBr, to unsymmetrically substituted alkenes, such as propene.

unsystematic name adj 非系統名 syn *common name, trivial name*

▶ The simple alkanoic acids have long been known by descriptive but unsystematic names that correspond variously to their properties, odors, or natural origin.

untenable adj 維持し得ない syn *unsupportable, unsustainable*

▶ The laws of classical physics are wholly untenable when the responses of particles to small transfers of energy are examined.

until conj …まで 文頭では till ではなく until を用いる.

▶ Until recently, the most successful O_2 carriers were cobalt(III) complexes.

untouched adj 手を付けていない syn *intact*

▶ As with all synthetic methods, emphasis is on the development of highly selective reagents, which will operate on only one functional group in a complex molecule and leave the other functional groups untouched.

unusable adj 役に立たない syn *impractical, ineffective, useless*

▶ Unusable by-products from many chemical and metal-processing operations often contain toxic or polluting materials that become environmental threats.

unused adj 使い残された syn *fresh, pristine, untouched*

▶ Because of the unused bonding capability of the atoms at the surface, chemistry here can be qualitatively different from that of the same reactants brought together in solution or in the gas phase.

unusual adj 異常な syn *atypical, exceptional, uncommon*

▶ The NiAs structure is unusual in that the anions and cations have the same coordination number but do not have the same coordination environment.

unwanted adj 望ましくない syn *undesirable, unwelcome*

▶ It is difficult to prevent the formation of unwanted impurity phases, including non-superconducting copper oxides, hydroxides, and carbonates due to reaction with air and products due to reactions with the crucibles used.

unwritten adj 文字に表していない

▶ The unwritten terms represented by the dots are assumed to be too small to be significant.

unyielding adj 堅い syn *hard, rigid*

▶ Phase transformations generally involve a change in the specific volume which, if the present phase is unyielding, will give rise to stresses and strains.

up adv **1 up and down** 上下に

▶ Molecules point alternately up and down from each hexagon, cages being left between hexagons.

up to (prep) 至るまで syn *equal to*

▶ Selenium forms a variety of halides: fluorides up to SeF_6, and chlorides and bromides up to SeX_4, but no iodides.

upcoming adj 近づく syn *forthcoming*

▶ Currently, new flu vaccines are generated each year based on the three main flu virus strains that are expected to be prevalent during the upcoming flu season.

upfield adj 高磁場への ☞ downfield

▶ In aromatic annulenes, the protons inside the ring experience an upfield shift and the outer protons experience a downfield shift.

uphill adj 上り坂の

▶ The autoprotolysis of water is a reaction which is uphill by some 20 kcal/mol, and this process, therefore, has a rate constant of the order of 10^{-5} sec^{-1} despite the fact that it is a simple proton transfer between two electronegative oxygen atoms.

upon prep **1** …の上に syn *on*

▶ Liquids are usually evaporated upon the water bath.

2 …によって syn *on*

▶ Upon intercalation, the graphite layers move apart somewhat, though less than expected as estimated from the diameter of the potassium ion.

upper adj 上のほうの syn *higher*

▶ The butyl bromide is in the upper layer since the aqueous solution of inorganic salts has a higher specific gravity.

upper atmosphere n 高層大気

▶ Our ability to model has allowed us to predict a set of photochemical reactions in the upper atmosphere.

upper limit n 上限 ☞ lower limit

▶ The recommended upper limit of dietary intake of cholesterol in an adult human is less than 400 mg per day.

uppermost adj 最上の syn *highest*

▶ Oxidation of the central metal atom will result in removal of electrons from the uppermost part of the d_{z^2} band and hence result in a net increase in bonding.

upset vt **1** ひっくり返す syn *invert*

▶ There were reactions in which this general inductive order seemed to be upset.

2 くつがえす syn *overcome, overwhelm*

▶ Supersaturation of the hot solution may be upset by rapid cooling, stirring, or even a slight motion of the flask.

upstream n 上流 ☞ downstream

▶ To transcribe a particular stretch of sequence, RNA polymerase binds to the DNA at a specific site, called a promoter, just upstream of the transcriptional start site.

uptake n 捕捉 syn *acquisition, gain*

▶ Obviously, the average valence state of copper changes upon the uptake or removal of oxygen.

urgently adv 緊急に syn *intensely, vigorously*

▶ This is an important matter that needs to be urgently addressed.

urinary adj 尿の

▶ The inherited disease phenylketonuria, which causes severe mental retardation, is characterized by the urinary excretion of phenyl pyruvate, lactate, and acetate.

usability n 有用性 syn *usefulness, utility, utilization*

▶ Several studies provide evidence for marked differences in usability of different types of tailor-made and commercial Q dots for live cell and animal imaging.

usable adj 使える syn *practical, useful*

▶ Tantalum is usable well in excess of 1200 ℃ and, with care in design and sealing, to at least 40-atm internal pressure.

usage n 用法 syn *convention, practice, use*
▶ Current usage tends to apply the term mole to an amount containing Avogadro's number of whatever units are being considered.

use n **1** 利用 syn *application, employment*
▶ Another aspect of stereochemistry concerns control of the molecular configuration of high polymer substances by use of appropriate catalysts.

2 make use of 利用する syn *take, use, utilize*
▶ In discussing the electrode reactions, we have made little use of the fact that the electrolyte is sodium chloride.

3 役に立つこと syn *usability, usefulness, utility*
▶ Conventional NMR measurements on solids give broad, featureless bands which are of little use for structural work.

vt **4** 使用する syn *employ, utilize*
▷ By using a flow technique, concentrations of reactants or products can be determined at various positions along the tube, corresponding to various times.

5 use up 使い切る syn *consume, exhaust, expend*
▶ When an animal eats more carbohydrate than it uses up, it stores the excess: some as the polysaccharide glycogen but most of it as fats.

used aux **used to＋不定詞** 以前は…であった syn *accustomed to*
▶ Moving down group 3, coordination numbers greater than 6 become the rule rather than the exception as used to be thought.

useful adj 有用な syn *advantageous, practical, usable*
▶ Nitroalkanes are a particularly useful source of stabilized carbanions for asymmetric addition to electron-donor alkenes.

usefully adv 有効に
▶ Because heavy atoms are most readily detected, EXAFS has been usefully employed to learn the immediate chemical environment of transition metal atoms as they occur in biologically important molecules.

usefulness n 有用性 syn *advantage, applicability, utility, value*
▶ Much of the usefulness of hydroboration-oxidation lies in the unusual orientation of the hydration.

useless adj 役に立たない syn *impractical, ineffective, unsuccessful*
▶ Methyl Orange which changes color at pH = 2.9–4.0 is useless for the titration of acetic acid with sodium hydroxide, but phenolphthalein which change color at pH = 8.3–10 should give a good result.

usher vt 先導する syn *direct, guide, lead*
▶ The demonstrated success of these studies has ushered in a new era of supermolecular chemistry aimed toward the design of organized molecular assemblies as artificial receptors to capture substrates by the use of intermolecular forces.

usual adj **1** 通常の syn *normal, ordinary, typical*
▶ The usual P−H bond is some 20 kcal weaker than the usual C−H bond.

2 as usual 例のとおり syn *usually*
▶ As usual, we are assuming that the vapor behaves ideally.

usually adv 一般に syn *as a rule, customarily, mainly, predominantly*
▶ Since arsenic is usually found in sulfide minerals, it is no surprise that a rich series of sulfides can be made in the laboratory.

utility n **1** 有用 syn *advantage, usefulness*
▶ The relation of the fugacity to the pressure is sometimes approximate but gives results of practical utility.

adj **2** 実用的な syn *practical*
▶ Where a general utility manometer is required, it is well to purchase a simple and inexpensive U-tube and fill and mount it.

utilization n 利用 syn *application, employment, use*
▶ Utilization of dynamic techniques, such as isotopic tracers, make it possible to determine the relationships between surface and other species.

utilize vt 利用する syn *apply, employ, use*
▶ An important method of synthesis of alkyl-

benzenes utilizes an alkyl halide as the alkylating agent together with a metal halide catalyst, usually aluminum chloride.

V

vacancy n **1** 空き syn *emptiness*

▶ The effective nuclear charge is greatest in elements which have a single vacancy in their valence shell, i.e., in the halogens.

2 空位 syn *vacant site*

▶ The predominant defects in AgCl are Frenkel defects, i.e., interstitial Ag^+ ions associated with Ag^+ ion vacancies.

vacant adj 空の syn *empty, unoccupied*

▶ Tungsten trioxide may be considered as having the perovskite structure of $CaTiO_3$ with all of the calcium sites vacant.

vacant orbital n 空軌道

▶ Vacant d orbitals permit some delocalization of electron density from the formally nonbonding pairs on neighboring atoms; thus, these nonbonding pairs become slightly bonding.

vacate vt 空位にする syn *depart, leave*

▷ The dissociation step of a bi- or multidentate ligand would usually be easily reversible, because the coordinating atom remains in the vicinity of the vacated spot in the metal atom coordination sphere.

vaccination n 予防接種

▶ The dosage of antigen can be rigorously controlled in vaccination but may vary widely in natural infections.

vaccine n ワクチン

▶ Commercial vaccines are composed of a polysaccharide extracted from an organism which is linked to a carrier protein.

vacuum n 真空 syn *emptiness*

▶ The X-rays pass out from the tube through windows which must be strong enough to hold a vacuum but which must absorb X-rays only slightly.

vacuum distillation n 減圧蒸留

▶ Vacuum distillation is not confined to the purification of substances liquid at ordinary temperatures but often can be used to advantage for solid substances.

vacuum sublimation n 真空昇華

▶ The Schiff bases were readily purified by vacuum sublimation, which if necessary was repeated until consistent and narrow phase transitions were obtained by differential scanning calorimetry.

vague adj 漠然とした syn *ambiguous, obscure*

▶ Hardness is a term ordinarily used with a rather vague meaning; as applied to crystals it refers to the resistance offered to the production of a scratch on a smooth surface.

vain n **in vain** むなしく syn *fruitlessly*

▶ It has been attempted repeatedly to separate ethylene from benzene but usually in vain.

valence n 原子価

▶ We define the valence of a bond (s_{ij}) between atoms i and j such that

$$V_i = \Sigma s_{ij}$$

for all j, where V_i is the valence of atom i.

valence angle n 原子価角 syn *bond angle*

▶ Formation of a four-membered ring of carbon atoms can only be achieved with substantial distortion of the normal valence angles of carbon regardless of whether the ring is planar or nonplanar.

valence band n 価電子帯 syn *filled band*

▶ The electrons in the valence band are not free to move because they are involved in chemical bonding.

valence bond n 原子価結合　複合形容詞として使用

▶ Resonance is used to describe or express the true chemical structure of certain compounds that cannot be accurately represented by any one valence-bond structure.

valence electron n 価電子

▶ The extra valence electron of phosphorus atom is easily ionized, because it is not needed to bond the phosphorus atom to its four neighboring silicon atoms in the tetrahedral geometry of the atoms in the crystal.

valence shell electron pair repulsion model, VSEPR model n 原子価殻電子対反発モデル

▶ The VSEPR model is based, explicitly or implicitly, on the assumption that the core beneath the valence shell of the central atom A is spherical and, therefore, has no influence on the geometry.

valence state n 原子価状態

▶ In the case of MnO_2, the original solid structure remains essentially unchanged during the reduction process as Mn^{4+} is reduced to a lower valence state.

valence tautomer n 原子価互変異性体

▶ Cycloheptatriene and norcaradiene are a fascinating pair of valence tautomers.

valid adj 確実な syn *certain, real, true*

▶ Even if the assumption of a linear variation of volume of mercury with temperature is valid at ordinary temperatures, it may be questioned whether the linear dependence is valid down to the absolute zero.

validate vt 確認する syn *confirm, justify, verify*

▶ We have made a considerable effort to validate molecular mechanical models on a wide variety of test cases, including conformational energies, barriers, and intermolecular interactions.

validation n 確認 syn *proof, verification*

▶ In addition to providing validation for the catalytic cycle shown in Scheme 1, the stereospecificity of epoxide-expansion carbonylation is vital to its application in organic synthesis.

validity n 妥当性 syn *effectiveness, significance, value*

▶ In order that the result of any measurement may be of scientific or technical value, it is desirable to have some numerical estimate or measure of its validity.

valley n 谷 syn *hollow*

▶ In a typical allowed concerted reaction, there is a continuous valley running from reactant to product via the transition state.

valuable adj 有益な syn *helpful, instrumental, profitable, useful*

▶ At present, the type of solid-state reaction most valuable for the synthetic organic chemist is the photochemical dimerization of unsaturated compounds.

value n **1** 価値 syn *merit, usefulness*

▶ The protective value of any film depends on its being 100% continuous, i.e., without holes or cracks, since it must form an efficient barrier to molecules of atmospheric water vapor, oxygen, etc.

2 of value 重要な syn *important, significant*

▶ The preparation of aldehydes, ketones, and carboxylic acids by the oxidation of alcohols is of great value in organic synthesis.

3 数値 syn *assessment*

▶ The conductivity values, e. g., 10^{-3} ohm^{-1} cm^{-1} for Na^+ ion migration in β-alumina at 25 ℃, are comparable to those observed for strong liquid electrolytes.

vt **4** 評価する syn *appreciate, find valuable*

▶ Talc is valued for the softness, smoothness, and dry lubricating properties.

van der Waals force n ファンデルワールス力 ☞ intermolecular force

▶ The forces between the chains in the crystallites of polyethylene are the so-called van der Waals forces, which are the same forces acting between hydrocarbon molecules in the liquid and solid states.

van der Waals radius n ファンデルワールス半径 ☞ atomic radius, covalent radius, ionic radius

▶ With respect to other atoms to which it is not bonded every atom has an effective size, called its van der Waals radius.

vanish vi 消え失せる syn *disappear, dissipate*

▶ If the ions of the electrolyte are infinitely far apart, then their interaction vanishes.

vanishing adj 消え失せそうな syn *disappearing*

▶ As the pressure is reduced at constant temperature, the dissociation of an atom into a positive ion and an electron becomes greater, until finally at vanishing pressure the dissociation can become complete, even at ordinary temperatures.

vanishingly adv 消え失せそうに

▶ The screening contribution equals the

number of electrons in the correlated orbital on F atom regardless of the symmetry of the orbital in this limit of vanishingly small forces.

vapor n 蒸気 syn *fog, mist, steam*

▶ Not only does mercury have a very low melting point but the liquid is very volatile, giving a monomeric vapor.

vapor deposition n 蒸着 複合形容詞として使用

▶ A vapor-deposition technique is used to form a tightly adherent aluminum coating from 0.2- to 1-mil thick on titanium and steel.

vaporization n 蒸発 syn *evaporation*

▶ To understand the vaporization process clearly, it is necessary to allow for the nature of the gaseous species.

vaporize vi **1** 蒸発する syn *evaporate*

▶ A more volatile part of the liquid vaporizes, and a less volatile part of the vapor condenses.

vt **2** 蒸発させる syn *evaporate*

▶ Only the difference between the binding energies of the species in the vapor and those in the crystal must be supplied in order to vaporize the solid.

vapor phase n 蒸気相 複合形容詞として使用

▶ In any vapor-phase deposition technique, it is necessary to provide a volatile species of each component required for a given compound.

vapor-phase chromatographic analysis n 蒸気相クロマトグラフ分析

▶ Vapor-phase chromatographic analysis is extraordinarily useful for detection of minute amounts of impurities when these are separated from the main peak.

vapor pressure n 蒸気圧

▶ The vapor pressure of a spherical droplet of a liquid may be very considerably larger than that of the bulk liquid.

variability n 変動性 syn *inconsistency, irregularity*

▶ The resonances of N–H protons are not so easily identifiable; considerable variability arises from differences in degree of hydrogen bonding.

variable n **1** 変数 記号にはイタリック体を用いる.

▶ In thermodynamics, a reversible change is one that can be reversed by an infinitesimal modification of a variable.

adj **2** 変動しうる syn *changeable, uncertain, unfixed*

▶ Since the structure can exist with incomplete filling of holes, the formulas of these clathrates are variable.

variance n **1** 可変度

▶ The number of degrees of freedom, or variance F, of a one-component system is given by $F = 3 - P$, where P is the number of phases in equilibrium.

2 at variance 不一致で syn *in conflict, in disagreement*

▶ Such an order is completely at variance with the experimental facts.

variant n 変形 syn *alternative, modification, variation*

▶ The synthesis of ethyl 1-naphthoate illustrates the preparation of a carboxylic ester by a variant of the Grignard carboxylation route to a carboxylic acid.

variation n **1** 変化 syn *change, difference*

▶ An electron density map is a plot of the variation of electron density throughout the unit cell.

2 変形 syn *modification, variant*

▶ Many variations of the original BET have been suggested in an endeavor to extend the range of applicability.

variation method n 変分法

▶ The variation method not only allows reliable calculations to be made on complex atoms, but it also provides considerable insight into the nature of atoms and of chemical bonds.

varied adj 種々の syn *diverse, diversified*

▶ A series of thymidine nonpolar analogues that have varied size and a series that have varied shapes have been studied.

variety n **1** 多様性

動詞が続くとき,集合を強調するときは単数形,個々のものを強調するときは複数形を用いる.

syn *diversity, multiplicity*

▶ A variety of chemical wastes such as detergents, cleaning fluids, paints, and vanishes make their way into the trash.

2 種類 syn *kind, species, type*

▶ HgO exists in a red and a yellow variety.

various adj **1 色々な** 多様性を強調しる表現で，これに続く名詞は複数形である． syn *a number of, many, numerous*

▶ Various spectroscopic and chemical investigations are now being planned in an attempt to confirm the existence of the postulated conformers.

2 それぞれの syn *different, distinct, individual*

▶ The free energy of a crystal will depend upon its volume and the numbers of molecules in its various surfaces, edges, and corners.

variously adv **様々に**

▶ The systems which contain hydrocarbon chains as major constituents include polyethylene, variously substituted alkanes, and biological membranes.

vary vi **1 変わる** syn *change, differ, range*

▶ Platinum alone forms all four tetrahalides, and these vary in color from the light-brown PtF_4 to the very dark-brown PtI_4.

vt **2 変える** syn *alter, change, modify*

▶ Each of the components of the simple Friedel-Crafts alkylation just given can be varied.

varying adj **連続的に変化する** syn *variable*

▶ The linear silicone polymers are liquids of varying viscosity depending on the chain length.

vast adj **膨大な** syn *enormous, huge, immense, tremendous*

▶ The vast majority of organic substances are compounds of carbon with hydrogen, oxygen, nitrogen, or the halogens.

vastly adv **大いに** syn *greatly, very much*

▶ Bicyclohexadiene is a known nonplanar molecule with vastly different bond angles and bond lengths than benzene.

vat dye n **建染め染料**

▶ Vat dyes are remarkably fast to washing, light, and chemicals.

vector n **1 ベクトル**

▶ A magnitude such as a velocity or force, which is not completely defined at any point unless its direction is given as well as its magnitude, is called a vector.

2 ベクター

▶ A vector is a DNA molecule into which foreign DNA molecules can be spliced and inserted into cells so that the recombinant DNA can be replicated.

vectorial adj **ベクトルの**

▶ The macroscopic dipole moment of a polyfunctional molecule is the resultant of the vectorial addition of all the group moments in the molecule.

vein n **1 鉱脈** syn *seam*

▶ Pyrite often occurs in massive lenses but may also appear in veins or in dispersed zones.

2 筋道 syn *course, manner, way*

▶ A polypyridyl ligand set was next pursued in hopes of accessing less electron-rich metal fragments. In this vein, a series of bpy- and terpy-based Re(II) complexes was synthesized.

velocity n **速度** syn *rapidity, speed*

▶ The average molecular velocity, being a measure of both the direction and the magnitude of the rate of molecular movement, is zero.

venture n **1 冒険的試み** syn *experiment, undertaking*

▶ The chances of success in this venture would have appeared slim indeed at that time to any reasonable chemist.

vt **2 意見を述べる** syn *advance, offer*

▶ The subject is a very complex one, and one may readily venture the opinion that much remains to be learned.

verifiable adj **立証しうる** syn *actual, factual, real, true*

▶ We then make a verifiable assumption, namely, that the reaction follows the simplest path from reactants to products.

verification n **確認** syn *demonstration, evidence, proof*

▶ As further verification, the addition of small

amounts of water to the reaction again generated the product in lower yield and higher enantioselectivity.

verify vt 確かめる syn *affirm, confirm, prove*

▶ To verify the quantum theory of magnetism, Stern and Gerlach passed narrow beams of vaporized atoms or molecules through a magnetic field of varying intensity.

versatile adj **1** 色々に使える syn *adjustable*

▶ Metal organic chemical vapor deposition is a versatile technique, and it is at various stages of development according to the materials being deposited.

2 変わりやすい syn *flexible, variable*

▶ Alkali metals form a fascinating variety of binary compounds with oxygen, the most versatile being Cs which forms nine compounds with stoichiometries ranging from Cs_7O to CsO_3.

versatility n 可能性 syn *adaptability, flexibility*

▶ Grignard reagents are probably the most extensively used of all organometallic reagents because of their easy preparation and synthetic versatility.

version n **1** 説明 syn *account, interpretation, understanding*

▶ We will give a more modern version, in terms of the concept of the activated complex.

2 変形 syn *variant, variation*

▶ The electron-pair domain version of the VSEPR model in which a lone-pair domain is assumed to be larger than a bond-pair domain is simpler and leads directly to an unambiguous prediction of the structures of AX_4E, AX_3E_2, and AX_2E_3 molecules.

versus prep …に対して

▶ Much more serious is the problem of coordination number 6 versus 8.

vertex n 頂点　複数形は vertexes, vertices syn *apex, peak*

▶ The cubic diamond structure contains carbon atoms which make use of four hybrid orbitals pointing to the vertices of a tetrahedron.

vertical adj 縦方向の syn *perpendicular, upright*

▶ Vertical size trends in the periodic table are easy to describe, and their explanation is for the most part simple and straightforward.

vertically adv 縦方向に syn *perpendicularly, upright* ☞ laterally

▶ The ratios among the dissolved components are quite constant throughout most of the oceans. However, the absolute values vary considerably from place to place, both laterally and vertically.

vertical transition n 垂直遷移

▶ Because nuclear motions are slow compared with electronic transition, the energy absorbed in going to the upper state is E_{op}, corresponding to a vertical transition.

very adj **1 the very** まさしくその syn *exact, particular, precise*

▶ The very assignment of definite partial charges to the individual atoms of so complex an electrical system as a molecule implies a physical picture that is doubtless a gross oversimplification.

adv **2** 大変に syn *exceedingly, extremely*

▶ A very important group of reactions of diazonium ions involve aromatic substitution by the diazonium salt acting as an electrophilic agent to yield azo compounds.

vesicle n ベシクル

▶ The vesicle has a limited life in the cell but acts as a transient store or waste collector as it crosses the cell.

vessel n 容器 syn *container*

▶ The decomposition of ozone usually occurs partly on the surface of the vessel and partly in the gas phase.

via prep **1** …を経て syn *through*

▶ Tungsten is obtained from the ore $CaWO_4$ via the trioxide, which can be reduced with hydrogen.

2 …によって syn *by means of*

▶ Alternative high-yield syntheses of these various boranes via hydride-ion abstraction from borane anions by BBr_3 and other Lewis acids have been devised.

viable adj 実行可能な syn *feasible, possible, practicable, tenable*

▶ Thin films of polythiophene are red in the doped state and deep blue in the undoped state, and they may provide viable alternatives to liquid crystal displays.

vibrate vi 振動する syn *oscillate*

▶ In all crystals, the atoms are vibrating with thermal motion, some more, some less, according to the temperature of the crystal and the strength of attraction of the atoms for each other.

vibration n 振動 syn *oscillation*

▶ Superimposed on the vibrations of the center of gravity of the molecules on a lattice are the vibrations of the atoms within the molecule relative to each other.

vibrational adj 振動の

▶ Absorption from about 5000 to 1250 cm^{-1} is generally associated with changes in the vibrational states of the various bonds and is relatively characteristic of the types of bonds present.

vibrational band n 振動バンド

▶ In contrast to pure rotational spectra, only one vibrational-absorption band has been treated for a given diatomic molecule.

vibrational energy n 振動エネルギー

▶ If the energy is transformed into vibrational energy, the probability that at some instant it will reside exclusively in some individual bond is so great that the molecule is almost certain to break up into radicals by rupture of that bond.

vibrational level n 振動準位

▶ Even when the vibrational heat capacity approaches the value Nk_B per vibration, most of the molecules are still in the $v = 0$ and $v = 1$ vibrational levels.

vibrationally adv 振動運動の上で

▶ In the chemical laser the H + Cl_2 reaction generates highly vibrationally excited HCl.

vibrational mode n 振動モード

▶ The principal selection rule for a vibrational mode to be Raman active is that the nuclear motions involved must produce a change in polarizability.

vibrational relaxation n 振動緩和

▶ At high pressures, vibrational relaxation will occur in the upper electronic state, and fluorescence will occur from the vibrational ground state.

vibrational spectroscopy n 振動分光法

▶ Vibrational spectroscopy is a sensitive, though not always reliable, diagnostic for the mode of bonding.

vibrational spectrum n 振動スペクトル

▶ The vibrational spectra of these aggregates contain a wealth of information about composition, structure, size, and shape.

vice versa adv 逆もまた同じ syn *conversely, the other way around*

▶ Most of the very intense bands found in the spectra of transition-metal complexes arise from the transfer of an electron from a ligand orbital to a metal orbital or vice versa.

vicinal adj 近接の

▶ Chlorine and bromine add to double bonds to form vicinal dihalides.

vicinity n 付近 syn *neighborhood*

▶ When an exciton is formed, the excited ion will interact differently with its neighbors, so that at first, the lattice in its vicinity is not in its configuration of minimum energy.

view n **1** 眺め syn *perspective, picture*

▶ Figure 1 is a view of the crystal structure of the yellow form in which the orientation has been changed so that the view is nearly normal to the $(\bar{1}0)$ face.

2 見解 syn *opinion, point of view*

▶ A direct corollary of this view was that the electron-repelling power of a methyl group should be increased by successive replacement of its hydrogen atoms by methyl groups.

3 in view of …の点から見て syn *because of*

▶ In view of the great spread in behavior of individual large crystals, it is surprising to find any simple kinetic equation applicable in the solid state.

vt **4** みなす syn *believe, consider, regard*

▶ The simple electrostatic treatment views

metal ions and ligands as rigid and undistorted.

5 調べる syn *examine, observe, regard*

▶ There is alternative way to view the stabilization of an ion by a solvent.

viewpoint n 観点 syn *point of view*

▶ The structural variation in the dimers has been rationalized from a purely electrostatic viewpoint also.

vigorous adj 激しい syn *energetic, strong*

▶ If a small amount of anhydrous aluminum chloride is added to a mixture of benzene and methyl chloride, a vigorous reaction occurs, hydrogen chloride gas is evolved, and toluene can be isolated from the reaction mixture.

vigorously adv 激しく syn *actively, strongly*

▶ When vigorously shaken with excess water, phopholipids form first lamellar phases and, eventually, single bilayer vesicles of diameter 25–50 nm.

violate vt 違反する syn *disregard, ignore*

▶ Unlike oxygen, nitrogen, and carbon, boron seems to violate the octet rule and be satisfied with a sextet of valence electrons, but this is not always the case.

violation n 違反 syn *disobeying*

▶ Unmixing does not involve a violation of the first law, but it would violate the second law.

violence n 猛烈さ syn *intensity, strength*

▶ If to the hydrogen–oxygen mixture we introduce a lighted match, the reaction will occur with explosive violence.

violent adj 激しい syn *acute, drastic, energetic, intense*

▶ Aqueous solutions of XeO_3 are quite stable but solid XeO_3, which is easily obtained from the solutions, is a sensitive and violent explosive.

violently adv 激しく syn *severely*

▶ $P(CN)_3$ is a highly reactive white crystalline solid which reacts violently with water to give mainly phosphorous acid and HCN.

viral adj ウイルス性の

▶ A wide range of viral nanoparticles have been utilized as templates for material fabrication

virial coefficient n ビリアル係数

▶ The second virial coefficients are due to molecular pair interactions; the other coefficients are due to higher-order interactions.

virial equation n ビリアルの式

▶ .A very convenient form of the equation of state, valid for all gases, is the power series known as the virial equation of state.

virion n ビリオン

▶ Fluorescent labeled tobacco mosaic virus particles have been used as templates for photoreduction of metals on the surface of the virion.

virtual adj 仮想の syn *calculated, effective*

▶ A virtual N–I transition temperature of 162° was obtained by extrapolation.

virtually adv 事実上 syn *almost, nearly, practically*

▶ An attractive feature common to virtually all polymer-supported reactions is that the species attached to the polymer can easily be separated from the other reaction products.

virtue n **1** 長所 syn *asset, strength*

▶ Classification according to functional groups has the virtue of providing rationale for the behavior of the vast numbers of organic compounds that are known.

2 by virtue of …によって syn *because of, on account of, owing to*

▶ Many reactions are irreversible by virtue of an insoluble or gaseous product.

virulent adj 病原性の

▶ Quinine is the longest known of the antimalarial drugs and is still today the most effective against some of the most virulent strains of malaria.

virus n ウイルス 複数形は viruses

▶ Viruses consist, in their simplest form, of a protein coat and nucleic acid.

virus-like particle n ウイルス様粒子

▶ Heterologous expression systems can give rise to high yields of virus-like particles.

viscoelastic adj 粘弾性の

▶ A material that is both elastic and plastic is said to exhibit viscoelastic properties.

viscosity n 1 粘性

▶ The macroscopic viscosity of silicone oils results from the tangling of the long polymer chains.

2 粘度 syn *viscosity coefficient*

▶ Water has a higher viscosity than benzene because its molecules bond together more strongly and this hinders the flow.

viscosity coefficient n 粘性率 syn *viscosity*

▶ From this rate of flow, and with the knowledge of the pressure acting and the dimensions of the tube, the viscosity coefficient can be calculated on the basis of a theory developed by Poiseuille.

viscous adj 粘性の syn *sticky, thick*

▶ These artificial tear solutions are based on viscous polymers, most commonly containing methyl cellulose, hydroxypropylmethylcellulose, or polyvinyl alcohol.

visibility n 目に見えること

▶ The phenomenon of fluorescence is shown by a small proportion of dyes, and such compounds are becoming increasingly important as high visibility colorants, as the emitting materials in dye lasers, and as the absorbing species in solar energy collectors.

visible n 1 可視領域

▶ The irradiation of KCl and NaCl with X-rays gives rise to an intense coloration in the crystal because of the production of strong absorption in the visible.

adj 2 可視的な syn *discernible, detectable, observable, perceptible*

▶ When more than one phase is formed from single crystals, the regions occupied by each may be macroscopically visible or they may range in size down to intimate mixtures in which the individual phases are detectable only by X-rays.

visible and ultraviolet light n 可視紫外光

▶ Absorption of visible and ultraviolet light produces changes in the electronic energy of molecules associated with excitation of an electron from a stable to unstable orbital.

visible spectrum n 可視スペクトル

▶ Azo compounds can be made with practically any color in the visible spectrum.

vista n 展望 syn *prospect, view*

▶ The answers to these questions will not only explain much of the bioinorganic chemistry of these processes but open a new vista in the chemistry of heterogeneous interface.

visual adj 目に見える syn *visible*

▶ The incorporation of up to 10 mol% LiCl led to a change of deposit morphology from microcrystalline to nanocrystalline and a change in visual appearance from metallic to black.

visualization n 視覚化

▶ Visualization was achieved by either UV fluorescence, acidic $KMnO_4$ solution and heat, ammonium molybdate solution and heat, or iodine vapor.

visualize vt 想像する syn *conceive, envisage, imagine*

▶ A type of catalyst-containing vesicle can be visualized in which the reactants would enter the vesicle and the product would be expelled into the bulk solution.

vital adj 肝要な syn *central, crucial, important*

▶ Besides inorganic substances, there are many organic catalysts that are vital in the life processes of plants and animals.

vitality n 活気 syn *activity*

▶ Nowhere is the vitality of materials chemistry more evident than in our emerging ability to control the arrangement of atoms from the nanoscale to the macroscale.

vitamin A n ビタミンA

▶ Biochemical reduction of the aldehyde carbonyl of retinal yields the important growth factor, vitamin A.

vitamin deficiency n ビタミン欠乏

▶ Vitamin D deficiency, leading to rickets, can be prevented in children by exposing them to sunshine or by feeding them cod liver oil which is rich in vitamin D.

vitreous adj ガラス状の

▶ Oxidation of silicon in air is not measurable

below 900°; between 950° and 1160° the rate of formation of vitreous SiO_2 rapidly increases, and at 1400° the N_2 in the air also reacts to give SiN and Si_3N_4.

vitreous state n ガラス状態

▶ An amorphous material, such as glass and a high polymer, changes from a brittle vitreous state to a plastic state at the glass transition temperature.

vivid adj 鮮やかな syn *clear, distinct, striking*

▶ A vivid example of the use of NMR in the study of motions within molecules is afforded by the fluorine resonance spectrum of 1,2-difluorotetrachloroethane.

void n **1** 空所 syn *space, vacancy*

▶ Sodium *p*-nitrophenolate penetrates the α-CD cavity to only a limited extent but is more deeply embedded in the larger β-CD void.

adj **2** 空の syn *empty, vacant*

▶ The unit cell is a pattern for the atoms as well as for the void spaces among the atoms.

volatile n **1** 揮発性物質

▶ Volatiles were evaporated in vacuum and the resulting solid was extracted with acetonitrile (20 mL).

adj **2** 揮発性の syn *evaporable, vaporizing*

▶ A volatile copper halide imparts a green color to the flame.

volatility n 揮発度

▶ Association of the molecules of primary and secondary amines through hydrogen bonding is significant and decreases their volatility relative to hydrocarbons of similar size, weight, and shape.

volatilization n 揮発

▶ The reduction of mixtures of metal oxides and boric oxide with carbon is not generally satisfactory owing to heavy losses of boric oxide by volatilization and contamination of the product with boron, carbon, and boron carbide.

volatilize vt 揮発させる

▶ Rhenium is extracted by roasting the flue dust of molybdenum plants to volatilize it as the oxide.

voltage n 電圧

▶ Piezoelectric materials are ceramic materials that will expand or contract when a voltage is applied to them.

voltammetric adj ボルタンメトリック

▶ Cyclic voltammetric studies on viologen-decorated mosaic virus nanoparticles showed the characteristic two successive, one electron, reversible steps of the methyl viologen moieties.

voltammetry n ボルタンメトリー ☞ cyclic voltammetry

▶ Voltammetry at the rotating platinum electrode was accomplished using conventional three-electrode cell design.

volume n **1** 容積 syn *size*

▶ Volume is a fundamental property of molecules that is important in understanding their structure, function, and interactions.

2 量 syn *amount, quantity*

▶ The largest volume use of perfumes is in soaps, lotions, shaving creams, and cosmetics.

3 本 syn *book*

▶ The volume is divided into four sections.

volume of activation n 活性化体積

▶ It follows that a rate constant increases with increasing pressure if the volume of activation is negative, that is, if the activated complex has a smaller volume than the reactants.

volumetric adj 容量測定の

▶ One of the most important factors governing the performance of a cell is the energy which can be stored per unit weight and volume, i.e., the gravimetric and volumetric energy density.

volumetric analysis n 容量分析

▶ Accurate results can be obtained by a volumetric analysis only when the instruments used are accurately calibrated.

voluminous adj 膨大な syn *enormous, substantial, tremendous, vast*

▶ In the last 20 years, the experimental work devoted to studies of molecular complexes has been voluminous.

vulcanization n 加硫

▶ A physicochemical change resulting from crosslinking of the unsaturated hydrocarbon chain of rubber with sulfur is called vulcaniza-

tion.

vulcanize vt 硬化させる

▶ Natural rubber can be vulcanized not only with sulfur but also with selenium, organic peroxides, and quinone derivatives, but these have limited industrial use.

W

wafer n ウェーファー syn *flake*
▶ A small wafer of single-crystal n-type Si is oxidized by heating it in O_2 or H_2O vapor to form a thin surface layer of SiO_2.

wall n 壁
▶ The solution is cooled at once in an ice bath, and the walls of the flask are scratched until the *p*-toluidine salt crystallizes.

walled adj 壁で囲った syn *enclosed*
▶ Single walled carbon nanotubes are emerging as ideal model systems for studying discrete and atomically regulated crystal growth.

want n **1** 欠乏 syn *deficiency, lack*
▶ The molecular orbitals are, for want of better models, approximated by linear combinations of atomic orbitals centered on the various nuclei.
vt **2** 欲する syn *desire*
▶ We want to know the factors influencing structural preferences and choices made not only in the dimer but also in higher order oligomers and the extended solids.

warm adj **1** 温かい syn *heated*
▶ The reaction mixture was filtered while still warm to remove salt by-product.
vi **2** 温まる
▶ The reaction mixture was stirred at −78 ℃ and allowed to warm slowly to room temperature overnight.
vt **3** 温める syn *heat*
▶ An unknown is warmed for a few minutes with alcoholic silver nitrate; halogen is indicated by formation of a precipitate that is insoluble in dilute nitric acid.

warming n 加熱
▶ If nickel is first saturated to atmospheric pressure with hydrogen at −196° and then allowed to warm up, the rate of warming is accelerated by the heat given off during the transition of physically adsorbed to chemisorbed hydrogen.

warn vt 注意する syn *advise, caution*
▶ Although these models are a useful framework for the interpretation of the experimental results, the reader is warned that they represent an oversimplified picture of micelle formation.

warning n 注意 syn *caution, notice*
▶ A word of warning should be given about experiments in which hydrogen is admitted rapidly to active nickel at low temperatures.

warrant vt 正当とする syn *approve, justify*
▶ Too often the physicist or physical chemist makes measurements of a precision that is not warranted by the purity of the sample under study.

wash n **1** 洗浄 syn *washing*
▶ No precipitate is absolutely insoluble, so that it is clear that every unnecessary excess of wash water causes harm by removing a fraction of the precipitate.
vt **2** 洗浄する syn *clean, cleanse*
▶ Blue-black needles separated almost immediately, and after 2 hours at room temperature the product was collected and washed on the filter with acetonitrile and ether.

washing n **1** 洗浄 syn *wash*
▶ Peroxides can be removed from ethers in a number of ways, including washing with solution of Fe^{2+} ion or distillation from concentrated H_2SO_4.
2 洗浄液
▶ About 100 mL of chloroform was shaken with 5 mL of concentrated sulfuric acid, washed with water until the washings were neutral to litmus, dried, and distilled in an all-glass apparatus through a Widmer column.

waste n **1** 浪費 syn *abuse*
▶ Any attempt to derive improved sets of van der Waals radii to predict or interpret intermolecular contact distances in condensed phases seems a waste of effort.
2 廃棄物 syn *garbage*
▶ Toxic chemical wastes are defined as those

which will cause death, disease, cancer, or genetic malfunctions in any organisms with which they come into contact.

vt **3** 浪費する syn *dissipate*

▶ Matching the external pressure to the internal pressure ensures that none of the pushing power of the system is wasted.

wasteful adj 不経済な syn *uneconomical*

▶ The furnace route is very wasteful of energy, and where sodium hydroxide is cheap, direct dissolution of sand in caustic soda is attractive.

wastewater n 廃水

▶ Everyday household materials such as cleaning agents, detergents, perfumes, caffeine, insect repellents, pesticides, fertilizers, and medicines appear in our wastewaters.

water content n 含水量

▶ Extraction with a water-immiscible solvent is useful for isolation of natural products that occur in animal and plant tissues having high water content.

water-in-oil emulsion n 油中水滴型エマルション ☞ oil-in-water emulsion

▶ When higher concentrations of soap are attempted, the rather oleophilic soap curd fibers remain undissolved and form the water-in-oil emulsion.

water-insoluble adj 水に不溶性の

▶ Starch contains about 20% of a water-soluble fraction called amylose and 80% of a water-insoluble fraction called amylopectin.

water of crystallization n 結晶水 ☞ anion water, lattice water

▶ Phloroglucinol contains two molecules of water of crystallization; these are removed by heating for 12 hours at 120 ℃.

waterproof adj 防水の syn *impermeable, sealed, tight*

▶ The lead wires are protected by waterproof plastic tubing that fits tightly onto the end of the glass tube.

water-resistant adj 水をはじく

▶ A calcium soap grease is water-resistant and can be made of any desired consistency.

water-soluble adj 水溶性の

▶ Naphthalene-picric acid complex is split by treatment with ammonia solution to form the water-soluble ammonium salt of picric acid.

waterway n 水路 syn *stream*

▶ High amounts of ammonium can affect the environment by causing algal blooms and diminishing oxygen concentrations in waterways.

wave n 波動

▶ Electrons have wave characteristics which allow them to be used for diffraction experiments.

wave equation n 波動方程式

▶ A wave equation has a series of solutions, called wave functions, each corresponding to a different energy level for the electron.

wave function n 波動関数

▶ The complete wave functions should actually be taken as a sum of individual product wave functions in which complete permutation of electrons and orbitals is established.

wavelength n 波長

▶ The short wavelengths of the visible spectrum give us the color violet.

wavenumber n 波数

▶ It is increasingly common to specify radiation by the wavenumber, the reciprocal of the wavelength and giving the number of waves per cm in vacuum.

wave packet n 波束

▶ The wave packet can represent the excitation energy also.

waxy adj ろう質の

▶ If large amounts of long-chain alkanes remained in the oil, they might crystallize to waxy solids in an oil line in cold weather.

way n **1** 方法 syn *approach, manner, method, practice, procedure*

▶ Substances that act in some ways like catalysts but which are consumed in the course of reaction should not be called catalysts; they can be called pseudocatalysts or activators.

2 点 syn *aspect, feature, point*

▶ Selenium behaves like sulfur in a number of ways.

3 方向 syn *course, direction*

▶ The unstable jelly is on its way not to coagulation but to crystallization or recrystallization.

4 やり方 syn *approach, character, manner, style*

▶ Chemists have realized two important facts about organic and molecular crystals which have transformed our way of thinking about these solids.

5 all the way はるばる syn *completely, entirely, thoroughly, totally*

▶ Triple bonds are easier to reduce than double bonds, so that a triple bond can be reduced to a double bond without carrying the reaction all the way to the alkane.

6 by way of (a) …として syn *as*

▶ By way of notation, the functional dependence is represented by $y=f(x)$, read "y is a function of x".

(b) …のために syn *by means of, through*

▶ Zero-valent metal atoms act as π donors by way of filled d orbitals.

7 give way to 屈する syn *break down, collapse*

▶ When large side groups are present, planar zigzag chains give way of necessity to severely deformed conformations.

8 in one way or another あれやこれやで syn *somehow*

▶ Most kinds of solid are thermally active in one way or another and may be profitably studied by thermal analysis.

9 no way 決して…でない syn *by no means*

▶ There appears to be no way to extend this approach to the simulation of shape or size effects in infrared spectra of much larger particles.

10 the other way round 逆に

▶ When we want to dilute sulfuric acid, we must always add acid to water and not the other way round.

11 under way 進行中 syn *in progress*

▶ The further application of this chemistry to specific targets is under way.

weak adj 弱い syn *feeble, low*

▶ The weak intermolecular $W_2\cdots O$ distance is 2.671 Å. This is barely a bond in terms of W−O distances of known compounds, but it is sufficient to cause the ordering in the solid state.

weak acid n 弱酸

▶ The titration of a weak acid with a weak base should be avoided whenever possible because of the difficulty in getting a sharp end point.

weak base n 弱塩基

▶ In a solution containing both a weak acid and its conjugate weak base, the weak acid becomes a buffer when alkali is added, and the weak base becomes a buffer when acid is added.

weak electrolyte n 弱電解質

▶ The weak electrolytes produce fewer ions and exhibit a much more pronounced fall of molar conductivity with increasing concentration.

weaken vt 弱める syn *diminish, lessen, lower, reduce*

▶ The bonding of hydrogen to the second oxygen weakens the O–H bond and lowers the energy and, hence, the frequency of O–H stretching vibration.

weakly adv 弱く syn *feebly*

▶ Signals from the methyl and methylene carbons are seen weakly upfield.

weakness n **1** 弱さ syn *fragility*

▶ The cadmium chloride and cadmium iodide structures reveal the weakness of the interlayer bonding, for they generally display excellent cleavage parallel to the layers.

2 欠点 syn *fault, shortcoming*

▶ Another weakness of the Debye–Hückel theory is that the ions are treated as point charges; no allowance is made for the fact they occupy a finite volume and cannot come close to each other.

wealth n **1** 富 syn *prosperity*

▶ A wealth of interesting science and technology remains to be explored, and doubtless new and unexpected phenomena will be discovered.

2 豊富 syn *abundance*

▶ Although there is a wealth of qualitative

information on the reactions of square complexes, only a limited amount of quantitative data is available.

wear n 磨耗 syn *abrasion, damage, deterioration*

▶ Formerly indium was much used to protect bearings against wear and corrosion.

wear-resistant adj 磨耗に耐える

▶ Tungsten carbide WC is extremely hard and wear-resistant.

weather n **1** 気象 syn *climate*

▶ Below 20 ℃, gray tin is formed, and objects made of tin often deteriorate in cold weather owing to the formation of this allotropic modification.

vi **2** 風化する

▶ Igneous minerals weather in temperate climates to give clay minerals such as kaolinite, montmorillonite, and vermiculite.

weathering n 風化 syn *erosion*

▶ In general, weathering selectively extracts the more basic ions and leaves the more acidic in the rock.

weave vt 考えを織り込む syn *blend, combine, merge*

▶ With more complex molecules, chemical, physical, and spectral data must be expertly and artfully woven together in order to arrive at a reasonable and consistent picture of molecular structure.

wedge n くさび

▶ More wedge-shaped molecules tend to favor the formation of micelles, while cylindrical molecules tend to form bilayers.

weigh vt **1** 目方を量る syn *gauge, measure*

▷ Good crystals of triglycine sulfate were grown weighing up to 360 g, by an automatic cooling from 52 to 34 ℃ over a period of 10 to 12 days.

2 weigh out 量り分ける

▶ As both solid tin(IV) chloride and its concentrated solution are very hygroscopic, it is necessary to weigh out the portion for analyses from a stoppered vessel.

weighing n 計量

▶ A set of weights as furnished by the maker cannot be relied upon closer to 0.5 mg for weighings of about 1 g.

weight n **1** 重量 syn *mass*

▶ The eutectic composition is 67 percent by weight of tin and 33 percent lead: this metal melts at 183 ℃, and it is used in electric solder.

2 重要さ syn *impact, importance, influence*

▶ The weight of evidence, which is mainly physical, favors at least some aromatic character in borazine and certain of its derivatives.

weight-average molecular weight n 重量平均分子量 ☞ number-average molecular weight

▶ Other methods of molecular weight determination, such as diffusion, sedimentation, and light scattering, involve terms related to the weights or volumes of molecules and give a molecular weight known as the weight-average molecular weight.

weighted average n 加重平均 ☞ arithmetic mean

▶ Since the proton-transfer reaction has been shown by the NMR results described above to be rapid on the X-ray time scale, the average bond lengths should be a weighted average of those of the two molecular structures in equilibrium.

weighting n 加重値

▶ Estimates of the abundance of Mg depend sensitively on the geochemical model used, particularly on the relative weightings given to the various igneous and sedimentary rock types.

weld vt 溶接する syn *bond, connect*

▷ Concentration gradients, particularly of oxygen, in the electrolyte or changes in the composition of the metal, for example, at bimetallic connections or near welded joints, can all cause the spatial separation of the cathodic and anodic processes.

well adv **1** 満足に 比較級・最上級は better, best syn *satisfactorily*

▶ $HOCH_2CH_2NH_2$ can be named almost equally well as 2-hydroxyethylamine or as 2-amino-ethanol.

2 かなり syn *clearly, obviously, plainly*

▶ There are well over a hundred compounds with the spinel structure reported to date.

3 as well そのうえ 文の最後に置く. syn *besides, moreover*

▶ The hydrogen-bonding relationships derived for nitroanilines are useful for understanding the hydrogen-bonding properties of other molecules as well.

4 as well as (prep) …だけでなく A and B とは異なり, A as well as B は A に重点を置く表現 syn *besides, in addition to, not only ... but also*

▶ As well as the two extreme types of behavior exhibited by the normal and inverse spinels, the complete range of intermediate cation distributions is possible, and in some cases, the distribution changes with temperature.

well-defined adj 明確な syn *clear, precise*

▶ 1,2-Naphthoquinone, highly sensitive and reactive, does not have a well-defined melting point but decomposes at about 145–147 ℃.

well-known adj よく知られた 比較級は better-known, 最上級は best-known syn *established, familiar*

▶ The turbidity exhibited by soap or detergent solutions is the best-known indication of micelle formation

well-established adj 確立した syn *actual, sound, well-known*

▶ Whatever the exact nature of mesomorphous state may be, it is a well-established fact that the transition from the mesomorphous to the amorphous liquid state takes place sharply at a definite temperature when the pressure is constant.

wet adj **1** 湿式の syn *moist*

▶ For purpose of qualitative analysis, only reactions in the wet way are applicable as are easily perceptible to our senses.

vt **2** 濡らす syn *moisten*

▶ Graphite is wetted and immediately penetrated by liquid potassium, rubidium or cesium.

wettability n 湿潤性

▶ Surface energy is a measure of extent and nature of the wettability of the surface in a general sense, i. e., without consideration of specific surface–adsorbate interactions.

wettable adj **1** 濡らせる

▶ Poorly wettable or slowly and sparingly soluble substances become more soluble when complexed with β-cyclodextrin, and consequently their bioavailability is improved, and smaller doses are effective.

2 湿気を吸う,

▶ The wettable surface with the carboxyl groups outward sorbed 138 times as much ammonia as the ordinary surface with the paraffin chains outward.

wetting n 濡れ syn *soaking*

▶ Wetting usually goes parallel with low surface tension and, in particular, with low interfacial tension.

what pron …するもの

▶ Much of what can be said about glucose can also be said about the other monosaccharides.

whatever adj どんな…でも

▶ Whatever other methods are applied, the study of a thermal transformation can hardly be regarded as complete until X-ray techniques have been used.

whatsoever adj どんな…でも whatever の強調形, 否定的に用いる.

▶ Most of the free-base porphyrins show no mesophase whatsoever, but on the introduction of one of several metals, a discotic phase is stabilized.

when pron **1** その時

▶ A very simple case is when the rate is proportional to the first power of the concentration of a single reactant.

adv **2** その時に

▶ If the substitution of silicon by aluminum proceeds to a greater extent, the added ions can be calcium ions when the layers are still more firmly bound to one another.

conj **3** …した時 特定の時に何かが起こると期待される場合に用いる. 不確実であれば, if を用いる.

▶ Iodine produces a blue color with starch only when hydriodic acid or a soluble iodide is present.
主文と主語が同じであれば，従属節の主語を省略することができる．
▶ The process called steam distillation is frequently used for substances that would decompose when boiled at atmospheric pressure.
節の形を取ることなく，when に形容詞，前置詞句が続く用い方もある．
▶ Modifications to operationally useful patterns of thinking should be introduced only when absolutely necessary and when the experimental evidence for these modifications can be impressively mustered.
▶ In general, although not always, the molecule is more polarizable when the electron is in the antibonding orbital since it is less tightly bound when in the excited state.

whenever conj **…する時にはいつでも** syn *at every time*
▶ Whenever a reaction occurs with a change in the number or kind of ions present so that the electric conductivity changes.

where adv **1** 制限的用法で，**…である場合に**
▶ Less severe distortions occur in cases where a cation is only slightly too small for its anion environment.
2 接続詞の働きを兼ね非制限的に，**そしてそこで**
▶ The most effective stabilizers for PVC plastics are R_2SnX_2, where R is an alkyl residue, typically *n*-octyl, and X is laurate, maleate, etc.
3 先行詞はなく名詞節を導いて，**…するところ**
▶ The most studied bilayer is where an enzyme layer is used over an electrochemical transducer, producing an enzyme electrode.

whereas conj **それに反して** 主節に対照的なことを述べる節を導く． syn *but, on the other hand*
▶ Potassium hexachloroplatinate(IV) is practically insoluble in absolute alcohol, whereas the corresponding sodium salt is soluble.

whereby adv **それによって**
▶ The most accurate method for estimating ozone consists in allowing the ozonized oxygen to act upon potassium iodide solution whereby free iodine is formed.

whereupon conj **そうすると**
▶ As a rule, the liquid to be dispersed is added in small quantities to the continuous phase whereupon it is spread into a thin, unstable film which spontaneously breaks up into droplets under the influence of surface tension.

wherever conj **…するところはどこでも**
▶ Iron occurs in smaller quantities wherever rocks and the soils derived from them are brown, yellow, orange, or red.

whether conj **1 …であろうとなかろうと** 通常，whether…or…と相関的に用いる．
▶ Most reactions involving organic compounds, whether they occur in the liquid or gas phase, take place in more than one step.
2 …かどうか 二者を選択するのに用いられる
▶ The charge of proteins is positive or negative according to whether the surrounding solution is acidic or basic.
3 whether or not …かどうか
▶ Whether a particular solid phase in a particular equilibrium system is to be considered as pure or not, or as invariable or variable, is a matter of experiment.

which pron **1 どちら**
▶ The positions of substitution may be rationalized by considering which of the various alternatives affords the most stable transition state or substitution intermediate.
2 …する（もの） 関係節の主語または目的語となる．意味を伝えるのに欠くことができない制限的な場合には，節の前にコンマを用いない．この場合，which ではなく that を用いるのが望ましい．
▶ Fracture of lead azide crystals must lead to the local production of atoms which are not thermally hot but which are exceptionally reactive because they no longer occupy normal lattice sites.
関係節の目的語として用いた which はこれを省略することができる．

▶ The chemical energy (which) a green plant stores by photosynthesis provides the total energy requirement of the plant.

そしてそれは 補足的な説明であって，なくても文の意味は失われない非制限的な場合には，which で始まる非制限的な節の前，あるいは前後にはコンマをおく．

▶ The words that are used in describing nature, which is itself complex, may not be capable of precise definition.

通常は，物を表す名詞を先行詞とするが，先行する全文や節を先行詞とすることもできる．意味が曖昧になる恐れがあるときは，this で始まる別の文にする．

▶ Cations fill the interstices between adjacent anion layers, which means that in these hcp or ccp packing schemes the coordination geometry of a centered cation is limited to octahedral or tetrahedral.

なお，which は above, along, at, below, by, during, for, from, in, into, of, on, through, to, with などの前置詞を伴う形で用いられる．

▶ Single-crystal X-ray studies are the usual way by which new solid-state products and structures are first characterized.

▶ The primary significance of bond energies lies in the calculation of the enthalpy of a reaction involving a compound for which no enthalpy data are available.

▶ The speed with which it relaxes can be measured, usually by spectrophotometry, and the rate can be obtained.

adj **3 そしてこの**

▶ The solution was exposed to direct sunlight for seven days during which time the solid product crystallized out of the solution.

while conj **1 …なのに比べて** syn *although, though*

▶ While electron-rich nucleophiles add to the carbonyl double bond, carbon–carbon double bonds of olefins are subject to attack only by electrophiles.

2 …するうち

▶ Using a sample of liquid carbon dioxide, T. Andrews gradually raised its temperature, while maintaining the pressure constant.

whisker n **ホイスカー**

▶ The whisker is able to grow without containing any dislocations, and such whiskers are orders of magnitude stronger than ordinary crystals.

white n **1 白身**

▶ The familiar, rapid and irreversible changes that occur when the white of an egg is heated provide a vivid example of the character of protein denaturation reactions.

adj **2 白い**

▶ Radium is a brilliant white solid but turns black on exposure to air.

white light n **白色光**

▶ If the light has all wavelengths of the visible spectrum, then it is white light.

whole n **1 全体** 一つのまとまったものとして考える． syn *sum total, total*

▶ For high frequencies, the whole of the dielectric constant of alkali halides arises from the polarization of ions.

2 as a whole 全体として syn *altogether*

▶ Since the system as a whole is electrically neutral, there are an equal number of free positive and negative charges present.

3 on the whole 概して syn *chiefly, largely, mainly*

▶ On the whole, the high-valence oxides of the first transition-metal series are more acidic than the corresponding high-valence oxides of the second and third series.

adj **4 すべての** syn *entire, total*

▶ For whole crystals, there are long induction periods which are very largely removed on grinding.

whole number n **整数** ☞ integral

▶ The ionic radii are r_A and r_B and the charges are $z_A e$ and $z_B e$; e is the electronic charge and z_A and z_B are whole numbers, positive or negative, that indicate the numbers of charges on the ion.

wholly adv **全く** syn *absolutely, entirely, fully*

▶ What is efficient and economical in a laboratory-scale synthesis may be wholly impractical in industrial production.

whose pron **その** syn *of which*

▶ Lanthanum combines energetically with hydrogen at 240 ℃, forming a black product approximating LaH_3, but whose exact composition depends on temperature and the pressure of the hydrogen.

why adv **1** なぜ

▶ Why some organisms utilize silica rather than calcium carbonate as a structural material is unknown.

2 reason を先行詞として，…との理由

▶ Apart from the fact that mercury is liquid at ordinary temperatures, there is no reason why its alloys should be treated as different from the alloys of any other metal.

wide adj **1** 幅の広い syn *considerable, extreme, sizable*

▶ The ease with which an optically active substance can be transformed into the corresponding racemic modification varies between wide limits.

2 範囲の広い syn *extensive, wide-ranging*

▶ Ferroelectric properties are found in a wide range of materials.

widely adv 広く syn *extensively, generally, thoroughly*

▶ It is widely accepted that the lithium surface is covered by a thin, compact, ionically conducting film.

widen vi 広くなる syn *broaden, expand, increase*

▷ It is important that the definitions of SI be continually re-examined in the light of widening experience and improved where possible.

wide-ranging adj 広範囲にわたる syn *broad, comprehensive, extensive*

▶ Catalysis and catalytic mechanisms permeate almost every aspect of chemistry and are of such wide-ranging importance that they have long been the subject of continuing research.

widespread adj 一般的な syn *common, general, worldwide*

▶ Zinc blende, ZnS, is the most widespread ore of zinc and the main source of the metal.

width n 幅 syn *breadth, span* ☞ line width

▶ The width of the bands decreases with decreasing temperature, indicating a reduced interaction with the surrounding lattice.

wildlife n 野生生物

▶ Oil spills have had a tremendous impact on the environment, specifically on large number of wildlife.

wildly adv **1** はなはだしく syn *excessively, extremely, intensely*

▶ The behavior of the metal is wildly variable and unpredictable.

2 広範囲に syn *amply*

▶ The most widely used detergents are sodium salts of alkylbenzenesulfonic acids.

wild type n 野生型

▶ The characteristics of the wild type reflect the usual or native form of a gene.

will n **1 at will** 意のままに syn *as wishes*

▶ The crossover $S_1 \rightarrow T_1$ can be essentially stopped or made 100% efficient at will by changing the mass of the solvent molecules.

aux **2** だろう

▶ If a compound has a hydroxyl group, there will be an absorption in the 3600 cm^{-1} region.

win vi **win out** 勝ち抜く syn *overcome, prevail*

▶ Strongly activating groups generally win out over deactivating or weakly activating groups.

wind vi **1** 巻きつく syn *coil, curl*

▶ The backbone of the α helix winds around the long axis.

vt **2** 巻く syn *coil, roll*

▶ In a chromosome, DNA is wound around a series of very basic proteins called histones.

winding n 巻いたもの syn *coil*

▶ The drop in temperature at the ends of a heated tube due to increased heat losses may be compensated by spacing the windings or by putting on additional windings at the ends.

window n **1** 窓

▶ The X-rays leave the tube through windows made of beryllium.

2 ウインドー

▶ Wavelengths of 1300 and 1550 nm corresponds to absorption windows in the silica core material and have been the focus of development efforts.

wipe vt ぬぐう syn *clean, cleans*
▶ On removal from the oil bath, the flask should be allowed to drain into the bath and then wiped.

wire n 針金 syn *cable*
▶ Commercially successful superconducting wires or tapes must be able to carry a current density of at least 10^4 A/cm^2, and, in some cases, they must retain this capacity when exposed to a strong magnetic field.

wire gauze n 金網
▶ Both sodium and magnesium sulfates can be obtained in anhydrous condition by heating hydrated or partially hydrated material in a casserole over a wire gauze.

with prep **1** 一緒に
▶ Methanol is a clear, colorless, mobile, highly polar liquid and is miscible with water, alcohols, and ether.

2 …につれて
▶ With increasing pressure, the absorption maxima of F centers in NaCl and KCl shift to higher energies as the crystal contracts.

3 同時に syn *accompanying*
▶ Acetylsalicylic acid melts with decomposition at temperatures reported from 128 to 137 ℃.

4 …について syn *concerning*
▶ With metals, the total energy cannot be obtained by summation of the interactions of the atoms since these are not additive.

5 …のある syn *having, possessing*
▶ A ketone such as cyclohexanone, with four α hydrogens, yields a mixture of halogenated products under haloform conditions.

6 …を用いて syn *by means of, using*
▶ Naphthalene can be sulfonated with concentrated sulfuric acid at 0–60 ℃.

withdraw vt 取り出す syn *remove*
▶ After quick stir the solution was withdrawn to another flask, concentrated to 1 mL, and left to stay overnight.

withdrawal n 引き上げ syn *removal*
▶ Because of powerful electron withdrawal by the fluorine atoms, 2,2,2-trifluoroethanol is much more acidic than ethanol.

within prep …の中に syn *in, inside*
▶ Graphite's strong covalent bonds lie entirely within planes, making this a two-dimensional layered solids.

without prep …なしに syn *not making use of*
▶ It is possible to synthesize ammonia from nitrogen and hydrogen without high temperatures and pressures by means of nitrogenase.

withstand vt 耐える syn *confront, resist*
▶ Silicones have high thermal stability and, in the absence of air, will withstand temperatures up to roughly 250–300°.

witness n **bear witness to** 証拠となる syn *testify*
▶ The remarkable resemblances among the lanthanides bear witness to the overwhelming influence of identical charge and similar size in these species.

wonder vi いぶかる syn *be curious, question*
▶ Faced with these numerous potential benefits one may wonder if ionic liquids have any problems associated with their use.

word n **1** 言葉 syn *expression, name, term*
▶ The use of the word chemisorption implies that bonding of the gas molecules to the solid occurs by ordinary chemical bonds, i.e., that the bonds can be described in terms of ionic and covalent character.

2 in other words 言い換えれば syn *namely*
▶ The Lowry–Brønsted definitions, like the classical ones, permit the existence of amphoteric substances, or, in other words, of substances which are both acids and bases at the same time.

work n **1** 熱力学的な仕事
▶ At least in principle, any process that tends to proceed spontaneously can be made to do useful work.

2 仕事 syn *effort*
▶ The purpose of this work was to develop a method for simultaneous measurement of gas adsorption and specific magnetization.

3 at work 働いて syn *functioning*
▶ There may be other factors at work here.

vt **4 put…to work 仕事をさせる** syn *employ, exert, use*

▶ As a first example of putting the last equation to work, we will consider the Gibbs function of mixing of liquids A and B.

vi **5 働く** syn *function, operate*

▶ There are many different spectroscopic techniques, but all work on the same basic principle.

6 work out 達成する syn *evolve, develop*

▷ Synthetic dyestuffs contributed to the development of the chemical industry primarily through the synthesis techniques that chemists worked out to make a wide variety of compounds.

vt **7 細工する** syn *effect, make, produce*

▶ Rhenium metal is hard and resistant to corrosion and must be worked by powder metallurgy.

8 work up 対処する syn *advance, rise*

▷ A frequently used method of working up a reaction mixture is to dilute the mixture with water and extract with ether in a separatory funnel.

workable adj **実際に使える** syn *feasible, practicable*

▶ Aluminum is widely distributed in igneous rocks, but the only workable ore is bauxite.

worker n **研究者** ☞ investigator

▶ A metastable red modification of tin(II) oxide has been reported by some workers.

work function n **仕事関数**

▶ In metals, the electron affinity equals the photoelectric work function.

work hardening n **加工硬化**

▶ Bending a piece of copper wire leads to the creation of new dislocations, which become pinned and then pin other dislocations in a process known as work hardening.

working adj **実際に役に立つ** syn *usable, workable*

▶ In many cases, the working concentration equates to 5 to 10 mol/L which although seemingly high is not overly different to many aqueous plating solutions.

working hypothesis n **作業仮説**

▶ In pharmaceutical research, scientific theories must be viewed as working hypotheses.

working temperature n **作業温度**

▶ Soda-lime glass has a high coefficient of expansion and must be brought to the working temperature slowly in a yellow flame to avoid cracking.

world n **区分として界** syn *area, sphere*

▶ The valence electrons of an atom are not very effective in shielding the outside world from the positive charge on the nucleus of the atom.

worldwide adj **世界中で** syn *extensively, internationally, universally*

▶ Needless to say, new techniques and approaches for purification of ionic liquids are currently being investigated in many laboratories worldwide.

worry vt **心配させる** syn *bother, concern, trouble*

▶ Most scientists are not in fact worried that theories cannot be logically proved.

worse adj **もっと悪い**

▶ In industry it is worse than useless to build a sophisticated plant if the overall reaction has a thermodynamic tendency to run in the wrong direction.

worth adj **…に値する** syn *meriting*

▶ It is worth noting that the as-synthesized nanofibers tend to disperse into the aqueous solution, while the product from a conventional slow mixing synthesis precipitates out after the stirring is stopped.

worthwhile adj **やりがいのある** syn *productive, profitable*

▶ Binuclear complexes of this type seemed very worthwhile objectives at that time because they had the potential to show new types of metal-metal interactions and new types of reactivity at the bimetallic site.

worthy adj **価値のある** syn *worthwhile, deserving*

▶ After all the time spent on the preparation,

the final product should be worthy of a carefully executed and secured label.

would aux 推定を表して，あろう

▶ Ions such as Cu^{+}, Ag^{+}, Au^{+}, and Hg^{2+} would be expected by any simple electrostatic theory to possess high coordination numbers.

wrap vt 包む syn *cover, enclose, envelop*

▷ The flash lamp is shown as being wrapped around the ruby rod.

wriggly line n 波状線

▶ The second step indicated in the diagrams by solid, wriggly lines is the collisional deactivation which brings the hot molecule into thermal equilibrium with its environment.

write vt **1** 書く syn *make up*

▶ If we try to write a conventional electron-pair structure for diborane, we see at once that there are not enough valence electron pairs for six normal B–H bonds and one B–B bond.

2 write down 書きとめる syn *record*

▶ It is not possible to write down a single, satisfactory, classical bonding diagram for S_4N_4.

writer n 著者 syn *author*

▶ The writer acknowledges with pleasure helpful discussion with Professor S. Kanda.

wrong adj 誤った syn *improper, inappropriate, unacceptable*

▶ The radius ratio is a simple and easily understood approximation for explaining the structures that are adopted by ionic compounds, but in many cases it makes the wrong predictions.

X

xerogel n キセロゲル

▶ If the solvent phase is removed by conventional drying from a gel, xerogels result.

X-ray n X線

▶ X-rays are scattered almost entirely by the electrons of an atom, and the intensity of the scattered radiation must depend on the distribution within the atom of all the electrons.

X-ray absorption fine structure n X線吸収微細構造

▶ X-ray absorption fine structure has one great advantage: by tuning in to the absorption edge of each element present in the material in turn, the local structure around each element may be determined.

X-ray diffraction n X線回折

▶ The technique of X-ray diffraction is the determination of the structures of crystalline solids by an analysis of the diffraction patterns produced when X-rays are passed through them.

X-ray emission spectrum n X線発光スペクトル

▶ Each element gives a characteristic X-ray emission spectrum composed of a set of sharp peaks.

X-ray photoelectron spectroscopy n X線光電子分光法 ☞ ultraviolet photoelectron spectroscopy

▶ In X-ray photoelectron spectroscopy, the energy of the incident photons is so great that electrons are ejected from the tightly bound atomic cores.

X-ray spectrum n X線スペクトル

▶ The tightly bound inner-shell electrons give rise to the X-ray spectrum.

Y

year n 年，複数形で長期間

▶ A mixture of hydrogen and oxygen can be kept for years without appreciable reaction.

yeast n 酵母

▶ The different types of yeast vary greatly in size, but a majority are larger than bacteria and spherical, elliptical, or cylindrical in shape.

yet adv **1** まだ syn *still*

▶ There is yet another most interesting group of boron compounds: the nitrides and their derivatives.

2 as yet 今までのところでは syn *so far*

▶ To begin with, there is as yet no theoretical explanation why the $T_1 \rightarrow S_0$ process should be viscosity-dependent.

3 not yet まだ…ない

▶ The mechanism of color formation is not yet definitely known.

conj **4** それなのに syn *despite, nevertheless*

▶ Diffusion is one of the simplest and yet most significant properties of substances in solution.

yield n **1** 収量 syn *production, output*

▶ 2-(4'-Methoxyphenyl)-1,4-benzoquinone was made in nearly quantitative yield by the method of Brassard and L'Écuyer.

vt **2** 与える syn *produce, supply*

▶ London's theory yields this force law and applies to cases where the valence shells of all atoms are filled.

yielding n 曲がり

▶ It is clear that a jelly must be elastic and rigid and yet show yielding or relaxation under persistent stress.

yield point n 降伏点

▶ The application of stresses above the yield point produced an immediate increase in the conductivity of alkali halide crystals, followed by a slow recovery.

yield strength n 降伏強度

▶ The low yield strengths of crystals are due to slip resulting from the movement of dislocations.

yield value n 降伏値

▶ There is a yield value in the loosely packed structure which resists sedimentation under gravity alone but gives under a larger force.

Young's modulus n ヤング率

▶ The ratio of stress to strain in the region of proportionality is known as Young's modulus.

Z

zenith n 絶頂 syn *high point, peak, summit*

▶ The amount incident on the earth's surface in clear weather with the sun at the zenith is close to 1 kW m^{-2}.

zeolite n ゼオライト

▶ Many zeolites occur naturally, but the majority of those used commercially have been synthesized and are designated by a letter or group of letters.

zero adj **1** ゼロ syn *nil*

▶ At equilibrium, the cell would be able to perform no useful work, and its emf must then be zero.

vt **2** ゼロに合わせる

▶ Circular dichroism spectra often need to be zeroed so it is essential that data at least 20 nm beyond the normal absorption envelope are available.

zero-dimensional adj ゼロ次元の

▶ Porous silicon has a highly complex nanoscale architecture made up of one-dimensional crystalline nanowires and zero-dimensional nanocrystallites.

zero-point energy n ゼロ点エネルギー

▶ The vapor pressure isotope effect has its physical origin in the fact that the zero-point energy of a light molecule is larger than that of a heavy molecule.

zero-point vibration n ゼロ点振動

▶ Even at the absolute zero of temperature, some broadening is to be expected, owing to the zero-point vibrations.

zeta potential n ζ電位

▶ If a liquid is forced through the pores of a membrane or through capillary tube, a potential difference is observed, its magnitude depending on the ζ potential.

zig-zag n **1** ジグザグ形

▶ Saturated acid chains are extended in a linear fashion with, of course, the zig-zag due to the tetrahedral bond angles and fit together rather well.

adj **2** ジグザグ形の

▶ Assuming a rigid, extended zig-zag alkyl chain, the odd series have more carbon–carbon bonds along the molecular axis giving the higher $\Delta\alpha$.

zinc dust n 亜鉛末 syn *zinc powder*

▶ Zinc dust may form explosive mixture with air and may heat and ignite spontaneously on exposure to air when damp.

zinc enzyme n 亜鉛酵素

▶ The importance of studying synthetic analogues of zinc enzymes resides with the fact that such species are more amenable to structural, spectroscopic, and mechanistic studies than the enzymes themselves.

Zintle phase n チントル相

▶ The structural requirements of Zintle phases are explained by assuming the presence of both ionic and covalent contributions to the bonding picture of the structure.

zone n **1** 晶帯

▶ All the faces that are parallel to a given direction are called a zone, and the direction to which they are parallel is the axis of the zone.

2 区域 syn *area, region*

▶ Often design works as intended up to a point, beyond which one moves into a zone of exploration.

zone-refine vt 帯域精製をする

▶ Samples as small as 0.1 g have been zone-refined by using a projection lamp and ellipsoidal mirror to focus a narrow spot of heat on the sample.

zone refining n 帯域精製

▶ The revolutionary advantages of zone refining over other methods of purification result from the fact that a heater or heaters may be passed over an ingot repeatedly without separation of the impurities from the bulk of the sample and without exposing the sample to solvent or to air.

zwitterion n 双生イオン

▶ An amino acid in approximately neutral

aqueous solution adopts a zwitterion form rather than a noncharged configuration.

zwitterionic adj 双生イオンの

▶ Detergents may have head groups which are ionic, zwitterionic, or nonionic.

索　引

数字

ζ

A

D

F

N

P

Q

R

S

X

あ

い

う

え

お

穏やかな
gentle 256
mild 364
穏やかに
gently 256
mildly 364
落ち着く settle down 527
負っている owe 416
オートクレーブ autoclave 47
落とし穴 pitfall 437
オートラジオグラフィー autoradiography 48
驚き surprise 577
驚くばかりに amazingly 24
驚くべき
amazing 24
startling 556
surprising 577
驚くほどに surprisingly 577
同じ
identical 284
same 514
（と）同じ the same as 515
同じことを意味する synonymous 581
（と）同じ程度に as much as 376
同じでない unlike 621
同じように analogously 27
（と）同じように as…so 40
オニウム化合物 onium compound 403
おのおの each 186
覚えておく bear in mind 54
重い heavy 271
思いがけなく unexpectedly 619
思い出す recall 481
思いもよらない unsuspected 624
思う believe 55
（と）思う deem 147
面白い amusing 26
主な
chief 88
principal 457
主に chiefly 88
（に）おもむく傾向がある lean toward 333
思わしくない discouraging 170
思われる
appear 34
seem 522
およそ
about 2
around 38
ca. 74
in the neighborhood of 384
および and 27
およびまたは and/or 28
及ぶ
range 476
span 547
及ぼす exert 218
（を）及ぼすこと imposition 290
折り返し報告する report back 496
折り重なる fold 241
オリゴ糖 oligosaccharide 402
オリゴマー oligomer 402
オリゴマー化 oligomerization 402
オリゴマーの oligomeric 402
織り込む weave 641
折りたたみ folding 241
折りたたみ構造 folded structure 241
折りたたみの folding 241
折りたたむ fold 241
折り目 fold 241
折り目を付ける crease 133
オール化 olation 402
オルガネラ organelle 409
オルガノシラン organosilane 410
オルトケイ酸塩 orthosilicate 411
オルト-パラ配向の ortho-para-directing 411
おろそかにする neglect 383
終わって over 414
終わり conclusion 111
終わりに臨んで in conclusion 111
終わる
result in 503
terminate 588
恩義 indebtedness 296
音響の acoustic 8
温室効果 greenhouse effect 263
温浸 digestion 166
温浸する digest 165
温度因子 temperature factor 586
温度係数 temperature coefficient 586
温度勾配 temperature gradient 586
温度ジャンプ法 temperature-jump method 586
音波 sound wave 546
音波処理 sonication 545
オンライン式の on-line 403
穏和な soft 541

か

が for 242
（である）が although 24
加圧する press 455
がある there is 591
かあるいはその逆で or otherwise 412
害 harm 269
界
kingdom 327
world 647
回 time 598
外圧 external pressure 224
外因性の exogenous 218
外因性の extrinsic 225
回映軸 axis of rotatory inversion 49
灰化 incineration 293
外界 surrounding 577
外界の ambient 24
外殻 outer shell 413
外殻電子 outer-shell electron 413
概観 overview 416
開環 ring opening 508
概観する survey 578
懐疑的な skeptical 537
解決する settle 527
外見 appearance 34
外圏機構 outer-sphere mechanism 413
会合 association 42
会合液体 associated liquid 42
解こう剤

き

け

こ

交互に
alternately 23
alternatively 24
交互に現れる alternate 23
交差
crossing 135
crossover 135
交差カップリング cross coupling 135
工作する engineer 204
交差した crossed 135
交差する
cross 135
cut 140
intersect 314
考察 consideration 118
考察する
consider 117
look at 344
交差分極マジック角スピニング cross-polarization magic angle spinning 135
交差分子線 crossed molecular beam 135
鉱酸 mineral acid 365
光散乱 light scattering 337
格子 lattice 331
光子 photon 434
格子エネルギー lattice energy 332
抗しがたいほどに compellingly 105
格子間原子
interstitial 314
interstitial atom 314
格子間の interstitial 314
公式 formula 244
公式でない unofficial 621
光軸 optic axis 407
格子欠陥 lattice defect 331
高次コイルの supercoiled 572
格子振動 lattice vibration 332
格子水 lattice water 332
行使する take 584
格子定数 lattice constant 331
格子点 lattice point 332
抗磁場 coercive field 97
高磁場への upfield 625
格子面 lattice plane 332
後者 latter 331
光重合性の photopolymeric 434
高重合体 high polymer 274
高周波の high-frequency 274
抗腫瘍性の antitumor 33
高純度の high-purity 275
向上
boost 65
improvement 291
工場 plant 438
孔食 pitting 437
高スピンの high-spin 275
構成 architecture 37
校正
calibration 74
proofreading 461
剛性 rigidity 507
合成 synthesis 581
構成員 member 358
合成ガス synthesis gas 581
構成原理 building-up principle 71
合成高分子 synthetic polymer 581
合成ゴム synthetic rubber 581
合成樹脂 synthetic resin 581
校正刷り proof 461
校正する calibrate 74
構成する
constituent 119
constitute 119
make up 350
（を）構成する compose 107
合成する synthesize 581
合成洗剤 synthetic detergent 581
合成染料 synthetic dye 581
構成単位 unit 620
構成的な constitutive 119
合成によって synthetically 581
剛性の rigid 507
合成の synthetic 581
高性能な smart 539
構成物 construct 120
抗生物質 antibiotic 31
抗生物質耐性 antibiotic resistance 31
校正補正 calibration correction 74
構成要素 building block 71
剛性率 modulus of rigidity 369
光線 ray 478
公然と openly 405
光線を平行にする collimate 99
酵素 enzyme 207
酵素-基質複合体 enzyme-substrate complex 207
構造 structure 566
構造異性 structural isomerism 566
構造異性体 structural isomer 566
構造因子 structure factor 566
構造解析 structure analysis 566
構造化された structured 566
構造決定 structure determination 566
構造式 structural formula 566
構造上の structural 566
高層大気 upper atmosphere 625
構造的に structurally 566
構造粘性 structural viscosity 566
構造のない structureless 566
構造敏感性の structure-sensitive 566
構造変換 structural transformation 566
構造用の structural 566
酵素活性 enzyme activity 207
拘束 containment 121
高速液体クロマトグラフィー high-performance liquid chromatography 274
拘束する bind 58
高速反応 fast reaction 229
酵素的な enzymic 207
酵素的に enzymatically 207
酵素電極 enzyme electrode 207
酵素による enzymatic 207
抗体 antibody 31
剛体球 hard sphere 268
剛体球ポテンシャル hard-sphere potential 268
広大な expansive 219
光沢

luster 346
polish 442
光沢剤 brightener 69
光沢のある lustrous 346
構築
building 71
construction 120
好都合に advantageously 16
肯定する affirm 17
肯定的な positive 446
肯定的に positively 447
交点 intersection 314
光電効果 photoelectric effect 433
光電子工学 optoelectronics 407
光電子工学の optoelectronic 407
光電子スペクトル photoelectron spectrum 433
光電子倍増管 photomultiplier tube 434
光電子分光 photoelectron spectroscopy 433
光電子分光の photoelectron spectroscopic 433
光電子放出 photoelectric emission 433
後天性の acquired 9
光電セル photocell 432
光伝導 photoconduction 432
光伝導性の photoconductive 432
光伝導体 photoconductor 432
光電流 photocurrent 432
高度 altitude 24
硬度 hardness 268
光度 luminosity 346
抗凍結の antifreeze 32
行動すること doing 178
行動中 in the act of 9
口頭で orally 408
高度な of high order 408
高度の high 274
光度の photometric 434
購入する purchase 467
購入できる obtainable 399
光二量体 photodimer 433
勾配
gradient 260
pitch 437
slope 538
勾配を付けた graded 260
勾配をなす ramp 476
広範囲な
extended 223
extensive 223
広範囲に
extensively 223
wildly 645
広範囲にわたる wide-ranging 645
広範囲の broad 70
高品質の premium 453
光付加 photoaddition 432
降伏強度 yield strength 650
降伏値 yield value 650
降伏点 yield point 650
鉱物 mineral 365
興奮 ferment 231
光分解 photolysis 433
光分解する photolyze 433
光分解の photolytic 433
高分解能分光学 high-resolution spectroscopy 275
興奮させる exciting 217
高分子 polymer 443
高分子電解質 polyelectrolyte 443
高分子の macromolecular 348
光変換 phototransformation 434
候補 candidate 75
酵母 yeast 650
後方散乱 backward scattering 50
抗マラリア剤 antimalarial 32
高密度化 densification 153
高密度に densely 153
高密度の dense 153
鉱脈 vein 631
巧妙さ ingenuity 301
巧妙な ingenious 301
こうむる
suffer 571
undergo 617
鉱油 mineral oil 365
合理性
rationality 478
reasonableness 481
効率 efficiency 187
効率のよい efficient 187
効率よく efficiently 187
合理的な
legitimate 334
rational 478
reasonable 481
合理的に rationally 478
合理的に説明する rationalize 478
向流抽出 countercurrent extraction 131
向流の countercurrent 131
香料 perfume 427
抗力 drag 181
効力
effectiveness 187
potency 448
効力のある potent 448
(を) 考慮して in the light of 337
考慮する revolve around 506
考慮中の under consideration 118
考慮に入れない leave aside 334
考慮に入れる
take account of 6
take into account 6
(を) 考慮に入れる
allow for 22
take into consideration 118
行路 path 423
(を) 越えて
above 2
beyond 57
(を) 超えて over 414
越える
cross 135
transcend 604
超える exceed 215
氷 ice 284
氷で冷やす ice 284
氷のような glacial 258
固化 solidification 542
誤解 misunderstanding 367
誤解させる
mislead 366
misleading 366
誤解された misunderstood 367
五角形 pentagon 425

さ

609
三角フラスコ　Erlenmeyer flask　210
酸化剤　oxidant　416
酸化させる　oxidize　417
酸化状態　oxidation state　417
酸加水分解　acid hydrolysis　8
酸化数　oxidation number　416
酸化する　oxidize　417
（に）参加する　take part in 421
三角形の　triangular　608
酸化的開裂　oxidative cleavage 417
酸化的カップリング　oxidative coupling　417
酸化的な　oxidative　417
酸化的付加　oxidative addition 417
酸化できる　oxidizable　417
三価の　trivalent　610
酸化物の　oxidic　417
酸化防止剤　antioxidant　33
酸化力のない　nonoxidizing 390
残基　residue　498
三脚配位子　tripod ligand　609
酸強度　acid strength　8
三形の　trimorphic　609
三原子を有する　triatomic　608
残光　afterglow　17
参考文献　reference　487
三座配位子　tridentate ligand 608
三軸の　triaxial　608
三次元構造　three-dimensional structure　596
三次構造　tertiary structure 589
三次の　cubic　138
三斜晶系　triclinic system　608
三斜晶系の　triclinic　608
三重結合　triple bond　609
三重項　triplet　609
三重項状態　triplet state　609
三重衝突　three-body collision 596
三重線　triplet　609
三重点　triple point　609
三重に　triply　609
算術的に　arithmetically　38
算術平均　arithmetic mean　38
参照　reference　487
参照する
consult　120
refer to　487
reference　487
参照点　reference point　488
参照電極　reference electrode 487
酸触媒　acid catalyst　8
酸触媒作用　acid catalysis　8
酸性雨　acid rain　8
酸性化　acidification　8
酸性酸化物　acidic oxide　8
酸性せっけん　acid soap　8
酸性度　acidity　8
酸性にする　acidify　8
酸性の
acid　7
acidic　8
三成分の　ternary　588
酸性溶液　acidic solution　8
酸素化　oxygenation　417
酸素飽和曲線　oxygen saturation curve　417
酸素を除く　deoxygenate　154
三段重ね　triple-decker　609
三中心結合　three-center bond 596
酸定数　acid constant　8
暫定的な
provisional　465
tentative　587
暫定的に　tentatively　588
酸滴定　acidimetry　8
サンドイッチ構造　sandwich structure　515
サンドイッチ状のもの　sandwich　515
算入　inclusion　294
残念な　regrettable　490
残念にも　regrettably　490
酸の解離　acid dissociation　8
酸敗　rancidity　476
三分子反応　trimolecular reaction　609
三分の二　two-thirds　613
三方晶　trigonal crystal　608
三方晶系　trigonal system　609
三方錐　trigonal pyramid　609
三方錐の　trigonal pyramidal 609
三方両錐　trigonal bipyramid 608
散漫　diffuseness　165
三面冠三角柱　tricapped trigonal prism　608
残余の　residual　498
散乱　scattering　517
散乱角　scattering angle　517
散乱させる　scatter　517
散乱された　scattered　517
残留磁化　residual magnetization　498
残留抵抗　residual resistance 498
残留物　residue　498
三量化する　trimerize　609
三稜形の　prismatic　457
三量体　trimer　609
三量体の　trimeric　609

し

仕上げ　finish　234
ジアステレオ異性化　diastereoisomerization　162
ジアステレオ異性体の　diastereoisomeric　162
ジアステレオ選択性　diastereospecificity　162
ジアステレオトピック　diastereotopic　162
ジアステレオマー　diastereomer　162
ジアステレオマー塩　diastereomeric salt　162
ジアステレオマーの　diastereomeric　162
ジアゾ化　diazotization　162
ジアゾ化する　diazotize　162
ジアゾカップリング　diazo coupling　162
ジアゾの　diazo　162
強いてさせる
constrain　119
force　243
強いる　enforce　203
ジエノフィル　dienophile　163
ジェミナル　geminal　255
シェーレルの式　Scherrer's formula　518

acquisition 9
gain 254
取得する acquire 9
主として
largely 330
mainly 349
primarily 456
principally 457
首尾一貫性 self-consistency 523
首尾よく successfully 570
手法 technique 586
寿命
life 336
life-span 336
lifetime 336
longevity 344
需要 demand 152
腫瘍 tumor 610
受容する accept 4
受容性 receptivity 482
受容性のある receptive 482
受容体 acceptor 4
主要な
central 82
foremost 243
leading 333
main 349
prime 456
授与する confer on 113
主流の mainstream 350
シュリーレン組織 Schlieren texture 518
種類
category 79
kind 327
sort 545
variety 631
シュレーディンガー方程式 Schrödinger equation 518
準安定性 metastability 361
準安定な metastable 361
準位 level 335
順位 ranking 476
巡回の cyclic 141
循環 circulation 90
瞬間
instant 305
moment 372
循環 round 511
循環過程 cyclic process 141
循環させる circulate 90
循環する
circulating 90
cycle 141
循環性 cyclability 141
瞬間的な
instantaneous 305
momentary 372
準結晶の paracrystalline 419
準周期的 quasiperiodic 471
順序
order 408
sequence 525
純粋な pure 467
純粋に purely 467
潤沢 richness 507
準直線形の quasilinear 471
純度 purity 467
純度が極めて高い ultrapure 614
順応させる accommodate 5
準備 provision 465
準備中 in preparation 453
商 quotient 472
仕様 specification 548
昇位 promotion 461
昇位させる promote 461
小員環 small ring 539
漿液 serum 526
常温 ordinary temperature 409
浄化
cleanup 93
purge 467
消化 digestion 166
昇華 sublimation 567
障害
bottleneck 66
obstacle 399
昇華エンタルピー enthalpy of sublimation 205
小角散乱 small-angle scattering 539
消化酵素 digestive enzyme 166
昇華させる sublime 567
浄化する
clean up 92
purge 467
消化する digest 166
昇華する
sublimable 567
sublime 567
消化できる digestible 166
昇華熱 heat of sublimation 271
消化の digestive 166
昇華物 sublimate 567
消化不良 indigestion 297
蒸気 vapor 630
蒸気圧 vapor pressure 630
蒸気相 vapor phase 630
蒸気相クロマトグラフ分析 vapor-phase chromatographic analysis 630
小規模の small-scale 539
小球 pellet 425
小球状にする pellet 425
小球状にすること pelletizing 425
状況 affair 16
商業上 commercially 102
商業上の commercial 102
状況的な circumstantial 91
蒸気浴 steam bath 558
消極的な negative 383
小区分
ramification 475
subsection 568
小区分される ramify 476
衝撃
bombardment 64
shock 530
衝撃波管 shock tube 530
衝撃を与える bombard 64
焼結 sintering 536
焼結させる sinter 536
焼結物 sinter 536
上下に up and down 625
条件
condition 112
qualification 470
stipulation 561
上限 upper limit 625
証拠 evidence 214
消光 quenching 472
照合する check 86
消光する quench 471
常光線 ordinary ray 409
証拠となる
testify 589
bear witness to 646

す

せ

そ

sheet 530
相 phase 430
槽 trough 610
創案する invent 317
相違
difference 163
divergence 177
gap 254
掃引 sweep 579
相応して correspondingly 130
増加
gain 254
increase 295
総括反応速度 overall reaction rate 415
相関 correlation 129
相間移動触媒作用 phase-transfer catalysis 431
層間隔 layer spacing 332
層間化合物 intercalation compound 309
増感光分解 sensitized photolysis 524
増感剤 sensitizer 524
増感されていない unsensitized 623
相関図 correlation diagram 129
相関長 correlation length 129
相関的な correlative 129
層間の interlamellar 311
増感ルミネセンス sensitized luminescence 524
操業 operation 405
双極イオン dipolar ion 168
双極構造 dipolar structure 168
双極子 dipole 168
双極子-双極子相互作用 dipole-dipole interaction 168
双極子の dipolar 167
双極子モーメント dipole moment 168
双曲線の hyperbolic 282
遭遇する
meet 357
see 521
総計 sum total 572
象牙質 dentin, dentine 153
相互依存 interdependence 310
相互依存の interdependent 310
総合的視野で in perspective 429
相互拡散
counterdiffusion 131
interdiffusion 310
相互関係
interrelation 314
interrelationship 314
相互貫入 interpenetration 313
相互凝結 mutual coagulation 379
相互作用
interaction 309
interplay 313
相互作用エネルギー interaction energy 309
相互作用をしていない noninteracting 389
相互晶出 intercrystallization 310
相互に mutually 379
相互に関連した interconnected 309
相互に作用する interact 309
相互に連結させる interconnect 309
相互に連結した interconnected 309
相互に連結する interlink 312
相互の reciprocal 482
相互変換 interconversion 309
相互変換する interconvert 310
相互変換性 interconvertibility 310
相互変換のできる interconvertible 310
操作
manipulation 351
operation 405
走査 scan 516
相殺 cancellation 75
相殺する
cancel 75
compensate 105
counterbalance 131
offset 401
走査型電子顕微鏡法 scanning electron microscopy 517
走査型トンネル顕微鏡 scanning tunneling microscope, STM 517
創作 creation 133
操作上 operationally 405
操作上の operational 405
操作する manipulate 351
走査する scan 516
操作の operating 405
喪失 loss 345
双晶 twin 612
双晶形成 twinning 612
相乗効果 synergy 581
層状格子 layer lattice 332
層状構造 layer structure 332
相乗的な synergic 580
相乗的に synergistically 581
双晶にする twin 612
双晶になった twinned 612
層状の
laminar 330
layered 332
双晶の twin 612
増殖する multiply 378
装飾品 artifact 39
そうすると whereupon 643
双生イオン zwitterion 651
双生イオンの zwitterionic 652
総説 review 505
想像 imagination 286
想像された supposed 575
想像する
envisage 207
imagine 286
visualize 635
想像的に imaginatively 286
想像できる imaginable 286
想像も及ばない inconceivable 294
想像力に富む imaginative 286
増大
enhancement 204
intensification 308
相対誤差 relative error 492
増大させる increase 295
増大する
growing 264
increase 295
rise 508

た

ち

つ

electro-optic 195
electro-optical 195
電気浸透 electroosmosis 195
電気抵抗 electric resistance 189
電気抵抗率 electric resistivity 189
電気的中性 electroneutrality 193
電気的中性の原理 electroneutrality principle 193
電気伝導率 electric conductivity 189
電気透析 electrodialysis 190
電気二重層 electric double layer 189
電気分解 electrolysis 191
電気ベクトル electric vector 189
電気めっき electroplating 196
電気めっきする electroplate 196
電気陽性の electropositive 196
電極 electrode 190
電極過程 electrode process 190
電極触媒反応 electrocatalysis 190
電極電位 electrode potential 190
電極反応 electrode reaction 190
典型 prototype 464
典型的な
archetypal 37
classic 92
prototypical 465
typical 613
点欠陥 point defect 441
点検 check 86
電弧 arc 37
電子移動 electron transfer 195
電子雲
cloud 95
electron cloud 192
電子雲膨張系列 nephelauxetic series 384
電子雲膨張効果 nephelauxetic effect 384
電子エネルギー準位 electronic energy level 193
電子エネルギー損失分光 electron energy loss spectroscopy 193
電子回折 electron diffraction 192
電子化合物 electron compound 192
電子過剰の electron-rich 194
電子間の interelectronic 310
電子間反発
electron-electron repulsion 193
electron repulsion 194
電子求引性の electron-withdrawing 195
電子吸収スペクトル electronic absorption spectrum 193
電子供与 electron donation 192
電子供与性基 electron-donating group 192
電子供与体 electron donor 192
電子供与体-受容体錯体 electron donor-acceptor complex 192
電子計算機 computer 109
電子欠損 electron deficiency 192
電子顕微鏡 electron microscope 194
電子顕微鏡による electron microscopic 194
電子顕微鏡法 electron microscopy 194
電子工学的に electronically 193
電子工学の electronic 193
電子交換 electron exchange 193
電子構造 electronic structure 194
電子散乱 electron scattering 194
電子写真 electrophotography 196
電子銃 electron gun 193
電子受容性の electron-accepting 191
電子受容体 electron acceptor 191
電子衝撃
electron bombardment 192
electron impact 194
電子状態 electronic state 194
電子親和力 electron affinity 191
電子スピン electron spin 194
電子スピン共鳴 electron spin resonance, ESR 195
電子スペクトル electronic spectrum 194
電子線 electron beam 191
電子遷移 electronic transition 194
電子対供与体 electron-pair donor 194
電子対受容体 electron-pair acceptor 194
電子的に electronically 193
電子伝達鎖 electron-transport chain 195
電子伝導率 electronic conductivity 193
電子導体 electronic conductor 193
電子の electronic 193
電磁波 electromagnetic wave 191
電子配置
electron configuration 192
electronic configuration 193
電子反発 electronic repulsion 194
電子非局在化 electron delocalization 192
電子不足の electron-deficient 192
電子分極 electronic polarization 193
電子分光法 electron spectroscopy 194
電子分布 electron distribution 192
電子放射 electron emission 193
電子密度 electron density 192
電子密度図 electron density map 192
転写 transcription 604

と

な

locity　356
二色性　dichroism　163
二色性の　dichromic　162
二進法の　binary　58
にすぎる　too　600
にする　render　494
に精通して　aquinted with　9
二成分系　binary system　58
二成分の　binary　58
偽の　spurious　553
二相の　biphasic　61
に属する　belong to　55
に沿って　along　23
に備えて　against　17
に備える　make provision for　465
に存する　reside　498
に対して
against　17
for　242
to　600
toward　602
versus　632
に対処する　contend with　121
に対する　for　242
に耐える
resist　499
resistant　499
に携わる
engage in　203
enter into　205
（に）似た　like　338
似たもの　like　338
に頼る
rely on　493
resort to　500
に近づく　approach　36
日常生活　everyday life　214
日常的な　routine　511
日常的に　routinely　511
にちなんで　after　17
に注目を払う　pay attention to　46
に直角に　at right angles to　507
について
as to　40
with　646
についで
following　242
next　386
に通じている　familiar with　228
につき　per　426
日光　sunlight　572
に続く　follow　242
につれて　with　646
似ていない　dissimilar　174
（に）似ている　resemble　498
二糖　disaccharide　168
二等分する　bisect　61
にとって　to　600
に伴う　attend　45
ニトロ化　nitration　386
ニトロ化剤　nitrating agent　386
ニトロ化する　nitrate　386
になる
become　55
come　102
に似た　like　338
に似ている　resemble　498
に乗り出す　embark upon　198
二倍に　twice　612
二倍にする　double　180
二倍になる　double　180
に入る　go into　260
に反対して　against　17
に反対の　contrary to　123
二番目の　second　520
に匹敵する　compare with　104
に等しい　amount to　25
鈍い　dull　184
二分子求核置換反応　bimolecular nucleophilic substitution　58
二分子層　bilayer　58
二分子脱離反応　bimolecular elimination　58
二分子の　bimolecular　58
に変じる　turn　611
二変の　bivariant　61
二方向の　bidirectional　58
にほかならない　nothing other than　393
二本鎖 DNA　double-stranded DNA　180
二本鎖の　double-stranded　180
にまたがって　across　9
二面角　dihedral angle　166
にもかかわらず
despite　157
notwithstanding　394
in spite of　551
even though　596
に用いる　apply　35
に基づいて
on the basis of　53
on　402
に基づく
base on　52
based on　52
rest upon　501
に役に立つ　lend itself to　334
乳化　emulsification　200
乳化剤　emulsifying agent　200
乳化する　emulsify　200
乳光　opalescence　404
入射　incidence　293
入射角　angle of incidence　28
入射光　incident light　293
入射する　incident　293
入射の　incoming　294
入手可能なこと　accessibility　5
入手しやすい　accessible　5
入手の可能性　availability　48
に優先して　in preference to　452
乳白色を発する　opalescent　404
乳鉢　mortar　375
乳棒　pestle　430
に有利に　in favor of　229
入力　input　304
尿の　urinary　625
によって
by means of　356
on　402
upon　625
via　632
by virtue of　634
により　through　597
によれば　according to　6
二量化　dimerization　167
二量化する　dimerize　167
二量体　dimer　167
二量体内の　intradimer　315
二量体の　dimeric　167
に類似の　analogous to　27
に分かれる　separate into　525
任意に　arbitrarily　37
任意の　arbitrary　37

ぬ

ね

は

反応物　reactant　478
反応分子数　molecularity　371
反発　repulsion　497
反発する
repel　495
repulsive　497
半波電位　half-wave potential　266
反復　iteration　324
反復していう　reiterate　491
反復の　iterative　324
反物質　antimatter　33
半分　half　266
半分の　half　266
半分満たされた　half-filled　266
ハンマーで打つ　hammer　267
判明する
prove　465
turn out　611
繁茂　overgrowth　415
反粒子　antiparticle　33
範例　paradigm　419
販路　outlet　413

ひ

比　specific　548
非圧縮性の　incompressible　294
非イオン性の　nonionic　389
緋色の　scarlet　517
微液滴　microdroplet　363
非会合性の　nonassociated　387
非会合の　unassociated　615
非回転的に　irrotationally　322
控えめな　modest　369
控えめに　modestly　369
控えめにいう　understate　617
控えめにいっても　to say the least　333
控えめの　conservative　117
非可逆波　irreversible wave　322
比較　comparison　104
比較して　by comparison　104
（と）比較して
as compared with　104
in comparison to　104
in comparison with　104
in relation to　491
relative to　492
（と）比較する　compare　104
比較的
comparatively　104
relatively　492
比較的重要でない　minor　365
比較できるほどに　comparably　104
比較にならないほど　incomparably　294
ヒ化ニッケル型構造　nickel arsenide structure　386
光　light　337
光イオン化　photoionization　433
光解離　photodissociation　433
光り輝く　brilliant　69
光ガルバニ電池　photogalvanic cell　433
光触媒作用　photocatalysis　432
光増感　photosensitization　434
光増感剤　photosensitizer　434
光毒性の　phototoxic　434
光ポンピング　optical pumping　407
光誘起の　light-induced　337
光レセプター　photoreceptor　434
微環境　microenvironment　363
非環式の　acyclic　11
非干渉性散乱　incoherent scattering　294
引き合いに出す　invoke　318
引き上げ　withdrawal　646
引き上げる　lift　336
引き起こす
breed　68
excite　217
induce　298
invoke　318
occasion　399
present　454
produce　458
provoke　465
引き裂き　tear　585
引き裂く　tear apart　586
引き算　subtraction　569
引きずっていく　drag　181
引き出す
bring out　69
draw　182
引き付ける　attract　46
引き続いて　successively　570
引き続きの　continued　122
引き止める
detain　158
hold back　276
引き抜き　abstraction　3
引き抜く　abstract　3
引き伸ばされた　elongated　198
引き伸ばされていない　unstretched　623
引き伸ばす　stretch　564
非吸湿性の　nonhygroscopic　389
非共有相互作用　noncovalent interaction　388
非共有電子対　unshared electron pair　623
非局在化　delocalization　151
非局在化エネルギー　delocalization energy　151
非局在化軌道　delocalized orbital　152
非局在化する　delocalize　151
非局在化電子　delocalized electron　151
非局在の　nonlocal　390
引き渡す　deliver　151
卑金属　base metal　53
非金属　nonmetal　390
非金属性の　nonmetallic　390
ピーク　peak　424
引く　subtract　569
低い　low　345
（より）低い　lower　345
低い位置にある　low-lying　346
低くする　lower　346
低く評価する　discount　170
非経験的な　nonempricial　388
非系統名　unsystematic name　624
非結合原子　nonbonded atom　387
非結合性電子対　nonbonding electron pair　387
非結合性の　nonbonding　387
非結合相互作用　nonbonded interaction　387

ふ

へ

ほ

ま

み

む

め

も

や

ゆ

よ

ら

る

れ

ろ

わ

を

編著者略歴

まつながよしお
松永義夫

1929年　岐阜県に生まれる
1952年　東京大学理学部化学科卒業
1966年　北海道大学理学部教授
現　在　北海道大学名誉教授
　　　　理学博士

主な著書　『物性化学』（裳華房）
　　　　　『現代の物理化学』（三共出版）
　　　　　『入門化学熱力学』（朝倉書店）
　　　　　『化学英語のスタイルガイド』（朝倉書店）

化学英語[精選]文例辞典　　定価はカバーに表示

2015年8月20日　初版第1刷

編著者　松　永　義　夫
発行者　朝　倉　邦　造
発行所　株式会社　朝　倉　書　店
東京都新宿区新小川町 6-29
郵便番号　162-8707
電話　03(3260)0141
FAX　03(3260)0180
http://www.asakura.co.jp

〈検印省略〉

真興社・牧製本

ISBN 978-4-254-14100-9　C 3543　Printed in Japan